Lecture Notes in Networks and Systems

Volume 401

The series "Lecture Notes in Networks and Systems" publishes the latest developments in Networks and Systems—quickly, informally and with high quality. Original research reported in proceedings and post-proceedings represents the core of LNNS.

Volumes published in LNNS embrace all aspects and subfields of, as well as new challenges in, Networks and Systems.

The series contains proceedings and edited volumes in systems and networks, spanning the areas of Cyber-Physical Systems, Autonomous Systems, Sensor Networks, Control Systems, Energy Systems, Automotive Systems, Biological Systems, Vehicular Networking and Connected Vehicles, Aerospace Systems, Automation, Manufacturing, Smart Grids, Nonlinear Systems, Power Systems, Robotics, Social Systems, Economic Systems and other. Of particular value to both the contributors and the readership are the short publication timeframe and the world-wide distribution and exposure which enable both a wide and rapid dissemination of research output.

The series covers the theory, applications, and perspectives on the state of the art and future developments relevant to systems and networks, decision making, control, complex processes and related areas, as embedded in the fields of interdisciplinary and applied sciences, engineering, computer science, physics, economics, social, and life sciences, as well as the paradigms and methodologies behind them.

Indexed by SCOPUS, INSPEC, WTI Frankfurt eG, zbMATH, SCImago.

All books published in the series are submitted for consideration in Web of Science.

For proposals from Asia please contact Aninda Bose (aninda.bose@springer.com).

More information about this series at https://link.springer.com/bookseries/15179

M. Shamim Kaiser · Juanying Xie ·
Vijay Singh Rathore
Editors

Information and Communication Technology for Competitive Strategies (ICTCS 2021)

Intelligent Strategies for ICT

Springer

Editors
M. Shamim Kaiser
Jahangirnagar University
Dhaka, Bangladesh

Vijay Singh Rathore
IIS University
Jaipur, India

Juanying Xie
School of Computer Science
Shaanxi Normal University
Xi'an, China

ISSN 2367-3370 ISSN 2367-3389 (electronic)
Lecture Notes in Networks and Systems
ISBN 978-981-19-0097-6 ISBN 978-981-19-0098-3 (eBook)
https://doi.org/10.1007/978-981-19-0098-3

This Springer imprint is published by the registered company Springer Nature Singapore Pte Ltd.
The registered company address is: 152 Beach Road, #21-01/04 Gateway East, Singapore 189721, Singapore

Preface

Sixth International Conference on Information and Communication Technology for Competitive Strategies (ICTCS-2021) targets state of the art as well as emerging topics pertaining to information and communication technologies (ICTs) and effective strategies for its implementation for engineering and intelligent applications.

The conference is anticipated to attract a large number of high-quality submissions, stimulate the cutting-edge research discussions among many academic pioneering researchers, scientists, industrial engineers, students from all around the world and provide a forum to researcher; propose new technologies, share their experiences and discuss future solutions for design infrastructure for ICT; provide a common platform for academic pioneering researchers, scientists, engineers and students to share their views and achievements; enrich technocrats and academicians by presenting their innovative and constructive ideas; focus on innovative issues at international level by bringing together the experts from different countries.

The conference was held during 17 and 18 December 2021, physically at Four Points by Sheraton Jaipur, India, and digitally on Zoom organized by Global Knowledge Research Foundation.

Research submissions in various advanced technology areas were received, and after a rigorous peer review process with the help of programme committee members and external reviewer, 150 papers were accepted with an acceptance rate of 19%. All 150 papers of the conference are accommodated in two volumes, and also, papers in the book comprise authors from eight countries.

This event success was possible only with the help and support of our team and organizations. With immense pleasure and honour, we would like to express our sincere thanks to the authors for their remarkable contributions, all the technical program committee members for their time and expertise in reviewing the papers within a very tight schedule, and the publisher Springer for their professional help.

We are overwhelmed by our distinguished scholars and appreciate them for accepting our invitation to join us through the virtual platform and deliver keynote speeches and technical session chairs for analysing the research work presented by the researchers. Most importantly, we are also grateful to our local support team for their hard work for the conference. This series has already been made a continuous series which will be hosted at different locations every year.

Dhaka, Bangladesh
Xi'an, China
Jaipur, India

M. Shamim Kaiser
Juanying Xie
Vijay Singh Rathore

Contents

Editors and Contributors

About the Editors

Dr. M. Shamim Kaiser is currently working as Professor at the Institute of Information Technology of Jahangirnagar University, Savar, Dhaka-1342, Bangladesh. He received his Bachelor's and Master's degrees in Applied Physics Electronics and Communication Engineering from the University of Dhaka, Bangladesh, in 2002 and 2004, respectively, and the Ph.D. degree in Telecommunication Engineering from the Asian Institute of Technology (AIT), Pathumthani, Thailand, in 2010. His current research interests include data analytics, machine learning, wireless network and signal processing, cognitive radio network, big data and cybersecurity, and renewable energy. He has authored more than 100 papers in different peer-reviewed journals and conferences. He is Associate Editor of the *IEEE Access Journal* and Guest Editor of *Brain Informatics Journal* and *Cognitive Computation Journal*. Dr. Kaiser is Life Member of Bangladesh Electronic Society, Bangladesh Physical Society. He is also Senior Member of IEEE, USA, and IEICE, Japan, and Active Volunteer of the IEEE Bangladesh Section. He is founding Chapter Chair of the IEEE Bangladesh Section Computer Society Chapter.

Dr. Juanying Xie is currently Professor at Shaanxi Normal University, Xi'an, China. I have been Full Professor at School of Computer Science of Shaanxi Normal University in PR China. My research interests include machine learning, data mining, and biomedical data analysis. I have published around 50 research papers and published 2 monograph books. I have been Associate Editor of *Health Information Science and Systems*. I have been Program Committee Member of several international conferences, such as the international conference on Health Information Science, etc. I have been Senior Member of China Computer Federation (CCF), Member of Chinese Association for Artificial Intelligence (CAAI), Member of Artificial Intelligence and Pattern Recognition Committee of CCF, and Member of Machine Learning Committee of CAAI, etc. I have been a peer reviewer for many journals, such as *Information Sciences* and *IEEE Transactions on Cybernetics*, etc. I was awarded my Ph.D. in signal and information processing from Xidian University. I cooperated with Prof. Xiaohui Liu at Brunel University in UK from 2010 to 2011 in machine learning and gene selection research. I received an engineering master degree in the application technology of computers at Xidian University and a bachelor degree of science in computer science at Shanxi Normal University.

Dr. Vijay Singh Rathore is presently working as Professor in the Department of CS and IT, IIS (Deemed to be) University, Jaipur (India). He received Ph.D. from the University of Rajasthan and has teaching experience of 20 years. He is Secretary, ACM Jaipur Chapter, Past Chairman, CSI Jaipur Chapter, got two patents published, Ph.D. Supervised (Awarded: 16, Under Supervision: 07), 80+ research papers, and 10+ books got published. He is handling international affairs of The IIS University, Jaipur. His research areas are Internet security, cloud computing, big data, and IoT.

Contributors

R. Yasir Abdullah Computer Science and Bussiness System, SKCE, Coimbatore, TamilNadu, India

Abhishek Electronics and Communication, NSUT East Campus, Delhi, Delhi, India

Aasakti Agarwal Department of Computer Science and Engineering, Amity University, Kolkata, India

Pooja Agarwal PES University EC Campus, Bangalore, India

Nisha Agrawal Centre for Development of Advanced Computing, Pune, India

Faiyaz Ahmed Department of Computer Science and Engineering, Independent University, Bangladesh, Dhaka, Bangladesh

G. Aishwarya KLE Technological University, Hubballi, India

Asha Ambhaikar Department of CSE, Kalinga University, Raipur, India

Aman Anand Department of Computer Science and Engineering, Amity University, Kolkata, India

Prateek Anand Electronics and Communication, GGSIPU, Dwarka, Delhi, India

K. Annapurani Department of Networking and Communication, School of Computing, SRM Institute of Science and Technology, Kattankulathur, India

B. U. Anu Barathi Department of CSE, Sathyabama Institute of Science and Technology, Chennai, Tamil Nadu, India

Maheriya Arpita Gujarat Technological University—Graduate School of Engineering and Technology, Ahmedabad, Gujarat, India

M. K. Arti Electronics and Communication, NSUT East Campus, Delhi, Delhi, India

G. Arun Prasath Department of Networking and Communication, School of Computing, SRM Institute of Science and Technology, Kattankulathur, India

Arzoo Guru Nanak Dev University, Amritsar, India

K. Ashwini KLE Technological University, Hubballi, India

Sudeep Balan Centre for Development of Advanced Computing (CDAC), Thiruvananthapuram, Kerala, India

Samir Bandyopadhyay Lincoln University College, Petaling Jaya, Selangor D.E, Malaysia

Saubhik Bandyopadhyay Department of Computer Science and Engineering, Amity University, Kolkata, India

Reenu Batra Department of Computer Science and Engineering, SGT University, Gurugram, India

Spurthi S. Bhat Vishwakarma Institute of Technology, Pune, Maharashtra, India

Sonali Bhatnagar Department of Physics and Computer Science, Dayalbagh Educational Institute, Agra, India

N. P. G. Bhavani EEE Department, Dr. MGR Educational and Research Institute, Chennai, India;
Department of ECE, Saveetha School of Engineering, SIMATS, Chennai, Tamil Nadu, India;
School of Electrical Engineering, Vels Institute of Science, Chennai, India;
ECE Department, Dr. MGR Educational and Research Institute, Chennai, India

Ankit Bhavsar GLS University, Ahmedabad, India

Bhoomika PES University EC Campus, Bangalore, India

Kiran Bhowmick Department of Computer Engineering, D. J. Sanghvi College of Engineering, Mumbai, India

Pravin Kumar Bhoyar Symbiosis Institute of Management Studies (SIMS), A Constituent of Symbiosis International Deemed University, Pune, India

Vineeta Bhujle KJ Somaiya College of Engineering, Mumbai, India

R. Bhuvaneshwari College of Engineering, Anna University, Chennai, India

Siddhartha Sankar Biswas Department of Computer Science and Engineering, School of Engineering Sciences and Technology, Jamia Hamdard, New Delhi, India

Payal Bose Lincoln University College, Petaling Jaya, Selangor D.E, Malaysia

Rajashree Vinod Bothale National Remote Sensing Centre, Hyderabad, Telangana, India

H. N. Champa University Visvesvaraya College of Engineering, Bengaluru, India

Arti Chandani Symbiosis Institute of Management Studies, Symbiosis International (Deemed University), Pune, India

Khyati Chavda Shantilal Shah Engineering College, Bhavnagar, Gujarat Technology University, Bhavnagar, Gujarat, India

Pradnya Chitrao Symbiosis Institute of Management Studies (SIMS), A Constituent of Symbiosis International Deemed University, Pune, India

Pradnya Vishwas Chitrao Symbiosis Institute of Management Studies (SIMS), Constituent of Symbiosis International Deemed University, Pune, India

Komal Chopra Symbiosis Institute of Management Studies (SIMS), Constituent of Symbiosis International Deemed University, Pune, India

Aryan Chouhan Dwarkadas J. Sanghvi College of Engineering, Mumbai, India

Aakrut Dabhi Parul University, Gujarat, India

Abhishek Das Centre for Development of Advanced Computing, Pune, India

Alaka Das IIIT Bhubaneswar, Bhubaneswar, Odisha, India

Maheshwari Das JSS Science and Technology University, Mysuru, India

Swapnil Das Department of Computer Science and Engineering, Amity University, Kolkata, India

Manoj Kumar Deka Bodoland University, Assam, India

Vaskar Deka Gauhati University, Guwahati, Assam, India

V. Devendran Computer Science and Engineering, Lovely Professional University, Phagwara, India

Rajiv Divekar Symbiosis Institute of Management Studies (SIMS), Constituent of Symbiosis International Deemed University, Pune, India

Faraz Doja Department of Computer Science and Engineering, SEST, New Delhi, India

Pankaj Dorlikar Centre for Development of Advanced Computing, Pune, India

Suyash Kumar Dubli PES University EC Campus, Bangalore, India

Shawni Dutta The Bhawanipur Education Society, Kolkata, India

S. Gavaskar Bharathiar University, Coimbatore, Tamilnadu, India

Ujwalla H. Gawande Yeshwantrao Chavan College of Engineering, Nagpur, India

S. S. Gharde School of Computer Sciences, Kavayitri Bahinabai Chaudhari North Maharashtra University, Jalgaon, India

Lakshmi S. Gopal Center for Wireless Networks and Applications (WNA), Amrita Vishwa Vidyapeetham, Amritapuri, India

V. S. S. N. Gopala Krishna Pendyala National Remote Sensing Centre, Hyderabad, Telangana, India

Lipichanda Goswami Bodoland University, Assam, India

Sanjay Gowdru PES University EC Campus, Bangalore, India

Vishal Goyal GLA University, Mathura, India

Jyostna Grandhi Centre for Development of Advanced Computing (C-DAC), Hyderabad, Telangana, India

Pooja Gupta Computer Science and Engineering, Maharaja Agrasen Institute of Technology, Delhi, India

Aravinda Harithsa JSS Science and Technology University, Mysuru, India

Indu Hariyale Computer Technology, YCCE, Nagpur, India

Mahady Hasan Department of Computer Science and Engineering, Independent University, Bangladesh, Dhaka, Bangladesh

Nikita R. Hatwar Yeshwantrao Chavan College of Engineering, Nagpur, India

Shrilalita Hegde KLE Technological University, Hubballi, India

Aishwarya Hugar KLE Technological University, Hubballi, India

Nalini C. Iyer KLE Technological University, Hubballi, India

Jeevan J. Jacob Department of Computer Science and Engineering, Christ University, Bangalore, Karnataka, India

Aarti Jain Electronics and Communication, NSUT East Campus, Delhi, Delhi, India

Ashu Jain Dr. Akhilesh Das Gupta Institute of Technology and Management, New Delhi, India

Dhyanendra Jain Dr. Akhilesh Das Gupta Institute of Technology and Management, Guru Gobind Singh Indraprastha University, New Delhi, India

M. Jansi Rani Mepco Schlenk Engineering College, Sivakasi, Tamil Nadu, India

Vidula S. Jape Department of Electrical Engineering, PES's Modern College of Engineering, Pune, India

U. Jayalatsumi EEE Department, Dr. MGR Educational and Research Institute, Chennai, India;
Department of ECE, Saveetha School of Engineering, SIMATS, Chennai, Tamil Nadu, India;
School of Electrical Engineering, Vels Institute of Science, Chennai, India;
ECE Department, Dr. MGR Educational and Research Institute, Chennai, India

Deepak Jayaswal St. Francis Institute of Technology, Mumbai, India

Malick Jeng Symbiosis Institute of Management Studies, Symbiosis International (Deemed University), Pune, India

Chandan Kumar Jha Department of Computer Science and Engineering, School of Engineering Sciences and Technology, Jamia Hamdard, New Delhi, India

Nisheeth Joshi Computer Science and Engineering, Banasthali University, Vanasthali, Rajasthan, India

Dinesh Kumar Kalal GLS University, Ahmedabad, India

Muhammad Abul Kalam Bharath Institute of Higher Education and Research, Chennai, Tamil Nadu, India

K. P. Kaliyamurthie Department of Computer Science and Engineering, Bharath Institute of Higher Education and Research, Chennai, India

Hemantha Kumar Kalluri Vignan's Foundation for Science Technology and Research, Vadlamudi, Andhra Pradesh, India

Mehul Kapoor Department of Information Technology, UIET, Panjab University, Chandigarh, India

M. S. Karthika Devi College of Engineering, Anna University, Chennai, India

M. Karuppasamy Kalasalingam Academy of Research and Education, Krishnankoil, Tamil Nadu, India

Puneet Jai Kaur Department of Information Technology, UIET, Panjab University, Chandigarh, India

Ramandeep Kaur Computer Science and Engineering, Lovely Professional University, Phagwara, India

K. Kavin Pragadeesh College of Engineering, Anna University, Chennai, India

T. Kavitha EEE Department, Dr. MGR Educational and Research Institute, Chennai, India;
Department of ECE, Saveetha School of Engineering, SIMATS, Chennai, Tamil Nadu, India;
School of Electrical Engineering, Vels Institute of Science, Chennai, India;
ECE Department, Dr. MGR Educational and Research Institute, Chennai, India

Khushi Khanchandani KJ Somaiya College of Engineering, Mumbai, India

Mohammed Kharodawala Parul University, Gujarat, India

Karan Khunti Parul University, Gujarat, India

K. Krishnakumar EEE Department, Dr. MGR Educational and Research Institute, Chennai, India;
Department of ECE, Saveetha School of Engineering, SIMATS, Chennai, Tamil Nadu, India;
School of Electrical Engineering, Vels Institute of Science, Chennai, India;
ECE Department, Dr. MGR Educational and Research Institute, Chennai, India

Haripriya H. Kulkarni Professor and HOD, MMCOE, Department of Electrical Engineering, PES's Modern College of Engineering, Pune, India

Milind V. Kulkarni Vishwakarma Institute of Technology, Pune, Maharashtra, India

Parag Kulkarni iKnowlation Research Laboratory, Shivajinagar, Pune, Maharashtra, India

Anil Kumar Guru Nanak Dev University, Amritsar, India

Ankit Kumar Department of Computer Science and Engineering, Siksha 'O' Anusandhan Deemed to be University, Bhubaneswar, Odisha, India

Atul Kumar Parul University, Gujarat, India

Balwant Kumar Goverment College for Girls, Hisar, Haryana, India

K. Senthil Kumar ECE Department, Dr. MGR Educational and Research Institute, Chennai, India;
Department of Civil Engineering, Dr. MGR Educational and Research Institute, Chennai, India;
EEE Department, Meenakshi College of Engineering, Chennai, India

Nilesh Kumar National Informatics Center Hyderabad, Hyderabad, Telangana, India

Amogh Kuruwatti JSS Science and Technology University, Mysuru, India

P. R. Lakshmi Eswari Centre for Development of Advanced Computing (C-DAC), Hyderabad, Telangana, India

G. Lalitha Devi Department of Mathematics, GITAM University, Visakhapatnam, India

Netra M. Lokhande HOD, IT Department, School of Electrical Engineering, MIT World Peace University, Pune, India

Jayashree Mallidu KLE Technological University, Hubballi, India

Nilesh Maltare Government Engineering College, Modasa, India

Venkatesh Mane KLE Technological University, Hubballi, India

Iti Mathur Computer Science and Engineering, Banasthali University, Vanasthali, Rajasthan, India

S. Mayukha Bharathiar University, Coimbatore, India

Mita Mehta Symbiosis Institute of Management Studies, Symbiosis International (Deemed University), Pune, India

Aishwarya Girish Menon REVA University, Bengaluru, Karnataka, India

Manish Modani NVIDIA, Pune, India

E. Mohana Department of CSE, Rajalakshmi Engineering College, Chennai, Tamil Nadu, India

Andrew Moitra Department of Computer Science and Engineering, Amity University, Kolkata, India

Neha Motagi Pimpri Chinchwad College of Engineering, Pune, India

Pooja Mudgil Computer Science and Engineering, Banasthali University, Vanasthali, Rajasthan, India

Debasish Mukherjee Department of Computer Science and Engineering, Christ University, Bangalore, Karnataka, India

M. P. R. Murthy Department of Mathematics, GITAM University, Visakhapatnam, India

Sunil Kumar Muttoo Department of Computer Science, University of Delhi, Delhi, India

Md. Tabrez Nafis Department of Computer Science and Engineering, School of Engineering Sciences and Technology, Jamia Hamdard, New Delhi, India

Komal Nair Pimpri Chinchwad College of Engineering, Pune, India

Rupali Narayankar Pimpri Chinchwad College of Engineering, Pune, India

Manali Narendra Department of Computer Science and Engineering, Siksha 'O' Anusandhan Deemed to be University, Bhubaneswar, Odisha, India

Rajender Nath Department of Computer Science and Applications, Kurukshetra University, Kurukshetra, India

J. Naveen John Nesamony Memorial Christian College Affiliated to Manonmaniam Sundaranar University, Abishekapatti, Tirunelveli, Tamil Nadu, India

P. Nikita KLE Technological University, Hubballi, India

Asmita Nirmal Datta Meghe College of Engineering, Navi Mumbai, India

U. Barakkath Nisha Information Technology, Sri Krishna College of Engineering, Coimbatore, India

Nisha Department of Computer Science, University of Delhi, Delhi, India

Shailesh Panchal Gujarat Technological University—Graduate School of Engineering and Technology, Ahmedabad, Gujarat, India

Amit Kumar Pandey Dr. Akhilesh Das Gupta Institute of Technology and Management, Guru Gobind Singh Indraprastha University, New Delhi, India

Ronakkumar Patel U & P U. Patel Department of Computer Engineering, Chandubhai S. Patel Institute of Technology (CSPIT), Charotar University of Science and Technology (CHARUSAT), Anand, India

Sachin Patel SAGE, Indore, India

Mohit Pathak Symbiosis International (Deemed University), Pune, India

Rishi Pathak Centre for Development of Advanced Computing, Pune, India

Mahesh Patil Centre for Development of Advanced Computing (C-DAC), Hyderabad, Telangana, India

Neha Patil KJ Somaiya College of Engineering, Mumbai, India

Manu Phogat SGT University, Gurugram, Haryana, India

Madhu Ponakala St. Francis College for Women, Hyderabad, India

S. Poonkuzhali Department of CSE, Rajalakshmi Engineering College, Chennai, Tamil Nadu, India

M. Prabha Velammal College of Engineering and Technology, Madurai, India

Rekha Prabha Center for Wireless Networks and Applications (WNA), Amrita Vishwa Vidyapeetham, Amritapuri, India

M. Prabhakar REVA University, Bengaluru, Karnataka, India

P. K. Preetha Department of Electrical and Electronics Engineering, Amrita Vishwa Vidyapeetham, Amritapuri, India

Preeti Rajiv Gandhi Goverment College for Women, Bhiwani, Haryana, India

R. Priya Bharathiar University, Coimbatore, Tamilnadu, India

Rupok Chowdhury Protik Department of Computer Science and Engineering, Independent University, Bangladesh, Dhaka, Bangladesh

M. M. Raghuwanshi Computer Engineering, GHRCEM, Pune, India

Devashri Raich Department of CSE, Kalinga University, Raipur, India

P. K. Rajani Pimpri Chinchwad College of Engineering, Pune, India

Kanchana Rajaram Department of Computer Science and Engineering, Sri Sivasubramaniya Nadar College of Engineering, Chennai, Tamil Nadu, India

Rakhee The University of West Indies, Mona Campus, Kingston, Jamaica

Baskaran Ramachandran College of Engineering, Anna University, Chennai, India

Maneesha Vinodini Ramesh Center for Wireless Networks and Applications (WNA), Amrita Vishwa Vidyapeetham, Amritapuri, India

R. J. Ramteke School of Computer Sciences, Kavayitri Bahinabai Chaudhari North Maharashtra University, Jalgaon, India

Jyoti Rani Department of Computer Science and Applications, Kurukshetra University, Kurukshetra, India

Smita Rath Department of Computer Science and Engineering, Siksha 'O' Anusandhan Deemed to be University, Bhubaneswar, Odisha, India

G. Ratnakanth Anna University, Chennai, India

P. Revathy Department of CSE, Rajalakshmi Engineering College, Chennai, Tamil Nadu, India

Sandeep Romana Centre for Development of Advanced Computing (C-DAC), Hyderabad, Telangana, India

Nader Salam Computer Science Engineering, Karunya Institute of Technology and Sciences, Coimbatore, Tamil Nadu, India

Shriya Salunkhe Department of Computer Engineering, D. J. Sanghvi College of Engineering, Mumbai, India

S. Sarika Naipunnya Institute of Management and Information Technology, Pongam, Kerala, India

V. S. Saroja School of ECE, KLE Technological University, Hubballi, Karnataka, India

S. Sarva Lakshmi Department of Mathematics, GITAM University, Visakhapatnam, India

A. K. Sarvaiya Government Engineering College Bhavnagar, Bhavnagar, Gujarat, India

S. Savitha Department of Computer Science and Engineering, Christ University, Bangalore, Karnataka, India

K. Preethi Naveena Selvi Department of Computer Science and Engineering, Sri Sivasubramaniya Nadar College of Engineering, Chennai, Tamil Nadu, India

Diptiman Senapati Department of Computer Science and Engineering, Siksha 'O' Anusandhan Deemed to be University, Bhubaneswar, Odisha, India

S. K. Shafiya ECE Department, Dr. MGR Educational and Research Institute, Chennai, India;
Nalla Narasimha Reddy Group of Institutions, Hyderabad, India

Manav Shah Dwarkadas J. Sanghvi College of Engineering, Mumbai, India

Shyamal Shah U & P U. Patel Department of Computer Engineering, Chandubhai S. Patel Institute of Technology (CSPIT), Charotar University of Science and Technology (CHARUSAT), Anand, India

Anupam Kumar Sharma Dr. Akhilesh Das Gupta Institute of Technology and Management, Guru Gobind Singh Indraprastha University, New Delhi, India

Dharmendra Sharma STME, NMIMS, Indore, India

K. S. Shashank JSS Science and Technology University, Mysuru, India

I. Shatheesh Sam Department of PG Computer Science, Nesamony Memorial Christian College Affiliated to Manonmaniam Sundaranar University, Tamil Nadu, Abishekapatti, Tirunelveli, Tamil Nadu, India

K. V. Shiny Department of Computer Science and Engineering, Noorul Islam Centre for Higher Education, Kumaracoil, India

M. Shobana Department of CSE, Rajalakshmi Engineering College, Chennai, Tamil Nadu, India

Saroja V. Siddamal KLE Technological University, Hubballi, India

P. Sijin University Visvesvaraya College of Engineering, Bangalore University, Bengaluru, India

Kavita R. Singh Computer Technology, YCCE, Nagpur, India

Archana Singhal IP College for Women, University of Delhi, Delhi, India

Prashant Singh Dr. Akhilesh Das Gupta Institute of Technology and Management, Guru Gobind Singh Indraprastha University, New Delhi, India

Sandeep Singh SGT University, Gurugram, Haryana, India

Yashpal Singh Department of CSE, Kalinga University, Raipur, India

L. R. Sreedhanya Centre for Development of Advanced Computing (CDAC), Thiruvananthapuram, Kerala, India

K. Sreeram Department of Electrical and Electronics Engineering, Amrita Vishwa Vidyapeetham, Amritapuri, India

Vandita Srivastava University of Lucknow, Lucknow, India

V. Srividhya ECE Department, Dr. MGR Educational and Research Institute, Chennai, India;
Department of Civil Engineering, Dr. MGR Educational and Research Institute, Chennai, India;
EEE Department, Meenakshi College of Engineering, Chennai, India

M. Sudhakar Department of Computer Science and Engineering, Bharath Institute of Higher Education and Research, Chennai, India

N. Sugitha Department of Information Technology, Noorul Islam Centre for Higher Education, Kumaracoil, India

K. Sujatha EEE Department, Dr. MGR Educational and Research Institute, Chennai, India;
Department of ECE, Saveetha School of Engineering, SIMATS, Chennai, Tamil Nadu, India;
School of Electrical Engineering, Vels Institute of Science, Chennai, India;
ECE Department, Dr. MGR Educational and Research Institute, Chennai, India

Sadiya Sulaiman Computer Science Engineering, Karunya Institute of Technology and Sciences, Coimbatore, Tamil Nadu, India

V. B. Suneeta KLE Technological University, Hubballi, India

V. Budihal Suneeta School of ECE, KLE Technological University, Hubballi, Karnataka, India

Ch. Suneetha Department of Mathematics, GITAM University, Visakhapatnam, India

Sreehari Surendran Department of Electrical and Electronics Engineering, Amrita Vishwa Vidyapeetham, Amritapuri, India

Prachi Tawde Dwarkadas J. Sanghvi College of Engineering, Mumbai, India

Sandeep Tayal Department of Computer Science and Engineering, Maharaja Agrasen Institute of Technology, GGSIP University, New Delhi, India

Mahesh Bapusaheb Toradmal Symbiosis Institute of Management Studies, Symbiosis International (Deemed University), Pune, India

Deepali Ubale Department of Electrical Engineering, PES's Modern College of Engineering, Pune, India

R. Udayakumar Bharath Institute of Higher Education and Research, Chennai, Tamil Nadu, India

R. Vadivel Bharathiar University, Coimbatore, India

Prashant Vats Dr. Akhilesh Das Gupta Institute of Technology and Management, Guru Gobind Singh Indraprastha University, New Delhi, India;
Fairfield Institute of Management and Technology, GGSIP University, New Delhi, India;
Department of Information Technology, Indira Gandhi Delhi Technical University for Women, New Delhi, India

Deepak A. Vidhate Dr. Vithalrao Vikhe Patil College of Engineering Ahmednagar, Ahmednagar, Maharashtra, India

A. B. Vijayalakshmi School of ECE, KLE Technological University, Hubballi, Karnataka, India

M. S. Vinay Prasad JSS Science and Technology University, Mysuru, India

S. Vinodhkumar Department of CSE, Rajalakshmi Engineering College, Chennai, Tamil Nadu, India

P. Vishal KLE Technological University, Hubballi, India

Varun Vora Dwarkadas J. Sanghvi College of Engineering, Mumbai, India

V. S. Yeshwanth School of ECE, KLE Technological University, Hubballi, Karnataka, India

Efficient Prediction of Fake News Using Novel Ensemble Technique Based on Machine Learning Algorithm

M. Sudhakar and K. P. Kaliyamurthie

Abstract It improves the detection rate of fake news from Twitter, Facebook, and other social media. We are using a novel machine learning algorithm to detect fake news. We proposed a novel algorithm for classifying phony information and actual news. This study deals with logistic regression, SVM, and novel ensemble approach based on machine learning algorithms. It is divided into sample size values of 620 per group. The experiment uses a dataset of 10,000 records with binary classes (fake news, real news). The result demonstrated that the proposed novel ensemble approach obtains a better accuracy value of 95% and a loss value of 05% compared with other algorithms. Thus, the obtained results prove that the proposed algorithm is an ensemble approach that combines decision tree techniques with AdaBoost by varying parameters and can get a significantly higher accuracy value.

Keywords Fake news · Ensemble · Decision tree · AdaBoost · KNN · SVM

1 Introduction

Recently, the increase in Internet usage makes most people share their articles, ideas, posts, videos across Twitter, Facebook, Instagram, etc. The increase in information sharing in social media leads to a rise in the spread of phony information. Therefore, authentication of news is also becoming very important. As technologies have developed, manipulation of authentic images and videos has become an easy job. Fake news plays a significant influence on the opinion of public people. Fake news refers to untrue information shared in any form, such as reports, articles, news, images, videos. The main goal of the spread of phony information is to manipulate the opinion of the public [1]. There are different types of fake news available: targeted fake news, fake headlines, viral posts, and Satire. Bubbles of information, manipulation of news, and

M. Sudhakar (✉) · K. P. Kaliyamurthie
Department of Computer Science and Engineering, Bharath Institute of Higher Education and Research, Chennai, India
e-mail: sudhakarmtech@gmail.com

M. S. Kaiser et al. (eds.), *Information and Communication Technology for Competitive Strategies (ICTCS 2021)*, Lecture Notes in Networks and Systems 401,
https://doi.org/10.1007/978-981-19-0098-3_1

distrust of the media are increasingly severe problems in our society. The knowledge is required for identifying the fake news and where it is coming from. AI, machine learning algorithms, natural language processing algorithms will help us in this situation [1–5].

2 Literature Review

Wang [9] The author introduces a new dataset, namely LIAR, that is used to detect fake news automatically. Compared with the previous dataset, LIAR is an order of magnitude higher, making it possible to develop statistical and computational methods to see fake news. The different backgrounds of different speakers also contribute to extensive research on the development of counterfeit news detectors. We show that combining metadata with text can significantly improve particle holiday news detection. With detailed analysis reports and links to source documents in the dataset, automatic fact-checking tasks based on redundant knowledge can also be explored in the future. Our corpus can also be used to classify NLP locations, extract parameters, model topics, etc. Rumor Investigation [3]. The data are collected from the WWW, which includes articles from various news. The main aim of the author is to determine the pattern of distinguishing misinformation and the real news. The author uses the LIWC tool to extract various text functions from the article and uses many functions as input to the model [6]. The model is trained, and parameters are fine-tuned to obtain the best accuracy. We use several performance indicators to compare the results of each algorithm. Cooperative students performed better than individual students on all performance indicators. Varshney and Vishwakarma [8] the author proposes the Hoax News Inspector technique that automatically collects manipulated news and distinguishes them into fake or actual binary classes, which can benefit further research to predict and understand phony news [6]. The author implements two techniques to detect fake news. The first stage is characterization, and disclosure is the second stage. The first stage will help us highlight the concepts and principles of fake news in social media [8].

The current methods are reviewed during the discovery stage to detect fake news using various algorithms [10]. From a wide range of heterogeneous social news networks, many explicit and implicit features can be extracted from the text information of newspaper articles, authors, or topics. The network model is proposed to integrate the information about the network structure into the learning model. In this article, we also introduce a new diffusion unit model, namely the Gated Diffusive unit. The Gated Diffusive unit model will accept multiple inputs simultaneously from different sources. Effectively to get the output, they combine these inputs. Content "forget" and "adjust" the door. Many experiments on the real fake news dataset PolitiFact show that the proposed model effectively detects phony news articles [9, 10].

3 Materials and Methods

This method involves three main steps,

Text processing, feature extraction, model construction.

3.1 Steps Involved in Text Processing.

- A text file is read as a string consisting of raw text.
- Number values are converted into words using normalization.
- All the texts are converted to the same format.
- Eliminate all non-words such as punctuations, white spaces, special characters.
- Break the sequence of strings into words, keywords, symbols, etc. This process is called tokenization.
- Closely redundant words are replaced with the same word called lemmatization or stemming.
- Frequently used stop words are removed.

3.2 Feature Extraction

Transforming text data to numerical data is called vectorization or feature extraction. This feature extraction can be done using various techniques, namely BoW, N-grams, term frequency. The BoW algorithm works in two sessions. In the first session, the entire corpus is created from a dictionary of words and in the second session as a vector of words called bags of words. In N-grams, contextual information is extracted to enrich data. The following technique is [4] term frequency which is otherwise called inverse document frequency (Tf-idf), where TF(x) is no. of times the term x appears in the specified document out of a total number of terms. Similarly, IDF(t) is a loge of an unlimited number of records processed divided by the document's total number of specific words [10].

3.3 Construction of a Model

Hamdaoui [4] Construction of a model can be done in two ways; the first is the PageMaker; it needs to upload the data to S3, consisting of three objects: model artifacts, training code, and inference code.

Step1: Split the dataset into training and testing specified in the 80% and 20% ratio, respectively.
Step2: Several transforms are to be chained together to form a pipeline.

Pipeline adopts two methods: fit transform, which transforms results to the next transformer, and the fit predicts, where the channel ends with an estimator.
Step2: Train the model using a training set.
Step3: Fine-tune the parameters for the decision tree algorithm to make the model for the dataset.
Step4: Repeat Step2 and Step3 until the model fits the given dataset.
Step5: Evaluate the model using a test set.

3.4 Support Vector Machine (SVM)

Ray [7] SVM is a supervised learning technique used for classifying the problem or regression problems, and it is mainly used for classification problems. It can handle linearly and nonlinearly datasets well. Even if the amount of data is limited, the SVM algorithms can continue to work wonders [7].

3.5 Linear Regression

("How to Conduct Linear Regression" 2010) Linear regression analysis does the job of fitting the linear line through a cloud of records points. The first step develops a model where the X variables are causally related to the Y variables and linear relationships. To fit the regression line, we can use regression analysis. It is the second step of regression analysis. Minimize unexplainable residues. The second stage does reading the correlation and directionality of the records, estimating the model [9].

3.6 K-Nearest Neighbor

Rithesh [8] The KNN supervised machine learning algorithm is used for classification and regression, mainly for classification problems. No particular learning stage uses all learning data in variety, so KNN is a lazy learning algorithm. It is also called a non-parametric learning algorithm because it does not make assumptions about the underlying data. The distance is calculated using various techniques such as Euclidean, Manhattan, or Hamming distance [6].

4 Description of Dataset

The open-source dataset used in this experiment can be obtained for free on the Internet. This data contain false and actual news articles collected from various sources. Some published news articles contain accurate information and descriptions of real-world events. Some fake news sites control statements that are not correct. Two datasets are accessed from Kaggle, namely Dataset 2, containing 20,386 articles for training and 5126 articles for testing. Many sources of the dataset were collected from the Internet. Individual domains serve as a guide because they contain real and fake articles from various other fields. The second dataset, named DS3, consisting of 3352 pieces, is included, either true or false. The actual articles come from trusted online sources. The fake news comes from untrustworthy news websites. This one covers different areas, including sports news, entertainment news, and political news. Dataset 1 is combined dataset from collected articles from all two datasets. The attributes are.

- Unique id for a news article: id
- The title of a news article: Title
- Author of the news article: Author
- The text of the paper could be incomplete: Text
- A label that marks the report as potentially unreliable and reliable: Label
- A testing training dataset with all the same attributes at train.csv without the title: test.csv (Fig. 1)

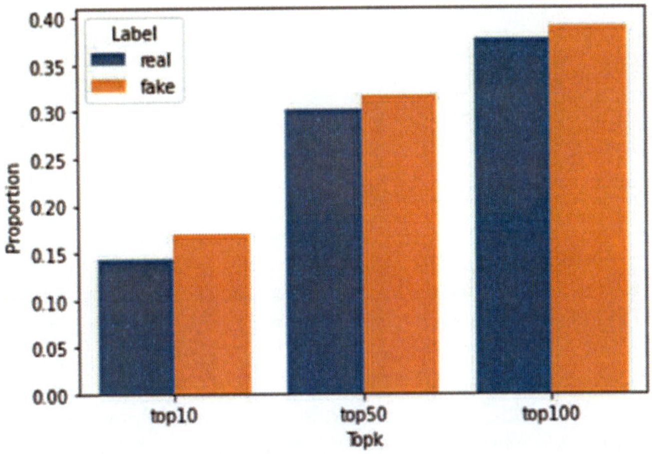

Fig. 1 Top frequency proposition of fake news and real news

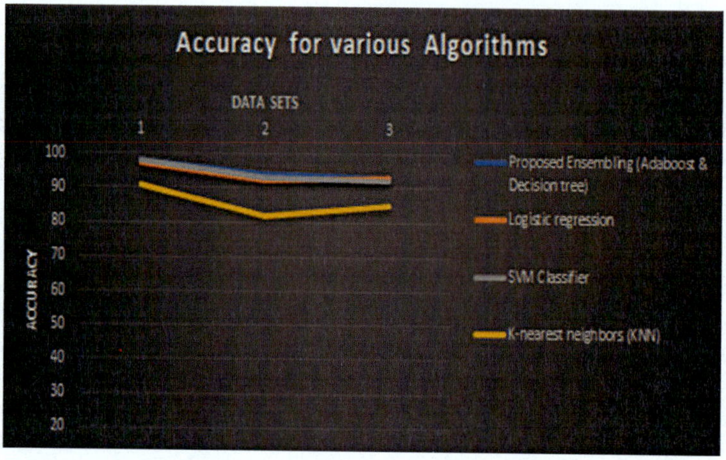

Fig. 2 Accuracy value of three datasets (DS1, DS2, and DS3) using the various algorithm

5 Results and Discussion

The result summarizes the accuracy achieved by each algorithm. This experiment is executed using the proposed algorithm ensemble technique and compared with three standard algorithms. Three datasets are considered for the executions, two datasets downloaded from Kaggle, named DS2 and DS3, and the third dataset DS1 is the combination of DS2 and DS3 [7–10].

Accuracy			
Proposed ensembling (AdaBoost and decision tree)	98	**94**	93
Logistic regression	97	92	93
SVM classifier	98	93	92
K-nearest neighbors (KNN)	91	82	85

From Fig. 2, it is a proof that the maximum accuracy achieved on DS1 is 98%, DS2 is 94%, and DS3 is 93% achieved by the proposed ensembling algorithm (AdaBoost and decision tree). SVM achieved an accuracy of 98%, 93%, and 93%, respectively, for DS1, DS2, and DS3. Logistic regression obtained 97%, 92%, and 93%, respectively, for DS1, DS2, and DS3 datasets, while K-nearest neighbors (KNN) algorithm achieved the least accuracy of 91%, 82%, and 85% for DS1, DS2, and DS3 datasets, respectively.

Precision			
Proposed ensembling (AdaBoost and decision tree)	95	91	92
SVM	93	72	70

(continued)

(continued)

Precision			
Logistic regression	98	74	95
K-nearest neighbors (KNN)	89	83	83

From Fig. 3, it is evident that the precision successfully achieved on DS1 is 95%, DS2 is 91%, and DS3 is 92% achieved by the proposed ensemble algorithm (AdaBoost and decision tree). SVM achieved 93%, 72%, and 70% accuracy, respectively, for DS1, DS2, and DS3. Logistic regression obtained 98%, 74%, and 95%, respectively, for DS1, DS2, and DS3 datasets, while K-nearest neighbors (KNN) algorithm achieved the least accuracy of 89%, 83%, and 83% for DS1, DS2, and DS3 datasets, respectively.

6 Conclusion

The problem of fake news is dangerous, and as information is made available to the public in various forms, it spreads like wildfire. We can control the spread of misinformation with the help of artificial intelligence techniques more efficiently. Data at each stage are analyzed, and then predictions are made. Many attempts were made to solve the misinformation problem. Because social media sites such as Facebook and Twitter collect many data, top models are getting better every day. This work has applied three standard algorithms KNN, SVM, and linear regression techniques and proposed an ensemble technique to detect fake news. The obtained result proves that the ensemble techniques get better results of 95% accuracy and

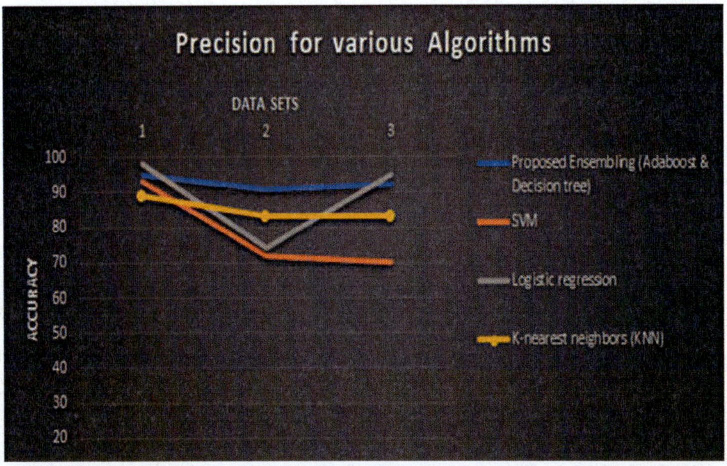

Fig. 3 Precision value of three datasets (DS1, DS2, and DS3) using the various algorithm

98% of precision. It shows that the proposed model can detect more accurately than any other method. The main challenge in fake news prediction may be incorrect data, which means that any predictive model may have anomalies and errors. In the future, various NLP techniques such as POS tags, word2vec can be used to improve the performance.

References

1. 1&1 IONOS Inc (2020) What is fake news? Definition, types, and how to detect them. 1&1 IONOS Inc, 27 July 2020. https://www.ionos.com/digitalguide/online-marketing/social-media/what-is-fake-news/
2. Agarwal V, Parveen Sultana H, Malhotra S, Sarkar A (2019) Analysis of classifiers for fake news detection. Proc Comput Sci. https://doi.org/10.1016/j.procs.2020.01.035
3. Ahmad I, Yousaf M, Yousaf S, Ahmad MO (2020) Fake news detection using machine learning ensemble methods. Complexity 2020 (October). https://doi.org/10.1155/2020/8885861
4. Hamdaoui Y (2019) TF(Term Frequency)-IDF(Inverse Document Frequency) from scratch in python. Towards data science, 10 Dec 2019. https://towardsdatascience.com/tf-term-frequency-idf-inverse-document-frequency-from-scratch-in-python-6c2b61b78558
5. How to Conduct Linear Regression (2010) 21 Dec 2010. https://www.statisticssolutions.com/free-resources/directory-of-statistical-analyses/how-to-conduct-linear-regression/
6. Khanam Z, Al Wasl IBN, Sirafi H, Rashid M (2021) Fake news detection using machine learning approaches. IOP Conf Ser: Mater Sci Eng. https://doi.org/10.1088/1757-899x/1099/1/012040
7. Ray S (2017) Understanding Support Vector Machine (SVM) algorithms from examples (along with code), 12 Sept 2017. https://www.analyticsvidhya.com/blog/2017/09/understaing-support-vector-machine-example-code/
8. Rithesh RN (2017) SVM-KNN: a novel approach to classification based on SVM and KNN. Int Res J Comput Sci. https://doi.org/10.26562/irjcs.2017.aucs10088
9. Varshney D, Vishwakarma DK (2020) Hoax news-inspector: a real-time prediction of fake news using content resemblance over web search results for authenticating the credibility of news articles. J Ambient Intell Humaniz Comput. https://doi.org/10.1007/s12652-020-02698-1
10. Wang WY (2017) Liar, liar pants on fire: a new benchmark dataset for fake news detection. In: Proceedings of the 55th annual meeting of the association for computational linguistics, vol 2, Short Papers. https://doi.org/10.18653/v1/p17-2067
11. Zhang J, Dong B, Yu PS (2019) Deep diffusive neural network-based fake news detection from heterogeneous social networks. In: 2019 IEEE international conference on big data (big data). https://doi.org/10.1109/bigdata47090.2019.9005556. IEEE 36th International Conference on Data Engineering (ICDE). https://doi.org/10.1109/icde48307.2020.00180

Hybrid Ant Particle Swarm Genetic Algorithm (APSGA) for Task Scheduling in Cloud Computing

Arzoo and Anil Kumar

Abstract In cloud computing technology, task scheduling is one of the research challenges. For these various algorithms, works such as particle swarm optimization (PSO), firefly algorithm, ant colony optimization (ACO) and genetic algorithm (GA). PSO is inspired by the bird's movement, and ACO is based on the behaviour of ants. GA works based on the natural evolution process. This paper presents the hybrid of PSO-ACO-GA for task scheduling on virtual machines of cloud computing known as ant particle swarm genetic algorithm (APSGA). Here, GA and PSO will perform iteration to get the task basis on fitness value and further ACO will distribute the task on specific virtual machines. This paper has achieved improved results for parameters such as CPU utilization, makespan and execution time. Our proposed algorithm has achieved makespan that is reduced by 27.1%, 19.45% and 21.24% with compare to PSO, ACO and GA, respectively. It has achieved maximum of CPU utilization and execution time.

Keywords Task scheduling · Cloud computing · Makespan · Nature-inspired algorithms · Resource utilization · Simulation · CloudSim

1 Introduction

In the research sector, cloud computing [1] is a famous technology. This technology is beneficial in various fields such as mobile systems, medical health care, resource scheduling, business and many more. Due to the requirement of research work, cloud computing technology is developing more and more. As this terminology has introduced new terms such as virtualization [2], with the assistance of this, multiple users distribute the resources of cloud computing to the various clients.

Cloud computing provides different services such as infrastructure as a service, security as a service, platform and software as a service [3]. These facilities make the cloud computing technology more valuable. These all services are charged from

Arzoo (✉) · A. Kumar
Guru Nanak Dev University, Amritsar, India
e-mail: arzoorana98@gmail.com

M. S. Kaiser et al. (eds.), *Information and Communication Technology for Competitive Strategies (ICTCS 2021)*, Lecture Notes in Networks and Systems 401,
https://doi.org/10.1007/978-981-19-0098-3_2

the user as per the usage. At present time, most of the industries are migrating on this technology.

To use this at the industry level, cloud computing technology has raised many issues and lightens up various research topics such as resource scheduling [4], task scheduling [5] and migration [6]. In this paper, we will propose a new technique to perform task scheduling.

Task scheduling [5] is considered as a major source to completely utilize the resources provided by the vendor. Task scheduling provides an efficient way to utilize the services of the cloud. This process mainly distributes the user requests to the various resources to maximize the CPU utilization [7] and to reduce the execution time [8]. Task scheduling is considered as of two types [9]: one is static and the second is dynamic. In static task scheduling, a scheduler knows the detail of task and resources in advance, whereas, in dynamic task scheduling, scheduler is unaware about the details that make the dynamic plans to distribute the tasks. In this paper, we proposed a method to dynamically schedule the tasks on virtual machines.

The main contributions of this paper are:

- This paper proposed ant particle swarm genetic-based task scheduler that helps to schedule the task.
- The main aim of this optimal scheduling algorithm to minimize the makespan, to complete the maximum number of tasks within the time quantum.
- The proposed method is implemented in CloudSim.
- The results are computed for makespan, execution time and CPU utilization.

2 Literature Survey

Kumar and Sharma [10], Lin et al. (2018). This paper has proposed a new algorithm that is named as PSO-COGENT. This algorithm is based on particle swarm optimization and is used to allocate resources to the working applications in efficient manner. This method has taken deadline as constraint, and it optimizes the time as well as cost. The author has proposed a mathematical framework that reduces the consumption of energy on cloud data centres. In this paper, the computed parameters are makespan, energy consumption, cost and throughput. The tool used is CloudSim to perform real-time simulation. In the last, they have compared the results with honeybee and traditional PSO.

Verma et al. [11] This paper reviewed traditional approaches and proposed a new method for balance the dynamic load on cloud datacentres that is based on ant colony optimization (ACO). In this paper, author has explained various issues, challenges and future scope of the technique used. For implementation, they have used CloudSim simulation kit that made the load balancing more efficient. In the end, advanced ACO is compared with traditional ACO.

Nilesh et al. [12], This paper has proposed improved ACO and compares this proposed algorithm with the traditional ACO. The simulation is performed on CloudSim simulation toolkit. The parameter computed here is makespan. This paper has performed the task scheduling along with balancing the load on cloud data centres.

Xiang et al. [13], This paper has proposed a new method for workflow scheduling. Here, ACO is used and named as Greedy ACO-based workflow scheduling. The author has implemented this algorithm to reduce the execution time. Heuristic information based on forwarding and backward dependency is adduced to build transition probability for ants to generate task priorities. At the same time, a greedy minimum strategy for machine allocation is incorporated to complete scheduling. Experiments on both synthetic graphs and real application graphs demonstrate the effectiveness of the Greedy-Ant algorithm. The results extracted that the proposed algorithm outperforms the state of the art up to 18% in the metric of speedup.

Chen [14], This paper has presented a fuzzy-based new algorithm. The author has proposed, two algorithms fuzzy geometric mean (FGM) and fuzzy weighted average, and these are based on alpha cut operations and ant colony optimization (ACO), respectively. Here, in this paper, the algorithm is applied to check the fuzzy level of the factor. In the end, results are compared and fuzzy weight term is derived.

Li and Wu [15], This paper focused on load balancing in system-wide information management (SWIM) task scheduling. The author has focused on large-scale network information method for completing the task initialized. The ACTS-LB algorithm improved the defect that pheromone updating can easily fall into the local optimum and ensures that the system's load balancing requirements were maximized while the scheduling task was effectively completed. The experimental setup is not performed in real. For implementation, only simulation is performed.

3 Nature-Inspired Task Scheduling Algorithm

3.1 Particle Swarm Optimization

Particle swarm optimization follows the normal conduct of birds rushing, and it is created in 1995 by Kennedy and Ederhart. Typically, the birds will run when they discover their food in a spot. This nature conduct of birds is utilized in cloud to track down the ideal VM to deal with the responsibility. It is accepted that the bird fly in the cloud space and their regular conduct of rushing while at the same time getting their food tracks down the best answer for getting the VM to deal with the responsibility. Each task tracks down its nearby best (Lbest) VM and furthermore keeps the data about worldwide best (Gbest) VM and the appropriate briefest way distinguished by the assignment around then. The undertakings are allocated with their position and speed with which they travel through a multidimensional cloud space [16]. Basic algorithm is shown in Fig. 1

Fig. 1 Particle swarm optimization [16]

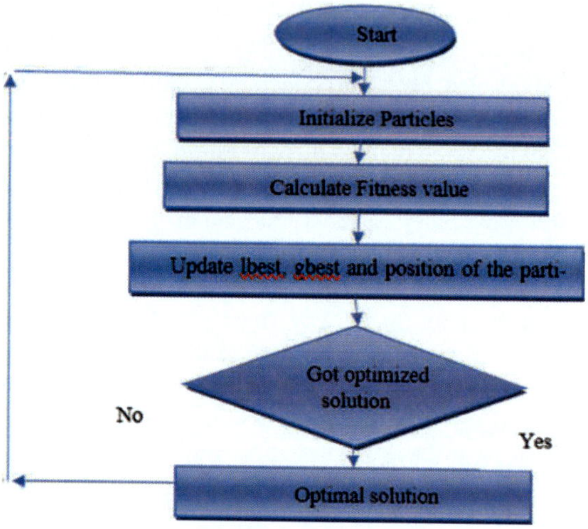

3.2 Ant Colony Optimization

Ant colony optimization is an algorithm inspired from the ant species those find out the shortest way to reach their homes from the food location [17]. In this method, at every iteration, ants find out the best move to go for completion of whole task and put the pheromone. Here, other ants follow the same path as shown in the algorithm steps in Fig. 2.

3.3 Genetic Algorithm

Genetic algorithm is used for simulation of problems basis on natural evolution [19]. This process follows the steps such as random generation of population, evaluation, crossover, selection, regeneration and mutation [20]. The algorithm works in multiple fields like task scheduling, travelling salesman problem, resource scheduling and many more. The main goal of this algorithm is to find out the fitness of each task on a virtual machine and further division of tasks basis on fitness functions. The main problem faced with this algorithm is need of high processing time. The flow of individual GA is shown in Fig. 3

3.4 Hybrid of PSO, ACO and GA

There are several works based on hybrid of two algorithms among PSO, ACO and GA. An author of [22] has used hybrid of GA and ACO for travelling salesman

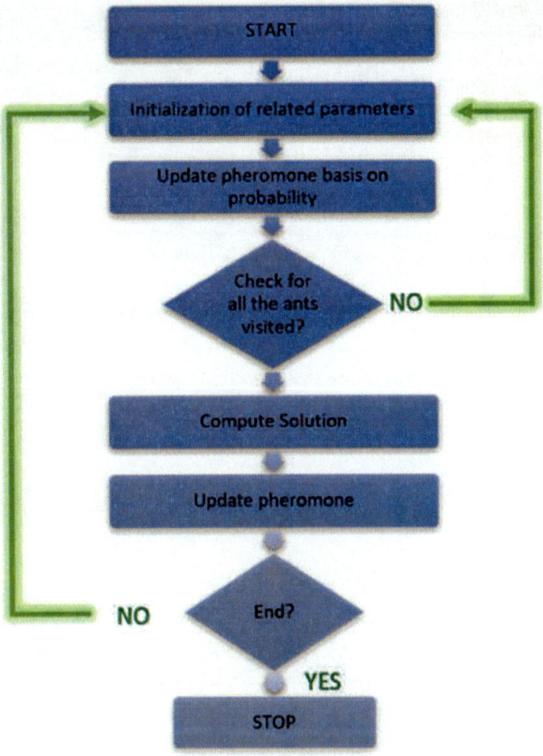

Fig. 2 Ant colony optimization [18]

problem. In [14], author has used genetic algorithm and bacterial foraging algorithm to minimize makespan and energy consumption. This section discusses the framework of proposed method shown in Fig. 4 that presents a scheduler named as ant particle swarm genetic algorithm (APSGA). Figure 5 discusses the detailed flow of the proposed work. First portion works as particle swarm optimization. The GA and PSO are used to divide the task, and ACO is used here to get the optimal solution for scheduling the tasks.

4 Methodology

This section shows the steps of the proposed algorithm.

Step1. INPUT: Task set $\{t1, t2, t3 \ldots. tn\}$.

Step2. OUTPUT: Arrange the task based on proposed algorithm.

1. Initialize all the tasks.
2. Repeat
   a. Calculate fitness value.

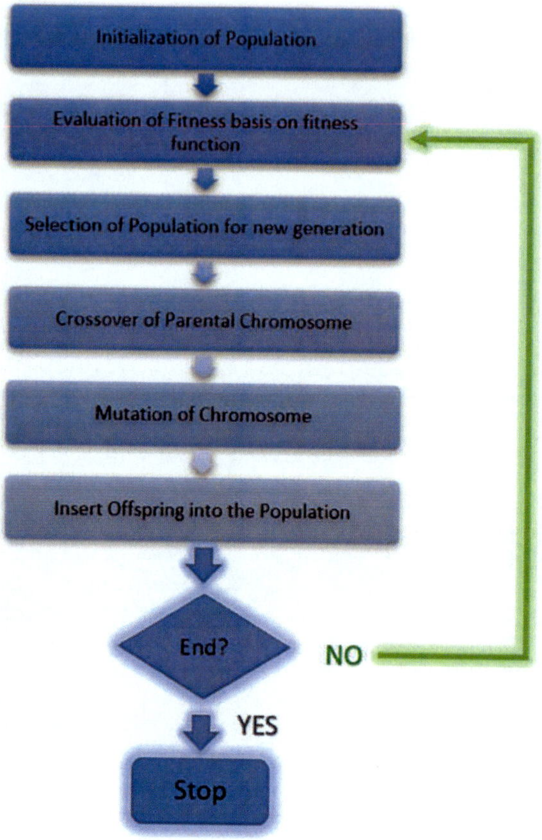

Fig. 3 Genetic algorithm [21]

   b. Determine optimal fitness value.
   c. Divide the task into half.

3. One set with best fitness value and other set with worst fitness value.
4. Find parameters such as selection, crossover and velocity.
   a. Generate the vectors of tasks.
5. To apply ant colony optimization
   a. Distribute tasks
      Repeat

      Perform probability on each ant.
         Perform probability on each ant.
      Create another vector.
6. Get arranged tasks.
7. Calculate makespan, CPU utilization and execution time.

Fig. 4 Framework of proposed work

8. Comparison with other algorithms such as ACO, PSO and GA.

5 Experimental Setup

For this experiment, CloudSim as cloud simulator is used to evaluate the proposed algorithm. This runs on Intel i5-1035G1 1.19 GHz. The proposed technique is used for computation and then compared with individual GA, PSO and ACO (Table 1).

6 Results

Simulation shows that our hybrid approach performed better than ACO, GA and PSO. In the following, Fig. 6 shows the simulation in CloudSim where various tasks are assigned to virtual machines.

The various parameters are computed in this paper:

Makespan: Makespan is the time required for processing all the input tasks.

$$\text{Makespan} = \max\{\text{Finish time of all the tasks}\} \tag{1}$$

Execution time: It is the time spent while executing all the tasks.

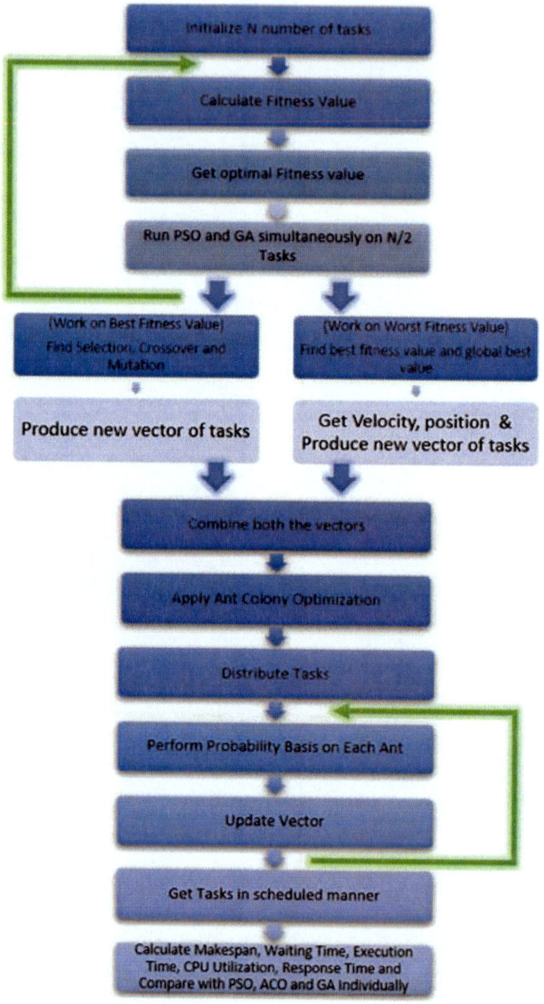

Fig. 5 Flowchart of proposed algorithm

$$\text{ETavg} = \frac{1}{n}\sum_{i=1}^{n} \text{ET}(i) \tag{2}$$

$$\text{ET(each virtual machine)} = \frac{\text{Total length of virtual machine operating the tasks}}{\text{Number of processors} * \text{speed}} \tag{3}$$

CPU utilization: It is defined as the amount of usage of processor while processing the tasks. CPU utilization has been computed in CloudSim using the function get UtilizationOfCpu(). It returns the CPU usage in percentage. The average makespan of the ACO, GA, PSO and proposed algorithms are shown in Fig. 7. With the increase

Table 1 Settings of cloud simulator

Data type	Parameters used	Their values
Virtual machine	Number of virtual machines	100
	Number of Pepper VM	1–3
	Mips of processing element	512–1024
	Bandwidth	500–1200
	Memory	512–2048
	Storage	100,000–800,000
	Unit cost of VM	1–10
Task	Number of task	100–500
	Length of each task	5000–100,000
	File size	300–5000
DataCenter	Number of datacenter	5
	Number of host	30

of the quantity task, proposed algorithm takes the timeless than other metaheuristic algorithms. So, here proposed algorithms work better. In Fig. 8, the CPU utilization is compared while scheduling the different number of tasks. It results that proposed algorithm utilizes the CPU in better manner. Figure 9 shows the execution time for scheduling the tasks using all the algorithms, and it shows proposed work requires less execution time and gives better throughput.

7 Conclusion and Future Scope

In this paper, hybrid of ACO, PSO and GA algorithm is performed for achieving task scheduling in cloud computing. Here the values of makespan, execution time and CPU utilization are compared for the proposed algorithm with the metaheuristic algorithms such as GA, ACO and PSO for the tasks varying from 100 to 500. Simulation provides the result that gives the proposed algorithm performed better than others. For the future work, we can enhance it for green computing and compare with other algorithms.

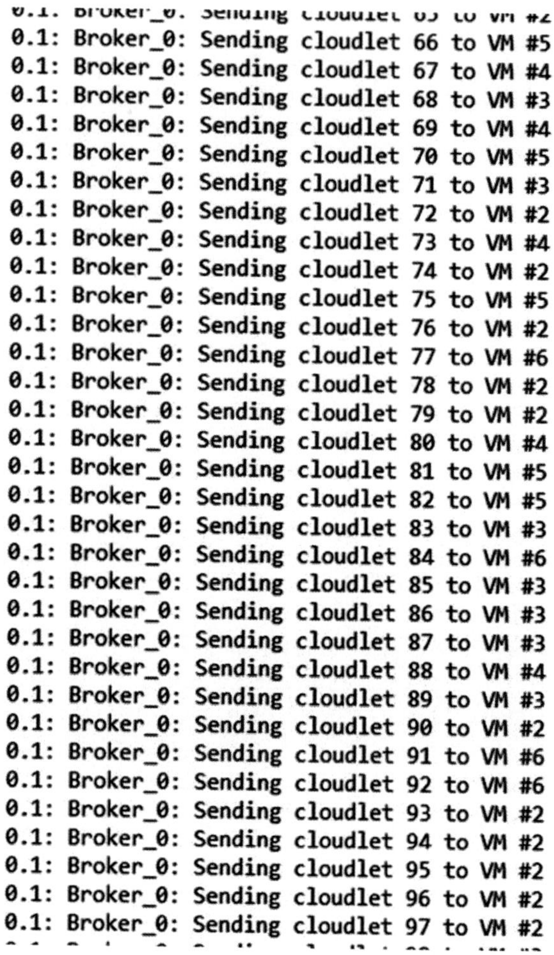

```
0.1: Broker_0: Sending cloudlet 66 to VM #5
0.1: Broker_0: Sending cloudlet 67 to VM #4
0.1: Broker_0: Sending cloudlet 68 to VM #3
0.1: Broker_0: Sending cloudlet 69 to VM #4
0.1: Broker_0: Sending cloudlet 70 to VM #5
0.1: Broker_0: Sending cloudlet 71 to VM #3
0.1: Broker_0: Sending cloudlet 72 to VM #2
0.1: Broker_0: Sending cloudlet 73 to VM #4
0.1: Broker_0: Sending cloudlet 74 to VM #2
0.1: Broker_0: Sending cloudlet 75 to VM #5
0.1: Broker_0: Sending cloudlet 76 to VM #2
0.1: Broker_0: Sending cloudlet 77 to VM #6
0.1: Broker_0: Sending cloudlet 78 to VM #2
0.1: Broker_0: Sending cloudlet 79 to VM #2
0.1: Broker_0: Sending cloudlet 80 to VM #4
0.1: Broker_0: Sending cloudlet 81 to VM #5
0.1: Broker_0: Sending cloudlet 82 to VM #5
0.1: Broker_0: Sending cloudlet 83 to VM #3
0.1: Broker_0: Sending cloudlet 84 to VM #6
0.1: Broker_0: Sending cloudlet 85 to VM #3
0.1: Broker_0: Sending cloudlet 86 to VM #3
0.1: Broker_0: Sending cloudlet 87 to VM #3
0.1: Broker_0: Sending cloudlet 88 to VM #4
0.1: Broker_0: Sending cloudlet 89 to VM #3
0.1: Broker_0: Sending cloudlet 90 to VM #2
0.1: Broker_0: Sending cloudlet 91 to VM #6
0.1: Broker_0: Sending cloudlet 92 to VM #6
0.1: Broker_0: Sending cloudlet 93 to VM #2
0.1: Broker_0: Sending cloudlet 94 to VM #2
0.1: Broker_0: Sending cloudlet 95 to VM #2
0.1: Broker_0: Sending cloudlet 96 to VM #2
0.1: Broker_0: Sending cloudlet 97 to VM #2
```

Fig. 6 Simulation of task to virtual machines

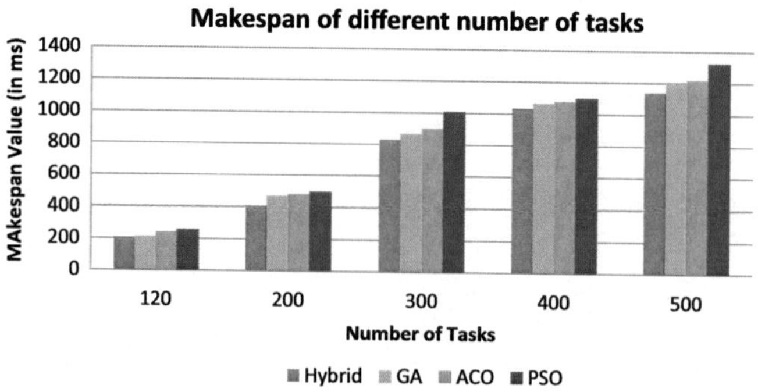

Fig. 7 Makespan of proposed algorithm and its comparison with GA, PSO and ACO basis on different range of tasks

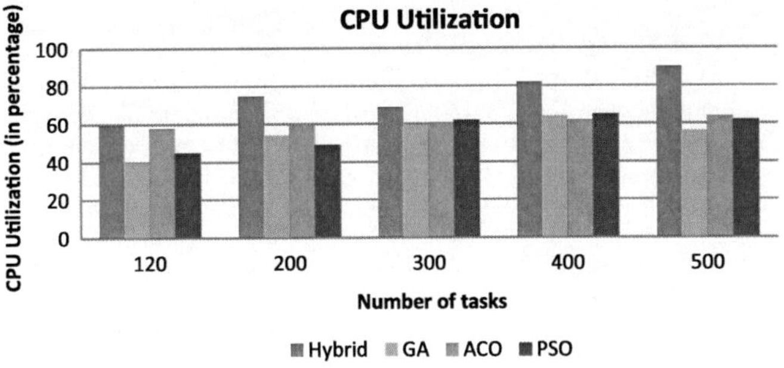

Fig. 8 CPU utilization with GA, ACO, PSO and proposed algorithm

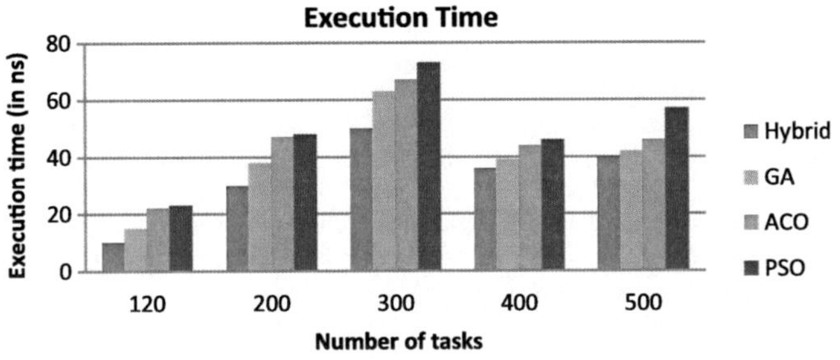

Fig. 9 Execution time with GA, ACO and PSO algorithm

References

1. Dillon T, Wu C, Chang E (2010) Cloud computing: issues and challenges. In: 2010 24th IEEE international conference on advanced information networking and applications. IEEE, pp 27–33
2. Campbell S, Jeronimo M (2006) An introduction to virtualization. Published in "Applied Virtualization", Intel, 1–15
3. Zhao G, Rong C, Jaatun MG, Sandnes FE (2012) Reference deployment models for eliminating user concerns on cloud security. J Supercomput 61(2):337–352
4. Singh S, Chana I (2016) A survey on resource scheduling in cloud computing: issues and challenges. J Grid Comput 14(2):217–264
5. Singh RM, Paul S, Kumar A (2014) Task scheduling in cloud computing. Int J Comput Sci Inf Technol (IJCSIT) 5(6):7940–7944
6. Boutaba R, Zhang Q, Zhani MF (2014) Virtual machine migration in cloud computing environments: benefits, challenges, and approaches. In: Communication infrastructures for cloud computing. IGI Global, pp 383–408
7. Ergu D, Kou G, Peng Y, Shi Y, Shi Y (2013) The analytic hierarchy process: task scheduling and resource allocation in cloud computing environment. J Supercomput 64(3):835–848
8. Jiao H, Zhang J, Li J, Shi J, Li J (2015) Immune optimization of task scheduling on multidimensional Qos constraints. Clust Comput 18(2):909–918
9. Arunarani AR, Manjula D, Sugumaran V (2019) Task scheduling techniques in cloud computing: a literature survey. Futur Gener Comput Syst 91:407–415

10. Kumar M, Sharma SC, Goel A, Singh SP (2019) A comprehensive survey for scheduling techniques in cloud computing. J Netw Comput Appl 143:1–33
11. Verma P, Shrivastava S, Pateriya RK (2017) Enhancing load balancing in cloud computing by ant colony optimization method. Int J Comput Eng Res Trends 4(6):277–284
12. Nilesh AM, Patel CA (2017) Load balancing in cloud computing using ant colony optimization. Int J Comput Eng Technol 8(6)
13. Xiang B, Zhang B, Zhang L (2017) Greedy-ant: ant colony system-inspired workflow scheduling for heterogeneous computing. IEEE Access 5:11404–11412
14. Srichandan S, Kumar TA, Bibhudatta S (2018) Task scheduling for cloud computing using multi-objective hybrid bacteria foraging algorithm. Future Comput Informatics J 3(2):210–230
15. Li G, Wu Z (2019) Ant colony optimization task scheduling algorithm for SWIM based on load balancing. Future Internet 11(4):90
16. Naseri A, Navimipour NJ (2019) A new agent-based method for QoSaware cloud service composition using particle swarm optimization algorithm. J Ambient Intell Hum Comput 10(5):1851–2186
17. Dorigo M, Maniezzo V, Colorni A (1996) The ant system: optimization by a colony of cooperating agents. IEEE J Syst Man Cybernetics 26(1):1–13
18. Al Salami NM (2009) Ant colony optimization algorithm. UbiCC J 4(3):823–826
19. Raza Z, Vidyarthi DP (2009) A computional grid scheduling model to minimize turnaround using modified GA. Int J Artif Intell 3(A9):86–106
20. Banga VK, Singh Y, Kumar R (2007) Simulation of robotic arm using genetic algorithm and AHP. J Eng Technol 25:95–101
21. Zhao C, Zhang S, Liu Q, Xie J, Hu J (2009) Independent tasks scheduling based on genetic algorithm in cloud computing. In: 2009 5th international conference on wireless communications, networking and mobile computing. IEEE, pp 1–4
22. Shang G, Xinzi J, Kezong T (2007) Hybrid algorithm combining ant colony optimization algorithm with genetic algorithm. In: Proceedings of the 26th Chinese control conference

Anonymization of Multi-relations Datasets Using Single Table Algorithms

Nisha, Archana Singhal, and Sunil Kumar Muttoo

Abstract Nowadays, immense educational data is available for analysis. To extract useful information from huge data, data analysis is required. For analysis, data has to be shared with a third party called data analyst. Sharing data with data analyst introduces the risk of privacy for individuals whose sensitive information is present in the shared data. Privacy is the right of an individual, so it needs to be maintained while sharing the data with a third party. There are many privacy-preserving algorithms available. k-anonymization is one of the most popular and traditional privacy-preserving methods. It modifies data to protect an individual's sensitive information by disconnecting the link between an individual and its sensitive information. k-anonymization uses generalization and suppression tools to anonymize the data. Generalization replaces specific values of attributes with general values, and suppression deletes unmatched values of attributes. There are many existing single table anonymization algorithms. The proposed work presents how a single table anonymization algorithm can be applied on multi-relation dataset. It also discusses the problems encountered when single table anonymization algorithms are used for the multi-relation dataset. Present work also discusses the limitations of k-anonymization and gives future research directions.

Keywords Learning analytics · Educational data · Privacy · Anonymization · Utility · Multi-relational

1 Introduction

Huge educational data is available in today's world which contains valuable information. Learning analytics is used to collect, measure and analyse educational data

Nisha (✉) · S. K. Muttoo
Department of Computer Science, University of Delhi, Delhi, India
e-mail: nisha1988.d@gmail.com

A. Singhal
IP College for Women, University of Delhi, Delhi, India
e-mail: asinghal@ip.du.ac.in

M. S. Kaiser et al. (eds.), *Information and Communication Technology for Competitive Strategies (ICTCS 2021)*, Lecture Notes in Networks and Systems 401,
https://doi.org/10.1007/978-981-19-0098-3_3

of students to enhance their educational practices. It extracts valuable information from educational data with the help of analytical tools that can benefit the stakeholder: family, educators and students to improve learning processes [1]. It also helps in developing a predictive model based on past educational data to improve future practices. The main goal of learning analytics is to increase the completion rate and drop the failure rate of students by guiding them in their learning processes as and when required.

Learning analytics also helps researchers to ground their research in learning on real educational data. For harvesting valuable information from educational data, it needs to be shared with a third party called a data analyst. Although sharing and using of data introduce privacy risks for data subjects, privacy is the right of an individual to secure his/her personal information from unauthorized access [1]. Data privacy is the main concern of sharing the sensitive information of students with a third party. There are many privacy-preserving techniques available for educational data. One of them is data publishing, in which the data curator anonymizes the dataset and shares the anonymized data with the data analyst [2]. The goal of anonymization is to transform the dataset in such a format that privacy is enforced. k-anonymization is one of the most popular privacy-preserving techniques available. It states that "each record in the published dataset needs to be indistinguishable from at least $k - 1$ other records concerning to quasi-identifiers (QI) values" [1]. Quasi-identifiers-wise groups of indistinguishability are called equivalence classes [2]. The main issue of k-anonymization is that it does not consider the distribution of sensitive values within the equivalence class.

The l-diversity is another privacy-preserving technique that is also required along with k-anonymization to provide privacy to educational data. L-diversity takes care of the distribution of sensitive attributes values within the equivalence class. In l-diversity, every equivalence class is required to contain l well-represented values for each sensitive attribute [2].

The utility of data is another important factor along with privacy. So, it is also very important to maintain the balance between utility and privacy of educational data.

Many algorithms have been proposed for k-anonymization with single relation [3–5]. These all algorithms assume that each entity is stored as a single tuple record in a single relation dataset. Existing algorithms fail when information of each entity does not store in single relation and require a multi-relation dataset [3].

Current work discusses single table algorithms for multi-relation databases. It also discusses problems and their solutions, when single table algorithms are applied on multi-relation databases.

This paper is divided into six sections. Section 2 describes the literature survey. Section 3 provides the introduction of k-anonymization. Section 4 discusses k-anonymization single table algorithms for the multi-relational dataset. Section 5 discusses proposed work. Section 6 presents the conclusion and the future work.

2 Related Work

In [2], detailed literature review has been provided and challenges of learning analytics also discussed. According to researchers, aim of learning analytics is to improve the learning experience and practices. Author also discussed different privacy-preserving techniques which can be applied on multi-relational dataset. In [1], k-anonymization technique has been discussed which anonymizes dataset to provide privacy. k-anonymization can be applied on single relational dataset as well as multi-relational dataset [6].

Data analysis is used to analyse data and extract useful information from it. After pre-processing, data can be stored in a single table or multi-relational dataset. Many anonymization algorithms exist that can be applied to a single table. In [3, 4], single table algorithms are used for anonymization. Existing algorithms work on the notion that every private entity is stored in a single tuple. When data related to an entity cannot be stored in a single table and require multiple tables, then existing algorithms do not perform adequately. Some algorithms also use data pruning for reducing the size of data and efficient k-anonymization [5, 7].

In a multi-relational dataset, search spaces are large and require some assumptions for the processing which reduce optimality. In [8, 9] multi-relational anonymity has been applied and showed that k-anonymity provides more flexibility in anonymization. Clustering-based algorithms are more appropriate for multi-relation anonymization [9].

The present work discusses problems, when single table anonymization algorithms are applied on multi-relation databases, and also provides the solutions to these problems.

3 k-anonymization

As discussed earlier, there are many benefits of collecting, utilizing and sharing educational data. Collected data also has sensitive information of individuals which raises privacy concerns. In k-anonymization, the data owners anonymize the dataset and share it with the data analyst. Anonymization provides privacy by modifying the dataset. It uses generalization and suppression to convert specific data into more generalized data. It states that each record needs to be indifferent from at least $k - 1$ records concerning to quasi-identifiers value in the published dataset. Indifferent groups of quasi-identifiers are called equivalence classes [1].

Table 1 shows the tabular form of educational dataset. Data is arranged in the table using different attributes. Quasi-identifiers are the linking attributes that can identify the individual with the help of external databases. For example, in Table 1, age, gender and course attributes are the quasi-identifiers. Sensitive attributes are those attributes which an individual does not want to share with others. For example, the grade is a sensitive attribute. Table 1b has three equivalence classes each represented

Table 1 Single table dataset, **a** private dataset, **b** 2-anaonymous dataset

(a)

Age	Gender	Course	Grade
21	Male	Computer Science	A+
22	Male	Computer Science	A+
18	Male	Electrical Engineering	B
18	Female	Mechanical Engineering	C
22	Female	English	A-
23	Female	History	B+

(b)

Age	Gender	Course	Grade
20-25	Male	Computer Science	A+
20-25	Male	Computer Science	A+
18	*	Engineering	B
18	*	Engineering	C
20-25	Female	Arts	A-
20-25	Female	Arts	B+

in a different colour. Table 1a shows the private original dataset, and Table 1b shows the 2-anonymous version of the original dataset. In Table 1b, generalization changes specific values of age, gender and course attributes into general values. For example, in tuples 1 and 2, value of the age attribute is generalized to range 20–25. Suppression replaces unmatched values of gender attribute for tuples 3 and 4 with "*".

Although generalized data provides privacy to students, it reduces the utility of data. If data is more generalized, no useful information can be extracted from it. It is very important to maintain a balance between utility and privacy of the data. Anonymization maintains the balance between utility and privacy of data using loss metrics.

The next section discusses some methods of single table algorithms on multi-relation databases.

4 Multi-relation Anonymization Using a Single Table

This section discusses methods for multi-relation anonymization using a single table. In every algorithm, multi-relation dataset is converted into one or more single tables before anonymization.

4.1 Universal Table Anonymization

In this method, multi-relation database is converted into a single universal table, and anonymization is applied on this universal table. Table 2 shows a multi-relational database. For anonymization, attributes are identified, which require to be modified. The solution is to choose quasi-identifiers. For example, Table 3b shows the anonymized results of Table 3a in which generalization and suppression are applied

Table 2 Original multi-relational database

Sid	GPA
S1	7.06
S2	6
S3	8.2
S4	8.9

SCid	Sid	Course	Grade
SC1	S1	Mathematics	93
SC2	S1	English	91
SC3	S1	Physics	85
SC4	S2	Computer Science	78
SC5	S2	Hindi	78
SC6	S2	Physics	62
SC7	S3	History	42
SC8	S3	Religion	80
SC9	S3	Mathematics	67
SC10	S4	History	68
SC11	S4	Religion	85

SCid	Book	Price
SC1	Discrete	$68
SC2	English Grammar	$78
SC3	Text Book	$67
SC4	Fundamentals	$65
SC5	Hindi Grammar	$68
SC6	Textbook	$78
SC7	System of Mughal	$67
SC8	AM. History	$65
SC9	Calculus	$80
SC10	System of Mughal	$67
SC11	AM. History	$65

on course attribute. Although Table 3b does not provide privacy because if an adversary knows that a student studies Mathematics, English and Physics, then it can easily be mapped to student S1 who study two science and one language course.

The reason for the failure of single universal table anonymization is that an entity has multiple entries in the table. For example, information of student S1 is stored in three rows in Table 3a which is treated as three different entities during anonymization. It is also possible that rows of the same student can get anonymized with each other. So, it does not provide privacy. One solution to this problem is to delete data of joining attributes like Sid, which leads to information loss. It can be observed from the table that student S2 studies three subjects Maths, Hindi and Physics. It

Table 3 **a** Single universal table, **b** anonymized table

(a)

Sid	Course	GPA
S1	Mathematics	7.06
S1	English	7.06
S1	Physics	7.06
S2	Computer Science	6
S2	Hindi	6
S2	Physics	6
S3	History	8.2
S3	Religion	8.2
S3	Mathematics	8.2
S4	History	8.9
S4	Religion	8.9

(b)

Sid	Course	GPA
S1	Science	7.06
S1	Language	7.06
S1	Physics	7.06
S2	Science	6
S2	Language	6
S2	Physics	6
S3	History	8.2
S3	Religion	8.2
*	*	*
S4	History	8.9
S4	Religion	8.9

means Sid has a 1-N relationship with the course attribute. If joining attribute Sid gets deleted, then these relationships would also be lost.

The next section presents another method of anonymization using bitmap.

4.2 *Bitmap Anonymization*

The bitmap method uses Boolean values to store the data. It stores data of every entity in a single tuple. It uses value "1" for an attribute that exists for an entity; otherwise, it stores "0". Table 4a represents the bitmap form of Table 2 dataset. Attribute possesses by the entity represented by "1" in Table 4a; otherwise, "0" is stored for an attribute that does not exist for an entity. Table 4b shows a 2-anonymized dataset of Table 4a.

Table 4 **a** Bitmap representation of Table 2, **b** 2-anonymization of Table 4a

(a)

Sid	Maths		English	Physics	Hindi	Computer Science	History	Religion	GPA
	Dis	Cal	Re	TB	HG	Fund	SM	AH	
S1	1	0	1	1	0	0	0	0	7.06
S2	0	0	0	1	1	0	1	0	6.0
S3	0	1	0	0	0	0	1	1	8.2
S4	0	0	0	0	1	0	0	0	8.9

(b)

Sid	Maths		English	Physics	Hindi	Computer Science	History	Religion	GPA
	Dis	Cal	Re	TB	HG	Fund	SM	AH	
S1	*	0	*	1	*	0	*	0	7.06
S2	*	0	*	1	*	0	*	0	6.0
S3	0	*	0	0	*	0	*	*	8.2
S4	0	*	0	0	*	0	*	*	8.9

It shows that k-anonymization can be applied to the bitmap representation of some databases.

So, multi-relation databases can be converted into bitmap representation, and k-anonymization can also be applied to bitmap representation. There are some shortfalls of this approach. First, generally, data is stored in relational databases and bitmap conversion adds additional cost, time and storage. Second, every multi-relation database cannot be converted into a bitmap representation. In databases where an entity maps to another entity uncertain number of times, it cannot be converted into bitmap representation. Third, in bitmap representation, suppression removes the certainty of tuples. For example, S1 studies Physics and S2 studies Chemistry that can be generalized into science courses which means both students are taking at least one science course. Fourth, it does not consider the similarity between entities of nested tables.

Next section presents in detail the solutions proposed for these limitations.

5 Proposed Work

This section proposes multi-relation anonymization to overcome the limitations of approaches discussed in the previous section. An anonymized database maintains the dependency preservation of the original database. It fails inference attack by an adversary. It also maintains the semantic of the database. Multi-relation anonymizations take some assumptions for the schema. This technique assumes that every table should have one foreign key and no primary and foreign key can be a quasi-identifier.

In multi-relation anonymization, data can be organized in hierarchical representation. Figure 1a shows the hierarchical representation of Table 2 for students S1 and S2. Figure 1b shows the anonymized result of Fig. 1a. In Fig. 1a, every level of the tree shows the nested relation of an entity.

As discussed, earlier generalization and suppression are used for anonymization. Generalization uses domain generalization hierarchy of attributes to change general values to specific values as shown in Fig. 2. For example, in Fig. 1a, S1 has three subjects {Mathematics, English, Physics}, and S2 has {Computer Science, Hindi, Physics}. In domain generalization hierarchy of course in Fig. 2a, Mathematics and Computer Science come under one parent named Science. So, values of Mathematics and Computer Science are replaced with science in Fig. 1b. Suppression deletes values of a node with unmatched nodes. For example, in Fig. 1a, Mathematics and Computer Science courses have different books. So, books for both subjects are replaced with "*" in Fig. 1b.

Table 5 shows the k-anonymized result of Table 2 for the value of $k = 2$. "*" represents suppressed values. Generation is did with the help of domain generalization hierarchy as explained above. Multi-relation anonymization algorithms provide privacy to data as well as maintain the utility of data. It also considers the nested relations of entities.

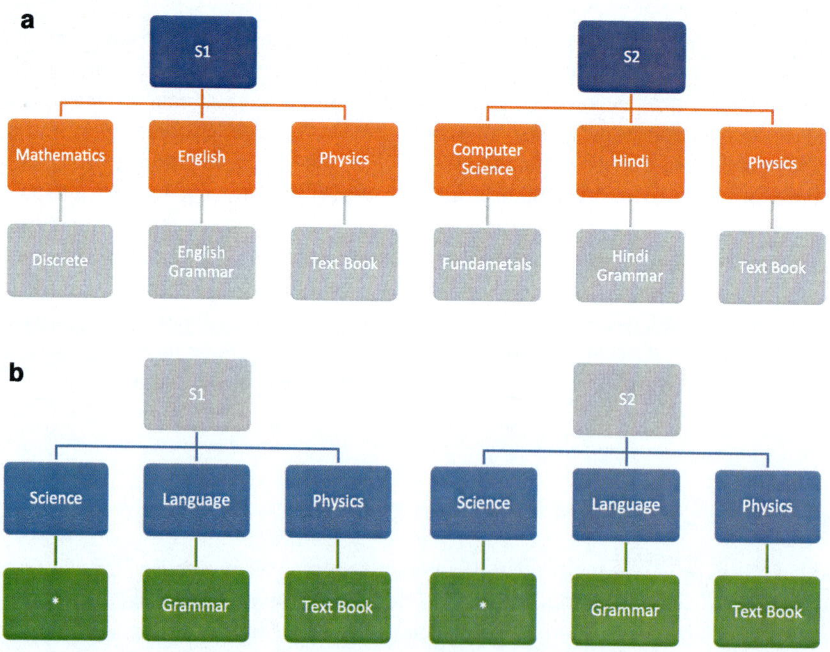

Fig. 1 **a** Hierarchal representation of Table 2 for students S1 and S2, **b** hierarchal representation of anonymized data of students S1 and S2

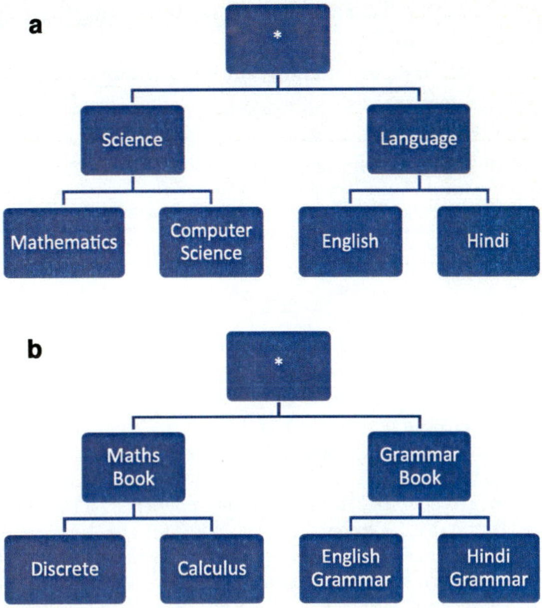

Fig. 2 **a** Domain generalization hierarchy for course, **b** domain generalization hierarchy for book

Table 5 Anonymized multi-relational database of Table 2

Sid	GPA
S1	7.06
S2	6.0
S3	8.2
S4	8.9

SCid	Sid	Course	Grade
SC1	S1	Science	93
SC2	S1	Language	91
SC3	S1	Physics	85
SC4	S2	Science	78
SC5	S2	Language	78
SC6	S2	Physics	62
SC7	S3	History	42
SC8	S3	Religion	80
*	*	*	*
SC10	S4	History	68
SC11	S4	Religion	85

SCid	Book	Price
*	*	*
SC2	Language Book	$78
SC3	Text Book	$67
*	*	*
SC5	Language Book	$68
SC6	Text Book	$78
SC7	System of Mughal	$67
SC8	AM. History	$65
*	*	*
SC10	System of Mughal	$67
SC11	AM. History	$65

6 Conclusion and Future Scope

To maintain the privacy of the educational data, it needs to be anonymized. Present work discusses how single table anonymization methods can be used for the multi-relation dataset. A hierarchical representation has been suggested to overcome the problems of single table anonymization methods when applied on the multi-relation dataset.

Although k-anonymization is a popular privacy-preserving method, it has some limitations. It does not consider the distribution of sensitive attributes values within the equivalence class. It is also vulnerable to homogeneity and background knowledge attacks.

L-diversity is another privacy-preserving algorithm that overcomes the drawbacks of k-anonymization. This work can be extended by applying l-diversity on the k-anonymized dataset. Arbitrary schema representations can also be considered in future in place of hierarchical representation [10].

References

1. Gursoy ME, Inan A, Nergiz ME, Saygin Y (2017) Privacy-preserving learning analytics: challenges and techniques. IEEE Trans Learn Technol 10(1)
2. Nisha AS, Muttoo SK (2020) Learning analytics: a literature review and its challenges. In: Proceedings of the 5th international conference on Information and Communication Technology for Competitive Strategies (ICTCS). Springers (Scopus Indexed)
3. Fung KWB, Yu P (2005) Top-down specialization for information and privacy preservation. In: Proceedings 21st International Conference on Data Engineering (ICDE)
4. Nergiz ME, Clifton C (2006) Thoughts on k-anonymization. In: Proceedings of 22nd International Conference on Data Engineering Workshops (ICDEW '06), p 96
5. LeFevre K, DeWitt D, Ramakrishnan R (2005) Incognito: efficient full-domain k-anonymity. In: Proceedings of ACM SIGMOD '05, June
6. Gasevic D, Siemens G, Rose CP (2017) Guest editorial: special section on learning analytics. IEEE Trans Learn Technol 10(1):3–5
7. Ozalp I, Gursoy ME, Nergiz ME, Saygin Y (2016) Privacy-preserving publishing of hierarchical data. ACM Trans Privacy Security 19(3), Article 7
8. Domingo-Ferrer J, Torra V (2005) Ordinal, continuous and heterogeneous k-anonymity through microaggregation. Data Min Knowl Disc 11(2):195–212
9. .LeFevre K, DeWitt DJ, Ramakrishnan R (2006) Mondrian multidimensional k-anonymity. In: Proceedings of 22nd International Conference on Data Engineering (ICDE'06), pp 25–35. https://doi.org/10.1109/ICDE.2006.101
10. Nergiz ME, Clifton C, Nergiz AE (2009) Multirelational k-anonymity. IEEE Trans Knowl Data Eng 21(8)

Analytical Performance of QO-STBC Transmission for Shadowed-Rician (SR) Fading Channels

Abhishek, M. K. Arti, Aarti Jain, and Prateek Anand

Abstract This paper presents investigation and analysis of the analytical performance of QO-STBCs transmitted over SR fading channels with M-PSK constellations. Specifically, we derive signal-to-noise ratio for four symbols transmitted by antennas employed at earth station (ES) for instantaneous and identically distributed SR fading channels. The basic approach behind this study is to evaluate the analytical performance of QO-STBC for land mobile satellite (LMS) communication system. Following this, by using the SNR, we further derive moment generating function (MGF) in terms of Meijer's G-function. Using derived MGF, we evaluate SER of the QO-STBC-based LMS communication systems under various shadowing conditions with M-PSK constellations.

Keywords Quasi-orthogonal space-time block code (QO-STBC) · Instantaneous and identically distributed (i.i.d) shadowed-Rician fading channels · land mobile satellite (LMS) · M-ary phase shift keying (M-PSK) · Multiple-input multiple-output (MIMO)

1 Introduction

In an era of digital revolution, the satellite situated communication system plays vital role in the field of wireless communication. Nowadays, satellite stationed terrestrial communication systems attract the attention of mobility service developers to exploits its serving capabilities of communication system in terms of large coverage and connectivity such as spanning the earth sensing arena, in military surveillance and in commercial sector, offers vital global connectivity and recognizing capabilities [1, 2]. For preliminary explanation of the advantages of multiple-input multiple-output (MIMO) revealed that increased transmit diversity, coverage arena

Abhishek (✉) · M. K. Arti · A. Jain
Electronics and Communication, NSUT East Campus, Geeta colony, Delhi, Delhi 110031, India
e-mail: abhichaudhary79915@gmail.com

P. Anand
Electronics and Communication, GGSIPU, Dwarka, Delhi 110078, India

M. S. Kaiser et al. (eds.), *Information and Communication Technology for Competitive Strategies (ICTCS 2021)*, Lecture Notes in Networks and Systems 401,
https://doi.org/10.1007/978-981-19-0098-3_4

and capacity accelerate the swiftness of developments in the existing technology that attracts researchers toward hybrid satellite-terrestrial communication systems which lead to its benefits in wireless communication system [3–6]. To reach effective data rates in advance telecommunication technologies based system, long-term evolution (LTE) standard motivates the developers to study and design the MIMO-based satellite communication system.This research introduces a large diversity-order blueprint design of MIMO with QO-STBC transmission scheme-based system that reaches better capacitive performance and low bit-error rate (BER) performances than the conventional QO-STBC configurations [7, 8]. However some former MIMO standards were design on an OSTBC, which restricted to two transmit antennas and lower data rate, whereas to manage the requirement for higher data rates by avoiding the channel state information (CSI) that encourages the consideration of higher antenna structural system using QO-STBC scheme under different shadowing conditions [9–11]. Space–time block coding is a novel technique that provides maximal diversity advantage over fading channels along with highly bandwidth efficient property which makes wireless communication more compatible. OSTBC designs provide full transmit diversity with lower decoding complexity of symbol detections [12–17]. A transmit diversity is an effective scheme for multiple antenna design communication system, and because of its effectiveness, it mitigates the effects of multipath fading, due to this mitigating effect of fading, the performance of the system is improved without additional bandwidth requirement [17–21]. However, with all above discussed advantages, OSTBC is not bandwidth efficient for four transmit antennas, and thus, OSTBC, i.e., Alamouti code, exists for two transmit antennas [22, 23]. Thus, to address this issue and achieve full rate transmission with more than two transmitting antenna, QO-STBC scheme has been introduced [23, 24].

The motivation behind this study is to evaluate the performance of QO-STBC for land mobile satellite link (LMS) while employing four transmit antennas at ES and single receive antenna at satellite. Fading model that is popular for satellite communication purpose is Shadowed-Rician (SR) model [26, 27] because of LOS randomness. The LOS fading comes from total or partial hindrance in the LOS by unlike objects operating on a direct transmitter–receiver line. To significantly overcome this severe effects of randomness issue, SR fading model is proposed [27]. The decoding complication of coding scheme QO-STBC is much lesser than a general non-orthogonal coding scheme STBC, Motivated by the advantages of high code rate and low decoding complexity of QO-STBC technology without channel state information (CSI) at the transmitter for depiction the diversity of transmission [22–25].

Basically in this paper, we particularly derive SNR for four conveyed symbols from their respective antennas employed at ES for i.i.d positioned SR fading channels [27–30]. Using this SNR, we derive MGF using the relation of Meijer's G-function. By using this derived MGF, we have analytically analyzed the performance of QO-STBC transmitted over shadowed-Rician (SR) fading channels. Further, we evaluate SER that defines performance analysis of the QO-STBC-based LMS communication systems under various shadowing conditions with M-PSK constellations over SR fading channels [15, 22, 23, 25]. It is shown that QO-STBC transmission is convenient and practicable approach for communication systems developed for satellites.

2 System Model

We consider a MIMO technology-enabled satellite communication system in which four antennas for transmission are deployed at ES and one antenna is placed at satellite for reception. The uplink transmission that is transmission from ES to satellite is assumed to follow SR fading [29]. QO-STBC is transmitted from ES in four time intervals. It is assumed that doppler shift is properly compensated in the considered communication system [9]. The QO-STBC is written as [20]:

$$\mathbf{X} = \begin{bmatrix} x_1 & x_2 & x_3 & x_4 \\ -x_2^* & x_1^* & -x_4^* & x_3^* \\ -x_3^* & -x_4^* & x_1^* & x_2^* \\ x_4 & -x_3 & -x_2 & x_1 \end{bmatrix}, \tag{1}$$

where (.*) denotes complex conjugate.
The signal received at the satellite is written as:

$$\begin{bmatrix} y_1 & y_2 & y_3 & y_4 \end{bmatrix} = \begin{bmatrix} h_1 & h_2 & h_3 & h_4 \end{bmatrix} \begin{bmatrix} x_1 & x_2 & x_3 & x_4 \\ -x_2^* & x_1^* & -x_4^* & x_3^* \\ -x_3^* & -x_4^* & x_1^* & x_2^* \\ x_4 & -x_3 & -x_2 & x_1 \end{bmatrix} + \begin{bmatrix} n_1 & n_2 & n_3 & n_4 \end{bmatrix}, \tag{2}$$

where n_1, n_1, n_3, n_4 are the complex valued additive white Gaussian noise (AWGN) with σ^2 variance, zero mean and h_1, h_2, h_3, h_4 are the SR random variable.

In order to decode the symbols at satellite, (2) can be rewritten as:

$$\mathbf{Y} = C_1 \begin{bmatrix} x_1 \\ x_4 \end{bmatrix} + C_2 \begin{bmatrix} x_2^* \\ x_3^* \end{bmatrix} + \begin{bmatrix} n_1 \\ n_2^* \\ n_3^* \\ n_4 \end{bmatrix}$$

here $$\mathrm{C}_1 = \begin{bmatrix} h_1 & h_4 \\ h_2^* & -h_3^* \\ h_3^* & -h_2^* \\ h_4 & h_1 \end{bmatrix}, \quad C_2 = \begin{bmatrix} -h_2 & -h_3 \\ h_1^* & -h_4^* \\ -h_4^* & h_1^* \\ h_3 & h_2 \end{bmatrix} \tag{3}$$

By left multiplication of (3) with C_1^H, we get :

$$\mathbf{Y_1} = C_1^H.C_1 \begin{bmatrix} x_1 \\ x_4 \end{bmatrix} + C_1^H.C_2 \begin{bmatrix} x_2^* \\ x_3^* \end{bmatrix} + n.C_1^H,$$

where $$C_1^H.C_2 = 0, \tag{4}$$

and by left multiplication of (3) with C_2^H, we obtain

$$\mathbf{Y_2} = C_2^H.C_2 \begin{bmatrix} x_2 \\ x_3 \end{bmatrix} + C_2^H.C_1 \begin{bmatrix} x_1 \\ x_4 \end{bmatrix} + n.C_2^H,$$

$$\text{where} \quad C_2^H.C_1 = 0, \tag{5}$$

The decoder decodes the symbols from Eq. (4) as

$$(C_1^H.C_1)^{-1} Y_1 = \begin{bmatrix} x_1 \\ x_4 \end{bmatrix} + (C_1^H.C_1)^{-1} C_1^H.n$$

$$\text{where} \quad (C_1^H.C_1)^{-1} = \frac{1}{a^2 - b^2} \begin{bmatrix} a & b \\ b & a \end{bmatrix}, \text{ and}$$

$$Tr.(C_1^H.C_1)^{-1} = \frac{2a}{a^2 - b^2}, \qquad \text{here}$$

$$a = (|h_1|^2 + |h_2|^2 + |h_3|^2 + |h_4|^2), b = (h_1^* h_4 + h_1 h_4^* - h_3^* h_2 - h_2^* h_3) \tag{6}$$

From (4) and (6), the received SNR of pair of symbol at satellite can be evaluated as:

$$\gamma = \frac{\left[E|x_1|^2 + E|x_4|^2\right]}{2Tr.(C_1^H.C_1)^{-1}.\sigma^2} \tag{7}$$

$$\gamma = \frac{\bar{\gamma}\left(a - \frac{b^2}{a}\right)}{4} \tag{8}$$

If a >> b then, instantaneous γ is rewritten as:

$$\gamma = \frac{\bar{\gamma}.a}{4}$$

where,

$$a = (|h_1|^2 + |h_2|^2 + |h_3|^2 + |h_4|^2), \ \bar{\gamma} = \frac{\left[E|x_1|^2 + E|x_4|^2\right]}{\sigma^2} \tag{9}$$

3 Formulation with Channel Model

We consider the i.i.d positioned SR fading channel model for ES-satellite communication link with approximate PDF of $\|h\|^2$, where $h = [h_1\ h_2\ h_3\ h_4]$ is given below as [29]:

$$f_{\|}h\|^2(\mathrm{p}) = \kappa^M \sum_{v=0}^{j} \binom{j}{v} \omega^{j-v} \left(\frac{p^{u-v-1}}{\Gamma(u-v)}{}_1 F_1\big(u; u-v; -(\omega-\delta)p\big) + \right.$$
$$\left. \frac{\psi \delta p^{u-v}}{\Gamma(u-v+1)}{}_1 F_1\big(u+v; u-v+1; -(\omega-\delta)p\big) \right), \tag{10}$$

where $M = 4, \kappa = 0.5(2bq/2(2bq + \Omega))^q/b, \omega{=}(0.5/b), \delta = 0.5\Omega/(2b^2q + b\Omega)$, b is the half of the average power of the multipath portion, $q \geq 0$ is the strength in respect to shadowing, Ω is the average power of the LOS portion, $a = (u - M)^+$, $\psi = qM - u, u = \max[M, (qM)]$, $[p]$ denotes maximal integer not exceeding than p; $(p)^+$ is that if $p \leq 0$, utilize $p = 0$; max is to choose between the greatest of two positive integers; and $\Gamma(.)$ represents the gamma function.

3.1 MGF Calculation

Through equation (10), we derived MGF [Eq. 15, [24]] in closed form as:

$$M_\gamma(s) = E\left[\mathbf{e}^{-s\gamma}\right] \tag{11}$$

By using (9) and (11), the MGF can be rewritten as:

$$M_\gamma(s) = \int_0^\infty \mathbf{e}^{boldsx\frac{\bar{\gamma}}{4}} f_\gamma(x)dx \tag{12}$$

By substituting the expression of $f_\gamma(x)$, we obtain

$$M_\gamma(s) = \frac{\kappa^4}{\bar{\gamma}} \sum_{v=0}^{j} \binom{j}{v} \omega^{j-v} \int_0^\infty \mathbf{e}^{-\mathbf{sx}\frac{\bar{\gamma}}{4}} \left(\frac{p^{u-v-1}}{\Gamma(u-v)} \right.$$
$$\times {}_1 F_1\big(u; u-v; -(\omega-\delta)\frac{p}{\bar{\gamma}}\big) + \frac{\psi \delta p^{u-v}}{\Gamma(u-v+1)}$$
$$\left. \times {}_1 F_1\big(u+1; u-v+1; -(\omega-\delta)\frac{p}{\bar{\gamma}}\big) \right) dx. \tag{13}$$

For solving the integral in Eq. (13), we use following relations:

$$e^{-\frac{sE_x p}{\sigma^2}} = G_{01}^{10}\left(\frac{sE_x p}{\sigma^2} \middle| \begin{matrix} \cdot \\ 0 \end{matrix} \right) \tag{14}$$

and

$$
{}_1F_1\left(u; u-1; -(\omega-\delta)\,y\right) = \frac{\Gamma(u-v)}{\Gamma(u)} \times G_{12}^{11}\left((\omega-\gamma)y\,\middle|\,\begin{matrix}1-u\\0, 1-u+v\end{matrix}\right) \tag{15}
$$

By using the relation in Eqs. (14) and (15), the function of integral in Eq. (13) may be expressed (closed-form expression) as:

$$
\begin{aligned}
M_\gamma(s) = \kappa^4 \sum_{v=0}^{j} \binom{j}{v} \omega^{j-v} \Bigg(\frac{(s\bar{\gamma})^{-u+v}}{\Gamma(u)} \\
\times G_{22}^{12}\left(\frac{(\omega-\delta)}{s\bar{\gamma}} \,\Bigg\|\, \begin{matrix}1-u, 1-u+v\\0, 1-u+v\end{matrix}\right) + \frac{\psi\delta(s\bar{\gamma})^{-u+v-1}}{\Gamma(u+1)} \\
\times G_{22}^{12}\left(\frac{(\omega-\delta)}{s\bar{\gamma}} \,\Bigg\|\, \begin{matrix}-u, -u+v\\0, -u+v\end{matrix}\right)\Bigg)
\end{aligned} \tag{16}
$$

where $\bar{\gamma} = \frac{\left[E\,|x_1|^2 + E\,|x_4|^2\right]}{\sigma^2}$. In Eq. (16), we seen that MGF of the introduced communication system can be represented in terms of finite summation of Meijer G-function. Thus, this derived MGF is easy to be implemented in MATLAB.

3.2 SER Calculation

By using the derived result of MGF, the SER of the QO-STBC scheme over LMS-based communication satellite links for M-PSK constellations [[11]; Eq. (21)] can be obtained using following equation given below :

$$
\mathbf{Q}_{MPSK} \approx \sum_{n=1}^{4} \eta_r \mathbf{M}_\gamma(s)\,(\chi_r)\,, \tag{17}
$$

Here, the SER of the proposed communication system can be evaluated by using Eqs. (16) and (17) in closed form as:

$$
\begin{aligned}
Q_{MPSK} \approx \sum_{n=1}^{4} \eta_r \kappa^M \sum_{v=0}^{j} \binom{j}{v} \omega^{j-v} \Bigg(\frac{(\chi_r\bar{\gamma})^{-u+v}}{\Gamma(u)} \\
\times G_{22}^{12}\left(\frac{(\omega-\delta)}{\chi_r\bar{\gamma}} \,\middle|\, \begin{matrix}1-u, 1-u+v\\0, 1-u+v\end{matrix}\right) + \frac{\psi\delta(\chi_r\bar{\gamma})^{-u+v-1}}{\Gamma(u+1)} \\
\times G_{22}^{12}\left(\frac{(\omega-\delta)}{\chi_r\bar{\gamma}} \,\middle|\, \begin{matrix}-u, -u+v\\0, -u+v\end{matrix}\right)\Bigg).
\end{aligned} \tag{18}
$$

4 Numerical Results

This paper has investigated QO-STBC scheme [22, 23] over a SR fading channel based on LMS communication system with four antennas at the ES, and satellite is equipped with one antenna with M-PSK constellation under different fading conditions, i.e., frequent-heavy-shadowing (FHS) , average-shadowing (AS) and

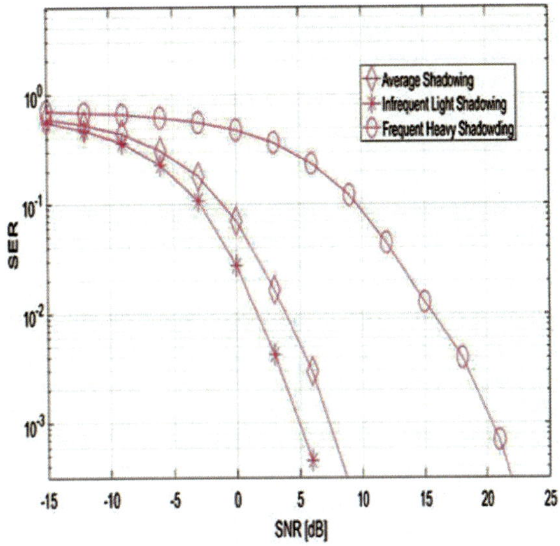

Fig. 1 Analytical versus simulated result of SER under different shadowing conditions with M-PSK over SR fading channels

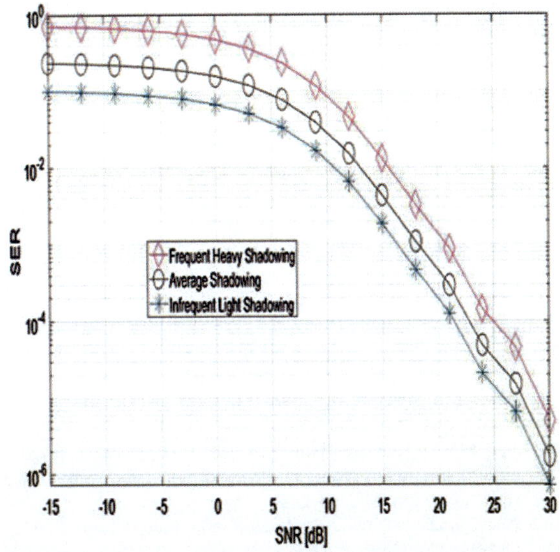

Fig. 2 Analytical versus simulated result of SER of QO-STBC-based LMS communication system under different fading conditions

infrequent-light-shadowing (ILS) for simulation and analysis. From Fig. 1, in simulation, we plot the simulated SER versus SER plots under different fading conditions for M-PSK (M = 4) over SR fading. It can be seen from figure that SNR is increased to a large extent with simultaneous decrease in SER in all shadowing conditions. Moreover, the analysis indicates that the performance of the considered system yields better for both ILS and AS shadowing conditions with loss of approximately 14 dB diversity gain as compared to FHS shadowing conditions. It can be observed from figure that with Q-PSK constellation, the SNR is increased with simultaneous decrease in SER in all shadowing conditions. Figure 2 shows the simulated plots of diversity order. It can be distinguished from the figure that the diversity order of the QO-STBC-based satellite communication system is less than four. It is not full diversity code as seen from Fig. 2.

5 Conclusion

We have presented the analytical study to observe the performance of QO-STBC transmission over SR fading channels. Particularly, we considered the system having four transmitting antennas has been deployed at ES and one antenna at the satellite with M-PSK constellation based on LMS communication system. Analytical evaluation is done by using MGF expression. By utilizing this closed-form MGF term, we derived the SER for the presented system. It can be concluded from the result that the performance of QO-STBC over SR fading is improved with increasing values of SNR. From all the observations, it can be concluded from the result that the execution of QO-STBC-based satellite stationed communication system is best under ILS conditions and worst in the presence of FHS conditions. Further, diversity order plot of the considered system is obtained by using simulations.

References

1. Minoli D (2015) Innovations in satellite communications and satellite technology, 1st edn. Wiley publications, New Jersey (USA)
2. Laneman JN, Tse DNC, Wornell GW (2004) Cooperative diversity in wireless networks: efficient protocols and outage behavior. IEEE Trans Inf Theory 50(12):3062–3080
3. Arti MK, Bhatnagar MR (2014) Beamforming and combining in hybrid satellite-terrestrial cooperative systems. IEEE Commun Lett 18(3):483–486
4. Bhatnagar MR, Arti MK (2013) Performance analysis of AF based hybrid satellite terrestrial co-operative network over generalized fading channels. IEEE Commun Lett 17(10):1912–1915
5. Bhatnagar MR, Arti MK (2013) Performance analysis of hybrid satellite-terrestrial FSO cooperative system. IEEE Photon Technol Lett 25(22):2197–2200
6. Arti MK (2015) Imperfect CSI based AF relaying in hybrid satellite terrestrial co-operative communication systems. In: Proceedings of the International Conference on Communications (ICC), London, UK, pp 1–6
7. Anoh K, Okorafor G, Adebisi B, Alabdullah A, Jones S, Abd-Alhameed R (2017) Full-diversity QO-STBC technique for large-antenna MIMO systems. Electronics 6(2):37

8. An K et al (2015) Performance analysis of multi-antenna hybrid terrestrial satellite relay networks in the presence of interference. IEEE Trans Commun 63(11):4390–4404
9. An K et al (2015) Outage analysis of multi-antenna cognitive hybrid satellite-terrestrial relay networks with beamforming. IEEE Commun Lett 19(7):1157–1160
10. Alamouti SM (1998) A simple transmit diversity technique for wireless communications. IEEE J Select Areas Commun 16(8):1451–1458
11. Arti MK, Jindal SK (2015) OSTBC transmission in shadowed-Rician land mobile satellite links. IEEE Trans Vehicular Technol 65(7):5771–5777
12. Foschini G, Gans M (1998) On limits of wireless communications in a fading environment when using multiple antennas. Wireless Personal Commun 6:311–355
13. Vishwanath R, Bhatnagar MR (2007) High data rate Alamouti code from field extension. Wireless Personal Commun 40(4):489–494
14. Chen S et al (2009) Performance analysis of OSTBC transmission in amplify-and-forward cooperative relay networks. IEEE Trans Vehicular Technol 59(1):105–113
15. Arti MK (2016) OSTBC transmission in large MIMO systems. IEEE Commun Lett 20(11):2308–2311
16. Arapoglou P et al (2011) MIMO over satellite: a review. IEEE Commun Surv Tut 13(1):27–51
17. Park U, Kim S, Lim K, Li J (2008) A novel QO-STBC scheme with linear decoding for three and four transmit antennas. IEEE Commun Lett 12(12):868–870
18. Larsson E, Stoica P (2003) Space-time block coding for wireless communications. Cambridge University Press, New York
19. Tarokh V, Jafarkhani H, Calderbank AR (1999) Space-time block codes from orthogonal designs. IEEE Trans Inf Theor 45:1456–1467
20. Papadias CB, Foschini GJ (2003) Capacity-approaching space-time codes for systems employing four transmitter antennas. IEEE Trans Inf Theor 49:726–733
21. Tirkkonen O (2003) Optimizing space-time block codes by constellation rotations. FWCW 2001, pp 59–60. Sharma N, Papadias CB (2003) Improved quasi-orthogonal codes through constellation rotation. IEEE Trans Commun 51:332–335
22. Jafarkhani H (2001) A quasi-orthogonal space-time block code. IEEE Trans Commun 49:1–4
23. Sharma N, Papadias CB (2003) Improved quasi-orthogonal codes through constellation rotation. IEEE Trans Commun 51(3):332–335
24. Arti MK, Jain A, Chaudhary A (2021) Performance analysis of QO-STBC transmission for Shadowed-Rician (SR) fading channels. TAES-202100211
25. Su W, Xia X (2002) Quasi-orthogonal space-time block codes with full diversity. Globecom 2002, pp 1098–1102
26. Bhatnagar MR, Hjørungnes A, Song L (2009) Interference cancellation by using quasi-orthogonal STBC in two-user MIMO system. In: 2009 15th Asia-Pacific conference on communications, IEEE
27. Bhatnagar MR, Arti MK (2014) On the closed-form performance analysis of maximal ratio combining in Shadowed-Rician fading LMS channels. IEEE Commun Lett 18(1):54–57
28. Arti MK (2015) Performance evaluation of maximal ratio combining in Shadowed-Rician fading LMS channels with estimated channel gains. IET Commun 1–12, in press
29. Abdi A, Lau WC, Alouini MS, Kaveh M (2003) A new simple model for land-mobile satellite channels: first-order and second-order statistics. IEEE Trans Wireless Commun 2(3):519–528
30. Tarokh V, Jafarkhani H, Robert Calderbank A (1999) Space-time block coding for wireless communication: performance results. IEEE Trans Selected Areas Commun 17:451–460
31. Yilmaz F, Alouini M-S (2012) A unified MGF-based capacity analysis of diversity combiners over generalized fading channels. IEEE Trans Commun 60(3):862–875

Smart Health and Cybersecurity in the Era of Artificial Intelligence

Maheriya Arpita and Shailesh Panchal

Abstract Since 2020, the world has been impacted badly by the pandemic situation that arose due to the coronavirus. Artificial intelligence plays a crucial role in the healthcare system, specifically identifying symptoms of disease with the help of various machine learning algorithms during the diagnosis stage. The identified symptoms in various diagnostic tests are used to predict the clinical outcome of early detection of diseases, which results in human life saving. Machine learning algorithms have been successfully used in automated interpretation. With the advanced technology of cybersecurity aspects, we can emphasize data protection for better results. Artificial intelligence can enhance the security of medical science data. Furthermore, they improvise cybersecurity techniques with machine learning technologies.

Keyword Artificial engineering · Medical data · Cyber-attacks · Secure healthcare data · Early disease prediction · Data privacy · Machine learning

1 Introduction

Nowadays, the healthcare sector is at the risk of exposing the privacy of patients. An effective and fruitful organization must not be vulnerable to security-related fundamental elements. Those elements are: confidentiality, authentication, integrity, and non-repudiation. Two points are the key things for any sector that are: availability of data and truthiness. But these might be influenced by hackers who seek vulnerable organizations. Very few studies have examined cybersecurity in the healthcare sector [1]. Confidentiality of data eventually exceeds trust in the healthcare system. Furthermore, the healthcare sector is falling behind other sectors in cybersecurity. The highest level of threat to health care is: healthcare services, medical equipment, and

M. Arpita (✉) · S. Panchal
Gujarat Technological University—Graduate School of Engineering and Technology, Ahmedabad, Gujrat, India
e-mail: arpita_maheriya@gtu.edu.in

S. Panchal
e-mail: sdpanchal@gtu.edu.in

M. S. Kaiser et al. (eds.), *Information and Communication Technology for Competitive Strategies (ICTCS 2021)*, Lecture Notes in Networks and Systems 401,
https://doi.org/10.1007/978-981-19-0098-3_5

infrastructure. Those could compromise the safety of patients. The healthcare sector is growing use of electronic record systems. That could increase the chance of cyber-attacks on healthcare record systems. Exchange data with third parties for research and analysis to improve the efficiency of the healthcare system. While sharing, the data that secures the privacy of patients, critical information is a necessary task. The healthcare sector is often affected by hackers due to the sensitivity of the data they hold. This literature study covers the security challenges in the healthcare sector. Artificial technology and machine learning techniques have sent their waves widely in many sectors. In addition, the intelligence of AI techniques could be used to unlock the hidden information from the massive amounts of healthcare data. With the use of this large set of data, it helps to create an intelligence system to take decisions and make predictions [2]. The incentive behind this survey is to discuss ML algorithms used in disease identification and cybersecurity aspects of the healthcare sector.

We discussed this study in brief in eight sections. The second section has a background history and related literature review on the use of modern techniques in the medical science field. The third section is about the effect of cybersecurity attacks on the healthcare system in the period of the COVID-19 pandemic. The fourth section is about the aims and objectives of the study. And then, section of the research gaps found in the literature review study. In the end, the conclusion of this review article.

2 Background Study

Argaw [1] study is based on cyber violence in healthcare sector. In modern years, rife of the cyberattacks has noticed. Health sector has rapidly become a target for cyberattacks. Literature is conducted based on cyberattacks on hospitals and recognizes the various health domains for further research. The review was extracted and analyzed from four databases: Web of Science, PubMed, Scopus, ProQuest, and ProQuest) published between 1997 and 2017. Cyberattacks on the health sector contain a diversity of threats. Like, Denial-of-Service and brute force attacks, malware phishing and social engineering techniques are to violate security. One report discloses that the health organization came in the top three domains, which more influence using ransomware attack globally. The healthcare field is most vulnerable to attackers due to the nature of the work. Moreover, data like personal information could risk patients' life safety and could be blackmailed. The stolen data could have devastating inferences such as criminal activities, terrorists, and retribution. Additionally, analysis of the selected reviews directs to discriminating the domains of research: (1) cybersecurity context and trends, (2) disseminating awareness, (3) connected medical devices and equipment, information systems of hospitals, (4) methodology of information security, and (5) types of specific attacks. This study is an up-to-date analysis of the latest research domains and determine recommendations on healthcare cybersecurity. It is an interesting research field, but at present, the papers published remain limited in number.

In [3], the study is highly focused on exploring various kinds of cybersecurity threats, and the current rise of cybersecurity breaches in patient records. Risk of being exposed, lost, stolen, displaced, and hacked information into the system. These violate patients' privacy and decrease safety level. Hackers are focusing on the healthcare sector because healthcare data is more valuable than data from other industries for illegal sales. There is a need to focus on a better understanding of the complex issue of cyberattacks in the healthcare sector. The study discusses the many cybersecurity threats/attacks and the incentives behind them. This study describes the vital roles of cyber attackers, cyber defenders, developers, and end-users. It also provides guidance for policymakers and healthcare organizations to make security strong in the health sector.

In [4], the main impulse is to determine and identify the reproducibility of subtypes of COPD. This is done with the help of generating cluster analysis techniques on large-scale electronic healthcare records. The training set includes risk factors. Performed dimensionality reduction with multiple correspondence analysis. Here, they used two methods of unsupervised clustering machine learning algorithms (k-means and hierarchical clustering) which results in a $k = 5$ discrete cluster. These discrete clusters have been labeled as subtypes of COPD. The five subgroups identified are (1) depression/anxiety; (2) severe airflow obstruction and frailty; (3) diabetes and cardiovascular disease (4) atopy/obesity; and (5) low prevalence of most comorbid conditions. To establish reproducibility, the decision tree classifier (DTC) is trained with an 80% predetermined labeled cluster. DTC results in 96% accuracy performance on test data. The Jaccard index was computed at above 89%. The Jaccard index result is based on patients that were reassigned to the original clusters from the test set results. Some limitations have also been disclosed. The first primary care concern is the high rates of misdiagnosis. And the other one is overlapping and not quite clear boundaries for some clusters. Due to this, it would be possible for complex patient data to be in more than one cluster.

The objective of the [5] paper is to early diagnose diabetes. It is extremely vital to preserve a healthy life. Machine learning (ML) is a process that rectifies performance and makes more accurate predictions by learning from past encounters. They have used a dataset of Pima Indian diabetes to discover patterns and trends along with risk factors. A total of 768 cases were differentiated into diabetic and non-diabetic classes. Along with eight divergent risk factors, the Boruta package is used for an unbiased and steady selection of useful features. With the least medical field knowledge, this method is preferable to manual selection of the attributes. Through the Boruta feature selection method, you can get appreciative precision and accuracy using the least number of parameters. They described how supervised algorithms are used to build models for diabetes detection. (1) Kernel support vector machine, (2) SVM-linear, (3) radial basis function (RBF), (4) k-NN, artificial-neural networks (ANNs), (5) multifactor-dimensionality reduction (MDR) algorithms are used to generate five discrete models. To avoid over- and under-fitting, tenfold cross-validation is applied. From this study, it could be determined that with the use of linear kernel support vector machine and k-nearest neighbors (k-NN), both models are the finest models.

Nonetheless, early detection of diabetes disease could be useful for improving the quality of health, comfort, and happiness of patients.

In [2], paper reviewed the current status of AI applications in health care. It also shows the future of AI in the healthcare sector. They discussed the purpose of AI in the healthcare sector. Various healthcare data are available to analyze and survey major disease types using AI. The study examined two major categories of AI devices: ML (SVM, neural network, and deepening techniques) and natural language (NLP). The review showed SVM has been the most used in medical research. Deep learning is also in the current trend. In particular, convolution neural networks (CNN) have been used most. The disease includes cardiology, neurology, stroke care, and cancer that are areas that use AI technology for early detection and diagnosis, treatment, and result prediction. Management is required to continue taking advantage of a good amount of information. Data exchange is a hurdle. Data supply has become a critical problem for further improvement of the healthcare system. The present-day healthcare environment does not provide incentives or rules for sharing data within the system (Fig. 1).

Masood et al. [6] The study describes how, in recent years, patient data privacy and security threats have increased. Specifically, in the distributed environment of sensor-cloud infrastructure (S-CI). This literature is focused on health sector information-related privacy and security in S-CI. Techniques are in use at present time and are classified into: attribute-based encryption, dynamic probability packet marking, and priority-based data, chaotic maps, multi-biometric key generation, pairwise key establishment, hybrid encryption, Number Theory Research Unit, tri-mode algorithms, and forwarding techniques, hash functions, according to their application areas. The pros and cons of these methods are also described in a particular order. The study also presented a generic framework of six-step for patient physiological parameters (PPPs) of privacy and security in S-CI. S-CI really needs proper attention

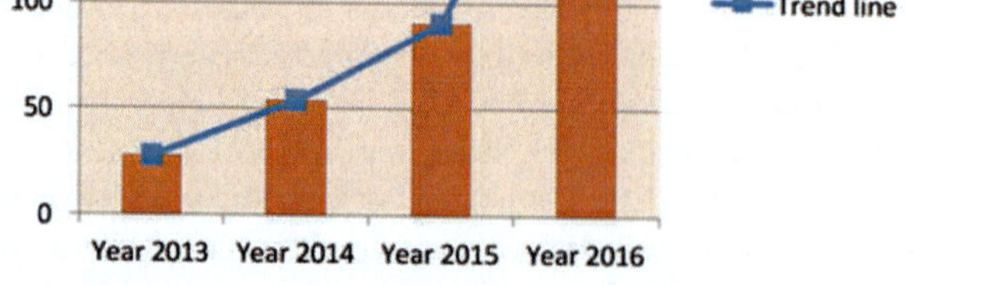

Fig. 1 Current trend for deep learning. The data is generated through searching the deep learning in health care and disease category on PubMed [2]

for effective access control, utilization quality of patient data access, network security, and user-friendly and real-time implementation of applications. The distributed nature of the S-CI system motivates a disparate set of problems for researchers to study in this thriving area. They have also explained 13 snapshots of this.

In [7], the author analyzed further scientific research, especially general transparency in patient-level data sharing. It is one of the rising areas for study. This article presents two recommendations: 1. Whenever data is shared with third-party researchers, better practices are to anonymize or de-identify the clinical trial record. 2. Control access and data sharing agreements (DSA). The aim of the article is to promote an approach that balances both data utility and privacy risk. The main motive is maintaining patient confidentiality, while sharing clinical experiment/trial data. Data need to be appropriately prepared and shared with third-party researchers. Additionally, research is required to identify and generate competing possibilities for regulations, attitudes to risk, and technologies evolving in sharing data.

In this current era, we have been facing many diseases due to environmental pollution and our living habits. So, at this stage, the prediction of disease in the starting stages is a very vital task. For doctors, predicting the precious and accurate results of the disease is a challenging problem. The medical care system has a large amount of data generated per year. Due to this, a large amount of data analysis is possible which helps with early diagnosis. This study [8] proposed a general prediction of disease on the basis of the symptoms of disease and patients. They used two machine learning approaches, k-nearest neighbor and convolutional neural network, for prediction accurately. The k-nearest neighbor was proved to be better than the convolutional neural network in this study in terms of accuracy and timing. The proposed system could be able to diagnose associated risks with general disease higher or lower.

3 Cybersecurity in Healthcare System During Pandemic Situation of the COVID-19

The healthcare system has been made more susceptible in the world to cyberattacks during COVID-19. Healthcare organizations worldwide are fighting the virus (COVID-19). They have been frontline workers in this critical situation. We should manage the security of sensitive sectors like healthcare organizations. In the last few years, healthcare organizations have reported many attacks on cybersecurity.

In this article [9], they express the attack that happened on the University of Vermont (UVM) healthcare system. On October 28, 2020, the system was halted for 40 days, along with the electronic healthcare records system. During the period of the attack, identification of the attack and resolving it were the main concerns. The healthcare sector of UVM had a cost of $1.5 million per. The financial amount was more than $63 million to regulate the system back to normal condition. On October 26, 2020, a cyberattack called Ryuk ransomware affected six hospital networks in

New York and California for a day. The hackers asked for a ransom, which is 10% of the hospital system's income. The uniform kind of ransomware attack was first distinguished in 2018. At that time, the ransom was $2.1. Many patients had to suffer. Important types of equipment in medical systems were shut down. Distributed Denial-of-Service happens in network overload situations. The vital reason for the DDOS attack is to gain access to the hospital's critical information and financial data. As such, crucial sectors like the healthcare sector are the targets of the attackers. DDoS attacks targeted Boston Children's Hospital. That is one of the remarkable attacks in year 2014. The hospital had a loss of $300,000 to recover from the DDOs cyberattacks in 2014 [10]. Moreover, the healthcare sector has some internal threats as well. These kinds of internal breaches are due to either employment contact or human error. An insider attack has three categories [11]: (1) insider malicious or criminal, (2) carelessness of employers, and (3) thief of credentials were disturbed by outsiders due to a data breach in 2019. Montpellier University Medical Center reported access by outsiders due to a data breach in 2019. More than 600 computers were affected. Health organizations have critical data such as patients' names, information, mobile numbers, DOB, insurance, bank details. Influencing any of the equipment used in medical care is very agonizing which can provide a wrong result that may have an impact on a person's life [12]. Medjacking is the process or practice in which hackers manipulate medical equipment and instruments to harm patients' health. This is to put the reputation of hospitals at risk. Artificial intelligence could be used to increase cybersecurity.

4 Aims and Objective

As we discussed above, the healthcare sector is facing challenges with security. Cybersecurity is important for every organization. However, in the health sector, the impact is far greater. It is the most vulnerable in the case of security. There has been a gradual increase in cyberattacks and threats to the healthcare sector. Make sure that patients' personal information and medical data stays safe. The security of sensitive information can save or destroy some lives. This is a very crucial subject for scientists. There is a need to enhance security with the latest technologies. There has been a gradual increment in cyberattacks and threats to the healthcare sector. There is a requirement to establish a framework with finer functions to reduce the number of attacks and protect the healthcare system.

Due to the availability of a vast amount of data in the medical field, it can be advantageous to utilize a wide range of research such as analysis of risk, exploring patterns, and implementing disease prediction and risk of chronic disease. There has been a continuous rise in air pollution all around the world. India is in the third position in the air pollution index. Air pollution is one of the causes of chronic obstructive pulmonary disease and other chronic inflammatory diseases. Applying modern technological tools, especially machine learning and artificial intelligence, to improve the accuracy of disease diagnosis quickly. AI can provide efficient human judgment and regulate AI tools to protect policy security and network architecture. By using intelligence and training systems along with AI and ML, we could extend the security of the healthcare system. Motivate the research scientist community to study in this flourishing area.

5 Research Gap Findings

In the current era, especially in a pandemic, cyberattacks have hit hospitals and healthcare systems globally. Cyberattacks are a dilemma influencing the health sector which creates vulnerabilities in medical devices, infrastructure, and security culture. That needs to be a significant consideration. Technologies are inducing substantial attention in medical research, but still facing obstacles in real-life implementation. Such as emerges from regulations and data exchange. Safety in data sharing could revive ideas for researchers. Today's regulations have leaner standards to evaluate efficient safety in healthcare systems. There is a lack of standard architecture in security policies and publications for medical data. The concern is ensuring patient data privacy and unreliability [6]. This study also proposes giving attention to designing secure public sharing and data publication [2]. As reviews of AI in healthcare, the research is mostly based on a few disease types, such as the nervous system, cancer, and cardiovascular disease. That would need to focus on another chronic disease that leads to death. Improving the analysis process is the strength of AI techniques. Research is primarily needed to improve performance in: prediction approaches of disease, securing data while sharing it with a third party for research, and early disease identification. These are bright fields for research.

6 Conclusion

Cybersecurity applications can be made easier and more efficient with the use of AI and machine learning algorithms. The comparison of algorithms across all papers we studied in an objective manner is challenging due to each study's performance being reported using variable methodologies on different populations with different sample distributions and characteristics. The distributed data across multiple sources needs to be collectively used in a particular order or format. Intrinsic difference between security challenges among those which are already addressed by the large medical dataset. The process of adaptation of clinical data for machine learning algorithms needs to be understood. Need to perform security of data by establishing or confirming claims made by or about the subject are true and authentic and provide functionality for securing access to any trusted networks, protecting the identities of users.

References

1. Argaw ST, Bempong NE, Eshaya-Chauvin B, Flahault A (2019) The state of research on cyberattacks against hospitals and available best practice recommendations: a scoping review. BMC Med Inf Decis Making 19(1):10
2. Jiang F, Jiang Y, Zhi H, Dong Y, Li H, Ma S, Wang Y, Dong Q, Shen H, Wang Y (2017) Artificial intelligence in healthcare: past, present and future. Stroke Vascular Neurol 4(2):230–243
3. Bhuyan SS, Kabir UY, Escareno JM, Ector K, Palakodeti S, Wyant D, Kumar S, Levy M, Kedia S, Dasgupta D, Dobalian A (2020) Transforming healthcare cybersecurity from reactive to proactive: current status and future recommendations. J Med Syst 44(5):98
4. Pikoula M, Quint JK, Nissen F, Hemingway H, Smeeth L, Denaxas S (2019) Identifying clinically important COPD sub-types using data-driven approaches in primary care population based electronic health records. BMC Med Inf Decis Making 19(1):86
5. Kaur H, Kumari V (2020) Predictive modelling and analytics for diabetes using a machine learning approach. Applied, ahead-of-print
6. Masood I, Wang Y, Daud A, Aljohani NR, Dawood H (2018) Towards smart healthcare: patient data privacy and security in sensor-cloud infrastructure. Wireless Commun Mobile Comput Article ID 2143897, 23 pages
7. Tucker K, Branson J, Dilleen M, Hollis S, Loughlin P, Nixon MJ, Williams Z (2016) Protecting patient privacy when sharing patient-level data from clinical trials. BMC Med Res Methodol
8. Dahiwade D, Patle DG, Meshram E (2019) Designing disease prediction model using machine learning approach. In: 3rd International Conference on Computing Methodologies and Communication (ICCMC), pp 1211–1215
9. The University of Vermont (UVM) Health Network Cyberattack. https://www.uvmhealth.org/uvm-healthnetwork-cyber-attack
10. Six hospital ransomware attacks in 24 hours prompts US advisory: 8 things to know. Laura Dyrda (Twitter)-Thursday, 29 Oct 2020
11. DDoS Case Study: Boston's Children's Hospital DDoS attack Mitigation. https://www.radware.com/security/ddos-expertsinsider/ert-casestudies/boston-childrens-hospital-ddosmitigation-case-study/ Cybersecurity &Healthcare During COVID-19
12. Cybersecurity and Healthcare During Covid-19 by Susan Alexandra on April 2020

A Comprehensive Framework for the IoT-Based Smart Home Automation Using Blynk

Faraz Doja, Reenu Batra, Sandeep Tayal, Prashant Vats, and Siddhartha Sankar Biswas

Abstract The phrase "IoT" mentions to any technology that makes it possible for a device to connect to the Internet. Data collecting is essential for such systems. The information is subsequently utilized for Internet-based control, monitoring, and transfer of data to other devices. Smart home technology is the future of domestic technology, aiming to offer and distribute a variety of services both inside and outside the home via networked devices in which all of the many applications and intelligence behind them are merged and interconnected. Given the constant availability of a broadband Internet connection, these smart gadgets have the potential to communicate information with one other. As a result, technology for the smart home is now a part of the Internet of Things (IoT). In this research paper, we have offered a framework for remote administration and monitoring IoT-enabled appliances and systems in this paper. Home automation provides homeowners with allowing them to have peace of mind to keep an eye on and defend their homes from afar.

Keywords Internet of Things (IoT) · Smart homes · Blynk · Automation · Sensors

F. Doja (✉) · S. S. Biswas
Department of Computer Science and Engineering, SEST, Jamia Hamdard, New Delhi, India
e-mail: farazdoja@jamiahamdard.ac.in

S. S. Biswas
e-mail: ssbiswas@jamiahamdard.ac.in

R. Batra
Department of Computer Science and Engineering, SGT University, Gurugram, India

S. Tayal
Department of Computer Science and Engineering, Maharaja Agrasen Institute of Technology, GGSIP University, New Delhi, India
e-mail: sandeeptayal@mait.ac.in

P. Vats
Department of Information Technology, Indira Gandhi Delhi Technical University for Women, New Delhi, India

M. S. Kaiser et al. (eds.), *Information and Communication Technology for Competitive Strategies (ICTCS 2021)*, Lecture Notes in Networks and Systems 401,
https://doi.org/10.1007/978-981-19-0098-3_6

1 Introduction

A smart home is a residence where appliances may be remotely monitored and controlled via Internet-connected devices. Smart home technology enables homeowners to control smart appliances from a smartphone or other connected device, resulting in increased security, comfort, convenience, and energy savings. The IoT includes devices and systems for smart homes, often collaborating, exchanging data on customer usage, and automating processes based on the homeowners' wishes. One of the most often stated advantages of home automation is the reduction of energy use, as a result of which, it offers homeowners with peace of mind by allowing them to keep an eye on their homes from afar and avert dangers such as an unprotected front door or a forgotten coffee maker. In smart homes, user preferences for convenience may be met. Home automation can help consumers be more productive. Rather than keeping the air conditioner running all day, it is possible that a smart home system will learn how people live, behave, and ensure that the house is cool when they get home from work. Using IoT-based home automation, grass will only be watered when it is required by a smart irrigation system, and with the precise quantity of water required. Home automation provides for more efficient use of energy, water, and other resources, from the conservation of both money and natural resources for customers. More than just smart devices and appliances collection, a smart home is a way of life; it is a system of networked items that can be controlled from anywhere. All of the gadgets are under your control by a master home automation controller, often referred to as a smart home hub. A smart home hub is a physical device that connects to the Internet and serves as the system's brain, sensing, processing, and transmitting wirelessly. It connects all of your smart home apps in one into a single app that consumer can use from anywhere.

2 Related Work

Menachem [1] has proposed their work for cloud-based enhanced smart house idea and implementation-based LAN technology, where the IoT, smart homes, and cloud computing are three components that are just loosely linked and have been integrated. They have used the cloud computing technology to integrate the IoT-based devices using LAN.

Cristina et al. [2] have proposed their work by using ESP8266 chips and Raspberry Pi boards to create smart home automation. In their work, the data from sensors, actuators, and other sources are interconnected to create a system that may be used for many home automation applications. Using the qtoggle system, users may control a number of home appliances and sensors.

Sandeep et al. [3] have proposed a low-power Wi-Fi-based IoT-based smart home using wireless sensor networks, where the WSNs are used to create the IoT (IoT), which connects smart home gadgets via the Internet Protocol (IP). The advantage of this system is the triangle-based security algorithm (TBSA)-based integrated WSN and Wi-Fi network. In their work, the suggested utilizing WSNs can be used to secure IoT-based home automation applications.

Adi et al. [4] have presented a smart grid system for the smart home-based on IoT. They have presented an overview of the IoT as it relates to smart grid and smart house technologies. The use of IoT in smart grids might help to cut energy usage.

Timothy et al. [5] have presented a smart home with IoT using FLIP Platform, where the Frugal Labs IoT Platform (FLIP) would serve as a foundation for the development of IoT-enabled smart homes. The suggested solution is extremely beneficial in terms of monitoring and regulating the smart home environment. The technology proposed in this study is extremely adaptable and extendable to meet the demands of users while maintaining security.

Mohammed et al. [6] have given an approach for monitoring our health in real time using IoT-based security systems to create a smart home security systems which is based on the Internet of Things in telemedicine architecture that allow for monitoring of health in real time. The suggested solution intends to provide a secure substructure that is up to date in IoT-based telemedicine infrastructures.

Bharat et al. [7] have given a Blynk framework for IoT-based smart homes by utilizing Wi-Fi as a communication channel and as a private server, which use a Raspberry Pi to manage and monitor appliances through smartphone. The proposed system can be used to solve many difficulties encountered by Nepalese people in their everyday lives, an IoT-based smart home using the Blynk.

Swapna et al. [8] have given a secure framework for IoT-based smart homes by which IoT-based smart home safety requirements are being improved. By using the proposed system, the IoT is being used to improve safety standards in smart homes.

Tran et al. [9] have given a real-time security monitoring systems based on the Internet of Things. The monitoring framework is built on the web application's strength, allowing for features for real-time storage, display, and warnings for administration of remote or local monitoring.

Mubeena et al. [10] have given a smart home monitoring system based on IoT and tri-level context making that offers an IoT-based monitoring system for smart home services based on a tri-level context making model. The suggested system may be utilized to update values or parameters in the house using various communication protocols through the Internet. As a future implementation, displaying data and information is required.

3 Proposed Framework for Smart Home Automation

In this research, we offer a framework for a smart house controller that is based on the IoT (IoT) in this article. The proposed methodology may be used to manage household electrical equipment including lights, fans, air conditioners, and water pumps using a smartphone. We have utilized a DHT-11 sensor to monitor room temperature and humidity, a soil moisture sensor with a garden motor pump, and an ultrasonic sensor (HC-SR04) to indicate the water level in an above water tank in this framework. By using smartphone, we could graphically check the status, which included light ON/OFF, room temperature, soil moisture level, and water level. The IoT-based smart home device controller's circuit design is depicted in Fig. 1. It is made up of a Node MCU (ESP1), two light bulbs (Load3 and Load2), four 12 V single-changeover relays (RL1 through RL4), a 5 V regulator 7805 (IC1), a temperature and humidity sensor (DHT11), a 5 V DC soil moisture sensor, an ultrasonic sensor (HC-SR04), and a few more parts.

In the proposed work, the room temperature and humidity are monitored using a DHT11 sensor in Fig. 1. To show humidity and temperature, we have utilized the Blynk app with virtual levels V5 and V6. In the proposed framework, the garden soil moisture level is detected using a soil moisture sensor. When the soil becomes dry, the garden motor pump automatically activates to provide water to the plant. The water level in the above water tank is monitored using an ultrasonic sensor

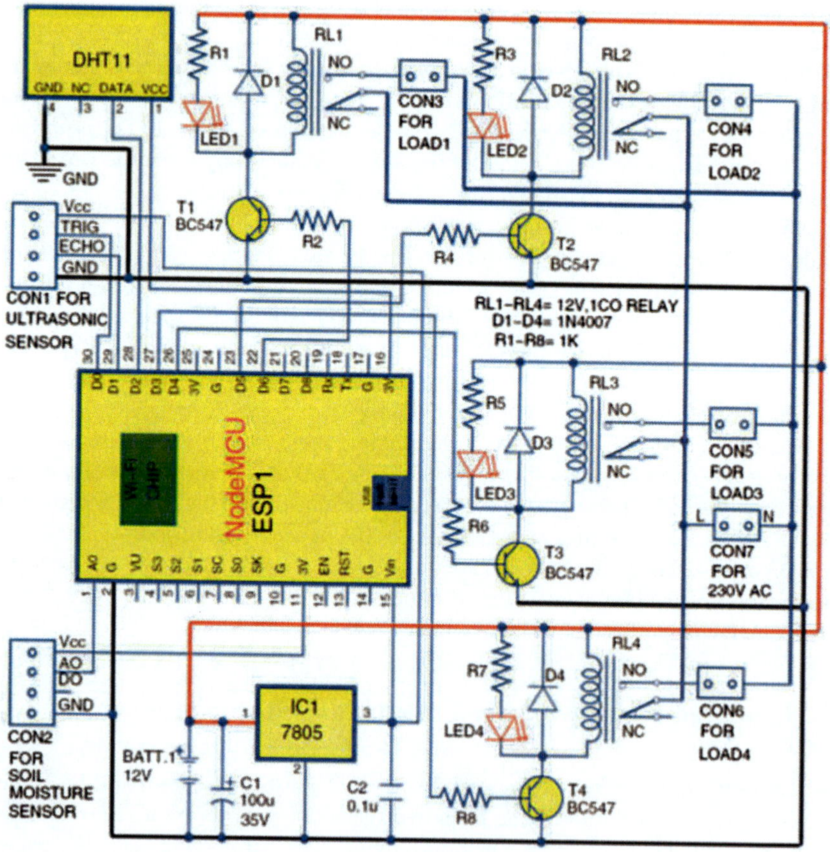

Fig. 1 IoT-based smart house controller's circuit diagram

(HC-SR04), as illustrated in Fig. 6. In the proposed framework, as illustrated in the circuit the water level is monitored using the HC-SR04 sensor, and the water motor pump can be controlled using RL1, which is linked to NodeMCU pin D6. The garden motor pump can be controlled using RL4, which is connected to NodeMCU pin D3. The NodeMCU connects to Wi-Fi network when power is switched ON by attaching a 12 V battery. In the proposed framework, we utilized the Blynk app on a smartphone cellular device with the needed NodeMCU software to operate the electrical appliances. The IoT is aided by the Blynk app. It is capable of remotely controlling hardware and displaying sensor data. The most frequently utilized IoT platform is Blynk and can be used for developing apps to manage and connecting devices to the cloud, and monitor them remotely, and managing them. It is a platform as a service (Paas) application. The code used for the Blynk is shown in Fig. 2.

```
#define BLYNK_PRINT Serial  // Comment
this out to disable prints and save space
#include <SPI.h>
#include <ESP8266WiFi.h>
#include <BlynkSimpleEsp8266.h>
#include <SimpleTimer.h>
#include <DHT.h>
#include <Adafruit_Sensor.h>
#include<NewPing.h>
// You should get Auth Token in the Blynk
App at the time of //project creation to
registered mail.
char auth[] = "";
// Your WiFi credentials.
// Set password to "" for open networks.
char ssid[] = "";
char pass[] = "";
#define DHTPIN D2        // What digital pin
we're connected to
#define TRIGGER 16
#define ECHO   5
// NodeMCU Pin D0 > TRIGGER | Pin D1 >
ECHO
#define DHTTYPE DHT11   // DHT 11
DHT dht(DHTPIN, DHTTYPE);
SimpleTimer timer;
int low=1024;
int high=390;
int m=19;
int ma=0;
void sendSensor()
{
  float h = dht.readHumidity();
  float t = dht.readTemperature();
  int r=analogRead(A0);
//Serial.println(r);
int per=map(r,low,high, 0,100);
//Serial.println(per);
long duration, distance;
  digitalWrite(TRIGGER, LOW);
  delayMicroseconds(2);
   digitalWrite(TRIGGER, HIGH);
  delayMicroseconds(10);

   digitalWrite(TRIGGER, LOW);
  duration = pulseIn(ECHO, HIGH);
  distance = (duration/2) / 29.1;
  Serial.print(distance);
  Serial.println("Centimeter:");
  int z=20;//taken height of tank is 20cm
 int y=z- distance;
 int l=y*5;
  if (isnan(h) || isnan(t)) {
   Serial.println("Failed to read from DHT
sensor!");
   return;
  }
  // You can send any value at any time.
  // Please don't send more that 10 values per
second.
  Blynk.virtualWrite(V5, h);
  Blynk.virtualWrite(V6, t);
  Blynk.virtualWrite(V7, per);//soil dry
percentage
   Blynk.virtualWrite(V8, l);// water level
percentage
  }

void setup()
{
  Serial.begin(9600); // See the connection
status in Serial Monitor
   Blynk.begin(auth, ssid, pass);
pinMode(TRIGGER, OUTPUT);
  pinMode(ECHO, INPUT);
  dht.begin();
  // Setup a function to be called every
second
  timer.setInterval(1000L, sendSensor);
}
void loop()
{
  Blynk.run(); // Initiates Blynk
  timer.run(); // Initiates SimpleTimer
  }
```

Fig. 2 To demonstrate the code required to create Paas IoT-based smart house controllers using Blynk

4 Working of the Proposed Framework

The system is made up of three separate subsystems: the first is a Blynk app module that displays the status of the equipment; the second is a DHT11 temperature sensor that measures temperature, a PIR sensor that detects motion, and an ultrasonic sensor that measures distance; the third is a master microcontroller that serves as the centrifuge. The master microcontrollers also controlled the appliance at the site through a relay module. The data from the different sensors is retrieved to the user interface, which is assisted by smart phones or tablets, using an ESP8266 as the private server. Essentially, the owner has authority over whether or not to turn ON or OFF the entire system. When the system is turned ON, it looks for the service established identifier (SSID) that has been set and connects to the Internet automatically; otherwise, it stays offline and does the automatic tasks and controlling task that does not require the owner's input. Sensors collect a variety of environmental conditions and send them to the microcontroller, which analyses each sensor's data individually before sending the collected data to the Web server in real time. Each sensor's readings may be viewed by the user at any time and from any location. Furthermore, all sensor data is recorded every second for future data processing. The system has two modes of operation: automated and manual. When it is set to automatic mode, all household appliances, such as fans and heaters, are programmed to function in accordance with the ambient conditions observed by the sensors. When it is set to manual mode, on the other hand, the user may monitor and manage each of her household appliances locally or remotely from her smartphone or office desktop PC. To summarize, she has complete authority over the surveillance and management of all household appliances at her fingertips.

Some of the activities go beyond the auto-manual mode of operation, monitoring the water level, for example (in the water-storage tank) and blaring sirens, as well as turning ON or OFF the water pump when a person is detected at the door, and putting motion detectors in your house or office to remotely notify the owner, which can also be used for security. Because these devices run constantly under the orders of the microcontroller, the user does not need to switch them ON and OFF at regular intervals. Similarly, various alarm signals and warnings, such as a home on fire, a gas leak, or the discovery of an intruder, are wirelessly sent and alerted to the house owner by e-mail and phone notification. The owner may track and locate her vehicle using GPS at any time and from any location. To summarize, the IoT offers a higher level of physical and psychological security.

5 Experimental Set-up and Results

We open widget and choose buttons as shown in Fig. 3 in the proposed framework using the Blynk software on a smartphone cellular device. We have included add label value to display temperature, humidity, super chart for graphical representation, and

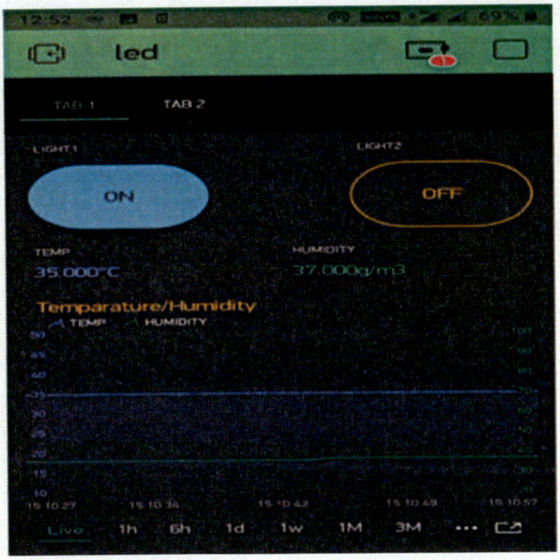

Fig. 3 Addition of control tabs

other features to the proposed framework with Blynk app as shown in Fig. 4. We will use the Arduino IDE to access the smart home.ino code. In the proposed framework, a smartphone is used to set-up Node MCU with Wi-Fi credentials, and then the authorization code issued by the Blynk app is used. After putting the circuit together,

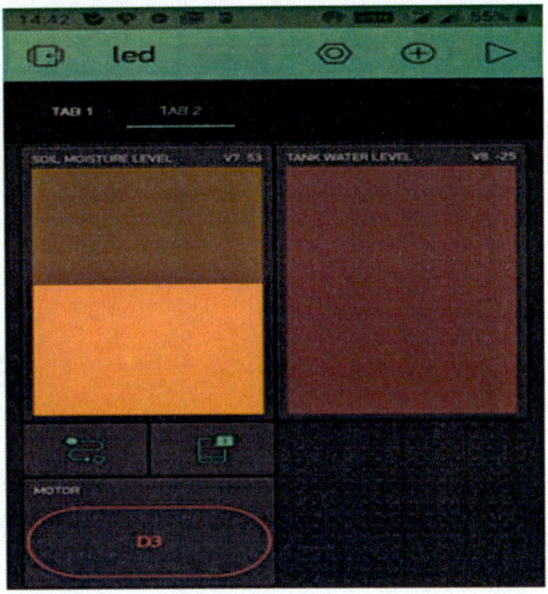

Fig. 4 Set Tab2 to the moisture level of the soil

using the Blynk app, we have linked all devices to the Internet in the suggested framework (Figs. 5 and 6).

The home appliances may be operated from any remote place by pushing a virtual button on the smart phone. One of the benefits of this software is that it may be shared with all members of the household's family. When one of the app's members turns ON or OFF an appliance, the activity is visible to all other app users. Similarly,

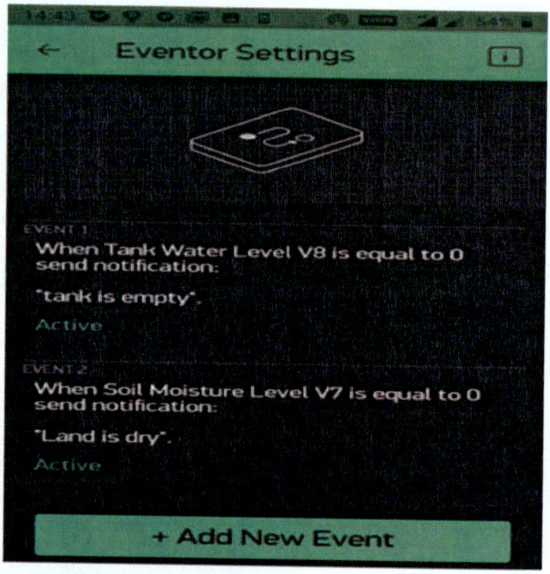

Fig. 5 Configuration of the eventer

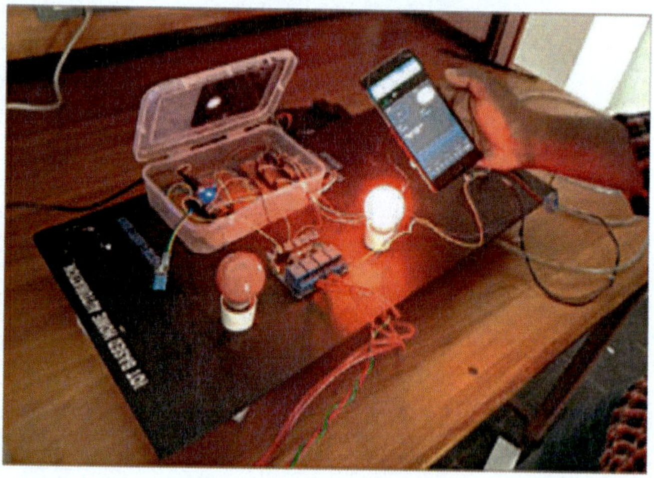

Fig. 6 IoT-based smart house controller prototype

the app may be used to get real-time and historical data on temperature, humidity among other things. This system may also be used in a variety of settings, including banks, hospitals, laboratories, traffic checkpoints, apartments, houses, highways, and greenhouses and poultry farms. In a nutshell, this method may be used to a variety of sectors and regions in order to improve their efficiency.

6 Conclusion and Future Work

This research work is primarily concerned with overcoming the challenges that people experience on a daily basis in a society, where recurrent power outages, uncontrolled urbanization, a shortage of labour in agriculture and farming, and other issues are all too obvious. Home security and automation in real time, remote system monitoring and control are all possible with our prototype system. In this paper, we have proposed a framework for remote monitoring and administration of appliances and systems utilizing the IoT in this article. The suggested framework may be used to monitor, manage, and transfer data to other devices from afar. This allows for the automated execution of defined activities whenever certain conditions arise. The proposed Internet of Things-based home automation solution has several benefits over wired systems, including ease-of-use, ease-of-installation, avoidance of the complexity of running through cables or weak electrical connections, problem detection and triggering, and, most all, mobility. As a future work, the system needed to be tested for large-scale systems.

References

1. Domb M et al (2019) Smart Home Systems based on Internet of Things (IoT) for automated and smart applications, Yasser Ismail. Intech Open. https://doi.org/10.5772/intechopen.84894
2. Stolojescu-Crisan et al (2021) An IoT-based smart home automation system. Sensors 21:3784
3. Pirbhulal S et al (2017) A novel secure IoT-based smart home automation system using a wireless sensor network. Sensors 17:69
4. Swastika AC et al (2017) IoT-based smart grid system design for smart home. pub. In: Proceedings of 3rd IEEE International Conference on Wireless and Telematics (ICWT), pp 49–53
5. Malche T et al (2017) Internet of Things (IoT) for building smart home system, pub. In: Proceedings of IEEE International Conference on I-SMAC (IoT in Social, Mobile, Analytics and Cloud)
6. Talal M et al (2019) Smart Home-based IoT for real-time and secure remote health monitoring of triage and priority system using body sensors: multi-driven systematic review, pub. Springer J Med System 43, 42
7. Bohara B et al (2016) IoT based Smart Home using Blynk framework. Zerone Scholar 1:26–30
8. Badabaji S et al (2018) An IoT based Smart Home service system. Int J Pure Appl Math 119(16):4659–4667

9. Khoa TA et al (2020) Designing efficient Smart Home Management with IoT Smart Lighting: a case study. In: Wireless communications and mobile computing, vol 2020
10. Begum M et al (2016) IoT-based monitoring system using tri-level context making for Smart Home Services. IOSR J Electrical Electronics Eng (IOSR-JEEE) 11(4):01–05 (Ver. I)

Predict Future Events Over Smart Environment Through Modified Apriori Algorithm

Lipichanda Goswami, Vaskar Deka, and Manoj Kumar Deka

Abstract In a smart environment, mining of patterns of occurrences of the past events leads to the prediction of the appropriate event to be occurred as next event. As such, this work is to identify the challenges appearing in smart environment research with respect to finding the pattern of occurrences, followed by a proposed solution. The proposed solution is designed and implemented over the existing Apriori data mining algorithm with certain required modifications to incorporate the time information of occurrences of events. During the implementation of the said modified algorithm, appropriate datasets related to the events of daily activities of the inhabitants of the given environments have been provided; and it is found that the implementation has given the appropriate results with respect to the requirements of the work. Moreover, one more modification over the Apriori algorithm is made for considering all possible subsets of a candidate item set; otherwise impact of the introduction of the time duration information is not up to the level as desired. From the experiment through implementation, it is claimed that the proposed mechanism can choose the next event from the list of predicted events more accurately after inclusion of time duration information between consecutive events.

Keywords Apriori algorithm · Smart environment · Internet of Things · Association rules · Frequent item set

1 Introduction

Smart environment has become an emerging research area nowadays. Smart environment generally refers to a world, where human can perform their day-to-day activities by man–machine interaction. The backbone of the transition of physical

L. Goswami (✉) · M. K. Deka
Bodoland University, Debargaon, Kokrajhar, BTR, Assam, India
e-mail: lipichanda93@gmail.com

V. Deka
Gauhati University, Gopinath Bordoloi Nagar, Guwahati, Assam, India

M. S. Kaiser et al. (eds.), *Information and Communication Technology for Competitive Strategies (ICTCS 2021)*, Lecture Notes in Networks and Systems 401,
https://doi.org/10.1007/978-981-19-0098-3_7

environment to smart environment concept is mainly because of the introduction and development of the sensor and actuator technologies, energy efficient communication medium, the new computing methodologies with learning capabilities, etc. One of the suitable definitions on smart environment, defined by Diane J Cook is "One that is able to acquire and apply knowledge about an environment and also to adapt to its inhabitants in order to improve their experience in that environment" [1]. As smart environment is a domain which is interdisciplinary in nature, there are several scopes for the researchers to work with different challenges of different domains at different levels. For making the daily living activities more efficient through smart environment, researchers of almost all parts of the globe are working with different issues like building a home automation as test bed, incorporation of data mining and machine learning as part of learning mechanism, compatibility issues of IoT platform, communication medium, computer vision, etc. Some of the projects as part of such works are as follows:

CASAS project [2]: The CASAS project designed in Washington State University is a smart home containing three different layers, namely physical layer, middleware, and application layer. In the physical layer of the architecture, hardware like sensors and actuators are used and a wireless mesh network is used for direct communication with those components. The middleware performs services like adding time span for the events, allocations of UUIDs, etc. Finally, the topmost layer, i.e., the application layer is involved in recognition and discovery of activities.

The Neural Network House [3]: Main aim of this project was to predict the need of individuals. The electrical appliances of the household such as light, fan, geezer, etc., are automatically turned on and off based on the needs of the individual before s/he does it manually.

The Kids Room [4]: This is a perceptual world, where with the help of Internet of Things, a child can gather knowledge and enrich their vocabulary on English language.

Georgia Tech Aware Home [5]: The aim of this project is to automate the day-to-day activities such as turning on the coffee maker in the morning, displaying the list of medicines, turning on the room heater on a rainy day, etc.

MavHome Project [6]: Another project on smart environment is the MavHome Project. It aims at helping the elderly and physically challenged persons by automating the day-to-day activities, so that they can avail all the facilities without any external assistance.

2 Apriori Algorithm for Association Rule Mining

Apriori algorithm is a data mining algorithm for finding frequent item sets and generating association rules from those item sets. Apriori property states that all the subsets of a frequent item set will be frequent and all the supersets of infrequent subset will be infrequent. The working principle of the Apriori algorithm can be better understood by the flow diagram given in the Fig. 1. In Fig. 1, Min _Sup is the

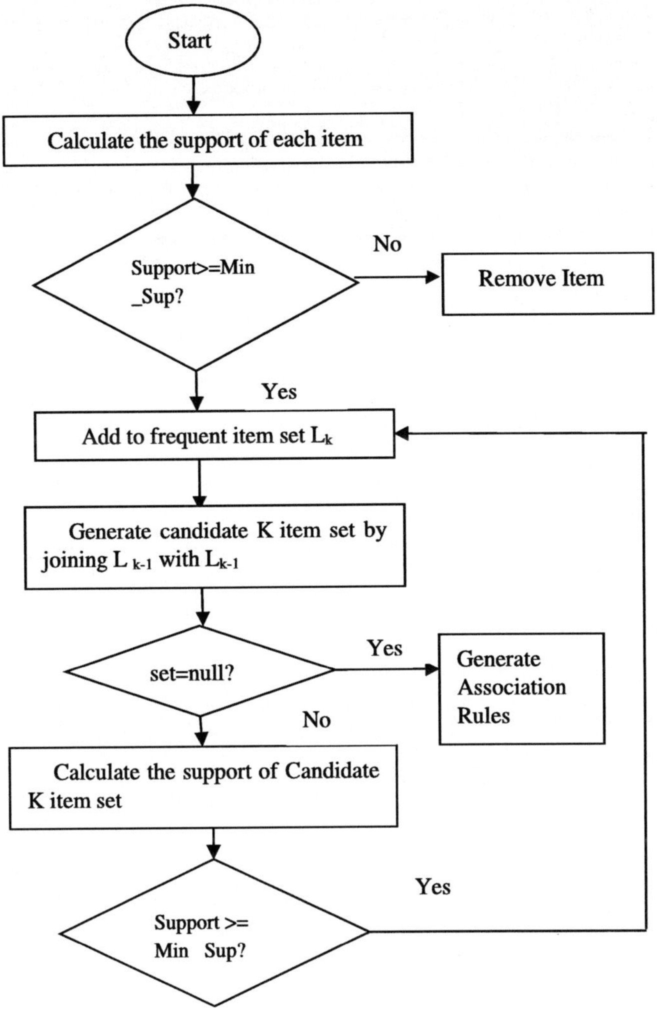

Fig. 1 Flow diagram of Apriori algorithm

threshold value named as "Minimum Support" for finding out the frequent item set; where support of any item set *X*, denoted by supp(*X*), is the number of transactions containing the item set X out of the total number of transactions. Moreover, once the frequent item sets are generated, the association rules are to be derived from those item sets using the metric confidence as defined below $\text{Conf}(X \to Y) = \text{Sup}(X \cup Y)/\ \text{Sup}(X)$ $\text{Conf}(X \to Y)$ means confidence of *X* to *Y*. *X* is called antecedent, and *Y* is called consequent. $\text{Conf}(X \to Y)$ gives the probability of the item *Y* to be present in dataset, in combination with the item *X*.

Working of the algorithm can be better understood from the following flow diagram:

3 Challenges and Proposed Mechanism:

While using the conventional Apriori algorithm in a dataset of smart environment, frequent item sets and association rules could be derived successfully. But there is no way to predict the future events based on the frequent item sets. In a smart environment, start time and end-time of an activity can be recorded by monitoring the sensor information. Thus, it opens up a possibility of adding time difference between consecutive events, which may lead to prediction of future events based on the events occurred at any given time.

While introducing the time parameter in Apriori algorithm, the basic structure of the algorithm is needed to be modified slightly so that the main purpose of adding the time value between consecutive events becomes fruitful.

The problem with classic Apriori algorithm is that it does not consider the order between two or more events. That means if three events occur in an order {1,2,3}, {2,3,1},{3,2,1}, then the algorithm will consider it as single item set. In other words, it considers the item set {1,2,3} as combination of the three and not the permutation.

Therefore, in this work, an attempt has been made to modify Apriori algorithm in such a way that it considers all possible subsets as a separate candidate set. Upon generating the candidate item sets, cumulative average time difference has been calculated between consecutive item sets.

Example below will show how the proposed algorithm works with secondary datasets (Fig. 2; Table 1).

Considering minimum support = 2.

In the above example, the item set {1,2} has come 4 times in the transaction. So the average difference is calculated as follows:

1(10:03:00) − 2(10:13:00) = 00:10:00
1(11:30:00) − 2(11:45:00) = 00:15:00
Avg = (00:10:00 + 00:15:00)/2 = 00:12:30
1(12:04:00) − 2(12:20:00) = 00:16:00
Avg = (00:12:30 + 00:16:00)/2 = 00:14:15
1(15:10:00) − 2(15:20:00) = 00:10:00
Avg = (00:14:15 + 00:10:00)/2 = 00:12:07

This is the average duration between 1 and 2 and is same for all the candidate-2 item set.

For candidate-3 item set, {1,2,1} we have calculated difference between {1,2} from those transactions which contain the item set {1,2,1}.

Thus, from 1(10:03:00) 2(10:13:00) 1(10:30:00) we get {1,2} = 00:10:00 {2,1} = 00:17:00.

1(11:30:00) 2(11:45:00) 1(11:50:00) gives {1,2} = 00:15:00 {2,1} = 00:05:00.

Considering minimum support=2

Sensor ID	Support
1	8
2	5
3	4

⇒ L2

Sensor ID	Support	Duration
{1,1}	2	00:23:30
{1,2}	4	00:12:07
{1,3}	3	00:17:00
{2,1}	3	00:09:15
{2,3}	2	00:8:30
{3,1}	2	00:10:00

L3

Sensor ID	Support	Duration
{1,2,1}	2	{1,2}=00:12:30 {2,1}=00:11:00
{1,2,3}	2	{1,2}=00:10:00 {2,3}=00:08:30

Fig. 2 Item sets with time duration in Modified Algorithm

Table 1 Respective transaction ID with sensor ID and start time

Transaction ID	Sensor ID with start time
1	1(10:03:00) 2(10:13:00) 3(10:20:00) 1(10:30:00)
2	2(11:20:00) 1(11:30:00) 4(11:40:00) 2(11:45:00) 1(11:50:00)
3	1(12:04:00)3(12:15:00)2(12:20:00)
4	1(13:03:00)5(13:10:00)
5	3(14:10:00) 1(14:20:00)

So average difference between {1,2} = 00:12:30 {2,1} = 00:11:00.

For candidate-3 item set {1,2,3}, the transaction 1(10:03:00) 2(10:13:00) 3(10:20:00) gives {1,2} = 00:10:00 and {2,3} = 00:07:00.

Next transaction 1(15:10:00) 2(15:20:00) 3(15:30:00) gives {1, 2} = 00:10:00 {2, 3} = 00:10:00. So average duration between {1, 2} = 00:10:00 {2, 3} = 00:08:30. It shows that duration between {1,2} in item set {1,2,3} and in item set{1,2,1} are different. Hence, by looking at the duration we can predict the next event to be occurred.

Association rules from the above item sets taking minimum confidence = 0.3 are as follows:

1 → 2 Support: 3, Confidence 0.375.
2 → 1 Support: 4, Confidence 1.0
2 → 3 Support 3, Confidence: 1.0
3 → 1 Support: 2, Confidence: 1.0
1, 2 → 1 Support: 3, Confidence: 1.0

1, 2 → 3 Support 2, Confidence: 1.0

3.1 The Modified Algorithm

<Forall Recordsets do> // Recordsets € T, a Record set with time information

<Generate Candidate-1 item set based on minimum support>

<for(k=2; L_{k-1}={ };k++) do begin> // until no more frequent item sets generate

<C_k= apriori-gen(L_{k-1})> // all possible item set of size k, considering the permute tions of the items

<C_t=subset(c_k,t)> // calculates support of each subset of C_k

<Forall candidates c€ C_t do>

<c.count++> // to get the support for each candidate frequent itemset

<End for>

<L_k={c € C_k | c.cout>=min-s}> // all frequent k item sets

<C_{ktime}=avg(L_m,L_{m+1},L_{m+2}) > // calculate cumulative average time difference between the consecutive items in an item set and store the time values

<Forall candidates c€ C_{time} do>

<c.count++> // to get the support for each candidate frequent item set

<End for>

<AlgoGenRules(F, minconf)> // Generate Association Rules and show the corresponding time difference

4 Experiments and Results

4.1 Dataset Details

The dataset used to carry out the experiment has two attributes: sensor IDs and start time of corresponding sensor. Dataset is a data file with .dat extension. Each odd numbered row contains transactions, and the even numbered row contains start time of the sensors in that specific transaction. Different datasets are available on daily living activities; one of the popular examples is the CASAS dataset. The dataset contains activities, sensors with start time and end-time, and object of the sensor for each transaction. All the other information is cleaned up from the dataset keeping sensor IDs and start times of the sensor and is made suitable to carry out this experiment.

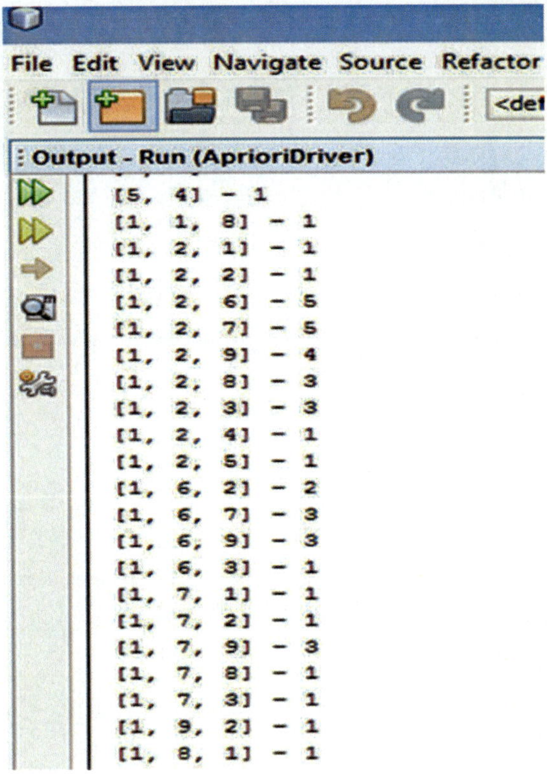

Fig. 3 Generation of candidate-2 item set

4.2 *Discussion on Results*

The enhanced Apriori algorithm has been implemented in Java platform in Windows 10 PC. Using the abovementioned dataset, all the frequent events occurred in day-to-day activities have been find out. After that, cumulative average time difference has been calculated between consecutive frequent events. Once the time value has been found according to the requirements, association rules are derived using minimum confidence value. Finally, a diagrammatic representation is made where the association rules contains the time difference between consecutive events.

Some of the snapshots are given below (Figs. 3, 4, 5, and 6):

5 Conclusion and Future Work

Smart environment has become an emerging research area in today's world. In the emerging technology, researchers are working on daily activity recognition in a home setting. In order to get the proper ordering of day-to-day activities, it is necessary

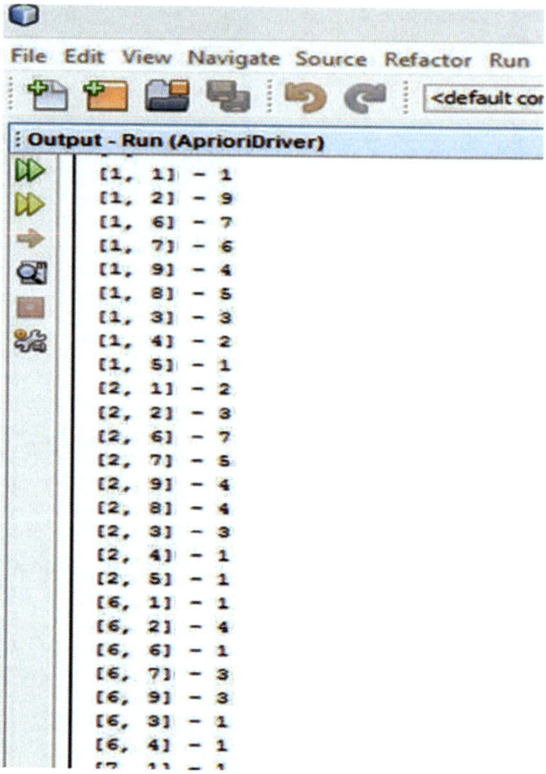

Fig. 4 Generation of candidate-3 item set

to include time information. The database that stores the time information can be named as temporal database. The classic association rule mining algorithm does not provide time duration between item sets.

In this work, the algorithm is modified to integrate the time information between consecutive sensor events. To fulfill this objective, a secondary dataset is used and necessary experiments are carried out using Java as a programming language.

As a future work, the end-time of the event can be considered, which will tell us the duration of time, for which the predicted event will occur.

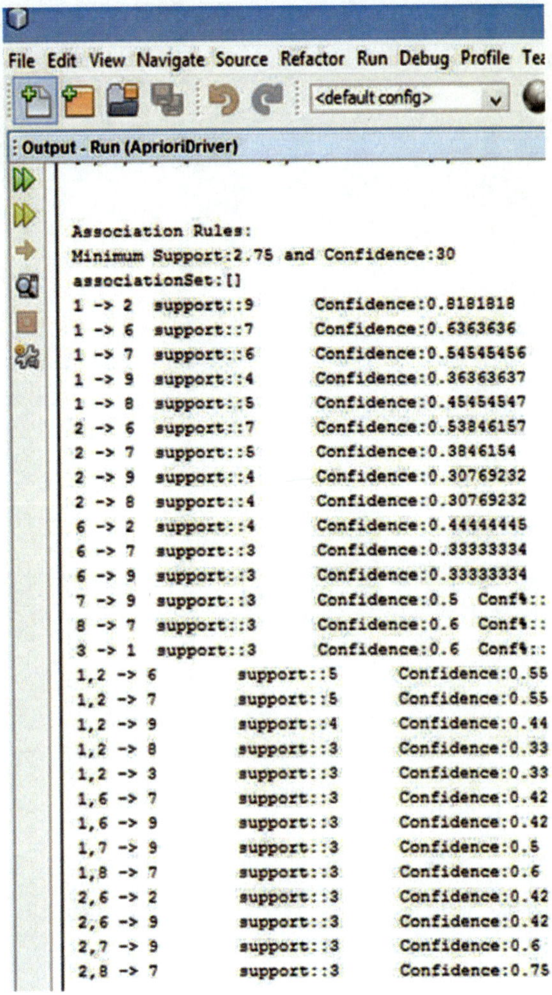

Fig. 5 Generation of association rules

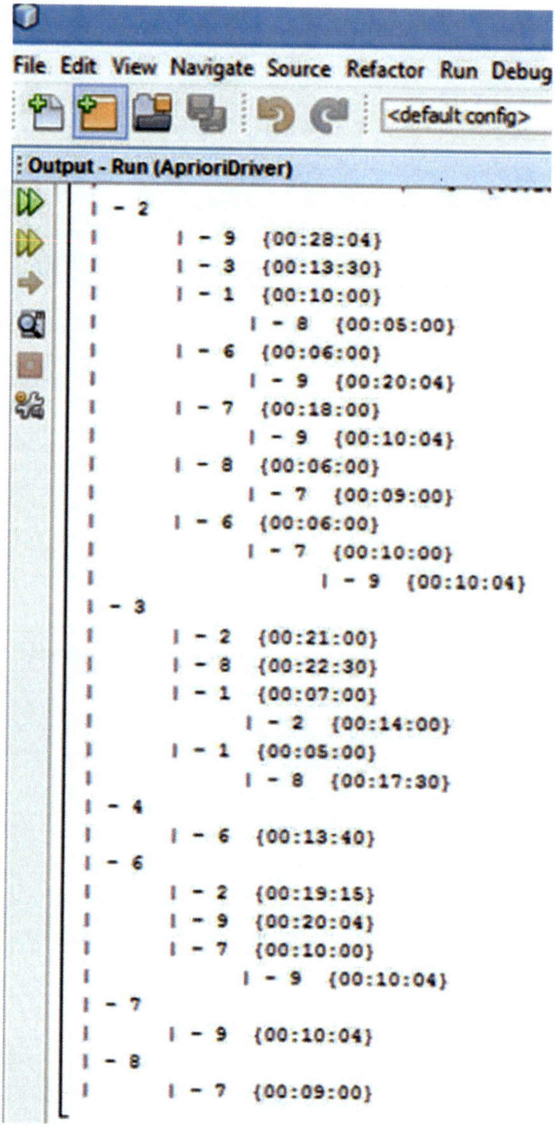

Fig. 6 Association rules with time duration

Acknowledgements This work is part of All India Council for Technical Education (AICTE) funded research project titled "Design and Development of a Prototype Model for Smart Environment" under RPS-NER scheme.

References

1. Cook DJ, Das SK (2005) Smart envrionments technologies, protocols and applications. John Willey and Sons, Inc., Hoboken, New Jersey
2. Cook DJ, Crandall AS, Thomas BL, Krishnan NC (2012) CASAS: a smart home in a box. Washington State University, Pullman, WA USA, pp 62–69. https://doi.org/10.1109/MC..328
3. Mozer MC (1998) The neural network house: an environment that adapts to its inhabitants. In: Proceedings of the American Association for Artificial Intelligence, pp 110–114
4. Bobick A, Intille S, Davis J, Baird F, Pinhanez C, Campbell L, Ivanov Y, Schutte A, Wilson A (1999) TheKidsRoom: a perceptually-based interactive and immersive story environment. Presence 8(4):369–393
5. Kidd CD, Orr R, Abowd GD, Atkeson CG, Essa IA, MacIntyre B, Newstetter W (1999) The aware home: a living laboratory for ubiquitous computing research. In: Cooperative buildings: integrating information, organizations and architecture. Springer, Heidelberg, pp 191–198
6. Cook D, Youngblood M, Heierman I, Gopalratnam EOK, Rao S, Litvin A, Khawaja F (2003) Mavhome: an agent-based smart home. In: Proceedings of the first IEEE international conference on Pervasive Computing and Communications. (PerCom 2003), pp 521–524

Imbibing Auxiliary Information Using Transfer Learning to Improve Collaborative Filtering-Based Recommender System

Shyamal Shah and Ronakkumar Patel

Abstract The available data on the Internet can be utilized to obtain valuable information on any given subject. But the data is massive and intricate, which produces many problems for users in finding information of their interests. Recommender Systems (RSs) are successful software techniques designed to solve this issue. The RSs have a broad range of uses, such as providing consumers with a suggested list of products for online shopping, recommending articles or books for reading online, film or music recommendations, media recommendations, and so on. The traditional Collaborative Filtering (CF) recommender systems use different item ratings as a foundation to recommend new items to various users. The sparsity of this user-item matrix affects the performance of RS. Also, most consumers on today's e-commerce websites would not purchase a product that they do not find appealing, even though it has especially favorable feedback. As a result, we can say that consumers give equal importance to product images and the item's ratings. This paper aims to research to improve the traditional collaborative filtering recommender system by imbibing visual characteristics with explicit feedback (ratings). The visual features are extracted using transfer learning with the convolutional neural network (CNN). The amazon product dataset for clothing, shoes, and jewelry is used to validate the proposed approach. The results are compared using standard evaluation measures like MAE and RMSE.

Keywords Collaborative filtering · Recommender system · Convolutional neural network · Transfer learning

S. Shah (✉) · R. Patel
U & P U. Patel Department of Computer Engineering, Chandubhai S. Patel Institute of Technology (CSPIT), Charotar University of Science and Technology (CHARUSAT), Anand, India
e-mail: d19ce152@charusat.edu.in

R. Patel
e-mail: ronakpatel.ce@charusat.ac.in

M. S. Kaiser et al. (eds.), *Information and Communication Technology for Competitive Strategies (ICTCS 2021)*, Lecture Notes in Networks and Systems 401,
https://doi.org/10.1007/978-981-19-0098-3_8

1 Introduction

The data generated are enormous and accelerating day by day. On YouTube, 500 h of videos are uploaded per minute [1]. With such an enormous amount of data on the Internet, information filtering becomes difficult and very much complicated. RSs play a vital role in helping users to filter useful information from this data. A variety of areas utilizes RSs as an integral part of their system to provide users a personalized experience. The RSs are suggesting video on YouTube, music on Spotify, products on Amazon, friends on Facebook, etc.

RSs usually make use of either one or both Content-based filtering (CB) and Collaborative filtering (CF). The CB utilizes the features of the items and users to find products having similar traits as users. The CF [2] takes into consideration the numerical ratings of the items provided by the users and finds either similar users or similar items. There are many applications nowadays combine both CB and CF in their recommendation engine. This technique is called the Hybrid recommender system [3].

The user-item matrix for CF in real-world applications has millions of records but is very sparse, resulting in low performance and challenging to identify correlations between users or items. The sparsity in the user-item matrix triggers one big problem, which is the key explanation for unsatisfactory outcomes. Since the amount of goods sold on the Internet is incredibly high and continuously increasing, and the number of products rated by active users is still limited. This results in having a large margin of error in calculating ratings of unrated items. The CF uses the rating profile in the recommendation process, but new users or products will not have enough ratings, resulting in a problem where it will not be recommended to any user at all. This situation is called a cold start problem. The auxiliary information like product image, review, metadata are available and RSs can take advantage of it. This paper focuses on improving the CF approach using transfer learning which uses the product's image. The pre-trained CNN on myntra fashion data [4] is used to extract the features from the amazon product images.

The remaining paper is organized as follows: Sect. 2 discusses the literature in the domain. The proposed approach is described in Sect. 3, which includes training of CNN on fashion dataset, use of this pre-trained model in amazon dataset, similarity measurement calculation, and a method for predicting rating. Experiments and result analyses are discussed in Sects. 4 and 5 concludes the paper.

2 Literature Review

In this section, we outline several methodological approaches, techniques, and algorithms put in place to manage the influence of data sparsity in CF RSs and image feature extraction.

The authors in [5] proposed Coordinate System Transfer (CST) which integrates the auxiliary data domains of both the user and item through a principled matrix-based transfer learning framework. The proposed method in [6] to reduce the sparsity problem using fuzzy inference rules that will increase the recommendation accuracy. The authors have developed 5 fuzzy inference rules to overcome the sparsity in the user rating matrix which grouped users into groups such that the prediction of the user-item preference becomes easy.

In the paper [7], authors came up with a method that reduces the sparsity problem and the cold start problem using Bi-separated clustering algorithm. This algorithm clusters the rating matrix simultaneously into users and items bi-clusters based on rating classification. Then used a Bi-Mean imputation algorithm to calculate and plug the missing ratings for the items in the bi-clusters. The authors used this newly created rating matrix and executed traditional CF to eliminate cold start and sparsity problems. The authors [8] discussed a hierarchical Bayesian framework that embodies metadata of items into the pair-wise ranking models for CF. The proposed method is called Collaborative Topic Ranking (CTRank) and it is based on the fact that items to be recommended have more information than the implicit feedback or ratings. The CTRank incorporates metadata like "bags-of-words" and implicit feedback simultaneously for a CF system.

In the paper [9], authors proposed a differentially private recommendation scheme that will reduce the sparsity problem along with protects users' preferences by protecting the user-rated items list and user rating information. The authors [10] presented a method for alleviating the sparsity problem in CF by using trust inferences. The trust inferences are transferable connections between users in the supporting social network. The trust inferences supplement the standard collaborative filtering method, which decreases sparsity and the cold start problem. Moreover, their model takes into account the subjective concept of trust and represents how it is raised in real-world social networks.

In the paper [11], authors discussed a method for mapping image content to low-level features. The author also identified some of the existing issues in the Content-Based Image Retrivel (CBIR) and Biometrics systems for image content and image feature extraction. The authors in [12], presented a comprehensive overview of recent advances in the domains of image feature extraction and image feature representation. The author expanded on the most common visual image features, such as color, shape, and texture. The author also investigated the efficacy of combining global and local features in image processing, employing various classic models and their illustrative examples from the literature.

The authors [13] provided a detailed identification technique of CNN. The author has discussed the fundamentals of CNN, including its structure, distinct layers, and operating method. In addition, the author has provided a practical example of how CNN models can be used to identify handwriting.

Takumi and Nobuyuki [14] proposed two techniques for extracting image features that use 2nd order statistics, namely, spatial and orientational auto-correlations of local image gradients. The authors were able to extract more information from images using the Gradient Local Auto Correlation (GLAC) and Normal Local Auto Correlation (NLAC) techniques than using typical histogram-based methods. In the paper [15], authors introduced a deep CNN for extracting image features for Hyperspectral Image Classification (HIS). To extract deep features from nonlinear, discriminant, and invariant HSIs, the proposed method utilizes multiple convolutional and pooling layers. More crucially, the authors presented a 3-D CNN-based feature extraction approach with coupled regularization for extracting effective spectral-spatial characteristics from hyperspectral images. The suggested deep feature extraction gives up new avenues for future research.

The authors of the study [16] used deep learning to compare the five distinct classifier models. Convolutional neural networks are used to extract visual features. On the CIFAR-10 dataset, the authors trained and evaluated K-nearest neighbors (KNN), support vector machines (SVM), softmax, Fully-connected (FC) neural networks and CNN classifiers. The results demonstrate that CNN significantly decreases the mistake rate.

In the study [17], authors proposed a local binary model method and CNN technique that are utilized to extract features from tumor CT images in the medical area. As a continuous optimization method, the algorithm and model presented in this study use the enlarged data set to train a new tumor image-sensitive to deep CNN and aim to include additional types of cost variables in the classification process to continue the algorithm improvement.

3 Proposed Approach

In this section, we describe the proposed approach which uses ratings and image features of the item for rating calculation. As shown in Fig. 1 the CNN model which is required to extract image features is trained on fashion product dataset. Once the model is trained, then this trained model is used to extract the visual features from the Amazon product images. When we combine the features, at that time with the image features of the product we also have the ratings of the users who rated both the items. This combined feature (rating + image feature) is used to find similarities between items and then for the rating prediction.

The CNN model architecture used for the transfer learning and then feature extraction is shown in Fig. 2. The CNN model is trained for the classification task on the 72 categories of fashion dataset [4]. Multiple architectures with different parameters and hyper parameters are tested and then the final CNN architecture is chosen.

3.1 Rating Calculation

The rating calculation depends on the similarity between users or items. In this RS, the IbCF has been used and the similarities between items are calculated based on

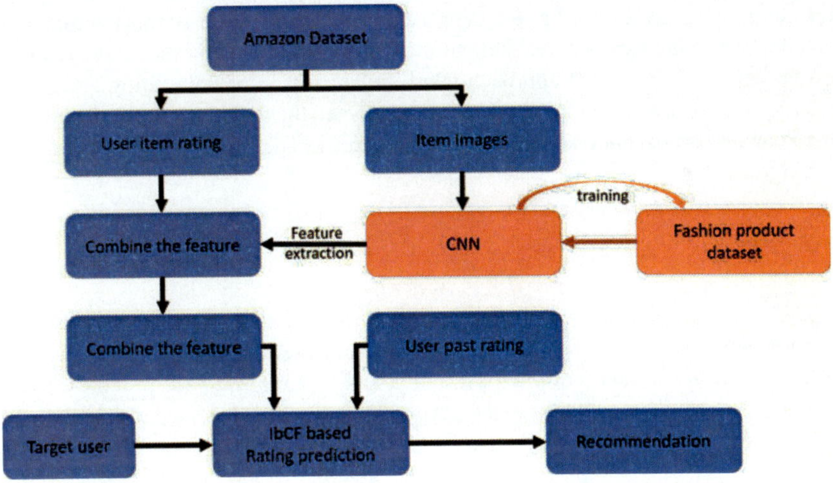

Fig. 1 Flowchart of the proposed approach

Layer (type)	Output Shape	Param #
conv2d_1 (Conv2D)	(None, 64, 64, 32)	896
leaky_re_lu_1 (LeakyReLU)	(None, 64, 64, 32)	0
max_pooling2d_1 (MaxPooling2	(None, 32, 32, 32)	0
dropout_1 (Dropout)	(None, 32, 32, 32)	0
conv2d_2 (Conv2D)	(None, 32, 32, 64)	18496
leaky_re_lu_2 (LeakyReLU)	(None, 32, 32, 64)	0
max_pooling2d_2 (MaxPooling2	(None, 16, 16, 64)	0
dropout_2 (Dropout)	(None, 16, 16, 64)	0
flatten_1 (Flatten)	(None, 16384)	0
dense_1 (Dense)	(None, 64)	1048640
leaky_re_lu_3 (LeakyReLU)	(None, 64)	0
dropout_3 (Dropout)	(None, 64)	0
dense_2 (Dense)	(None, 72)	4680

Total params: 1,072,712
Trainable params: 1,072,712
Non-trainable params: 0

Fig. 2 Convolutional neural network (CNN) architecture for image feature extraction the combined features (rating + image feature). The advantage of this combination is when the ratings are not available in that case it will focus on the image features of the item. The proposed approach takes the advantage of both rating and image features if both are available. The similarity between the two items is calculated with the Pearson correlation coefficient as given in Eq. 1. The $\overline{x} = \frac{1}{n}\sum_{i=1}^{n} x_i$ (mean) and analogously for $\overline{y}$.

$$\text{sim}(x, y) = \frac{\sum_{i=1}^{n}(x_i - \overline{x})(y_i - \overline{y})}{\sqrt{\sum_{i=1}^{n}(x_i - \overline{x})^2}\sqrt{\sum_{i=1}^{n}(y_i - \overline{y})^2}} \tag{1}$$

Here sim (x, y) stands for the similarity between product x and y. Here, n is a set of users who have rated both items x and y, and r (u, x) and r (u, y) represent the rating given by user u to item x and y, respectively. The Eq. 2 is used to calculate the ratings of the items for the target user.

$$\hat{r}(u, x) = \frac{\sum_{j}^{J} \text{sim}(x, y) * r(u, y)}{\sum_{j}^{J} |\text{sim}(x, y)|} \tag{2}$$

where u denotes the user, and x and y represent the product. Here, we are trying to predict a rating for user u and product x which is represented by $\hat{r}(u, x)$. J denotes a set of products rated by user u and sim (x, y) is the similarity between product i and j.

4 Experiment Setup and Result Analysis

4.1 *Dataset and Evaluation Measures*

Deep learning models are more powerful than traditional machine learning models. However, these models are also very data-hungry: to achieve reliable performance, a large amount of training data is needed. The CNN model is trained on the myntra fashion product dataset [4] which consists of data of over 44,000 products. The dataset includes an image of each product, type of the product, for which gender the product is, name of the product, base color, master category, the year it was released, the season in which it is used, etc. This dataset does not contain the ratings and reviews of the products. The traditional CF approach uses the ratings to find the most relevant products to the users.

The Amazon Data (2018)'s Clothing, shoes, and jewelry 5 core Ranting dataset [18] are used as the base for applying a collaborative filtering algorithm and to evaluate the proposed approach. This data set also contains images of the products and some metadata as well but does not have a specific category that can be used as the base to classify products. But this dataset contains ratings to these products in the range of 1 to 5 by different users. So, transfer learning is used with the help of CNN to extract image features. The dataset is divided into 70%:30% for training, and testing. The proposed approach is validated using Mean Absolute Error (MAE), and Root Mean Square Error (RMSE) as given in Eqs. 3 and 4, respectively.

$$\mathrm{RMSE} = \sqrt{\frac{\sum_{i=1}^{n} \left(\hat{r}_i - r_i\right)^2}{n}} \tag{3}$$

$$\mathrm{MAE} = \frac{\sum_{i=1}^{n} \left(\hat{r}_i - r_i\right)}{n} \tag{4}$$

where n is the test dataset size, r_i, is the actual rating of the item indexed with i, $\hat{r}_i$, is the predicted rating of the item indexed with i.

4.2 *Result Analysis*

The proposed approach is compared with other approaches on given evaluation metrics. Method-1 is a traditional CF system that relies on ratings only, Method-2 is finding similarity based on the image features only and Method-3 is combining the rating and image features to find similarity.

The results of the 5-fold cross-validation of three methods prove that both RMSE and MAE in calculating rating are high in Method-1. The results in Tables 1 and 2 indicate the improvement in the proposed methods (Method-3). There is drastic improvement compare to Method-1 and Method-2. So, it indicates the more insight into the user and item you get, the performance of the RS will be enhanced. The proper fusion of such features are required. The visual features and ratings are complementing each other.

Table 1 Comparison of three methods with MAE

	Method-1	Method-2	Method-3
Fold-1	1.408	1.340	0.818
Fold-2	1.417	1.347	0.820
Fold-3	1.412	1.343	0.817
Fold-4	1.403	1.344	0.823
Fold-5	1.408	1.346	0.819
Average	1.410	1.344	0.819

Table 2 Comparison of three methods with RMSE

	Method-1	Method-2	Method-3
Fold-1	2.166	1.906	1.369
Fold-2	2.173	1.913	1.369
Fold-3	2.169	1.909	1.368
Fold-4	2.161	1.909	1.377
Fold-5	2.163	1.910	1.366
Average	2.166	1.909	1.370

5 Conclusion and Future Work

Major factors which increase the errors in rating prediction in the CF RSs are the cold start problem and data sparsity. Most of the RSs having other auxiliary information as well which can be incorporated to enhance the performance of the RSs. The use of image features is explored in this paper and the experiment indicated the improvement in RS when image features are used. As we include image features in the rating prediction process, the new items will now have the possibility to make correlations based on image features with other items and hence will not suffer from the cold start problem and the newly added item also has a chance to get recommended. In future work, other information like review, review score, metadata, etc. can be used to enhance the RSs performance.

References

1. Susan Wojcicki: https://blog.youtube/news-and-events/youtube-at-15-my-personal-journey
2. He X, Jin X (2019) Collaborative filtering recommendation algorithm considering users' preferences for item attributes. In: 2019 International conference on big data and computational intelligence (ICBDCI), IEEE, pp 1–6
3. Jung K-Y, Park D-H, Lee J-H (2004) Hybrid collaborative filtering and content-based filtering for improved recommender system. In: Computational Science—ICCS 2004. Lecture Notes in Computer Science, vol 3036
4. Aggarwal P (2019) Fashion product images (small). https://www.kaggle.com/paramaggarwal/fashion-product-images-small
5. Pan W, Xiang E, Liu N, Yang Q (2010) Transfer learning in collaborative filtering for sparsity reduction. In: Proceedings of the AAAI conference on artificial intelligence, vol 24, No 1.
6. Sachan A, Richhariya V (2013) Reduction of data sparsity in collaborative filtering based on fuzzy inference rules. Int J Adv Comput Res 3(2):101
7. Roko A, Almu A, Mohammed A, Saidu I (2020) An enhanced data sparsity reduction method for effective collaborative filtering recommendations. Int J Educ Manage Eng 10(1):27
8. Yao W, He J, Wang H, Zhang Y, Cao J (2015) Collaborative topic ranking: Leveraging item meta-data for sparsity reduction. In: Proceedings of the AAAI conference on artificial intelligence, vol 29, No 1
9. Kim T, Jung K, Park S (2020) Sparsity reduction technique using grouping method for matrix factorization in differentially private recommendation systems. IEICE Trans Inf Syst 103(7):1683–1692

10. Papagelis M, Plexousakis D, Kutsuras T (2005) Alleviating the sparsity problem of collaborative filtering using trust inferences. In: International conference on trust management. Springer, Berlin, Heidelberg, pp 224–239
11. Choras RS (2007) Image feature extraction techniques and their applications for CBIR and biometrics systems. Int J Biol Biomed Eng 1(1):6–16
12. Ping Tian D (2013) A review on image feature extraction and representation techniques. Int J Multimedia Ubiquitous Eng 8(4):385–396
13. Liu YH (2018) Feature extraction and image recognition with convolutional neural networks. J Phys Conf Ser 1087(6): 062032
14. Kobayashi T, Otsu N (2008) Image feature extraction using gradient local auto-correlations. In: European conference on computer vision. Springer, Berlin, Heidelberg, pp 346–358
15. Chen Y, Jiang H, Li C, Jia X, Ghamisi P (2016) Deep feature extraction and classification of hyperspectral images based on convolutional neural networks. IEEE Trans Geosci Remote Sens 54(10):6232–6251
16. Jogin M, Madhulika MS, Divya GD, Meghana RK, Apoorva S (2018) Feature extraction using convolution neural networks (CNN) and deep learning. In: 2018 3rd IEEE international conference on Recent Trends in Electronics, Information and Communication technology (RTEICT). IEEE, pp 2319–2323
17. Yang A, Yang X, Wu W, Liu H, Zhuansun Y (2019) Research on feature extraction of tumor image based on convolutional neural network. IEEE Access 7:24204–24213
18. Ni J, Li J, McAuley J (2019) Justifying recommendations using distantly-labeled reviews and fine-grained aspects. In: Proceedings of the 2019 conference on empirical methods in natural language processing and the 9th international joint conference on natural language processing (EMNLP-IJCNLP), pp 188–197

An Efficient Resource Allocation Mechanism Using Intelligent Scheduling for Managing Energy in Cloud Computing Infrastructure

M. Karuppasamy, M. Jansi Rani, and M. Prabha

Abstract We hit the issue of conservation of energy in cloud data centers with the effective utilization of resources used for the completion of tasks in cloud. Here, we consider the virtual machine requirements and task completion time of process. We model the system based on the process priority and task completion with the availability of resources for intelligent scheduling. Based on the scheduling algorithm, the energy efficiency and performance got increased. Then, we can reduce the power and cost of maintenance using the proposed methodology.

Keywords Cloud computing · Resource allocation · Green computing

1 Introduction

Clouds play a vital role in storing data, providing services based on the customer needs. Many of the cloud service providers like Google, Amazon, Microsoft Office, and Salesforce are providing the quality of services based on the customer needs. Cloud architecture is built with hardware and network devices. In last decades, the number of data centers is increasing to support the online services and processing big data in digital world [1]. The performance varies from place to place. While accounting a single system, the performance is based on how fast it completes the execution of the process in time. While accounting a data center, the performance is measured with how many users' task is completed in time. The storage and energy efficiency are more important than shear speed performance [2]. The data centers are

M. Karuppasamy (✉)
Kalasalingam Academy of Research and Education, Krishnankoil, Tamil Nadu, India
e-mail: karuppasamy.m1987@gmail.com

M. Jansi Rani
Mepco Schlenk Engineering College, Sivakasi, Tamil Nadu, India
e-mail: jansirani@mepcoeng.ac.in

M. Prabha
Velammal College of Engineering and Technology, Madurai, India
e-mail: mpr@vcet.ac.in

M. S. Kaiser et al. (eds.), *Information and Communication Technology for Competitive Strategies (ICTCS 2021)*, Lecture Notes in Networks and Systems 401,
https://doi.org/10.1007/978-981-19-0098-3_9

built with thousands of servers. The cost of building the data centers are increased day by day [3]. Only, 30% of data center costs in purchasing the equipment needed. The remaining 33% is assigned to chiller; 18% amount tends toward uninterruptible power supply (UPS), 9% amount to computer room air conditioning (CRAC), and 7% to power distribution, lighting, and transformer costs. From this, we can come to know that 60% of amount spent on maintenance and management.

The challenges in managing cloud data centers are as follows: (1) scalability, (2) reliability, (3) multiuser manageability, (4) green information technology [4]. The power consumption by the data centers is more problematic. In 2018, the report says that US data centers used greater than 90 billion kilowatt-hours of electricity a year. Global data centers used roughly 416 terawatts (4.16×10^{14} watts) (or about 3% of the total electricity), nearly 40% more than the entire United Kingdom, and this consumption will be doubled every four years. So, it leads to the increase in demand of electricity and also leads to the emission of CO_2 and naturally the problem of global warming [5–8].

For the issue, if we schedule the user's task wisely, we can manage the data centers efficiently. The paper in detail discusses about the scheduling which tends to manage the power consumption by the resources in cloud data centers. The Sect. 2 explains the related work; Section 3 discusses the proposed scheduling techniques in energy conservation; Section 4 deals with conclusion.

2 Related Work

First, we could analyze the service requested by users. The scientific workflows execution with efficient energy utilization becomes necessary. The efficient scheduling of the scientific workflows will reduce the amount consumption of energy [9]. Cloud architecture consists of large number of users, and the users' service will be executed on different virtual machines (VMs). It assumes the user that their services are handled by infinite machines [10, 11]. Scheduling should consider the dependencies and communications of the tasks. Scientific workflow needs a large amount of computational resources for their execution. It is very important to schedule the resources in managing the energy with less consumption. If we fail in the sense, it is a total waste of energy. Here, they also consider the threshold temperature to reduce the energy consumption. Lin et al. [12] propose the SHEFT workflow scheduling algorithm to schedule a workflow elastically on a cloud computing environment. By this, they organize the resources with same capability of computation is considered as a cluster, and tasks are ranked; the resources are allocated based on their availability. Green computing is a terminology for designing, managing, using, and disposing the computing resources to reduce the power, cost, and emissions of carbon dioxide (CO_2). With the help of green scheduler and using renewable resources, we can utilize the energy efficiently [13].

For the process, scheduling HEFT algorithm was proposed to increase the performance. The algorithm assigns the task to the processor having the finishing time fast

[14]. Considering the data centers, it is growing by its infrastructure. We cannot stop because there is always demand in servicing tasks by the users. The cloud service providers (CSPs) are also in competition to provide the quality of service (QoS). Data centers energy consumption has raised speeder and substantial to the operation costs and impacts in environment [15, 16]. There are more number of researches are going on in considering the issue of data centers to go into the technology adapts green [17, 18]. Resource allocation is performed as sequential manner, and every methodology should fail at intervals of time. Since this fashion continues, the cloud environment will be in a place to switch over from time to time [19].

3 Proposed Scheduling Techniques in Energy Conservation

The customer's request in cloud must be satisfied by the service providers based on intelligent scheduling with effective utilization of energy. For that we can categorize the customers into periodical and non-periodical customers (Fig. 1). They are further divided according to their needs in cloud. The periodical customers are the one having touch with the cloud in a regular manner. The non-periodical customers are not in regular touch with the cloud, but they will contact the cloud in random basis. Considering the periodical and non-periodic customer's analysis, the data centers are going to switch on the resources used such as virtual machines, switches, and network. In case of any dynamic need of resources, some resources may put in sleep mode. Through this, we can effectively use the resources. We can avoid the wastage of energy. While moving to the scenario of users' task execution, the intelligent scheduler should segregate the users' task based on their needs. The computation time and communication costs are to be calculated and prioritize accordingly.

On the other hand, the resources in data centers are also switched on according to the historic. We have to check for the availability of resources is sufficient to service the request. If it is not sufficient, then we can switch on the additional resources. The resources available (R_{avail}) are exceeding the requested tasks (R_{req}); we cannot

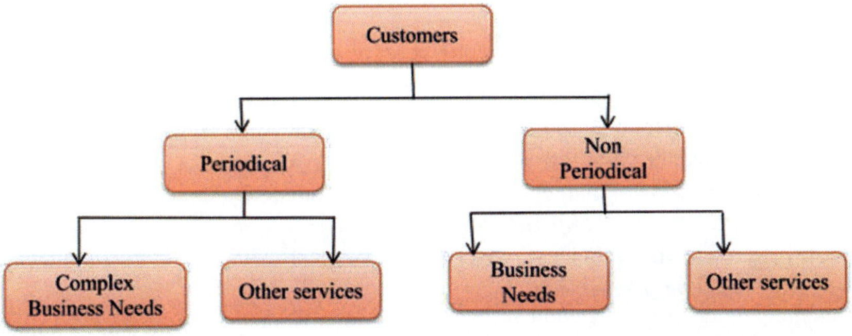

Fig. 1 Categories of customers

switch off the additional resources. We should put the additional resources in sleep mode. This is the better methodology to reduce the power consumption.

Intelligent Scheduling Algorithm

Step 1 Identify the customers

Step 2 Split the tasks based on needs and calculate the execution time cost and communication cost

Step 3 Prioritize the tasks

Step 4 If ($R_{req} > R_{avail}$)

Step 5 Switch on additional resources

Step 6 Else if ($R_{avail} > R_{req}$)

Step 7 Extra resources will be in sleep mode

Step 8 End if

Step 9 Arrange the tasks in ascending order by considering the priority

Step 10 While there are any incomplete task in the list do

Step 11 Allocate the resources to the task which is best fit

Step 12 End while

$$\text{Generally, the energy usage is related to power and time :}$$
$$E = P * T$$

where **power** P is in watts, **Power** $P =$ work/time (W/T). **Energy** E is in joules, and time T is in seconds. For every resource in data centers, like virtual machines, routers, nodes, storage, network, cooling unit consume some amount of energy [20]. If the resources are in idle condition, they are using some amount of energy. According to the proposed algorithm, the resources are switched on whenever there is a necessity. So, the amount of energy consumed by the resources in idle condition is avoided. We can preserve the energy using the methodology. The emissions are also reduced.

The intelligent resource allocation algorithm will tend to save the energy 30%. The graphical representation shows the relation that before applying the methodology and after applying the methodology in data centers (Fig. 2).

4 Conclusion

The effective usage of power is enhanced through the intelligent scheduling is proposed in this paper. Based on the history and customers' needs, the resources are allocated, and the resources that will be used for critical period are being in sleep mode. The remaining resources are in switched off mode. We can control the usage of power by applying this methodology. The economy is improved; the emissions of CO_2 are reduced. By this way, we can satisfy many users per day. This optimizes the performance of data centers.

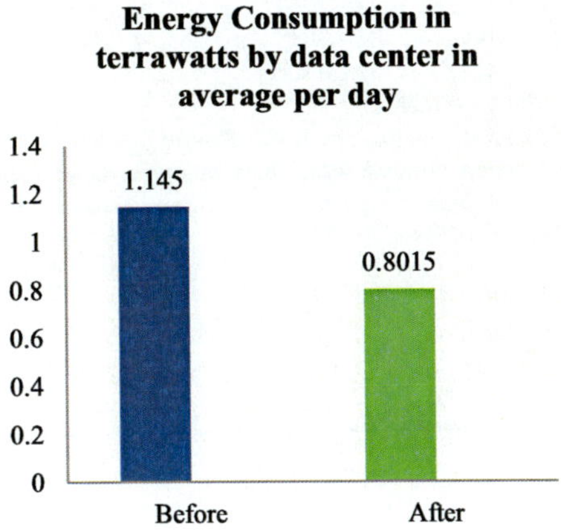

Fig. 2 Energy consumption graph for before versus after applying the methodology

References

1. Koomey JG (2008) Worldwide electricity used in data centers. Environ Res Lett 3 (3)
2. Hwang K, Fox GC, Dongarra JJ (2012) Distributed and cloud computing: from parallel processing to the internet of things, Elsevier Inc. 2012 ISBN: 978-0-12-385880-1
3. Dayarathna M, Wen Y, Fan R (2016) Data center energy consumption modeling: a survey. IEEE Commun Surveys Tutor 18(1)
4. Bahari HI, Shariff SSM (2016) Review on data center issues and challenges: towards the green data center. In: 2016 6th IEEE international conference on control system, computing and engineering, 25–27 November 2016, Penang, Malaysia
5. Doyle J, Shorten R, O'Mahony D (2013) Stratus: load balancing the cloud for carbon emissions control. IEEE Trans Cloud Comput 1(1)
6. Karuppasamy M, Suprakash S, Balakannan SP (2018) Energy-aware resource allocation for an unceasing green cloud environment. In: IEEE proceedings of 2017 international conference on intelligent computing and control, I2C2 2017, pp 1–4
7. Karuppasamy M, Suprakash S, Balakannan SP (2018) Energy efficient utilization of cloud resources using hybrid ant colony genetic algorithm for a sustainable green cloud environment. In: IEEE proceedings of 2017 international conference on intelligent computing and control, I2C2 2017, pp 11–14
8. Karuppasamy M, Suprakash S, Balakannan SP (2013) Energy efficient cloud network towards a sustainable green environment. Int J Res Eng Adv Technol 1(1):1–3
9. Ghose M, Verma P, Karmakar S, Sahu A (2017) Energy efficient scheduling of scientific workflows in cloud environment. In: 2017 IEEE 19th international conference on high performance computing and communications; IEEE 15th international conference on smart city; IEEE 3rd international conference on data science and systems
10. Armbrust et al M (2009) Above the clouds: a berkeley view of cloud computing
11. Buyya R et al (2009) Cloud computing and emerging it platforms: vision, hype, and reality for delivering computing as the 5th utility. Futu Gen Com Sys 25(6):599–616
12. Lin C, Lu S, Scheduling scientific workflows elastically for cloud computing. In: 2011 IEEE 4th international conference on cloud computing
13. Gu C, Liu C, Zhang J, Huang H, Jia X (2015) Green scheduling for cloud data centers using renewable resources. In: IEEE INFOCOM 2015 workshop on mobile cloud and virtualization
14. Topcuoglu H, Hariri S, Wu M-Y (2002) Performance-effective and low-complexity task scheduling for heterogeneous computing. In: IEEE transactions on parallel and distributed systems 13(3)

15. Chait K, Juiz C (2013) Research line on improving energy efficiency in web servers
16. Koomey J, Belady C, Patterson M (2009) assessing trends over time in performance, costs, and energy use for servers, Lawrence Berkeley Natl. Lab. Stanford Univ. Microsoft Corp. Intel Corp. Tech. Rep.
17. Ricciardi S, Careglio D, Santos-Boada G, Sole-Pareta J, Fiore U, Palmieri F (2011) Saving energy in data center infrastructures, data compression. In: 2011 First international conference on Data Compression, Communications and Processing (CCP), pp 265–270
18. Pendelberry SL, Thurston M, Strasenburgh J, Stein R (2012) Case study—the making of a green data center. In: 2012 IEEE international symposium on sustainable systems and technology, pp 1–6
19. Fiandrino C, Kliazovich D, Bouvry P, Zomaya AY (2015) Performance and energy efficiency metrics for communication systems of cloud computing data centers. IEEE Trans Cloud Comput
20. Karuppasamy M, Balakannan SP (2018) Energy saving from cloud resources for a sustainable green cloud computing environment. J Cyber Secur Mobil 7(2):95–108

Online Learning App

Spurthi S. Bhat and Milind V. Kulkarni

Abstract Nowadays, many students are using online learning apps due to the surge of people having Internet connectivity as well as due to ongoing pandemic where getting an education in a traditional way is not possible. The students can learn and practice at their own pace as well as educators can monitor student's progress and teach them easily through online learning platforms. Online learning app is developed using the Java language which can be used by students to learn different coding concepts, data structures, and algorithms. Students can test their knowledge by attempting quiz regarding the same. The platform being used is Android Studio. The backend used is Firebase Realtime Database where the assets required by the app are stored and email authentication is performed using the authentication service provided by Firebase.

Keywords Online learning · Online education · App development · Java · Android Studio · Firebase realtime database · Authentication · Coding · Data structures · Algorithms

1 Introduction

Online learning can be defined as a process of obtaining education or knowledge without being physically present in the classroom or in front of the teacher and through a web page or mobile app. The students have access to study material throughout the day and all days of the week, and teachers can monitor the progress of students easily. Education has enormously changed over the last few years with increased awareness about distance learning and the ongoing pandemic in the world. Obtaining certifications as well as performing guided projects through online learning

S. S. Bhat (✉) · M. V. Kulkarni
Vishwakarma Institute of Technology, Pune, Maharashtra 411037, India
e-mail: spurthi.bhat19@vit.edu

M. V. Kulkarni
e-mail: milind.kulkarni@vit.edu

M. S. Kaiser et al. (eds.), *Information and Communication Technology for Competitive Strategies (ICTCS 2021)*, Lecture Notes in Networks and Systems 401,
https://doi.org/10.1007/978-981-19-0098-3_10

platforms has become very common these days with the increasing number of smartphones and online learning apps. Hence, developing an online learning app has many benefits. This online learning app has been developed using Java which makes it compatible with Android operating system. This app has the backend as Firebase Realtime database which is very powerful in terms of authentication and can be used to create real-time databases very easily. Online learning app has been developed on the Android Studio platform which has many tools for the development of Android apps. The user interface of the app has many features which give the user a smooth experience throughout the app. The app has many screens such as home screen, code screen, profile screen, login screen, registration screen, leaderboard screen, quiz screen, contact details of the developer, and eBooks screen.

2 Literature Review

E-learning makes the learning process more effective, and due to the resources available at all the times, students are able to learn at their own pace and are able to remember the concepts due to colorful pictures and animations. The flexibility provided through online learning helped the students to develop their individual skills more efficiently as well as to make use of variety of tools available on the Internet suitable as per their needs. According to the survey conducted by the authors, the major disadvantage of e-Learning was the need to buy equipment such as computers and the cost of Internet access. Also, the students missed out on interacting with their peers during the learning process on online platforms which hindered their social skills [1]. The major impact of online learning actually depends on a number of factors such as the quality of content, interaction between instructors and learners. The instructors on online platforms must be well-prepared and must provide good support to the students to create a sense of collaborative learning. This has been supported by rapid advancement in technology in terms of audio and video tools such as Google Hangout and Skype. Online learning is expected to develop further by incorporating many interesting technologies such as virtual reality, social network, and video conferencing [2]. The authors conducted a study based on Transactional Distance Theory and Bloom's Taxonomy Theory to throw light on utilization of online learning platforms in students' improvement in academics. The results revealed that students use online learning platforms only if it meets their study requirements and if it is better than traditional learning. Also, the prior knowledge of such online learning platforms plays an important role in student's preference of such platforms. Instructors formed an important part of communication with learners. Online learning platforms gave students' a sense of autonomy such as students' by setting their own goals and taking their own decision in learning process. Students have also shown to have better understanding, applying of knowledge, analyzing, and remembering skills in their courses on online learning platforms [3]. The author has studied the impact of COVID-19 pandemic on education in India and found that the students and teachers have been encouraged to be technological savvy. There has been a great demand of online learning management systems, and students have been accustomed to make use of soft copy of materials due to the unavailability of hard copy of study materials because of lockdown in many cities in India. Also, teachers and students have been forced to learn and use digital technologies which have led to worldwide exposure of resources and instructors. Students have become better with time management due to online learning. But the negative impact of pandemic is that many teachers have been conducting lectures on Google Meet, Zoom, etc., which is not true online learning without any devoted online learning platform. Numerous students have limited or no Internet students which has increased the gap between rural and urban. The author has suggested some measures to improve online learning process in India like establishing quality assurance mechanisms for ensuring standard of contents on Internet and provisioning maintainable access to Internet and learning resources to all the students and to create a safe learning space [4]. The authors have conducted a survey for performing methodical research of e-learning platforms to solve different problems faced by Indian engineering students during the pandemic. Google Meet and Zoom have been top preference by engineering students. Still there have been security concerns on these platforms such as video hijacking and spreading hateful comments. The educational institutions have been conducting exams on the same platform without no proctoring resulting in large number of malpractices by students. The lack of integration of tools for various purposes on one platform is one of the major drawbacks of such platforms [5]. The authors have developed an app which is based on Android operating system and can be installed on a smartphone with Internet access provides learning material in one part of the app as well as chat room for queries in the other part of the app. The learning material can be accessed offline and free of cost. The queries in the chat room are resolved in real time [6].

3 Methodology

Online learning app is a tool designed for students to gain knowledge about data structures and algorithms in a simplified manner. This app has been developed to provide a single platform for providing learning material from the best resources on the Internet and to test one's knowledge regarding the same with the help of quiz. It also provides free eBooks which can be downloaded by students to refer it offline. Live leaderboard updates motivate the students to perform better each time and to enhance their performance in academics. It basically promotes mobile learning which can take place through smartphones which is affordable to most of the learners with Internet connection which is much better than existing learning systems which usually requires computers. Since it is not limited to any particular educational platform company, it can be used to include resources which are totally free of cost and which promotes self-learning among students.

The steps involved in the development of online learning app are as follows:

Design

The online learning app is divided into three sections such as home section, code section, and quiz section. The different widgets required for these sections are designed accordingly. The individual widgets of each screen are as follows:

- Splash Screen: It displays the logo of the app with animation for about two seconds.
- Home Screen: All the categories of different data structures and profile icon are displayed. On clicking any topic leads to its subtopics which in turn lead to webpages from where information about the particular topic is available. The profile icon leads to the profile screen. A drawer has been provided where users can access leaderboard, eBooks, and about screens.
- Leaderboard Screen: The total scores of all the users of the app are displayed along with their user names and rankings.
- Ebooks Screen: This screen displays all the books that are available, and on clicking any book title, it opens the book in pdf readable format. The option to download the book is also present.
- About Page: This screen displays the main motivation and idea of the app along with the contact details of the developer.
- Profile Screen: This screen displays the name and email id of the logged in user along with a button to logout.
- Code Screen*:* All the different data structures and algorithms are displayed, wherein when the user clicks on any topic it will show the problem statements within that particular topic. On clicking any problem statement, the code in C/Java language will be displayed.
- Quiz Screen: This screen displays different quiz questions about the concepts included in the app to test the knowledge of the users. A question along with the four options are displayed. On clicking an option, it will display the correct answer. There is also a button to share the question with others, and the score will be displayed at the end.
- Registration Screen: This screen has the fields to enter the username, email, and password to create an account of new user.
- Login Screen: Email and password authentication of the user takes place in this screen.

Figure 1 displays an overview of all the different screens and widgets in the online learning app.

Code

Code for the different user interface components and the screens are connected to each other through various widgets like navigation bar and buttons. The navigation drawer on the home screen enables smooth navigation for the users within these screens. Code is written on the platform of Android Studio. The main part of the code is the use of different libraries and packages. Android Pdf Viewer is a library for displaying PDF documents on Android, with animations, gestures, zoom, and double tap support. It is based on Pdfium Android for decoding PDF files, and it

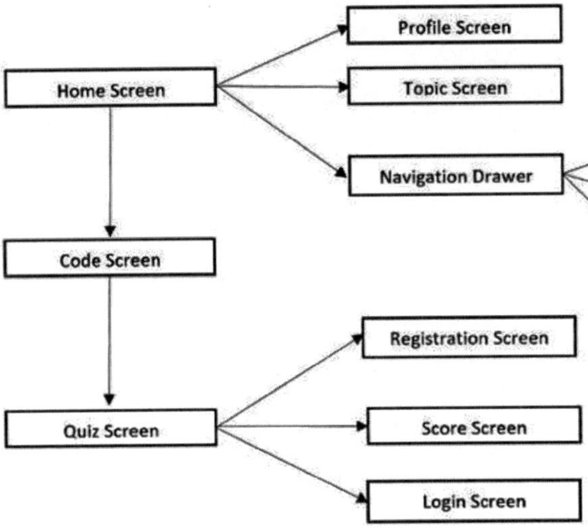

Fig. 1 Various screens and widgets of online learning

works on API 11 (Android 3.0) and higher. It has been used for displaying the eBooks within the app. Another important library is CodeView which is an Android widget which can show code and highlight it. It can automatically recognize the dominant languages such as Java, C++ , C#, python, and bash. This library has been used for displaying the codes for different problem statements, and these programs have been hardcoded into the app.

Creation of Database

Creation of a database in Firebase Realtime database allows to create a unique json file along with key value pairs. All the assets required for the app are stored in Google Cloud Platform in a bucket which allots the space for storing the assets in the cloud. Authentication of the user is performed by email authentication feature of Firebase. The database stores the scores, quiz questions and Ebooks for easy access and retrieval. For storing the quiz questions, the json of the database is downloaded and the required questions for the quiz are edited into the json and reuploaded. The structure of the database is that it has quiz collection with correct answer, question and options as the subfields. The scores collection have the user's name, id, and scores stored as subfields. The eBooks collection has the name of the book and download URL as the subfields.

Connecting the database to the App

To connect the Firebase Realtime Database with the app, the dependencies of firebase_auth and firebase_core are installed. The Firebase Realtime database is connected to the app, and errors regarding the same are handled.

Adding Features

Different features are added to the app to enhance the user's experience throughout the app. Lastly, the app can be tested by downloading it on different smartphones to test its performance.

4 Results

All the screens of the app have been interconnected with each other successfully. The images of the different sections in the home page are retrieved from the database with the help of database reference of the storage in the Google Cloud Platform. The user's name and email in the profile section are also retrieved from the Firebase Realtime database successfully. The user authentication is done with the help of Firebase authentication service. All the different parts of the app such as quiz and code screens have been coded in the form of fragments so that it can be easily accessed through horizontal navigation bar. The widgets and common functions used in all the screens have been coded separately into different Java files to maintain code quality and to increase code reusability. Some screens like leaderboard, about, eBooks can be accessed using the drawer in the home page which saves a lot of screen space instead of having a navigation bar.

Figure 2 displays the screenshots of home screen, code screen, and quiz screen from left to right.

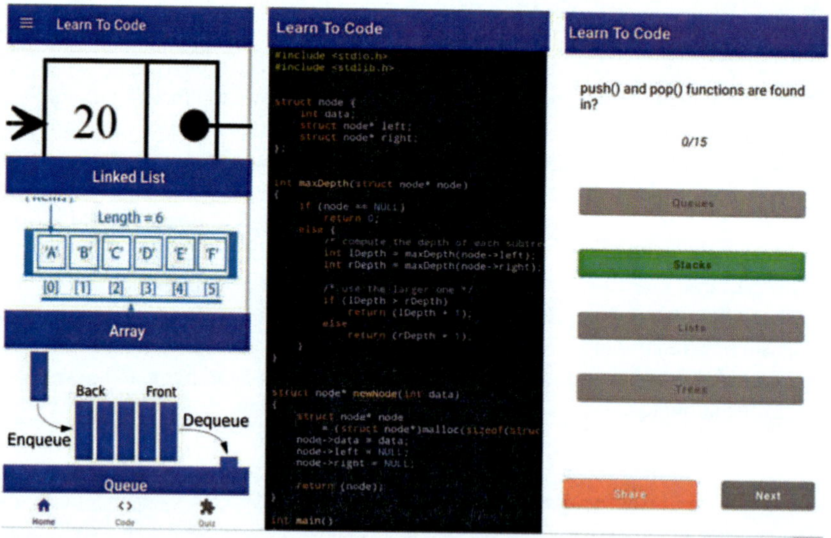

Fig. 2 Screenshots of home screen, code screen, and quiz screen (from left to right)

5 Limitations

There are some limitations of Android Development through Java as well as the app development which are as follows:

- It is compatible with only Android Operating System which makes it incompatible with iOS.
- Most of the codes presented in the code view are hardcoded which makes it restricted to limited content.
- The compilation and running time are more as compared to other languages of app development like Flutter and Kotlin.
- The app has massive file size of about 30 Megabytes which can lead to increased use of RAM on mobile phones.

6 Future Scope

- App can be enhanced by features such as live quizzes and integrating many more content.
- Chat feature can be added to communicate with other fellow learners as well as a discussion forum can be integrated within the app.
- Admin side of the online learning app can be integrated within the same app.
- Profile of the users can be improved to display their current progress in terms of learning a particular concept.
- By integrating a live coding environment, the users can practice questions and problem statements within the app itself.
- Video conferencing can be included to conduct direct interaction among the learners.

7 Conclusion

- The importance of online learning has increased by many leaps due to increasing use of Internet and the ongoing pandemic.
- While new and innovative online learning tools are coming up and education-related startups are rising at a very fast pace, India is beginning to see the drift from traditional education system to online learning.
- Also, with the help of Java, creating online learning apps have become very fast and cost-effective investment.
- The users have the benefit of having the choice of courses, ease of learning, convenience of having education at home and at reduced cost.
- Online learning app is a tool for learners to study and gain more knowledge regarding programming free of cost and to analyze their capabilities in the form of quiz.

References

1. Luaran JE, Samsuri NN, Nadzri FA, Mohamad Rom KB (2014) A study on the student's perspective on the effectiveness of using e-learning. Proced Soc Behav Sci 123:139–144
2. Sun A, Chen X (2016) Online education and its effective practice: a research review. J Inf Technol Educ Res 15:157–190
3. Abuhassna H, Al-Rahmi WM, Yahya N et al (2020) Development of a new model on utilizing online learning platforms to improve students' academic achievements and satisfaction. Int J Educ Technol High Educ 17:38
4. Jena PK (2020) Impact of pandemic COVID-19 on education in India. Int J Current Res 12(7):12582–12586
5. Thakker SV, Parab J, Kaisare S (2021) Systematic research of e-learning platforms for solving challenges faced by Indian engineering students. Asian Assoc Open Univ J 16(1):1–19
6. Tamhane KD, Khan WT, Tribhuwan SR, Burke AP, Take SB (2015) Mobile learning application. Int J Sci Res Publ 5(3). ISSN 2250-3153

An Efficient Underwater Image Restoration Model for Digital Image Processing

Jeevan J. Jacob, S. Savitha, and Debasish Mukherjee

Abstract Digital image processing (DIP) is showing a massive growth in today's trending world particularly, in the field of biological research. Underwater image analysis plays a vital role, where the images are easily prone to attenuation and haziness. Capturing underwater images has always been a challenging job due to dispersion and scattering of light inside water on a high scale. Several image enhancement and restoration methodologies are currently available to address these issues, where hazing and color diffusion are viewed as a common phenomenon in it. Such procedures normally includes two basic methodologies in it, namely dehazing and contrast or color enhancement, which improves the overall output of the degraded image. However, the quality and processing time of the images can still be enhanced with additional techniques incorporated to it. This work is intended toward proposing one such channel called improvised bright channel prior for dehazing the underwater images. The technique further improves on the existing methodologies by estimating the atmospheric light and refining the transmittance of the image along with image restoration. The experimental results show that the improvised bright channel prior methodology is found to perform better in dehazing underwater images with a balanced intensity in terms of dark and white patches obtained from it. When comparing and contrasting the processing time of the proposed methodology with the existing techniques, it is found that improvised bright channel prior performs better. Also, the quality of the dehazed underwater image obtained from the proposed channel is found to be effective when compared with the existing channels.

Keywords Digital image processing · Underwater image restoration · Dehazing · Improved bright channel prior · Atmospheric light · Transmittance image

J. J. Jacob · S. Savitha (✉) · D. Mukherjee
Department of Computer Science and Engineering, Christ University, Bangalore, Karnataka, India
e-mail: savitha.s@christuniversity.in

J. J. Jacob
e-mail: jeevan.jacob@btech.christuniversity.in

D. Mukherjee
e-mail: debasish.mukherjee@christuniversity.in

M. S. Kaiser et al. (eds.), *Information and Communication Technology for Competitive Strategies (ICTCS 2021)*, Lecture Notes in Networks and Systems 401,
https://doi.org/10.1007/978-981-19-0098-3_11

1 Introduction

Image processing is a method to perform some computations on an image, to obtain an enhanced image, or to deduce some useful information out of it. These days, it is viewed as rapidly growing technology and is considered a core research area in engineering and computer science disciplines. The importance of digital image processing (DIP) can be understood from the two levels of applications. One is improving the pictorial representation of the data for better human interpretation, and the other is to improvise a machine's perception of a real-time scenario to obtain the maximum data possible on actual scale. DIP applications range from inspection of underwater structures for the offshore industry to the identification and counting of fishes for biological research. However, no matter how big the impact of this technology can be on industry and research, it still is in the early stages of development compared to a traditional computer vision sector.

In underwater computer vision domain, image preprocessing is the most eminent procedure for object detection. The images obtained by the underwater vision system show the characteristics of uneven illumination, low contrast, and heavy noise when compared to terrestrial imaging [1]. Underwater images are highly degraded because of the presence of medium around it. Most of the images encounter poor visibility because of the absorption and scattering of light through the medium resulting in different wavelengths of light that gets randomly attenuated. After a certain depth, the natural light is not even visible underwater; thereby, artificial lights are used to improve the lighting for the image. The particles suspended in water also contribute to absorption and deflection. This leads to degraded images with color diffusion, hazing, and blurring resulting in an unclear perception of the scene under observation.

The existing methodologies use dark channel prior (DCP) as the dehazing method [2, 3] which was initially developed for terrestrial images. Further, the technique was incorporated in underwater imaging and was considered as the most optimal technique for underwater image enhancement. Even though the algorithm gave optimum results in terrestrial imaging, the technique was found to show errors for underwater images due to noise amplification in some images. The algorithm was expected to provide better results in terms of quality, intensity, and processing time, which are viewed as a major shortcoming.

Enhancements were made to the DCP, and a new technique was proposed as the bright channel prior (BCP) [4] which worked on a different approach. Here, the channel dilates each color channel of the image and finds the maximum intensity found in the window of the degraded image. This technique was not found to be efficient enough for irregularly illuminated images where artificial lights do not cover the entire image appropriately. Also, there are chances where water is contaminated often with a lot of suspended particles that could result in uneven brightness of the underwater images. Such bright noises in the image were amplified in the bright channel prior, leading to bright patches.

Improved bright channel prior is proposed to encounter these problems in DCP and BCP. The contribution to the proposed system is as follows,

- To use improved bright channel prior to estimate a channel which preserves both the high and low intensity details of an image.
- To estimate atmospheric light and to produce an enhanced transmittance image by using soft matting technique.
- To improve the overall computational time when dehazing the underwater image.
- To improve the overall contrast for better visualization of underwater images.

2 Related Works

The quality improvement methods for underwater images can be classified into two, namely [1], image formation model—free and image formation model—based methods. These two methods were classified based on whether it used an optical imaging physical model or not. IFM free image enhancement methods focused on the improvement of color and contrast features of the image. One of the most widely used deep learning models, convolutional neural networks (CNN), had also been used in such image enhancement techniques.

Li C., Guo, C, et al. proposed an underwater image enhancement benchmark (UIEB) [5] that used a dataset of around 950 images for image enhancement. Out of these, 890 images were used as reference images to train the baseline model and further employed the information for testing the remaining 60 number of images. Further to access the image quality assessment, this technique analyzed the image enhancement network named Water-Net with the existing proposition model.

In 2017, Wang Y., Zhang J, et al. proposed a CNN approach [6] for underwater image enhancement named Underwater Image Enhancement-Net (UIE-Net), where the model used 2,00,000 images for training. A better design of the algorithm was expected to address the individual patches of the images and to obtain the maximum features from it, for which, a pixel disrupting strategy was deployed that resulted in increased convergence and accuracy.

Another deep model named the joint residual model [7] was introduced in 2018 by Hou M., Liu R., et al. The model described about knowledge-driven scene residual formulation based on image domains and their transmittance. Along with this, an adaptive exposure map for color correction was also found to be deployed for image enhancement. Another CNN-based image enhancement is discussed [8], where Liu P., Wang G., et al. employed a technique through a deep residual framework where a cycle-consistent adversarial network (CycleGAN) was used to generate the synthetic underwater images for the trained CNN model.

Additionally, the very-deep super-resolution reconstruction model (VDSR) along with the Underwater Resnet model (UResnet) was proposed for underwater resolution applications that also served as a residual learning model for underwater image enhancement procedures. However, compiling a massive amount of trained dataset was found to be tough, hence a generative adversarial network (GAN). Guo et al. [9] were employed that used a complex GAN with a multi-scale approach, dense concatenation, and residual learning. The algorithm was also improved with nonsaturating GAN loss function and gradient loss to focus more on the image features of the percept.

As observed in the [3, 10], the histogram of underwater images were observed to obtain a clean concentrated distribution of its pixels when compared to terrestrial images. Spatial-domain enhancements were used to enhance the histogram distributions, deployed on the three basic color models namely, RGB, HSL, and HSV. It was also observed that these models redistributed the pixels based on the histogram and their respective color models. Guo et al. [3] used a Contrast Limited Adaptive Histogram Equalization (CLAHE) method to increase the global contrast of the image. This color model was later changed from RGB to HSV, and the V channel was adjusted to improve the brightness in images, along with a dehazing technique.

Hitam et al. [10] used a similar approach, which employed the CLAHE method on both color models along with the dehazing technique. Similarly, in 2014, Guo et al. [2] used a histogram stretching and redistribution technique in their respective models where equalizing the mean of the color values in each color channel had been the major focus along with the dehazing technique.

One more commonly used spatial image enhancement technique is white balancing (WB). This technique was found to be adopted across various areas [11] where Li C., and Zhang X. combined the background light estimation and white balancing that replaced the classical DCP as aforementioned under IMF methodologies. The inclusion of WB was referred to as automatic white balancing that helped reduce the color distortion and enhanced the color correction of the restored image.

An alteration to DCP was observed in [12] by Wen H., Tian Y., et al. named underwater DCP that employed an optical model with a combination of underwater DCP transmission, scattering light, and background light estimation. All the estimations values were utilized in proposed formula to replace the standard dehazing algorithms.

Another technique named gamma correction also described about a new spatial image enhancement technique. Ancuti et al. [13] introduced gamma correction by applying WB to images. Further, two versions of the original image were formed based on color and white balancing. An elaborate qualitative and quantitative evaluation revealed that the results obtained were found to be varying based on various aspects. But, the result validation proved that the algorithms were found to be independent of the instrument configuration and improved the accuracy of many image processing applications.

Multiple color model spatial image enhancement technique [14] called the Markov random field (MRF) was found to be useful in removing the fogginess of images by correlating the images before and after distortion, which further added a factorial factor to obtain an effective image quality.

An alternative approach was taken by Fan et al. [15] on proposing blind deconvolution combined with Well's small-angle approximation theory. The model used a pulsed laser imaging system to get their initial perception of the environment. It was found to be a frequency-based imaging model, where high frequency corresponds to edges in the percept region. These high frequencies represented the changes showing edges. On the other hand, low frequency represented the flat regions in the percept.

Although the algorithms provided efficient results, complexity in the algorithm was not taken into consideration.

So, a straightforward approach was observed in simple linear transformation for color and contrast improvement by Fu et al. [16] that deployed a simple approach considering pixel-wise operations that resulted in implementation on real-time application. Adopting image blurriness with an image formation model, Peng et al. in [17] proposed an approach to estimate depth between scene points of the percept. The results of the experiment indicated that the algorithm works better than other IFM based models, even in challenging lighting conditions.

Another approach of enhancement technique was introduced in transform-domain image enhancement method [18] by Khan et al. as homomorphic filter, high-boost filter, wavelet transform that incorporated wavelet-based fusion to enhance hazy underwater images. The algorithm also addressed the low contrast and color alteration issues in underwater images.

IMF-based methodologies or image restoration methods established degradation models based on the physical properties of propagation of light in water. This is one of the widely used techniques in dark channel prior (DCP) [19, 20] with improvised forms of DCP where the thickness of the haze is directly estimated to recover a high-quality haze-free image. However, it was found that of the haze-free terrestrial images contained few pixels with low brightness among all the other channels in the color model.

An adaptation of DCP by Chiang J. Y., and Chen Y. C. is again found in [21] as wavelength compensation and image dehazing (WCID) model. The algorithm proposed here incorporated the classic DCP with artificial light removal for better lighting in deeper environments. Among these prior-based image restoration techniques, it is observed that there are plenty such algorithms like DCP and WCID exists, which were not widely used but still worked effectively. One such technique can be observed in [22] that used an attenuation curve prior. The algorithm initially estimated smoothness properties and various other attenuation coefficients of light for each color channel. Later, the transmission map was estimated, which restored the original image from the degraded image.

A simple approach similar to this was found in [23] by Peng Y. T., & Cosman P. C., where attenuation and light absorption values were used to estimate the depth of the image. The proposed method was found to be much more efficient in estimating the depth of the scene in the percept compared to a DCP approach. On similar par, DCP was used by Gao et al. [4, 24] with a different dehazing technique called bright channel prior (BCP). This prior channel used an approach where the objects of percept were indicated with bright patches. This also resulted in local patches due to dilation. Also, most of the images contained some pixels of high intensities in at least one color channel. The resultant channel was found to be deployed in further evaluations in finding dehazed image.

A very recent development was made in 2020 with a new algorithm named [25] adaptive color compensation and dual transmission estimation (ACDTE). The proposed methodology was found to use color-tone adaption technique for hue estimation that estimated the global light for the degraded image (due to non-uniform turbid water) using quad-tree decomposition. Further, transmission maps were estimated for each color channel (R/G/B in RGB) for image restoration. This methodology has been found to be highly promising and hence has been employed in marine exploration and environmental monitoring. To enhance the contrast and color, quad-tree methodology was found to be combined with the ocean optical properties in [26]. Also, the depth in the image has been improved by using a multi-channel guided image filter that resulted in an efficient visual quality of underwater images.

An alternative to histogram distribution was introduced by Wang et al. [27] to increase the contrast and brightness of an underwater image. The algorithm yielded two types of output. The first one was an actual image of the perception based on relatively genuine color to depict the natural appearance of the image. The other one was the high contrast and high brightness version of the percept to discover more details and extract additional information from it. On testing the algorithm, results were observed to be better in yielding a good image quality.

Another improvement was done by using a bilateral filter or trilateral filter [28] to remove the luminance and improve the contrast enhancement. Lu et al. adopted trigonometric bilateral filters to overcome scattering and color distortion of underwater images. The algorithm deployed was found have an automated color enhancement feature that resulted in better exposure to the scene, improved global contrast, and fine details on edges to enhance objects in the scene.

A completely different approach was proposed very recently in 2019 by Gao et al. [29], named adaptive retinal mechanism or fish retina model to collect feedback from color-sensitive horizontal cells to cones to correct the non-uniform color bias in underwater images. This technique enabled collection of feedback from other cells like amacrine, interplexiform, and ganglion for edge correction, contrast improvement, and color enhancement. The methodology was found to provide efficient results in terms of accurate transmission map.

Another unique approach is taken by Petit et al. [30] where the inversion of light attenuation is used along with quaternions for color space contraction. In the general percept, the background water area is converted to gray and the objects in the percept stay colored. This improves the overall contrast of the scene under observation. It also creates a strong differentiation between the background and the objects in the percept, providing an easier way to extract features from the image.

3 Methodology

3.1 Dark Channel Prior (DCP)

On analysis of haze-free terrestrial images, it is observed that there are pixels with low intensity in at least one color channel within an observed local patch of the image.

$$J^{\mathrm{dark}}(x) = \min_{y \in O(x)} \left(\min_{c \in r,g,b} RGB^{c}(y) \right) \quad (1)$$

The dark channel of the image is evaluated based on the Eq. (1). Here, RGB represents the original image, $O(x)$ is a local patch with x as its center, c represents a color channel among R, G, B, and Jdark represent the dark channel. Using the dark channel, an estimate of the atmospheric light in each color channel and the transmittance of the image is obtained. Later, the value of atmospheric light and transmittance is estimated as shown in Eq. (2).

$$RGB(x) = J^{\text{dark}}(x).T(x) + A(1 - T(x)) \tag{2}$$

3.2 Bright Channel Prior (BCP)

Unlike DCP, the bright regions in a degraded image tend to have a very high intensity in at least one color channel of its color model.

$$J^{\text{bright}}(x) = \min_{y \in O(x)} \left(\min_{c \in r,g,b} RGB^c(y) \right) \tag{3}$$

Equation (3) shows the implementation to obtain the bright channel of the observed hazy image. Again, RGB represents the original image, $O(x)$ is a local patch with x as its center, c represents a color channel among R, G, B, and Jbright represent the bright channel. Now again using the bright channel, an estimate of the atmospheric light in each color channel and the transmittance of the image is obtained. Later, the value of atmospheric light and transmittance is used in a similar degradation model as the DCP as shown in Eq. (2), to finally obtain the dehaze image. Both the models are implemented to understand the working of how the image is affected.

3.3 Improved Bright Channel Prior Setup

The models explained above are the most commonly used dehazing technique in underwater images and terrestrial images. Even though both the techniques have given efficient results, still there are some limitations and a wider scope of improvement to be addressed. Particularly, in haze removal, the underwater images projected the overlapped regions of two images like, water and sky on it. Often times, these regions were hard to be differentiated. Also, flickering patches are observed in some cases with poor dehazing, and though bright regions are well preserved, still noise amplification is observed. To overcome these issues, the proposed improved bright channel prior approach combines the functionalities of both DCP and BCP. The algorithm starts by making a copy of the original image as depicted in Fig. 1.

After which, a general sliding-neighborhood operation called nlfilter is used to create the new channel. Now, the newly evaluated channel is used similarly to DCP, to estimate the atmospheric light of each color channel. Further, the new channel

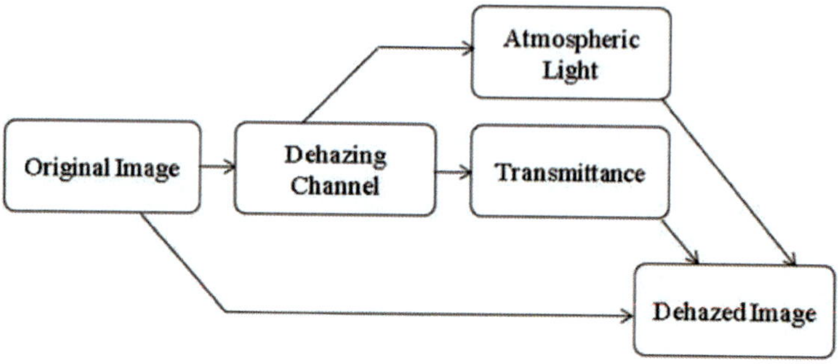

Fig. 1 Proposed System Model

is also used to find the transmittance image of the original hazy image. Here, the transmittance of the image is further improved by using a soft matting technique to estimate the foreground of the image. Matting is a widely used technique to extract or produce a matte that separates the foreground and background from a given image.

Finally, the atmospheric light for each channel is obtained and added along with the transmittance of the image to obtain the final dehazed image. The functions namely atmospheric light estimation, finding the transmittance of the image, estimating the foreground of the image, and removing the haze are executed in an incremental manner, where the inputs needed for each function are previously generated or initially declared. During this, the functions are also individually tested in unit testing. The output of each function is evaluated and compared at the end with the expected result from theory and actual result. All these functions are integrated and tested finally for the use.

4 Results and Analysis

The primary principle of improved bright channel prior as an underwater image restoration technique is to provide a better haze removal method. The method is expected to provide a distinctive channel image resulting in a better evaluation of atmospheric light and transmittance image. Following the work of DCP and BCP, the improved bright channel prior is used on a single degraded underwater image, and the results are compared.

As shown in Fig. 2b, it is observed that the dark channel resulted in dark patchy boundary when the image was analyzed. The output of the bright channel as observed in Fig. 2c indicates that the distance of the scene from the instrument of observation is indicated by the intensity, which also concludes the results as the closer an object is the brighter the patches appear in that specific region.

The proposed channel works more appropriate in object recognition as shown in Fig. 2d where the objects in the percept are identified by brighter intensity patches

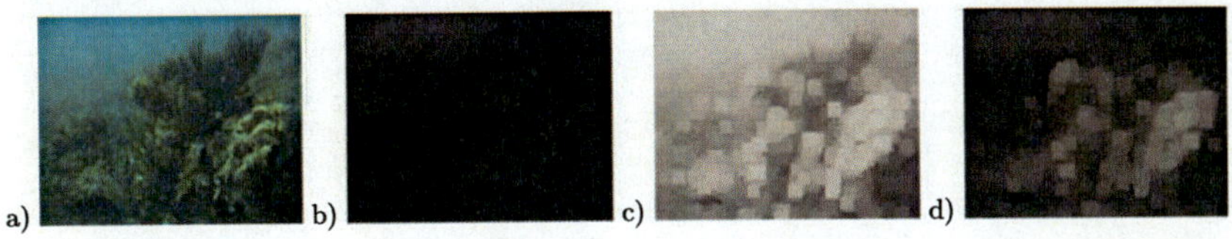

Fig. 2 **a** Original image as per the instrument's percept, **b** dark channel of the image, **c** bright channel of the image, **d** the proposed channel

in which the area of the object in the percept is observed clearly. That means objects closer to the instrument of observance are highlighted with bright patches. A higher scale of brightness indicates how close the image is in the percept. Also, it is found that the improved bright channel enhances the images on preserving the bright as well as the dark ends of the image as the background pixels are preserved with the darkest patches. The overall computational time of the improved bright channel is found to be reduced when compared to the existing systems as shown in the Fig. 3.

The comparative analysis of the algorithm with respect to the varying intensities between the dark channel prior, bright channel prior, and improved bright channel prior is shown in Fig. 4. Figure 4a shows the histogram of dark channel prior, where most of the pixels are observed in the lower intensities within the range of 0.05–0.3. Here, the higher intensity details of the image are lost in the process.

A similar observation is seen in Fig. 4 b) where most of the intensities are covered in the higher intensities within the range of 0.5–0.85. Though bright channel prior is

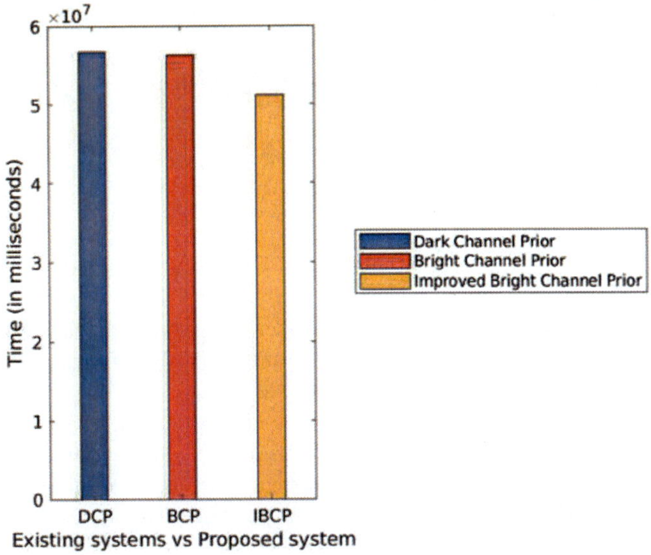

Fig. 3 Analysis of Computational Time

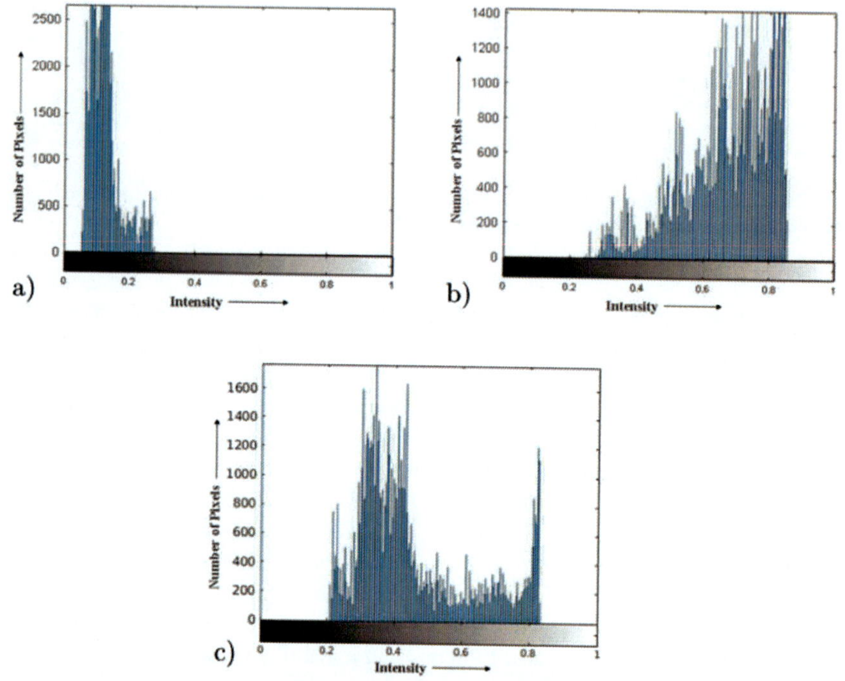

Fig. 4 Histogram analysis: **a** dark channel prior, **b** bright channel prior, **c** improved channel prior

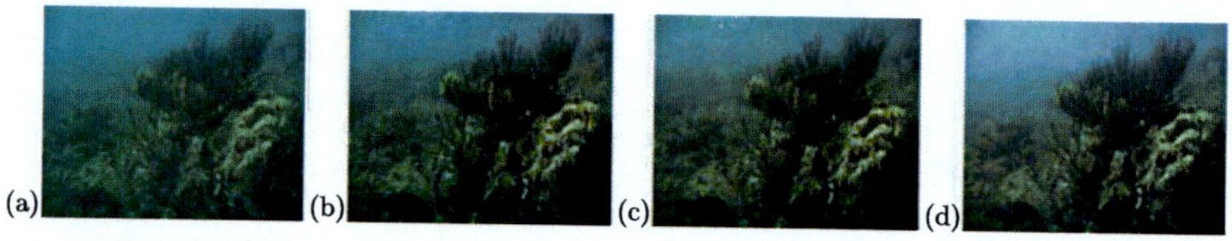

Fig. 5 Dehazed image: **a** original image, **b** dark channel prior, **c** bright channel prior, **d** improved channel prior

able to indicate some of the low intensity details, the number of pixels highlighting them is decreasing with lower intensities. On the other hand, Fig. 4c depicts the improved bright channel prior, covering a wider region with a reasonable amount of pixels highlighting each area. The range is observed to be between 0.2 and 0.8+ intensities, highlighting most of the details of the image.

The final dehazed image on deploying each channel can be observed in Fig. 5. The images b and c remove the haze, but patchiness is observed in it, whereas in BCP, lesser patchiness is observed when compared to DCP. Also, the improved bright channel prior is found to generate haze-free, no patchiness with an improvement on the overall contrast of the underwater image.

5 Conclusion and Future Work

The paper proposed an image dehazing technique called the improved bright channel prior that starts with the basic implementation of DCP. The method dehazed the hazy underwater images in terms of fogginess, patchiness, and contrast. Improved bright channel prior implemented a better channel prior compared to existing channels with respect to the intensities and processing time. In terms of the overall working principle of the system, it provided an enhanced channel for estimation of light and the transmittance for underwater image. This eventually helped in better haze removal with a high-quality underwater image restoration process. Also, incorporating soft matting technique has built the improved bright channel prior in such a way that the foreground estimation as well as increasing the transmittance of the image was easier to be achieved. Further, the performance of the channel as a dehazing method was compared with existing methodologies on factors like individual channel output, pixel preservation, computational time, and the final output of the channel. The proposed system is found to perform better on these aspects of underwater image restoration.

Although the final output was better in terms of intensity and processing time, the computational time was still high for a single underwater image. So the future work can be directed toward replacing the foreground estimation algorithm with a better algorithm that can greatly reduce the computational time. Also, as the improved bright channel prior focuses on haze removal, the algorithm could be combined with image enhancement techniques to various others fields of image processing and analysis. Doing so could improve the overall saturation of the image by using color enhancement or by improving the global contrast using contrast enhancements. The algorithm could also be designed in such a way, where a specific region of image is considered rather than deploying dilation of pixels to the entire image. This could further preserve finer details of the images processed.

References

1. Wang Y, Song W, Fortino G, Qi LZ, Zhang W, Liotta A (2019) An experimental-based review of image enhancement and image restoration methods for underwater imaging. IEEE Access 7:140233–140251
2. Guo JK, Sung CC, Chang HH (2014) Improving visibility and fidelity of underwater images using an adaptive restoration algorithm. In: OCEANS 2014-TAIPEI. IEEE, pp 1–6
3. Guo JK, Sung CC, Chang HH (2014) Restoration of underwater vision using a two-phase regularization mechanism. In: 2014 7th international congress on image and signal processing. IEEE, pp 243–247
4. Gao Y, Li H, Wen S (2016) Restoration and enhancement of underwater images based on bright channel prior. In: Mathematical problems in engineering
5. Li C, Guo C, Ren W, Cong R, Hou J, Kwong S, Tao D (2019) An underwater image enhancement benchmark dataset and beyond. IEEE Trans Image Process 29:4376–4389

6. Wang Y, Zhang J, Cao Y, Wang Z (2017) A deep CNN method for underwater image enhancement. In: 2017 IEEE International Conference on Image Processing (ICIP). IEEE, pp 1382–1386
7. Hou M, Liu R, Fan X, Luo Z (2018) Joint residual learning for underwater image enhancement. In: 2018 25th IEEE International Conference on Image Processing (ICIP). IEEE, pp 4043–4047
8. Liu P, Wang G, Qi H, Zhang C, Zheng H, Yu Z (2019) Underwater image enhancement with a deep residual framework. IEEE Access 7:94614–94629
9. Guo Y, Li H, Zhuang P (2019) Underwater image enhancement using a multiscale dense generative adversarial network. IEEE J Oceanic Eng 45(3):862–870
10. Hitam MS, Awalludin EA, Yussof WNJHW, Bachok Z (2013) Mixture contrast limited adaptive histogram equalization for underwater image enhancement. In: 2013 International Conference on Computer Applications Technology (ICCAT). IEEE, pp 1–5
11. Li C, Zhang X (2018) Underwater image restoration based on improved background light estimation and automatic white balance. In: 2018 11th international Congress on Image and Signal Processing, BioMedical Engineering and Informatics (CISP-BMEI). IEEE, pp 1–5
12. Wen H, Tian Y, Huang T, Gao W (2013) Single underwater image enhancement with a new optical model. In: 2013 IEEE International Symposium on Circuits and Systems (ISCAS). IEEE, pp 753–756
13. Ancuti CO, Ancuti C, De Vleeschouwer C, Bekaert P (2017) Color balance and fusion for underwater image enhancement. IEEE Trans Image Process 27(1):379–393
14. Yang HY, Chen PY, Huang CC, Zhuang YZ, Shiau YH (2011) Low complexity underwater image enhancement based on dark channel prior. In: 2011 second international conference on innovations in bio-inspired computing and applications. IEEE, pp 17–20
15. Fan F, Yang K, Fu B, Xia M, Zhang W (2010) Application of blind deconvolution approach with image quality metric in underwater image restoration. In: 2010 international conference on image analysis and signal processing. IEEE, pp 236–239
16. Fu X, Fan Z, Ling M, Huang Y, Ding X (2017) Two-step approach for single underwater image enhancement. In: 2017 international symposium on Intelligent Signal Processing and Communication Systems (ISPACS). IEEE, pp 789–794
17. Peng YT, Zhao X, Cosman PC (2015) Single underwater image enhancement using depth estimation based on blurriness. In: 2015 IEEE International Conference on Image Processing (ICIP). IEEE, pp 4952–4956
18. Khan A, Ali SSA, Malik AS, Anwer A, Meriaudeau F (2016) Underwater image enhancement by wavelet based fusion. In: 2016 IEEE international conference on Underwater System Technology: Theory and Applications (USYS). IEEE, pp 83–88
19. Tang A (2019) A restoration of underwater polarized images based on DCP. In: 2019 international conference on Sensing, Diagnostics, Prognostics, and Control (SDPC). IEEE, pp 675–678
20. Codevilla FM, da Costa Botelho SS, Drews P, Duarte Filho N, de Oliveira Gaya JF (2014) Underwater single image restoration using dark channel prior. In: 2014 symposium on Automation and Computation for Naval, Offshore and Subsea (NAVCOMP). IEEE, pp 18–21
21. Chiang JY, Chen YC (2011) Underwater image enhancement by wavelength compensation and dehazing. IEEE Trans Image Process 2121(4):1756–1769
22. Wang Y, Liu H, Chau LP (2017) Single underwater image restoration using attenuation-curve prior. In: 2017 IEEE International Symposium on Circuits and Systems (ISCAS). IEEE, pp 1–4
23. Peng YT, Cosman PC (2017) Underwater image restoration based on image blurriness and light absorption. IEEE Trans Image Process 26(4):1579–1594
24. Gao Y, Li H, Wen S (2016) Restoration and enhancement of underwater images based on bright channel prior. In: Mathematical problems in engineering
25. Wang Y, Cao J, Rizvi S, Hao Q, Fang Y (2020) Underwater image restoration based on adaptive colour compensation and dual transmission estimation. IEEE Access 8:207834–207843
26. Wu M, Luo K, Dang J, Li D (2017) Underwater image restoration using colour correction and non-local prior. In: OCEANS 2017-Aberdeen. IEEE, pp 1–5

27. Li CY, Guo JC, Cong RM, Pang YW, Wang B (2016) Underwater image enhancement by dehazing with minimum information loss and histogram distribution prior. IEEE Trans Image Process 25(12):5664–5677
28. Lu H, Li Y, Serikawa S (2013) Underwater image enhancement using guided trigonometric bilateral filter and fast automatic color correction. In: 2013 IEEE international conference on image processing. IEEE, pp 3412–3416
29. Gao SB, Zhang M, Zhao Q, Zhang XS, Li YJ (2019) Underwater image enhancement using adaptive retinal mechanisms. IEEE Trans Image Process 28(11):5580–5595
30. Petit F, Capelle-Laizé AS, Carré P (2009) Underwater image enhancement by attenuation inversionwith quaternions. In: 2009 IEEE international conference on acoustics, speech and signal processing. IEEE, pp 1177–1180

Molecular Dynamics Simulations Accelerate on Elastic Multi-GPU Architecture Build with FP64/TF32 Latest Streaming Multiprocessor Ampere Architecture

Nisha Agrawal, Abhishek Das, Rishi Pathak, Pankaj Dorlikar, and Manish Modani

Abstract Newtonian equations of motion for systems with millions of particles and bonded complex interactions describe the behavior of a physical system in terms of its motion as a function of time. Simulating such ordinary differential equations of second order with molecular dynamics application GROMACS for biochemical molecules is being used extensively. GROMACS's performance is optimized over the years. This paper focuses on the study of the behavior of molecular dynamics (MD) simulations using GROMACS on the PARAM Siddhi-AI system. The application performance is analyzed with the number of latest NVIDIA A100 GPUs, MPI tasks, and OpenMP threads. It is observed that 12 and 16 OpenMP threads give better performance for PME-GPU and PME-CPU runs, respectively. GPU-enabled GROMACS demonstrated ~18× and ~3× of relative speedup on single-node with 8× NVIDIA A100 GPUs for PME-GPU and PME-CPU, respectively, when compared to single-node CPU only performance. The GPU-enabled GROMACS multiple nodes each node with 8× A100 NVIDIA GPUs performance with CPU multi-node performance were also analyzed. Relative speedup shows the performance gain with GPU-enabled GROMACS is 3.4×, 9×, 8.9×, 7.6×, and 6× for 1, 2, 3, 4, and 5 nodes, respectively. NVIDIA Ampere architecture using recent CUDA version 11×, OpenMPI version 4.0×, multiples of millions of transistors fabricated on a TSMCs 7 nm N7 manufacturing technology, new third generation NVLink with lower latencies, and HPC software complexity. The introduction of a new sparsity feature accelerates mathematical operations while the faster and larger L1 cache and shared memory in Ampere architecture (192 KB per streaming multiprocessor (SM)) delivers additional speedups for high-performance computing (HPC) workloads.

N. Agrawal · A. Das · R. Pathak · P. Dorlikar
Centre for Development of Advanced Computing, Pune, India
e-mail: anisha@cdac.in

M. Modani (✉)
NVIDIA, Pune, India
e-mail: mmodani@nvidia.com

M. S. Kaiser et al. (eds.), *Information and Communication Technology for Competitive Strategies (ICTCS 2021)*, Lecture Notes in Networks and Systems 401,
https://doi.org/10.1007/978-981-19-0098-3_12

Keywords NVIDIA Ampere architecture · A100 · Accelerated computing · GPU · Molecular dynamics · Simulation · GROMACS · High-performance computing

1 Introduction

Recently, PARAM Siddhi-AI, the supercomputer established under the National Supercomputing Mission (NSM) at Center for Development of Advanced Computing (C-DAC) Pune, has achieved the global ranking of 62 in TOP 500 most powerful non-distributed computer systems in the world, released on 16th November 2020.

The system has 42 racks of NVIDIA DGX A100 servers, connected with HDR 200 Gbps InfiniBand Network and 10.5 PB of storage sub system. Each rack consists of 8 powerful NVIDIA A100 GPUs, Dual AMD EPYC Rome 7742 CPU (128 CPU cores) results in 19.5 Teraflops, double-precision peak performance per node. The PARAM Siddhi-AI system can deliver 210 AI Petaflops and 6.3 Petaflops of double-precision peak performances. The majority of the scientific applications are now capable to exploit the heterogeneity of the modern high-performance computing architecture, however, to extract, optimal performance on these state-of-the-art systems (e.g., PARAM Siddhi-AI), applications require further tuning and optimizations.

Molecular dynamics (MD) simulations used extensively to investigate and understand biomolecular function in atomic detail [1]. Almost all the MD applications (e.g., GROMACS, NAMD, LAMMPS, CHARMM, GENESIS, ACEMD, AMBER, DESMOND, OPENMM) are GPU-enabled and exploiting the performance of heterogeneous architecture. Turner et al. [2] discussed the performance of variety of scientific workloads, when executed only on CPUs and when GPUs are used. Groningen machine for chemical simulations (GROMACS) is one of the fastest MD engines, freely available, and widely used applications [1]. GROMACS CPU performance provided a challenge [3] as the optimized CPU SIMD kernels can achieve iteration rates of around 1 ms per step. To improve performance further, GROMACS was run on heterogeneous (GPU-enabled) architecture. The relative speedup on accelerators (GPUs), without extensive tuning, becomes less impressive.

Kutzner et al. [1] benchmark GROMACS 2018 performance on a diverse set of compute nodes with various generations of GPUs (GTX, RTX, K40, Quadro, and V100). Pall et al. [3] shared the GROMACS parallelization approach, GPU enablement, and performance. Konke et al. [1] implemented a performance-optimized, highly efficient GPU-enabled FMM and integrated it into the GROMACS code base as an alternative to particle mesh Ewald (PME). The performance advantage of FMM over PME is demonstrated over V100 GPUs. However, GROMACS performance is not discussed, on the latest NVIDIA GPU Ampere (A100) architecture.

This paper focuses on the performance analysis of GROMACS on the PARAM Siddhi-AI system. In the next section, the system architecture is discussed. GROMACS and its setup are discussed in the third section. GROMACS performance analysis on one node as well as on multi-node is discussed in the results and discussion section. GROMACS performance is analyzed with the number of OpenMP threads, number of GPUs, PME execution on CPU as well as when PME offloads on GPUs. GROMACS performance improvement over A100 GPUs in comparison with the previous generation GPUs (V100) was also demonstrated. The findings are summarized in conclusions section.

2 PARAM Siddhi-AI

The PARAM Siddhi-AI system is used to study GROMACS behavior on elastic multi-GPU architecture. Each node of the PARAM Siddhi-AI system consists of 8 NVIDIA A100 GPUs and Dual AMD EPYC Rome 7442 CPU (128 total CPU cores).

2.1 AMD EPYC Rome 7742 CPU

The NVIDIA DGX A100 system consists of the latest Dual AMD EPYC Rome 7742 with 128 total CPU cores. Apart from the higher core count, system consists of an L3 configuration which may also impact the performance of the CPU-based applications. The 64 cores in EPYC are organized into sixteen core complex (CCX), groups of four cores with each of the cores having private L1 and L2 cache, with L3 cache shared by all four cores in CCX. The L3 cache size of AMD EPYC 7742 is relatively on the larger side in comparison with the other CPU architectures.

2.2 NVIDIA A100 GPU

The newly released NVIDIA A100 GPU based on NVIDIA Ampere architecture is used for this application behavior analysis. A100 GPU has a third generation tensor core which deliver 300 teraFlops deep learning performance; next generation NVLink when connected with NVSwitch deliver up to 600 Gbps GPU-to-GPU bandwidth to unleash the highest application performance; 40 GB of high-bandwidth memory (HBM2) provides bandwidth of up to 1.6 TB/sec which is 1.7$\times$ higher than the previous generation of NVIDIA Volta GPU; the structural sparsity feature provides 2$\times$ higher performance for the sparse model.

The fives nodes of the PARAM Siddhi-AI system are used to perform the study of the behavior of GROMACS on elastic multi-GPU architecture. Each node consists of 8 NVIDIA A100 GPUs and Dual AMD EPYC Rome 7442 CPU (128 total CPU cores).

3 GROMACS

PARAM Siddhi-AI is designed to provide National Supercomputing Facility for HPC, AI, and HPC-AI applications. The goal of this work is to understand the behavior (scaling efficiency) of the widely used molecular dynamics application GROAMCS on a multi-GPU multi-node system. GROMACS is one of the most highly used HPC applications globally for biomolecular systems.

It evolves systems of particles using the Newtonian equations of motion, forces between particles dictate their movement (e.g., two positively charged ions will repel).

The force (F) on any atom

$$F_i = -\frac{\partial V}{\partial r_i} \tag{1}$$

is computed by calculating the force between non-bonded atom pairs:

$$F_i = \sum_j F_{ij} \tag{2}$$

plus the forces due to bonded interactions (which may depend on 1, 2, 3, 4, or 5 atoms), restricting and/or external forces. Energies (potential and kinetic) and pressure tensor are computed. Here, v represents bond potential.

Calculating forces is the most expensive part of the simulation as—all pairs of particles in the simulation can potentially interact. Force calculations typically fall into three major classes in GROMACS:

- Bonded forces: captures the specific behavior of bonds between particles, e.g., the harmonic potential when two covalently bonded atoms are stretched
- Non-bonded (short range) forces: These help to capture the behavior of particles within a certain cutoff range interact directly
- PME (long range) forces: Accounted through a "particle mesh Ewald" scheme, where fast Fourier transforms (FFTs) are used to perform calculations in Fourier-space. It is much cheaper than calculating all interactions directly in real-space.

PME implementations in molecular dynamics (MD) packages (e.g., GROMACS and NAMD) are highly optimized. For GROMACS, typical MD systems reach iteration rates of O (1000) steps per second, and the computation of all the forces takes less than a millisecond. However, with the increasing demand to study the larger system size (system of 108–109 particles), on high-performance systems [4, 5], the performance of PME runs into a communication bottleneck.

FFT needs to communicate with all (all-to-all communication) processes, implies quadratic scaling with the number of processes. Accordingly, PME scaling breaks down at an intermediate number of processes. At the same time, the FFT grid becomes memory intensive, particularly if high accuracy is required or for highly inhomogeneous charge distributions [6].

Bonded, non-bonded forces, and PME are offloaded on GPUs. However, the performance of FFTs in PME becomes critical in the case of single-node run (all-to-all communication remains in the node) and multi-node run (communication over a network). Therefore, in the present study, PME performance on CPU as well as on GPUs is analyzed in detail. The user community has put a lot of effort to port GROMACS on accelerators. All three types of force calculation can now be executed on the accelerator, but PME performance is a bottleneck as PME run on one MPI rank only.

It is not decomposed to run on many MPI ranks. For the performance study, GROMACS version 2021-dev is compiled with CUDA Version: 11.0, OpenMPI-4.0.5, and ucx-1.9.0. Satellite tobacco mosaic virus (STMV) proteins which consist of 1066 K atoms have been selected due to their relatively large problem sizes. The performance is measured in nanoseconds per day (ns/days).

4 Results and Discussion

GROMACS performance depends on several factors, such as the usage of GPUs, the number of MPI tasks and OpenMP threads, the loading balance algorithm, the ratio between the number of particle–particle (PP) ranks and particle mesh Ewald (PME) ranks, and the type of simulation being performed, force field used, etc.

Performance analysis is divided into two major sections such as (A) single-node multi-GPU and (B) multi-node multi-GPU. GROMACS performance is analyzed with the number of OpenMP threads, MPI tasks, and compared with CPU only performance as well as with the previous generation (V100) GPUs.

4.1 Single-Node Multi-GPU

A number of experiments were performed to study GROMACS single-node behaviors on PARAM Siddhi—AI system. These include the OpenMP thread and GPUs scalability analysis.

4.1.1 OpenMP Thread Scalability

This analysis was carried out to study the impact of the number of OpenMP threads on GROMACS performance. The experiment was performed by varying the number of OpenMP threads for the fixed number of GPU and MPI processes. The GPU-enabled GROMACS spans one MPI task per GPU. Therefore, we used 8 MPI ranks, which spans job on all the 8 GPUs in one node. Two kinds of experiments were performed:

- Case 1: PME, non-bonded, and bonded all the three types of force calculation are offloaded to GPUs. This case will be referred to as PME-GPU.
- Case 2: PME calculations performance on CPU whereas non-bonded and bonded force calculation is offloaded to GPU. It will be known as PME-CPU.

Figure 1 shows the variations in performance for OpenMP threads 4, 8, 10, 12, and 16. Here, a maximum of 16 threads per MPI ranks (total $16 \times 8 = 128$) is used as the CPU node has 128 cores.

Figure 1 also shows that for PME-GPU, GROMACS performance is increasing with the increase in OpenMP threads up to 12 threads. For 16 OpenMP threads, GROMACS performance decreased slightly. The performance variation is mainly because of the run time spent in waiting due to PP/PME imbalance, PME wait time for PP increased in the case of 16 OpenMP threads per MPI process when compared to 12 OpenMP threads per MPI process performance (Table 1).

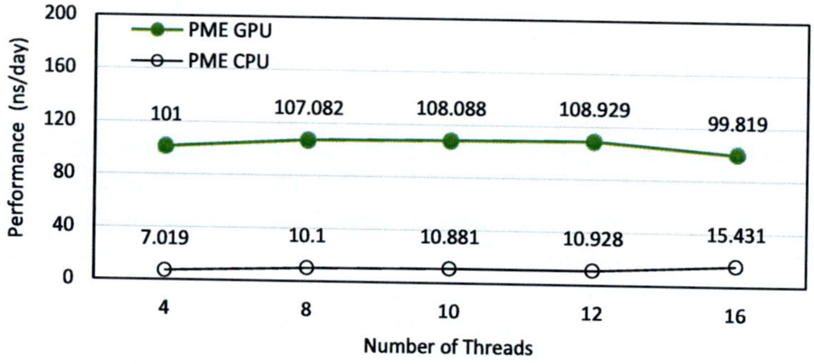

Fig. 1 Single-node PME-GPU and PME-CPU performance for varying number of OpenMP threads

Table 1 Summarize the hardware and software environment used for experiments

Node configuration	
CPU	2× AMD EPYC 7742 64C 2.25 GHz; 128 cores total
System memory (RAM)	1 TB
Accelerator	8× NVIDIA A100 –SXM4
Accelerator memory	40 GB/accelerator card
Storage	1.8 TB; raid:14 TB
OS	Ubuntu 18.04.5 (Kernel: 5.4.0–48-generic)
NVIDIA software	Cuda: 11.0, NVIDIA Driver Version:450.51.06
Message passing interface	OpenMPI-4.0.x

Table 2 Key functions/subroutines time with the change in OpenMP threads for the run on one node and 8 GPUs

Functions or subroutines	OpenMP threads				
	4	8	10	12	16
Domain decomp.	0.74	0.58	0.55	0.52	0.57
Send X to PME	1.1	0.00	1.09	1.12	1.15
Neighbor search	0.47	0.28	0.23	0.21	0.18
Force	0.14	0.08	0.08	0.07	0.07
PME mesh	0.21	0.21	0.21	0.21	0.21
PME wait for particle-particle	8.32	7.85	7.78	7.72	8.44
Wait + Recv. PME F	0.4	0.39	0.34	0.34	0.29
Wait PME-GPU gather	0.0	0.0	0.0	0.0	0.0
Wait bonded GPU	0	0	0	0	0
Wait GPU NB nonlocal	0.02	0.02	0.02	0.02	0.02
Wait GPU NB local	0.0	0.0	0.0	0.0	0.0
Wait GPU state copy	0.32	0.31	0.32	0.33	0.34
NV X/F/buffer ops	0.01	0.01	0.01	0.01	0.0
Comm. energies	0.12	0.12	0.12	0.13	0.18
Rest	0.53	0.42	0.40	0.39	0.4
Total	8.53	8.07	7.99	7.93	8.66

For PME-CPU, when bonded and non-bonded forces are offloaded to GPU and PME computed on CPU, the GROMACS performance increasing linearly with the increase in the OpenMP threads, and maximum performance is observed with 16 OpenMP threads (Fig. 1). It shows PME-CPU performance improves significantly with the increase in OpenMP threads (Table 2).

Figure 1 reveals the single-node performance of GPU enablement of PME (PME-GPU) is 7× better than that obtained by PME-CPU. It also shows GPU enablement of PME improved GROMACS performance significantly.

4.1.2 GPU Scaling

Here, the GROMACS single-node performance is analyzed with the increasing number of GPUs (up to 8). The OpenMP threads are fixed to 12 and 16 for PME-GPU and PME-CPU, respectively, to have optimum performance as discussed in the previous sections.

Figure 2 shows single-node GROMACS scalability with the variation in the number of GPUs. In the case of PME-GPU, 4× GROMACS performance is increased when the number of GPUs increased from 2 to 4. The performance further increased by 2× when GPUs increased from 4 to 8. In the case of PME-CPU, there is a linear

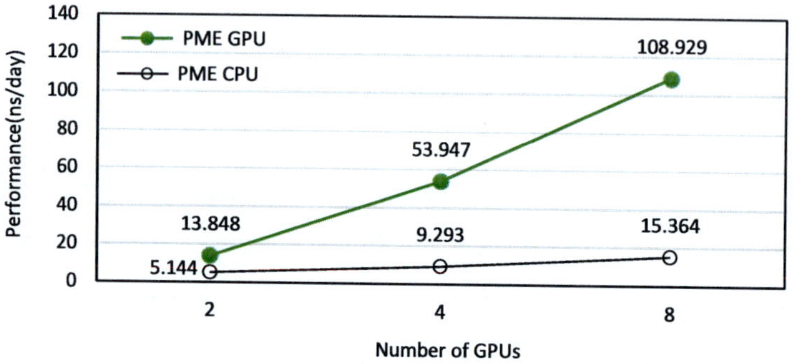

Fig. 2 Single-node PME-GPU and PME-CPU performance with the varying number of GPUs

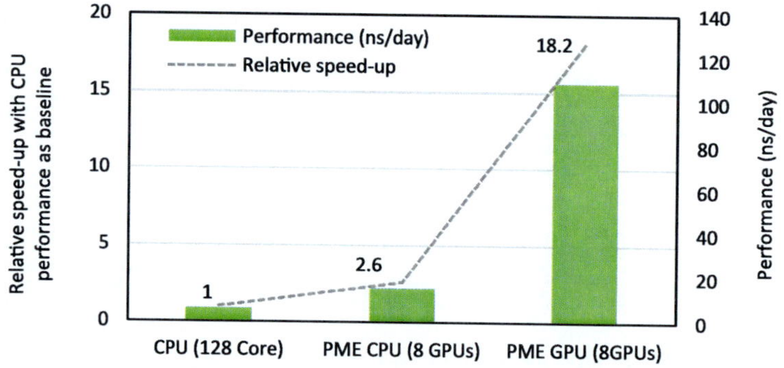

Fig. 3 Single-node CPU, PME-CPU, and PME-GPU performance and speedup

performance increase in GROMACS with 2, 4, and 8 GPUs. PME-GPU enablement (PME-GPU) shows approximately 8× improvement over PME-CPU (Fig. 2).

Figure 3 shows the relative speedup of PME-CPU and PME-GPU with CPU performance as a baseline for the performance. PME-CPU runs 2.6× faster, whereas PME-GPU runs 18.2× faster when compared with CPU runs.

For CPU only run, all the forces of GROMACS remain on CPU whereas in the case of GPUs, bonded and non-bonded force calculations are offloaded to GPUs. CPU runs performed using 16 MPI processes and 8 threads on each node as each node having 128 cores and for the configuration, 16 processes and 8 threads per process achieved maximum single-node CPU performance with GROMACS default parameters.

4.2 Multi-node Multi-GPU Scaling

In this section, GROMACS scalability with the number of GPUs and nodes is analyzed. In addition, GROMACS multi-node performance is also compared with CPU only run is performance is analyzed.

4.2.1 Scaling

As demonstrated above, PME is critical to have GROMACS optimum performance. PME is offloaded to GPUs; however, PME is not decomposed further. In other words, PME runs on the single MPI rank on GPUs. The single-node performance detailed in the previous section shows the PME-GPU performed better than PME-CPUs. For multi-node, multi-GPUs also, GROMACS run is performed for PME-GPU as well as PME-CPU.

In the case of PME-GPU two nodes run, the performance decreased is observed for 16 GPUs run (approximately ~ $-10\times$). Run time logs show the execution time taken by domain decomposition statistics (e.g., Domain decomp., DD comm. Load, Send X to PME, Wait + Recv. PME_F, etc.) is increased significantly for 16 GPUs (2 nodes) run in comparison with 8 GPUs (1 node) run. It again illustrates for PME-GPU scalability over 2 nodes is limited by PME decomposition.

In the case of PME-CPU, GROMACS scalability is observed with the increase of the number of nodes. Figure 4 shows the performance increase with the increase in the number of GPUs and nodes. The performance increases up to 4 nodes (32 GPUs) and remains the same for 5 nodes (40 GPUs) run. It should also be noticed, the performance gain with PME-CPU multi-node run is there; however, observed performance is still less than PME-GPU single-node run (~$3\times$). In other words, PME offload to GPU increases GROMACS performance significantly.

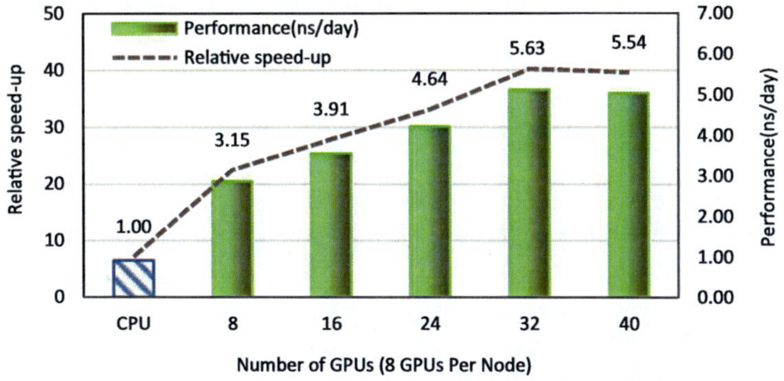

Fig. 4 GROMACS PME-CPU, multi-node multi-GPU scaling

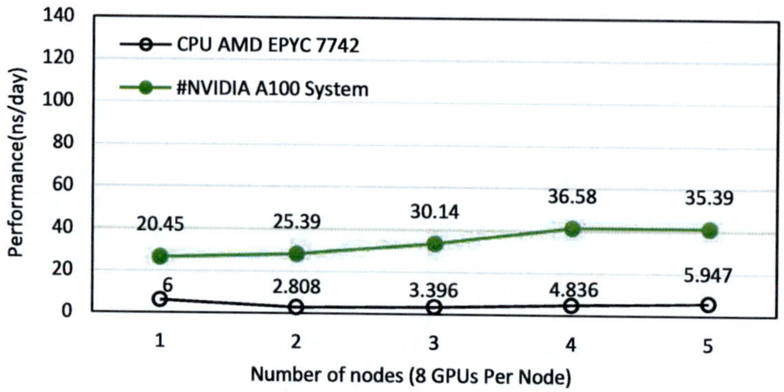

Fig. 5 GROMACS single-node multi-GPU and multi-node CPU scaling

4.2.2 Multi-node CPU and GPU Performance

In this section, GROMACS CPU only multi-node performance is compared with those observed with GPU-enabled, PME-CPU, multi-node performance.

For CPU only run, all the forces of GROMACS remain on CPU, whereas in the case of GPUs, bonded and non-bonded force calculations are offloaded to GPUs. CPU runs performed using 16 MPI processes and 8 threads on each node as each node having 128 cores and for the configuration, 16 processes and 8 threads per process achieved maximum single-node CPU performance with GROMACS default parameters.

In Fig. 5, GROMACS multi-node relative speedup is shown for CPU only and GPU-enabled (PME-CPU). In the case of CPU only runs, the performance decreases with the increase of CPU node. However, in the case of the PME-CPU run, GROMACS performance increases with the increase in the number of nodes. A significant performance gain is observed with the GROMACS GPU enablement. On 5 nodes, GROMACS ~7× performance gain is observed in comparison with CPU only run.

4.3 GPU-Enabled GROMACS Performance on Current Generation (NVIDIA A100) and Previous Generation (NVIDIA V100) GPUs

In this section, the performance of GROMACS is analyzed, on the current generation of GPUs (NVIDIA A100) versus the previous generation GPUs (NVIDIA V100) for 8 GPUs (single node) run. GROMACS V100 performance is considered from the results provided on the NVIDIA developer site [https://developer.nvidia.com/hpc-application-performance] for the STMV input dataset.

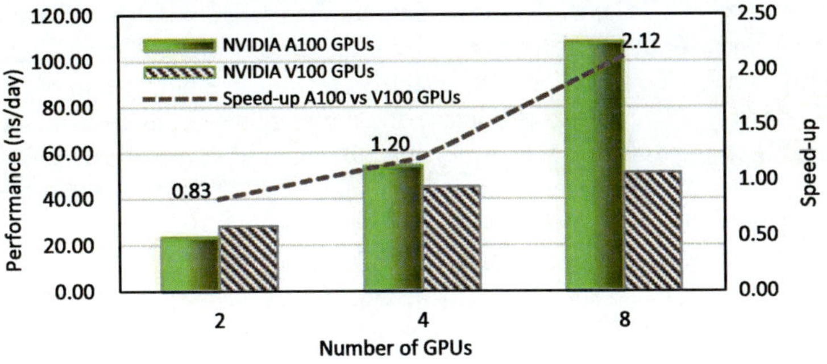

Fig. 6 GPU-enabled GROMACS performance on A100 and V100 GPUs

As shown in Fig. 6, in both cases, GROMACS is scaling well up to 8 GPUs. GROMACS on NVIDIA A100 GPU performed ~2× times faster in comparison with the performance observed on NVIDIA V100 GPUs. It is in sync with the capability difference of A100 and V100.

In the present work, GROMACS performance is analyzed on the PARAM Siddhi-AI cluster. The performance of GROMACS is analyzed for the single node as well as for multi-node runs (up to 5 nodes) on the latest GPU architecture. It is demonstrated that GROMACS scale well on multi-node. The significant performance gain is obtained with the GPU enablement of GROMACS forces. There is a need to further analyze the GPU performance and OpenMP threads behavior for further performance improvement. PME-GPU enablement is critical to have multi-node, multi-GPU performance gain. Other methods to PME, e.g., FMM [6] need to be analyzed on A100 GPUs. In the present, GROMACS GPU enablement, only 1 MPI rank spawns the job on 1 GPU. Therefore, a maximum of up to 8 MPI ranks is used for one node performance analysis. However, GPUs memory is not fully utilized. To further optimize, the multi-instance GPU (MIG) and multi-process services (MPS) need to be used to spawn jobs on GPUs from more than one MPI rank. GROMACS performance needs to be further analyzed on more than 5 nodes of the PARAM Siddhi-AI system.

5 Conclusions

GROMACS is one of the most popular and freely available codes for molecular dynamics (MD) simulations. Over the years, GROMACS performance improved with the advancement of GPU architectures and generations. In this article, we analyzed the characteristic of molecular dynamics application GROMACS on the new and most powerful supercomputer PARAM Siddhi-AI. The system has the latest generation AMPHERE (A100) GPUs. Single node, as well as multi-node GROMACS performance, is demonstrated. Detailed scalability analysis for OpenMP thread, number of GPUs, number of nodes, etc., is carried out. It is observed that 12 and 16 OpenMP threads give better performance for PME-GPU and PME-CPU runs, respectively. GROMACS GPU-enabled performance is also compared with the GROMACS CPU only performance for the single node as well as multiple nodes. GPU-enabled GROMACS demonstrated ~18× and ~3× performance gain on PARAM Siddhi-AI system for one and five nodes. The GROMACS multiple nodes performances with CPU multiple nodes performance were also compared. Relative speedup shows the performance gain with GPU-enabled GROMACS is 3.4×, 9.04× , 8.88×, 7.56×, and 5.95× when 1, 2, 3, 4, and 5 nodes are used to run GROMACS, respectively. In end, limitations of the present work and future work are discussed.

Acknowledgements The support and the resources provided by the Center for Development of Advanced Computing (C-DAC) and the National Supercomputing Mission (NSM), Government of India are gratefully acknowledged. We acknowledge the use of computing resources at PARAM Siddhi-AI (C-DAC, Pune).

References

1. Kutzner C, Pall S, Fechner M, Esztermann A, Groot BL, Grubmuller H (2020) Improved use of GPU nodes for GROMACS 2018. arXiv: 1903.05918v2
2. Turner D, Andresen D, Hutson K, Tygart A (2018) Application performance on the newest processors and GPUs. In: Proceedings of the practice and experience on advanced research computing July 2018 Article No. 37, pp 1–7
3. Pall S, Zhmurov A, Bauer P, Abraham M, Lundborg M, Gray A, Hess B, Lindahl E (2020) Heterogeneous parallelization and acceleration of molecular dynamics simulations in GROMACS. Chem Phys 153:134110. https://doi.org/10.1063/5.0018516
4. Berendsend HJC, van der Spoel D, van Drunen R (1995) GROMACS: a message-passing parallel molecular dynamics implementation. Comput Phys Commun 91(1-3):43–56
5. Abraham MJ, Murtola T, Schutz R, Pall S, Smith JC, Hess B, Linkahl E (2015) GROMACS: high performance molecular simulations through multi-level parallelism from laptops to supercomputers. SoftwareX 1–2:19–25
6. Bartosz Kohnke B, Kutzner C, Grubmüller H (2020) A GPU-accelerated fast multipole method for GROMACS: performance and accuracy. J Chem Theory Comput 16:6938–6949

Observations and Analysis of Power Quality Indices Using Custom Power Devices in Power Distribution Network

Vidula S. Jape, Haripriya H. Kulkarni, Netra M. Lokhande, and Deepali Ubale

Abstract The overall performance of the electricity distribution network is determined from various parameters. Power quality indices such as drop in voltage, rise in voltage, and harmonics play an essential role in this context. The article introduces a new index called the power quality distortion (DI). This index gives information about the contribution of non-linear load to the harmonic distortion in power. This effect has been observed due to the presence of non-linear component of current drawn by the load. Changes in the supply voltage in the form of dips, over voltages, harmonics, transients, etc., are considered major power quality issues. This has led to deterioration in the quality of the energy delivered to end customers and generates disruptions on the utility side. Load presents at the consumer end also explains the degradation in the quality of the power supply. Initially, penalties were charged for low p.f. consumers, and then, corrective measures were initiated by the consumer. In the same way to control the power quality, penalty is proposed to the consumers who are hampering the quality of power by application of non-linear load without using any mitigation technique. One of the mitigation techniques is custom power devices (CPDs). It is more effective as compared to traditional methods. This article provides various power quality indices such as distortion index as a power quality calculation parameter and the results obtained by application of dynamic voltage restorer (DVR) and hybrid-STATCOM for reducing the effect of the voltage sag and other indices of the quality of power in the electrical network. The proposed concept allows operation in non-linear conditions on the consumer side. The design of the

V. S. Jape (✉) · D. Ubale
Department of Electrical Engineering, PES's Modern College of Engineering, Pune, India
e-mail: jape.swati@moderncoe.edu.in

D. Ubale
e-mail: deepali.choudaki@moderncoe.edu.in

H. H. Kulkarni
Professor and HOD, MMCOE, Department of Electrical Engineering, PES's Modern College of Engineering, Pune, India
e-mail: haripriyakulkarni@mmcoe.edu.in

N. M. Lokhande
HOD, IT Department, School of Electrical Engineering, MIT World Peace University, Pune, India

M. S. Kaiser et al. (eds.), *Information and Communication Technology for Competitive Strategies (ICTCS 2021)*, Lecture Notes in Networks and Systems 401,
https://doi.org/10.1007/978-981-19-0098-3_13

system parameters is proposed based on the compensation parameter as an index of distortion of the distribution network. The simulation is carried out by the MATLAB model, and the results are compared with the analyzed performance of the power distribution system with regards to the index of distortion with injection of DVR and hybrid-STATCOM.

Keywords Power quality (PQ) · Custom power devices (CPDs) · Dynamic voltage restorer (DVR) · Hybrid-STATCOM · Distortion index (DI) · Total harmonic distortion in voltage (VTHD) · Total harmonic distortion in current (ITHD) · Active power filters (APFs) · Battery energy storage systems (BESSs)

1 Introduction

In deregulated era, layout of the electrical system became more complicated every day. It includes many power plants and loads interconnected by numerous transmission and distribution stations. Losses occurring in the electrical distribution system and consumption of the energy by consumers are a decisive factor for the utility to investigate the performance of the electrical system; in addition, the heavy use of power electronics circuits leads to growth of non-linearity. These types of loads are more immune to hamper power quality parameters. Deviation in voltage, current, or fundamental frequency shows remarkable existence of harmonic distortion in the system. Hence its continuous monitoring, analysis is required to take the mitigation actions, and as a result, improving the quality of power. Many traditional devices are available to reduce the power quality issues, but they are not sufficient to keep the power quality limits as per the prescribed standards. Due to application of electronic devices, it has been observed that it gives rise to PQ problems. The various mitigation techniques include CPDs, D-STATCOM, APF, DVR, BESS for improving the quality of the power supply. For voltage sag compensation-dynamic voltage restorer (DVR) and for improving the quality of power and reducing the index of distortion, hybrid-STATCOM is found to be more suitable [1].

2 Control Strategies of CPD and a Real-time Study

CPD control is one of the essential procedures which integrate the specific disturbances in voltage with appropriate recognition strategies. The DVR is considered to be an efficient custom power device (CPD), therefore the power supply to voltage converter is considered an important element of the DVR depicted in Fig. 1 [2].

11 kV/440 V, 200 KVA distribution system which supplies an educational institute is considered to evaluate the effect of power quality parameters. Simulink diagram for the same is depicted in Fig. 2, and subsystem of dynamic voltage restorer is presented by Fig. 3. The DVR is a custom device, which can be injected in between

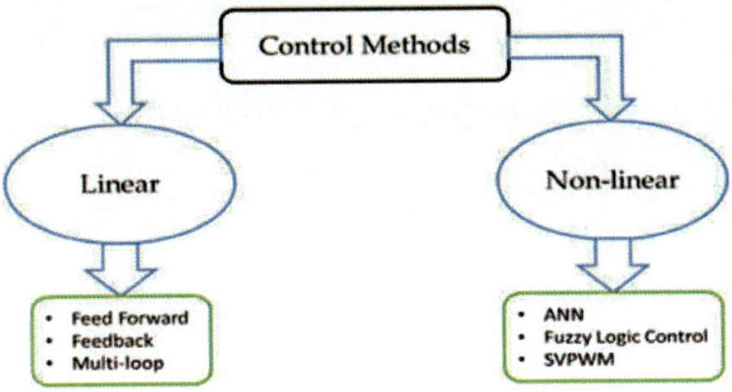

Fig. 1 Different control strategies for DVR

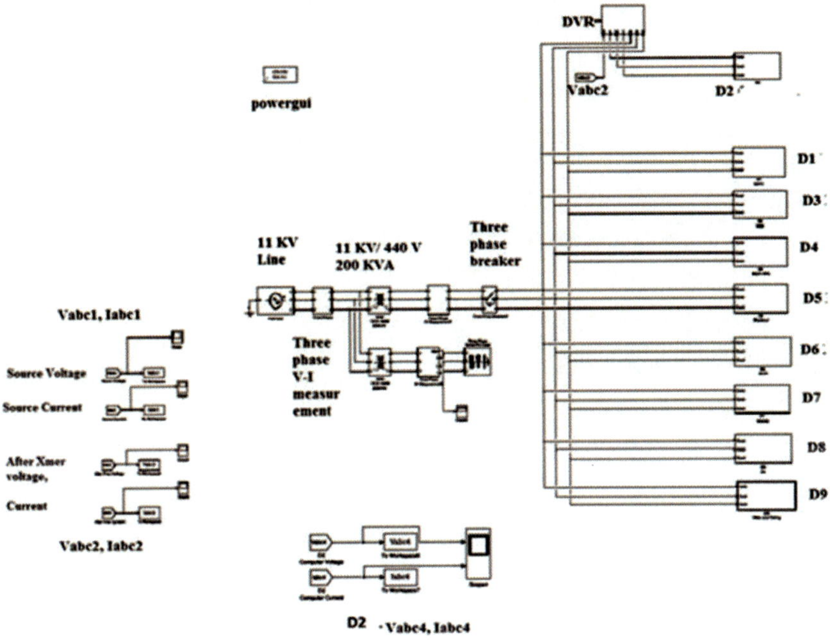

Fig. 2 Simulation diagram for institution with injection of DVR

the utility grid and the customer side [3]. Injection of compensation voltage required to alleviate power quality issues is the basic feature of DVR. Performance of DVR is identified by injecting DVR into the different departments of the institute [4]. The total system load is 600.68 kW. The DVR and the hybrid-STATCOM are connected in between the supply side and the different departments at various load conditions. Figure 4 represents the waveform of observations collected for voltage, in which the drop in voltage is determined between 0 and 1 s.

Figure 5 illustrates the waveforms of the department for voltage and current with

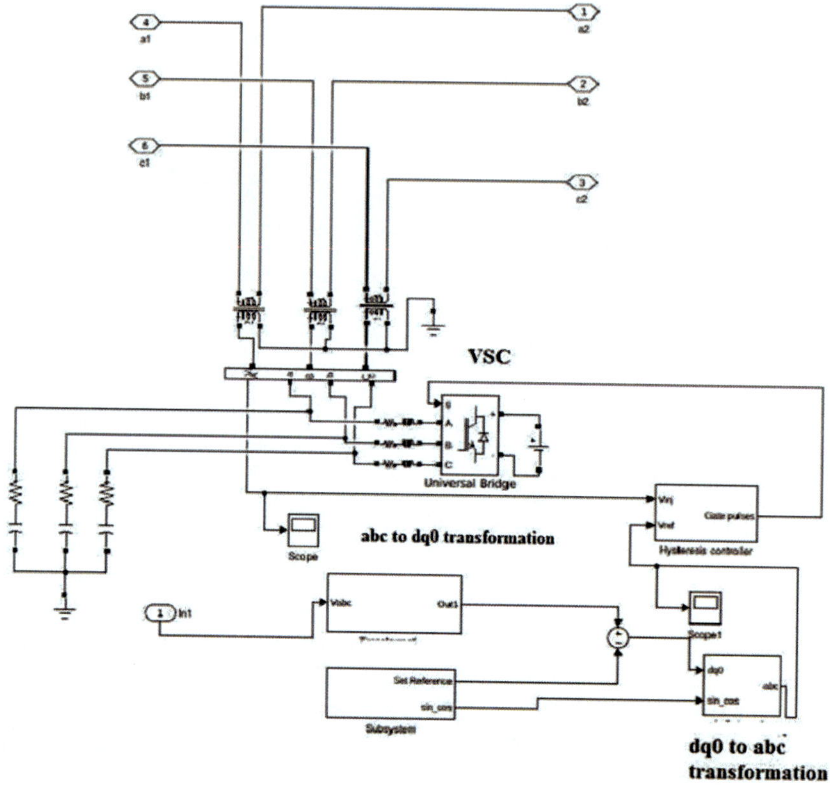

Fig. 3 Subsystem of dynamic voltage restorer

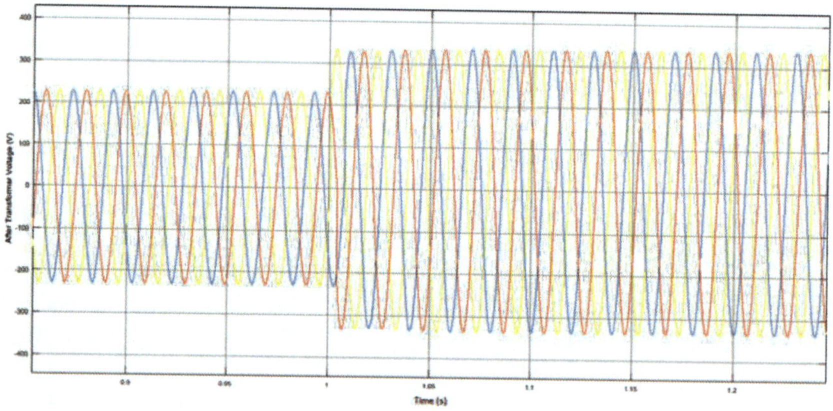

Fig. 4 Waveform of supply voltage

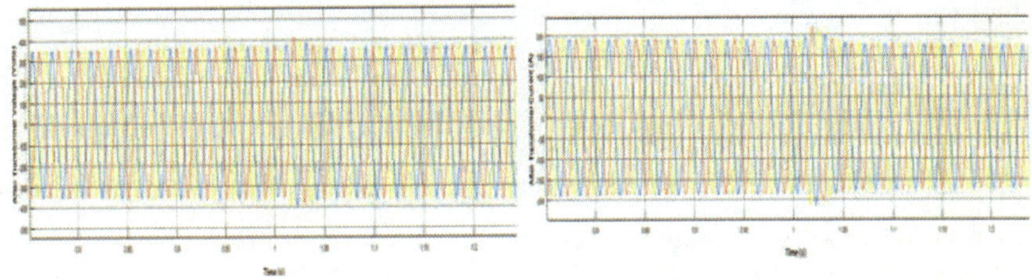

Fig. 5 Waveforms for voltage and current in the department, respectively

non-linear loads by injecting a DVR into the system. Harmonic distortion is observed in different departments. Total harmonic distortion in voltage values on the power supply side without DVR and with DVR for linear and non-linear load is shown in Figs. 6 and 7

The simulation diagram with injection of hybrid-STATCOM and the subsystem of hybrid-STATCOM are shown in Fig. 8 and in Fig. 9. Figures 10 and 11 indicate the source voltage at the hybrid-STATCOM terminals and the load voltage and current waveforms for D2 (with hybrid-STATCOM).

The THD voltage values before the hybrid-STATCOM connection and after the hybrid-STATCOM connection for linear and non-linear loads are shown in Fig. 12

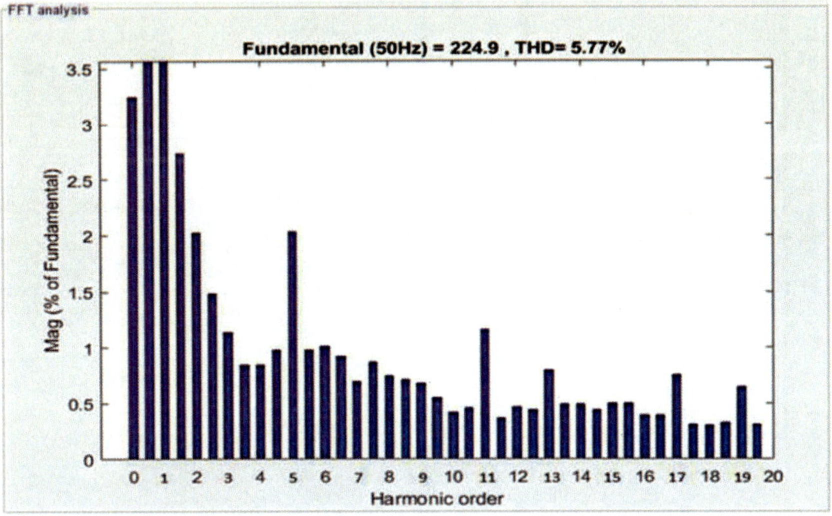

Fig. 6 Total harmonic distortion of voltage (before injection of DVR in the circuit)

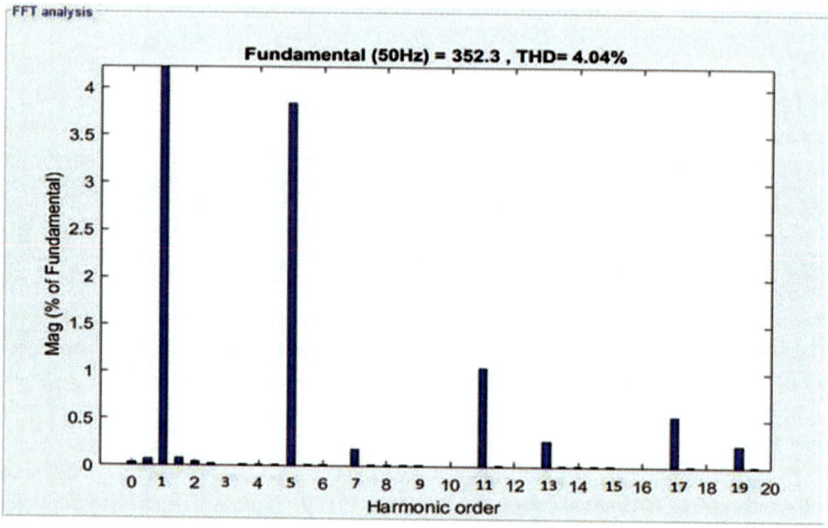

Fig. 7 Total harmonic distortion of voltage (after injection of DVR in the circuit)

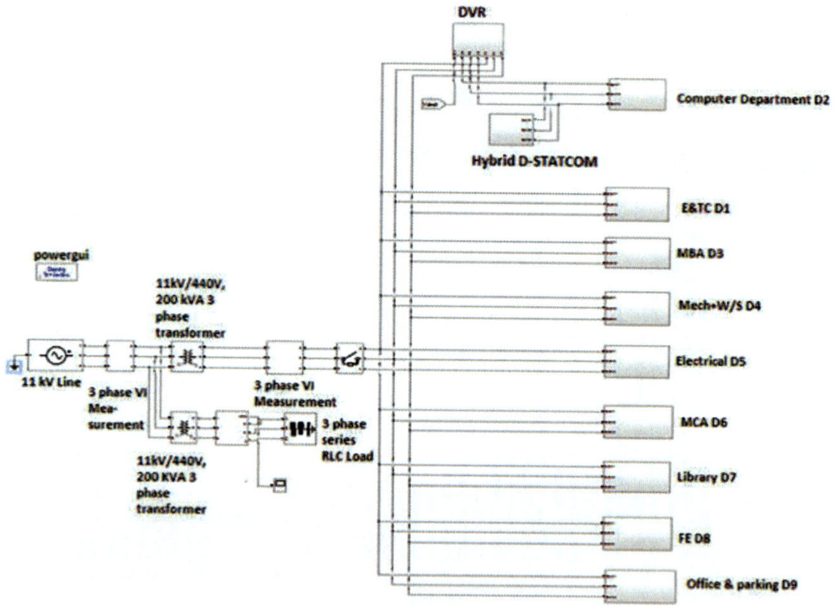

Fig. 8 Simulink diagram for institution with hybrid-STATCOM

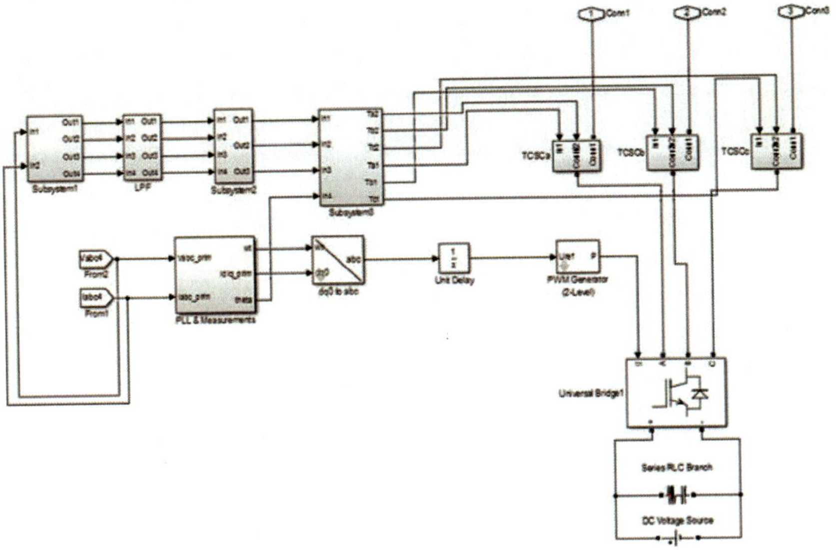

Fig. 9 Hybrid-STATCOM subsystem

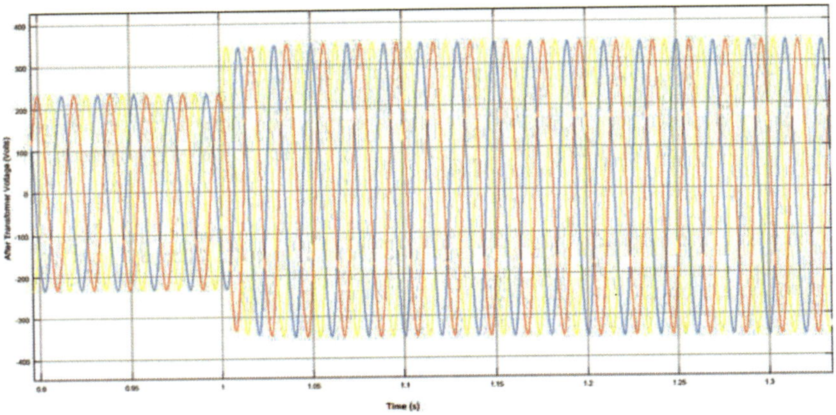

Fig. 10 Source voltage waveform with hybrid-STATCOM in the circuit

Sample calculations for Distortion Index

Distortion index and some other important parameter with THD are calculated for the example taken. It is evaluated through results obtained from software analysis. Equations are considered for calculation of distortion index [5].

V_1, I_1: Fundamental frequency voltage and current.

$V_3, V_5, V_7, \ldots V_n$: Order of harmonic voltages

$I_3, I_5, I_7, \ldots I_n$: Order of harmonic currents

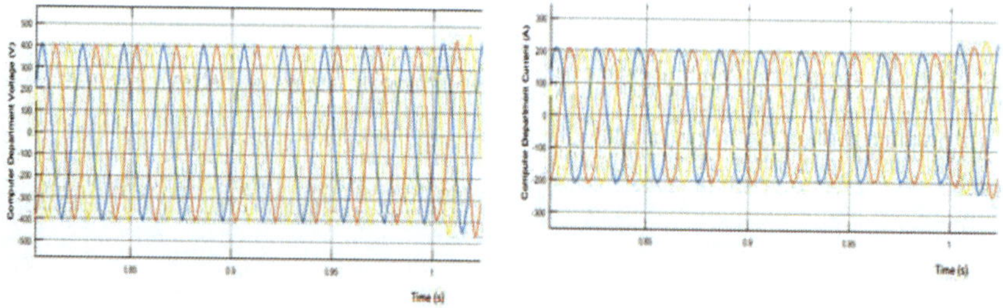

Fig. 11 Load voltage and current waveform with hybrid-STATCOM in the circuit

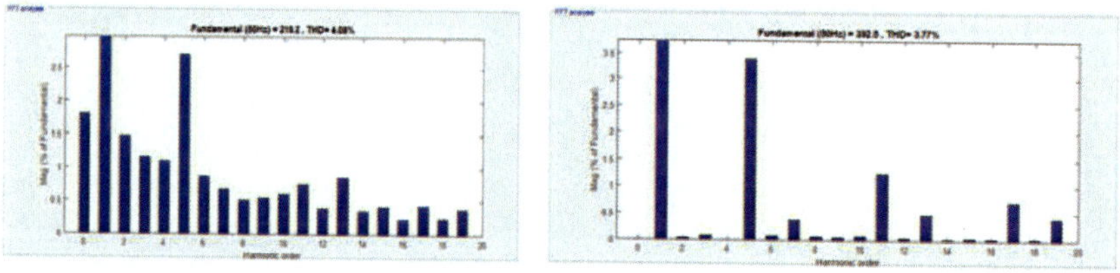

Fig. 12 Voltage THD before and after injection of hybrid D-STATCOM in the circuit

$$I_H = \sqrt{I_3^2 + I_5^2 + I_7^2} \tag{1}$$

$$V_H = \sqrt{V_3^2 + V_5^2 + V_7^2} \tag{2}$$

$$\text{VTHD} = V_H / V_1 \tag{3}$$

$$\text{ITHD} = I_H / I_1 \tag{4}$$

$$\text{Fundamental Apparent Power FAP} = V_1 I_1 \tag{5}$$

$$\text{Current Distortion Power CDP} = I_H * V_1 \tag{6}$$

$$\text{Voltage Distortion Power VDP} = V_H * I_1 \tag{7}$$

$$\text{Harmonic Distortion Power HDP} = V_H * I_H \tag{8}$$

$$\text{Non-linear Apparent power (NAP)}; = \sqrt{\text{CDP}^2 + \text{VDP}^2 + \text{HDP}^2} \tag{9}$$

$$\text{Total Apparent Power TAP} = \text{FAP}^2 + \text{NAP}^2 \tag{10}$$

Table 1 Distortion index

Departments	Distortion index in (%)		
	Distortion index in (%) without DVR	Distortion index in (%)With DVR circuit	Distortion index in (%) with hybrid -STATCOM
Source side	2.6507	1.9964	1.950
Load side	44.9758	30.632	2.8192
D1	34.995	22.232	3.2734
D2	45.9388	31.591	3.6303
D3	48.5182	25.3123	1.4027
D4	38.0075	24.672	3.2731
D5	36.9939	24.321	2.2728
D6	52.4725	24.288	2.4053
D7	42.992	24.108	3.2718
D8	45.9906	23.192	2.123
D9	44.9951	23.0889	2.3334

$$\text{Distortion Index DI} = \text{NAP/FAP} \tag{11}$$

Table 1 indicates distortion index values for different departments without injection of DVR in the circuit, with injection of DVR in the circuit and with injection of hybrid-STATCOM in the circuit.

3 Conclusions

From the case study considered here, it can be proved that the quality of the power can be improved by mitigating the harmonics and DI can be maintained by introducing the DVR in the system. In connection to this, a novel indices named DI are expressed newly in this paper which reflects the effect of non-linearity present in the load and its mitigation using DVR and without using DVR, for load with non-linearity connected to the power distribution network. With the application of the DVR in the presence of non-linear load, the distortion index can be mitigated, and therefore, the power quality can be enhanced.

From all above-mentioned results, THD shows the most effective VTHD or ITHD. The distortion index (DI) relates to the distortion of power in the distribution network. Present day power quality meters available in market provides total harmonic distortion in voltage or current. To read power distortion, DI is presented in this article as an innovative main power quality index for the identification and evaluation of power quality levels, present in the distribution network. The regulator can consider this index while billing the consumer on the basis of distortion. To overcome the shortcomings of the DVR as attenuation of voltage dips for distribution networks, a hybrid-STATCOM is offered. It includes a (TCLC) thyristor-based LC control acts as an active inverter [6]. The contributions in this article are briefly presented in following points.

- A hybrid-STATCOM offers property of voltage sag attenuation and other indices with respect to power like ITHD as well as VTHD is presented [7]. Significant changes in the value of the distortion index with hybrid-STATCOM are observed.
- From the extensive analysis carried out, it has been found that hybrid-STATCOM has better performance over traditional custom power devices with consideration of power quality.

References

1. Jape VS, Bharadwaj DG, Sarwade T (2018) Power quality problems mitigation using dynamic voltage restorer (DVR) using pi controller and fuzzy logic controller. IJET 2(12):214–218
2. Raval CH, Doorwar A (2014) An overview of various control methods of DVR. ICCPCT, p 53
3. Omar R, Rahim NA (2009) Implementation and control of a dynamic voltage restorer using space vector pulse width modulation (SVPWM) for voltage sag mitigation. In: IEEE-TECHPOS, p 1
4. Jape VS, Bharadwaj DG (2017) Substation monitoring system (SMS) for remote power quality monitoring and analysis of power distribution network using object oriented web enabled method. IJCTA (International Science Press), 10(9):533–539
5. Kulkarni HH, Jape VS, Bankar DS, Borkar RV (2019) Analysis of power quality indices and harmonics in power distribution network with reference to IEEE standard 519–1992. Int J Sci Technol Res 8(10):1065–1067
6. IEEE recommended practices and requirements for harmonic control in electrical power systems, IEEE Std 519–1992
7. Wang L, Lam C-S, Wong M-C (2015) A Hybrid-STATCOM with wide compensation range and low DC-link voltage. IEEE Trans Ind Electron 0278–0046

Redefining Text-to-SQL Task as a Machine Translation Problem

Alaka Das

Abstract Natural language interfaces are gaining popularity as an alternative interface for non-technical users. Natural language interface to database (NLIDB) systems have been attracting considerable interest recently that are being developed to accept user's query in natural language (NL), and then converting this NL query to an SQL query, the SQL query is executed to extract the resultant data from the database. This Text-to-SQL task is a long-standing, open problem, and towards solving the problem, the standard approach that is followed is to implement a sequence-to-sequence model. In this paper, I recast the Text-to-SQL task as a machine translation problem using sequence-to-sequence-style neural network models. To this end, I have introduced a parallel corpus that I have developed using the WikiSQL dataset. Though there are a lot of work done in this area using sequence-to-sequence-style models, most of the state-of-the-art models use semantic parsing or a variation of it. None of these models' accuracy exceeds 90%. In contrast to it, my model is based on a very simple architecture as it uses an open-source neural machine translation toolkit OpenNMT, that implements a standard SEQ2SEQ model, and though my model's performance is not better than the said models in predicting on test and development datasets, its training accuracy is higher than any existing NLIDB system to the best of my knowledge.

Keywords Text-to-SQL · SQL · NLIDB · Machine translation · OpenNMT · WikiSQL · SEQ2SEQ model · NLP

1 Introduction

Natural language interfaces are gaining popularity recently to be used as tools for non-expert users or common users to interact with computers. Natural language interface to databases (NLIDB) is such a tool that would allow non-technical users who does not necessarily have knowledge of specific database query languages to

A. Das (✉)
IIIT Bhubaneswar, Bhubaneswar, Odisha 751003, India
e-mail: c116002@iiit-bh.ac.in

M. S. Kaiser et al. (eds.), *Information and Communication Technology for Competitive Strategies (ICTCS 2021)*, Lecture Notes in Networks and Systems 401,
https://doi.org/10.1007/978-981-19-0098-3_14

retrieve information from databases. Many research works have been carried out to design and develop such interfaces using which these users would be able to formulate complex database queries conveniently and effectively. However, NLIDBs have not yet made their way into practical tools because of the difficulty in understanding natural language questions and translating it accurately to a formal query. Research in this area has started decades ago, but despite all these efforts, no commercial solutions are available at present.

The most important task of an NLIDB system is: Text-to-SQL task, i.e., to map natural language sentences to corresponding SQL queries. Dependency parsing and semantic parsing in natural language processing (NLP) are the two main alternatives used to carry out this Text-to-SQL task. The recent developments in deep learning algorithms have contributed a lot and seem to bring back into the foreground the NLIDB. In this approach, both the NL query and the SQL query is treated as sequences and a sequence-to-sequence model [28] or its variants [8] can be trained, which can be used as the parser. In recent years, some state-of-the-art methods with Seq2Seq encoder–decoder architectures can obtain more than 80% exact matching accuracy [12, 32, 33, 35, 39]. Though a lot of work has been done in this area with Seq2Seq encoder–decoder architectures, most of these systems have used semantic parsing technique and are trained using semantic parsing datasets that provide schema information along with natural language questions and corresponding SQL statements. Till date no NLIDB system is developed along with a dataset that gives more than 90% training accuracy, for which I have addressed this text-to-SQL problem in a different perspective: I treated this task as a neural machine translation problem and trained a Seq2Seq model with a parallel corpora that I have developed from the WikiSQL dataset [39]. The training accuracy is 99.8% which to the best of my knowledge no existing NLIDB models have achieved so far. Hence, this work may be considered as a state of the art in this research area.

The remaining parts of the paper are organized as follows: Sect. 2 describes in short the overview of related work in NL interfaces for databases; Sect. 3 contains a detailed description of the model I have used, the parallel corpora that I have developed from the WikiSQL dataset, implementation Details, the matrices to evaluate the correctness of the generated SQL queries and their results. In Sect. 4, I have discussed the performance of my model in comparison to some recent start-of-the-art models in this research area. In Sect. 5, I have concluded the paper by highlighting the key insights and future work.

2 Related Work

First attempts of language interfaces to databases appeared around the sixties. Those NLIDBs were basically NL interfaces for expert systems, specifically designed for a fixed domain and a particular DB. As a result of which, it was very difficult to modify them and to port them to other application domains. Early systems used dictionaries and dialogue systems to guide the user [6, 10]. In the mid-eighties, the primary focus was to mitigate portability issues and to develop generic interfaces that could be easily adapted to various databases. In the 90s, primary development focus was on learning approaches. There is a large literature describing systems that are purely rule based, semi-supervised, and unsupervised. Excellent overviews of such systems can be found in [1, 7, 17]. Most systems developed in 90s followed various approaches like pattern matching, syntactic parsing, semantic grammar systems, and intermediate representation languages for generating the query from text [1]. Most of the earlier work focused on specific databases, and to port the system to new database, additional customization was required. Recent work incorporates users' guidance [12, 17] to overcome this issue. Some literatures describe the data in the table as an additional input [20, 21]; but privacy issues may arise if table data would be used. Again if tables are huge in size, scalability could be another issue.

Many recent works use semantic parsing to parse a natural language to SQL queries. In most of the cases, a NL description is parsed into a logical form [2, 3, 5, 15, 16, 24, 25, 31, 36–38]. Most of these approaches focused on specific domain of interest, and it may be difficult to generalize them.

SQLizer [33] is a related work that uses a semantic parser [4, 18]. In this case, a NL question is translated into a query sketch, and then employing programming language techniques such as type-directed synthesis and automatic repair, the sketch is refined into the final query. Here, the user decides which of the top k queries is the right one. This approach is not restricted to any specific database. Sqlizer uses sequence-to-sequence neural semantic parsing and Word2Vec [19] for generating the query sketch. Query sketch includes clauses, but no database schema information is included(e.g., table and column names). That is why it is database independent. SQLizer does not work well for domain specific terms for which similarities cannot be computed using Word2Vec.

Dong and Lapata [8] provide a generic approach, an attention sequence-to-sequence neural semantic parser that achieves state-of-the-art performance on a host of semantic parsing datasets.

Seq2SQL [39] automatically synthesizes SQL code from English queries without user interaction. Deep learning method is used for translating NL questions to corresponding SQL queries using pointer-based generation akin to [29]. It achieves higher performance than [8], especially in generating queries with rare words and column names. This model takes as input a NL question and the columns of a table. The generated SQL query is executed, and the result of the execution is utilized to train the reinforcement learning algorithm. It uses WikiSQL dataset [39], GloVe word embedding [22] and character n-gram embedding [9]. It is trained using ADAM [13] and regularized using dropout[26]. Bi-directional LSTM [11] is used for encoding and decoding.

SQLNet [32]approach has all the merit points of Seq2SQL, and it also overcomes the drawbacks of a sequence-to-sequence model by replacing it with a sequence-to-set model with column attention approach. These techniques enable SQLNet to outperform Seq2SQL. It uses WikiSQL dataset [39], Stanford CoreNLP tokenizer [18], GloVe word embedding [22], and bi-directional LSTM [11].

TYPESQL [35] is build by modifying the state-of-the-art SQLNet [32]. TYPESQL impliments a sketch-based approach and considers the task as a slot filling problem. Here, different slots are grouped based on reasoning and relationships between attributes are captured, TYPESQL outperforms SQLNet in half of the original training time. Here, pre-trained Glove [22] and paraphrase [30] embedding are concatenated. Input questions are first prepossessed by type recognition. Using two bi-directional LSTMs, the words in the question are encoded to predict the values for the slots in the SQL sketch. The hidden states of LSTMs are used. In this model, accessing the content of databases for ill-formed users' queries can significantly improve the performance.

3 Model and Experiments

3.1 Matrices to Evaluate the Query

3.1.1 Logical-Form Accuracy

The synthesized SQL query and the ground truth query are compared to find out whether they match each other or not [39].

Logical-form accuracy = number of predicted queries that exactly matches with their corresponding gold queries/number of gold queries

3.1.2 Execution Accuracy

The synthesized query is executed and the ground truth query is also executed. Then their results are compared to check whether they match or not. [39].

Execution accuracy = number of predicted queries output matches that of gold queries/number of gold queries executed

3.2 Dataset

I have developed a large parallel corpora from the WikiSQL dataset proposed in [39]. In this work, the WikiSQL dataset is considered because it is different from most previous NL2SQL datasets [17, 21, 23, 27, 34], as Wiki SQL is a collection of 80,654 examples of questions, corresponding SQL queries, and database tables built from 24,241 table Schema. In order of magnitude, it is the largest hand-annotated semantic parsing dataset that have logical forms. It contains realistic data extracted from the Web by crowd sourcing the Amazon Mechanical Turk site. This large-scale dataset helps to train a neural network effectively.The WikiSQL dataset can be downloaded from its site https://github.com/salesforce/WikiSQL.

It is obvious that to correctly write an SQL statement, schema of the database must be known But as the training parameters of OpenNMT does not provide a scope for it presently, explicitly database schema information can not be provided to the model. So, I modified the datasets to include schema information in such a manner that the model can infer it during training .

I collected the natural language (NL) questions from the WikiSQL dataset and created the corresponding train, dev, and test source text files. I collected the corresponding SQL queries from the WikiSQL dataset, and with the help of the json files that store the details of the tables schema, I modified the SQL statements and created the corresponding train, dev, and test destination text files. I did these operations by writing python codes for them. As the model OpenNMT [14] that I am using only needs train and dev text files for training, I trained the model accordingly.

3.3 Implementation Details

I have used OpenNMT http://opennmt.net/ [14] which implements a standard SEQ2SEQ model. OpenNMT is a neural machine translation toolkit that can be freely downloaded from its site. After installation , while executing it, an user can set various parameters as per its requirements. I have set the following values for the parameters I have used: global-attention function logsoftmax is used; both the encoder and the decoder are long short-term memory (LSTM) cells with two hidden layers and 500 neurons with dropout of 0.3; the word embedding layer has 500 neurons; I have initialized encoder and decoder embeddings with GloVe's [22] values; I have used batches of maximum size 64 and trained the model with a learning rate of 1.0 and a learning decay of 0.5 till the perplexity did not decrease on the validation set. The code is available at https://github.com/Ali-Das/Text_to_SQL_using_OpenNMT

3.4 Evaluation of Result

The **training accuracy** is **99.8%** with **validation perplexity** of **5.11** and **validation accuracy** of **88.02%**. I have used the model to predict on the **Test** dataset. For this translation, I have used the trained model with a beam search of size 5. SQL queries were generated from the predicted text file after some modifications. I have also modified the test text file to generate the gold queries. All these queries are then compared to find out *logical-form accuracy* and executed to compare the results to check *execution accuracy*. The results are as shown below.

Logical-form accuracy = 0.33 , Execution accuracy = 0.49

Table 1 logical-form and the execution accuracy of models on the WikiSQL task

S. No.	Model	Dev dataset		Test dataset	
		($Accuracy_{lf}$) (%)	($Accuracy_{ex}$) (%)	($Accuracy_{lf}$) (%)	($Accuracy_{ex}$) (%)
1	Seq2SQL [39]	49.5	60.8	48.3	59.4
2	SQLNet [32]	63.2	69.8	61.3	68.0
3	TypeSQL [35]	68.0	74.5	66.7	73.5

Note $Accuracy_{lf}$ and $Accuracy_{ex}$ represent the logical-form accuracy and the execution accuracy

The accuracy is also not very good on **Dev** dataset. It was observed that most of the time the system generates the correct SQL query but when it fails, it generates a query based on the words used in the NL query. This generated SQL query may not be the exact match of the corresponding ground truth SQL query. As the model is not explicitly provided with the database schema information, more complex approaches that integrate schema information can produce better results.

4 Discussion

The Table 1 presents the logical-form accuracy and the execution accuracy of Seq2SQL, SQLNet and TypeSQL on the **Dev** and **Test** datasets of WikiSQL. My model's performance is not better than these previous state-of-the-art models on the dev and test datasets. But the **training accuracy of my model is far ahead** of all these models.

5 Conclusion

In this paper, I have described a NLIDB system that uses the open-source neural machine translation toolkit OpenNMT that implements a standard SEQ2SEQ model. I have trained this model with a large parallel corpora I have developed from the WikiSQL dataset. This system outperforms a strong baseline Seq2SQL [39] on training accuracy. As the training accuracy of my model is on the higher side, my model can be used in real life applications like: Chat-bots/ dialogue systems/ artificial intelligence powered personal digital assistants if it could be trained with sufficient real data. The most efficient further directions for improvement could be to integrate more schema information in the training corpora so that a significant increase in the accuracy could be achieved to produce better results.

References

1. Androutsopoulos I, Ritchie GD, Thanisch P (1995) Natural language interfaces to databases-an introduction arXiv preprint cmp-lg/9503016
2. Artzi Y, Zettlemoyer L (2011) Bootstrapping semantic parsers from conversations. In: Proceedings of the 2011 conference on empirical methods in natural language processing, pp 421–432
3. Artzi Y, Zettlemoyer L (2013) Weakly supervised learning of semantic parsers for mapping instructions to actions. In: Transactions of the association for computational linguistics, vol 1. MIT Press, pp 49–62
4. Berant J, Chou A, Frostig R, Liang P (2013) Semantic parsing on freebase from question-answer pairs. In: Proceedings of the 2013 conference on empirical methods in natural language processing, pp 1533–1544
5. Cai, Q, Yates A (2013) Large-scale semantic parsing via schema matching and lexicon extension. In: Proceedings of the 51st Annual Meeting of the Association for Computational Linguistics (Volume 1: Long Papers), pp 423–433
6. Codd EF (1974) Seven steps to rendezvous with the casual user. IBM Corporation (1974)
7. Copestake A, Jones KS (1990) Natural language interfaces to databases. Citeseer, 187
8. Dong L, Lapata M (2016) Language to logical form with neural attention, arXiv preprint arXiv:1601.01280
9. Hashimoto K, Xiong C, Tsuruoka Y, Socher R (2016) A joint many-task model: growing a neural network for multiple NLP tasks, arXiv preprint arXiv:1611.01587
10. Gary G, Sacerdoti ED, Sagalowicz D, Slocum J (1978) Developing a natural language interface to complex data. ACM Trans Database Syst (TODS) 3(2):105–147
11. Hochreiter S, Schmidhuber J (1997) Long short-term memory. Neural Comput 9(8):1735–1780
12. Iyer S, Konstas I, Cheung A Krishnamurthy J, Zettlemoyer L (2017) Learning a neural semantic parser from user feedback, arXiv preprint arXiv:1704.08760
13. Kingma DP, Ba J (2014) Adam: a method for stochastic optimization, arXiv preprint arXiv:1412.6980
14. Klein G, Kim Y, Deng Y, Senellart J, Rush AM (2017) OpenNMT: open-source toolkit for neural machine translation. Assoc Comput Linguist 67–72. arXiv preprint arXiv:1701.02810
15. Liu C, Chen X, Shin EC, Chen M, Song D (2016) Latent attention for if-then program synthesis. Adv Neural Inf Proc Syst 4574–4582
16. Liang P, Jordan MI, Klein D (2013) Learning dependency-based compositional semantics. Comput Linguist 39(2):389–446
17. Li F, Jagadish HV (2014) Constructing an interactive natural language interface for relational databases. Proc VLDB Endowment 8(1):73–84
18. Manning CD, Surdeanu M, Bauer J, Finkel JR, Bethard S, McClosky D (2014) The stanford CoreNLP natural language processing toolkit. In: Proceedings of 52nd annual meeting of the association for computational linguistics: system demonstrations, pp 55–60
19. Mikolov T, Sutskever I, Chen K, Corrado GS, Jeff D (2013) Distributed representations of words and phrases and their compositionality. Adv Neural Inf Proc Syst 3111–3119
20. Mou L, Lu Z, Li H, Jin Z (2017) Coupling distributed and symbolic execution for natural language queries. Int Conf Mach Learn 2518–2526
21. Pasupat P, Liang P, Compositional semantic parsing on semi-structured tables. ArXiv preprint arXiv:1508.00305
22. Pennington J, Socher R, Manning CD (2014) Glove: global vectors for word representation. In: Proceedings of the 2014 conference on empirical methods in natural language processing (EMNLP), pp 1532–1543
23. Price P (1990) Evaluation of spoken language systems: the ATIS domain. Speech and Natural Language: Proceedings of a Workshop Held at Hidden Valley, Pennsylvania
24. Quirk C, Mooney R Galley M (2015) Language to code: learning semantic parsers for if-this-then-that recipes. In: Proceedings of the 53rd annual meeting of the association for computational linguistics and the 7th international joint conference on natural language processing (Volume 1: Long Papers), pp 878–888

25. Reddy S, Lapata M, Steedman M (2014) Large-scale semantic parsing without question-answer pairs. In: Transactions of the Association for Computational Linguistics, vol 2. MIT Press, pp 377–392
26. Srivastava N, Hinton G, Krizhevsky A, Sutskever I, Salakhutdinov R (2014) Dropout: a simple way to prevent neural networks from overfitting. J Mach Learn Res 15(1):1929–1958
27. Tang LR, Mooney, RJ (2001)Using multiple clause constructors in inductive logic programming for semantic parsing. In: European Conference on Machine Learning, organization: Springer, pp 466–477
28. Vinyals O, Bengio S, Kudlur M (2015) Order matters: sequence to sequence for sets. ArXiv preprint arXiv:1511.06391
29. Vinyals O, Fortunato M, Jaitly N (2015) Pointer networks. Adv Neural Inf Proc Syst 2692–2700
30. Wieting J, Gimpel K (2017) Paranmt-50m: pushing the limits of paraphrastic sentence embeddings with millions of machine translations, arXiv preprint arXiv:1711.05732
31. Wong YW, Mooney R (2007) Learning synchronous grammars for semantic parsing with lambda calculus. In: Proceedings of the 45th annual meeting of the association of computational linguistics, pp 960–967
32. Xu X, Liu C, Song D(2017) Sqlnet: Generating structured queries from natural language without reinforcement learning. ArXiv preprint arXiv:1711.04436
33. Yaghmazadeh N, Wang Y, Dillig I, Dillig T (2017) SQLizer: query synthesis from natural language. In: Proceedings of the ACM on programming languages, vol 1, number: OOPSLA,ACM New York, NY, USA, pp 1–26
34. Yin P, Lu Z, Li H, Kao B (2015) Neural enquirer: learning to query tables with natural language. ArXiv preprint arXiv:1512.00965
35. Yu T, Li Z, Zhang Z, Zhang R, Radev D (2018) Typesql: knowledge-based type-aware neural text-to-sql generation. ArXiv preprint arXiv:1804.09769
36. Zelle JM, Mooney RJ (1996) Learning to parse database queries using inductive logic programming. In: Proceedings of the national conference on artificial intelligence, pp 1050–1055
37. Zettlemoyer L, Collins M (2007) Online learning of relaxed CCG grammars for parsing to logical form. In: Proceedings of the 2007 joint conference on empirical methods in natural language processing and computational natural language learning (EMNLP-CoNLL) pp 678–687
38. Zettlemoyer LS, Collins M (2012) Learning to map sentences to logical form: Structured classification with probabilistic categorial grammars. ArXiv preprint arXiv:1207.1420
39. Zhong V, Xiong C, Socher R (2017) Seq2sql: generating structured queries from natural language using reinforcement learning. ArXiv preprint arXiv:1709.00103

Rainfall Data-Based Time Series Forecasting Using Rolling Forecasting Model for Indian Geographic Area

Nilesh Maltare, Dharmendra Sharma, and Sachin Patel

Abstract In today's era, the agriculture industry plays an important role for the Indian economy. In India, more than 80% of households depend on the agriculture industry. These agriculture industries are majorly affected by unusual rainfall. A lot of crops are going into loss due to unwanted rainfall. Thus, there is a need to forecast the time-based rainfall data for different geographical areas. In this paper, we are using the rainfall-based time series data of Madhya Pradesh and Rajasthan (from 1901 to 2015) for forecasting rainfall data. We had implemented a rolling forecasting based supervised machine learning model with the assumption that rainfall data broadly affected by its geographical attributes like distance from sea, height of location from sea, etc. From the result, we found that the 12% amount of rainfall data is affected by its geographical location. We achieved an RMSE of 134.65 mm for Madhya Pradesh and 173.43 mm for Rajasthan state.

Keywords Machine learning · Rainfall prediction

1 Introduction

India is observing adverse effects of climate change. The increase in temperature is evident; according to the report [1], the average temperature of India has risen to 0.7 °C during 1901 to 2018. The significant change in pattern of rainfall is also noticeable. The monsoon rainfall (from June to September) in India has gone down by 6% from 1951 to 2015 [2]. The Indian economy heavily depends upon agriculture,

N. Maltare (✉)
Government Engineering College, Modasa, India
e-mail: nilesh.maltare@gecmodasa.org

D. Sharma
STME, NMIMS, Indore, India
e-mail: dharmendra.sharma@nmims.edu

S. Patel
SAGE, Indore, India
e-mail: drsachin.patel@sageuniversity.in

M. S. Kaiser et al. (eds.), *Information and Communication Technology for Competitive Strategies (ICTCS 2021)*, Lecture Notes in Networks and Systems 401,
https://doi.org/10.1007/978-981-19-0098-3_15

and increase in temperature and irregular rainfall affect crop yields badly. The change in rainfall patterns will impact in two severe forms, one is drought, and another is flood.

Global warming is affecting the whole world. The average temperature of surface, air, and oceans is increasing, glaciers are melting, sea level is rising, and frequent flooding and increasing areas of drought, etc. are noticeable.

In coming years, Himalayas will get more rainfall [3], while Indo-Gangetic plains and Western Ghats will observe decrease in precipitation. The state of Uttarakhand has faced the worst natural disaster in June 2013 [4]. It occurred during peak tourism, and in it, huge loss of human life and infrastructure happened. The disaster could not be predicted by any government agency and any scientific organization. The extreme rainfall in Kedarnath resulted in landslides. The roads, bridge, communication infrastructure, and everything were destroyed. The rescue work could not be done due to loss of infrastructure.

The accurate rainfall prediction model can help government agencies to prepare for such floods. Rainfall prediction is a tough task and in focus of scientific and government organizations. The computer scientists and engineers have been developing accurate models for rainfall predictions. There has been lots of work done for accurate prediction of rainfall.

Machine learning arises as one of the promising technologies in accurate rainfall prediction. In [5], regression on the historical rainfall data was applied to predict rainfall. It uses different algorithms such as boosted decision tree regression (BDTR), Bayesian linear regression (BLR), decision forest regression (DFR), and neural network regression (NNR) to predict rainfall. The study also claims acceptable accuracy of rainfall prediction.

The ability to handle a great amount of data encourages application of deep learning in rainfall prediction. Deep learning approach based on multilayer perceptron (MLP) and autoencoders was used to predict rainfall [6].

In this study, rainfall data was collected from the official website of the Indian government. The dataset consists of rainfall information of the last 100 years. In India, declining groundwater level, low yield of crops and frequent floods and droughts can be handled in a better way if we can predict rainfall accurately. In this paper, we tested the model and attempted for accurate prediction of rainfall for the states of Madhya Pradesh and Rajasthan.

The paper experimented rainfall prediction over rainfall data of two states: Madhya Pradesh and Rajasthan. Section. 2 of the paper discusses previous work on rainfall prediction. Section 3 is about methodology used in prediction. Section 4 talks about results, and Sect. 5 covers conclusion, limitation, and future work.

2 Literature Survey

The rainfall prediction is done using various machine learning techniques. The selection of machine learning algorithms is very important and depends on the availability of dataset and nature of prediction. It can predict short term or long term. The regression-based forecasting gives prediction, while approaches like decision tree, KNN, and SVM provide classification. The accurate prediction requires more features, but it can lead to complexity and requires more computational resources.

Linear regression is used [7] by establishing relationships between previous rainfall data and crop seasons. It uses standard deviation and mean in prediction of crop seasons. The comprehensive survey [8] discussed applications of machine learning in handling big data. Machine learning and big data analytics (BDA) are evolving with each other. The state-of-the-art developments in ML algorithms for BDA can handle huge amounts of data and provide better prediction. The paper also discusses key challenges associated with application of ML-based approaches. Machine learning-based big data processing is important to handle large datasets efficiently. The new methods and approaches are emerging for efficient data processing in order to discover interestingness for decision making.

Abhisek et al. [9] used layer recurrent network (LRN), artificial neural network in rainfall prediction, in which backpropagation algorithm (BPA) and cascaded backpropagation (CBP) algorithms were utilized for prediction. Niharika Solanki et al. [10] presented machine learning-based approach for rainfall prediction which employs artificial neural network and genetic algorithm.

The soft computing techniques have also been applied in prediction of rainfall [11] self-organizing map (SOM), backpropagation neural networks (BPNN), and fuzzy rule systems. In [12], all climatic factors that cause rainfall were analyzed, and a fuzzy rule-based K-nearest neighbor (KNN) approach was proposed for rainfall forecasting. The study uses fuzzy membership functions on historical rainfall data of Bangladesh. In [13], soft computing techniques like artificial neural networks, genetic algorithms, and fuzzy logic and their applications are reviewed in rainfall prediction. The study recommends the use of hybrid approach due to the harmony among soft computing approaches. Hybrid approaches can utilize the best feature of each technique and reduce shortcoming of individual technique.

Data-driven deep learning methods are gaining popularity in every domain. They enable faster learning by correlating features without being explicitly told to do so. Deep learning-based weather forecasting is expected to replace conventional methods of weather forecasting. In [14], a comprehensive survey of deep learning-based methods of weather forecasting is done. It analyzes different methods with respect to spatial scale (local, regional, and global), temporal scales (small, large), and also on different datasets and benchmarks. Deep learning neural network architecture encompasses convolutional neural network (CNN), and recurrent neural network (RNN) was proposed [15] for predicting and alarming flash floods. The relation between groundwater level and rainfall is demonstrated in [16]. The rainfall prediction and groundwater relation can be used to help decision makers to create and implement new policies for increasing groundwater resources.

3 Methodology

In this research work, we have used time series-based rainfall data for rainfall forecasting of Madhya Pradesh and Rajasthan states. We used a hybrid rolling forecasting model for forecasting rainfall data. The hybrid rolling forecasting model assumes that rainfall data is broadly affected by its geographical attributes like distance from sea, height of location from sea, etc. Our model is using six-year windows to forecast coming year rainfall data. We divided our task into four parts:

1. Data collection and missing value imputation.
2. Trend detection for time series data.
3. Training of forecasting model.
4. Testing with cross-validation dataset.

3.1 Data Collection and Missing Value Imputation

The study experimented with a real dataset collected from a government organization [rainfall data]. The data is cleaned, and for missing value imputation, we used random value imputation as it has minimal effect on the distribution of data points. The following steps are used for missing value imputation:

1. Create alias of feature.
2. Count number of NaN values.
3. Select random values equal to count in Step 2.
4. Fill NaN location with selected random values.

3.2 Trend Detection

There are various methods available to predict the trends of time series data such as differencing, shifting, evaluation of mean, standard deviation. The trend in time series data defines seasonal effect. The time series which is not affected by trends is considered as a stationary time series. We evaluated rolling mean and rolling average with every six consecutive samples to test the trend of series. The below Fig. 1 represents the comparative data of rainfall between Madhya Pradesh and Rajasthan states. Figure 2 represents rolling mean and standard deviation for the time series. From Fig. 2, we can observe that the rolling mean and standard deviation remain constant for every year of the time series.

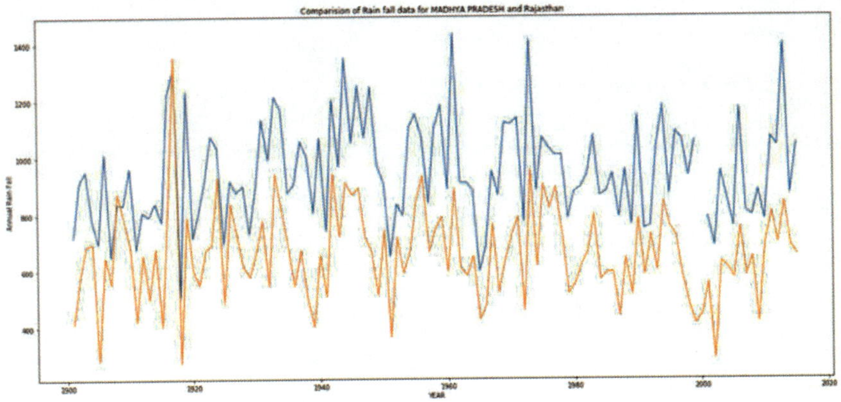

Fig. 1 Comparison of rainfall data for Madhya Pradesh and Rajasthan

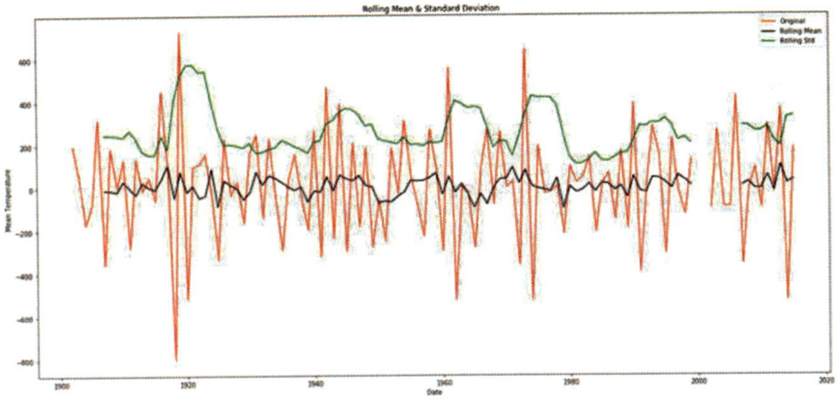

Fig. 2 Rolling mean and standard deviation

3.3 *Training and Testing of Forecasting Model*

The entire dataset is divided into two parts, first part used for training of model and second part used for validation of model. Model evaluation involved two components: performance measure and test strategy. To evaluate the performance of the model, we used root mean square error (RMSE). This is the baseline for our model that the model we make should at least have less or equal to RMSE.

A model is developed using a training dataset, and the forecasted value is compared with the expected value. The observation (forecasted value) is added to the training dataset, and the same process is iterated again and again. The trained model was evaluated using a walk-forward validation method. In the walk-forward validation method, we enumerate each time step from the test dataset. Walk-forward validation

```
#------------- code structure ---------------------- #
## ---------- ---------------- #

history = [x for x in train]

predictions = list()

for i in range(len(test)):
    # predicting test[i] and storing it in yhat
    yhat = ...
    predictions.append(yhat)

    # observation
    obs = test[i] #storing actual test[i] in observation
    history.append(obs)
    print('>Predicted=%.3f, Expected=%3.f' % (yhat, obs))

# Note : The actual observation from the test dataset will be added
# to the training dataset for the next iteration.
```

Fig. 3 Code structure

is a real way to evaluate any time series data because it updates the model as new forecasted values are available. This model is best for time series data. The model works as follows.

The walk-forward validation will work as follows:

1. Fifty percent of the dataset used for training of model.
2. Fifty percent of the data iterated for testing of model.

For each iteration of test dataset, we execute following steps:

1. Train the model using 50% of the dataset.
2. Store predicted value for later evaluation.
3. Add actual value of observation to the test dataset for the next iteration.
4. The predicted value obtained during iteration of test data evaluated and root mean square value are recorded.

Figure 3 presents the code structure for the model.

4 Results and Discussions

We used Python Jupyter Notebook with Pandas, NumPy, seaborn, matplotlib, and sklearn library for implementation of our model. We used 4116 records of rainfall data for training and testing of our model. Tables 1 and 2 represent the expected and predicted values of rainfall for Madhya Pradesh and Rajasthan states, respectively. With this model, we achieved an RMSE of 134.65 for Madhya Pradesh and 173.43 for Rajasthan state. The percentage difference between RMSE for Rajasthan and Madhya Pradesh is 12%. This means that on average, the model was wrong by

Table 1 Expected and predicted values of rainfall for Madhya Pradesh

S. no.	Year	Predicted	Expected
1	1945	832.9	1099.7
2	1946	1099.7	1181.4
3	1947	1181.4	882.3
4	1948	882.3	1433.7
5	1949	1433.7	905.9
6	1950	905.9	905.4
7	1951	905.4	880.1
8	1952	880.1	594.7

Table 2 Expected and predicted values of rainfall for Rajasthan

S. no.	Year	Predicted	Expected
1	1971	1003.3	781.9
2	1972	781.9	873.9
3	1973	873.9	890.2
4	1974	890.2	934.1
5	1975	934.1	1071.9
6	1976	1071.9	859.5
7	1977	859.5	874.7
8	1978	874.7	937.1

about 134.65 mm rainfall per year for each prediction made for Madhya Pradesh and 173.43 mm rainfall per year for each prediction made for Rajasthan.

5 Conclusion

In this research work, we used the rainfall-based time series data of Madhya Pradesh and Rajasthan (from 1901 to 2015) for forecasting rainfall data. We had implemented a rolling forecasting-based supervised machine learning model. We assume that rainfall data highly depends on geographical attributes like distance from sea, height of city from the sea, etc. From the result, we found that 12% of rainfall data is affected by its geographical location for Madhya Pradesh and Rajasthan states. Our model achieved root mean square error of 134.65 mm for Madhya Pradesh and 173.43 mm for Rajasthan state. In the future work, we can use groundwater level data to establish the relation between rainfall data and groundwater data. These relations can be used to optimize the groundwater level for different geographical areas.

References

1. Sanjay J et al (2020) Temperature changes in India. In: Krishnan R, Sanjay J, Gnanaseelan C, Mujumdar M, Kulkarni A, Chakraborty S (eds) Assessment of climate change over the Indian Region. Springer, Singapore. https://doi.org/10.1007/978-981-15-4327-2_2
2. Kulkarni A et al (2020) Precipitation changes in India. In: Krishnan R, Sanjay J, Gnanaseelan C, Mujumdar M, Kulkarni A, Chakraborty S (eds) Assessment of climate Change over the Indian Region. Springer, Singapore. https://doi.org/10.1007/978-981-15-4327-2_3
3. Pant GB, Kumar PP, Revadekar JV, Singh N (2018) Climate change: central himalayan perspective. In: Climate change in the Himalayas. Springer, Cham. https://doi.org/10.1007/978-3-319-61654-4_5
4. Uttarakhand Disaster 2013 (2018) National institute of disaster management, ministry of home affairs, Government of India. 2013. p 41. Retrieved 14 December 2018
5. Ridwan WM, Sapitang M, Aziz A, Kushiar KF, Ahmed AN, El-Shafie A (2021) Rainfall forecasting model using machine learning methods: case study Terengganu, Malaysia. Ain Shams Eng J 12(2):1651–1663. ISSN 2090–4479. https://doi.org/10.1016/j.asej.2020.09.011
6. Basha CZ, Bhavana N, Bhavya P, Sowmya V (2020) Rainfall prediction using machine learning and deep learning techniques. In: 2020 International Conference on Electronics and Sustainable Communication Systems (ICESC), pp 92–97. https://doi.org/10.1109/ICESC48915.2020.9155896
7. Thirumalai C et al (2017) Heuristic prediction of rainfall using machine learning techniques. In: 2017 International Conference on Trends in Electronics and Informatics (ICEI). IEEE
8. Bhatnagar R (2018) Machine learning and big data processing: a technological perspective and review. In: Hassanien A, Tolba M, Elhoseny M, Mostafa M (eds) The international conference on Advanced Machine Learning Technologies and Applications (AMLTA2018). AMLTA 2018. Advances in Intelligent Systems and Computing, vol 723. Springer, Cham. https://doi.org/10.1007/978-3-319-74690-6_46
9. Abhishek K, Kumar A, Ranjan R, Kumar S (2012) A rainfall prediction model using artificial neural network. In: 2012 IEEE Control and System Graduate Research Colloquium (ICSGRC), Shah Alam, Selangor, pp 82–87
10. Solanki N, Panchal G (2018) A novel machine learning based approach for rainfall prediction. In: Satapathy S, Joshi A (eds) Information and communication technology for intelligent systems (ICTIS 2017) - Volume 1. ICTIS 2017. Smart Innovation, Systems and Technologies, vol 83. Springer, Cham. https://doi.org/10.1007/978-3-319-63673-3_38
11. Wong K, Wong P, Gedeon T et al (2003) Rainfall prediction model using soft computing technique. Soft Comput 7:434–438. https://doi.org/10.1007/s00500-002-0232-4
12. Zahid Hasan M, Hossain S, Zubair Hasan KM, Uddin MS, Ehteshamul Alam M (2021) Fuzzy rule-based KNN for rainfall prediction: a case study in Bangladesh. In: Uddin MS, Bansal JC (eds) Proceedings of international joint conference on advances in computational intelligence. Algorithms for Intelligent Systems. Springer, Singapore. https://doi.org/10.1007/978-981-16-0586-4_41
13. Chandwani V, Vyas SK, Agrawal V, Sharma G (2015) Soft computing approach for rainfall-runoff modelling: a review. Aquatic Procedia 4:1054–1061: ISSN 2214-241X. https://doi.org/10.1016/j.aqpro.2015.02.133
14. Ren X, Li X, Ren K, Song J, Xu Z, Deng K, Wang X (2021) Deep learning-based weather prediction: a survey. Big Data Res 23:100178 ISSN 2214-5796. https://doi.org/10.1016/j.bdr.2020.100178
15. Panahi M, Jaafari A, Shirzadi A, Shahabi H, Rahmati O, Omidvar E, Lee S, Bui DT (2021) Deep learning neural networks for spatially explicit prediction of flash flood probability. Geoscience Frontiers 12(3):101076. ISSN 1674-9871. https://doi.org/10.1016/j.gsf.2020.09.007
16. Prajapati R, Upadhyay S, Talchabhadel R, Thapa BR, Ertis B, Silwal P, Davids JC (2021) Investigating the nexus of groundwater levels, rainfall and land-use in the Kathmandu Valley, Nepal. Groundwater for Sustainable Development 14:100584. ISSN 2352-801X. https://doi.org/10.1016/j.gsd.2021.100584

Reforming Title Deeds via Blockchain and Challenges—Indian Perspective

Jyostna Grandhi, Sandeep Romana, Mahesh Patil, P. R. Lakshmi Eswari, and Nilesh Kumar

Abstract It is crucial to have accurate title records that assure the property's ownership. In India, the challenges in uniquely identifying property, verifying the title ownership, syncing database of different departments, no implicit linkage to the property registrations lead to the creation of fake registration documents and double registration. For citizens to trust the system, it has to be transparent and with accurate title records. We leverage the Hyperledger Fabric blockchain framework to improve the existing workflow's limitations. For this, the existing workflow followed for the registration of property in the state of Telangana is studied comprehensively. The designed system integrates fully with the existing workflow without disturbing the current system. We present example-based scenarios to illustrate the benefits of being able to track (a) fraudulent transactions via the creation of fake registration documents, (b) insider attacks, (c) double registrations, (d) typing errors and avoiding manual verification of data, and (e) reliable encumbrance search for encumbrance certificates. We present the challenges faced during the system design. A pilot deployment at the sub-registrar office (SRO) in the Shamshabad district of Telangana state shows the practical usability of the designed system. Migration of the actual data from existing registration records makes the designed system work on real-time data.

J. Grandhi (✉) · S. Romana · M. Patil · P. R. Lakshmi Eswari
Centre for Development of Advanced Computing (C-DAC), Hyderabad, Telangana, India
e-mail: gjyostna@cdac.in
URL: https://www.cdac.in

S. Romana
e-mail: sandeepr@cdac.in

M. Patil
e-mail: maheshp@cdac.in

P. R. Lakshmi Eswari
e-mail: prleswari@cdac.in

N. Kumar
National Informatics Center Hyderabad, Hyderabad, Telangana, India
e-mail: nilesh.kumar18@nic.in
URL: https://ts.nic.in

M. S. Kaiser et al. (eds.), *Information and Communication Technology for Competitive Strategies (ICTCS 2021)*, Lecture Notes in Networks and Systems 401,
https://doi.org/10.1007/978-981-19-0098-3_16

The performance benchmarking data of the developed system using the Hyperledger Caliper framework shows its scalability to the entire state of Telangana.

Keywords Reforms · India · Blockchain · Title deed · SRO · Proof-of-existence · Hyperledger Fabric · Hyperledger Caliper

1 Introduction

There are many limitations to the existing registration process followed by the states of India. Firstly, the ways to identify a property uniquely are not in place. Multiple departments such as Survey & Settlement, Stamps & Registration and Revenue get involved during the property title registration. Secondly, the Revenue and Stamp & Registration departments are independent entities and maintain their separate database that does not regularly sync. Thirdly, the legal authority for property registration, the Sub-Registrar's Office (SRO), has no legal power and mandate to ask the parties to prove their property ownership. There is no mandate to verify the identity of the buyer and the seller. In India, the challenges in uniquely identifying property, verifying the title ownership, syncing database of different departments, no implicit linkage to the property registrations lead to the creation of fake registration documents and double registration.

Globally, the authors at [1] lists the problems of document forgery and document availability in the Integrated Land Management Information System (ILMIS) of Tanzania. They propose a model that integrates the ILMIS with *factom* and *bitcoin* blockchains to solve the listed problems. Meghna Bal [2] mentions the deficiencies in the existing system of land records in India despite the introduction of the Digital India Land Records Modernization Programme (DILRMP). The author discusses the underlying problems and suggests that blockchain technology is a possible solution for record-keeping. Further, the report from Chromaway lists the benefits of implementing a blockchain-based system in the Indian scenario [3].

It is required to have a system with features that enable transparency across the departments, build trust across the stakeholders, maintain a single source of truth, keep entrenched records, and identify the potential irregularities during registration. ADYTY (blockchAin baseD propertY record managemenT sYstem) incorporates the above-listed features to resolve land registration problems. The Indian Registration Act is adopted with minor modifications by the various states of India [4]. Thus, the developed system can be extended to India's other states property registration process with minor modifications. The presented work demonstrates leveraging blockchain for the property registration process followed by the Telangana State Government.

1.1 About Blockchain

Blockchain technology is suitable for decentralised and transactional data sharing across an extensive network of untrusted participants. Using blockchain, one can maintain a continuously growing list of records called blocks and link them in a distributed manner, ensuring their security against tampering. It allows distributed software architecture to build trust between peers without any trusted central authority. Further, it allows for pseudonymity, cryptographic guarantees, immutability, shared read and write, accountability, transparency, and distributed ownership [5].

Cryptocurrencies such as *bitcoin* use blockchain as the underlying technology. Any user can connect to the blockchain network, add new and verify transactions and take part in the appending blocks of transactions via a process known as *mining*. Besides cryptocurrencies, blockchain has potential in governance, health care, cybersecurity, automobiles, media, travel, hospitality, energy, smart cities. Few specific applications of blockchain are tracking taxpayer money, direct benefit transfer (DBT), online voting, tracking database records for different governance applications, smart contracts, smart payments, music payment and licencing, decentralised management of Internet of Things (IoT), the authenticity of a review, notary, supply chain communications, proof-of-provenance, rural payments, corporate audits, patient data management, and global wallets [6]. We have developed a property record management application leveraging blockchain technology. The subsequent sections provide the details of the evolved system.

2 Related Work

Blockchain has brought forward the potential of a paradigm shift in building trusts from using trusted centralised systems to untrusted decentralised systems. Globally, many countries have started working on property registration applications to bring transparency to property records. The Republic of Georgia launched the first-ever blockchain land registry system to strengthen property owners' rights, enhance citizens' trust in Government and reinforce data security. They have registered more than 1.5 million land titles, with an average of 3 min registration time [7]. The Swedish land registry is testing real estate and other multi-party transactions on a blockchain [8]. UK Government department responsible for recording 25 million ownership titles of lands and properties valued at over $7 trillion ($8.7 trillion) partnered with ConSenSys Codefi to explore the use of blockchain and smart contract technology to aid speed, simplicity and transparency in the real estate market [9].

Upwards of 78% of the land in Ghana is unregistered. In partnership with Bitland, the Government of Ghana launched a pilot project to register land title ownership on a blockchain and carried out a trial in 20 communities in Kumasi, Ghana [10]. The Rwandan Government is utilising blockchain to architect a paperless, secure, corruption-proof system. There is a plan to centre the first pilot around the land registry [11]. Uganda government partnered with CryptoSavannah to create a proof of concept for land titles registry [12]. The Center for Affordable Housing Finance in Africa (CAHF) has piloted a working example of a blockchain-based property registry [13]. Most of these projects and some implementations in this direction are still in the PoC stage [14–16]. In this work, we present a fully functional pilot system deployed in real time for overcoming the existing issues. For demonstration and evaluation, the developed system is integrated to the existing property registration workflow followed by the state of Telangana in India.

3 Design Details

3.1 Existing Workflow

The existing workflow of the registration process followed by the Telangana State Government is a four-step process described in Fig. 1.
(a) Pre-processing the data: The Sub-Registrar's Office (SRO) verifies the data from the document provided by the document writer or public data entry service.
(b) Check slip report generation: The existing software application used by the SRO takes the following as an input to generate a check slip number:

1. All the buyer(s) and seller(s) details, including the Aadhaar number. Aadhaar is a 12 digit individual identification number issued by the Unique Identification Authority of India on behalf of the Government of India. This number serves as a proof of identity and address anywhere in India [17]. It is similar to the Social Security Number of the USA.
2. Property details such as survey number and area of extent. A survey number is a unique number assigned to a specific piece of land provided by the Survey and Settlement Department to maintain records.

The generated check slip no. is for inter-departmental usage. SRO checks whether the property is on the abeyance list and prohibited list and verifies the receipts of the taxes paid.

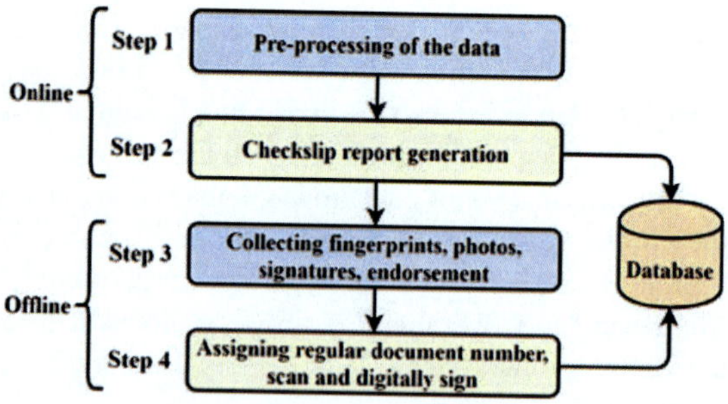

Fig. 1 Existing workflow

(c) Endorsing the registration document: This step includes capturing the following data and getting it endorsed on the registration document's backside to generate a new regular document number: (a) buyer, seller and witness data such as fingerprints, signatures, photographs, (b) the execution date and registration date.
(d) Generation of the final regular document: The endorsed document is scanned and digitally signed by the SRO before handing it over to the buyer. The existing workflow requires adding the property details into the local database at two different instances. Once during the check slip number generation (i.e. Step 2) and a second time during the final regular document generation (i.e. Step 4), as depicted in Fig. 1. The existing application of Telangana state has digitised all the records from 2008 onwards. The steps, pre-processing the data and check slip report generation are currently online and automated, whereas verifying the person's identity, collecting fingerprints and photographs, assigning regular document numbers, scanning the sale deed and digitally signing the document are offline activities.

3.2 *Challenges*

There were several challenges in designing a property management system based on blockchain. The following text highlights some of the most important of these challenges.
(a) Unique identification of the property: Evolving a unique identification for a property from the property attributes is crucial for the blockchain-based system to work correctly. The unique property ID (UPID) is evolved based on—survey number, district, mandal, village, few other attributes.
(b) Blockchain platform selection: Various blockchain frameworks are available in open source, catering to different target applications' needs. A survey on the capabilities of different blockchain frameworks helped to select the right framework for this application. The selection criteria were: maturity, active development, community support, technical capability, modularity, type of platform, i.e. public, private or permissioned. Based on the above parameters, Hyperledger Fabric, a permissioned blockchain, is identified for implementing ADYTY.
(c) Identifying interfaces: It is vital to identify interfaces between them to integrate a new system (ADYTY) into the existing system. An in-depth understanding of the existing system and blockchain leads to identifying two interfaces—check slip and regular document transactions.
(d) Data migration: The existing property records had many attributes attached to each data record. Understanding each attribute's meaning and selecting attributes necessary for storing records on blockchain was a challenge.
(e) Encumbrance search: The encumbrance search provided by the existing system is complex and requires manual intervention. To simplify and fully automate this search feature, ADYTY relied on UPID, PPID and a few other property attributes.
(f) Selection of design approach: There are legal challenges in using blockchain for a property record management system developed from scratch. It involves stamp

papers and manual signatures of buyers, sellers and witnesses. To overcome these challenges, we preferred to integrate ADYTY into the existing workflow, followed by the Telangana State Government for one of the SRO's at Shamshabad as a parallel system.

3.3 *Design*

ADYTY integrates to the Telangana Government's current property registration application, pulls the near real-time data of check slip and regular document transactions and appends it to its ledger. The check slip transaction verifies the property's current owner details with the blockchain ledger and cautions SRO if there is a mismatch. It also provides a reliable encumbrance certificate (EC) search without any manual intervention. The paring of the ADYTY with the Proof-of-Existence (PoE) ledger allows for storing a timestamped hash of the sale deed document for later verification. PoE calculates any digital artefact's cryptographic digest and stores it on the blockchain along with the timestamp. Later, searches based on the hash of the digital artefact are possible. A successful search proves the existence of stored hash on the blockchain at a specific time. Figure 2 provides an overview of ADYTY's architecture, and Fig. 3 depicts its components.

Through an external interface, its REST server periodically pulls the data from Telangana's property registration system. It exposes REST APIs and thus acts as a middleware between the web application and the blockchain network. The exposed REST APIs allows for creating channel, joining peers to the channel, installing smart contract on the channel, authenticating and authorising the users for invoking and

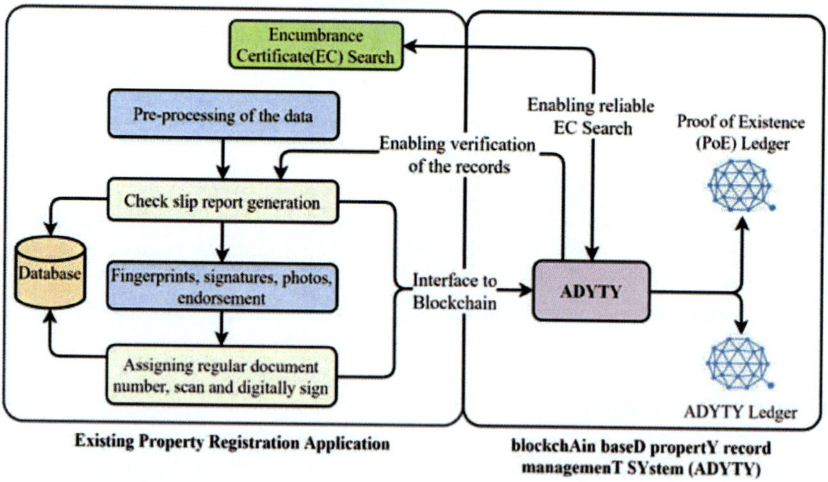

Fig. 2 Architecture

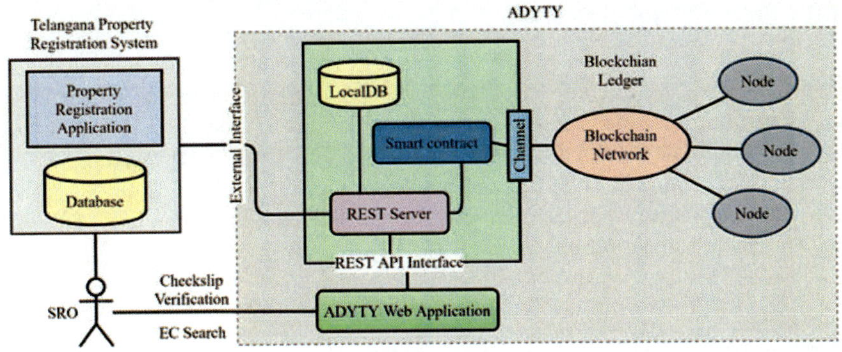

Fig. 3 Components

querying a transaction on the blockchain. The web application allows SRO to verify the check slip details and perform an encumbrance search.

4 Implementation Details

This section focuses on implementation details of ADYTY's components. We leveraged the Hyperledger Fabric blockchain platform for the development of ADYTY.

4.1 Blockchain Network

The blockchain network is composed of *organisations*, *peer nodes* (endorsement peers, anchor peers and committing peers), *Membership Service Provider* (MSP), *ordering service nodes* (Orderers), and *communication channel* for communicating with trusted entities in a network. The endorsing peers simulate and endorse transactions, and the committing peers verify these endorsements and validate the transaction results before committing transactions to the blockchain. *Orderer* service node accepts endorsed transactions, orders them into a block and delivers the blocks to the committing peers [18].

MSP is used to create a digital identity for each node in the network and builds trust across all the actors participating in the blockchain network. All the peers view the current state of the network by maintaining a copy of the ledger. The configured system uses the Raft protocol that provides a crash fault-tolerance ordering service. Raft selects one of the *orderer* nodes as a leader and the other *orderer* nodes as followers. A leader can replicate the messages to the follower nodes, who send them to the nodes.

4.2 Smart Contract

Hyperledger Fabric refers to the smart contracts as a chaincode, and the endorsing peers must have the chaincode installed on them. The chaincode initialises the ledger, manages its state through transactions and executes it in a separate and secure Docker container. ADYTY's chaincode implements code for recording the check slip and regular document transactions, querying check slip information and performs encumbrance searches on the property.

4.3 REST Server

The REST server is implemented using NodeJS and exposes REST APIs for easy integration with third-party applications. It uses Hyperledger Fabric SDK to communicate with the blockchain network. It provides the APIs for registering users through the authentication & authorisation component, recording the check slip and regular document details, verifying the check slip data and performing encumbrance searches. The designed system pulls the check slip and regular document details periodically from the existing application through the REST APIs. It then sends them to the chaincode for recording on the blockchain. It receives the check slip in JavaScript Object Notation (JSON) format, containing data about check slip number, year, property details, seller, buyer and witness details.

ADYTY creates a unique property ID (UPID) from the given property details by considering the property's essential attributes. The designed system can handle multiple schedules containing more than one property registered to the single entity and bifurcation of the property, i.e. parts of a property sold to different entities. During the check slip transaction, via the chaincode, the property is recorded on the blockchain using the UPID, check slip number and year of registration. Simultaneously, the system verifies the property's current owner data from the blockchain and shows it to the SRO via the web application. Regular document transaction involves assigning a new regular document number and year if the transaction is successful. REST server provides encumbrance search results to SRO, citizens and various departments based on the regular document number and the year of registration. Figure 5 shows the sample encumbrance search results for a specific property. The figure shows how the ownership changes are tracked by the chain of transactions on a specific property.

4.4 Web Application

The web application is developed using Angular and provides a web interface to the SRO for verifying the property's current owner details and performs encumbrance search on the blockchain data, as shown in Figs. 4 and 5, respectively. When provided

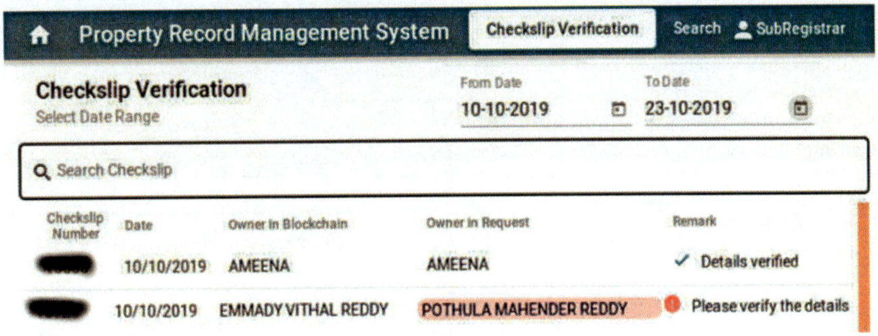

Fig. 4 Check slip verification

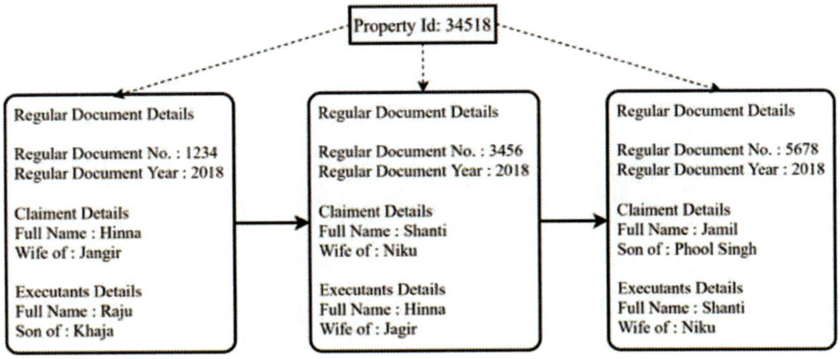

Fig. 5 Encumbrance search

with the input of the regular document number and year of registration, the system shows the list of transactions on that property by considering the UPID and parent property ID (PPID). It uses PPID to trace to the first transaction on any property for reliable encumbrance search and certificate.

5 Benefits of ADYTY over Existing Workflow

This section describes five different scenarios to illustrate the benefits of ADYTY. The developed system denies double registration, prevents fraudulent transactions via creating fake registration documents, prevents insider attacks, eliminates typo errors during manual data entry and provides a reliable encumbrance search for encumbrance certificate (EC).

An illustration of improved workflow with the integration of ADYTY to the existing system is as follows. For illustration, consider Ramesh and Rajesh in the role of seller and buyer, respectively. Rajesh can verify a property's linked documents to

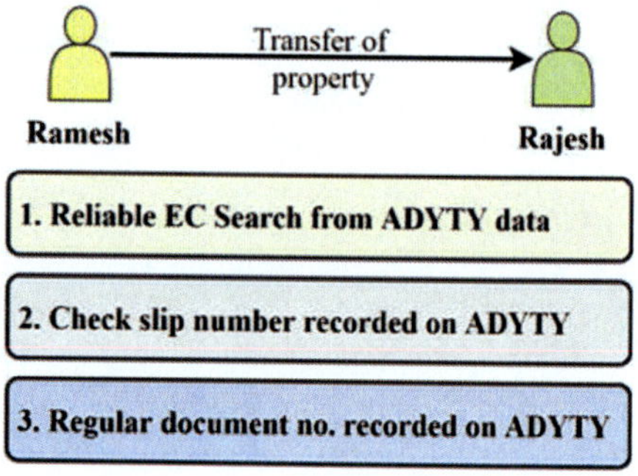

Fig. 6 Improved workflow with ADYTY

know the ownership changes through encumbrance search on the blockchain-based system. If found valid, he proceeds to provide the required details to the SRO. The SRO verifies the parties' identities from the Aadhaar database and ownership data from the blockchain. The SRO generates a unique check slip number and then subsequently assigns a regular document number. Both these transactions get recorded on the blockchain. A new owner gets a copy of the sale deed after uploading it to the existing system's database. Figure 6 describes this example scenario.

5.1 *Scenario 1—Preventing Double Registration*

As described above, the existing workflow saves data to the database during check slip and final regular document generation. There is a manual effort to enter the details of fingerprints of the buyer, seller, witnesses and other data, which causes a time delay between the two transactions in the existing workflow. The current application allows any attempt to sell the same property to a third person during this delay. It happens because the encumbrance search does not show the status of the newly initiated check slip transaction. ADYTY resolves this issue by notifying the SRO that the property is under the transaction and waiting for the final regular document generation.

For illustration, say Suresh is buying property P1 from Ramesh and waiting for final regular document generation after check slip transaction. During this time, if Ramesh tries to sell property P1 to Krishna, the existing system cannot notify Krishna that the property is already under transaction during the encumbrance search. ADYTY prevents the generation of two check slips simultaneously on property P1. Figure 7 describes this scenario.

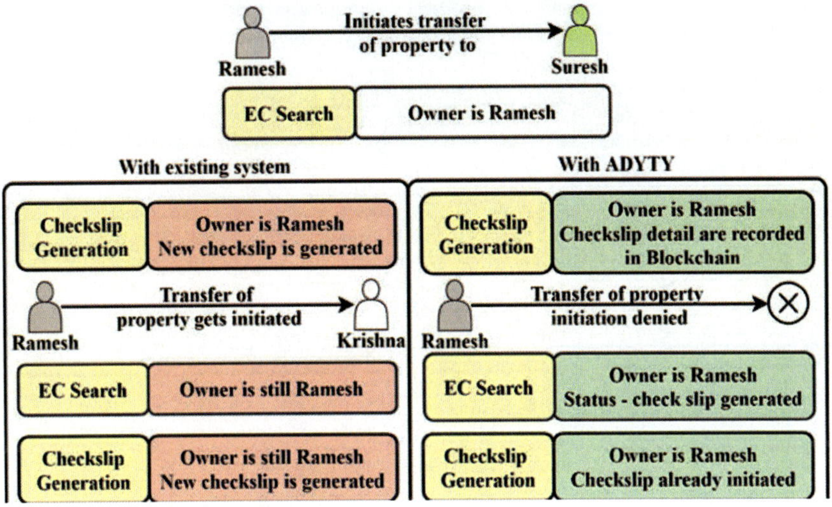

Fig. 7 Preventing double registration

5.2 *Scenario 2—Handling Fake Registration Documents*

Assume a hypothetical scenario where Ramesh is the genuine owner of the property P1. Let us assume that Rajesh fraudulently creates fake registration documents of Ramesh's property P1 showing the owner as Rajesh. If Rajesh tries to sell property P1 to Vinod, the existing system fails to identify this as an invalid transaction as ownership data is not verified. In this case, the ADYTY generates an alert for the SRO, and the buyer with the owner verification failed messages. One must note that though Aadhaar data is collected, it helps verify only the individual identity of the parties involved in the transaction, not the linkage of the identity to the ownership of the property. With ADYTY, the attacker has to modify the blockchain ledger's immutable records. Figure 8 illustrates this scenario.

5.3 *Scenario 3—Preventing Insider Attack*

To understand this scenario, consider a hypothetical scenario where say fake seller named Rajesh is trying to sell the owner Ramesh's property with the help of an insider having access to the registration database. For this, say the insider helps Rajesh modify property's ownership details from Ramesh to Rajesh. With the property's modified owner, ADYTY generates an alert for the SRO and the buyer during the encumbrance search and check slip transaction, thus preventing insider attacks. Figure 9 illustrates this scenario for pictorial understanding.

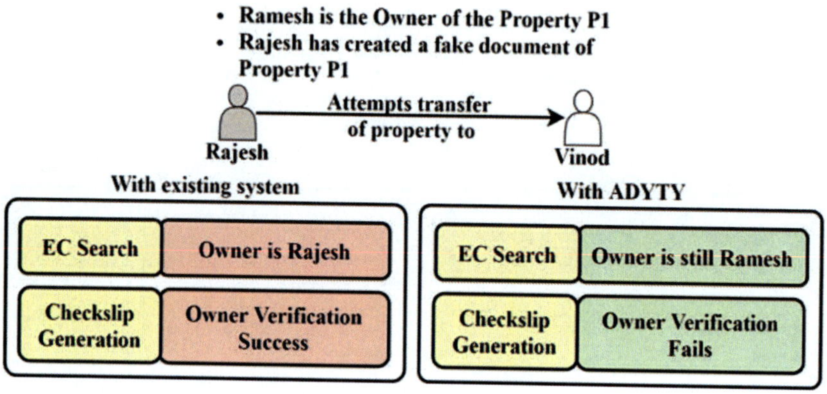

Fig. 8 Handling fake registration documents

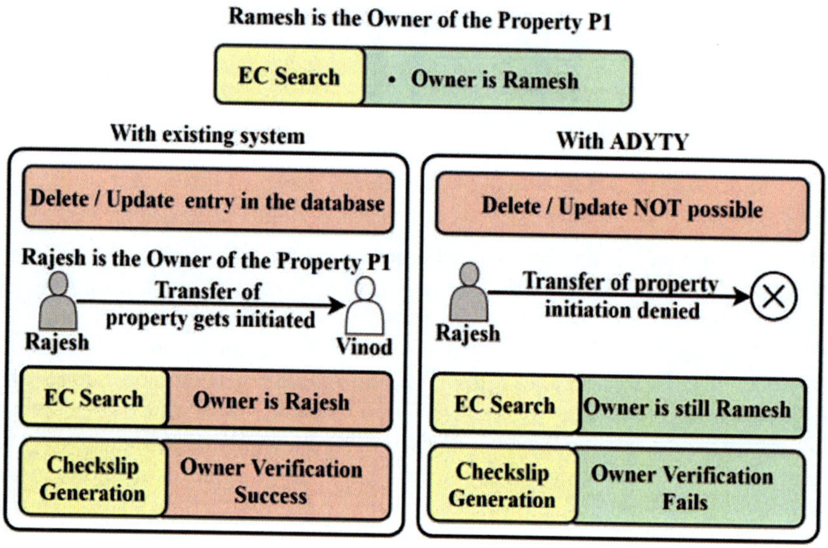

Fig. 9 Preventing inside attack

5.4 *Scenario 4—Eliminating Typo Errors and Manual Verification*

The existing workflow of property registration requires manually entering data into the registration portal during the pre-registration phase. There is no inbuilt mechanism for verifying the manually entered data, which increases the probability of typo errors. With ADYTY, auto-fetching data from the blockchain during the pre-registration phase improves the existing system preventing unnecessary typo errors and verification during entry into the registration portal.

5.5 Scenario 5—Tracking Property to the Origin

A blockchain is a ledger of cryptographically linked blocks such that each block contains a link to the hash of previous block data. This inherent property facilitates reliable encumbrance search and tracking of the property to its origin using UPID and PPID.

6 Deployment Setup and Evaluation

A pilot deployment of ADYTY at Shamshabad district of Telangana state uses ten nodes. Eight nodes out of ten are part of the blockchain network; one is a REST server, and one is configured with a front-end application, as shown in Fig. 10.

The client with the transaction proposal provides the details to the REST server through the front-end application. The REST server routes the transaction based on the client's access role. If the client is mapped to the writer role, it will send the transaction through the peer node of *RegOrg* for further processing. On the other hand, if the client is a reader, then the transaction can be process through any of the peer nodes. After the transaction is endorsed by other organisations endorsing peers, the transaction along with the endorsement details is sent to the leader of the orderers. The leader is elected as per the rules of the consensus algorithm (RAFT in this case). The *orderer* then orders all transactions and creates a block. This block is sent to the *committing* peers of all the organisations, gets verified by these committing peers and then updated in the ledger. As this is a permissioned blockchain, it is assumed that the addition of the nodes into the network can be achieved only via the administrator of the network.

The pilot deployment of ADYTY uses the https://www.cdacchain.in/pmapp/. We migrated property records from the year 2008 till 2018 for the Shamshabad district to ADYTY. Besides this, we regularly pull the near real-time transactions from January 2019 onwards using Telangana state's web services. The data showed that, on average, approximately 50 registrations happen per day in Shamshabad SRO. A separately obtained data for the year 2019 showed that, on average, nine registrations per minute (*tpm*) happened in the state of Telangana for the year 2019.

We used a server with a processor configuration of Intel Xeon (2.40GHZ/20-core/27.50 MB/150W), 512GB RAM, and SAS 19*2.4TB hard disc for performance benchmarking. The blockchain network had 11 nodes, out of which six were peer nodes and five were *orderer* nodes. For nodes, eight virtual machines were used, out of which three nodes were organisations, and each organisation had two peers, and five nodes were ordering nodes.

Table 1 presents the performance results for five rounds of testing. Each round shows the data for the *recordProperty* transaction with a varying number of transactions from 200 to 600. The table data shows that none of the transactions failed, and we observed a maximum throughput of approximately 47 transactions per second

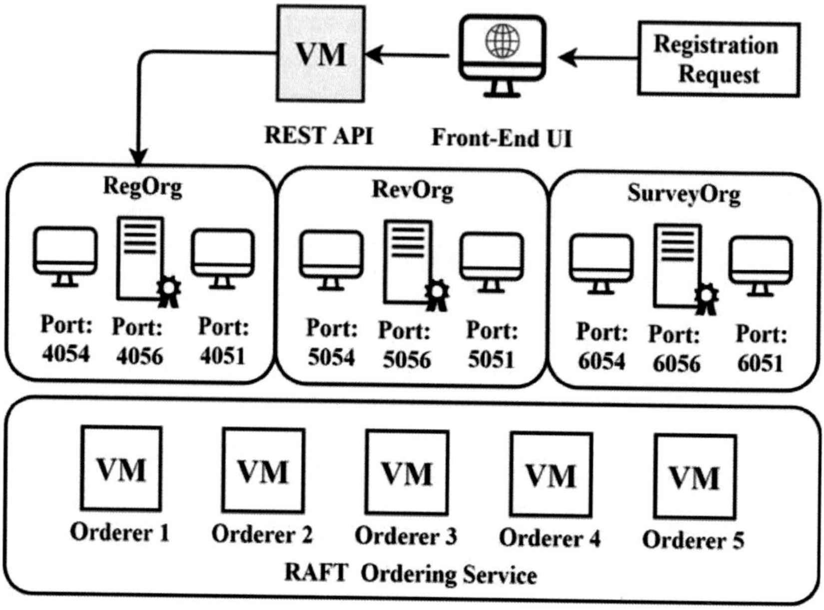

Fig. 10 Deployment set-up

Table 1 Performance benchmarking—*recordProperty* transaction

Test	Name	Success	Fail	SendRate	Throughput
1	*recordProperty*	200	0	10.2	9.6
2	*recordProperty*	300	0	25.3	19.6
3	*recordProperty*	400	0	49.6	33.8
4	*recordProperty*	500	0	75.5	44.9
5	*recordProperty*	600	0	100.4	47.3

(*tps*) with a send rate of 100 *tps*. Throughput is the average number of transactions processed per second.

There are 141 SROs in Telangana state, and as per the registration data taken from the year 2019, nine registrations per minute happened in the entire state of Telangana in 2019. With a measured throughput of 47 *tps*, the tested system can easily handle the nine *tpm* requirements for the entire state of Telangana.

7 Conclusion and Future Plan

The scope of ADYTY is to store the registration details, verify the title ownership and provide a reliable encumbrance search. The pilot system deployed at Shamshabad district of Hyderabad leverages blockchain with the existing registration workflow followed by the Telangana State Government. The system ensures that the fraudulent overtaking is significantly tricky and leaves an audit trail in case of any such attempt. The ability to track property history and a single source of information across departments ensures reliable encumbrance. The developed system is citizen-friendly and ensures privacy as it equips the citizens to view data relevant to them. Due to the decentralised nature of the blockchain networks, the developed system is highly available and resilient to malware attacks. The system is easily scalable for other districts of the Telangana state.

The derivation of all Indian states' registration processes from a single-parent source makes the presented system customisable for India's other states, whereas fully leveraging the power of ADYTY may require an amendment to existing laws and regulations. The presented system is evaluated on VMs hosted on a single machine, as network delay plays a significant role; the results may differ in real-life scenarios. Only the direct transfer property transactions are considered from Shamshabad SRO data for evaluation purposes. The case of transfer by will, gift or other scenarios is not considered for the pilot implementation. Moreover, if legal registration occurs and then a court order changes the legality of the registration, a new transaction has to be performed to transfer the property back to the owner as directed by the court order.

We plan to extend the Proof-of-Storage (PoS) system, providing secure and efficient file storage for sale deed documents. Unlike the traditional database, PoS is content-addressable storage (CAS), a mechanism for storing information that can be retrieved based on its content, not its storage location. The stored information in a CAS system uses a unique identifier in the form of the hash of the content and permanently link it to the content itself [19]. Interplanetary File System (IPFS), a peer-to-peer, distributed file-sharing system, can be leveraged for its implementation. PoS is a decentralised method of storing, sharing and retrieving the files and supports downloading the file/data from multiple nodes simultaneously.

Acknowledgements This work is funded by the Ministry of Electronics and Information Technology (MeitY) Govt. of India. Efforts are supported by Information Technology, Electronics & Communications department, National Informatics Centre and Stamps & Registration departments of Telangana State Government. Any views, opinions and findings made in this paper are only of the authors and do not reflect the views of MeitY and other departments mentioned above.

References

1. Manyilizu M, Kombe C, Mvuma A (2017) Design of land administration and title registration model based on blockchain technology
2. Bal M (2017) Securing property rights in India through distributed ledger technology. Obs Res Found
3. Mizrahi A (2017) A blockchain-based property ownership recording system. http://chromaway.com/papers/A-blockchain-based-property-registry.pdf Online
4. The registration act (1908). https://www.indiacode.nic.in/bitstream/123456789/13236/1/the_registration_act
5. Zheng Z, Xie S, Dai H, Chen X, Wang H (2017) An overview of blockchain technology: architecture, consensus, and future trends. In: 2017 IEEE International Congress on Big Data (BigData Congress), pp. 557–564
6. George Foroglou A-LT Further applications of the blockchain. https://www.researchgate.net/publication/276304843
7. Shang Q, Price A (2018) A blockchain-based land titling project in the republic of Georgia. https://doi.org/10.1162/inova00276
8. Kempe M (2017) The land registry in the blockchain-testbed. A Development Project with Lantma-teriet, Landshypotek Bank, SBAB, Telia Company, ChromaWay and Kairos Future
9. consenSys (2020) Hmlr case study-real estate asset tokenisation in the uk. https://codefi.consensys.net/hmlr
10. Allison I (2016) Bitland is revolutionising african land registry in partnership with ccedk2. https://landportal.org/news/2016/09/bitland-revolutionising-african-land-registry-partnership-ccedk
11. Thompson S (2018) Mlg, rwanda government partner on blockchain property rights platform. https://coinrivet.com/mlg-rwanda-government-partner-on-blockchain-property-rights-platform/
12. Thompson S (2019) The cryptosavannah.com website. https://cryptosavannah.com/projects.php
13. africanews (2019) south-africas-first- ever-blockchain-based-property-register-pilot. [Online]. Available: https://www.africanews.com/2019/11/06/south-africas-first-ever-blockchain-based-property-register-pilot/
14. Mukne H, Pai P, Raut S, Ambawade D (2019) Land record management using hyperledger fabric and ipfs. In: 2019 10th International Conference on Computing, Communication and Networking Technologies (ICCCNT), 2019, pp. 1–8
15. Aquib M, Dhomeja LD, Dahri K, Malkani YA (2020) Blockchain-based land record management in pakistan. In: 2020 3rd International Conference on Computing, Mathematics and Engineering Technologies (iCoMET), pp 1–5
16. Nandi M, Bhattacharjee RK, Jha A, Barbhuiya FA (2020) A secured land registration framework on blockchain. In: 2020 Third ISEA Conference on Security and Privacy (ISEA-ISAP), pp 130–138
17. https://uidai.gov.in/what-is-aadhaar.html
18. Androulaki E et al (2018) Hyperledger fabric: a distributed operating system for permissioned blockchains. Proceedings of the thirteenth EuroSys conference
19. Benet J (2014) Ipfs-content addressed, versioned, p2p file system." arXiv preprint arXiv:1407.3561

Analysis of Various Security Schemes of Internet of Things

Mehul Kapoor and Puneet Jai Kaur

Abstract The Internet of Things (IoT) is a decentralized network in which sensor nodes can join and depart at any time. As a result of the network's design, hostile nodes infiltrate the network, triggering a variety of active and passive assaults. Each layer in the network has various security issues which includes authentication, multiparty computation, identity authentication, and encryption. The various authors proposed schemes to enhance security of IoT. The light weight cryptography schemes are required which can improve security of Internet of Things. In this paper, we have studied the detailed security schemes of a IoT layers and also classify the different solutions based upon deep learning, machine learning, and encryption techniques to overcome the security threats and attacks associated with these layers. In application layer, we have detailed explore the attacks and some protocols in this like data authentication, data access, phishing attacks, malwares attacks, etc., so in the future, we have to do some analysis of the solutions and the algorithms based upon the layer.

Keywords IoT security · Encryption · Authentication · Multiparty · UDP · TCP

1 Introduction

The Internet of Things is a concept of M2M connectivity between devices embedded in physical objects. The devices of Internet of Things are small in size, embedded, and endowed with only a few hundred KB's of read only memory for avoiding the interference with the utility of the thing [1]. The emergence of IoT in many application sectors such as home automation, smart transportation systems, wearable technology for healthcare, and industrial process management has changed the fundamental way,

M. Kapoor (✉) · P. J. Kaur
Department of Information Technology, UIET, Panjab University, Chandigarh, India
e-mail: mehulkhatri89@gmail.com

P. J. Kaur
e-mail: puneet@pu.ac.in

M. S. Kaiser et al. (eds.), *Information and Communication Technology for Competitive Strategies (ICTCS 2021)*, Lecture Notes in Networks and Systems 401,
https://doi.org/10.1007/978-981-19-0098-3_17

and we view and govern the physical environment. According to the estimations, the amount of IoT devices will reach about 35 billion by 2021 [2].

Almost all the devices based on wireless communication technology having finite potentials with regard to computation and storage and less expensive. The issues regarding the security and consistency of the transmission of data to and from the Internet of Things devices have increased rapidly due to the enlargement of IoT systems in sensing and managing highly complex eco-systems. Various studies have stated that a number of security issues are occurred in these networks such as revealing of information, privacy, authentication, eavesdropping.

To deal with the application layer security, strong authentication is required for the privacy reason so that only recipients can read the sender's message. There are different kinds of technologies available for the protection of the application layer like installing strong antivirus for avoiding malwares attacks, secure algorithms like configuring data collection algorithm, SDN-based data transfer security model named also as M-G security model, Deffie-Hellman security model, etc. The various applications protocols like EDHOC, MQTT, SMQTT, SSI, LWM2M, M3DA, SOAP, JavaScript IoT, etc., are also required for better communication.

The standards like TLS, PKI addresses the different principles of a security like confidentiality, integrity, and authentication where TLS provides the security over the internet through IP address and public key infrastructure (PKI) is set of a different kinds of a technology, standards, protocols associated with the cryptographic techniques for the achievement of the better communication.

2 IoT Security

Security is a procedure carried out for protecting sources from physical damage, unauthorized access, or theft. To achieve this goal, high levels of secrecy and integrity are maintained for asset information, as well as information about that item that may be executed as needed [2]. The enhancement of various applications is facilitated in several sectors through IoT. However, various security problems have arisen as a result of the use of limited IoT devices and technology in such critical applications.

Internet of Things is depending upon connectivity of numerous devices in order to execute its operation. The depiction of a security assault is thus possible as much as possible. An assault in IT is an effort to damage, modify, and rob the illegal access to an asset. The cryptographic security protocols are, to demonstrate, an important part of security services to create network-wide communications.

The demonstration of a protocol default is known as an attack on a protocol and is frequently regarded an acts sequence done by a dishonest person using any hardware or software instrument to undermine the security of the protocol. An attack on the Internet of Things is not unique to an attack on an IT asset. The new thing is the scalability and relevance of the easiness of IoT assaults. Figure 1 shows the different types of securities associated with each layer.

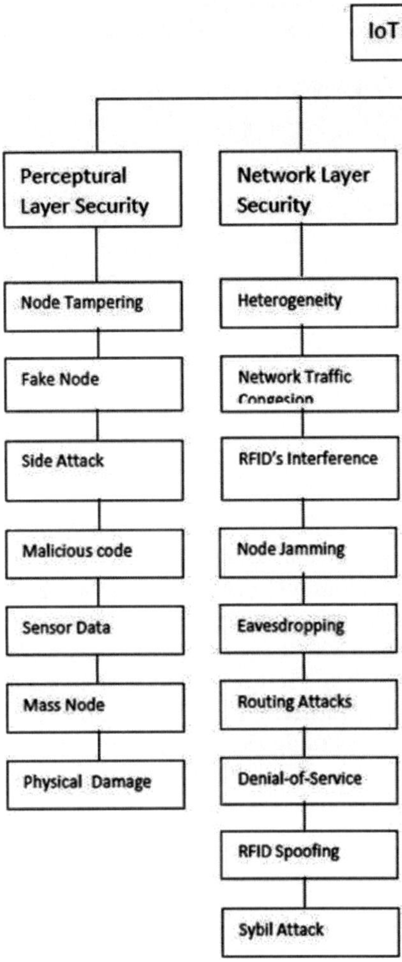

Fig. 1 IoT security

3 IoT Security Layers

This means that millions and billions of devices are potential targets of traditional cyber assaults yet are widely accessible and frequently limited or non-protected.

3.1 Perception Layer

The devices belonging to the perception layer are performed in LLN where energy, memory, and processing power are restricted compared to the location of network nodes on conventional Internet platforms [3]. Eavesdropping is considered as an unauthorized attack of real time at which an attacker interrupts the private communications like phone calls, text messages, or video conferences. An attacker can control the entire key node. It can reveal all the information about the communication, a key which is utilized for secure communication and information of kept memory [4]. An intruder can spy the communication between sender and receiver, and authentic information is sent to the victim from intruder.

3.2 Network Layer

The database connection permits the perception of layer devices in the network layer in order to perform various applications inside the application layer. Several probable security drawbacks in which the operations of the entire Internet of Things architecture have cooperated are present as this layer provides the connectivity to other layers [5]. DoS, MITM attack, etc., are such a security threats which are take place in this layer. The DoS is an attack of preventing authentic users from trying to access the devices. The communication between sender and recipient who believe that they are directly contacted by the attacker is stopped during the main-in-the-middle attack. As an attacker has fully control over the conversation and able to change the messages.

3.3 Application Layer

Which the utilization of Internet of Things technology can be done is known as application layer. Generally, there are two kinds of feasible attacks appears in application layer [6]. These attacks are based on encryption and software. The authentic credentials of the user are disclosed through an attacker. For this purpose, those credentials are imitated as a trusted authority. Malware, worms, adware, spyware, and Trojans are very likely to attack owing to the heterogeneity of IoT applications and cryptanalytic attacks, and cipher text are some of the illustrations of these kinds of attacks.

3.4 Support Layer

A fourth layer is generated for the security purpose in the architecture of Internet of Things. The direct transmission of information to the network layer leads to maximize the chances of threats occurrence [7]. A novel layer is suggested because of the limitation of three-layer model. There are two tasks of support layer. Firstly, the information that the authentic users forwarded is validated and protected from threats. The pre-shared secrets, keys, and passwords are employed for its deployment. The second task of the support layer is to send the information to the network layer [8].

4 Literature Review

Roukounaki et al. [9] recommended scalable and configurable data collection architecture in the security Internet of Things that was data driven. The major goal of this approach was to gather the security data from various components of systems of IoT such as smart objects and nodes. The existing technologies were incorporated for creating the recommended architecture scalable in order to gather, stream, and store the enormous amount of data. On the other hand, the potential of its configuration was based on an extensible approach to model the security of data.

Shin and Seto [10] described that the IoT devices had acquired a lot of popularity due to which various incidents had taken place, and susceptibility counter measures were proved insufficient. Thus, the significance of security information regarding devices of Internet of Things was enlarged. But, the large amount of burden of introduction and operation costs led to restrict the deployment of CyExec that had contained a virtual environment.

Liu et al. [11] established a SDN-based data transfer security model named M-G model whose major intend was to mitigate the network latency. Initially, dataflow abstraction and a heuristic algorithm had employed for placing the middle boxes regarding the defined secure policies at the exact locations [11]. Subsequently, an offline ILP pruning algorithm was suggested for dealing with the constraints of switch volume so that any conversion of middle box to a hotspot was neglected. At last, secure systems were presented for addressing a number of attacks and enhancing the security performance.

Cheng and Huang [12] stated that the production of huge volume of data in IoT became an issue because of the finite resources of sensor nodes. Thus, the data aggregation was intended considering the energy consumption and security. This approach assisted in alleviating the data redundancy, saving energy consumption, and other resources because of aggregating multiple data into one data. In this way, network became secure.

Wang and Wei [13] focused on developing a platform of power IoT on the basis of SDN technology and expanding the open flow protocol. For this, a few kinds of

Table 1 Surveys related to IoT security

Author	Year	Description	Outcome
Aikaterini Rou-kounaki, Sofoklis Efremidis, John Soldatos, Juergen Neises, Thomas Walloschke, Nikos Kefalakis	2019	Recommended scalable and configurable data collection architecture in the security of Internet of Things that was data driven	This approach allowed the execution of efficient security analytics algorithms in order to recognize the threats and related pattern attacks
Sanggyu Shin, Yoichi Seto	2020	The significance of security information regarding devices of IoT was enlarged	This system was less expensive and flexible security system that was utilized to disseminate the security
Yanbing Liu, Yao Kuang, Yunpeng Xiao, Guangxia Xu	2018	Established a SDN-based M-G model whose major intend was to mitigate the network latency	The outcomes of experiment revealed that the established model had potential for enhancing security performance in IoT system
Yunli Cheng, Min Huang	2019	The data aggregation technology was intended considering the energy consumption and security	Delay while transmitting the data was decreased and network became reliable, secure, and stable
Yang Wang, Xiaogang Wei	2020	The DFI approach was implemented for gathering and analyzing the traffic in the IoT	This assisted in creating a security prevention and control strategy model that was proved suitable for power IoT without detecting any payload

actions and meters were inserted in order to perform on-demand monitoring and dynamic defense. The DFI approach was implemented for gathering and analyzing the traffic in the IoT.

Pérez et al. [14] suggested to accomplish a procedure on the basis of application layer schemes. Generally, the EDHOC protocol was computed for OSCORE. A compact ephemeral Diffie-Hellman over COSE was presented as a lightweight substitute on the basis of suggested procedure [14] (Table 1).

5 Application Layer Security and Protocols

This layer defines all the applications of the Internet of Things like smart homes, healthcare, education, and smart cities projects. The main aim of application layer is to provide services to the applications of the IoT. The various security issues come in this layers and consist of data access and authentication, phishing attacks, malicious active x scripts, malware attacks, etc. (Table 2).

Table 2 Application layer protocols

Protocols	Standards	Functions	Architecture	TCP protocols
MQTT	OASIS	Messaging	Centralized	TCP
COAP	IETF	Messaging	Centralized	UDP
AMQP	OASIS	Messaging	Centralized	TCP
DDS	OMG	Messaging	Decentralized	TCP/UDP
XMPP	IETF	Messaging	Centralized	TCP

Communication protocols on the application layer are an essential part of the IoT since they provide the foundation of all IoT and cloud/edge interactions. The application layer has five messaging protocols and two service discovery protocols (i.e., MQTT, CoAP, AMQP, DDS, and XMPP) (i.e., mDNS and SSDP). Some protocols utilize centralized while others are based on distributed architectures.

Message queue telemetry transportation (MQTT) is a standard protocol accessible to messaging. This protocol is now extensively used in the IoT environment and is simple, lightweight, and suitable for IoT situations. MQTT offers several authentication and TLS-based encryption methods.

CoAP is an upcoming open protocol of web transmission. Although CoAP has many features under the HTTP protocol, it was especially intended for confined devices with limited energy, capacity for processing, storage, and transmission. CoAP allows the use of the DTL security and offers safeguards for UDP implementations of the TLS.

Advanced message queuing system (AMQP) is an open commercial communications protocol (OASIS standard). The protocol provides advanced functionality and is now extensively utilized in various situations that need dependable asynchronous communication between endpoints. As far as security is concerned, AMQP supports the client authentication and security layer (SASL) framework and TLS to ensure secrecy.

Data distribution service (DDS) is an object management group-defined data-centric standard protocol. The protocol is typically used for the management of data interchange between lightweight devices and massive networks of high-performance sensors. In industrial implementations, such as air traffic control, smart grid management, transport, and healthcare services DDS does not constitute a conventional IoT solution.

Extensible messaging and presence protocol (XMPP) is an open XML technology for asynchronous real-time communication between two or more organizations. The XMPP protocol offers strong security features by supporting SASL for authentication and the TLS for data secrecy. However, the absence of end-to-end support for encryption renders the protocol susceptible to many kinds of attacks.

The multicast domain name system (mDNS) is an open protocol that is now extensively used for service discovery and name resolution on local connections. This protocol, combined with DNS-based service discovery (DNS-SD), provides the flexibility that environments need to integrate new devices automatically and conduct DNS-like operation without the need of a traditional DNS server.

Simple service discovery protocol (SSDP) is an open protocol commonly used in home or small business networks for service discovery and advertising (UPnP forum). The protocol contained in the UPnP architecture enables plug-and-play devices to be transparent without the need for any human setup. Finally, it is imperative that users be aware of the possible safety hazards connected with IoT device ownership and usage.

6 Conclusion

The choice of the application protocol relies on the type and needs of the IoT systems. MQTT and CoAP are especially suited for services in restricted settings needing data gathering. On the contrary, AMQP, DDS, and XMPP respond to particular service needs such as corporate messaging, online presence detection, and real-time exchanges. In service discovery, the protocols for IoT are mDNS and SSDP. Many gadgets are regarded by designers, producers, which may easily be used by hackers to cause all sorts of harms. The issues regarding the security and consistency of the transmission of data to and from the IoT has increased rapidly due to the enlargement of IoT systems in sensing and managing highly complex eco-systems. To accomplish a security, a high confidentiality and integrity of the information is maintained and information related to that object which can be executed when needed is also created. The light weight and block encryption scheme is required to improve security and more research efforts should be focused on privacy of protocols. In addition, solutions for end-to-end systems security.

References

1. Ali I, Sabir S, Ullah Z (2016) Internet of things security, device authentication and access control: a review. Int J Comput Sci Inf Secur 14(8)
2. Razzaq MA, Qureshi MA, Gill SH, Ullah S (2017) Security issues in the internet of things (IoT): a comprehensive study. Int J Adv Comput Sci Appl 8(6)
3. Vignesh R, Samydurai A (2017) Security on internet of things (IOT) with challenges and countermeasures. IJEDR 5
4. Okul S, Aydın MA (2017) Security attacks on IoT. In: Proceedings of international conference on computer science and engineering
5. Vishwakarma R, Jain AK (2019) A honeypot with machine learning based detection framework for defending IoT based Botnet DDoS attacks. In: Proceedings of 3rd international conference on trends in electronics and informatics
6. Waragaa AO, Bettayebb M, Nasir Q, Talib AM (2019) Design and implementation of automated IoT security testbed. Elsevier
7. Karande JB, Joshi SA (2018) Comprehensive assessment of security attack detection algorithms in internet of things. In: Proceedings of fourth international conference on computing communication control and automation

8. Liang L, Zheng K, Sheng Q, Huang X (2016) A denial of service attack method for an IoT system. In: Proceedings of 8th international conference on information technology in medicine and education
9. Roukounaki A, Efremidis S, Soldatos J, Neises J, Walloschke T, Kefalakis N (2019) Scalable and configurable end-to-end collection and analysis of IoT security data: towards end-to-end security in IoT systems. Global IoT summit
10. Shin S, Seto Y (2020) Development of IoT security exercise contents for cyber security exercise system. In: Proceedings of 13th international conference on human system interaction
11. Liu Y, Kuang Y, Xiao Y, Xu G (2018) SDN-based data transfer security for internet of things. IEEE Internet Things J
12. Cheng Y, Huang M (2019) Based on the internet of things environment data aggregation security requirements and key technology research. In: Proceedings of international conference on information technology and computer application
13. Wang Y, Wei X (2020) A security model of ubiquitous power internet of things based on SDN and DFU. In: Proceedings of information communication technologies conference
14. Pérez S, Hernández-Ramos JL, Raza S, Skarmeta A (2020) Application layer key establishment for end-to-end security in IoT. IEEE Internet Things J

Detection of Solar Panel Farms Using Machine Learning

Vandita Srivastava

Abstract There has been an increase in the number of solar energy farms all over India in an effort to move toward more sustainable sources of energy. The installation of these is widely unregistered, and thus, they are popping up everywhere without any official record. So, we can use satellite images taken over remote villages to map and find out the number of solar farms. We can train our machine learning algorithm over these images to correctly identify solar farms. This can then be used in the future to keep a tab if and when the number of solar farms changes.

Keywords Solar panel farms · Satellite images · Remote sensing · Object detection · Semantic segmentation

1 Introduction

For as long as we have been using fossil fuels as sources of power, there has been an ongoing discussion about its harmful effects on the environment. The damage they cause is growing over the years.

There has been lots of discussion and ideas about ways to reduce the pollution. One of the ways to tackle environmental problems is the use of solar panels. They are used all over the world for alternative power generation.

An example of alternative generation is the use of solar power. It has a small environmental impact and is sustainable.

Solar panels are cropping up everywhere for this reason all over India, including the rural regions, where it is difficult to keep track of them. They are unregistered and may or may not be up to standard. One way to keep track of the growing numbers of solar panel farms is with the use of object detection using satellite images. This would also help in accurately gaging the economic conditions of an area. Since it is very difficult to provide electricity in the traditional way in many parts of India, they do not develop economically. These solar panel farms give them the resources

V. Srivastava (✉)
University of Lucknow, Lucknow, India
e-mail: vanditasrivasatva19@gmail.com

M. S. Kaiser et al. (eds.), *Information and Communication Technology for Competitive Strategies (ICTCS 2021)*, Lecture Notes in Networks and Systems 401,
https://doi.org/10.1007/978-981-19-0098-3_18

to do so, and they can then have better opportunities to advance and use modern technologies.

2 Solar Power, Satellite Images, and Remote Sensing

2.1 Solar Power

Solar power is made using photovoltaic cells to convert it into electricity. There are other indirect methods of use as well, such as using the heat to warm up a building.

Most importantly, it is a sustainable and non-polluting energy source.

2.2 Satellite Images

Satellite images are a vast and easily accessible resource of data that allow the earth to be viewed from a wider perspective. This information can be obtained for any particular location on earth. They also allow for the ability to track the progress of any area by comparing its images over a period of time. This can then be used to predict how the area is likely to change in the future, based on its past trends.

2.3 Remote Sensing

Special cameras collect remotely sensed images, which help researchers "sense" things about the earth [1]. It is a very useful tool in implementation of spatial planning.

Working with satellite imagery is quite a challenging task. Various algorithms have been implemented over the past 70 years to interpret the images; however, their success has only been abysmal over even simple tasks. Routine tasks like identifying building footprints or distinguishing tree canopy in urban areas were laborious sub-specialties of the field.

The advent of deep learning meant that useful algorithms could now be applied to satellite images with much greater success. Machine learning has made object detection, semantic segmentation, and image classification easier.

2.4 Satellite Image Used in Object Detection and Deep Learning

Object detection using satellite images has led to a greater understanding and extraction of large-scale areas of the earth. Object detection is useful in finding out the number of objects belonging to the target classes that we want to identify. It then creates a bounding box around the object, helping to locate it.

It helps in many applications such as rural area planning and geographic information system (GIS) updating [2–9].

3 Working

3.1 Detection of Solar Panels

Solar panels and solar panel farms are often of irregular shapes and sizes. They do not have a set pattern which could be used to easily identify them. This makes them hard to work with.

Machine learning-based approaches, especially those of convolutional neural networks, can be used to overcome these obstacles. When combined with satellite imagery, they show accuracy as high as 0.94, which is promising and leaves room for improvement for further research and better algorithms [10].

3.2 Data Preparation

Data preparation is an important step for neural networks. It improves the model quality and significantly speeds up the training process. Data preparation can sometimes be allocated about 50–70% of the time spent working with neural networks [11].

Solar panel farms cannot be encapsulated into a simple polygon of rectangular bound boxes. They need to be bounded in a multisided polygon with each polygon being of different shapes—different number of edges—for each solar panel farm in each image.

This approach, therefore, requires the use of image segmentation instead of object detection. Image segmentation would isolate and label each pixel inside the mapped area as that labels and works best for irregular shapes.

For the purpose of image segmentation being done on the solar panel farms, semantic segmentation should be enough as the need is to distinguish the entire solar panel farm from the background, as opposed to also identifying the individual solar panels present in the solar panel farm as separate instances. Semantic image segmentation aims to label individual pixels of an image with a class. This is also known as dense prediction.

3.3 Comparison

Various approaches for semantic segmentation are available, and they have been compared to find the best method for this purpose:

1. Region-based Segmentation
   (a) Threshold segmentation
      (i) It works when the target can be easily distinguished from the background due to high contrast between the two.
      (ii) However, without high contrast between background and target, the results are poor.
2. Edge Detection Segmentation
   (a) It detects the edge of the target object by finding the discontinuities between the gray values in adjacent regions of the image.
3. Segmentation based on clustering
   (a) Clustering algorithms, such as k-means clustering, work by clustering the samples according to distances between them.
   (b) It may, however, fail to cluster targets into the same class if they happen to lie on different regions, far apart in the image.
4. Segmentation based on weakly supervised learning in CNN
   (a) CNN model is used to frame the ROI and then segmented by non-machine learning segmentation method to improve the segmentation effect.
   (b) Faster R-CNN is widely used for object detection tasks. For a given image, it returns the class label and bounding box coordinates for each object in the image.

After comparison, Faster R-CNN would be the best method to get the highest possible accuracy for this task.

4 Implementation

First, two datasets are required

1. Training dataset.
2. Testing dataset.

For the datasets, satellite images from freely available sources can be taken from the internet.

To make the training dataset, the images need to be annotated by manually creating multisided polygons over the solar panel farms and labeling them. This is a tedious process. Various online tools such as labelme and labelimg can be used for this task.

It is important to have a high number of training images to not cause the model to overfit. The number of images should ideally be in thousands. If the images acquired are not enough, various transformations can be applied to the existing images to create more images. Once all the images have been properly annotated, they can be used as the training dataset.

A small percentage of the images is used as the testing dataset and does not need manual annotation.

The Faster R-CNN model is then applied to the training images.

1. The input image is first passed through convolutional neural network to get the feature map.
2. This is then passed though the region proposal network to get the object proposals and objectiveness score.
3. An RoI pooling layer is applied so that proposals can be all of the same size.
4. Finally, it goes through a softmax and a linear regression layer.
5. The outputs are classified bounding boxes of all the object present in the image.

The training should continue till the loss function becomes sufficiently low, at which point, it can be stopped.

Then, the accuracy of the model is to be tested using the testing dataset. Higher accuracy means that the model has been successful.

This model can now be implemented over the rural and remote places of India to record the number of solar panel farms that have cropped up everywhere. This will help establish their numbers as well as allow us to see the progression of them over time. It can also then help correlate progress with the advancement in these places. It will be easier to keep a tab on these solar panel farms.

5 Applications

This system can be used to keep track of the solar panel farms in the rural areas of India. They do not have easy access to the traditional methods of electricity generation and transportation. Furthermore, there is a need to move toward more sustainable sources of energy. Therefore, more and more places are getting equipped with either individual solar panels or entire solar panel farms to provide power to those places. India is developing fast, and it is very important for us to make sure that everyone can have access to electricity. This is a way of keeping track of that progress.

6 Challenges

The main challenge that presents itself is that the solar panel farms are of irregular shapes. Therefore, simple rectangular bound boxes cannot be used to capture them. During data preparation, a lot of images need to be labeled, and since the farms require different sided polygons to mask them, this process cannot be easily automated.

Further, the training images need to be very diverse. They need to have a distinct and versatile topography to train the model for different locations—such as grasslands and deserts. While satellite images are easy to acquire, it is difficult to find a wide range of variety in them.

This method of object detection can further be developed and expanded to also take into account other areas of sustainable energy, such as windmills for wind-powered electricity generation and watermills for water-powered electricity generation. If we are able to effectively keep track on these sources of power, we will then be able to more efficiently help these communities.

7 Concluding Remarks

There is an increasing need all over the world to switch to more sustainable and environmentally friendly sources of power. India must not remain behind in this respect. More and more areas are starting to build solar panel farms for their power needs, and this has also greatly helped with the development of the nation. Their numbers are increasing day by day, and so satellite images taken over the region can be very helpful to keep a tab on these new developments.

References

1. USGS. Retrieved 10 Aug 2021, from https://www.usgs.gov/faqs/what-remote-sensing-and-what-it-used?qt-news_science_products=0#qt-news_science_products
2. Cheng G, Han J, Zhou P, Guo L (2014) Multi-class geospatial object detection and geographic image classification based on collection of part detectors. ISPRS J Photogramm Remote Sens 98:119–132
3. Cheng G, Han J, Zhou P, Xu D (2019) Learning rotation-invariant and fisher discriminative convolutional neural networks for object detection. IEEE Trans Image Process 28:265–278
4. Li K, Cheng G, Bu S, You X (2018) Rotation-insensitive and context-augmented object detection in remote sensing images. IEEE Trans Geosci Remote Sens 56:2337–2348
5. Felzenszwalb PF, Girshick RB, Mcallester D, Ramanan D (2010) Object detection with discriminatively trained part-based models. IEEE Trans Pattern Anal Mach Intell 32(9):1627
6. Jia Y, Shelhamer E, Donahue J, Karayev S, Long J, Girshick R, Guadarrama S, Darrell T (2014) Caffe: convolutional architecture for fast feature embedding. In: ACM MM
7. Krizhevsky A, Sutskever I, Hinton GE (2012) Imagenet classification with deep convolutional neural networks. In: NIPS
8. Girshick R, Donahue J, Darrell T, Malik J (2014) Rich feature hierarchies for accurate object detection and semantic segmentation. In: CVPR

9. LeCun Y, Bengio Y, Hinton G (2015) Deep learning. Nature 521(7553):436–444
10. Castello R, Roquette S, Esguerra M, Guerra A, Scartezzini J-L (2019) Deep learning in the built environment: automatic detection of rooftop solar panels using convolutional neural networks. J Phys Conf Ser 1343:012034. https://doi.org/10.1088/1742-6596/1343/1/012034
11. Yu L, Wang S, Lai KK (2006) An integrated data preparation scheme for neural network data Analysis. Proc IEEE 18(2):217–230

Design and Analysis of Band Stop Filter Using Hexagonal Split Ring Resonator for Wireless Application

Khyati Chavda and A. K. Sarvaiya

Abstract In wireless application, the microwave filter is the most considerable component to pass the required band and remove the unwanted frequency band. Metamaterials have artificial characteristics that *minimize the size* of a unit cell for active and passive design in the field of microwaves. In this proposed work, design, simulate, and fabricate a band stop filter using a microstrip line loaded with pair of hexagonal-shaped split ring resonator for single and dual frequencies. The simulated and fabricated results of a band stop filter for WLAN applications with desired results are like insertion loss and return loss at 2.4 and 3.6 GHz. The proposed band stop filter design provides good isolation in MIMO applications, accurate rejection with good selectivity in RF transmitter receiver designs for WLAN applications.

Keywords Hexagonal split ring resonator (HSRR) · Band stop filter · Metamaterial

1 Introduction

1.1 Microwave Band Stop Filter

The most important parameter in wireless communication systems is the filter design. Microwave planer filters will allow the required frequency band while rejecting the unwanted frequency band and will be used for isolation in MIMO system applications. Filters have characteristics such as allowing some frequencies to pass through while rejecting others, thereby blocking the flow of electrical current through it [1–3].

The most important aspects of microwave filter design are miniaturization and electrical parameter performance. A metamaterial is an artificial material which has

K. Chavda (✉)
Shantilal Shah Engineering College, Bhavnagar, Gujarat Technology University, Bhavnagar, Gujrat, India
e-mail: khyati.chavda@gmail.com

A. K. Sarvaiya
Government Engineering College Bhavnagar, Bhavnagar, Gujrat, India

M. S. Kaiser et al. (eds.), *Information and Communication Technology for Competitive Strategies (ICTCS 2021)*, Lecture Notes in Networks and Systems 401,
https://doi.org/10.1007/978-981-19-0098-3_19

a negative refractive index. Veselago [4–7] introduces the first theoretical prediction. The fundamental properties of metamaterial are DNG and SNG, which stand for double negative and single negative material properties, respectively [4–10]. The majority of left-handed transmission designs are used for active and passive RF structures, trying to design planar microstrip line-based filters with notches at desired frequency using a single negative material property. The left-handed metamaterial transmission line was created to demonstrate a new efficient approach for planar structures with small dimensions [8–17]. Square split ring resonator (SRR), complementary split ring resonator (CSRR), and hexagonal split ring resonator (HSRR) have all been used in the past to create this type of filter [8, 10, 11, 16, 18]. The scattering and absorption properties of MTM are mostly appropriate for this type of band stop filter design. Either permeability or permittivity is negative when a single negative metamaterial attribute is present [4–7]. The negative permeability correlates with the magnetic field, which is designed using a split ring resonator or n-side polygon-shaped ring resonator [8–17, 19] that is called a sub-wavelength resonator loaded with left-hand material transmission line and is dependent on effective inductance and capacitive gap of the ring resonator, and artificial permeability [4–7, 20] has been produced.

In wireless communication systems, filter design is critical, such as in MIMO-multiple i/p and multiple o/p applications to reduce frequency interference and offer isolation between two MIMO components [21, 22]. The fundamental requirement for a MIMO antenna in a WLAN application is sufficient isolation between the different feeding lines. This type of band stop filter is good for those with HSRR, which create one notch according to resonance and absorb that specific frequency. Band stop filters are crucial in multiservice radio applications for selecting suitable isolation between two channels. The proposed work is based on different research papers and literature [8–17, 19, 20, 23, 24].

In this paper's design structure, lumped design elements such as inductor and capacitor different valued resonators are used in filter resonance at specific frequencies. Nowadays, planar structures are in high demand for all microwave components, such as surface-integrated waveguide (SIW), as well as planar filters for satellite applications and other wireless applications as per IEEE 802.11 standards [19]. In this work, the left-hand microstrip transmission line is loaded with single and multiple HSRR arrays, and various parameters including such return loss and insertion loss are analyzed [25–29].

The simulated work and fabrication of a planar microstrip metamaterial band stop filter is presented in this research. There are two HSRRs loaded on a microstrip line, which create resonance at single and dual frequencies with low insertion and return loss. The hexagonal split ring resonator generated an artificially magnetic characteristic. When magnetic flux penetrates the hexagonal-shaped metal rings, it induces rotating currents in the ring, which produce their own flux to oppose the incident field, and the dielectric material used is RT duroid 5880 with a dielectric constant of 2.2.

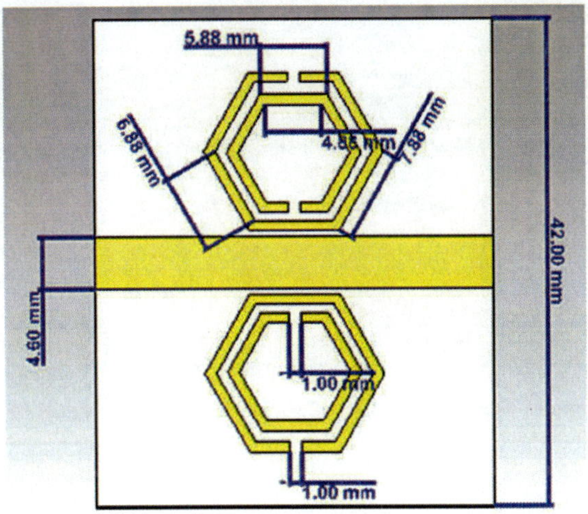

Fig. 1 Simulated result for the single stop band at 2.45 GHz using a hexagonal split ring resonator loaded with microstrip line

1.2 Hexagonal Split Ring Resonator Band Stop Filter for Single Frequency in S Band

The hexagonal split ring resonator features a negative refractive index metamaterial structure with negative permeability. The material used in the design of a single-ring HSRR band stop filter is RT duroid 5880 with a dielectric constant of 2.2, a substrate height of 1.57 mm, and a copper thickness of 0.035 mm. The additional parameters for HSRR are the spacing between two rings (d_1), ring radius (r_1), copper thickness (t), width of the filter (w), length of the filter (l), width of the feedline (w_f), and width 0.8 mm.

The design of a single pair of hexagonal split ring resonators loaded with a 50 Ω impedance microstrip line is shown in Fig. 1, which was simulated using CST software. This results in the resonance frequency shown in Fig. 2. The simulated result is at 2.45 GHz center frequency with 52 MHz (at −3 dB) bandwidth and fractional bandwidth 2.1 at 2.43 GHz with lower cutoff and 2.48 GHz with higher cutoff frequency.

Figure. 3 shows a hexagonal split ring resonator band stop filter with dual frequency in the S band. The material used in the design of a dual-frequency HSRR band stop filter is RT duroid 5880 with a dielectric constant of 2.2, a substrate height of 1.57 mm, and a copper thickness of 0.035 mm. The other parameters for HSRR are the thickness of the split ring resonator with ring-1 which resonates at 2.45 GHz frequency, the gap between two rings (d_1) value is 1 mm, the ring radius (r_1) value is 7.83 mm, the copper thickness (t) = 0.035 mm, the width of the filter (w) = 42 mm, the length of the filter (l) = 42 mm, and the width of the feedline (w_f) = 4.6 mm. For the second HSRR, which resonates at 3.6 GHz, the radius of ring-2 r_2 = 5.35 mm, the gap between two rings d_2 = 0.8 mm, and the split cut width = 0.4 mm.

Figure. 4a depicts the simulated design result of a dual pair of hexagonal split ring

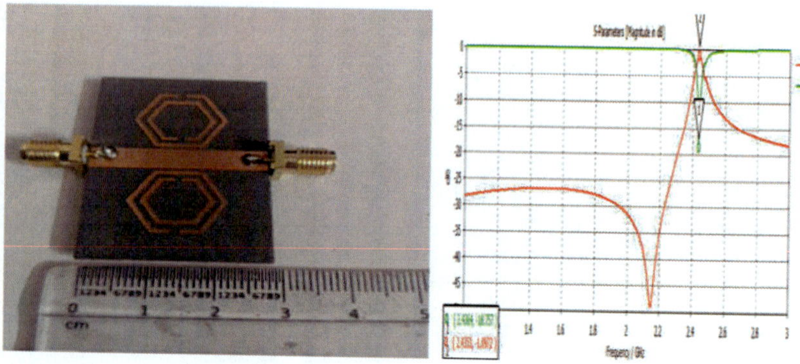

Fig. 2 Fabrication design and simulated result for the single frequency using hexagonal split ring resonator

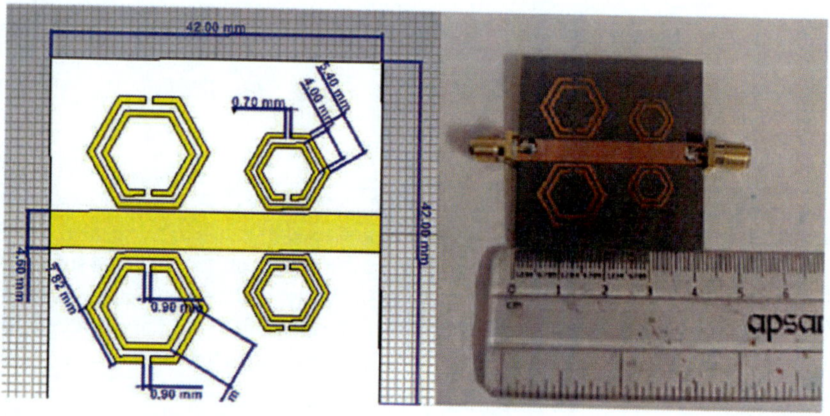

Fig. 3 Simulated and fabricated a Dual-frequency band stop filter with a hexagonal split ring resonator loaded with microstrip line

resonators loaded with 50 Ω impedance microstrip lines in CST software. Figure 4b depicts the constructed result using a vector network analyzer, which generates two resonances depicted in the pictures. The simulated results are at 2.45 GHz and 3.65 GHz center frequencies, respectively, with bandwidths of 52 MHz and 70 MHz. Due to fabrication and soldering loss, there is some difference between the simulated and measured results.

Figure. 5 shows that the negative permeability of each pair of HSRR provides a magnetic coupling with the microstrip line, which is initiated to resonate at multiple frequencies and behaves as a single negative metamaterial. The simulated design functions as a band stop filter, rejecting two separate frequencies: 2.45 GHz with a rejection level close to −17.854 dB and 3.6 GHz with a rejection level near to 18.40 dB. Additionally, the insertion loss is less than −2 dB.

Figure. 6 shows the design of the dual stop band line impedance value. The

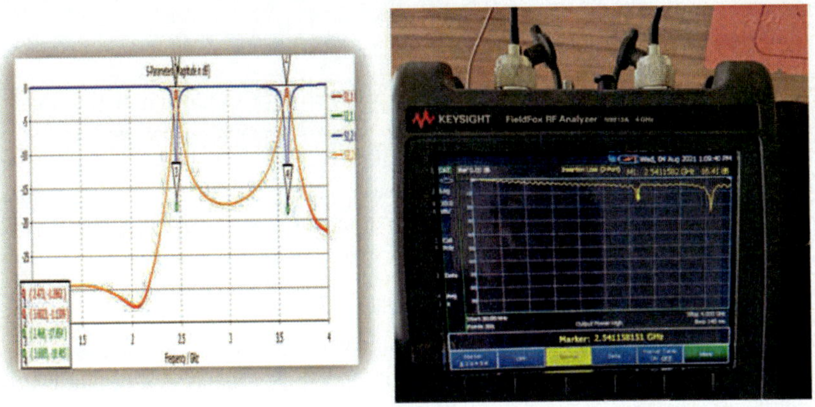

(a) (b)

Fig. 4 **a** Simulated result and **b** fabricated result measured with vector network analyzer (Keysight FieldFox RF analyzer N9913A 4 GHz, August 4th, 2021, Government Engineering College, Bhavnagar) for dual-frequency band stop filter using dual-ring pair hexagonal split ring resonator loaded with microstrip line

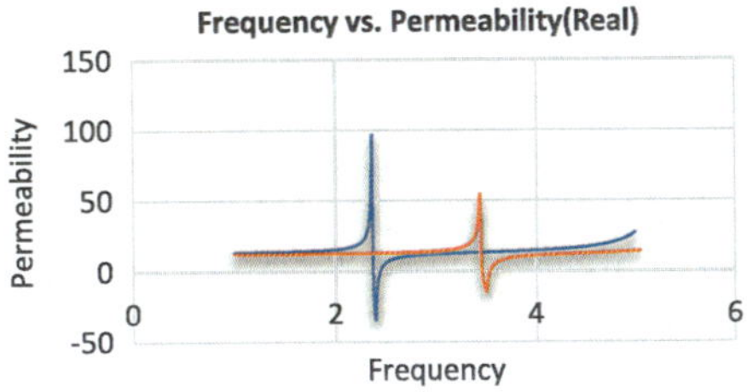

Fig. 5 Frequency versus permeability for HSRR

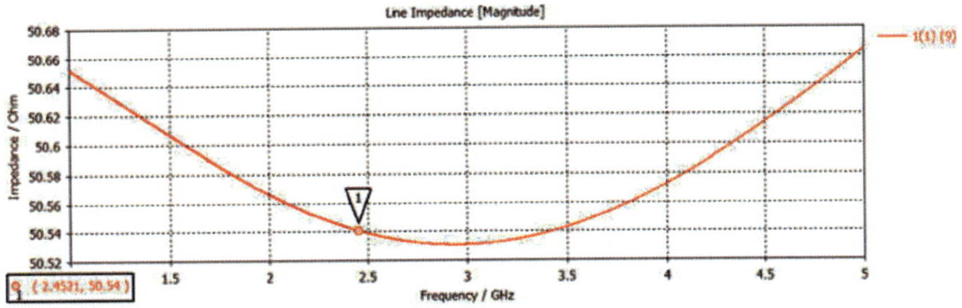

Fig. 6 Simulated result of line impedance for band stop filter

proposed design consists of a 50 Ω microstrip line along with two pairs of HSRRs of different sizes that are placed across the feedline. The two pairs of HSRRs are arranged so as to have effective coupling with the feedline.

2 Result and Discussion

For the simulation of a hexagonal split ring resonator for single and dual frequency, some important dimension parameters are affected by electric parameters such as return loss and insertion loss. The simulated result reveals several critical details about the hexagonal split ring resonator's dimensions. The hexagonal split ring resonator structure takes into account the following parameters: ring width, spacing between two rings, and coupling between strip line and HSRR. In the hexagonal split ring configuration, increasing the number of rings improves return loss but increases insertion loss. From single to multi-ring, single-ring return loss for band stop filter (S21) is −13.87 dB, dual ring is −18 dB, and triple ring is −22 dB (Table 1).

The HSRR is a metamaterial unit cell with inductive capacitive resonating structural features. The gap between the two rings of the HSRR is connecting different rings via a strong distributed capacitance. Through a capacitive gap between two rings, an induced current passes from one ring to the other. Some of the most critical factors influence HSRR's magnetic resonance. Total capacitance is created in the HSRR by two parameters: one from the split and the other from the space between the two concentric rings. The capacitance caused by splits prevents current passage around rings. The design of Inductive effects was affected by the size of the conducting ring. The metamaterial unit cell HSRR changes physical dimension and also physical resonant frequency shifting. The capacitance will be directly affected if the HSRR split width is increased. When the split width is increased, capacitance decreases, causing the frequency response shift on the higher frequency side to increase. When the spacing between two rings is extended, the resonant frequency moves to the higher frequency side. The ring's metal width reduces mutual inductance and capacitance, shifting the frequency to the lower frequency side. If an additional capacitance split is required in HSRR, the resonance frequency will be affected as well. As capacitance rises, the resonance frequency shifts to the lower frequency side. As a response, these are some important parameters to consider when designing a HSRR structure.

Table 1 Simulated result of HSRR-based band stop for single- and dual-frequency filter

S. no.	Band stop filter parameters				
	Parameter	S21 (dB)	S11 (dB)	BW	Size
1	Single frequency—HSRR BSF				
	2.4 GHz	−17.854	−1.2862	52 MHz	30 × 35 × 1.57 mm^3
2	Dual frequency—HSRR BSF				
	2.45 GHz	−17.854	−1.2862	52 MHz	40 × 40 × 1.57 mm^3
	3.6 GHz	−18.40	−1.1309	70 MHz	

3 Conclusion

In this paper, a hexagonal split ring resonator loaded with a microstrip line-based microwave band stop filter is designed for wireless applications. The band stop is developed for isolation in MIMO and selection and rejection of desired frequency in multiservice radio wireless applications at frequencies of 2.4 and 3.6 GHz. Increase the number of HSRR in the future to get better results in terms of selectivity and rejections, but insertion loss will rise. The filter dimensions are $30 \times 35 \times 1.57$ mm^3 at 2.4 GHz for a single frequency and $40 \times 40 \times 1.57$ mm^3 for dual frequencies at 2.4 and 3.6 GHz. The band stop filter has good agreement between measurement and simulation results.

References

1. Reja AH, Ahmad SN (n.d.) Studying the effects of metamaterial components on microwave filters 193–204. https://doi.org/10.1007/978-3-319-11218-3
2. Chavda KD, Sarvaiya AK (2020) A survey on microwave planar filter design using metamaterial properties—research design & development. Solid State Technol 63(2s)
3. Chavda K, Sarvaiya AK (2012) Design of passive RF beam forming network for wireless LAN application. In: 1st International conference on emerging technology trends in electronics, communication & networking
4. Veselago VG (1967) The electrodynamics of substances with simultaneously negative values of ε and μ. Usp Fiz Nauk 92:517–526
5. Smith DR, Padilla WJ, Vier DC, Nemat Nasser SC, Schultz S (2000) Composite medium with simultaneously negative permeability and permittivity. Phys Rev Lett 84(18):4184–4187
6. Pendry JB (2000) Negative refraction makes a perfect lens. Phys Rev Lett 85(18):3966–3969
7. Caloz C, Chang CC, Itoh T (2001) Full-wave verification of the fundamental properties of left-handed materials in waveguide configurations. J Appl Phys 90(11):5483–5486
8. Ayadi H, Machac J, Beldi S, Latrach L (2021) Planar hexagonal antenna with dual reconfigurable notched bands for wireless communication devices. Radioengineering 30(1):25–33
9. Nasiri B, Ennajih A, Errkik A, Zbitou J, Derri M (2020) Band-pass filter based on complementary split ring resonator. Telkomnika Telecommun Comput Electron Control 18(3):1145–1149
10. Rajalakshmi P, Gunavathi N (2021) Hexagonal split ring resonator enclosed circular split ring resonator inspired dual-band antenna for sub-6 GHz 5G NR and IEEE 802.11ba/be applications. Prog Electromagn Res C 115(August):1–15
11. Zhang FL, Zhao Q, Liu YH, Luo CR, Zhao XP (2004) Behaviour of hexagon split ring resonators and left-handed metamaterials. Chin Phys Lett 21(7):1330–1332
12. Christydass SPJ, Gunavathi N (2021) Octa-band metamaterial inspired multiband monopole antenna for wireless application. Prog Electromagn Res C 113(June):97–110
13. Faruque MRI, Siddiky AM, Ahamed E, Islam MT, Abdullah S (2021) Parallel LC shaped metamaterial resonator for C and X band satellite applications with wider bandwidth. Sci Rep 11(1):1–15
14. Gaurav Y, Chauhan RK (2017) Design of single narrow band bandpass filter using cascaded open loop triangular ring resonators embedded with rectangular ring. J Telecommun Electron Comput Eng 9(2):119–123
15. Hinojosa J, Martínez-Viviente FL, Alvarez-Melcon A (2021) Compact double notch coplanar and microstrip bandstop filters using metamaterial—inspired open ring resonators. Electronics 10(3):1–15

16. George B, Bhuvana Nair B, Sreedevi Menon K (2018) Mathematical modeling and validation of a hexagonal split ring resonator. In: Proceedings of the 2018 2nd International conference on advances in electronics, computers and communications, ICAECC 2018, no 1, pp 3–6
17. Hossain A, Islam MT, Misran N, Islam MS, Samsuzzaman M (2021) A mutual coupled spider net-shaped triple split ring resonator based epsilon-negative metamaterials with high effective medium ratio for quad-band microwave applications. Results Phys 22(January):103902
18. Afsar SU, Faruque MRI, Hossain MJ, Khandaker MU, Osman H, Alamri S (2021) Modified hexagonal split ring resonator based on an epsilon-negative metamaterial for triple-band satellite communication. Micromachines 12(8):878. https://doi.org/10.3390/mi12080878
19. Islam MR, Islam MT, Moniruzzaman M, Samsuzzaman M, Arshad H (2021) Penta band single negative meta-atom absorber designed on square enclosed star-shaped modified split ring resonator for S-, C-, X- and Ku-bands microwave applications. Sci Rep 11(1):1–22
20. Iyer AK, Eleftheriades GV (2002) Negative refractive index metamaterials supporting 2-D waves. In: 2002 IEEE MTT International microwave symposium (IMS) digest, Seattle, WA, 2–7 June 2002, pp 1067–1070
21. Christydass SPJ, Gunavathi N (2021) Octa-band metamaterial inspired multiband monopole antenna for wireless application. Prog Electromagnet Res C 113(June):97–110. https://doi.org/10.2528/PIERC21041102
22. Çinar A, Biçer S (2020) Band-stop filter design based on split ring resonators loaded on the microstrip transmission line for GSM-900 and 2.4 GHz ISM band. Int Adv Res Eng J 04(01):29–33. https://doi.org/10.35860/iarej.641459
23. Karthikeyan SS, Kshetrimayum RS (2015) Compact and wide stopband lowpass filter using open complementary split ring resonator and defected ground structure. Radioengineering 24(3):708–711
24. Sorace RE, Reinhardt VS, Vaughn SA (1997) High-speed digital-to-RF converter. U.S. patent 5 668 842, 16 Sept 1997
25. Perhirin S, Auffret Y (2013) A low consumption electronic system developed for a 10 km long all-optical extension dedicated to sea floor observatories using power-over-fiber technology and SPI protocol. Microwave Opt Technol Lett 55(11):2562–2568. https://doi.org/10.1002/mop.28085
26. Eleftheriades GV, Iyer AK, Kremer PC (2002) Planar negative refractive index media using periodically L-C loaded transmission lines. IEEE Trans Microwave Theory Tech 50:2702–2712. https://doi.org/10.1109/TMTT.2002.805197
27. Choudhary DK, Mishra n, Kumar R, Chaudhary RK (2017) Compact two pole metamaterial bandpass filter using inverted IDC, meander line and rectangular stub for WiMAX application. In: Progress in electromagnetics research symposium, November 2017, pp 128–131. https://doi.org/10.1109/PIERS-FALL.2017.8293124
28. Asci C, Sadeqi A, Wang W, Nejad HR, Sonkusale S (2020) Design and implementation of magnetically–tunable quad–band filter utilizing split–ring resonators at microwave frequencies. Sci Rep 10(1):1–8. https://doi.org/10.1038/s41598-020-57773-6
29. Berka M, Azzeddine HA, Bendaoudi A, Mahdjoub Z, Rouabhi AY (2021) Dual-band bandpass filter based on electromagnetic coupling of twin square metamaterial resonators (SRRs) and complementary resonator (CSRR) for wireless communications. J Electron Mater 0123456789. https://doi.org/10.1007/s11664-021-09024-1

Implementation of Generic Medical Device Software Integration with Multiple Protocol Support

Devashri Raich, Yashpal Singh, and Asha Ambhaikar

Abstract Pathology laboratory testing is helping in the medical industry for the diagnosis of almost every disease these days. Advance laboratory testing equipment is helping laboratory technicians and doctors to get more accurate results every time using standard procedure and automation every time. Technicians record the result extracted from the testing machine and enter the result in hospital or laboratory management software to generate the final patient result. These transferring records from machine to report generation software process are still manual in many pathology laboratories and clinics. The proposed research work mainly targeted at study different pathology testing analyzers and the communication protocols supported by the respective analyzers. This study also helps in understanding the different physical connection media and their significance. The system was implemented on medical ASTM protocol processing and data exchange. Using systematic parser as per the protocols and rules defined in every respective machine system understands and generates the result required by the report generation software. This approach will avoid manual transfer of results into software and improve the accuracy of the medical report, which helps to understand the quality of the result and performing automatic result interpretation easy.

Keywords LIS · Analyzer · HL7 · ASTM · HOST · Control characters

1 Introduction

Due to changes in lifestyle, eating habits, and lack of physical stress or exercise can cause thousands of problems in human health [1]. In the last couple of decade, along with the life of a human body, many diseases, including cancer and skin disease,

D. Raich (✉) · Y. Singh · A. Ambhaikar
Department of CSE, Kalinga University, Raipur, India
e-mail: devashriraich@gmail.com

Y. Singh
e-mail: yashpal.singh@kalingauniversity.ac.in

M. S. Kaiser et al. (eds.), *Information and Communication Technology for Competitive Strategies (ICTCS 2021)*, Lecture Notes in Networks and Systems 401,
https://doi.org/10.1007/978-981-19-0098-3_20

heart attacks, blood sugar, have been common in many societies and cultures that are killing many people every year. Data mining and machine learning are based on captured data and existing historical transactions to create a learning model that can be effectively predicting the new cases of a disease, whether they are safe or risky [2]. In order to achieve higher efficiency in algorithm, there must be higher level of accuracy in the prediction or the recognition of a given pattern or model [3]. In many diseases, few symptoms are common, and in few diseases, no symptoms found in day to day [4] life until it reaches the higher risk stages [5]. Right direction analysis in different components of in human bodies can lead to detection of different disease. On other hand, pathology laboratories are becoming more automated with the help of advance medical equipment's and machineries. Interfacing these machines and understanding different test mythology and components extracted from test are the major challenge in information technology or electronics industries [6]. This research work is mainly aimed at study about different laboratories machineries and finding out the multiple symptoms through laboratory test and human day to day life activities and predicting the possible chances of different disease and providing possible solutions to patients with the help of expert practitioners.

In the medical field, machine learning is used for prediction, detection, and diagnosis of disease [6]. The main goal of these techniques is to earlier detection of heart disease so that it may lead to earlier diagnosis of disease with good treatment on the heart disease. In past years, the medically based research done automatically extracted large amount of feature from structure data. In the database, there are three types of data that consist of structured data, unstructured data, and semi-structured data. Structured data means well-formed data which consists of actual patient's record like EHR, laboratory records, some medical tests reports data, etc. [2].

2 Proposed System

Different researchers are developing the system to predict and identify the disease, but through single logic and only direct relative information cannot help in predicting the disease. The main objective of this research work is to study about different laboratories machineries and finding out the multiple symptoms through laboratory test and human day to day life activities and predicting the possible chances of different disease and providing possible solutions to patients with the help of expert practitioners. This study involves different parameters like daily activities, eating habit, and the disease symptoms.

Figure 1 describes the basic architecture of the proposed system where multiple components and modules are presented. Proposed research work will be carried out in multiple stages, and result from any stage may change the action plan in consecutive stages. Following are the steps involved in the research work, and those are,

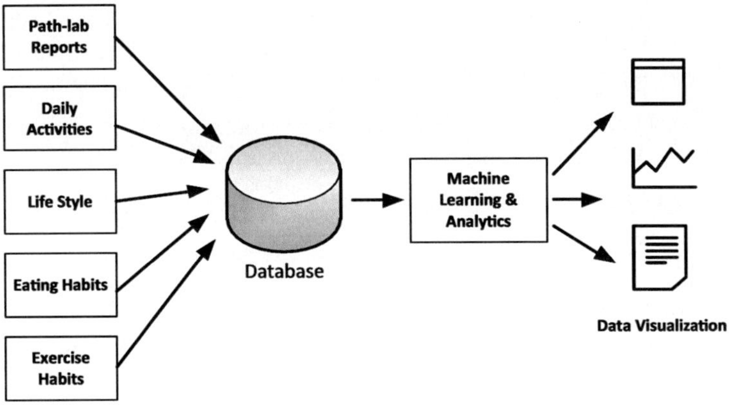

Fig. 1 System overview

- Data gathering stage where first we start with the collection volunteers or patients in formation like his/her, lifestyle, eating habits, exercise habits, daily activities, or the path-laboratory report in case of patients.
- Building a structure to find out the inter-relation between collected data and statistics.
- Developing and executing machine learning and data analytics methods on collected data and build relation modeling.
- Developing the presentation layer in order to generate visual presentation of present statistics and relatively predicted disease.

3 Implementation

In order to start analysis and prediction of different disease, process starts with the understanding of different body component and parameter mainly considered by pathology laboratories for investigation of particular disease. Maximum laboratories machines are having standard methods for interfacing with computers and supporting software system. These methods either work on continuous data reading or work on-demand data response mode. Every component or parameters is having standard value sets which may rarely differ between different patients and different scenarios. In the first phase, we will work on different research available on pathology laboratories machineries and process for investigations. Here, we will compare different communication protocols available in laboratory machines and data capturing methods. Secondly, listing the different parameter involved in computerized investigation.

Once we are ready with study of different laboratories communication protocol, in second phase, we will develop the interfacing software to read the desired component from machines and store in database for further analysis. This interface software will also describe the comparison with standard or normal value. Studying different symptoms and differential value in investigation component with respect to different disease, this will help us identifying which disease is more predictable and what are the values or point of considerations in prediction of disease or its stages.

- Creating interface for the pathology laboratory machines.
- Analysis of common and distinct symptoms of various diseases.

Challenges with proposed system are like,

- Since the medical device supports standard communication media, yet not every medical device supports or works on standard protocols for data exchange.
- Globally, ASTM- and HL7-like data exchange protocols are gaining the popularity among many medical instruments and vendors, and still many local or small-scale companies are going with proprietary communication protocols due to cost-effectiveness and ease of communication with basic data exchange requirement fulfillment.
- In order to develop a generic or maximum possible medical device supporting medical device integration system, it needs to understand standard interfaces and the communication protocols.

3.1 Programming Class Structure

Figure 2 describes the implemented class structure and interassociations between every classes with each other. Communicator is the main class, which connects to analyzer and maintains the logical connection for data exchange. Data incoming source for communicator class would be (a) LAN server class, (b) LAN client class, or (c) serial COM link. These three interfaces are supported by different analyzer, and every method has different way of handling the data exchange, but the data format might be the same in any other the interface. These formats could be an ASTM, HL7, or proprietary protocol.

Communicator class employees the data parser class which plays major role in conversion of data in to tokens or the subfields. Data parser class deals with database handler class in order to push incoming data to associated database tables and pull the sample, orders or the patient records to push data in to analyzer as a worklist in case of bidirectional communication. Finally, logger class is utilized to record the errors and the application activity in plain text file for debugging and troubleshooting purpose.

3.2 Parser Function Set

LIS host machine or integration software is primarily responsible to capture or demand records from analyzer and push these records to Hospital Information System

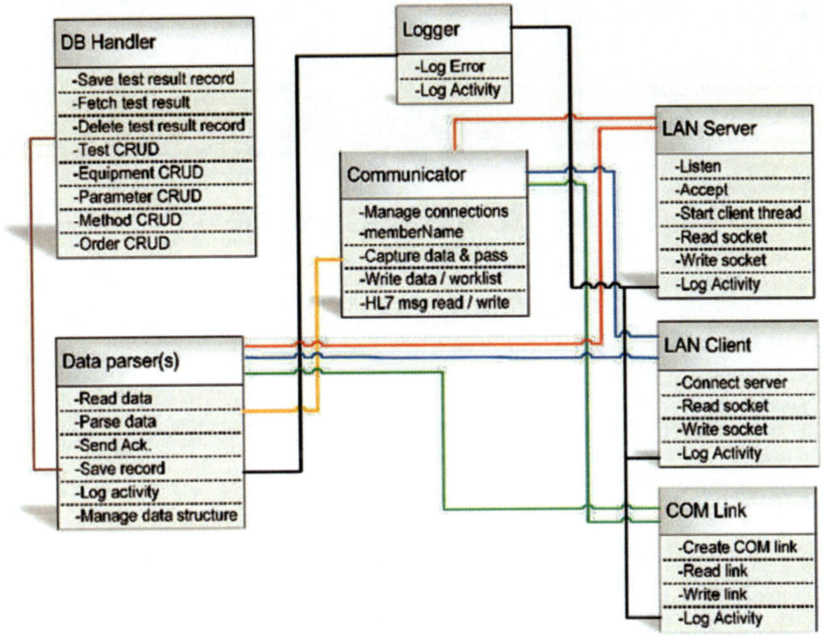

Fig. 2 Class diagram

(HIS) or Laboratory Information management System (LIMS) for reporting and storage purpose. LIS integration could be a part of the HIS or LIMS.

This section primarily discusses the programmatic functionality required for implementation of the LIS integration. This section could directly help LIS integration developer in software implementation. LIS implementation may consist following submodule and functionality. Few of the functions may vary analyzer to analyzer, and these are considerable functions as a guideline.

ASTM of HL7 data is formatted in predefined structure. Analyzer sends the data in bulk or in serial sequence or line by line. This data needs to interpret and tokenize as per the requirement. These formats could be seen in example section of this paper. ASTM and HL7 formatted data is constructed using multiple data fields, group of fields, and the special purpose special character. These fields are connected using predefined delimiters, and it needs right way of tokenization or parsing. Parsing class consists line splitting, field splitting, block splitting, and subfields parsing. Table 1 describes the different possible classes.

Development and implementation of any ASTM or HL7 protocol would be a challenging task as it needs detailed understanding of every block, field, and the subfield of supportable ATSM or HL7 protocols, and these protocols are very much strict about the syntax. Apart from programming expertise, understanding of the every meaningful keyword in analyzer provided with LIS host interface manual is

Table 1 Parser class function set

Parameter	Return	Description
Parser: SplitIncomingData()		
• Compare delimiter • Incoming string	Parsed string array	Parse incoming data with given delimiter
Parser: SplitBlocks()		
• Compare delimiter • Input string	Parsed string array	Parse input string block with given delimiter
Parser: SplitField()		
• Compare delimiter • Input string	Parsed string array	Parse input string field with given delimiter
Parser: SubString()		
• Input string • Start index • Length	Extracted string	Get mid string or substring from given input

most important. Upon that knowledge, a developer can refer and implement functions set mentioned in Table 1.

3.3 *Graphical User Interface*

ASTM or the HL7 communication could be a background process that is running continuously to receive the data and parse the data as per the defined structure. This application can be a console application or the desktop application, and we choose to implement using desktop application methodology. Figure 3 describes the main analyzer communicator window, where different subcomponent presents the execution methods. Very first textbox presents the every activity station like connection status, client connection request, client connected, data arrived, COM port opened for data receiving, and error occurred while execution. Next three textbox presents every data in raw format, which is received over logical connection link between analyzer and the LIS host. This could be serial link or the network socket link. With right experience of understanding the ASTM or HL7 data, one can read the information from raw data. Finally, since this is to be background running application, so we provided the hide button component, by clicking which user can hide the application but running in background. Using system-tray icon, user can again open the window.

Once the parsing process is completed, the parsed record needs to push in the database tables or supportable format from where LIS, HIS, or report generation software can fetch the data. Figure 4 presents the result data after successful parsing. Typically, from any analyzer, important data is patient test result data. Here, implementation system is presenting the different fields like,

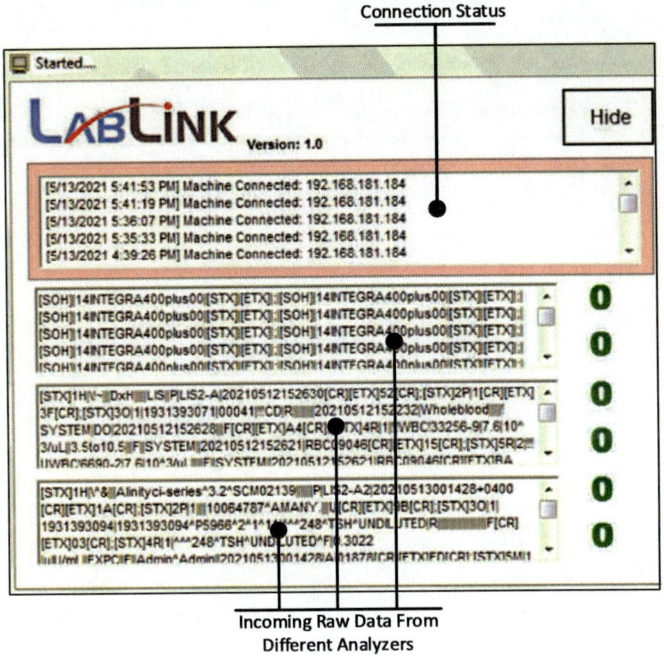

Fig. 3 Parser communication module

- **Sample ID**: This is and must be unique identification number to every sample used for test.
- **Parameter name**: Every test has some parameters, and it has some alias like WBC, PLT, GLU, etc.
- **Result value**: This is the actual result value given by analyzer after successful test. This is the main part of result.
- **Parameter result unit**: Every parameter is represented in different unit like mg, mg/l, u/l, gl/d, etc. This module also allows user to search data from specific analyzer, specific date, and for specific sample ID.

4 Conclusion

This study is made upon finding out the literature from medical automation research paper; at the same time, this requires detailed manuals from every individual medial testing analyzer machine from different manufacturer. Here, we tried to cover as much as possible aspect about the ASTM and HL7 protocol parser. Result shows the successful implementation of the ASTM and HL7 parser, which can extract the requested information from incoming raw data. In the near future, this research can be taken further for data analytics and the understanding of the pathology laboratory

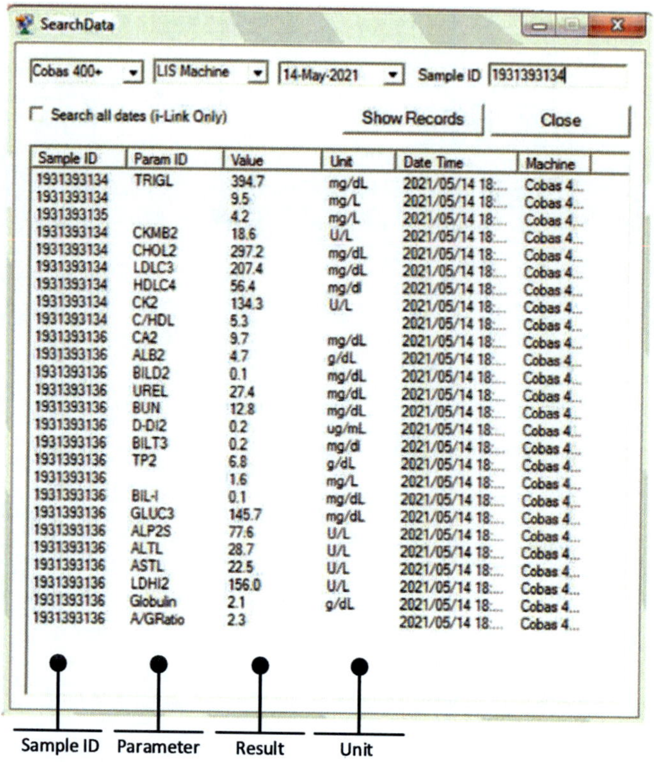

Fig. 4 Parser result figure

patient test data for diseases ratio analysis in society. In the near future, the proposed system can be enhanced to serve better service and the support to multiple communication protocols and analyzers. Following are some expected enhancements in the near future.

- System will support HL7 event based by directional communication with work list upload support.
- System will work in hybrid mode to support serial, USB and LAN connectivity simultaneously.

References

1. Kilintzis V, Kosvyra A, Beredimas N, Natsiavas P, Maglaveras N, Chouvarda I (2019) A sustainable HL7 FHIR based ontology for PHR data. In: 2019 41st Annual international conference of the IEEE Engineering in Medicine and Biology Society (EMBC). https://doi.org/10.1109/embc.2019.8856415

2. Houta S, Ameler T, Surges R (2019) Use of HL7 FHIR to structure data in epilepsy self-management applications. In: 2019 International conference on wireless and mobile computing, networking and communications (WiMob). https://doi.org/10.1109/wimob.2019.8923179
3. Gomes F, Freitas R, Ribeiro M, Moura C, Andrade O, Oliveira M (2019) GIRLS, a gateway for interoperability of electronic health record in low-cost system. Interoperability between FHIR and OpenEHR standards. In: 2019 IEEE International conference on e-health networking, application & services (HealthCom). https://doi.org/10.1109/healthcom46333.2019.9009602
4. Mukhiya SK, Rabbi F, Pun KI, Lamo Y (2019) An architectural design for self-reporting e-health systems. In: 2019 IEEE/ACM 1st International workshop on software engineering for healthcare (SEH). https://doi.org/10.1109/seh.2019.00008
5. Corici AA, Olaf R, Kraufmann B, Billig A, Caumanns J, Deglmann M et al (2020) Interoperable and discrete eHealth data exchange between hospital and patient. In: 2020 23rd Conference on innovation in clouds, internet and networks and workshops (ICIN). https://doi.org/10.1109/icin48450.2020.9059335
6. Khalique F, Khan SA, Nosheen I (2019) PHF-A framework for public health monitoring, analytics and research. IEEE Access 1–1. https://doi.org/10.1109/access.2019.2930730

Smart Water Management: A Learning-Based Approach

V. S. Yeshwanth, A. B. Vijayalakshmi, V. Budihal Suneeta, and V. S. Saroja

Abstract To ensure efficient management of resources, i.e., water was a constraint due to lack of supportive technology. The latest advancements in Internet of Things (IoT) and artificial intelligence (AI) have provided the opportunity to build real-time resource monitoring and management systems. We propose to build a pipeline to predict every day water usage. It can be integrated to any IoT device. The pipeline established can predict the requirements of water with 90.14% accuracy. The system is also designed to learn from every day's data; thus, ensuring the model is improving everyday without any human intervention. The data collected through sensors from various departments of the University was used. The data was preprocessed and used to train various ML and deep learning models out of which the RF model proved to be the best. AWS is used to deploy the model over the cloud and connect it to IoT devices which monitor the water levels and update the model.

Keywords Smart water management · AI · IoT · Random forest · ML

1 Introduction

Water is one of the priceless natural resources with a fixed amount of availability, and usable water is insufficient. Rapid industrialization and urbanization have improved standard of living that has increased the freshwater demand; hence, managing water usage is more important. With the evolution of IoT and AI, the ease of availability of resources through cloud and the ease of access to computing power have generated new possibilities to solve complex real-time water issues. In the proposed work, we explore the possibilities of predicting daily water usage using AI, ML, and IoT. The authors aim to establish and prove that the AI models clubbed with IoT systems can be used to monitor and efficiently manage water resources in real-life environments. The IoT devices are designed to continuously monitor the water levels in the tanks

V. S. Yeshwanth · A. B. Vijayalakshmi · V. B. Suneeta (✉) · V. S. Saroja
School of ECE, KLE Technological University, Vidyanagar, Hubballi, Karnataka, India
e-mail: suneeta_vb@kletech.ac.in

M. S. Kaiser et al. (eds.), *Information and Communication Technology for Competitive Strategies (ICTCS 2021)*, Lecture Notes in Networks and Systems 401,
https://doi.org/10.1007/978-981-19-0098-3_21

and transmit the information to AWS IoT Core. The IoT core would then record the data and suggest appropriate action to the IoT device such as toggling the water supply of the tank. The models developed reside in the AWS EC2 instance which is in constant communication with the AWS IoT Core. Everyday, the model predicts a value of water consumption. This predicted value will be sent to the IoT device through AWS to fill the tanks appropriately. At the end of day with the updated consumption values of the particular day, the model retrains itself.

The remaining portion of the paper is divided in to several sections. Section 2 discusses existing related works in this domain. Section 3 briefly explains the data preprocessing. Section. 4 discusses the methodology of implementation and fine-tuning. In Sect. 5, the results are discussed, while Sect. 6 provides the conclusions.

2 Related Works

The authors in paper [1] compare various classical and adapted ML algorithms such as ARIMA, SVR used for water predictions. To improve performance of water demand prediction, human mobility is considered. Around 51 days of readings on water consumption over 7 Million Geolocated mobility records from various urban areas are taken. Candelieri et al. [2] discussed the use of two-stage learning model, where in the first stage, the hourly data for daily water demand time series patterns is analyzed for their behaviors. SVM regression is used in the second stage, to get a specific forecasting model for each cluster, based on outcome of first stage.

In [3], forecasting models such as adaptive heuristic model, a transfer/noise model, and a multiple linear regression model are analyzed. Kwon et al. [4] aim to develop a nonlinear ensemble model to forecast hourly water demand which allows to uncertainty estimation across different model classes. The first model is a multi-model scheme based on a combination of independent prediction models. The second model is a cross-validation scheme named the bagging approach to derive weighting factors corresponding to individual models. Ji et al. [5] discussed to reduce the usage of non-revenue water (NRW), and they collected three sets hourly basis of data/day from each water meter. It uses sensors to detect leakage automatically and automatic checking of sensor error system. The linear regression, ARIMA model, and additive regression model are used to analyze the given data and implement automated alarms for abnormal water usage prediction. Prediction models for total revenue water and NRW have been studied in [6, 7], but research on alarm system design for the NRW of household is inefficient and insufficient. In some cases, the solution is provided to entire regions rather to individual household [8].

Paper [9] proposes continuous deep belief echo state network (CDBESN) for the prediction of urban water demand on an hourly basis. Continuous deep belief network (CDBN) is the feature extraction algorithm, and an echo state network (ESN) is the regression algorithm. This architecture is modeled for actual data on water demand with quick convergence and had global optimization ability. Other previous studies

Algorithm 1: Algorithm for preprocessing the data

1 Merge the year-wise data of the particular department
2 Convert timestamp to date-time format
3 Apply rolling mean averaging with window 3 to smoothen the 'TotalConsumption' column
4 Convert categorical data into numerical data
5 Fill NaN values using FFill
Result: Preprocessed dataset of the particular department

detected the fault measurement of urban water networks [10–12]. In [13, 14] propose the real-time algorithms and Kalman filters to detect the leakage in pipeline. Smolak et al. [1] water prediction is improved with human mobility data. The best forecasting algorithm employing a RF method achieved 90.4% accuracy and is better by 1% than the same algorithm using only water data, while the classic ARIMA approach achieved 90.0%. The blind prediction achieved 85.1% of accuracy. Candelieri et al. [2] the models showed that when using weather input, the largest forecasting errors can be reduced by 11% and the average errors by 7%.

3 Data Preparation

To get weekly data, strength of each department and estimated number of students is determined. This estimation was done considering the events during the week such as exams, co-curricular activities, project submissions. Upon arriving at the week-wise strength, we added weather data to the dataset. The weather dataset obtained by open-source sites was used to establish the amount of precipitation, temperature, and cloud cover which would directly affect the water consumption. The authors increased the water consumption by 5% for every 2 °C rise in temperature with 28 considered as datum. The vice-versa applied when the temperature decreased. The authors also made changes to the gardening consumption based on the rate of precipitation. Adding the obtained values to the gardening consumption and multiplying it with 7 provided the authors with the entire weeks water consumption. The data in the same process was generated separately for each department for a period of 4 year from 2016 to 2020. Figure 1 shows the representation of data before preprocessing, and Fig. 2 displays the representation of data after preprocessing. The authors further performed feature extraction using XGBoost feature extractor where the criterion used was F-score. The model discovered that only 'AvgTemp', 'Timestamp', and 'AvgCloudCover' are the important features of the dataset. The algorithm illustrates the preprocessing performed on the data.

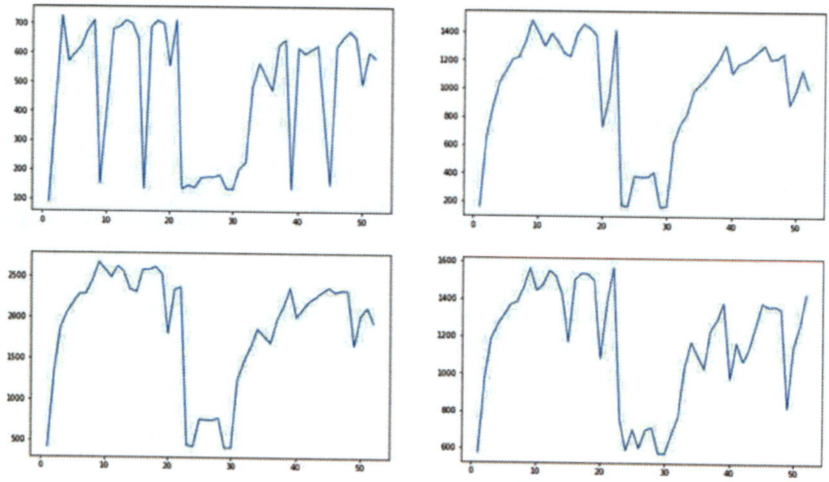

Fig. 1 Graphical representation of year-wise data of a department before preprocessing

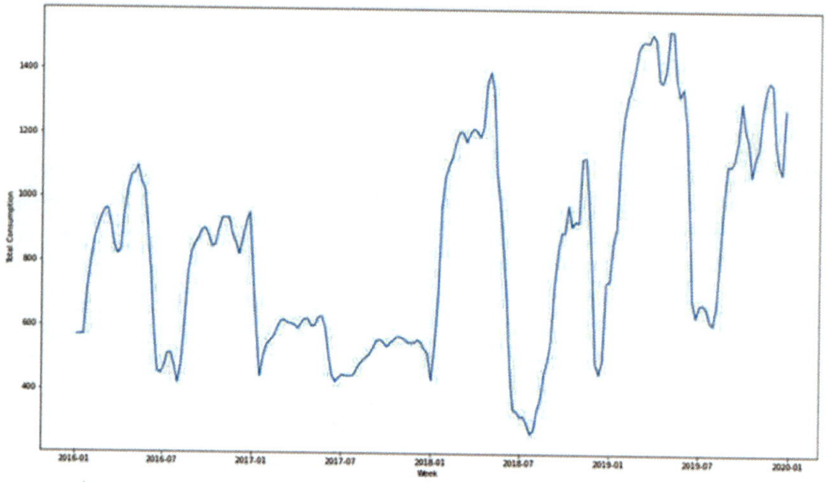

Fig. 2 Graphical representation of data of a department after preprocessing

4 Methodology

In this section, the authors discuss the methodology used to implement and integrate the pipelines of IoT device, AWS, and the AI model. Figure 3 displays the system architecture proposed by the authors for the implementation of the pipeline. Figure 4 demonstrates the activity diagram of the proposed system architecture. The subsections of this section discuss in detail the various components of the proposed system architecture.

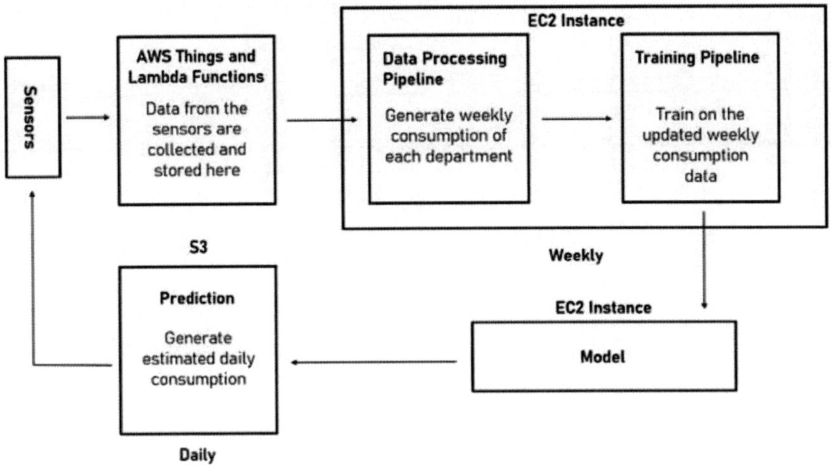

Fig. 3 Proposed system architecture for the implementation of the pipeline

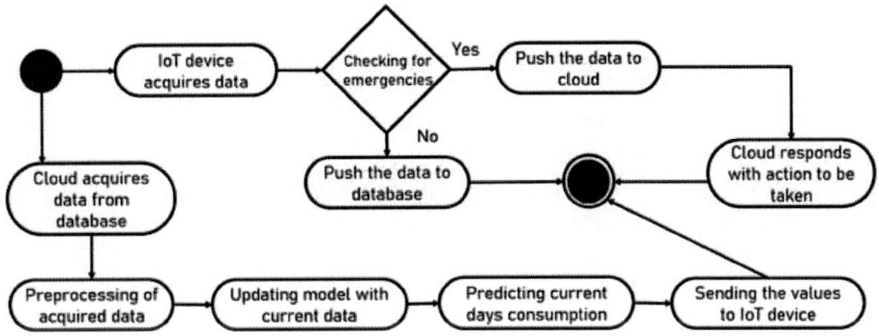

Fig. 4 Activity diagram of the proposed system architecture

4.1 Model Preparation

We experimented with several ML models such as XGBoost, decision tree, KNN, linear regression, among others. The best performing model among all the experimented models is RF regressor. The results of the experiments conducted using various models are discussed. The authors also experimented using neural network (NN) but due to the small dataset, the NN model overfitted very quickly and proved useless for further predictions.

RF is a supervised learning method where forest, it builds, is an ensemble of decision trees trained with the bagging method. A combination of learning models increases the overall result which is the idea of bagging method. In order to get a more accurate and stable prediction, RF builds multiple decision trees by merging them. Figure 5 contains the flowchart of RF. The authors performed hyperparameter

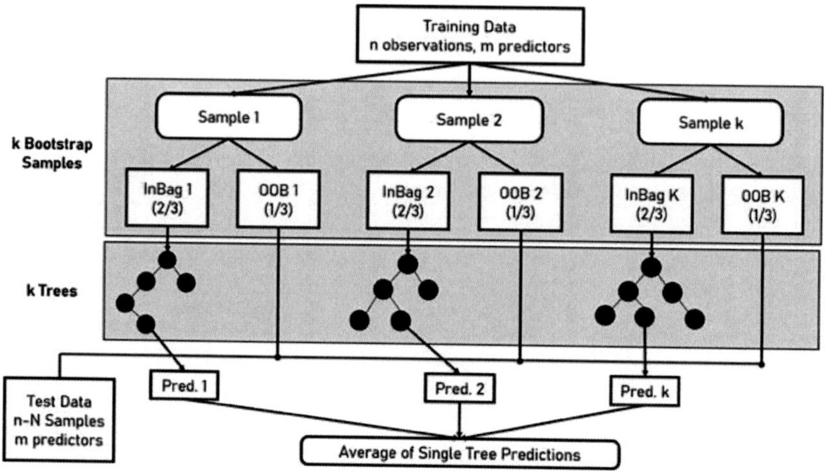

Fig. 5 Flowchart of random forest algorithm

tuning on the RF model. To find the optimal hyperparameters, the authors used random search algorithm. The results obtained after hyperparameter optimization are as follows: 'n_estimators': 1400, 'min_samples_split': 2, 'min_samples_leaf': 1, 'max_features': 'auto', 'max_depth': 100, 'bootstrap': True. All the models were trained using Google Colab. The authors used the free GPU provided by Colab for training the models. The data was uploaded onto the drive, and it was accessed through the Colab during training.

4.2 *Cloud Infrastructure*

Due to the limited computing power at the hardware end, the hard is to resort to cloud deployments for hosting the model. The authors used AWS as the cloud provider due to the availability of free tier systems. The authors used EC2 instance to host, train, and generate predictions from the model. S3 buckets were used as a medium of information exchange between the IoT devices and the cloud. The authors used an Ubuntu 18.x EC2 instance. This is a free tier eligible instance provided by AWS. The authors used this instance to run the ML model and generate predictions every day. The trained machine models were exported as pickle files. Every day as and when the data arrives, the pickle file would be loaded into the memory and will be trained on the newly obtained data. Once the weights of the models are updated, they are again saved as a pickle file which will be further used predictions. This ensured that the learning is propagated every day. The new data obtained by the sensors will be pushed to S3 buckets at a particular time. While training, the pushed file will be downloaded, preprocessed, and fed to the training function. Once the training is done,

Table 1 Comparison of three best performing algorithms for mechanical department

Algorithm	Training accuracy (%)	Validation accuracy (%)
Decision tree	31.78	31.61
XGBoost	41.05	44.82
Random forest	**96.02**	**91.90**

a CSV file is generated and will be later pushed to an S3 bucket. The automation of the entire process was done using shell scripts and crontab (Table 1).

The data transfer between the EC2 instance and the sensors was done using S3 buckets. At an agreed time, every day the sensor would upload the day's data to the S3 bucket. The newly updated data will then be used for training the model. Later at an agreed time, the model then predicts the water consumption and then stores the output file in S3. This file is then used by the sensor to read the required water and fill the tanks accordingly. In S3, each object will have a URI which can be used by anyone with access to read and access the files. The authors used AWS-CLI to push the data from EC2 to the S3 instance. To download the files using the URI, the authors used WGET.

5 Results and Discussion

The authors experimented with various ML algorithms for each department, namely decision tree regressor, XGBoost, KNN, linear regressor, RF, etc. The results of the top three performing models are presented department-wise in the following subsections. The accuracy mentioned in the subsections is defined as accuracy $= (100 - (100 * (\text{abs}(\text{predictions} - \text{testlabels})/\text{testlabels})))$.

The fit of RF is illustrated in Fig. 6; from the figure, we can observe that due to smooth trends in the dataset, the model performed well relative to architecture department model. The strength of the department is around 1000+ including faculty

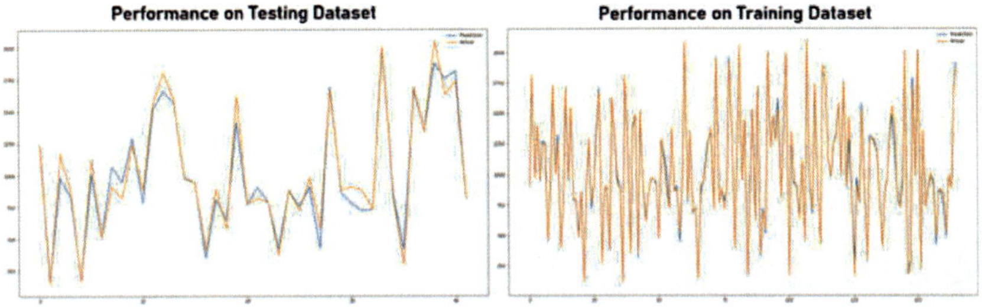

Fig. 6 Fit of RF model on training and testing datasets of mechanical department

Table 2 Comparison of the top three best performing algorithms for EC department

Algorithm	Training accuracy (%)	Validation accuracy
Decision tree	66.48	62.57%
XGBoost	66.85	68.46%
Random forest	**96.25**	**88.86**% height

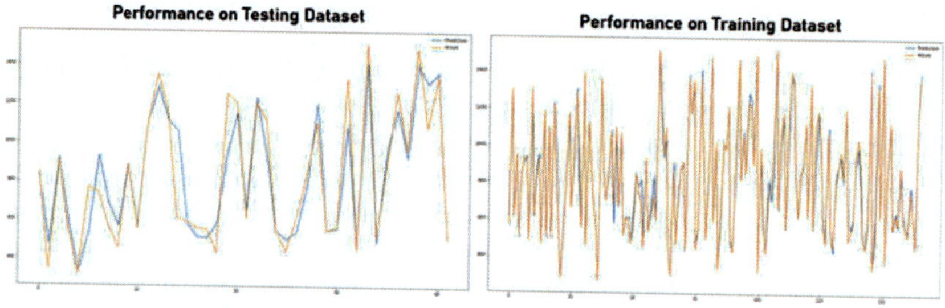

Fig. 7 Fit of RF model on training and testing datasets of EC department

Table 3 Department-wise RF model performance

Department	Training accuracy (%)	Validation accuracy (%)
Architecture	95.76	88.61
Mechanical	96.02	91.90
Main building	96.87	91.29
Electronics and communication	96.25	88.86

along with students. The real-time data of water utilization is very much tuned with the strength of the department. The deviation from the actual prediction is very less indicating that the proposed framework suits the gathered dataset. The fit of RF is illustrated in Fig. 7. The general observation here is that the model is not able to accommodate sharp variations in data. The strength of the students is more, and the number of visitors is also more as it has a small infrastructure made available to industry. Table 5 summarizes all the use cases of the proposed framework for water prediction in four different scenarios and strengths of the departments. The proposed methodology outperforms the other two techniques and provides stable results with good accuracy (Table 2).

6 Conclusion

On each department, the RF algorithm performed the best. On an average, the RF model produced an accuracy of 96.22% on training sets and 90.14% on testing sets. From Table 3, it can be inferred that the model performs consistently well over various departments, and we can understand that the model can perform well with any type of environment. This pipeline can be used on any other building with any number of people, and the model would still perform consistently well. With this, we would like to conclude that it is feasible to predict daily water consumption of any establishment given sufficient real-time data.

References

1. Smolak K et al (2020) Applying human mobility and water consumption data for short-term water demand forecasting using classical and ML models. Urban Water J 17(1):32–42
2. Candelieri A et al (2015) Short-term forecasting of hourly water consumption by using automatic metering readers data. Procedia Eng 119:844–853
3. Bakker M et al (2014) Improving the performance of water demand forecasting models by using weather input. Procedia Eng 70:93–102
4. Kwon H et al (2014) Development of ensemble model based water demand forecasting model. In: EGU General assembly conference abstracts, pp 3711–3721
5. Ji M et al (2019) Central prediction system for time series comparison and analysis of water usage data. IEEE Access 8:10342–10351
6. Liang C et al (2015) Water supply network monitoring based on demand reverse deduction technology. Procedia Eng 119:19–27
7. Candelieri A (2017) Clustering and support vector regression for water demand forecasting and anomaly detection. J Water 9(3):134–224
8. Patil A et al (2020) Smart security and surveillance system in laboratories using machine learning. In: Thampi SM, Piramuthu S, Li KC, Berretti S, Wozniak M, Singh D (eds) Machine learning and metaheuristics algorithms, and applications. SoMMA 2020. Communications in computer and information science, vol 1366. Springer, Singapore
9. Xu Y et al (2019) Hourly urban water demand forecasting using the continuous deep belief echo state network. J Water 11(2):351–361
10. Ragot J et al (2006) Fault measurement detection in an urban water supply network. J Process Control 16(9):887–902
11. Abbas et al (2018) Improving river flow simulation using a coupled surface-groundwater model for integrated water resources management. In: EPiC series in engineering, vol 3, pp 1–9
12. Siddamal SV et al (2020) Framework for a green Campus-LoRa^{TM}-based low-power smart water management for campus. In: Information and communication technology for competitive strategies, vol 191, Singapore
13. Suneeta VB et al, Facial expression recognition using supervised learning. In: Advances in intelligent systems and computing, vol 1108. Springer, Cham. https://doi.org/10.1007/978-3-030-37218-7-32
14. Delgado-Aguiñaga JA, Besancon G, Begovich O, Carvajal JE (2016) Multi-leak diagnosis in pipelines based on extended Kalman filter. Control Eng Pract 49:139–148

Bank Management System Using Blockchain Technology

Shrilalita Hegde, G. Aishwarya, Aishwarya Hugar, and V. B. Suneeta

Abstract A blockchain is a decentralized, ordered, and immutable ledger that permits a recording of transactions over a network, wherein digital information is stored in a public shared database. Bitcoin utilizes blockchain technology to maintain the records securely. The banks are stimulated with the aid of using economic and digital transformation and financial innovations. Blockchain technology is fundamental technology with promising application in the financial area. The paper includes the architecture and different features of blockchain technology and how it functions using a consensus algorithm. An angular frontend application is built to visualize and interact with JavaScript blockchain implementation.

Keywords Blockchain · Bitcoin · Cryptocurrency · Banking system

1 Introduction

A blockchain is a decentralized distributed technology with the advanced features that help securing the data and the system as a whole. In blockchain, decentralization refers to the transfer of control and decision-making from a centralized entity (individual, organization, or groups) to a distributed network. It is peer-to-peer technology where each peer indicates a node having copies of every transaction over the network which is called distributed ledger. That is all nodes have access to the same ledger. Therefore, it is also called as "Distributed ledger technology." A distributed ledger is a database that is consensually shared and synchronized across multiple sites, institutions, geographies accessible by multiple people that are connected over the network. It has the feature called immutability where the data remains unchanged. One cannot alter the package information. For such security, it uses cryptographic hash concept. Many sectors are using this technology that requires high security such as voting, banking, trading, and e-commerce.

S. Hegde · G. Aishwarya · A. Hugar · V. B. Suneeta (✉)
KLE Technological University, Hubballi, India
e-mail: suneeta_vb@kletech.ac.in

M. S. Kaiser et al. (eds.), *Information and Communication Technology for Competitive Strategies (ICTCS 2021)*, Lecture Notes in Networks and Systems 401,
https://doi.org/10.1007/978-981-19-0098-3_22

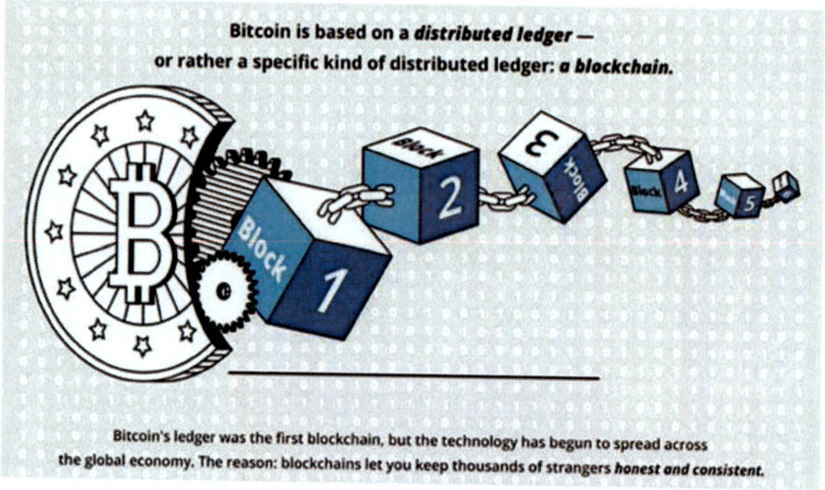

Fig. 1 Blockchain structure with example of the bitcoin [1]

For enhanced customer service, banking system is exploring different ways to make transaction. Enhanced customer service should promise the transparency as well as cost-effectiveness. This will help to change the financial business and makes the interaction more equitable, productive, and transparent secure. Blockchain joins a few technologies like encryption algorithms, point-to-point transmission, consensus mechanism, and distributed data storage. This technology acts as a decentralized ledger that monitors transactions between two parties. The system is essentially difficult to hack as these parties have simultaneous access to update digital ledger (record) constant. Figure 1 shows the blockchain system by taking the example of bitcoin.

2 Literature Survey

The following section presents a brief review of state-of-the-art work carried out in the similar domain. [1] introduced the idea of blockchain technology that transforms the underlying technology of the credit information systems and payment clearing in banks and accordingly transmutes and upgrades them. Additionally develop the "multi-center" formation, which will improve the effectiveness of the financial business. In this paper [2], the authors explore how blockchain will affect the banking industry and also about consensus algorithm. In this paper, advantages and limitation of the blockchain technology are discussed.

In paper [3], they have proposed a technology in order to eliminate the trusted third party to make transactions, to securely store the banking transactions at the lower cost. [4] present the blockchain technology which is immutable, distributed record that can be executed via blockchain. The blockchain technology is a distributed peer-to-peer linked structure which can take care of the issue of maintaining and recording transactions in a financial framework. The paper [5] talks about application of blockchain technology in banking sector. It describes the types of blockchain: implementation of the blockchain in banking and the limitations of the blockchain. Features of the blockchain are described, and this paper also talks about digital currency, supply chain financing, know your customer, and how COVID-19 in affecting the digital currency.

The paper [6] talks about banking transaction without using tokens. Blockchain technology is used for protection. In the beginning, it talks about need for blockchain in banking industry and drawbacks of the centralized system. This paper also talks about how decentralized system will help to reduce attacks and also talks about how blockchain will help to reduce such attacks and helps to build a secure system. The paper [7] talks about implementation of the blockchain in financial sector in order to enhance stability. This paper has conducted a research about how blockchain will affect the banking or financial industry, also talks about secure transactions can be done using blockchain technology, and how financial sector is benefited from the blockchain technology. The survey revealed that there is a need to develop the most secured system for banking that can provide complete security as well as secrecy. The objective is to develop a banking system using blockchain technology for secured transactions.

3 Blockchain Architecture

This section provides complete details about the architecture of the blockchain and its functionalities. Each block in the blockchain can be identified with the help of hash which is generated using SHA-256 algorithm. The header block consists of timestamp, hash, previous hash, data, index, and nonce. Genesis block is the first block in the blockchain. This block contains the unique values of hash. There will be no transaction data in the genesis block. The transactions' details can be seen from the next block on chain. In this way, the transaction is safe and secured from tampering revision. Figure 2 shows blockchain which contains multiple blocks, hash, and previous hash value of the blocks are linked. As shown in Fig. 3, transaction can be started by signing with private key. In a block, multiple transactions can be stored. This block is broadcasted to the peer-to-peer network. Once the block is verified by the peer-to-peer network, it will added to the blockchain.

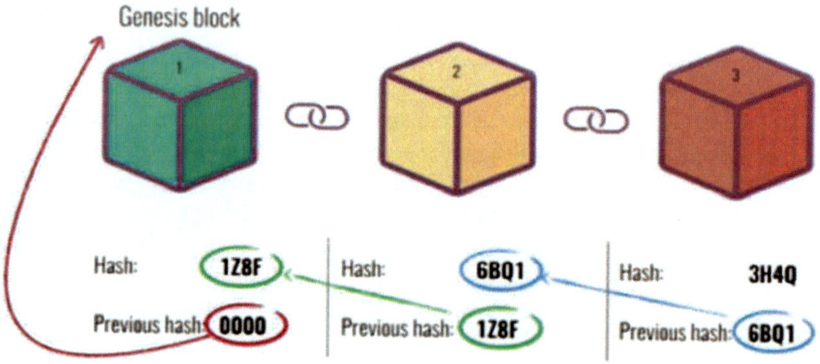

Fig. 2 Previous hash and hash value are linked in blockchain [2]

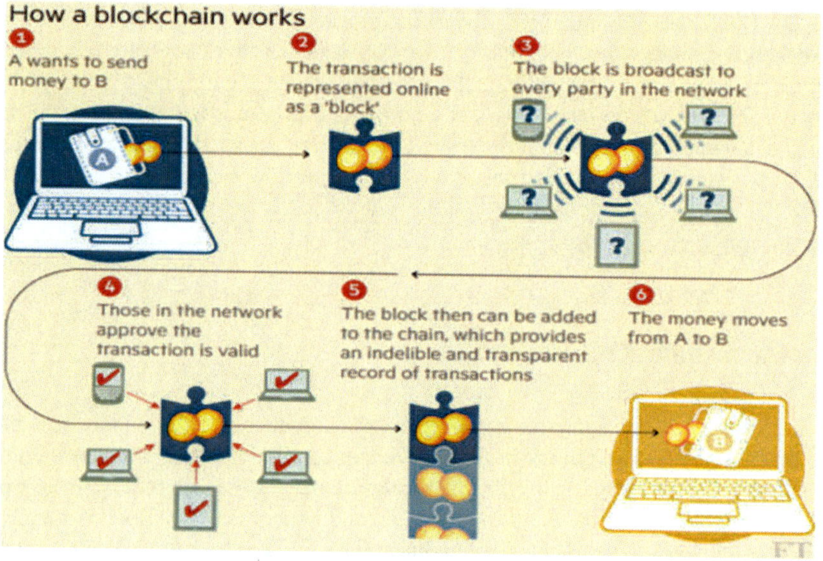

Fig. 3 A blockchain working in a transaction [3]

3.1 *Anatomy of a Blockchain Transaction*

The blockchain architecture includes three concepts, and they are decentralization, digital signature, and mining. The detailed anatomy of the blockchain. includes decentralization: Blockchain will give control to all the peers on the chain, and it also creates shared infrastructure. Digital signature: Owner knows the private key, and code for the description is known by all the members in the network. Mining: The users in the system mines and burrows the information are assessed by the cryptographic rules. Furthermore, it recognizes miners for confirmation of the transactions and affirmation.

The key characteristics of blockchain are consensus-driven: All the other parties must agree up for the validity of a transactions. Immutability: No member can alter the transaction after it has been recorded. The blocks cannot be changed whenever a block is added to the chain. This makes trust in the transaction record. Decentralized: A blockchain in bitcoin is stored in a file that can be copied by any node on the network and accessed with consistency. Transparent: Due to the property of open, a blockchain document can be accessed and evaluated by any party. This makes provenance under which resource lifetime can be determined. Participants know that where the asset came from and its ownership has changed after some time.

4 Methodology

In this section, we introduce and give details of the algorithms that are being proposed. Creating a blockchain consists of timestamp, data, previous hash. The previous hash helps in maintaining the integrity of the system. It shows the blockchain. Creation of the genesis block and the transactions, timestamp, nonce, and how the individual blocks are connected to the blockchain. The SHA-256 algorithm uses padding and dividing procedures to secure the block data with cryptographic hashing. (1) Padding: Pad the input string to the multiple of 512-bit length. Uses ASCII chart to convert letters and signs to number. Appends 1 to the converted string of 1's and 0's. Later appends bunch of zeroes that total whole bits to 448-bit length. Finally appends 11000 (24) making it to a total of 512-bit length string. (2) Dividing: Divides the padded binary into 512-bit chunks. Divides each chunk into 32-bit words (16 words per chunk). SHA-256 algorithm hashes the 32-bit words into 64 times. Then bought together to generate the result which is final. That is SHA-256 uses 64 rounds of encryption and can even specify the number of rounds manually.

4.1 *Proposed Framework*

New blocks can be added very quickly. All it has to do is create a transaction and compute its hash and add it to the blockchain. A modern computer can do that very quickly like 100–1000 blocks per second. But the purpose is not to create that many blocks per second. Since data in the block can be modified and hashes of all the blocks after that can be recalculated. It will end up as an invalid chain, even though we have tampered with it. So there is a security issue also. To solve this issue, blockchain has a property called PoW. Proof of work is the consensus algorithm that requires a lot of computation power. This implementation is also called mining. Computation power is also called difficulty level. An increase in the difficulty will increase the complexity. So mining block is created where the difficulty level will be set. The consensus algorithm is utilized to authenticate the transaction and creates a new block to the chain. This algorithm involves the mining process where miners

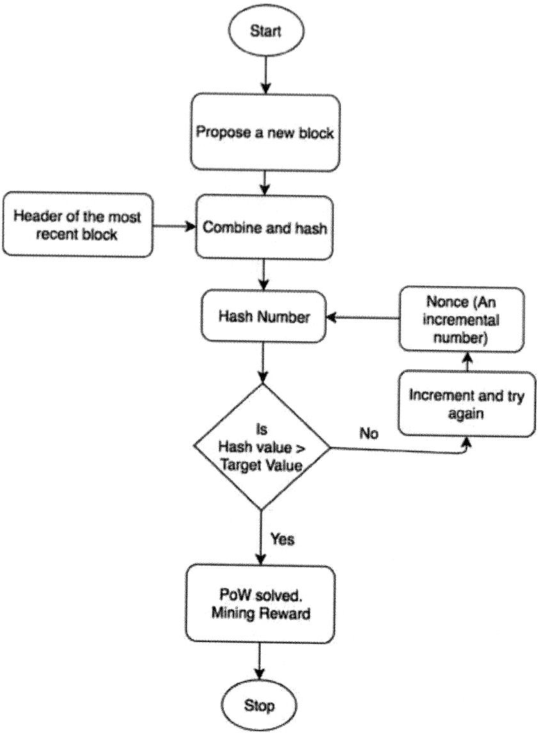

Fig. 4 Flowchart indicating the proposed framework

compete with one another in order to complete the transactions in the network. As soon as miners successfully created a valid block, he gets rewarded. From Fig. 4, the block is created using a block class. The miners calculate the hash until the result of it becomes less or equal to the target value. The nonce is a counter that handles to change of the value of the hash to meet the hash criteria. If the hash value is not greater than the target value the hash number is incremented. Once the correct hash value is calculated by miners it immediately broadcasts the block to the network. The miner who solved the first proof of work and added the block to the blockchain will be rewarded with bitcoin.

4.2 *Mining Rewards and Signing Transactions*

Now the blocks are secured with proof of work. In a block, multiple transactions are added and adding rewards to miners. When the cryptocurrency is started, it needs to have virtual money or coin. Transaction class contains from address and to address and the amount of money that needs to be transferred. Pending transaction class is also created since it has created only one block every ten minutes in the bitcoins case. The problem in the current system is anyone can make any transactions. So anyone can make a transaction that does not belong to them. So to solve this issue, it is mandatory to sign with public and private keys. So that only that person can spend the money who is having private keys. The private key is used to create a signature for each blockchain transaction a client sends out. The signature authenticates the transactions that have come from the client. Furthermore, it also blocks the transactions from being altered by anyone once it has been issued. So, using the private key, the client can sign the cryptocurrencies to send to others. Figure 5 explains about signing process. Private key which is only visible to the user is given to algorithm transformation. The private key is used to mathematically infer the public key. To create the address that others can see, the key is changed with a hash function. The cryptocurrencies are received when others send coins to our wallet address (which is a consequence of the hash of your public key and some extra information). The whole process of blockchain implementation is depicted in a flowchart as shown in Fig. 6. The flowchart starts with creating a block which consists of timestamp, hash value, nounce, and previous hash. This is implemented in visual studio code in Java. After creating a block, SHA-256 algorithm is used to calculate the hash value in bitcoin. In the next step, the generated blocks are stored. At the last step, the blockchain validation is performed to check whether it is valid or not and also to verify it by signing transactions where the user has the private and public keys to sign the transactions.

Fig. 5 Process of signing [6]

Fig. 6 Flowchart of blockchain implementation in the proposed approach

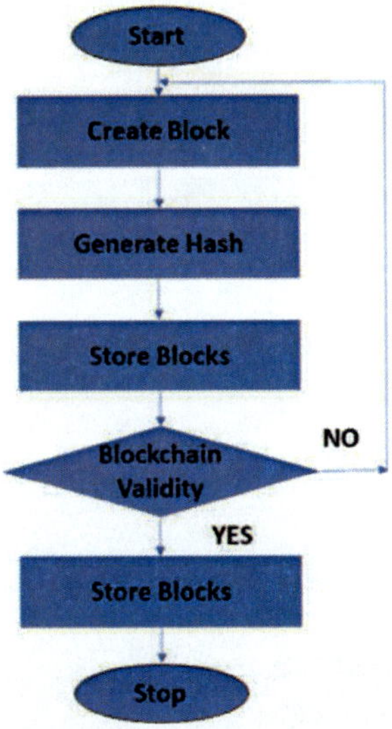

5 Results and Discussions

We built an angular frontend application that allows us to interact and visualize with blockchain implemented in JavaScript. As shown in Fig. 7, this is the homepage of the frontend blockchain that allows us to see blocks on the chain. It has two blocks: first block is the Genesis block. The block consists of the hash, the hash of the previous block, nonce, and timestamp. There will be no transactions stored in the first block. In the second block, we can see it contains one transaction, and this transaction came from the system to wallet address with 100 coins. The chain also contains whether the chain is valid or not. As shown in Fig. 8, we can create new transactions to any wallet for any amount (no validation). We can sign and create transaction. New transactions will be added to the list of "pending transactions," ready to be included in the next block. As shown in Fig. 9, list of all pending transactions. These are the

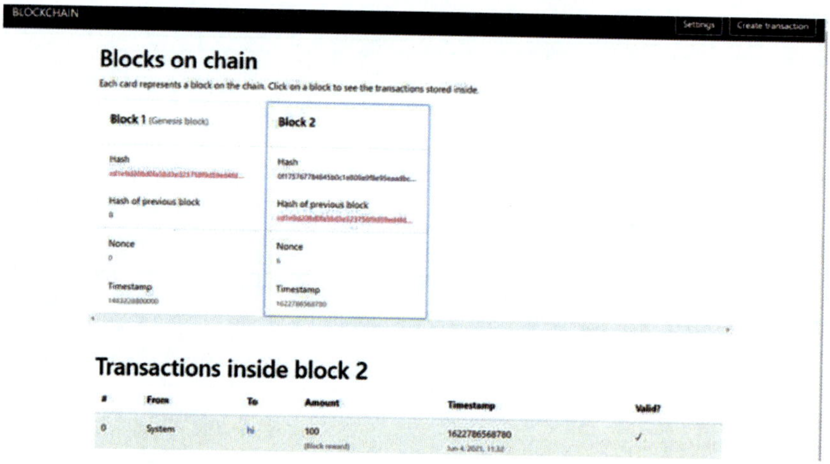

Fig. 7 This is the homepage of the frontend blockchain

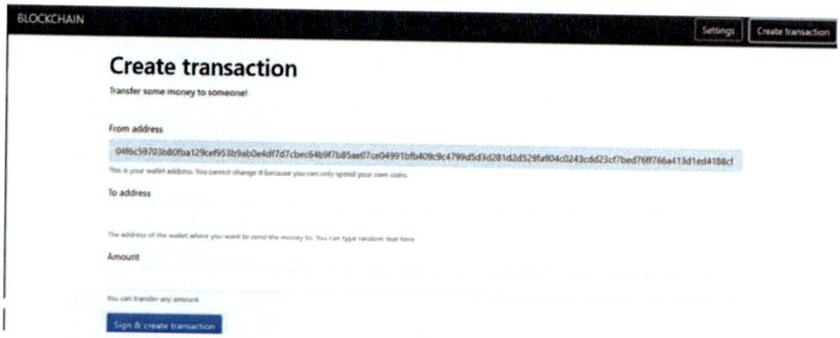

Fig. 8 Creating new transactions

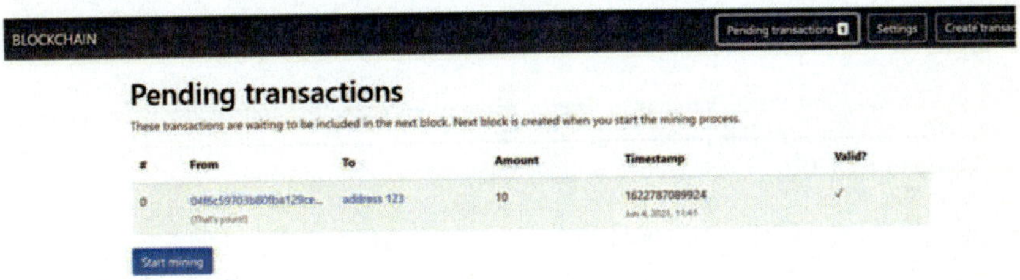

Fig. 9 Pending transactions

Fig. 10 Wallet details

transactions that are waiting to be included in the next block. So for that we have to start the mining process. We can keep adding transactions, and it will mine blocks and add them to blockchain. As shown in Fig. 10, by clicking on any wallet address and see an overview of that wallet: its current balance and all transaction to/from that wallet. There is also a setting option, where difficulty is set and configured the mining reward.

6 Conclusion

In this paper, architecture of the blockchain and its key characteristics are presented. The blockchain implementation includes the basic blockchain, the typical consensus algorithm (proof of work), mining rewards and transactions, and signing transactions. Furthermore, we built an angular frontend application to visualize and interact with JavaScript blockchain implementation. Blockchain technology in banking will do a lot of functionalities like consensus algorithm, mining rewards and transactions, and signing transactions.

References

1. Ye G, Chen L (2017) Blockchain application and outlook in the banking industry. Financ Innov 2. https://doi.org/10.1186/s40854-016-0034-9
2. Palihapitiya T (2020) Blockchain in banking industry
3. Sakho S et al (2019) Improving banking transactions using blockchain technology
4. Akhilesh NS et al (2020) Implementation of blockchain for secure bank transactions, pp 1–10. https://doi.org/10.23919/ICOMBI48604.2020.9203095
5. Sharma D (2020) Application of blockchain in banking sector. Glob Sci J 8(5):526–530
6. Pravin NP et al (2020) Blockchain technology for protecting the banking transaction without using tokens. In: Second international conference on inventive research in computing applications (ICIRCA), pp 801–807
7. Sneha P et al (2017) Real-time vehicle-type categorization and character extraction from the license plates. In: International conference on cognetive informtics and soft computing-2017, Dec 20th–21st 2017, Hyderabad, pp 557–565

Designing an Ensemble Classifier for Classifying Multi-label Data

Shriya Salunkhe and Kiran Bhowmick

Abstract In today's categorization challenges, it is critical to organize the ever-increasing amount of available data, and numerous labels are commonly connected with single occurrence. Multi-label classification is real-world problem that can be found in a variety of settings. Most traditional learning methods are developed for single-label problems because they involve more than one class, multi-label classification challenges. To address multi-label problems, this paper proposes a new model based on the binary relevance (BR) methodology and classifier chain (CC) techniques, which employs traditional classifiers such as k-nearest neighbor (KNN), Gaussian Naïve Bayes (GNB), and decision trees (DT) as well as ensemble classifiers such as random forest (RF) and stacking classifiers (SC). In this research, it is demonstrated that the two ensemble classifiers outperform the three traditional classifiers.

Keywords Multi-label classification · Binary relevance · Classifier chain · Traditional classifiers · Ensemble classifiers · Accuracy score

1 Introduction

Machine learning is a multidisciplinary field that includes Artificial Intelligence, Mathematics, Statistics, and Probability [1]. The dataset contains a significant quantity of data, with each specimen belonging to a distinct type and belonging to multiple classes, creating a classification challenge [2]. The labels are related [3]. These label associations provide additional information that can be explored during the learning process [4]. Traditional algorithms cannot handle multi-labels because they are used

S. Salunkhe (✉) · K. Bhowmick
Department of Computer Engineering, D. J. Sanghvi College of Engineering, Mumbai, India
e-mail: shriyas051996@gmail.com

K. Bhowmick
e-mail: kiran.bhowmick@djsce.ac.in

M. S. Kaiser et al. (eds.), *Information and Communication Technology for Competitive Strategies (ICTCS 2021)*, Lecture Notes in Networks and Systems 401,
https://doi.org/10.1007/978-981-19-0098-3_23

to classify single label [5]. This paper presents a novel model for dealing with multi-label situations. Binary relevance (BR) and classifier chain (CC) are two approaches used in this model. Traditional classifiers are easier to implement, interpret, and very efficient to train. So, traditional classifiers include *k*-nearest neighbor (KNN), Gaussian Naïve Bayes (GNB), and decision tree (DT). Ensembles of classifiers (EN) are a type of classifier that combines several learning algorithms to obtain better predictive performance. Ensemble classifiers include random forest (RF) and stacking classifier (SC). We compared it with the *k*-nearest neighbor, Gaussian Naïve Bayes, and decision tree. They are employed with the binary relevance and classifier chain techniques in this study. Section 2 shows the paper which we referred and the review of the papers in literature survey. Section 3, classification method, has a comprehensive overview of the model, methodology, and classifiers.

2 Literature Survey

A sample is associated with a number of labels in multi-label learning. Each sample may have a different label cardinality, as well as complete or incomplete label information [6]. Multi-label learning has become a hot issue in the machine learning field in recent years, and it has been successfully used to a range of real-world applications, such as picture identification, protein prediction, and text categorization [7]. The [8] study employs a multi-label categorization methodology. This algorithm combines LP with a rule-based evolutionary machine learning approach. As a result, MLRBC (Multi-Label Rule-Based Classifier) develops as a multi-label rule-based evolutionary learner. The LMMAL model is proposed in this paper. In [9] paper, the impact of noisy characteristics during the training phase is considered. It starts by selecting a set of representative characteristics and recording their relationships using an affinity propagation mechanism. Finally, a multi-label support vector machine is employed. Our main contribution is a novel multi-label learning method that we call the representative multi-label learning method (RMLLA) [10]. In a variety of real-world scenarios, multi-label classification comes in handy. As a result, multi-label classification is divided into binary and multi-class classifications. The evaluation process took into account Hamming loss, accuracy, precision, F_1 measure, and recall [11]. We could want to classify multi-label data in a variety of situations. Each data instance in any multi-label classification (MLC) issue has a high number of class labels. In this study, we solved MLC using the problem transformation method (MLC-LR) [12].

Understanding and problems to be focused: In this review of the literature, several ways to classifying multi-label data were discussed. The datasets used have many labels. The exponential numbers of possible label sets, capturing label relationships, the unknown volume of data, limited time and processing resources, missing labels, and other challenges have all been addressed. Several methods for classifying multi-label data were discussed in this study of the literature.

Table 1 Example of binary relevance (BR) technique

X	Y_1	Y_2	Y_3	Y_4
X_1	1	1	1	1
X_2	1	0	0	0
X_3	1	1	1	0
X_4	1	0	0	1
X_5	0	1	1	1

⇨

X	Y_1
X_1	1
X_2	1
X_3	1
X_4	1
X_5	0

X	Y_2
X_1	1
X_2	0
X_3	1
X_4	0
X_5	1

X	Y_3
X_1	1
X_2	0
X_3	1
X_4	0
X_5	1

X	Y_4
X_1	1
X_2	0
X_3	0
X_4	1
X_5	1

3 Method for Classification

3.1 Classification Techniques

3.1.1 Binary Relevance Technique (BR)

The simplest method is to train one binary classifier for each class label, with the binary classifier's prediction determining the relevance of each class label for the unseen instance. Binary relevance breaks down the multi-label learning problem into q independent binary learning problems. Let us have a look at the illustration below. The following is the dataset: The independent variable is X, and the target variable is Y (label). This problem is divided into four single-class classification problems having binary significance, as shown below. As a result, the labeling has the disadvantage of being identical, which means labels can be misplaced (Table 1).

3.1.2 Classifier Chain Technique (CC)

CC is a multi-label classification approach that takes advantage of label relationships. It generates a set of binary classifiers, each of which predicts whether or not a specific label is present. C_i uses the predictions of all the classifiers C_j, where $j <= i$. A sequence of binary classifiers C_0, C_1, C_2, …, C_n; where C_i uses the predictions of all the classifiers C_j, where $j < i$. As a result, the classifier chain can account for label correlations. Let us take a look at an example. The input space in the tuple below is X, and the labels are Y. Using classifier chains, this problem would be divided into four independent single-label issues, as shown below. The input space is represented by the yellow part of the picture, while the target variable is represented by the white part (label). This is similar to BR, but it uses chains to keep label correlation (Table 2).

3.2 Traditional Classifiers

The following three are the classic classifiers used:

Table 2 Example of classifier chain (CC) technique

X	Y_1	Y_2	Y_3	Y_4
X_1	1	1	1	0
X_2	1	1	0	1
X_3	0	1	1	1

⇨

X	Y_1
X_1	1
X_2	1
X_3	0

X	Y_1	Y_2
X_1	1	1
X_2	1	1
X_3	0	1

X	Y_1	Y_2	Y_3
X_1	1	1	1
X_2	1	1	0
X_3	0	1	1

X	Y_1	Y_2	Y_3	Y_4
X_1	1	1	1	0
X_2	1	1	0	1
X_3	0	1	1	1

3.2.1 *K*-Nearest Neighbor (KNN)

K-nearest neighbor (KNN) is a machine learning technique that is basic, easy to grasp, and versatile. KNN has a wide range of applications. KNN stands for the number of closest neighbors. The quantity of neighbors is the most important consideration. If there are two courses, K is usually an odd number. The procedure is known as the nearest neighbor algorithm when $K = 1$. This is the most basic scenario. Distance measures are used to determine the distance between points in order to discover the closest similar points [4].

3.2.2 Gaussian Naïve Bayes (GNB)

Gaussian Naïve Bayes is a Naïve Bayes variation that allows continuous data and follows the Gaussian normal distribution. Gaussian NB is a classification programed that uses the Gaussian Naïve Bayes technique. Gaussian NB is a classification programed that uses the Gaussian Naïve Bayes technique. The features' likelihood is considered to be Gaussian. Gaussian Naïve Bayes accepts continuous valued features and models them all as Gaussian (normal) distributions [5].

3.2.3 Decision Tree (DT)

A decision tree is a simple yet effective way of learning. It is a procedure for making decisions that follows a set of steps. Each branch node indicates a choice among numerous options, whereas each leaf node indicates a decision. An internal or test node is a node with outgoing edges. The leaves are the property of the other nodes. The divisions are determined by the leaves (labels) [6].

3.3 Ensemble Classifiers

The following two are the classic classifiers used:

3.3.1 Random Forest (RF)

Random forest produces a "forest" out of a collection of decision trees that are often trained using the "bagging" method. The main idea of the bagging method is that combining many learning models enhances total output. Make a dataset for each tree that is bootstrapped. Make a decision tree with the dataset, but use a random subset of variables or features to divide on at each node. To create a huge forest with a diverse range of trees, repeat these three stages hundreds of times. It is simple to make predictions with a random forest [8].

3.3.2 Stacking Classifier (SC)

Stacking generalization is another name for a stacking classifier. Each estimate's predictions are stacked and supplied into a final estimator, which computes the result. The estimator's parameter is a list of estimators that are applied in parallel to the input data. The projections of the estimators will be fed into the final estimator. To train the estimators and final estimator, the fit technique must be named on the training data. To solve the problem, this classifier is used. Logistic regression, KNN, and SVM are used as Base classifiers, and random forest is used as a Meta classifier. If we use numerous algorithms to tackle a problem, the strengths of one method may potentially disguise the shortcomings of the other, and vice versa. Stacking is based on this concept.

4 Proposed Model

In this paper, a new model is proposed. The five datasets have been taken into consideration for multi-label classification. Here, first the datasets are divided into train and test datasets with 70–30%. Then, we first apply BR technique on the training set. It transforms multi-label problems into single-label problems. Secondly, the CC technique is used on the training set. It transforms multi-label problems into a chain of single-label problems. Now, we move on to the classifiers which are used after BR and CC techniques. The five classifiers KNN, GNB, DT, RF, and SC are used to classify the label problems. Here, KNN, GNB, and DT are traditional classifiers, and RF and SC are ensemble classifiers. In SC, logistic regression, KNN, and SVM are used as Base estimators (BE) and random forest is used as Meta estimator (ME). The prediction is evaluated using accuracy scores. Next, we calculate the accuracy score for the classifiers:

$$\text{Accuracy} = \frac{\text{Number of correctly predicted labels}}{\text{Total number of labels}} \tag{1}$$

The result is shown using bar graph for BR technique and CC technique. Figure 1 shows the flow diagram.

5 Experiments and Results

5.1 Dataset

To conduct the studies, we used five multi-label datasets [13] from distinct learning domains. These datasets cover a variety of application domains, including picture,

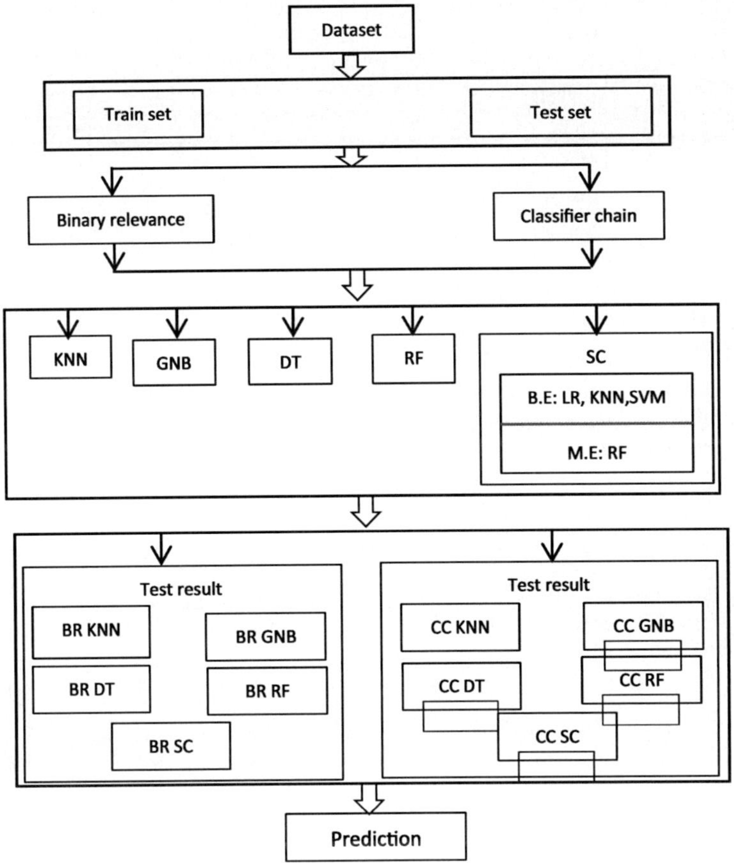

Fig. 1 Flow diagram of the model

audio, biology, and music. The Yeast dataset is made up of 2417 microarray expression data and phylogenetic profiles. Each gene has 103 descriptive properties. There are 14 functional classes in this version of the dataset, and a gene can be connected with any number of them. Scene dataset offers information on image qualities and classes and has six different labels. Emotions dataset is made of 72 music features for 593 songs categorized into one or more out of 6 classes of emotions. Flags dataset is made of 194 instances with 19 features and 7 labels. Bird dataset is made of 645 instances with 260 features and 19 labels. Table 3 contains the specifics of the selected datasets.

Table 3 Dataset description

Dataset	Domain	Instances	Attributes	Labels
Yeast	Biology	2417	103	14
Scene	Image	2407	294	6
Emotions	Music	593	72	6
Flags	Image	194	19	7
Birds	Audio	645	260	19

5.2 *Result*

The result here is predicted using Accuracy score. Accuracy score is defined as the ratio of the predicted correct labels to the total number (predicted and actual) of labels for that instance. Overall accuracy is the average across all instances. Figures 2, 3, 4, 5, and 6 represent the accuracy scores of the five datasets.

Table 4 shows the percentage increase.

Fig. 2 Comparison of accuracy w.r.t BR versus CC for yeast dataset

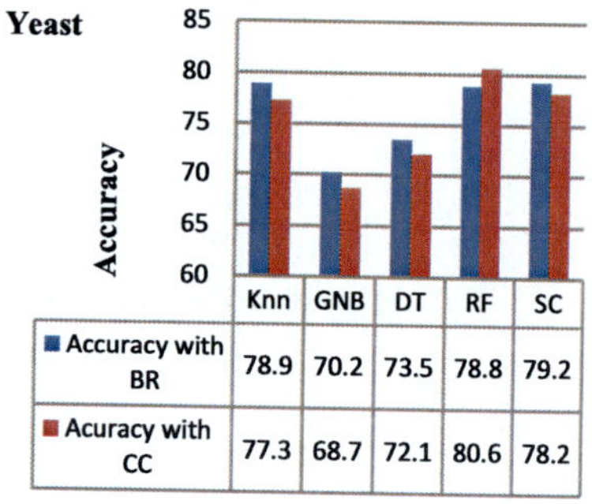

Fig. 3 Comparison of accuracy w.r.t BR versus CC for scene dataset

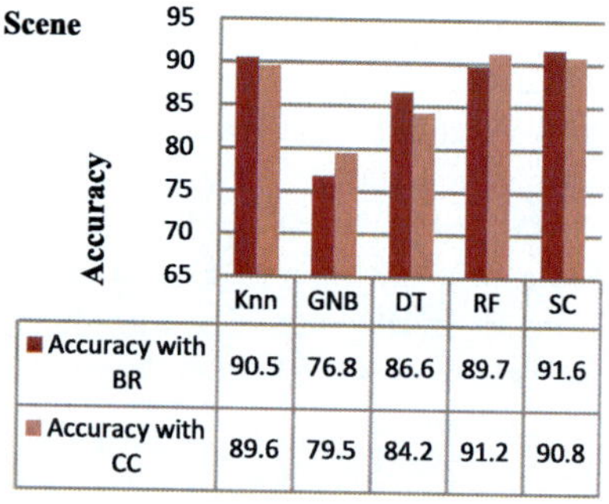

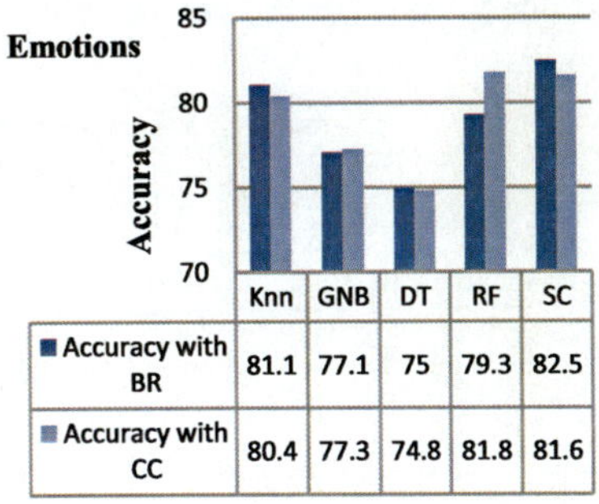

Fig. 4 Comparison of accuracy w.r.t BR versus CC for emotions dataset

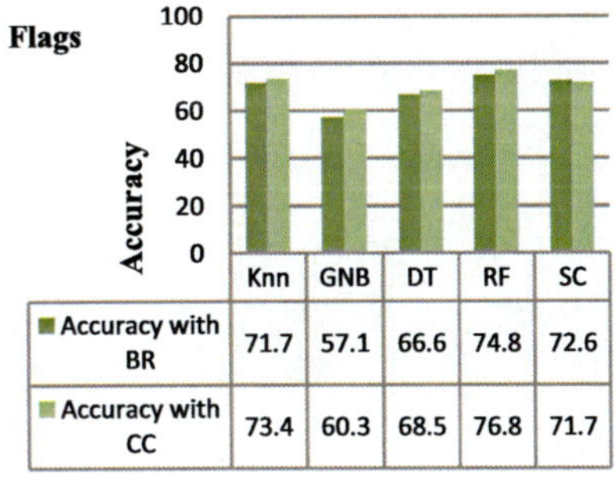

Fig. 5 Comparison of accuracy w.r.t BR versus CC for flags dataset

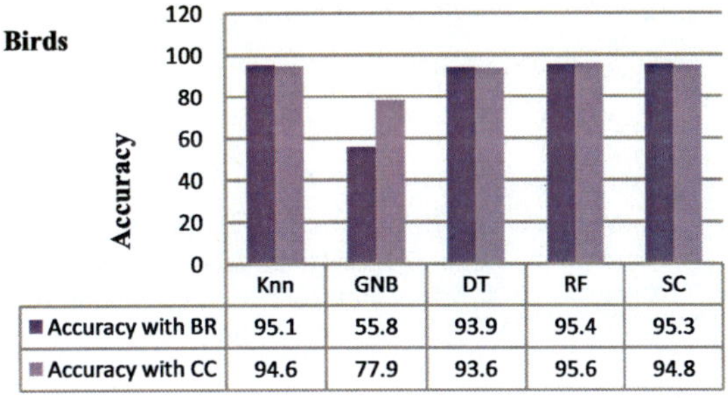

Fig. 6 Comparison of accuracy w.r.t BR versus CC for bird dataset

6 Conclusion and Future Scope

It is noteworthy that, while there are different methods of data classification, various challenges and issues affect the accuracy of the model performance. Considering the output, RF and SC classifiers show better performance, i.e., accuracy score as well as solving the classification of multiple labels of the given data. So, the ensemble classifiers are having more accuracy score than the three other traditional classifiers. Experimental results show that the random forest (RF) and stacking classifier (SC) well solved the classification problem of multi-labels in the given data as compared to the other three classifiers.

Table 4 Percentage (%) increase

	Yeast		Scene		Emotion	
	% Increase BR	% Increase CC	% Increase BR	% Increase CC	% Increase BR	% Increase CC
RF-KNN	−0.1	3.3	−0.8	1.6	−1.8	1.4
RF-GNB	8.6	11.9	12.9	11.7	2.2	4.5
RF-DT	5.3	8.5	3.1	7	4.3	7
SC-KNN	0.3	0.9	1.1	1.2	1.4	1.2
SC-GNB	9	9.5	14.8	11.3	5.4	4.3
SC-DT	5.7	6.1	5	6.6	7.5	6.8

	Flags		Birds	
	% Increase BR	% Increase CC	% Increase BR	% Increase CC
RF-KNN	3.1	3.4	0.3	1
RF-GNB	17.7	16.5	39.6	17.7
RF-DT	8.2	8.3	1.5	2
SC-KNN	0.9	−1.7	0.2	0.2
SC-GNB	15.5	11.4	39.5	16.9
SC-DT	6	3.2	1.4	1.2

Deep learning neural networks might be useful in the future. Deep learning neural networks are an example of a multi-label classification system that is built-in.

References

1. Nguyen TTT, Nguyen TT, Liew AW-C, Wang S-L, Liang T, Hu Y (2019) An online variational ınference and ensemble based multi-label classifier for data streams. In: Eleventh ınternational conference on advanced computational ıntelligence (ICACI), pp 302–307
2. Chu Z, Li P, Hu X (2019) Co-training based on semisupervised ensemble classification approach for multi-label data stream. In: IEEE International conference on big knowledge (ICBK), pp 58–65
3. Alattas AM (2018) Adaptive model over a multi-label streaming data. IEEE, pp 18–22
4. Huang J, Qin F, Zheng X, Cheng Z, Yuan Z, Zhang W (2018) Learning label-specific features for multi-label classification with missing labels. IEEE, pp 1–5
5. Luta RBG, Baldovino RG, Bugtai NT (2018) Multi-label classification of pH levels using support vector machines. IEEE, pp 45–48
6. Budhiraja M (2018) Multi label text classification for un-trained data through supervised learning. In: International conference on intelligent computing and control (I2C2). IEEE, pp 32–34
7. Wang R, Kwong S, Jia Y, Huang Z, Wu L (2018) Mutual ınformation based K-labelsets ensemble for multi-label classification. In: IEEE International conference on fuzzy systems (FUZZ-IEEE), pp 17–23
8. Nazmi S, Yan X, Homaifar A (2018) Multi-label classification using genetic-based machine learning. In: IEEE International conference on systems, man, and cybernetics (SMC), pp 675–680

9. Guo A, Wu J, Sheng VS, Zhao P, Cui Z (2017) Multi-label active learning with low-rank mapping for image classification. IEEE, pp 259–264
10. Li J-J, Alzami F, Gong Y-J, Yu Z (2017) A multi-label learning method using affinity propagation and support vector machine. IEEE Access 5:2955–2966
11. Venkatesan R, Wu S, Pratama M (2016) A novel online real-time classifier for multi-label data streams. In: 2016 International joint conference on neural networks (IJCNN). IEEE, pp 1833–1840
12. Prathibhamol CP, Jyothy KV, Noora B (2016) Multi label classification based on logistic regression (MLC-LR). In: International conference on advances in computing, communications and informatics (ICACCI). IEEE, pp 2708–2712
13. Data repository. http://www.uco.es/kdis/mllresources/. Accessed: 1st Aug 2021

Ultrasonic-Based Two-Factor Authentication System for Web Applications

K. S. Shashank, Aravinda Harithsa, Amogh Kuruwatti, Maheshwari Das, and M. S. Vinay Prasad

Abstract Authentication is the first step to access a resource, so it is of vital importance in a system. Many two-factor systems currently fail to address attacks such as eavesdropping or man-in-the-middle attacks. This paper presents a secure two-factor authentication method using sound. The proposed system authenticates a user without any physical contact at no additional cost of hardware by leveraging existing mobile phone technology. This idea is an extension of mobile connectivity-based audio communication in near-ultrasonic spectrum which is inaudible to the human ear.

Keywords Two-factor authentication · Sound · Ultrasonic · Security

1 Introduction

In today's world, the most widely used authentication mechanisms are one-factor authentication based on password and two-factor authentication methods which require a password and another factor like biometric or OTP. Passwords might suffer misuse by others; second-factor authentication needs extra procedures and incurs long delay, decreasing usability and deploy-ability of the system. This work overcomes the security vulnerabilities of a SMS system and also the hardware limitations of a biometric system. It is also more secure than a biometric system as some biometric systems can be fooled by a picture instead of a face or by a fake fingerprint, etc.

K. S. Shashank (✉) · A. Harithsa · A. Kuruwatti · M. Das · M. S. Vinay Prasad
JSS Science and Technology University, Mysuru, India
e-mail: shashankks0987@gmail.com

M. S. Vinay Prasad
e-mail: vpms@sjce.ac.in

M. S. Kaiser et al. (eds.), *Information and Communication Technology for Competitive Strategies (ICTCS 2021)*, Lecture Notes in Networks and Systems 401,
https://doi.org/10.1007/978-981-19-0098-3_24

2 Prior Study

We have surveyed and reviewed multiple papers on topics similar to ours. Most systems achieved sound-based authentication by comparing ambient sound of the two devices. All the reviewed methods employ a variation of ambient sound to achieve their goals. The system that we wanted to improvise on is a patent "System and method for using ultrasonic communication" [1] that discusses the potential of ultrasonic communication. The quest to replace passwords [2] this is a paper that compares and contrasts various means of authentication in use. QuickAuth [3], SonicAuth [4], Soundproof [5] rely only on ambient sound. A zero-effort two-factor authentication scheme [6] also considers the proximity of users "Two-factor user authentication with SMS and voiceprint challenge response" [7] is a paper that aims to tackle man-in-the-middle attack and socially engineered attacks. "Secure communication based on ambient audio" [8] employs fuzzy cryptography for better security. Implementation of a near-ultrasonic communication protocol in the 18.5–20 kHz band [9]. The protocol described in this paper is a component of Google's nearby platform used to localize the devices.

3 System Architecture and Methodology

3.1 Encoding

Encoding of sonic message will be done utilizing two different entities. We will use an UUID assigned to each user and stored securely in firebase. We will also add the current epoch time to the message payload with delimiter symbols. Then each character of the message will be further encoded second time with 8-bit binary digits allocated to each which is even more secure. Then we feed this input stream of bits into our audio synthesis function which will tune and generate the required ultrasonic message with the help of FSK in the predetermined frequency. Epoch timestamp will be obtained from the system to append to our payload to protect man-in-the-middle attacks and allows only 120 s of buffer time for authentication process. Figure 1 explains the end-to-end data flow during encoding process.

3.2 Decoding

Wave library was utilized to decode the recorded wave file with PCM encoding to establish an input audio stream for the system to process the data using a buffer size of 2048 samples. We read total of 512 samples at a time, here each sample is represented by 16-bit signed integer. Hamming window was applied to prevent the sound data from distorting frequency space transforms. The block diagram for audio decoder is shown in Fig. 2.

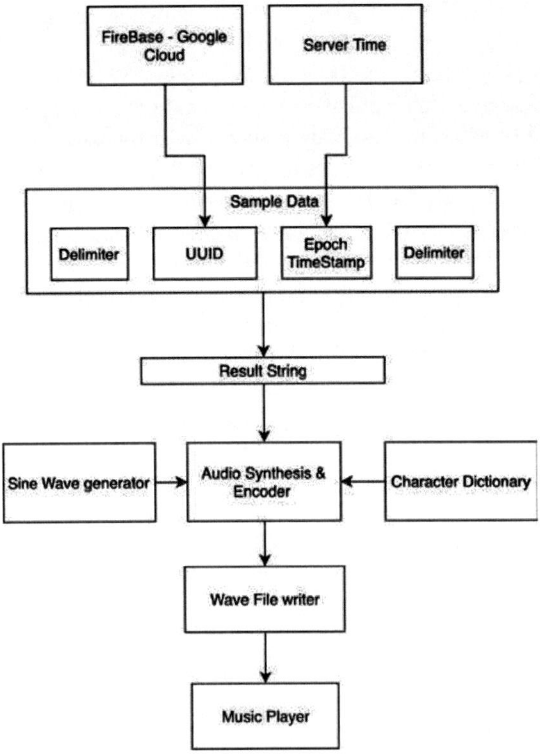

Fig. 1 Data flow during encoding

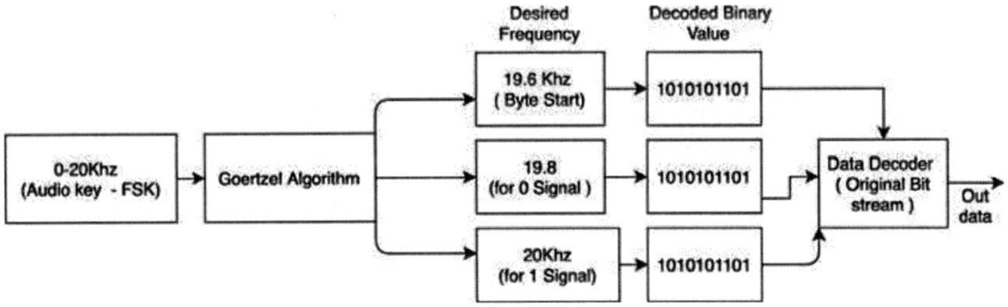

Fig. 2 Audio decoder

We have determined signal power at each particular frequency of interest by implementing the Goertzel algorithm for a baseline frequency (19,400 Hz), the byte start frequency (19,600 Hz), the 0 signal frequency (19,800 Hz) and the 1 signal frequency (20,000 Hz), which efficiently returns only the required component of the discrete Fourier transform. Frequency powers are normalized and then divided by the baseline frequency power in order to cancel out a significant amount of noise in the signal, since noise at 19,400 Hz tends to be similar to the noise at the three other frequencies of interest. We experimentally determined thresholds for which, if any, signal was present during each chunk. As the power of each signal rose dramatically when that signal was on, the threshold was generally highly effective; however, noisy environments would sometimes cause the recorded signal at high frequencies to partially drop out. The result is 0, 1, byte start or no signal for each chunk of 512 samples. Data flow during decoding is described in Fig. 3.

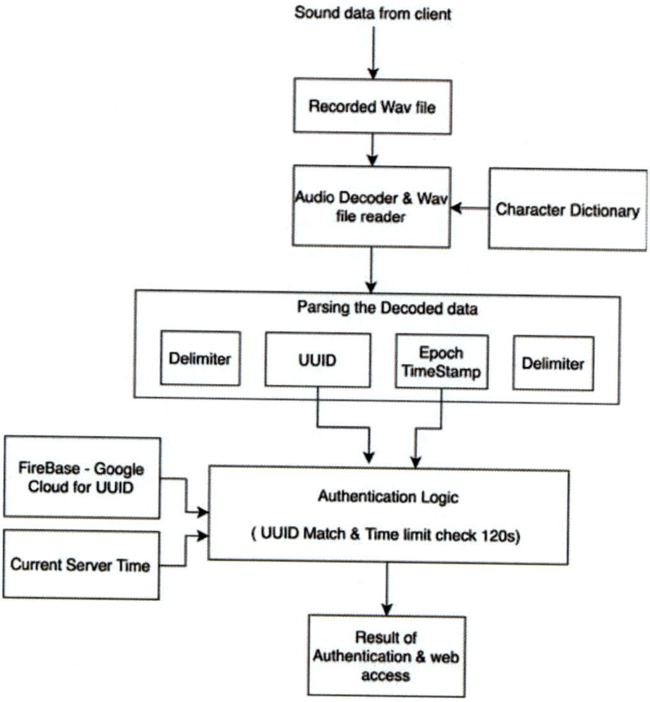

Fig. 3 Data flow during decoding

3.3 The Goertzel Algorithm

Infinite impulse response (IIR) filter structure is used to perform spectrum analysis for detection and measurement of sinusoidal tones which are sent as part of encoded audio key. The standard method for spectral energy is the discrete Fourier transform (DFT), typically implemented using a fast Fourier transform (FFT) algorithm. As our current application requires spectrum analysis only over a subset of the N bin-centre frequencies of an N-point DFT we will be using Goertzel algorithm for computing sparse FFT results using an IIR filter implementation. Goertzel algorithm is more efficient than the fast Fourier transform (FFT) in computing a point DFT if less than DFT coefficients are required.

$$y[n] = \mathrm{e}^{jw_0 n} \sum_{k=0}^{n} x[k]\mathrm{e}^{-jw_0 k}$$

- $x[k]$ is the k-th input sample in the filter window—a real number. The filter window contains previous samples, numbered $0 \ldots n$ in reverse order. In particular, $k = 0$ corresponds to the current sample.
- $y[n]$ is the output filter value at the current time point, a complex number.
- w_0 is the target frequency of the filter (the frequency you are trying to measure) in radians per second.

For our application, we do not require phase of the output. Only the magnitude, which is a real number, is enough to get the power level at a particular frequency.

$$|y[n]| = \left| \mathrm{e}^{jw_0 n} \sum_{k=0}^{n} x[k]\mathrm{e}^{-jw_0 k} \right|$$

FFT can be used to detect the presence of a single sinusoidal tone in a time-domain sequence $x(n)$. In our application, we want to detect 3 tones at 19.6, 19.8 and 20 KHz in time-domain. If we start performing a 64-point FFT, then we would examine the magnitude of the $X(n)$ sample to validate it exceeds some predefined threshold to be high bit. This method is very inefficient. In our example, we'd be performing 192 complex multiplication operations to obtain the 64-point complex $X(n)$ in order to compute the three different points $X(n)$ of interest. We would have to discard 95% of our computations results. The Goertzel algorithm is implemented in the form of a second-order IIR filter.

3.4 System Model

The database is connected to both the mobile device and the backend server. Both these devices use the user schema in the database for first-factor authentication. The desktop trying to authenticate will speak via REST APIs to the server which is the authenticator too. Sound data is recorded by the desktop from the trusted device and sent via a POST request to the server. There are three main fundamental entities associated:

- Mobile Phone: Used as the trusted device for second-factor authentication
- Computer: Device which wants to authenticate to access resources
- Remote Server: which is the authenticator

The server authenticates by comparing the ultrasonic message key captured by microphones of the computer. If the received information is decrypted and it matches with user key by the server, then the authorization is successful. This will verify that cellphone and the computer are in the same geographic location as sound needs to be transferred from a maximum distance of 2 m. Figure 4 shows the block diagram of the system architecture in detail.

Whenever a new user signs up, an UUID is created for that user along with storing his username and password. The username and password are used for the first-factor authentication, and a part of the UUID is used for the second-factor authentication.

A user enters an email and password which is authenticated against the stored email and password in firebase. In case of a match, the user is sent a view to record audio from his second trusted device. This recorded sound is sent to the two-factor authentication flow. If the recorded sound is as expected, we authorize the user to use the website. If authentication fails at any stage the user is redirected back to the login page.

Two-factor flow: After successful first-factor authentication, the user sends audio that is recorded using the mobile application. The post-request is received by the server and stored as a temporary file until the audio processing is done. The audio is processed using the decoder. The retrieved/decoded sound data is compared as per our protocol which is explained in another section. If the data matches, the user is authorized and hence can access the website or else is redirected to the signin page.

On successful authentication, the user gets access to the website.

For security reasons, the audio clip generated at time of login will be valid for a limited time. When generating the audio clip, we add a timestamp to the clip so that someone else does not record the clip and use it for their personal gain.

When any user logs in based on their unique UUID assigned by firebase the timestamp based on epoch time are combined and used as a key. Each bit has been encoded according to a hidden database so hackers won't be able to recreate the audio clips that have been generated without the knowledge of how it has been encoded. The generated clip is new every session and stored separately for every user.

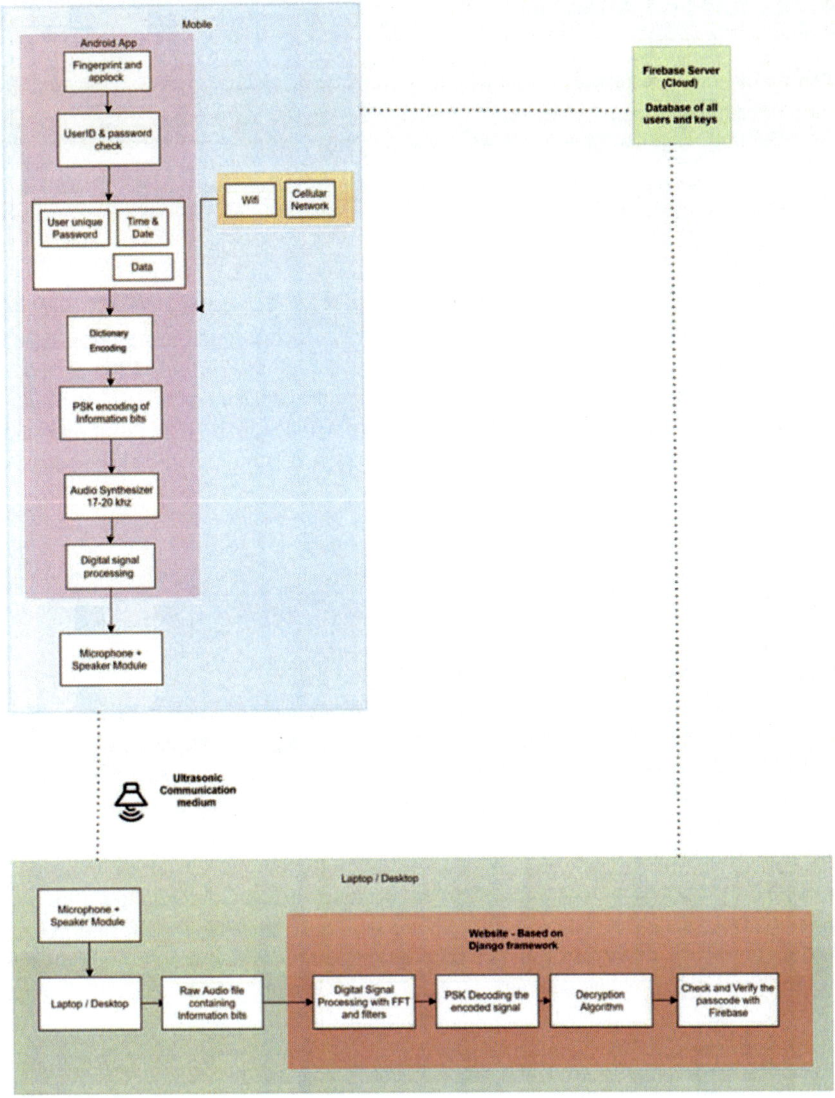

Fig. 4 Block diagram of 2FA system

4 Results and Analysis

The main intention of this paper was to provide a more secure, foolproof way to authenticate a user using ultrasound transmission. Below we have analysed the various problems that were overcome and the ones that persist.

4.1 Security Analysis

To prevent man-in-the-middle attacks, we have incorporated timestamp-based approach during encryption of the data. We have set a hard limit of 120 s for each unique audio key to be valid for. If any attacker records the sound and tries to impersonate an legit user, then the timestamp difference will be greater than the above-said limit and the attacker will not be able to use this type of attack. A log of the same is shown in Fig. 5.

4.2 Spectrum Analysis

This section shows the spectrum analysis of sound data.

Image shows both 2-D and 1-D spectrum. We can clearly see the data stream being sent across 19.6 , 19.8 and 20 KHz in the 1-D image, and we can also see train of purple marks which are constantly spaced as captured by the spectrum which translates to our data bitstream encoded using FSK technique. The noise level at 19+ KHz will be minimal which is also visible from the analysis using spectrum analyzer. There are a few factors which will cause the current design to be limited to about a byte per second of throughput are accuracy of the Goertzel algorithm and sampling rate of the recording stream. When we try to increase the bit rate higher then there will be degradation in the received signal accuracy. Figure 6 shows the plot of spectrum of received signal.

4.3 Current Limitations

We have also explored the limitations of this medium, which are quite extensive; we are limited in our range of usable frequencies as well as by the sampling rate of our devices, causing a low transmission bit rate, non-negligible error rates and non-ideal multiparty communication mechanisms despite fairly good multicast performance by its nature. Thus, while some improvements can be made, particularly

```
2021-07-17 19:03:53 Starting new HTTPS connection (1): www.googleapis.com
2021-07-17 19:04:25 testtest@gmail.com
2021-07-17 19:04:25 .
2021-07-17 19:04:25 $
2021-07-17 19:04:25 P
2021-07-17 19:04:25 W
2021-07-17 19:04:25 A
2021-07-17 19:04:25 g
2021-07-17 19:04:25 d
2021-07-17 19:04:25 1
2021-07-17 19:04:25 6
2021-07-17 19:04:25 2
2021-07-17 19:04:25 6
2021-07-17 19:04:25 5
2021-07-17 19:04:25 2
2021-07-17 19:04:25 8
2021-07-17 19:04:25 3
2021-07-17 19:04:25 7
2021-07-17 19:04:25 7
2021-07-17 19:04:26 $
2021-07-17 19:04:26 .
2021-07-17 19:04:26 .$PWAgd1626528377$.
2021-07-17 19:04:26 ----------------------------------------------------
2021-07-17 19:04:26 THE RECIEVED SOUND DATA IS = .$PWAgd1626528377$.
2021-07-17 19:04:26 ----------------------------------------------------
2021-07-17 19:04:26 THE DELIMITED SOUND DATA IS = [1, 17]
2021-07-17 19:04:26 SUCCESS DATA RECIEVE
2021-07-17 19:04:26 ---------LOG OF DATA ---------------------
2021-07-17 19:04:26 Index Delimeter is =  [1, 17]
2021-07-17 19:04:26 THe raw recieved Sound data is =  .$PWAgd1626528377$.
2021-07-17 19:04:26 The parsed sound data is =  PWAgd1626528377
2021-07-17 19:04:26 Auth key generated Epoch time  1626528377
2021-07-17 19:04:26 Current Epoch time =  1626528866
2021-07-17 19:04:26 Time difference = (limit is 120)  489
2021-07-17 19:04:26 Unique userID:  PWAgd
2021-07-17 19:04:26 Decoded uuid is =  PWAgd
2021-07-17 19:04:26 ----------------------------------------
2021-07-17 19:04:26 FAIL--Please check if Data conditions were matched
2021-07-17 19:04:26 Starting new HTTPS connection (1): www.googleapis.com
2021-07-17 19:04:26 Starting new HTTPS connection (1): www.googleapis.com
```

Fig. 5 Back-end server logs when (time gap greater than 120 s)

to the reconstruction algorithms used by our decoder, we believe that it is unlikely that long-range, high throughput, low-error ultrasonic communication using common consumer electronics with current mobile speaker and microphone technology will be useful.

5 Conclusion

We have successfully demonstrated that standard consumer electronics like laptops, computers and mobiles are capable of both transmitting and receiving data encoded at ultrasonic frequencies, at least up until around 20,000 Hz. We have demonstrated this feature of ultrasonic data transfer can be effectively used for two-factor authentication. We created a web-based hosting of our website and demonstrated the end-to-end flow of the product. We have created a new ultrasonic communication protocol with secure features like UUID-based passcode and timestamp-based protection for audio

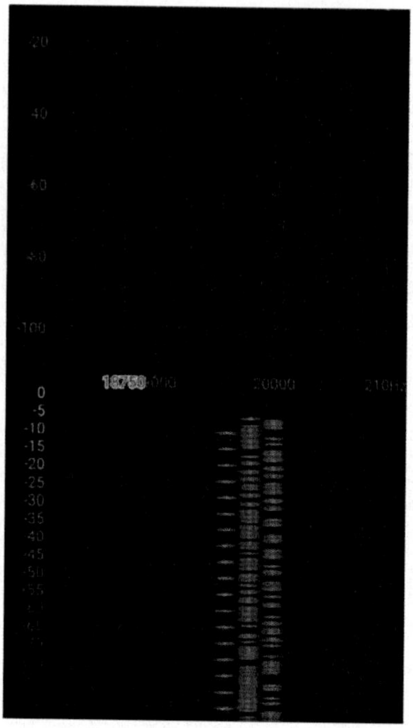

Fig. 6 Image of 2-D spectrometer

key. While our analysis is by no means exhaustive, we have demonstrated that ultrasonic communication in consumer devices is feasible and thus presents an interesting and dynamic problem space for future commercial and academic research.

References

1. Tsaty Y (2012) System and method for using ultrasonic communication. US patent 8854985B2 filed on 5th July 2012
2. Bonneau J, Herley C, Van Oorschot PC, Stajano F (2012) The quest to replace passwords: a framework for comparative evaluation of web authentication schemes. In: 2012 IEEE Symposium on security and privacy (SP), 20–23 May 2012
3. Zhu X, Yu S, Pei Q (2016) QuickAuth: two-factor quick authentication based on ambient sound. In: Proceedings of the IEEE global communications conference (GLOBECOM), Dec 2016, pp 1–6
4. Wang M, Zhu W-T, Yan S, Wang Q (2018) SoundAuth: secure zero-effort two-factor authentication based on audio signals. In: Proceedings of the IEEE conference on communications and network security (CNS), May 2018, pp 1–9
5. Karapanos N, Marforio C, Soriente C, Capkun S (2015) Sound-proof: usable two-factor authentication based on ambient sound. In: Proceedings of the 24rd USENIX security symposium (USENIX security), pp 483–498
6. AlQahtani AAS, Alamleh H, Gourd J (2020) 0EISUA: zero effort indoor secure user authentication. Access IEEE 8:79069–79078

7. Fujii H, Tsuruoka Y (2013) SV-2FA: two-factor user authentication with sms and voiceprint challenge response. In: 2013 8th International conference for internet technology and secured transactions (ICITST), 9–12 Dec 2013
8. Schürmann D, Sigg S (2013) Secure communication based on ambient audio. IEEE Trans Mob Comput 12:358–370
9. Getreuer P, Gnegy C, Lyon RF, Saurous RA (2018) Ultrasonic communication using consumer hardware. IEEE Trans Multimedia 20(6):1277–1290

Long Short-Term Memory (LSTM) Layers as a Proposed Learning Algorithm for Rainfall Prediction

Diptiman Senapati, Manali Narendra, Ankit Kumar, and Smita Rath

Abstract Rain is an important component of the overall cycle since it is capable of accumulating the majority of the freshwater on the planet. Hence, all variations in the weather have an impact on the agriculture as well as on the economy of the country. India being a farmer's-dependent economy indirectly depends on good rainfall. India gets a yearly precipitation of roughly 4000 billion cubic meter (BCM) which include snowfall, with an average rainfall of about 1170 mm. In the current times, climate change is a matter of major concern. Autoregressive integrated moving average (ARIMA) and point spread function (PSF) were used. But traditional methods have some limitations, are expensive and require large data set. Predictions are created by mathematical model defined by a set of equations that predict future rainfall with the help of a dynamical strategy. The paper performs a comparison between multiple linear regression (MLR), K-nearest neighbour (KNN), backpropagation with ANN and long short-term memory (LSTM) on the prediction of rainfall. Statistical methods are performed such as MAE, RMSE, MSE to compare the different prediction model and to choose the best suitable one. The test cases provided a lower MAE, RMSE and MSE value in case of LSTM model as it can memorize how elements in a series are ordered in a time series data. Instead of having the information which was before and unchangeable, LSTMs have the capacity to identify the context necessary to make forecasts in time series forecasting scenario.

Keywords Multiple linear regression · Artificial neural network · K-nearest neighbour · Long short-term memory · Precipitation

---

D. Senapati (✉) · M. Narendra · A. Kumar · S. Rath
Department of Computer Science and Engineering, Siksha 'O' Anusandhan Deemed to be University, Bhubaneswar, Odisha, India
e-mail: diptiman.sen1999@gmail.com

S. Rath
e-mail: smitarath@soa.ac.in

M. S. Kaiser et al. (eds.), *Information and Communication Technology for Competitive Strategies (ICTCS 2021)*, Lecture Notes in Networks and Systems 401,
https://doi.org/10.1007/978-981-19-0098-3_25

1 Introduction

Rainfall is essential for both plant and animal existence. It is responsible for bringing groundwater to the earth's surface. Water scarcity occurs when rainfall is low, which can lead to a drought-like situation. Rainfall in India is largely limited to the summer season. During the monsoons, India receives over 70% of its average rainfall, which has an impact on the productivity of summer crops including rice and pulses. India receives a yearly precipitation of roughly 4000 billion cubic meter (BCM) including snowfall, with an average rainfall of roughly 1170 mm. In the current times, climate change is also a matter of major concern. To keep droughts, floods and farmers yearly routine task in check, it is important to predict rainfall accurately, take precautions and plan the actions accordingly.

Earlier, statistical methods like autoregressive integrated moving average (ARIMA), point spread function (PSF) were used. Traditional methods have some erroneous disadvantages, are expensive and require large data set. As a result, scientists today employ techniques to ensure precision and accuracy. Regression, artificial neural networks and other empirical approaches to rainfall prediction are the most extensively utilized. Methodologies based on a series of formulas that forecast future rains through a dynamical approach provide predictions with the help of algorithms such as fuzzy logic and cluster data management technique. Earlier statistical methods like autoregressive integrated moving average (ARIMA), point spread function (PSF) require large data set and may give erroneous result if there is even a little disturbance or noise in data. Hence, there is a limit up to which traditional methods can predict accurate result. Parmar et al., suggested artificial neural network as better method to predict rainfall [1]. Rainfall data is very volatile and can have noise which may result in erroneous data. Cramer et al. suggest that machine learning validation provides more accurate outcomes of rainfall predictions after testing with extended Markov chain model and six other widely known machine learning techniques: genetic programming, support vector regression, etc. [2]. Basha et al. propose machine learning algorithms like artificial neural network, logistic regression, self-organizing maps and support vector machine, instead of traditional methods to get accurate predictions [3]. Diez-Sierra et al, investigated eight statistical and machine learning approaches for daily rainfall prediction based on weather patterns [4]. To protect the economy and human life from the destructions caused by draughts and floods, rainfall prediction is very crucial. Refonaa et al. suggests that due to dynamic nature of rainfall data statistical methods of prediction fail. Hence, he suggests artificial neural network-based prediction using Keras package in python [5].

Mosavi et al., suggest that hybridization of machine learning algorithms can increase the outcomes of the forecasting. According to them, the most effective tactics for improving ML systems for predicting rainfall more reliably are hybridization, data decomposition, algorithm ensemble and model optimization [6]. Thirumalai et al. recommended linear regression to help farmers in harvesting crops [7].

Hernández et al. suggest that deep learning techniques give better results than artificial intelligence methods [8]. Rainfall data is nonlinear in nature and hence may give erroneous results while predicting. Abdel-Kader et al. suggest PSO and multilayer perceptron (MLP) [9]. Dash et al. suggested that artificial intelligence methods like KNN and ANN give more accurate result with minimal prediction error score [10]. To predict rainfall, Poornima and Pushpalatha propose using an intensified long short-term memory (intensified LSTM)-based recurrent neural network (RNN) [11] Hasan et al. recommended fuzzy rule-based KNN for rainfall prediction as this gives less erroneous, accurate and precise result when compared to other algorithms [12].

The major objective of this study is to examine and investigate several machine learning methods for predicting rainfall that provides the most accurate and exact results. RMSE, MSE value are calculated to compare the different prediction model and to choose the best suitable one. The following sections provide a detail discussion on MLR, KNN and hybrid model of ANN, LSTM methods and give a graphical pictorial view of the actual value and target value closest to the actual value.

2 Machine Learning Approaches in Prediction of Rainfall

As stated above the idea behind this research-driven project is to test various machine learning algorithms with rainfall data and select the most suitable one which gives least error and most accurate prediction. The tested algorithms are.

2.1 Multiple Linear Regression

Multiple linear regression is a popular regression approach that models the linear connection between a single continuous dependent variable and multiple independent variables. In regression analysis, there are just two parameters, one of which is the dependent factor and the other of which is the independent factor and the relationship between them is as follows. The predictable model can be stated as follows:

$$Y_t = S + D(X_t) \tag{1}$$

Here Y_t = dependent variable, X_t = independent variable, $SandD$ = regression factors.

Usually, there are more than two parameters in multiple regressions, one of which is the dependent factor and the other are independent factors and the formula looks like this (2):

$$Y_t = \beta_{1i} + \beta_2 x_{i1} + \beta_3 x_{i2} + \beta_4 x_{13} + \cdots + \beta_p x_{ip} \tag{2}$$

The attribute is collected from the training data, and the variable is retrieved from the database using correlation to build the correlation matrix. The value r is called the degree and direction of the association among the two factors are measured by the correlation coefficient, where the equations are as follows:

$$r = \frac{n\sum xy - (\sum x)(\sum y)}{\sqrt{n(\sum x^2(\sum x)^2} - \sqrt{n(\sum y^2(\sum y)^2}} \quad (3)$$

If the regression line goes through every point on the dispersed plot, it will be able to describe all of the variance.

$$\text{R-squared} = \text{Variation defined/Total variation} \quad (4)$$

A higher R-squared indicates that the two variables have a linear connection. If R-squared = 1, the link among the two factors is flawless. This provides an indicator of variance of the forecast.

2.2 *K-Nearest Neighbour*

K-nearest neighbour is a monitored machine learning approach that uses a classification algorithm. The KNN method saves all available data and classifies a new data point based on its similarity to the past data. This means that utilizing the KNN technique, new data may be swiftly sorted into a well-defined group. This algorithm is a nonparametric method, which means it makes no assumptions about the underlying data.

2.3 *Artificial Neural Network Using Backpropagation*

The multi-layer feedforward network, which has one input layer, hidden layers, and one output layer, is used to create backpropagation with neural network (BPNN). Figure 1 shows a BPNN design with one hidden layer. The following steps are performed on the data sets.

(a) For input training, feedforward neural network (FFNN) is used. In this step, weights and nodes are adjusted.
(b) Find the error within the given rainfall value and forecasted value, and
(c) Updating of weights in backward phase.

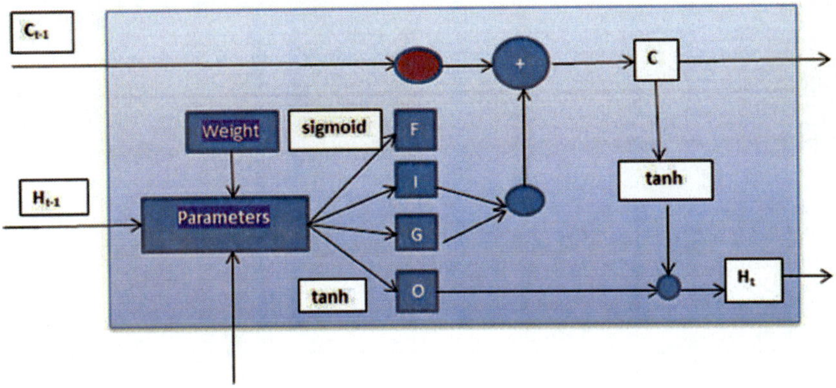

Fig. 1 Long short-term memory layout structure

2.4 *Long Short-Term Memory*

Long short-term memory (LSTM) is a deep learning model based on a recurrent neural network (RNN). Because there might be delays of undetermined duration between critical occurrences in a time series, LSTM networks seem to be well in classifying, analysing and generate forecasting based on time series data. The architecture depicts the working of LSTM in predicting rainfall in Fig. 1.

LSTM follows four sequences of different gates as defined below:

(1) Forget Gate (Fi): It finds how much amount of previous data is needed to forget.
(2) Input Gate (Inp): It finds the amount of data flowing onto the internal cell state.
(3) Modulation of Input Gate (Gi): It is frequently seen as a sub-component of the input gate, and much LSTM literature ignores it entirely, assuming that it is included inside the input gate. It is utilized to provide nonlinearity to the knowledge that the input gate will transfer onto the internal state unit, rendering it zero-mean. This is intended to shorten the learning period since zero-mean input converges faster.
(4) Output Gate (Oi): It decides what outcome (next hidden state) the present internal control action should produce.

This research has been done with the intention to test various machine learning algorithm and propose the one model that predicts rainfall with least error and most precision as shown in Fig. 2.

3 Results and Discussion

Data set contains the weather data which has been taken from Australia for the year 2012–2018 as well as from India for the year 2019. The various columnwise

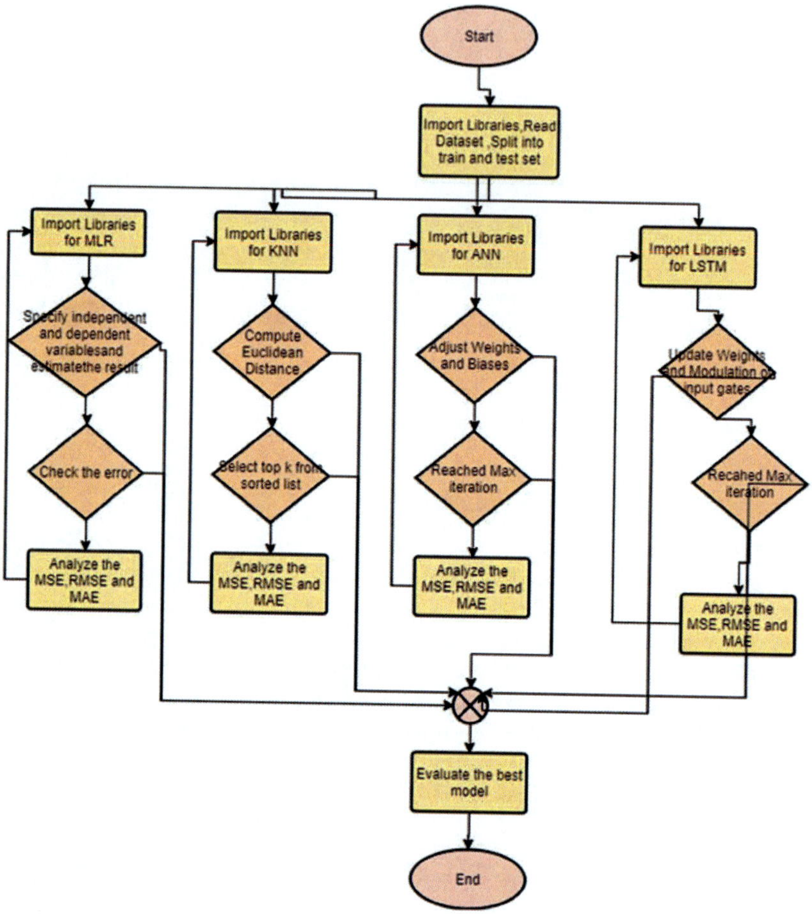

Fig. 2 Contrast study of machine learning techniques for prediction of rainfall

information or the inputs used from the data set are Name, Date time, Maximum Temperature, Minimum Temperature, Temperature, Heat Index, Snow, Snow Depth, Wind Speed, Wind Direction, Visibility, Cloud Cover, Relative Humidity and Precipitation. The displayed results with calculated value of MAE, MSE and RMSE prove the correctness of the program.

3.1 Long Short-Term Memory (LSTM) Layers as a Proposed Learning Algorithm

The suggested model is built on LSTM networks, and it analyses temporal weather data to find trends and forecasts. By adding memory units, the LSTM's fundamental ideas have the capacity to develop long-term dependencies. The network can learn, forgets previously hidden states and updates hiding states using the memory units.

3.1.1 Working of Long Short-Term Memory

The long short-term memory (LSTM) is a machine learning method capable of simulating problems with several independent factors in a relatively smooth manner. The recurrent neural network is made up of internal gates. The internal gates of the system, unlike those of other recurrent neural networks, enable the model to be effectively trained using Backpropagation over time, or BPTT, and escape the diminishing gradients problem.

The LSTM () class in the Keras deep learning package may be used to generate LSTM layers. It defines the number of memory units inside a layer of LSTM memory units by establishing a layer of LSTM memory units. Every component or cell in the layer has an inner cell state (abbreviated as "c") and an output hidden state (abbreviated as "h"). At each input, each LSTM unit will produce one hidden state h. The hidden state is the result of a LSTM layer or set of cells. The cell state, or c, is an internal state retained by each LSTM cell unit. The looping conditions are also checked with varying constraints during the execution of the program to find the lowest RMSE and MSE values to calculate the gap between real and projected values of rainfall in both the regions. Tables 1 and 2 provide a brief description about the accuracy of the model in terms of statistical measures.

Figures 3, 4, 5 and 6 provide a pictorial view of the actual and predicted values of random days taken from the test set that shows the amount of rainfall received.

Table 1 Results of accuracy tested on Australia data set using MLR, KNN, ANN and LSTM

Models used	Mean absolute error (MAE)	Mean squared error (MSE)	Root mean square error (RMSE)
Multiple linear regression	5.61798	110.7308	10.5228
K-nearest neighbours (KNN)	3.2280	107.4227	10.3644
Neural network with backpropagation	7.1207	971.5191	31.1692
Long short-term memory (LSTM)	2.0877	29.13104	5.39731

Table 2 Results of accuracy tested on Indian data set using MLR, KNN, ANN and LSTM

Models used	Mean absolute error (MAE)	Mean squared error (MSE)	Root mean square error (RMSE)
Multiple linear regression	8.634408	176.906907	13.3006356
K-nearest neighbours (KNN)	6.033420	149.752596	12.237344
Neural network with backpropagation	9.661610	281.076238	16.765328
Long short-term memory (LSTM)	4.426213	115.11469	10.729152

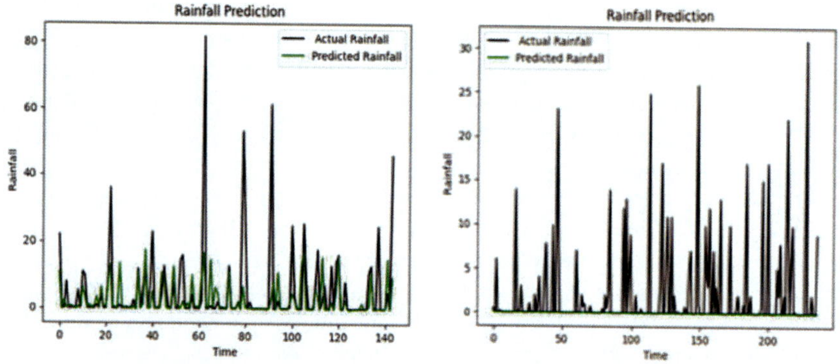

Fig. 3 Actual and projected graph of India and Australia (LSTM proposed model)

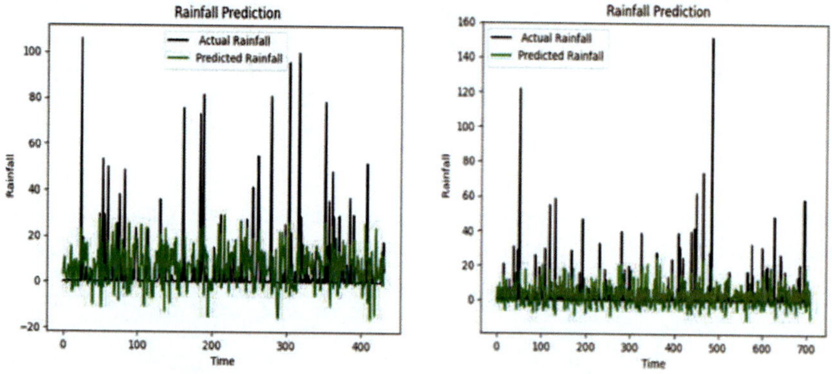

Fig. 4 Actual and projected graph of India and Australia (MLR algorithm)

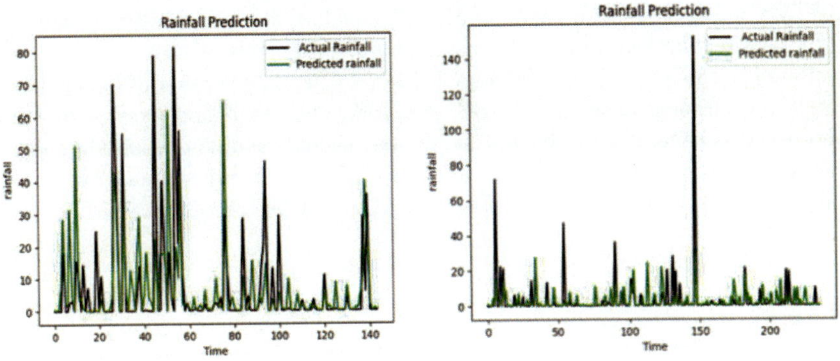

Fig. 5 Actual and projected graph of India and Australia (KNN algorithm)

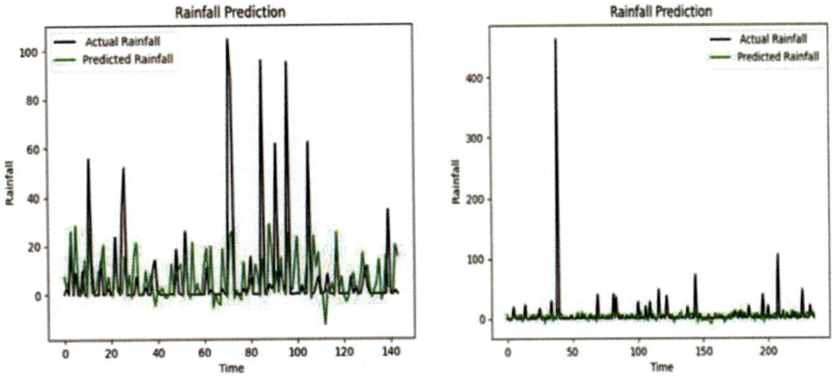

Fig. 6 Actual and projected graph of India and Australia (ANN with backpropagation)

The MAE, RMSE and MSE values are lower in case of LSTM model as compared to MLR, KNN and ANN with backpropagation model.

4 Conclusion and Future Work

The goal of this study is to evaluate and present a more accurate method to forecast rainfall. This will help in the above-stated scenarios that is the farmers and the NDRF to protect human life and economy by taking necessary precautions. LSTM overcomes the limitations of short-term memory of RNN and vanishing gradient problem in neural network. After testing Australia (2012–2018) and India (2019), rainfall data using multiple linear regression (MLR), K-nearest neighbour (KNN), artificial neural network using backpropagation (ANN) and long short-term memory

(LSTM), and results of MSE, MAE and RMSE conclude that long short-term memory (LSTM) gives the best result with least error and most precision.

Based on the research methodology used in this work, extension of the methods can be done through ensemble of LSTM network. The weight and biases within the network also affect the accuracy of the model. Therefore, optimization algorithms can be used to enhance the effectiveness of the model. For improved future outcomes, a hybrid network of neural learning strategies can also be considered.

References

1. Parmar A, Mistree K, Sompura M (2017) Machine learning techniques for rainfall prediction: a review. In: International conference on innovations in information embedded and communication systems, vol 3
2. Cramer S, Kampouridis M, Freitas AA, Alexandridis AK (2017) An extensive evaluation of seven machine learning methods for rainfall prediction in weather derivatives. Exp Syst Appl 85:169–181
3. Basha CZ, Bhavana N, Bhavya P, Sowmya V (2020) Rainfall prediction using machine learning and deep learning techniques. In: 2020 international conference on electronics and sustainable communication systems (ICESC), July 2020. IEEE, pp 92–97
4. Diez-Sierra J, del Jesus M (2020) Long-term rainfall prediction using atmospheric synoptic patterns in semi-arid climates with statistical and machine learning methods. J Hydrol 586:124789
5. Refonaa J, Lakshmi M, Dhamodaran S, Teja S, Pradeep TNM (2019) Machine learning techniques for rainfall prediction using neural network. J Comput Theor Nanosci 16(8):3319–3323
6. Mosavi A, Ozturk P, Chau KW (2018) Flood prediction using machine learning models: literature review. Water 10(11):1536–1539
7. Thirumalai C, Harsha KS, Deepak ML, Krishna KC (2017) Heuristic prediction of rainfall using machine learning techniques. In: 2017 international conference on trends in electronics and informatics (ICEI). IEEE, pp 1114–1117
8. Hernández E, Sanchez-Anguix V, Julian V, Palanca J, Duque N (2016) Rainfall prediction: a deep learning approach. In: International conference on hybrid artificial intelligence systems, April 2020. Springer, Cham, pp 151–162
9. Kader A, Hatem SM, Mohamed M (2021) Hybrid machine learning model for rainfall forecasting. J Intell Syst Internet Things 1(1):5–12
10. Dash Y, Mishra SK, Panigrahi BK (2018) Rainfall prediction for the Kerala state of India using artificial intelligence approaches. Comput Electr Eng 70:66–73
11. Poornima S, Pushpalatha M (2019) Prediction of rainfall using intensified LSTM based recurrent neural network with weighted linear units. Atmosphere 10(11):668–669
12. Hasan MZ, Hossain S, Hasan KZ, Uddin MS, Alam ME (2021) Fuzzy rule-based KNN for rainfall prediction: a case study in Bangladesh. In: Proceedings of international joint conference on advances in computational intelligence. Springer, Singapore, pp 509–524

Virtual Stocks: Stock Market Simulator

Khushi Khanchandani, Neha Patil, and Vineeta Bhujle

Abstract In today's world, investment is a must skill in anyone's repertoire, and unfortunately, it is also a costly and difficult skill to master. Neophytes need a platform where they can acquire their investment skills without losing their money. Virtual Stocks is a web application that provides a modern and easy-to-use solution to this problem. This paper describes the need and the development process of this application. The colossal volume of financial news available in the form of unstructured data was harnessed which yielded the market sentiment that assisted users in making prudent decisions. It was achieved using VADER which is an efficient sentiment analysis algorithm. The application was successfully developed with extensive study, comparison and testing, thus rendering an effective solution for a sanguine learning of investment skills.

Keywords Investment · Virtual · Modern web application · Financial news · VADER · Market sentiment analysis · Testing

1 Introduction

Stock markets serve the purpose of trading companies publicly which raises capital. Burgeoning businesses, abound jobs and expanding operations are effects of stock market trading. Investment is a skill now to be mastered because it lets you trade stocks at any point of time and possibly gain profits. It is a popular secondary source of income in today's world. It takes a lot of experience to be skilled, and unfortunately, innocuous mistakes here cost people their money. Callow investors are prone to

K. Khanchandani (✉) · N. Patil · V. Bhujle
KJ Somaiya College of Engineering, Vidyavihar, Mumbai, India
e-mail: khushikhanchandani@somaiya.edu

N. Patil
e-mail: patil.nv@somaiya.edu

V. Bhujle
e-mail: vineeta.b@somaiya.edu

M. S. Kaiser et al. (eds.), *Information and Communication Technology for Competitive Strategies (ICTCS 2021)*, Lecture Notes in Networks and Systems 401,
https://doi.org/10.1007/978-981-19-0098-3_26

irrational decision making due to emotional, familiarity and attention biases, loss aversion, the disposition effect, etc.

Our solution is meant to be a panacea to this prevailing problem. Virtual Stocks, a web application is a stock market simulator that allows users to trade with virtual money. It uses real stock data to give a fair experience helping users develop their skills without losing any money. It will help users gain confidence and experiment different strategies leading to prowess in investment. This mobile responsive web application also assists users by providing them a market sentiment of the stock, thus leading to a more deliberate decision making. The users can also analyze their progress with graphs, their portfolio and a leaderboard. We have focused on keeping the application modern and user friendly.

Market sentiment reflects the standpoint of investors toward the stock market. If the sentiment about a stock is mostly positive, then more people buy the stock and thus its price soars. Likewise, if the sentiment of the stock is mostly negative, then more people sell the stock, and thus its price dwindles [6]. This crowd psychology is dictated by the price movement and the news of the moment. This news is dispersed across many unstructured resources such as articles, tweets, and other social media. Winnowing through such a huge volume of data is complex and resource intensive and requires expertise [9]. Sentiment analysis automates this arduous process and yields vital information. We have used VADER algorithm for market sentiment analysis in our application. It is an excellent classifier which is easy to use and does not require any training data.

2 Background Work

Arul Agarwal in "Sentiment Analysis of Financial News" [1] has suggested a way to show how sentiment analysis has an effect on the stock prices. It utilizes VADER library in NLTK package in Python. VADER uses a lexicon-based approach to classify news sentiments. News headlines are classified into negative, positive and neutral sentiment. A compound score is generated based on the percentage of each sentiment in the text. The author finds a strong correlation between news sentiment and change in stock price. The drawback of the paper is that it has not used live stream data.

Zenon Chaczko in "A Virtual Stock Exchange Simulator for Education and Training" [10] has made an application, StockEt, which can be used to educate people on investment. It uses Yahoo finance and Google stock API to get near real-time data. The application has features like maintaining a portfolio for a user, leaderboard and provision to download stock market data. It is built on php, C# and MySQL. The disadvantage of this model is that it uses relative strength index (RSI) to analyze the stock, but RSI alone is not a good indicator for the users to make predictions. Their application is also not mobile responsive.

Laszl ´ o Nemes and A. Kiss in "Prediction of stock values ´ changes using sentiment analysis of stock news headlines" [5] has compared VADER, BERT, Textblob and recurrent neural network for sentiment analysis. The main focus was to reduce the neutral values as it increases the inaccuracy of stock market comparisons. The authors proceed to conclude that VADER and RNN gave finer results than TextBlob and BERT. However, in RNN training, the model is difficult and computing the results takes time.

Satyabrata Aich in "Analyzing stock price changes using event related Twitter feeds" [8] has investigated the relationship between event-related Twitter feeds and change in stock price. They have used Tweepy (a Python library) for accessing Twitter API and TextBlob (Python library) for sentiment analysis. They targeted the company Samsung, and the keyword that showed up the most was "Galaxy Note 7" and "Samsung". They found a positive correlation between sentiment score and change in stock prices. The only disadvantage of this study is that only 5 days Twitter data was considered which is inadequate for in-depth analysis.

Dev Shah in "Predicting the Effects of News Sentiments on the Stock Market" [2] has developed a dictionary-based sentiment analysis model with an accuracy of 70.59% in predicting the effect of news sentiment on stock market prices. This is carried out on pharmaceutical stocks. They have scraped news from sites like money control and converted data into ngram format. They have manually made a dictionary which contains words relevant to pharmaceutical industries. These words were mapped to positive, negative and neutral sentiment. The sentiment score was then calculated based on the frequency of positive and negative words. If the score lies above a particular threshold, then it is classified as sell.

Jessica in "Sentiment Analysis on Official News Accounts of Twitter Media in Predicting Facebook Stock" [7] has tried to implement sentiment analysis on tweets taken from Twitter accounts of official news media channels for predicting Facebook stock price. Random tweets are not taken into consideration as it might be misleading. The moving average method of 5, 10 and 15 days is applied to compute the sentiment score. The main focus of this paper is to appraise the contribution of news media channels content to predicting the stock price. The result shows that sentiment results of the CNBC channel are giving more accurate scores than others.

3 Scope

Our project is designed to provide the user an edifying experience by simulating a stock trading environment. It aims to educate the user using an interactive UI. It also provides a market sentiment analysis to assist users to determine the stock behavior.

3.1 Functional Requirements

- User can register himself and get an initial amount of 50 k at the start.
- User can add a stock to the wish list to observe it.
- User can search for a stock, view its details and buy or sell it.

- User can view the leaderboard which ranks all the users on maximum net profit.
- User can keep a tab on his account and view graphs to analyze his/her performance.
- User can view all the transactions made by him/her.
- User can view the top gainers and top losers in the market.
- User can see the details of all stocks he has bought in portfolio section.
- User can see the market sentiment graph in the stock details page.

3.2 *Non-functional Requirements*

- To ensure confidentiality, the passwords are hashed by Django and is not visible to any of the admins higher in the hierarchy.
- Token authentication is implemented.
- All input fields are validated to avoid any form of unsanitary input.
- Leveraged single page application feature of Angular to not reload entire application on every URL change, thus improving performance.
- Minimalistic UI for ease of use.

4 System Architecture

Figure 1 delineates the system architecture of our application. We have used JavaScript development framework, Angular10 for front end and Python development framework, Django for back end. The data is transferred between the two using

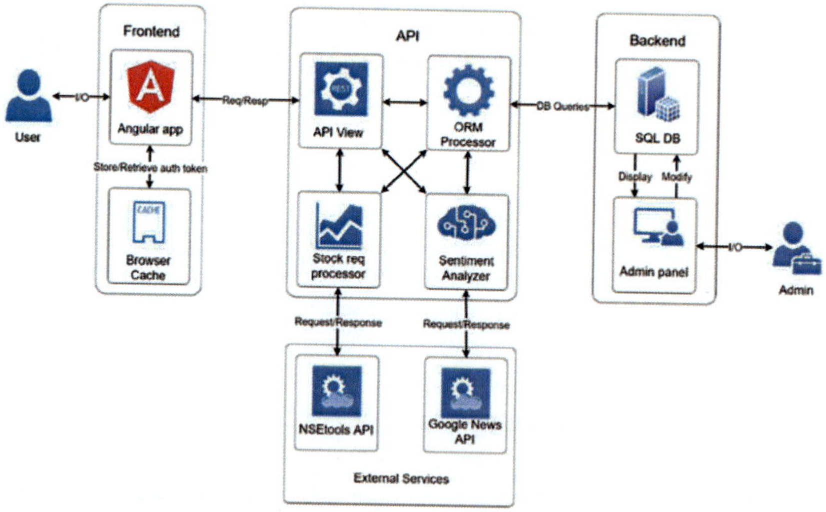

Fig. 1 System architecture

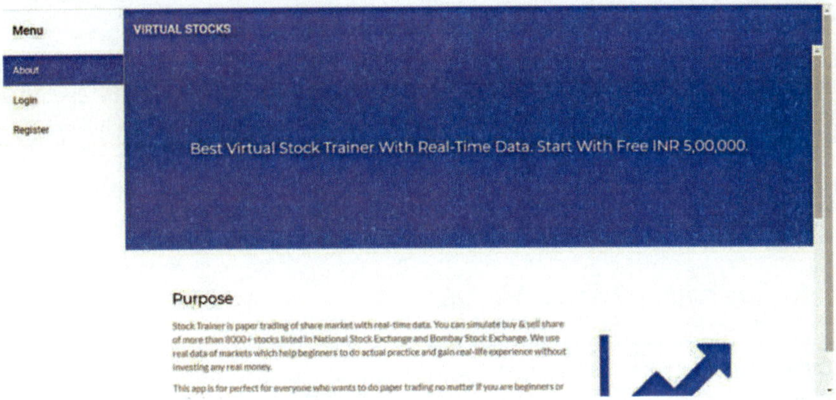

Fig. 2 Main page

REST API. We have used a relational database—SQLite. Two external services have been consumed: NSEtools API for the stocks data and Google News API for the market sentiment analysis. These modern technologies support modular application, rapidity and constancy.

5 System Modules

5.1 Main Screen

The user lands on this screen which contains information about the application. If the user is not logged in, he can see the login and register button. If the user is logged in, he can see the button to navigate to other modules in the application in the side navigation bar. Figure 2 is a screenshot of this screen.

5.2 User Registration/Login

The user has to register if he has visited the portal for the first time. He is asked for email and password for registration purposes, and both these fields are validated. User has to use the same credentials to log in to his account. A cookie is stored in the browser cache to save further sessions.

5.3 *Search Stock and Stock Detail Module*

The user can search any BSE stock and he can select a stock, and all the details of the stock are displayed using NSEtools API. The user can also view an informative graph to delineate the stock over time. The user can see the market sentiment graph of the stock for a stipulated period. Figure 8 is the screenshot of this module.

5.4 *Buy and Sell Stock Module*

The user can buy and sell stocks. The user cannot buy if he does not have enough funds in his portfolio or he cannot sell the stocks if he has not bought the required number of stocks already.

5.5 *Wish List*

The user can add stocks he is interested in to a wish list and observe it. He can also delete the stocks from the wish list. Figure 4 is a screenshot of this module.

5.6 *Leaderboard*

The user can see where he stands among all the users. The ranks are calculated according to the total net profit one has made from the start. Figure 5 is a screenshot of this module.

5.7 *Losers and Gainers*

This module displays the top ten losers and gainers of the BSE. Figure 6 is a screenshot of this module.

5.8 *Profile Module*

This module displays the account details like the net profit of the individual, the amount of funds left and the amount invested. It also displays graphs to analyze the user's progress. Figure 7 is a screenshot of this module.

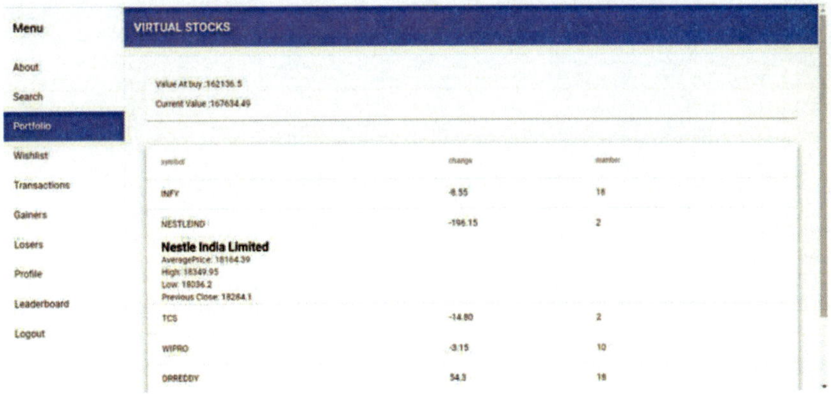

Fig. 3 Portfolio

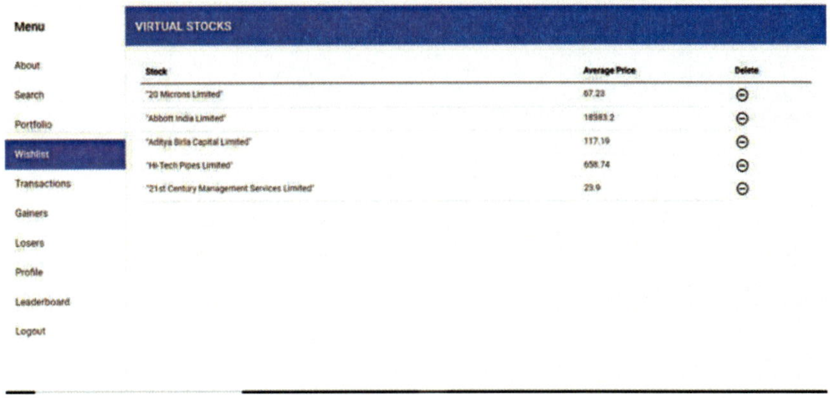

Fig. 4 Wish list

5.9 *Portfolio Module*

This module displays all the shares the user has bought, the quantity of shares, details of each share and the net profit or loss the user has on the share. Figure 3 is a screenshot of this module.

5.10 *Transaction Module*

This module displays details of all the transactions carried out like the time at which a share was bought, type of transaction, etc.

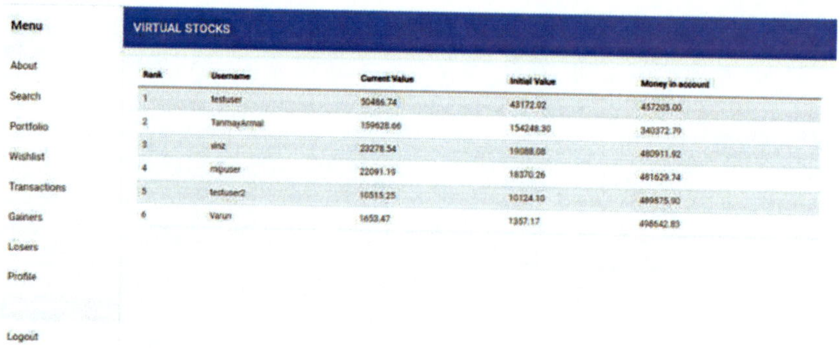

Fig. 5 Leaderboard

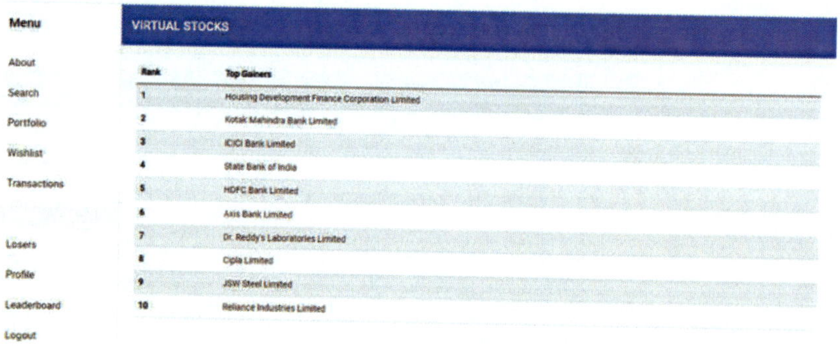

Fig. 6 Gainers

6 Algorithm

VADER algorithm has been used for market sentiment analysis to assist users make prudent decisions about their next action [3]. Valence aware dictionary for sentiment reasoning (VADER) is a NLTK model that recognizes the polarity (positive/negative/neutral) as well as the intensity of the sentiment. It uses a dictionary that maps lexical features to emotional intensities known as sentiment scores. The sentiment score of a text is calculated by adding the intensities of all words in the text [4]. Figure 9 depicts the output of this algorithm. Following are the advantages of using this model:

- It does not require any training data, thus working perfectly on real unlabeled data like the news source.
- It is attuned to social media texts because of its capability to recognize emoticons, slangs, punctuations, etc.

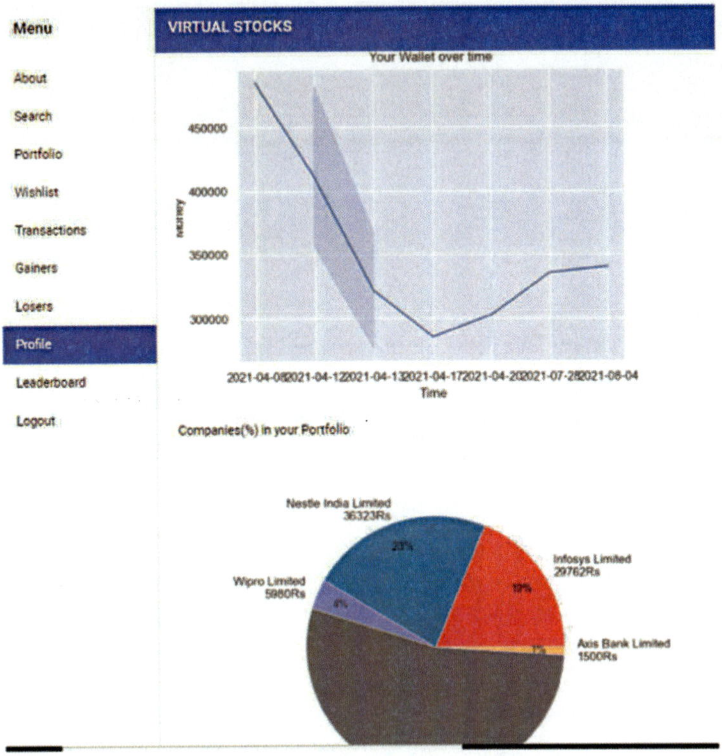

Fig. 7 Profile

- VADER can work with multiple domains.

7 Results

Testing of our application for performance was done using Apache JMeter. Twenty threads with a loop of 4 times giving up to 80 samples of each test were created. Figure 12 is a tabular summary report of the testing. We can understand from this graph that we have an average throughput of 22 request per minute and a zero percent error rate in all the modules. Figure 10 is an aggregate graph which is a visual representation of average response time (MS) versus different actions. Figure 11 is a response time graph which indicates the time taken by each test case which indicates that maximum time taken to load a module is 16 s.

The accuracy of VADER and TextBlob has been tested. Results showed VADER finished computation in 3.1 s, while TextBlob did the same in 6.5 s which verifies that VADER is considerably faster than TextBlob. The output is presented in Fig. 13.

Fig. 8 Stock details

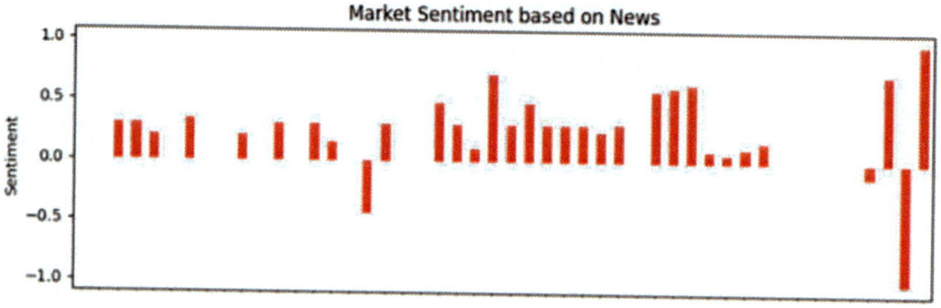

Fig. 9 Market sentiment graph

Our project has graphs showing market sentiment related to each stock which will help users make a prudent decision as market sentiment is a good indicator of stock price change. It also has graphs that helps the user analyze his trading history.

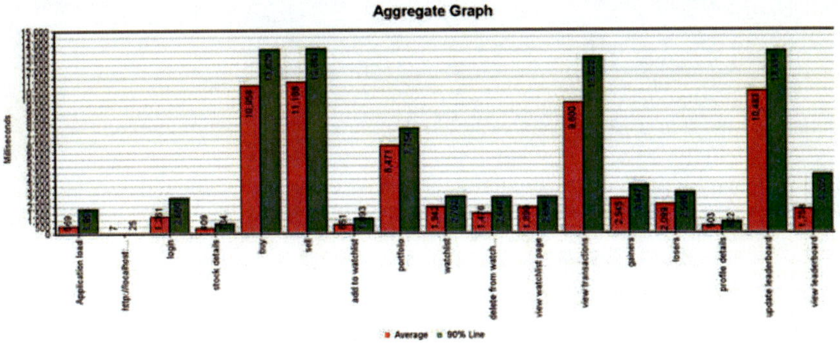

Fig. 10 Aggregate graph

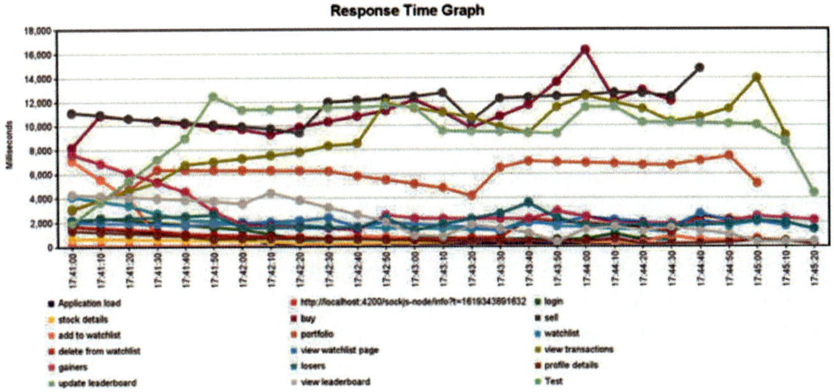

Fig. 11 Response time graph

Label	# Samples	Average	Min	Max	Std. Dev.	Error %	Throughput	Received KB/s	Sent KB/sec	Avg. Bytes
Application load	80	569	73	2521	714.21	0.00%	0.38617	2146.46	2.45	5691803.9
http://localhost:4200	80	7	0	29	10.02	0.00%	0.39408	0.14	0.13	367.8
login	80	1261	462	3427	853.42	0.00%	0.39307	0.12	0.17	320
stock details	80	409	133	1621	255.72	0.00%	0.39634	0.16	0.16	415
buy	80	10958	7606	16244	1933.73	0.00%	0.37694	0.12	0.18	333
sell	80	11188	7016	15967	1989.06	0.00%	0.36519	0.11	0.18	314
add to watchlist	80	651	259	1463	280.16	0.00%	0.38277	0.13	0.19	339.3
portfolio	80	6471	4140	9651	1086.47	0.00%	0.37503	0.63	0.15	1709.6
watchlist	80	1942	1058	3653	571.86	0.00%	0.38143	0.38	0.15	1029
delete from watchlist	80	1476	260	3681	883.58	0.00%	0.38155	20.34	0.19	54581.5
view watchlist page	80	1890	710	3464	602.77	0.00%	0.38035	0.37	0.15	1004.5
view transactions	80	9600	4514	17291	2760.54	0.00%	0.36701	8.12	0.15	22654.5
gainers	80	2545	797	5177	810.74	0.00%	0.37056	0.65	0.15	1799
losers	80	2099	878	3883	728.38	0.00%	0.37207	0.56	0.15	1530
profile details	80	503	132	1584	273.61	0.00%	0.37681	0.21	0.15	579
update leaderboard	80	10483	2591	16005	2672.44	0.00%	0.37086	0.11	0.18	308
view leaderboard	80	1758	57	7422	1783.42	0.00%	0.39517	1.37	0.16	3562.9

Fig. 12 Summary report

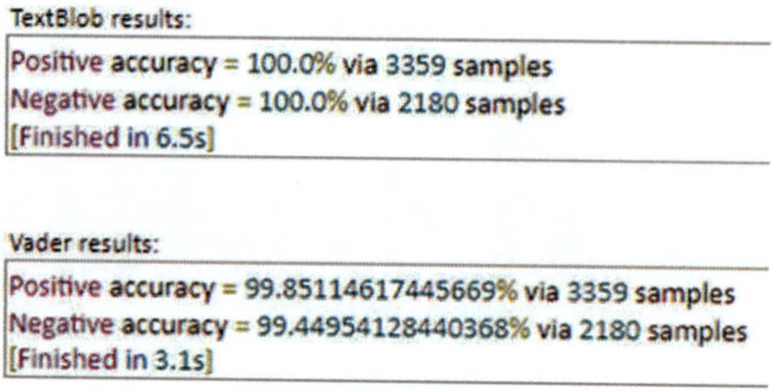

Fig. 13 TextBlob versus VADER

8 Conclusion and Future Scope

Virtual Stocks was successfully developed with a modern minimalist design that will assist users practice and learn, dabble in trading and remove any fear attached to the stock market without losing any real money. Market sentiment analysis aided users to make wise decisions. User data was analyzed and presented to users to evaluate their progress. Existing systems were compared as seen in Table 1. Extensive testing corroborates a stable and reliable application. Through VADER, we have achieved a 99% accuracy score and at least double the speed compared to the other sentiment analysis algorithms. Load testing of the application produced an error rate of 0% and an average throughput of 22 requests/min. Virtual Stocks surpasses the existing system due to the modern tech stack, mobile responsiveness, market sentiment analysis and extensive testing.

Table 1 Comparison with existing system

	Existing system	Virtual stocks
Stock analysis	They have calculated RIS which is an indictor whether a stock purchase should be made or not. However alone RIS is not a very accurate tool	We have calculated the market sentiment for the stock over a span of time which indicates the positive or negative sentiment among in the news related to the stock
Tech stack	PHP, MySQL, .NET and C#A	Angular, Django and SQLite
Testing	Testing of application is not done	Automated testing is done using JMeter with 0% error rate. Load testing was also carried out, and the average throughput came out to be 22 request per second which is admirable
Mobile responsive	No	Yes
API used	Yahoo finance and Google Stock API	Google News, Nsetools and Stockdio

Further aim is to continue enhancing the application and including more modules. An algorithm can be developed to recommend similar stocks to the ones user bought. This recommendation will be based on past interaction of user using cosine similarity or similar descriptions of a company using text vectorization. A module of social feed can be included, where users can interact and uplift each other. Performance can be improved by focusing on a potent way to deal with such huge volumes of data and increasing security as well as interactivity.

References

1. Agarwal A (2020) Sentiment analysis of financial news. In: 12th international conference on computational intelligence and communication. IEEE
2. Dev Shah HI (2018) Predicting the effects of news sentiments on the stock market. In: IEEE international conference on big data (Big Data). IEEE
3. Gilbert CH (2014) Vader: a parsimonious rule-based model for sentiment analysis of social media text. In: Proceedings of the international AAAI conference on web and social media
4. Janardhan VB (n.d.) A comprehensive study on lexicon based approaches for sentiment analysis. Asian J Comput Sci Technol
5. Kiss L´ (2021) Prediction of stock values changes using sentiment analysis of stock news headlines. J Inform Telecommun
6. Michael Graham JN (2003) Relative importance of scheduled macroeconomic news for stock market investors. J Econ Finan
7. Oetama RS (2019) Sentiment analysis on official news accounts of twitter media in predicting Facebook stock. In: 5th international conference on new media studies (CONMEDIA). IEEE
8. Satyabrata Aich H-CK (2017) Analyzing stock price changes using event related twitter feeds. In: 19th international conference on advanced communication technology (ICACT). IEEE
9. Walaa Medhat AH (2014) Sentiment analysis algorithms and applications: a survey. Ain Shams Eng J
10. Zenon Chaczko RB (2013) A virtual stock exchange simulator for education and training. In: 12th international conference on information technology based higher education and training (ITHET). IEEE

Real-Time Application for Deaf and Dumb: Bidirectional Communication Using Learning Methods

G. Arun Prasath and K. Annapurani

Abstract In this world, we have more than 466 million deaf dumb peoples. Normal peoples are communicating with voice, likewise deaf and dumb people are communicating by the use of sign language. Generally, without knowing the language, one cannot talk to other people in the world, similarly without knowing the sign language normal people cannot talk to deaf and dumb people. Within themselves, deaf and dumb people will communicate using sign language, which normal people cannot understand, but still hearing is one way to understand the people, so expression and gesture are also a way to understand thoughts of deaf and dumb people. To find a solution for this problem, either normal person needs to learn the sign language or we need to create a platform which will convert the sign language to voice and voice to sign language. Though we have many different types platform, still it not efficient so we need to use the advanced method which recognizes the input much efficiently. Since we are using gesture and voice as an input, this will act as bidirectional way communication. The focus of this work is to create a bidirectional-based system to identify sign language gestures from the sign language images and converting that to voice output. Since we are developing the application to computer and android, it should have two-way communication so that both can act as input and output node. For efficient voice input and output, we are using android mobile.

Keywords Gesture · Sign language · Bidirectional · Deaf and dumb · Two-way communication system · Computer vision · SIFT algorithm

G. Arun Prasath (✉) · K. Annapurani
Department of Networking and Communication, School of Computing, SRM Institute of Science and Technology, Kattankulathur 603203, India
e-mail: ag7678@srmist.edu.in

K. Annapurani
e-mail: annapook@srmist.edu.in

M. S. Kaiser et al. (eds.), *Information and Communication Technology for Competitive Strategies (ICTCS 2021)*, Lecture Notes in Networks and Systems 401,
https://doi.org/10.1007/978-981-19-0098-3_27

1 Introduction

In this current trend, gesture recognition is one of hottest topics for the research in the computer technology. Here research is focused on image processing domain [1]. Even though we have lots of techniques for deaf and dumb communication, still we need an efficient way for this deaf and dumb problem. Here we are trying to find a solution which is based on software because of the coast effectiveness and high accuracy [2]. The hardware coast may affect usage of the platform, then it will become a disadvantage for the system, it also has its own disadvantage like carrying the hardware, and this will limit the user. Since we are making the platform for deaf and dumb people, we need to make it in a user-friendly way. Here we are taking the input in image format, the accuracy will be defined by the model outcome, so the image recognition percent is very important. Because if the input accuracy is less, obviously the output accuracy also will be less so we need to take the accurate sign image from the user, and then we need to do preprocessing steps to improve the quality of image [3]. Since we are only familiar with voice-based communication among the normal humans, it is very tough to understand the deaf and dumb sign language so here we are developing the translator to convert the deaf and dumb sign language to the normal voice so our platform should also convert the normal voice data to the sign image which is understandable by deaf and dumb people [4].

It allows the human to interact with lots of hardware-based devices like robot and virtual play. In the future handling, the hardware-based tools are going to use this kind of human computer interaction platforms, it can be a vision-based technology or using image processing or the sensor, there are n number of ways available to communicate with the hardware, nowadays most of this kind of technology uses the sensor input, and the image processing tools will get the input from the user [2]. Compared to other kinds of approaches, hardware-based sensor data will give more accurate data. But still managing the hardware, carrying the hardware and the cost of hardware are all the major drawbacks in the hardware-based devices. According to our proposal, we are going to focusing on image processing domain which is the best way to enable or create the deaf and dumb communication. In image processing, we have different types of algorithms to extract the feature from the image; here in our work, we are using the shift algorithm which is a nonparametric approach [3]. Input will be taken from the user and will detect the object in the image, and it processes the image and then it compares the data with the existing trained data in the database. Gesture recognition will help computer to understand the human motions. It helps the human to interact with the machine without any external hardware; it has many applications like facial recognition and detecting and preventing the disease in the human body [5]. All over the world, the deaf and dump people are communicating through sign language, but they have been using different types of sign language based on the location. Since we are training the data set, we can also change the training data according to the user need [6].

2 Related Literature

In the previous research, different types of technology and method were used to recognize gesture and processing the input, but the main thing in this kind of system is user friendly so we need to develop a platform with better user-friendly system at the same time accuracy which is also very important. For that, we are using the MATLAB-based application. Some of recent works are done using the sensor gloves and input devices and sensors. But it is not possible to carry the kit everywhere so we need to fix that problem, and it is also not possible to buy the kit for normal person so we need to reduce the cost without compromising input quality. Since we are using the camera input system, we need to define the input image color and the angle. Here they have used geodesics shapes to find the need part of image [4]. And there are some methods which will trace the finger joints, and for recognition of the sign language, they have used fiber and flexible data gloves which consist of Arduino board and fiber sensing unit. They have proposed hidden Markov model, the proposed methods have used inertial motion sensor, and every finger has Acc sensor which will help to track even small movement of the finger [3].

3 Solving the Problem

The main objective of this paper is to aid a user-friendly and speech recognition feature with voice output. We are going to achieve that using the MATLAB with SIFT algorithm for better accuracy in the output. Even though we have lots of other methods, MATLAB with SIFT algorithm will give more accuracy. Here we are using the key points to find the matches between the input images and data set image; based on the key points counts, the input images can be identified and respective voice output will be played.

4 Types of Sign Language

Deaf and dump people around the world are using sign language to communicate in day-to-day life by their facial expression. In-between deaf and dumb people, sign language is enough to communicate. While expressing the sign language, deaf and dumb people will use their hands and the face parts [7]. But the most widely used body part of the expression is hands. As like deaf and dumb community, we are also going to talk to deaf and dumb people using this platform. And the sign language is not same around the world; it is different according to the country and the state they live. Like the language in the universe, deaf and dumb sign language is also different in the world. In India, we will use some sort of Indian sign language, and American people will use different kinds of sign language method called ASL [1]. In this kind of input method, camera input is one of the best ways to get the input from the user though we have different types of method to get the input from the user; vision-based camera input is the best way for image processing techniques because of the hardware handling problems [3]. This kind of sensor data will give more accurate result compared to other types of input data. But in the camera-based input, we only require laptop camera, no need of other kinds of hardware tools. Based on this kind of input method, we can achieve the real-time problem. Camera images data can be processed in different way but the objectives is to convert the sign image into voice [8].

Using the web camera in computer, we are going to capture the sign language image and then the input is processed; in our system, we already have some sample data set to compare the input with the dataset; from the data set, it will identify the input and will send the corresponding voice data set.

In gesture capture method, we are having different types of input method like camera and Kinect sensor. We can take input from the hardware device also.

Types of input for gesture input

a. Camera image
b. Sensor gloves
c. Hardware device.

4.1 Using Camera Image

In this proposed work, we are using camera input technique because other kinds of hardware are not easy to handle for deaf, and when the user is outside and we are using image processing tools to process the image and compare other ways of input technique, this kind of camera input data set is best to process [9]. Hardware-based input method will also give best result compared to image-based input, but major disadvantage is carrying out the hardware and handling the hardware is also not easy every time so the camera input is best way when it comes to easy way process with accuracy [10].

4.2 Sensor Gloves

In this kind of sensor-based gloves, input will be taken from sensor gloves, the accuracy is also best, here the gloves consist of couple of sensor and transmitter to transmit the data whenever we need to use the system, and we need the gloves. So, it is not such an easy to wear input glove everywhere, but the accuracy is much better in camera input system. It can translate the complete alphabets using the data set we have trained in our machine; according to our data set, it captures very sensitive movement also [7].

4.3 Hardware Device

Hardware-based input tools have sensor, and this kit captures the input and translates the sign language to voice-based information [4], but it requires hardware to process it so it is not possible to carry the hardware to everywhere so it is not so successful in deaf and dumb communication. However it gives the accurate result compare to other image processing techniques because it even captures very sensitive movement of the object and it will capture the complete frame from the full image it need to processed before the process the image for capture the needed frame the image need to be pre-processed, so the accuracy is more accurate compare to image processing methods but major drawback here is hardware-based input method and the input method will affect the output so if we need a better result we need proper input method and it depends on the input method but in our work we are using camera image and to process the image we are using SIFT algorithm in MATLAB it has lots of image processing tools to process the image [8].

5 System Application

MATLAB-based system application is deployed for deaf and dumb person communication purpose. The main model is developed to obtain their gesture inputs, and it will be processed. In this proposed work, camera is automatically initiated and captures the gesture input image provided by the deaf and dumb people. We also stored pre-recorded voices [2]. Normal person will speak through the MATLAB application, which recognizes the voice input, and it converts into hand gesture image [10]. Automatically, it will detect the motion, then camera will start capturing the input, and it transfers the image to the data set area for processing the image [11].

Basically, this consist of three parts.

1. Capture the input image
2. Process the input image in image processing
3. Display the corresponding language
4. SIFT algorithm-based conversion.

Capturing the input image

In our system application, we are tacking the input from the camera instant of hardware tools. Because of the hardware problems, the camera in the one system will capture the input since it has two nodes and the system both will act as input and output nodes [2]. The node one and the node two both can take input from the user since it is a two-way bidirectional communication system. Since the platform is used for deaf and dumb people, one node should get the sign language input from deaf peoples and another node should take the input from the normal people [10].

Processing the input image

In this image processing section, we have processed the input image, and here we are applying SIFT algorithm to get the better object of the image. The SIFT algorithm will plot point all over the object to identify the object in the complete image, and for comparing the input image, we need some previous data set to compare. It should be real-time sign language, then only we can compare our image point count with data set image count to get the corresponding result; it can be anything whether it is voice or the sign language. Since we have two types of nodes in this communication, we have to use the two-way communication, so here we have two types of input methods. This can be possible in MATLAB environment. Since we have lots of image processing tools in the MATLAB application for example, we need SIFT algorithm for processing and comparing the image, and it provides more accurate output [5]. Node 1: sign language input processing and Node 2: voice input processing in the sign language input processing, we need to process the sign language in one node, it has to be converted to normal voice-based message, and this can process the only the sign language. Receiving the sign language and displaying the sign language in the same node, so in this node, we cannot process the voice it can only process the sign language and can transmit to next end as a voice-based message. In the another end, we can transmit only the voice data from the normal user, and it will be converted to sign language which is understandable by the deaf and dumb people, so it will take voice-based data and the data will be converted to the sign language, for that it needs to compare the key word with our previous data set [6]. It will send the sign language to the next node; this node can only process the voice-based input and send the corresponding voice data, and it cannot process sign language input. This can only process voice-based input data only. For image acquisition, webcam is used, after that frame is sent to the server and edge detection of the video is done using SIFT algorithm, and tokens are being created from thinning image after tokens are fetched. The paper briefly describes the schemes of capturing the image from device, image detection, processing the image to recognize the gestures as well as voice result. In this, we can process the image based on the data set it sends corresponding to the voice (Fig. 1).

SIFT *algorithm-based conversion*

For recognizing the input object, we need to detect motion on the camera, and then the camera should capture the input. For that purpose, we are using SIFT algorithm. And matching the input image with our data set is the main work because common problem in this kind of work. It will not find the minor differences between the input image and database image, sometimes input image and database image might be almost same but meaning of the image may be different, so there is a chance of wrong output. To rectify this kind of issue, the algorithm should have the capacity of finding the difference between the similar inputs. Image scale may be same, and the construction will be same so SIFT algorithm will construct the scale space. It is a kind of representation of internal image [3]. And another thing is finding key point from image, it may be minimum or maximum different, but we need to find the count calculation. And one more thing is eliminating the bad key point from the image,

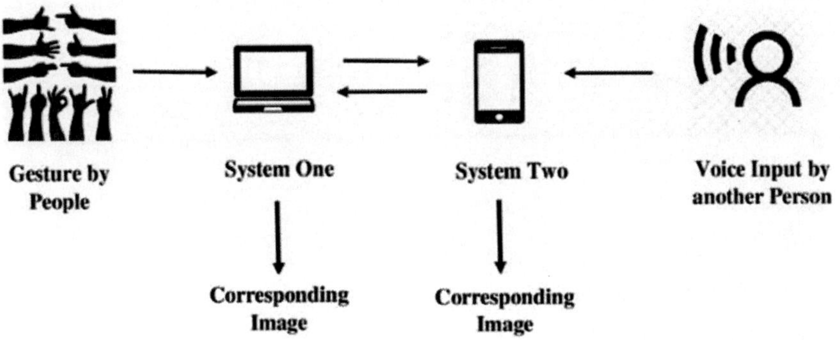

Fig. 1 Two-way communication

sometimes the background will also get allocated with key point, but it is not good when we are comparing sign images. When we have this kind of data in the data set, it will give the different output so it needs to be removed from the input image key point count; we can assign the different orientation to the key point, sometimes rotation of the image gives the different result but it may change the result totally different dimension [12]. The key points given by the SIFT algorithm will create the representation of original image [10]. It will generate points for that particular sign image; this will be used to compare the input image with data set to give a proper output.

6 Result

In this module, the camera in laptop will capture the input for processing the image; so while taking the input from the user, we need to be very careful about the background location since we are using SIFT algorithm to find the object in the input image; if the background is not empty algorithm, it will consider that also as an object. Then it will point full image location with key points so we need to be very careful in the trained data set; if the trained data set is having the proper key point count, then we can easily find the proper output for input voice; since we have two nodes in our work, one side will take the voice as input and another side will take sign image as an input. Then SIFT algorithm in this one side is to compare the input image and train the data set only. We have two buttons on the platform: One is to get the input from the user and another one is for training the machine. As shown in Fig. 2, it will capture some key points from the image like dot point.

Then the pattern will be stored in the database with key point count and the corresponding meaning of the image. The key points will be assigned to the moving objects in the camera So, the comparison of the image is easy to compare with other kinds of tools. It will show the result based on the key points count. If the key point

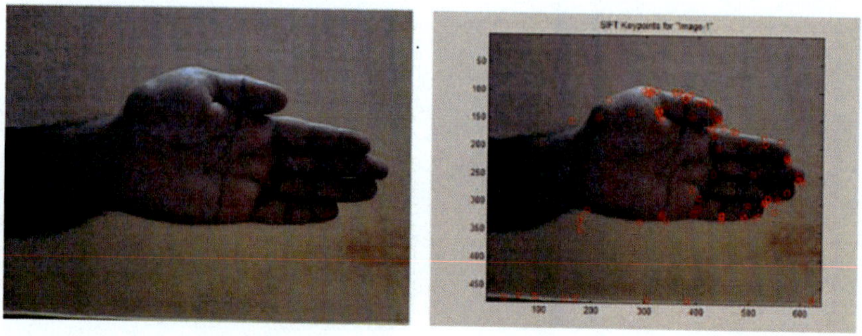

Fig. 2 Camera input vs SIFT key point finding

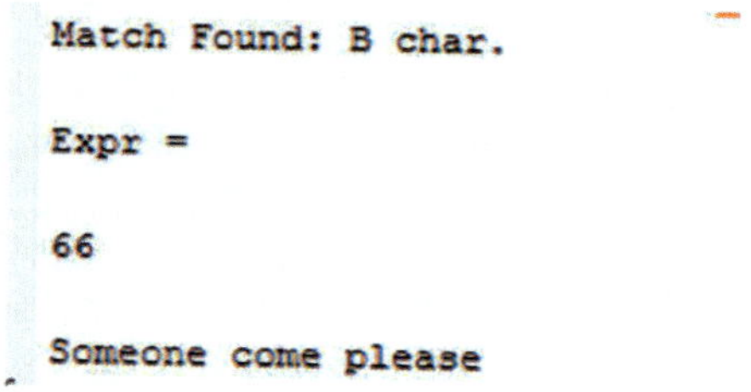

Fig. 3 Message to normal user

count is more, then it gives much better result. If the same image has minimum difference, then it has to check the other images too (Fig. 3).

7 Conclusion

In this work, we are using two-way communication system which means we can give the input in both sides. Every node will act as input and output node, so the deaf and dumb peoples can send sign language image in one machine, they can receive the output in the same screen, and it will convert to voice also. Hope this work for the deaf and dumb will remove the gap between the normal and deaf people. It may help the deaf and dumb people to communicate with normal world, and it will also help in the education purpose of the deaf and dumb people's life.

References

1. Jimenez J, Martin A, Uc V, Espinosa A (2017) Mexican sign language alphanumerical gestures recognition using 3D Haar-like features. IEEE Latin Am Trans 15(10)

2. Kishore et al (2018) Motionlets matching with adaptive kernels for 3-D Indian sign language recognition. IEEE Sens J 18(8)
3. Huang et al (2019) Tracing the motion of finger joints for gesture recognition. IEEE Sens J 19(20)
4. Nasreddine K, Benzinou A (2019) Shape geodesics for robust sign language recognition. IET Image Process 13(5):825–832
5. Deriche et al (2019) Intelligent ArSL recognition system using a pair of LMCs. IEEE Sens J 19(18)
6. Guo et al (2020) Hierarchical recurrent deep fusion using ACS for SLT. IEEE Trans Image Process 29
7. Tubaiz N, Shanableh T (2015) Glove-based continuous Arabic sign language recognition in user-dependent mode. IEEE Trans Human-Mach Syst 45(4)
8. Xiao Q et al Multimodal fusion based on LSTM and a couple conditional hidden Markov model for CSL recognition. Dig Obj Identifier. https://doi.org/10.1109/ACCESS.2019.2925654
9. Lee BG, Lee SM (2018) Smart wearable hand device for sign language interpretation system with sensors fusion. IEEE Sens J 18(3)
10. Gałka et al (206) Inertial motion sensing glove for sign language gesture acquisition and recognition. IEEE Sens J 16(16)
11. Bui D, Nguyen LT (2007) Recognizing postures in Vietnamese sign language with MEMS accelerometers. IEEE Sens J 7(5)
12. Aly W et al User-independent American sign language alphabet recognition based on depth image and PCANet features. Dig Object Identifier. https://doi.org/10.1109/ACCESS.2019.2938829

The Selection of Electroencephalogram System for Stress Detection

Nikita R. Hatwar and Ujwalla H. Gawande

Abstract Nowadays, mental stress is a social concern and powerful reason behind your performance of routine work. Stress suffered for a long period of time becomes chronic in nature. Chronic stress is the stress in which the retaliation to emotional distress suffered for a protracted amount of duration, within which an individual discern they have little or no control. Corticosteroids are released by an endocrine system in response to stress. Chronic stress may involve many psycho physiological illnesses. For instance, stress will increase the probability of depression, stroke, cardiac arrest, a heart attack, etc. The most recent neurobiology discloses that the mental stress primary target is human brain, as a result of the perception of the human brain determines a state of affairs that is threatening and nerve-racking. Stress detection is one of the dominant research areas and has become a trending topic in the study of neural engineering, rehabilitation, and brain science. Montreal imaging stress test (MIST) is used to instigate mild stress. MIST is a computer-based tool with different arithmetic tasks. MIST is extensively used to induce mild stress in a stress detection system. MIST is preferred over natural stressors, such as real-time examination, a questionnaire, public Stroop test, speaking task, and driving simulator. EEG signals are recorded in stress and control condition. In this paper, a brief review is given regarding the various EEG systems and a selection of device for stress detection.

Keywords EEG device · EEG signal · EEG channels · EEG dataset · Stress detection · MIST · EEG bands · Brain sensing locations · Sampling rate · PSD · Fourier transform

1 Introduction

In today's world, stress is a common thing due to different types of work pressure, money problem, or health issue. Hans Selye defined stress as "the non-specific

N. R. Hatwar (✉) · U. H. Gawande
Yeshwantrao Chavan College of Engineering, Nagpur, India
e-mail: npvanjari@gmail.com

M. S. Kaiser et al. (eds.), *Information and Communication Technology for Competitive Strategies (ICTCS 2021)*, Lecture Notes in Networks and Systems 401,
https://doi.org/10.1007/978-981-19-0098-3_28

response of body to any demand to change" [1]. Mental stress is a result of an exorbitant workload underneath a time restriction to satisfy the expectation. It can be said that stress = expectation – ability. Many physiological and psychological ailments are created if stress is not detected in time. It makes an individual become unequipped for adapting to an apparent danger to their physical, enthusiastic, or mental prosperity. Nervous tension, i.e. stress and its associated well-being conditions can represent a genuine danger to the personal satisfaction, and brokenness in day-by-day life can meddle with the public activity and bodily soundness of a person. Stress is related with a scope of conduct, intellectual, neurovascular, cardio vascular, and atomic impacts. The intellectual initiation hypothesis of stress (CATS) depicts the relationship of the cerebrum with mental pressure. There exists proof of pressure-related psychological impedance that prompts memory shortcoming and the obliteration of mental and physiological prosperity [2]. Stress has been accounted for to be related with the shrinkage of the hippocampus and prefrontal cortex just as the disability of prefrontal systems. In chronicity, stress can transform into pressure-related ailments with genuine effects, including burnout, sorrow, and post-horrible pressure issue [2].

Stress can be for all intents and purposes evaluated by polls strategies, for example, Perceived Stress Scale a psychological instrument [3], The Life Events and Coping Inventory (LECI) [4], The Ardell Wellness Stress Test, Stress Response Inventory (SRI) [5], and so forth to survey apparent pressure. But these methods are not reliable as person may pretend not to be stressed. It detects stress after a person gets stressed. So, a reliable method is required which will assess and quantify stress using neuroimaging modalities. Other methodologies to evaluate stress are facial appearance, physiological, and invasive approach such as hormonal analysis. These methodologies are uncertain and invasive. Hence, electroencephalogram (EEG) signal-based stress detection has been introduced, which is non-invasive, reliable, precise, and accurate. EEG is a most suitable non-invasive tool to detect stress by detecting different brain waves. Human nervous system's normal function is affected due to stress. To maintain the normal activity of nervous systems, brain reacts to these malformations by releasing sequence of chemical reaction (i.e. stress hormones) such as cortisol.

Different techniques are available to induce stress. The various techniques used are natural stressors, such as real-time examination, questionnaire [8], public Stroop test, speaking task [9], a driving simulator [10], and mental arithmetic task (MAT) [11–13]. MAT is extensively utilized to draw out stress in mental stress investigations. It is challenging to design a stress detection system which detects stress. The first challenge is to have such an EEG device that will record the brain signals from those brain locations, where we can detect stress. Another challenge is to create a dataset as per requirement. In literature, datasets such as DEAP [14], MAHNOB [15], and BNCI are also available for experimentation.

One of the challenges is to select an EEG device for research that will classify stress and normal data. For this proper EEG, device must be selected which should not have fixed channel electrodes. Many EEG devices are available that have fixed channels they have a cap and electrodes need to be placed at fixed locations. Otherwise, if EEG device has 12 channels, for example, first electrode needs to be kept at say FP1, so all remaining electrodes have fixed locations. Such types of device are not fully used for classification of stress and normal because we do not know exact channel which will generate stress signals or control signals. So, it is necessary to change position from one channel to another channel for the research purpose. To detect the area of the brain associated with stress and channel which will generate stress and normal EEG data.

After selection of device, next challenge is to have a dataset and what should be the format of dataset. Next is to select proper and sufficient subjects to create real-time dataset. Virtual dataset will not be helpful in the research. Subjects can vary from age, gender, profession whether it will work or not? Mild stress is induced using mental arithmetic test (MAT) tool that does not have medical impact on subjects. Subjects are chosen that does not have any diseases.

In the proposed work, the real-time stress detection system needs to be build that requires brain signals to be acquired using EEG device. Thus, an appropriate EEG device is acquired, and dataset is created as per the requirement of stress detection system. EEG is used to collect electrical activity of brain, which is a non-invasive medical imaging technique and is cost effective [7]. An EEG signal also provides an insight of how an information is handled in the brain. The selection of the EEG signal for stress detection is due to its technological advancement in recording ability. In this work, Montreal imaging stress test (MIST) [16] will be used to induce stress. Though the EEG signal for detecting the stress and MIST technique to induce the stress is not new, the testing will be done on our own database.

2 Literature Review

Stress detection has become very essential in today's scenario. Till date, many researchers have experimented with different EEG devices. In this section, a brief introduction to type of stressors, EEG device, and locations of the brain that are used for stress detection is given. Mind band dual is one of the most popular EEG devices which consists of two channels FP2 and FP1 to detect mental stress induced by a cognitive workload [17]. In [18], stress has been induced through—ODDBALL task. Natural stressor, i.e. construction site induces stress and EMOTIV EPOC + is an EEG device with 14 channel has also been used to monitor stress [7]. A 5-electrode EEG device is used in the OPTIMI project which uses FP2, FPz, and FP1 two ear lobes (A2 and A1) channels with natural stressors, i.e. workplace [21]. Arithmetic operation, lexical decision task, mental rotation, and finger tapping are four cognitive, and motor tasks have been used to induce stress, and it has been monitored with a wireless single channel, one electrode FP1 NeuroSky Mindwave device [25]. B-Alert X10® may be a wireless hardware system for EEG and ECG acquisition with 9 channel EEG (C4, C3, POz, P4, P3, F4, F3, Fz, and Cz) and single channel ECG. The stressor used is tasks, first 5 min of eyes closed (EC), then 5 min of eyes open focussing vision on a red colour dot on a dark screen (DOT) and 5 min of eyes open with attention on a bright screen (EO) [27].

Stressors mentioned above such as cognitive workload and Stroop test have been used in general to monitor the stress, but mostly a computer programme known as arithmetic task (MAT) has been employed to instigate stress that is based on the prototype of the Montreal imaging stress task (MIST). The EEG devices used are (1) a brain master 24E system with seven active electrodes [19]. (2) A 128 channel EEG electrode in which brain locations are connected FP2, FP1, F4, F3, F7, F8, FPz, temporal T4, T3, T6, T5, parietal P4, P5, P8, P7, occipital O2, O1, and central C4, C3 [22]. (3) An Emotiv Epoch + 14 electrodes EEG device with F7, AF3, F3, FC5, T7, P7, O2, O1, P8, T8, FC6, F4, F8, and AF4 [23]. (4) A RABio w8 is a portable, wireless, low-cost hardware–software system for the acquisition, and processing of multiple bio signals such as ECG, EEG, and EMG 4 FP2, FP1, F4, and F3 [24]. As MIST is the effective technique to induce stress, and it has been used in the proposed work.

After finalization of the technique to induce stress, the next is to select the device that can detect the stress in a more appropriate manner. Hence, it is a challenging task to select an EEG device, either it should be having wet electrode or dry electrode, how many channels should be there, and also it should be cost effective. Along with it, the aim is to select the device that is user friendly. After going through the specification of various devices such as EMOTIV EPOC, NeuroSky, MUSE headband, and many more, it has been concluded to use the devices suggested in [29] that gave details of the top 14 EEG devices.

3 Selected EEG Device

After rigorous survey, an EEG device with 6 + 1 channels which is made in India has been selected. Six channels are used to record EEG signals from brain locations, and one is used as a reference channel, as shown in Fig. 1. It has both wet and dry electrodes. Wet electrodes are preferred as it gives good conductivity between subject's skin and electrode. It is generally used where skin is hairy, as hair may

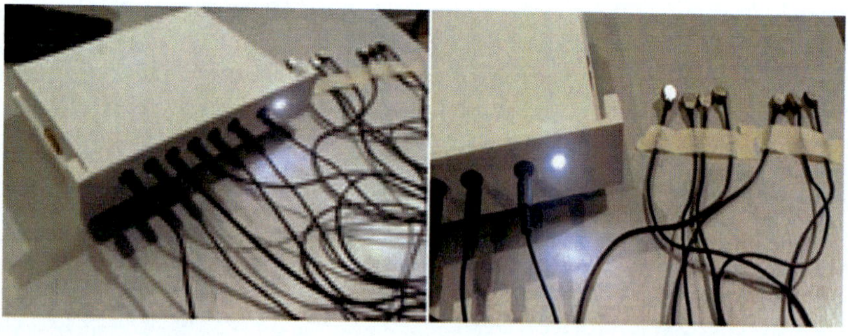

Fig. 1 EEG device with 6 + 1 channels

Table 1 Specifications of selected EEG device

EEG sensors	EEG signals	Supported platforms	Power
6 channels: Their brain locations are not fixed, and electrode positioning can be any locations such as FP1, FP2, F7, F8, F3, F4, etc.	Sampling rate: 128 samples/s	Windows and Linux	Power supply: 5 V DC USB support: Yes
One reference channel: left or right ear lobe			Dimensions: Length = 290 mm Breadth = 190 mm Height = 70 mm
Sensor material: silver metal			Weight: 950 g
Connectivity: wired (approx. 1.2 m)			

Table 2 Phase table

Phase name	Phase ID
REST	0
LEVEL1	1
LEVEL2	2
LEVEL3	3
LEVEL4	4
RELAX	5

cause conductivity issues. Dry electrode is subject friendly, and it requires less time to set up as compared to wet electrode. On hairy skin, dry electrodes with have less conductivity than wet electrode. For hairy locations, wet electrodes are used by applying gel. In this research project, brain locations such as FP1, FP2, AF7, AF8, F7, and F8 are used located at non-hairy positions. Mostly, in all EEG devices, brain locations are fixed, but in selected EEG device, brain locations are not fixed, and electrode positioning can be any locations such as FP1, FP2, F7, F8, F3, and F4. Electrodes are made of silver, the length of wire is approximately 1.2 m, and it is connected through USB of the system.

The specifications of the selected EEG device shown in Fig. 1, are depicted in Table 1.

3.1 Brain Sensing Locations

The location of the brain to detect the stress is discussed. In Fig. 2, the 23 brain locations are shown. Each brain location gives different signals such as emotions,

Table 3 Sample data of recorded file (recording.csv)

#CHANNEL = FP1
#DATE = THU MAR 15 19:37:58 IST 2021

#ID	TIME	CHANNEL_ID	PHASE_ID	VALUE
1	1,588,256,934,129	1	0	42
2	1,588,256,934,140	1	0	4
3	1,588,256,934,140	1	0	8
4	1,588,256,934,140	1	0	24
5	1,588,256,934,150	1	0	24
6	1,588,256,934,160	1	0	49
7	1,588,256,934,160	1	0	13
8	1,588,256,934,160	1	0	29
9	1,588,256,934,171	1	0	25
10	1,588,256,934,181	1	0	27

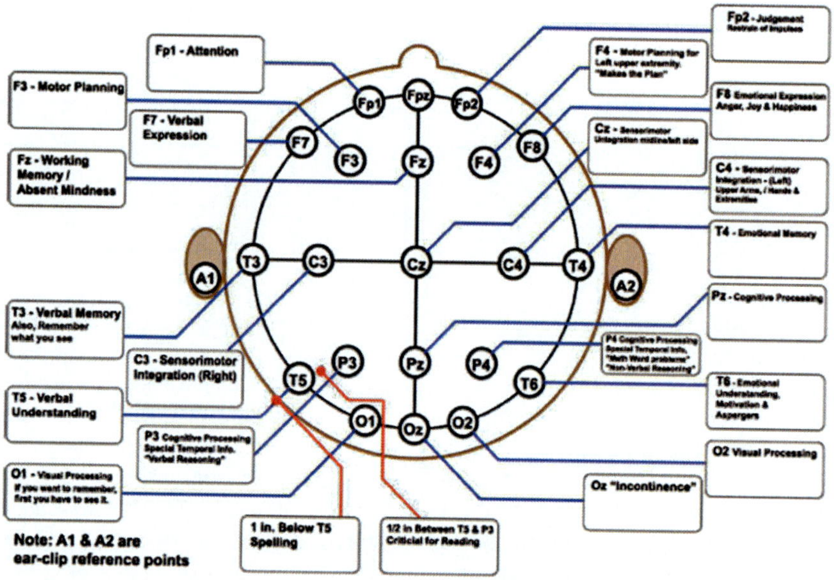

Fig. 2 Brain sensing locations

motor planning, eyesight, verbal understanding, and attention. Whilst recording the data from brain locations, it is important to focus on the application perspective of the project. Depending on it, the correct brain locations need to be selected.

For stress management, it is required to record the data from more than one brain locations such as FP1, FP2, F7, and F8.

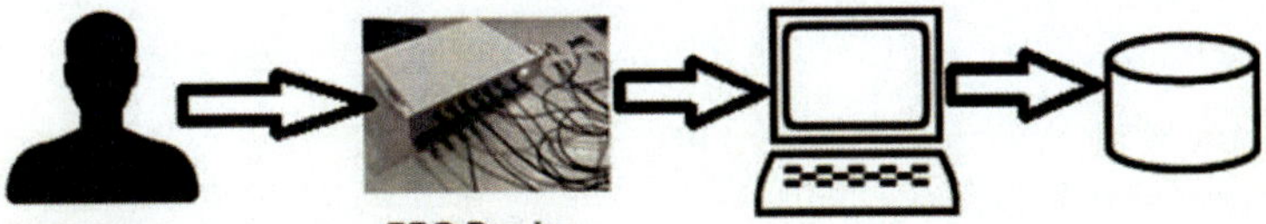

Fig. 3 Process of recording EEG signal

3.2 Recording of EEG Signals

Raw data are generated from 6 channels from various brain locations using a dry electrode. Dry electrodes are preferable over wet, because dry electrodes are subject friendly, and it is convenient to record data from the channel locations. The electrodes can also be used as wet electrode by applying alcohol-based sanitizer or aloevera gel. Figure 3 shows the block diagram of recording EEG signals from a subject and creating the raw database of EEG signals recording, from various brain locations.

MIST tool is developed to induce a mild stress in the subject which consists of two states, viz. normal and stress, which further consists of five phases.

As per requirement, a dataset is created, as shown in Fig. 4.

The root folder of file system is **dataset;** it contains the list of subject folders and a file **subject_model.dat**. The system is designed to store basic information of subject, i.e. ID, name, age, and gender as well as recorded data. Records of subjects are stored in a file **subject_model.dat,** and this file is a binary file and stored in a system. ID of each subject is auto generated by the system, where the value of ID varies from 1, 2, 3, etc. For every subject, a folder is created inside the root folder, and the name of the subject folder is same as subject ID. The system defines two types of condition, i.e. normal and stress. The system stores recorded brain signals under normal condition, in **normal** folder and signals under stress condition, in the **stress** folder for a corresponding subject.

(1) *Channel folder:* EEG device is supporting 6 channels; each channel is assigned unique ID from 1 to 6. User can record brain signals of any location such as FP1, FP2, F7, F8, F3, and F4 using any channel. If user is recording brain signals of FP1 location using channel 1 and FP2 location using channel 2, then the name of channel folder will be **1-FP1** and **2- FP2**. The name of the channel folder indicates channel ID and name of brain location from which the signals are recorded.

(2) *Format of recording.csv file*: The recorded file is a comma separated file (CSV). This file is divided into two parts, i.e. header and data part. Header part of recording file stores channel name and date, channel name indicates brain location, from which the signals are recorded and date indicates the recording date and time. Data part contains 5 columns, i.e. ID, TIME, CHANNEL_ID, PHASE_ID, and VALUE.

   (a) *The first column (ID)*—stores the record ID number.

```
[Dataset]
|
|—<Subject ID1>
|   |
|   |——— [NORMAL]
|   |    |
|   |    |—<Channel Folder1>
|   |    |  |
|   |    |  |——[recording.csv]
|   |    |  |——[RAW.csv]
|   |    |  |——[FILTERED.csv]
|   |    |  |——<Plot files>
|   |    |
|   |    |—<Channel Folder2>
|   |       |
|   |       |——[recording.csv]
|   |       |——[RAW.csv]
|   |       |——[FILTERED.csv]
|   |       |——<Plot files>
|   |
|   |———[STRESS]
|        |
|        |—<Channel Folder1>
|        |  |
|        |  |——[recording.csv]
|        |  |——[RAW.csv]
|        |  |——[FILTERED.csv]
|        |  |——<Plot files>
|        |
|        |—<Channel Folder2>
|           |
|           |——[recording.csv]
|           |——[RAW.csv]
|           |——[FILTERED.csv]
|           |——<Plot files>
```

Fig. 4 Structure of file system of dataset

(b) *The second column (TIME)*—stores the date and time of recording. Long value of the TIME column is the number of seconds measured between midnight say, 1 Jan 1970 and recording time.

(c) *The third column (Channel ID)*—stores the channel ID number through which the signals are recorded.

(d) *The fourth column (Phase ID)*—stores the ID number of phase.

It stores the phase of EEG signals recorded. Phases such as REST (0), LEVEL1 (1), LEVEL2 (2), LEVEL3 (3), LEVEL4 (4), and RELAX (5) (Table 2).

(e) *The fifth column (VALUE)*—stores a recorded brain signal value. The signal value (integer number) varies from 0 to 1023, where 0 corresponds to 0 mv and 1023 corresponds to 1100 mv. The electrical signal recorded from the brain is converted into digital numbers (0–1023). Here, the amplified signal represents the voltage value in millivolts, which is converted into digital numbers (Table 3).

A. *Sampling rate Hz*

The sampling rate is also known as sampling frequency. It is the number of recorded samples in 1 s. This is calculated as

$$\text{SR} = (N_r * 1000)/(T_l - T_f)$$

where

SR = Sampling rate (sampling frequency) in Hz
N_r = Number of records
T_l= Time of last record in milliseconds
T_f = Time of a first record in milliseconds.

Data are recorded in time domain, i.e. *y*-axis voltage in millivolts, and *x*-axis has time in seconds as shown in Fig. 5. This information is not sufficient to process EEG signal. The data recorded are having signals in thousands, even though the recording

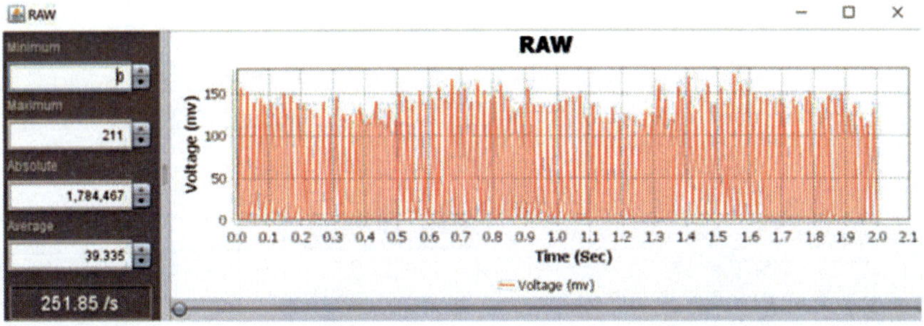

Fig. 5 Raw data waveform

of signals is done only for 2 min. It is difficult to identify the change in the state of brain signals in time domain; hence, they are transformed into frequency domain. This is done with the help of Fourier transform. In this, x-axis has frequency in Hertz, and y-axis has a power spectral density (PSD). PSD gives the power of each brain signals.

B. *Frequency Bands*

EEG data can be analyzed by decomposing it into functionally different frequency bands, i.e. delta (0–4 Hz), theta (4–7 Hz), alpha (8–12 Hz), beta (12–30 Hz), and gamma (30–100 Hz). A delta wave is activated in deep sleep and is generally present in small children. Theta wave is also found in sleep. Alpha waves are active whilst you are walking or when you are mentally active. When a person is alert, engaged in problem solving or decision-making and attentive than beta waves are present. Gamma waves are activated when you are in a state of universal love, high virtues, and altruism [30]. Figure 6 shows the different frequency bands.

Open, calm and balanced, psychological state is emulated by alpha waves, but alpha activity lowers in stress condition. Whereas, beta activity emulates emotional and cognitive processes which increases with mental workload and with stress [31]. Fourier transforms help in decomposition of the EEG signal into frequency components. Fourier transform returns, for every frequency bin, a complex number from which one can then easily extract the amplitude and phase of the signal, at that specific frequency. Thus, it is essential to convert time domain signals into frequency domain signals.

Fig. 6 Different frequency band

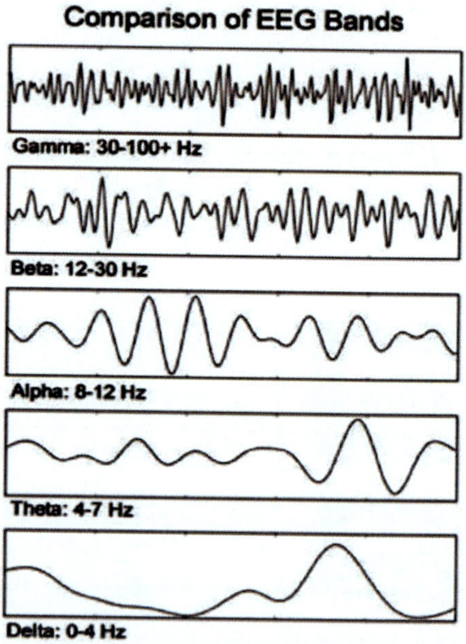

4 Conclusion

Detecting stress is a challenging task. Though there are various devices, mechanism, and technology available, EEG provides an economical solution that is feasible for remote applications in healthcare facilities and small clinics. From the study, it can be concluded from that stress affects a prefrontal and frontal region of brain, thus the signals acquired from these regions will give appropriate data for stress detection. The limitation of the device selected is that it is not portable.

Acknowledgements We would like to thank all those who voluntarily accepted to be part of our study. We are rigid to our commitment of not to use this study for any unethical purpose.

References

1. Seyle H (1956) The stress of life. Migra-Hill Edition, New York
2. Xia L, Malik AS, Subhani AR (2018) A physiological signal-based method for early mental-stress detection. Elsevier J Biomed Sig Process Control 46:18–32
3. Cohen S, Tyrrell DAJ, Smith AP (1993) Negative life events, perceived stress, negative affect, and susceptibility to the common cold. J Pers Soc Psychol 64:131–140
4. Dise-Lewis JE (1988) The life events and coping inventory: an assessment of stress in children. Psychosom Med 50:484–499
5. Koh KB, Park JK, Kim CH, Cho S (2001) Development of the stress response inventory and its application in clinical practice. Psychosom Med 63:668–678
6. Aftanas L, Reva N, Varlamov A, Pavlov S, Makhnev V (2004) Analysis of evoked EEG synchronization and desynchronization in conditions of emotional activation in humans: temporal and topographic characteristics. Neurosci Behav Physiol 34:859–867
7. Jebelli H, Hwang S, Sang Hyun Lee M (2017) EEG signal-processing frame work to obtain high-quality brain waves from an off-the-shelf wearable EEG device. ASCE J Comput Civil Eng 1–12
8. Hassellund SS, Flaa A, Sandvik L, Kjeldsen SE, Rostrup M (2010) Long term stability of cardiovascular and catecholamine responses to stress tests an 18-year follow-up study. J Am Heart Assoc 55:131–136
9. Karthikeyan P, Murugappan M, Yaacob S (2011) A review on stress inducement stimuli for assessing human stress using physiological signals. Blood Press 446–451
10. Schier MA (2000) Changes in EEG alpha power during simulated driving: a demonstration. Int J Psychophysiol Official J Int Organ Psychophysiol 37(2):155–62. Int J Pure Appl Math Special Issue 640
11. Karthikeyan P, Murugappan M, Yaacob S (2012) A study on mental arithmetic task based human stress level classification using discrete wavelet transform, pp 77–81
12. Liu X, Iwanaga K, Koda S (2011) Circulatory and central nervous system responses to different types of mental stress. Ind Health 49(3):265–273
13. Yu X, Zhang J (2012) Biomedical signal processing and control estimating the cortex and autonomic nervous activity during a mental arithmetic task. Biomed Sig Process Control 73:303–308
14. Koelstra S, Muehl C, Soleymani M, Lee J-S, Yazdani A, Ebrahimi T, Pun T, Nijholt A, Patras I (2012) DEAP: a database for emotion analysis using physiological signals. IEEE Trans Affect Comput 3(1):18–31

15. Al-Nafjan A, Hosny M, Al-Ohali Y, Al-Wabil A (2017) Review and classification of emotion recognition based on EEG brain- computer interface system research: a systematic review. J Appl Sci 1–34
16. Dedovic K, Renwick R, Mahani NK, Engert V, Lupien SJ, Pruessner JC (2005) The montreal imaging stress task: using functional imaging to investigate the effects of perceiving and processing psychosocial stress in the human brain. J Psychiatry Neurosci 30:319–325
17. Xu Q, Nwe TL, Guan C (2014) Cluster-based analysis for personalized stress evaluation using physiological signals. IEEE J Biomed Health Inform 1–16
18. Tsuda M, Lang Y, Wu H (2014) Analysis and identification of the EEG signals from visual stimulation. Procedia Comput Sci 35:1292–1299. ELSEVIER ScienceDirect
19. Al-sharpie F, Tang TB, Baudin N, Kiguchi M (2015) Mental stress quantification using EEG signals. In: International conference for innovation in biomedical engineering and life sciences, 18-12-2015, pp 1–5
20. Quesada-Tabares R, Molina-Cantero AJ, Gómez- Gonźalez IM, Merino-Monge M, Castro-Garćıa JA, Cabrera-Cabrera R (2017) Emotions detection based on a single-electrode EEG device. In: Proceedings of the 4th international conference on physiological computing systems (PhyCS 2017), pp 89–95
21. Hu B, Peng H, Zhao Q, Hu B, Majoe D, Zheng F, Moore P (2015) Signal quality assessment model for wearable EEG sensor on prediction of mental stress. IEEE Trans Nanobiosci 14(5):553–561
22. Subhani AR, Mumtaz W, Saad MNBM, Kamel N, Malik AS (2017) Machine learning framework for the detection of mental stress at multiple levels. Multi IEEE Open Access J 13545–13556
23. Vanitha V, Krishnan P (2016) Real time stress detection system based on EEG signals. Biomed Res S271–S275
24. Minguillon J, Perez E, Lopez-Gordo MA, Pelayo F, Sanchez-Carrion MJ (2018) Portable system for real-time detection of stress level. J Appl Sci
25. So WKY, Wong SWH, Mak JN, Chan RHM (2017)An evaluation of mental workload with frontal EEG. PLOS One 1–17
26. Jebelli H, Khalil MM, Lee S A continuous update, computationally efficient stress recognition framework using electroencephalogram (EEG) by Appien online multi-task learning algorithms (OMTL). IEEE J Biomed Heath Inform 1–12
27. Gaurav, Anand RS, Kumar V (2017) Non-invasive EEG-metric based Stress Detection. In: 4th IEEE international conference on signal processing, computing and control (ISPCC 2k17), 21–23 September 2017, pp 516–520
28. Liao C-Y, Chen R-C, Tai S-K (2018) Emotion stress detection using EEG signal and deep learning technologies. In: Meen, Prior, Lam (eds) Proceedings of IEEE international conference on applied system innovation 2018 IEEE ICASI 2018, pp 90–93
29. https://imotions.com/blog/top-14-eeg-hardware-companies-ranked/
30. Wasim M, Sajjad M, Ramzan F, Khan UG, Mahmood W (2018) A review and classification of widely used offline brain datasets. (IJACSA) Int J Adv Comput Sci Appl 9(2):399–408
31. Alberdi A, Aztiria A, Basarab A (2016) Towards an automatic early stress recognition system for office environments based on multimodal measurements: a review. J Biomed Inform 59:49–75

Comparative Analysis on Ontology-Based Contextual Senses Approaches for Search Engine

Pooja Mudgil, Pooja Gupta, Iti Mathur, and Nisheeth Joshi

Abstract Computers and the Internet have changed how we live our lives in ways that are often difficult to grasp, therefore it is extremely helpful to utilise computers and the Internet for the purposes of making computers and the Internet easier to use. An ontology-based search engine will assist in more accurate searches by utilising artificial intelligence to improve natural language processing and analysing keywords to more precisely target data (NLP). A general evaluation of some of the existing search engine design methodologies was introduced in this document, including with some information about the difficult areas of development. Text analysis with artificial intelligence NLP techniques could be used to build an engine that generates intelligent responses based on user requests. This research is all about intelligent ontology-based semantic search engines and evaluating their research in terms of creating methodologies. Furthermore, elements of artificial intelligence technology that are increasingly difficult to deal with, including such deep learning and machine learning, are investigated to influence development of search engines. These publications demonstrate considerable progress in the decade-long drive towards AI-infused ontology-driven search engine via use of ontologies.

Keywords Context · Search engine · Ontology · Text mining · Information retrieval · Artificial intelligence

P. Mudgil (✉) · I. Mathur · N. Joshi
Computer Science and Engineering, Banasthali University, Vanasthali, Rajasthan, India
e-mail: engineer.pooja90@gmail.com

N. Joshi
e-mail: jnisheeth@banasthali.in

P. Gupta
Computer Science and Engineering, Maharaja Agrasen Institute of Technology, Delhi, India

M. S. Kaiser et al. (eds.), *Information and Communication Technology for Competitive Strategies (ICTCS 2021)*, Lecture Notes in Networks and Systems 401,
https://doi.org/10.1007/978-981-19-0098-3_29

1 Introduction

The Internet allows many entities, such as private enterprises, colleges, and government agencies, to create computer networks and to talk to one another. A user can get services and information from any location due to the increased mobility of a rapidly growing organisation [1]. However, user queries for information in the Web database are particularly difficult. In general, Web databases deliver tremendous amounts of information. However, if people suffer from information overload, they are often forced to cope with very significant challenges like locating important information and also getting the opportunity to gather knowledge from many sites at once. For this, the end user must come up with a functional search mechanism (19) to enhance results based on what the customer needs. The searching technique will rely on the well resources and user queries matching. People that use designed search engines, like the kind used to search for information, should apply more sophisticated processes to increase the quality of their search results. So, for consumers, the alternative of using an ontology-based search engine (as shown in Fig. 1) is superior because it is more efficient [2].

Critical in knowledge-intensive jobs that require snap decisions, having up-to-date and correct information readily available is imperative. In this research, we provide a summary of current techniques for contextual searches in search engines. Human–machine interaction over the Internet has led to a simple way for users to obtain information and services. We understand that the Internet is a colossal, semi-structured database [17] in which data, according to the demands of the user, is

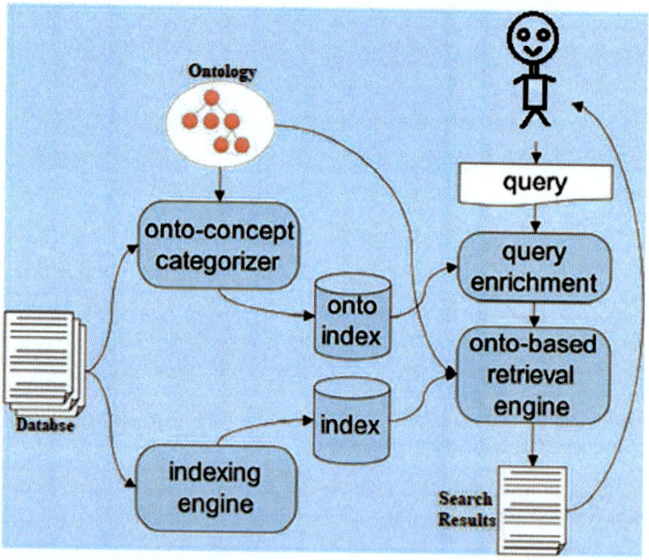

Fig. 1 Architecture of efficient search engine

plentiful. Because of the information increase, researchers confront two concerns: the increasing information overload makes finding information difficult, and the problems with retrieving and accessing multiple information sources are growing [1]. The researchers and consumers who do searches encounter several common challenges:

- Web databases are generally unstructured in their display of information, which makes them harder to find through search engines.
- There is poor interconnectedness because to data uncertainty.
- Content retrieval processes that do not provide security and are also not equipped to handle the growing community.
- Because of the lack of a common format, information offered by humans is not being understood by machines.

A significant shortcoming of present systems is the inability to retrieve and transfer data automatically. The search engine accurately finds the best material for its users, as it is ontology-based and provides efficient searches. In Fig. 2, the classic methodologies are demonstrated, and many ways to construct a good search engine to receive the correct information [3].

In this study, you can see how keyword searches are not sufficient for modern search engines. Since most people do not really have any formal background in linguistics, the algorithm that search engines use is, at best, rather simplistic and inaccurate, so it does not match their needs very well. This becomes a problem when search engines fail to provide the most relevant information because they are not very good at using complex linguistic terms. For this research, they looked at three ontology-based semantic search engines (DuckDukGo, Hakia, and Bing) in addition

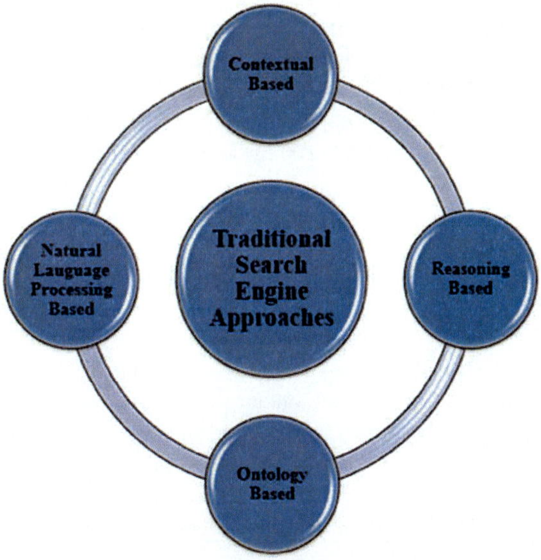

Fig. 2 Traditional search engine approach

to Google and Yahoo. Despite having increased their strategies for discovering what a search user actually wanted, search engines are falling short when it comes to understanding the search term's exact meaning. Many search engines provide the wrong URL on the first page of search results, making them less efficient.

Polysemic words and other factors suffer from an inadequate ability to return relevant search results. By searching using an ontology-based semantic context, a user will find results which align with their purpose, since the search engine detects their query and delivers results to them as they would expect. It is hard to debate the merits of Web searches without Google's search engine. However, there are techniques to search the Web without using ontology-based contextual information retrieval engines [21]. When people use ontology-based contextual semantic search engines, they are shown more intelligent and relevant results [4]. The powers of ontology-based contextual search engines come from their capacities to examine and compare data to offer the best results for queries. Ontology-based semantic Web technology enables one to implement knowledge-based online services by adding a layer of maximum accuracy, which is simply ontology-based contextual semantic Web. There are several other computer languages, such as RDF, OIL, DAML + OIL, and OWL, that provide context and meaning to the Web (OWL). You may find explanations of the search engines in the paragraphs below.

2 Ontology-Based Search Engine

2.1 RDF

It is a common ontology-based semantic search engine used for transferring complex data [8]. In the context of ontology-based RDF semantic search engine, each statement is separated into three parts; the first is the resource and is identified by a uniform resource locator, the second is a label, and the third is the information content (URI). Information may be better recognised, disambiguated, and linked by artificial intelligence (AI) systems with ontology-based contextual RDF semantic search engine.

2.2 OIL

It is a semantics-based search engine that caters exclusively to human readers, since its goal is to represent content to be easier to read. Computers cannot perceive the content of Websites, so there is a need for semantic representation, or standardised representations of data on the Web. As machines cannot detect the semantic differences amongst information types such as news, scientific papers, personal Websites, and corporate Websites, a standard representation of the semantic structure is needed to serve as a baseline. This inability to digest information significantly hinders the existing Web's operation. Computers cannot assist humans in digesting information by transmitting it to the Internet or presenting it. The overall objective of the semantic web is to design a Web where content can be processed by both people and machines. Of course, for information to be presented in an understandable fashion, it must be portrayed in a way that makes its meaning clear.

"Ontology" is included in every single definition of the acronym "OIL". Some of the definitions are ontology inference layer or ontology interchange language, but no matter what you pick, you get "ontology". Ontologies describe a community's common and agreed-upon definitions of certain terms in a specific subject. Ontologies often establish class hierarchies of things that are vital to a domain. Every class is defined by attributes that are shared by all of its members. Classes and their components have a close relationship in ontologies, thus important links between classes and the elements inside them are covered by ontologies. Now, in knowledge representation, natural language processing, information retrieval, databases, knowledge management, multi-agent systems, and more, the concept of ontologies is significant. A consensus is emerging that views them as a vital piece of the semantic Web's architecture. A semantic search engine that utilises information in the OIL ontology to aggregate information from many groups' work has been developed to enable general purpose markup language:

- Systems based on frames
- Description logics such as those that appear in discourse analysis tools
- Web coding guidelines.

2.3 DAML

DAML is an acronym for DARPA agent markup language, and it is a markup language, therefore it is also known as DAML, and it is based on extensible markup language (XML). In a nutshell, it aims to be better at expressing searchable things than current XMLs. By improving object connections, communicating meanings, and enabling a greater degree of interoperability with respect to user queries, it assists in Website development. DARPA is making DAML to incorporate an artificial intelligence system that allows the program to understand information on its own and to communicate with other programmers and agents independently. One of its features is a query language that allows it to uncover and process information relevant to its task. This means it can search and gather material on the Internet and compile it into a report that pulls together what the many online sources say.

2.4 OWL

It is furthermore a language for the semantic Web that is ontology-based and sophisticated, incorporating ways to describe ideas related to objects, groups of things, and their interrelationships. OWL is a computational logic-based language that is specifically designed to allow computer systems to use the information that is presented to uncover further insights. To publish ontologies on the Web, refer to the previously published ontologies, which are available to the world at large. To locate relevant sources and info, one has to execute well-defined searches. To help users who are looking for relevant information, improving the quality of information retrieval (IR) by using ontologies for query enrichment is a good way to get good results. By tailoring the system based on the user's priorities, we create an adaptive ontology for integration with typical vector-space retrieval systems, to give the user with relevant results and a full explanation of why they are relevant. To fit in with the vocabulary of the domain, the ontology ideas are represented as feature vectors. A feature vector helps make an existing query more informative. The ontology and retrieval system are being built as part of a broader effort to standardise the Norwegian oil and gas industry's semantic Web efforts.

Most semantic search engines give relevant results, but the responses are often bloated with irrelevant data, whilst the ontology-based contextual semantic offers exact replies with just relevant data. Artificial intelligence has spurred an increased adoption of the semantic Web, which primarily aims to make Web information intelligible not only to people, but also to machines [7]. The three different forms of AI depicted in Fig. 3 will be employed in ontology-based contextual semantic search engines.

This article discusses how various search engine strategies have developed. Section 3 specifically describes how ontology-based search engines have benefits and limitations in their working problems. Section 4 includes the coverage of the information review in the form of a survey of current research, whilst Sect. 5 lays out the general findings of the report and presents options for further action.

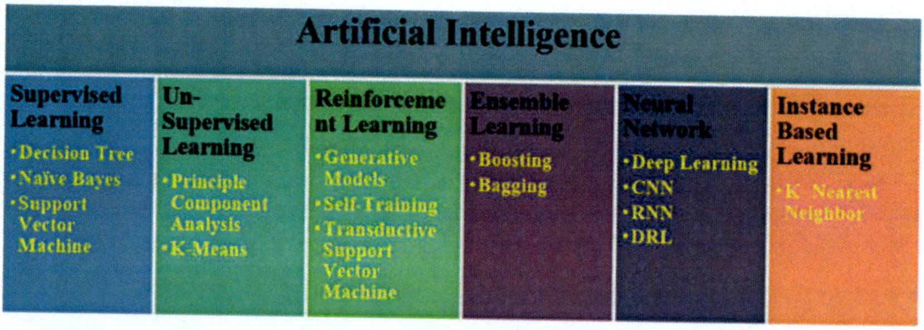

Fig. 3 Types of artificial intelligence in search engines

3 Background Survey

The results of the research appear in this part, and they are specific to the inquiry into how various artificial intelligence systems work whilst doing ontology-based contextual semantic search. A research done by B. Selvalakshmi and M. Subramaniam [5] found a semantic information retrieval model utilising feature selection and classification approaches that might help in developing an intelligent ontology. The authors employ a model for semantic information retrieval in their research, which provides a semantic retrieval facility more helpful than other accessible semantic networks and frames because of its usage of ontology for knowledge representation. The current data and information that is available through Web applications, such as Facebook and Twitter, grows very rapidly, which is too fast for older information retrieval or Web searching systems to handle. Additionally, the researchers also utilise a variety of search engines which are based mostly on syntactic information retrieval and use algorithms for ranking pages according to relevancy but are unable to give more accurate findings with regard to relevancy. To boost the system's relevance score, the researchers employed an artificial intelligence system to pick and classify features in their new semantic information retrieval system. This research had intelligent feature selection and a semantic information retrieval system that utilised fuzzy rough sets and an ontology. The suggested models and hybrid algorithms are known for their major benefits of providing a higher relevance whilst also allowing large data and speedy retrieval.

Following that, Nguyen et al. [6] designed a paradigm for an ontology-based semantic search of natural resources and environment for national databases. This study focusses on the semantic search to help users by helping them comprehend what they are searching for. In order to determine the semantic and the interrelationships between the user's search queries, ontology plays a key role in this context. Using the nation's resource and environment database is rather extensive. In the environmental and natural resources fields, it is critical to develop software with high accuracy and rapid speeds for exchanging data. Authors used semantic search software (VnNRESS) to help national databases in environmental and natural resource-related information search and discovery, resulting in a more efficient method of searching and accessing said information (NRE-search). Case-based retrieval of argument graphs uses a semantic textual similarity measure developed by M. Lenz et al. [7]. The application of computer methods for reasoning and decision-making based on argumentative frameworks is employed in argumentation, a notion of artificial intelligence. This research improves the ability to retrieve documents with similar structures to the problem by refining the matching phase of the method and explores a variety of innovative mechanisms for semantic textual similarity measures (both supervised and unsupervised mechanism) to improve the mechanism's performance for argument graphs. A range of tests was conducted to prove the reliability of an ontology-based semantic similarity measure that authors utilised to develop a system of argumentation schemes. They have shown how well the new technique outperforms old systems and what it has to offer.

Jiang et al. [8] suggested a method of using the concepts of hybridization of semantic and machine learning technologies to an ontology-based semantic search for open government data. The semantic search they described depends on metadata to define data's semantics. This research explores a reduced ontology for the transport domain to present and test the authors' ideas. Dataset NLP links, indexes, and enables the natural language used to describe it to be independent. Using domain experts and data publishers, human data connecting to ontology concepts are established, whilst automatic data linking is based on information already supplied by the data publisher. Linking to ideas manually is both laborious and dependent on domain expertise. The use of automated concepts linkage will relieve these burdens. This research shows that semantic search (with the use of ontologies) may be used to improve search quality and speed when open data are being searched.

In this study, several ontology-based contextual semantic search engine algorithms were covered, with artificial intelligence used. Nonetheless, it seems that the most successful way is to use artificial intelligence as a classifier and feature selection. The development of an ontology-based contextual semantic search engine system, tailored to different applications and reliant on diverse datasets, is possible. Table 1 summarises the research in existing work done to learn more about the process of building an ontology-based contextual semantic search engine.

After considering the results of studies on various approaches for ontology-enabled semantic search engines, the following are the key findings to bear in mind:

- Due to the more data in the system, the complexity has risen as data effects the search engine and causes the need to address system complexity.
- In the prior work, the model classification accuracy changes owing to an issue with unexpected database material, and so fluctuation in output is larger, and the likelihood of irrelevancy is higher.
- In order to get information relevant to user needs, the feature selection process must be integrated into current artificial intelligence models, which are constrained by a limited technical approach.
- The search takes longer, and the final results are worse quality, as irrelevant material exists and requires more effort to filter.

The results of such studies are implemented using artificial intelligence systems which offer categorization but are quite slow in providing feedback. Artificial intelligence algorithms help deal with inaccurate searches by working to enhance the system's categorization accuracy. Table 2 presents an overview of how current annotation contextual semantic search engine works.

4 Discussions

For the creation of an ontology-based system, we first studied the types of search engines that categorise their products based on information content, indexing

Table 1 Analysis of existing ontology-based contextual semantic search engine

References	Year	Description of model	Techniques	Drawbacks
[1]	2018	An intelligent ontology-based semantic information retrieval model	A fuzzy rough set and latent Dirichlet allocation (LDA) for semantic information retrieval were employed with intelligent feature selection	The response time increased because of the usage of irrelevancy, and it did not contain any semantic resources for users. A semantic resource with intelligent ontology would give the most relevant data possible in answer to a user's query
[2]	2019	A semantic search using an electric agents searching mechanism was developed	An electric agents searching mechanism	To address the limitations of the existing design, our VnNRESS model must use a new architecture for larger datasets. This new design needs to provide ontology-based semantic search in larger datasets
[3]	2019	A semantic textual similarity measures for case-based retrieval of argument graphs were proposed	The argumentation method (AT) has text similarity and another approach	The usage of a semantic similarity measure based on ontologies failed since the model did not handle all types of input
[4]	2019	An ontology-based semantic search for open government data was introduced	Use of NLP and machine learning technologies for semantic hybridization	The automated link-up significantly lowers manual models' burden

approach, and organisation method. Basically, search engines may be divided into two categories:

- Legacy search engines
- Search engines that provide results based on definitions.

4.1 *Traditional Search Engines*

The conventional search engines use decades-old practises and methods of recommendation and retrieval of beneficial information tailored to the user. Figure 4 illustrates the overall system design of traditional search engines, which is composed of several modules, each performing particular tasks.

Table 2 Main features of ontology-based contextual semantic search engine

Name of search engine module	Description and features
Google	This is the top search engine right now, and it is one of Google's most popular offerings (Software Company). Over 70% of the Web search market is controlled by Google. The engine's major function is to look for text on Web servers, whilst not searching for additional information, such as photos or data stored in databases
Bing	Bing was created in 2009 by Microsoft as a response to Google. Bing is the Microsoft Web browser's default search engine. They work tirelessly to improve Bing's search engine, but it still has a considerable way to go to catch up to Google
Yahoo	Google is better known, so Yahoo and Bing have a higher profile. According to netmarketshare.com, the market share of Yahoo has reached 7.68%. This also highlights the fact that despite their current leader is like a free email service, this is decreasing drastically with their recent disclosure that their user details and passwords were hacked a year ago
Baidu	It is a very popular search engine in China, with over double the market share as the runner-up in the country. It was established in 2000 by Chinese entrepreneur Eric Xu. Results for Websites, audio files, and pictures will be delivered via this online search. It provides other offerings including mapping, news, cloud storage, and several other services
AOL (America online)	And in addition to being a huge search engine, it was first known as Control Video Corporation, which began in 1983. AOL has two names over the years, with the second starting in 1991. It was America Online, then AOL in 2009. AOL is an American corporation headquartered in New York that produces multimedia content
Ask.com	They wanted search results to be structured as answers to basic questions. It is a question-and-answer site that includes plenty of archived data to give you an answer. The findings are less timely as they rely on archive and active user inputs
Excite	Whilst it is not something people know about, it nevertheless often ends up in the top 10. A basic service portal like Excite (a subsidiary of NetScape) delivers email, search engine, news, instant messaging, and weather updates to internet users
DuckDuckGo	A well-known search engine that many turn to for privacy. Their search results are generated in partnership with Yahoo, Bing, and Yummly, unlike Ask.com which is known for hiding how it actually delivers search results. It was created in California in 2008 by Gabriel Weinberg, and it gets money via search alliance network with Yahoo and Bing and through affiliate partners

(continued)

Table 2 (continued)

Name of search engine module	Description and features
Wolfram alpha	This is a computational knowledge search engine that does not deliver search results as a list of documents or Web pages. The outcomes were based on info about the search. The goal of their company is to make all information computable and available to everyone. They have a Pro option currently, which was launched in 2009 and focusses on student and educator pricing
Yandex	Russia's most popular search engine, Yandex, launched in 1997. Yandex has major exposure in Ukraine, Kazakhstan, Belarus, and Turkey, amongst other countries. It offers services such as Yandex Maps, Yandex Music, Yandex Money, online translators, and many others
Lycos	It has established itself as a leader in the search engine business. It offers email, Web hosting, social networking, and entertainment sites as its main areas of service
Chacha.com	Human-guided search is what it does, and it came into being in 2006. You may type in any question in the search box and have your response almost immediately. Additionally, it delivers mobile search and marketing. For top of installing the mobile applications on iOS and Android, you can get them on the Apple App Store as well

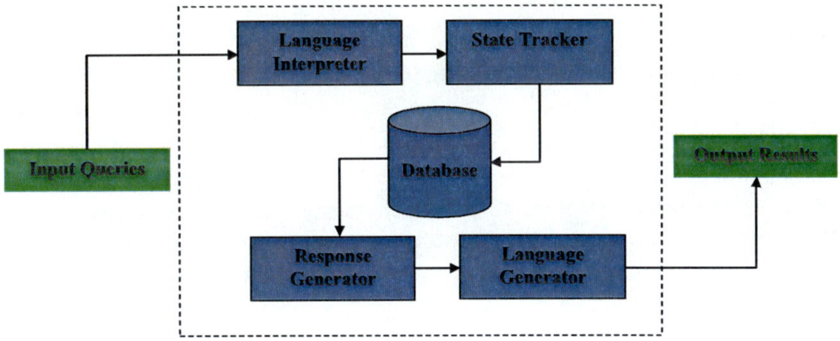

Fig. 4 Traditional search engines

4.2 *Semantic Search Engines*

Semantic search engines provide a powerful paradigm, especially when compared to standard search systems. Many search engine modules are simplified and more user-friendly. Because of this, all search engine modules are now being centralised into a single module as seen in Fig. 5.

Each type of search engines can be divided into two basic categories:

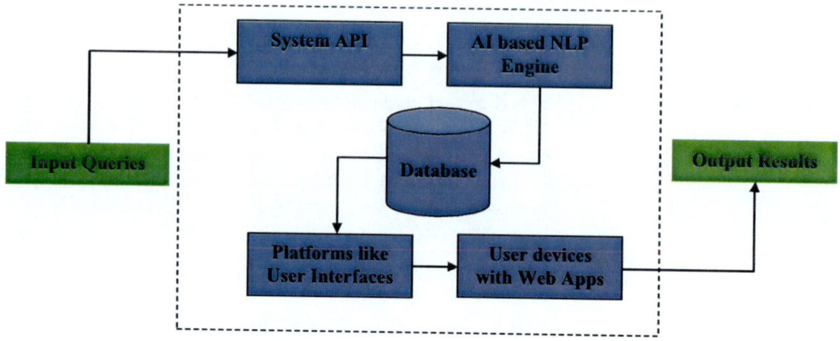

Fig. 5 Semantic search engines

4.3 Text or Keywords-Based Search Engines

Some search engines use keywords to help accomplish a certain aim. These kind of search engines can assist in finding a certain piece of information that has been entered using precise keywords. If the user completes his or her tasks, the iteration will be finished. There are lots of issues and challenges faced by the search engines which are discussed in Table 3.

4.4 Ontology-Based Search Engines

Tasks-free search engines do not have a defined objective and can continue to progress as long as it is believed that user and cyber-side agent iteration will continue to create improvements. Ontology based search engines considered the mapping of user query concepts and concepts of documents to retrieve the relevant information based on closeness of documents in vector space along with need of the user based on a particular domain.

Based on research and the current code, it is believed that artificial intelligence (AI) concepts will benefit from an ontology-based contextual sense approach for search engine usage. A feature selection strategy such as the hybrid method cooperation can even be used to increase search engine security.

5 Conclusion and Future Work

This survey article discusses how ontology-based contextual sense methods, using various searching parameters, may make search engines more efficient. It offers a

Table 3 Search engines challenges

Parameters	Challenges and issues
Database limitations	Searching through huge databases in order to locate information relevant to searches has always been quite difficult
Users way of searching	Each customer has their own unique means of contacting us (short sentences, long sentences, really long sentence in a search engines, multiple very short sentences). Understanding the requests made by search engine users is a tough challenge for AI-based search engines
Limitations of NLP	In the current level of NLP, many things have been handled with extraction of entities, synonyms, and blending of local language being dealt with. What remains are slang, words, and phrases, though, and the speed at which they are being introduced to our lexicon. Though it takes time, it is exciting and you can always change the level it is at now. This is a process that is evolving as needs emerge. Do you think we can utilise ontology-based contextual sense methods for search engines with the software?
Security	Some search engines have hacking problems with users' data, therefore focussing on the security of search engines beyond the various other factors would be wise. To increase search engine security, we will talk about a few important guidelines below • To safeguard your user's data, make sure you are utilising a protocol with your search engine • And you must make sure that the search engines will not provide answers to inquiries, for example, that might give important information about the company, that would let others access customer credit card numbers, or about other types of information • Display how and where the search engine is keeping user data to build transparency and confidence

more thorough perspective on the many types of artificial intelligence and their associated obstacles whilst attempting to construct the ontology-based semantic search engine. This study includes traditional and semantic search engines, as well as the many other innovative search engines, of which traditional and semantic search engines have the disadvantage of having incomplete or inaccurate data due to a lack of coverage on the accessible data. Additionally, it highlights the fact that there are several artificial intelligence strategies or tactics available for obtaining an effective outcome whilst employing ontology-based semantic search engine. To build a semantic search engine, selection of features and artificially intelligent architectures may need to combine in novel ways. To provide better information retrieval, future integration of artificial intelligence on twitter data set or on a particular data set is expected to use a soft computing algorithm for context based analysis of sentiments of the users.

References

1. Finkelstein L, Gabrilovich E, Matias Y, Rivlin E, Solan Z, Wolfman G, Ruppin E (2001) Placing search in context: the concept revisited. In: Proceedings of the 10th international conference on World Wide Web, pp 406–414
2. Tomassen SL, Gulla JA, Strasunskas D (2006) Document space adapted ontology: application in query enrichment. In: International conference on application of natural language to information systems. Springer, Berlin, Heidelberg, pp 46–57
3. Jiang JJ, Conrath DW (1997) Semantic similarity based on corpus statistics and lexical taxonomy. arXiv preprint cmp-lg/9709008
4. Sitthisarn S, Lau L, Dew PM (2013) A survey of semantic keyword search approaches. Int J Appl Comput Technol Inform Syst 2(2):29–37
5. Selvalakshmi B, Subramaniam M (2019) Intelligent ontology based semantic information retrieval using feature selection and classification. Clust Comput 22(5):12871–12881
6. Nguyen NV, Bui HS, Ha QT (2019) Ontology-based semantic search for national database of natural resources and environment. In: International conference on industrial networks and intelligent systems. Springer, Cham, pp 155–164
7. Lenz M, Ollinger S, Sahitaj P, Bergmann R (2019) Semantic textual similarity measures for case-based retrieval of argument graphs. In: International conference on case-based reasoning. Springer, Cham, pp 219–234
8. Jiang S, Hagelien TF, Natvig M, Li J (2019) Ontology-based semantic search for open government data. In: 2019 IEEE 13th international conference on semantic computing (ICSC). IEEE, pp 7–15
9. Ivanisenko TV, Saik OV, Demenkov PS, Ivanisenko NV, Savostianov AN, Ivanisenko VA (2020) ANDDigest: a new web-based module of ANDSystem for the search of knowledge in the scientific literature. BMC Bioinform 21(11):1–21
10. Gutierrez R (2018) A requirement ontology to guide the analysis of system life cycle processes. Doctoral dissertation, Concordia University
11. González CB (2018) Big data optimization: algorithmic framework for data analysis guided by semantics. Doctoral dissertation, Universidad de Málaga
12. Sayed A, Al Muqrishi A (2017) IBRI-CASONTO: ontology-based semantic search engine. Egypt Inform J 18(3):181–192
13. Singh R, Gonsalves TA (2018) Using ontologies to share access control information in rural business process outsourcing. IETE J Res
14. Hagelien TF (2018) A framework for ontology based semantic search. Master's thesis, NTNU
15. Rachman A, Ratnayake RC (2018) Ontology-based semantic modeling for automated identification of damage mechanisms in process plants. In: Working conference on virtual enterprises. Springer, Cham, pp 457–466
16. Tulasi RL, Rao MS, Ankita K, Hgoudar R (2017) Ontology-based automatic annotation: an approach for efficient retrieval of semantic results of web documents. In: Proceedings of the first international conference on computational intelligence and informatics. Springer, Singapore, pp 331–339
17. Chatterjee N, Kaushik N, Bansal B (2018) Inter sub-domain relation extraction for agriculture domain. IETE J Res
18. Zhang J (2021) Ontology based semantic engineering framework and tool for reconfigurable automation systems integration. Doctoral dissertation, University of Warwick
19. Chen X, Zhang X (2008) HAWK: a focused crawler with content and link analysis. In: Proceedings of IEEE international conference on e-business engineering, pp 677–680
20. Jain A, Mittal K, Tayal DK (2014) Automatically incorporating context meaning for query expansion using graph connectivity measures. Prog Artif Intell 2:129–139
21. Kumar S, Rana RK, Singh P (2012) Ontology based semantic indexing approach for information retrieval system. Int J Comput Appl 49(12)
22. Zhang R, Ying W, Wang J (2008) Research on ontology matching in semantic web. In: International conference on internet computing in science and engineering

23. Charles TM (2000) Text information retrieval systems, 2nd ed. Academic Press, San Diego, pp 131–132
24. Laurence AFP, Ramamohanarao K (2009) Efficient storage and retrieval of probabilistic latent semantic information for information retrieval. VLDB J Int J Very Large Data Bases Arch 18-1:141–155
25. Calegari S, Pasi G (2008) Personalized ontology-based query expansion. In: International conference on web intelligence and intelligent agent technology, vol 3, pp 256–259

Static GPS Surveys Using GAGAN SBAS Receiver for HR Satellite Photogrammetric Applications

V. S. S. N. Gopala Krishna Pendyala, Hemantha Kumar Kalluri, and Rajashree Vinod Bothale

Abstract Communication technology is used to transmit the error correction signal to the GPS receiver from geostationary satellites to improve the accuracy of position information provided by the receiver. This study uses GPS-Aided GEO Augmented Navigation (GAGAN) Satellite-Based Augmentation System (SBAS) GPS receiver in stand-alone mode to provide a simple and cost-effective solution with improved positional accuracy to collect the ground control points. In this study, the GAGAN receiver is operated in static mode for longer durations at International GNSS Service (IGS)—HYDE reference station, whose position is known very accurately. It is found that the GAGAN receiver can achieve accuracies of 0.3–0.6 m in position and up to 1.0 m in elevation depending on the duration of the observation meeting the requirements of many survey and engineering applications. Using the GCPs obtained by this process, orthophoto having a planimetric accuracy of 1.0 m can be generated from CARTOSAT-2E High-Resolution (HR) satellite images. The typical height accuracies of the order of 2.0 m (3–4 pixels) could be achieved from the Digital Surface Model (DSM) derived from this process.

Keywords GAGAN · Global Positioning System · Ground Control Point · High-Resolution Satellite · Photogrammetry

1 Introduction

Information and Communication Technology (ICT) has accelerated the development by helping to interpret and manage data about an area's resources and infrastructure. The tools that are responsible for this revolution are (i) devices that measure geographic location such as Global Positioning System (GPS), (ii) geographically

V. S. S. N. Gopala Krishna Pendyala (✉) · R. V. Bothale
National Remote Sensing Centre, Hyderabad, Telangana 500037, India
e-mail: gopalakrishna_pvssn@nrsc.gov.in

H. K. Kalluri
Vignan's Foundation for Science Technology and Research, Vadlamudi, Andhra Pradesh 522213, India

M. S. Kaiser et al. (eds.), *Information and Communication Technology for Competitive Strategies (ICTCS 2021)*, Lecture Notes in Networks and Systems 401,
https://doi.org/10.1007/978-981-19-0098-3_30

referenced data; and (iii) the data collection systems that provide periodic land use, land cover, and other thematic information.

The GPS is a space-based navigation system used for obtaining position, velocity, and time. GPS receiver estimates the position of a particular location by measuring the ranges from the satellites and applying trilateration principle. Based on the accuracy requirements, GPS is operated in 3 different modes (i) stand-alone mode, (ii) Differential GPS mode (DGPS), and (iii) augmented mode.

The communication revolution and its effective use for DGPS error correction have paved the way for the satellite-based augmentation systems (SBAS) which provide improved positional accuracy using the C/A code to suit real-time requirements. Systems of this type use a single-frequency, low-cost receiver that receives the augmented error correction from a geostationary communication satellite, similar to GPS satellite carrier signal. SBAS receiver can be deployed rapidly compared to DGPS [1, 2] since it is a single-receiver operation. The error budget of different operational modes of the GPS receiver is given in (Table 1).

The topographic maps derived using photogrammetry technique on High-Resolution (HR) satellite images find many applications in civil and other engineering domains to generate image products with improved positional accuracies. Photogrammetry uses the GPS measured Ground Control Points (GCPs) to geo reference the images/data as well as derive 2D and 3D maps and other resource details to help in increased accuracy which leads to productivity gains and cost reductions.

Traditionally, GCPs are derived by operating the GPS receiver in DGPS static mode for longer durations. DGPS involves 2 receivers of geodetic class operation and post-processing of data. This involves operating high cost receivers at 2 places, viz. base and rover requiring costly receivers and more logistic requirements.

Hence, experiments are conducted by operating GPS-Aided GEO Augmented Navigation (GAGAN) SBAS receiver, in static mode over long periods that offer a simple and cost-effective solution to meet the GCP accuracy requirements. This paper provides an analysis on the achievable accuracies using the GAGAN receiver in static mode. To demonstrate the usefulness of the study, photogrammetric product is

Table 1 Error budget of GPS measurements [1–3]

Error source		Typical positional error on the ground (using C/A code)		
		Standalone mode (m)	DGPS (m)	SBAS (m)
Ephemeris data		2	0	0.4
Clock model		2	0.1	0.2
Ionospheric delay	Single frequency	0.2		0.5 (using model)
	Dual frequency	0	0	–
Tropospheric delay		2.3–2.5	0.2	0.2
Multipath		0.5–1	0.15	0.1
Receiver noise		0.25–0.5	0.2	0.2
Typical error		8–10	0.5–1	1–2

generated using the CARTOSAT-2E of Indian Space Research Organisation (ISRO) High-Resolution Satellite stereo-images and checked for achieved accuracies.

2 GAGAN Configuration

GAGAN is the SBAS system available for Indian region developed and operated by ISRO along with the Airport Authority of India (AAI) used for real-time navigation applications in aviation. It reduces the errors in the GPS system using the satellite augmented corrections in real-time by utilizing modern communication technology.

Suryanarayana and Mrutyunjaya [1, 3] provided detailed information about GAGAN and its configuration as shown in Fig. 1. Fourteen reference stations collect data from GPS satellites, master control station estimates the clock and ephemeris correction for each GPS satellite, and vertical ionosphere delays on a grid model is named as ISRO GIVE Model-Multi-Layer Data Fusion (IGM-MLDF) [1, 3, 4]. Communication technology is the heart of GAGAN system. The master station sends these corrections and error bounds to the user receiver through GEO communication satellites [1].

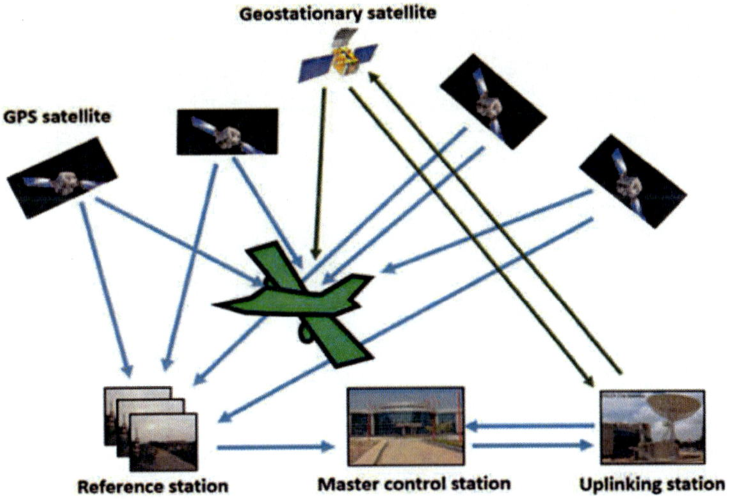

Fig. 1 Configuration of GAGAN [3]

3 Literature Study

Feasibility studies for using GAGAN receiver in static survey mode was carried out by Dammalge and Rani [5, 6]. For static observation of 10 min duration, Dammalge [5] found that the GAGAN active L1 receiver has significant accuracy improvement, and 79% of observations showing 3D positional accuracy of less than 1.0 m. Planimetric accuracy of 0.5–0.6 m was achieved by Rani [6] for observations of 15 min duration. These studies suggest that GAGAN receiver has the potential to provide cheaper solution with low-cost single-receiver deployment for GCPs.

4 Materials

This study used (i) GAGAN receiver for collecting the GCPs, (ii) International GNSS Service (IGS) reference station, and (iii) HR satellite stereo pair images for photogrammetric adjustment and on field accuracy assessment.

4.1 GAGAN User Receiver

General-purpose off-the-shelf U-Blox Neo-6 series GPS receiver configured for receiving GAGAN active signals with data logging capability is used for this study.

4.2 IGS Reference Station

IGS reference station is a network of 500 stations operate globally, permanently, and continuously having well-established coordinates, accurate to 3 mm in planimetry and 6 mm in vertical [7]. The GAGAN user receiver is operated for estimating the deviations for each epoch at HYDE station located at Hyderabad, India.

4.3 Dataset and Study area

CARTOSAT-2E HR satellite data having a Ground Sample Distance (GSD) of 0.6 m (PAN) and 1.6 m (MS) is used to generate photogrammetric products. The study area is in the Western part of Hyderabad with moderate undulations and urban built up.

5 Method

The sequence of operations followed for data collection, processing, and estimation of accuracy is listed below:

- Operating GAGAN user receiver at IGS station for longer durations
- Discard erroneous observations and outliers
- Divide the data into sessions of various durations and remove biases if any
- Estimation of RMS error for each session
- Estimation of combined RMS error for multiple sessions of each duration
- Analysis of data
- Photogrammetric process and estimation of product accuracy.

5.1 GAGAN Receiver Operations at IGS Station

The GAGAN used receiver operated in static mode with 2 s epoch at IGS station (Fig. 2) for three days by logging the output position coordinates data in NMEA format. The duration of observations is as follows:

- DAY 1: For 19 h from 12:03 on 20-12-2018 to 06:56 on 21-12-2018
- DAY 2: For 100 min from 14:32 to 16:12 on 03-01-2019
- DAY 3: For 8 h from 09:44 to 17:45 on 26-04-2019.

The observations were split into multiple sessions for every 15, 30, 45, 60, 120, 180, 240, 360, and 480 min. The data epochs that were discarded from data analysis (i) include the first 2 min data, (ii) are unable to record GAGAN signal (GSAT-8 or GSAT-11), and (iii) are having GDOP of more than two and (iv) deviations more than 3σ from the average.

Fig. 2 GAGAN GPS receiver operation at IGS station

Each epoch was compared with IGS HYDE station coordinates. Since low cost, the general-purpose receiver is used for this study; there will be biases due to the receiver clock as well as errors in orientation and datum [8]. Hence, the X, Y, and Z biases were estimated and applied to the observed data.

5.2 *Accuracy Estimation*

Session wise, RMS error was estimated, and then multiple sessions were grouped/binned for single duration of observation. Duration wise, the error is quantified as given in (Table 2). These results show that there is an association between the duration of observation and the RMS error. As the duration of observation increases, the RMS error reduces as seen from (Fig. 3).

5.3 *Photogrammetric Process*

GCPs: 6 Ground Control Points are acquired for 30 min duration each, using GAGAN receiver to provide a GCP accuracy of 0.5 m.

Block Triangulation: CARTOSAT-2E high-resolution images and the GCPs collected by GAGAN receiver are used for block adjustment. Detailed explanation of the block triangulation procedure using stereo-images, associated RPCs, and GCPs is provided by Pendyala [8–10]. ERDAS Imagine software is used for automatic tie point generation, orientation, and block triangulation. The triangulation accuracy measured as the total image unit weight RMSE is 0.54 pixels.

DSM and Orthophoto generation: The DSM extraction is carried out using the Semi-Global Matching (SGM) method [11]. Per pixel DSM [9] is derived having a point density of 2.75 points/sq. mt. Orthophoto is generated using the block adjusted orientation parameters, and the generated DSM [9] having a GSD of 0.6 m (Fig. 4).

6 Discussion

To assess the planimetric and height accuracy, a small subset of an area of 1.69 km^2. was extracted from the study area. Well-distributed 10 Independent check points (ICPs) were collected in this area (Fig. 4), using the traditional DGPS geodetic survey with 30 min observation duration having an RMS error of the order of 0.1 m in planimetry and 0.15 m in height.

The planimetry and height accuracy are assessed by the way of comparing the check points with the locations measured from orthophotos [12, 13]. The deviations

Table 2 GAGAN static surveys—accuracy achieved

Observation duration (s)	Number of sessions	X			Y			Positional error (m)	Z		
		Deviation		RMS	Deviation		RMS		Deviation		RMS
		Min	Max		Min	Max			Min	Max	
15	112	−1.08	1.29	0.49	−1.00	1.17	0.36	0.61	−2.67	3.19	1.11
30	55	−0.82	−0.77	0.42	−2.62	0.42	0.32	0.53	0.32	0.99	0.99
45	36	−0.85	−0.71	0.40	−1.98	0.40	0.29	0.49	0.29	0.95	0.95
60	28	−0.79	−0.65	0.38	−1.98	0.38	0.27	0.47	0.27	0.91	0.91
90	18	−0.70	−0.57	0.36	−1.49	0.36	0.23	0.43	0.23	0.84	0.84
120	12	−0.52	−0.50	0.36	−1.74	0.36	0.23	0.43	0.23	0.79	0.79
180	8	−0.42	−0.38	0.28	−0.79	0.28	0.21	0.35	0.21	0.57	0.57
240	6	−0.40	−0.30	0.26	−0.65	0.26	0.21	0.33	0.19	0.58	0.47
360	4	−0.20	−0.21	0.26	−0.50	0.17	0.20	0.33	0.20	0.30	0.40
480	3	−0.14	−0.08	0.20	−0.16	0.17	0.20	0.28	0.06	0.19	0.30

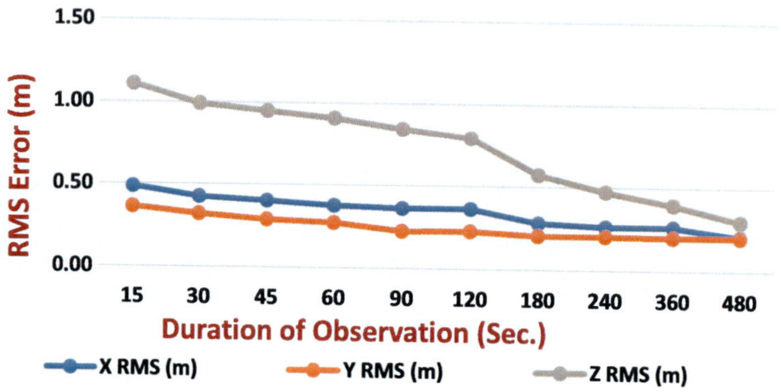

Fig. 3 Duration of observation versus control point accuracy

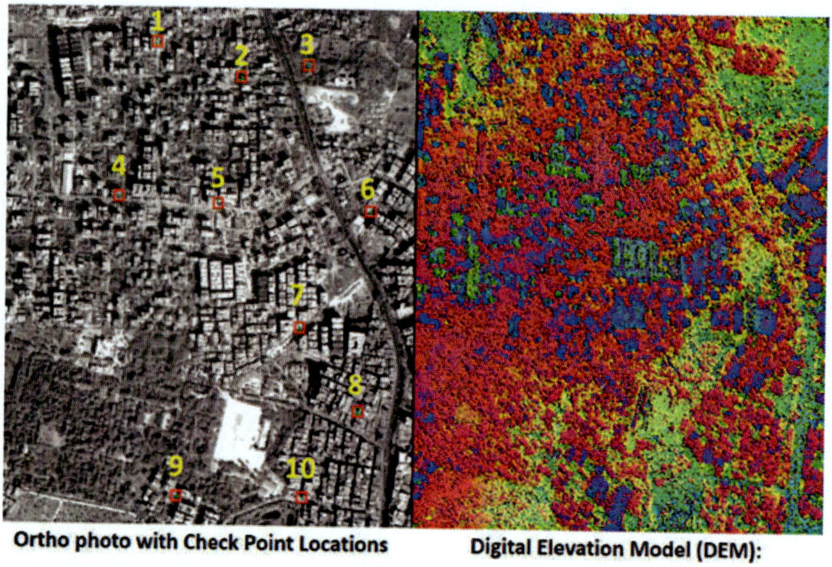

Fig. 4 Orthophoto and DSM

in planimetry (*X*, *Y*) between the ICPs and the same location in orthophotos are reported. A similar comparison was carried out with the DSM for deviations in elevation. It is observed that an RMS error of 1.03 m in planimetry and 1.97 m in height is observed (Table 3).

Table 3 Planimetric and elevation accuracy

Point id	*X* deviation (m)	*Y* deviation (m)	Planimetry deviation (m)	Z deviation (m)
1	1.59	0.84	1.27	2.97
2	0.11	2.37	1.68	−2.93
3	1.22	−0.47	0.92	−0.43
4	0.14	1.01	0.72	1.55
5	−1.29	−0.78	1.07	−0.45
6	−0.81	−0.81	0.81	−2.81
7	−0.37	−0.59	0.49	2.66
8	1.87	−0.07	1.32	−1.86
9	−0.8	−0.37	0.62	−0.25
10	1.02	0.48	0.80	−0.5
RMS error	1.08	0.98	**1.03**	**1.97**

7 Conclusions

From the study, we can conclude that GAGAN static GPS surveys present a viable alternative to DGPS control point and checkpoint surveys for the projects requiring a positional accuracy of 0.5 m and above. Using these GCPs and ICPs, orthophotos having a planimetric accuracy of 1.0 m (1–2 pixel accuracy) can be produced using CARTOSAT-2E High-Resolution satellite image data of 0.6 m GSD. The typical height accuracies of the order of 2.0 m (3–4 pixel accuracy) could be achieved from the DSM derived using these images. This method offers a cost-effective and faster solution for researchers as an alternative to DGPS surveys. However, it is not a complete replacement to the DGPS survey, and limitations such as multipath, receiver noise, and other biases need to be accounted.

Acknowledgements The authors thank Director, NRSC, for permitting to undertake this work and also for providing the satellite data set and the GAGAN receiver. The authors also acknowledge the support and help provided by the National Geophysical Research Institute (NGRI) by providing access to the IGS station.

References

1. Suryanarayana Rao KN (2007) GAGAN-Indian satellite based augmentation system. Indian J Radio Space Phys 36:293–302
2. Michael MA (2004) Signal quality monitoring for GPS augmentation systems. PhD dissertation, Stanford University
3. Mrutyunjaya L (2019) Overview of GAGAN and IRNSS. Retrieved 07 October 2019 from https://www.narl.gov.in › PDFs ›_Mruthunjaya_IRNSS_GAGAN
4. Balamurugan (2014) The land based positional accuracy of GPS and GPS augmented by GAGAN over the Indian geographical region. Int J Sci Technol Res 3(4):61–67

5. Dammalage T, De Silva DN, Satirapod C (2017) Performance analysis of GPS aided geo augmented navigation (GAGAN) over Sri Lanka. Eng J 21(5):305–315. https://doi.org/10.4186/ej.2017.21.5.305
6. Rani TE, Krishnaiah P, Muralikrishnan S, Sreedhar M (2013) Performance analysis of GCPs derived from GAGAN-SBAS for satellite data geo-referencing and mapping applications, vol XXXIII. Indian Cartographer
7. Craddock A, Gary J (2016) Innovation: the international GNSS service 25 years on the path to multi-GNSS. GPS World, September 2016
8. Jacobsen K (2013) DEM generation from high resolution satellite imagery. German Soc Photogrammetry Rem Sens Geoinform. https://doi.org/10.1127/1432-8364/2013/0194
9. Pendyala VSSNGK, Kalluri HK, Rao CV (2019) Dense DSM and DTM point cloud generation using CARTOSAT-2E satellite images for high resolution applications. J Indian Soc Rem Sens. https://doi.org/10.1007/s12524-019-01051-0
10. Jacobsen K (2013) DEM generation from high resolution satellite imagery. Photogrammetrie Fernerkundung Geoinformation 2013(5):483–493. https://doi.org/10.1127/1432-8364/2013/0194
11. Hirschmüller H (2008) Stereo processing by semi-global matching and mutual information. IEEE Trans Pattern Anal Mach Intell 30(2)
12. Aguilar MA, Aguera F, Aguilr FJ, Carvajal F (2008) Geometric accuracy assessment of the ortho rectification process from very high resolution satellite imagery for common agricultural policy purposes. Int J Rem Sens 29(24):7181–7197
13. Aksu O, Onder M, Iscan L, Müh Y, Atak VO (2019) Accuracy assessment of high resolution satellite images. Retrieved 08 October 2019 from https://www.harita.gov.tr/images/egitim/65b25dec393a9d8.pdf

COVID-19 Risk Prediction with Regularized Discriminant Analysis and Lasso Regression Using Booster Tree

Sadiya Sulaiman, Nader Salam, U. Barakkath Nisha, and R. Yasir Abdullah

Abstract The severe acute respiratory syndrome coronavirus which gives rise to coronavirus disease undergone transformation since the start of the pandemic, resulting in distinct virus types. The delta variation is one of these options. The CDC has designated the delta coronavirus as a different form of anxiety since it is highly contagious. Since the virus's outbreak, emergency rooms have been flooded with COVID-19 patients who needed to make a quick decision about whether they should be admitted or treated at home [1]. Many scientific articles have been released in order to better understand the sickness and assist humanity in overcoming the pandemic. The prediction method of COVID-19 is illustrated below using a mix of the regularized discriminant analysis feature selected (FE) technique, Lasso regression feature extraction (FS) algorithm, and a boosting tree model. The suggested analytical approach of ML could be readily installed and utilized in medical crisis unit to estimate people having coronavirus disease with high threat quickly and accurately. This article uses a dataset from Kaggle. All collected results demonstrate that RDA and Lasso R with booster tree outperform other approaches in predicting the risky people of coronavirus delta variant with efficiency 84%.

Keywords COVID-19 · Delta variant · Classification · Feature selection · Machine learning · Feature extraction · Lasso regression · Boosting tree · Regularized discriminant analysis

S. Sulaiman (✉) · N. Salam
Computer Science Engineering, Karunya Institute of Technology and Sciences, Coimbatore, Tamil Nadu, India
e-mail: sadiyasulaiman@karunya.edu.in

N. Salam
e-mail: nadersalam@karunya.edu.in

U. B. Nisha
Information Technology, Sri Krishna College of Engineering, Coimbatore, India

R. Y. Abdullah
Computer Science and Bussiness System, SKCE, Coimbatore, TamilNadu, India

M. S. Kaiser et al. (eds.), *Information and Communication Technology for Competitive Strategies (ICTCS 2021)*, Lecture Notes in Networks and Systems 401,
https://doi.org/10.1007/978-981-19-0098-3_31

1 Introduction

The emergence of the delta variety is both unwanted and unexpected to a world that has grown tired of playing defense against SARS-CoV-2 which was originally discovered in India in December 2020 [2]. Cases and hospitalizations are on the rise again around the world [2]. The established method can assist hospitals and medical institutions in determining who is at high risk and admitted to the hospital, triaging patients at the time of overload, and eliminating procrastination in delivering needed care. The algorithm forecasts death probability from victims biological states, indications, and demography features, allowing for correct care to be provided to those who require it. A series of methods for data mining for feature extraction (RDA), dealing with omitted features, removing unnecessary features, and feature selection are included in our proposed system (Lasso R). We employ boosting tree ML methods for the construction of prediction system that classifies the features, predicts the clinical case, and calculates its likelihood of death after preprocessing of data. The mined feature was made publicly available in the hopes of assisting the scientific community 1. The next part is divided as follows: Part 2 introduces various related literature survey, part 3 discusses each approach in depth, including the model, data pretreatment, feature selection, and feature extraction, as well as the obstacles we faced and the actions we took to overcome them. The results are presented in part 4 using a variety of methodologies and measurements. The conclusion is found in part 5.

2 Related Work

Many studies of ML to find coronavirus disease risk prediction have been published in the literature. Table 1 contains full information on the studies, including dates of completion, classification methods, feature extraction methods employed, and results obtained.

3 Proposed System

Dealing with data features containing tons of characteristics is becoming increasingly prevalent. When the total features in a DS approach (or exceeds!) the total finding contained in the DS, a machine learning model is likely to suffer from overfitting and then use one or the other: regularization or reduce dimension to avoid this type of difficulty (feature extraction). When regularized, it can lessen the risk of overfitting while also providing a variety of additional benefits like improvements in accuracy, overfitting, risk reduction, training speed, and improved data visualization are just a

Table 1 Literature survey

Study	Year	Method	Accuracy
Elena Casiraghi et al. [3]	2020	Associative tree, Boruta feature selection, random forest classifier	74
Pourhomayoun et al. [4]	2020	Filter feature selection and wrapper feature selection, neural net	89%
Hamza Turabieh et al. [5]	2021	A\Binary genetic algorithm (BGA) and CNN	76
Furqan Rustam et al. [6]	2020	LinearR•Lasso R•SVM•ES	
Norah Alballa et al. [7]	2021	LR	
Batista et al. [8]	2020	SVM and RF	84
Barbosa et al. [9]	2020	PSO, EA, and selection	
Celestine Iwendi et al. [10]	2020	Boosting, random forest classification	86
Muhammad et al. [11]	2020	Machine learning, decision tree	95

few examples. Feature extraction is a technique for reducing the amount of features in a dataset [12].

Feature selection is another method for reducing the amount of features in a dataset. The difference of FS and FE is that selecting features seeks to categorize values of current features in the DS and eliminate that are not as important (no new features are created). So regularized discriminant analysis is used to extract the features, while Lasso regression is useful for choosing features of the model proposed, after which boosted tree is used to predict the risky patients. The model proposed is depicted in Fig. 1.

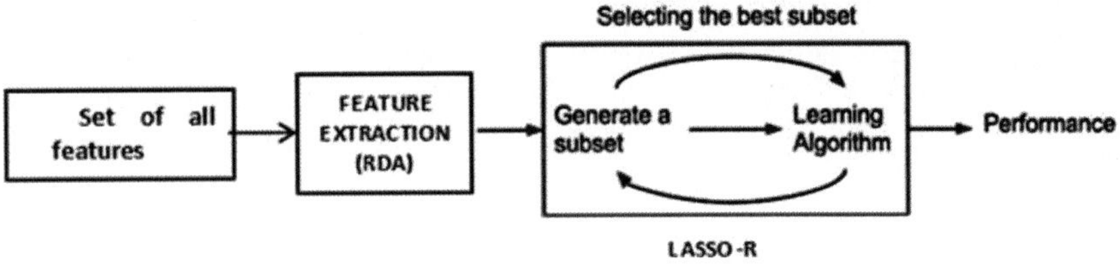

Fig. 1 Proposed model

3.1 Database Used

The dataset used in this study was accessed from Kaggle. Features are patientid, sex, patient_type: Outpatient, Inpatient, entry_date, date_symptomdate_died, pneumonia, age, pregnancy, diabetes, asthma, Hypertension, other_disease, cardiovascular, obesity, renal_chronic, tobacco, contact_other_covid, Awaiting Results, and icu.

3.2 Regularized Discriminant Analysis Feature Extraction

When the number of data exceeds the number of predictors ($n >$ pr), linear discriminant analysis and QDA are uncomplicated. We will make some changes to the LDA and QDA, namely we will use a tuning parameter α to create a new covariance matrix that combines the covariance matrix of LDA ($\widehat{\sum}$) and QDA ($\widehat{\sum}_p$) [13].

$$\sum_P(\alpha) = (1-\alpha)\sum_P + \alpha\sum_P \tag{1}$$

However, some version of regularized discriminant analysis uses another parameter (β) with the following equation:

$$\sum_P(\alpha,\beta) = (1-\beta)\sum_P(\alpha) + \beta\frac{1}{q}\mathrm{tr}\left(\sum_p(\alpha)\right)L \tag{2}$$

RDA limits the discrete variance of QDA toward the common variance of LDA that can enhance the estimates of variance matrix, where total predictions are more than the total set of trained data leads to the enhancement in system efficiency. In above Eqs. (1) and (2), the equation alpha and beta both have values between 0 and 1. Now, for all the four boundary values, it produces a special equation case for each one. Let us look at these special cases:

($\alpha = 0, \beta = 0$) the QDA covariance, i.e., each group's individual covariance
($\alpha = 0, \beta = 1$) LDA's covariance matrix, also known as a common covariance matrix.
($\alpha = 1, \beta = 0$) conditional independent variance.
($\alpha = 1, \beta = 1$) similar to the previous scenario, classification using Euclidean distance, but variances are the same for all groups [13].

3.3 Lasso Regression Feature Selection

The concern with linear regression is that the model's predicted coefficients might grow enormous, making it sensitive to inputs and potentially unstable. This is especially true for issues, where the number of observations (samples) or samples (*m*) is greater than the number of input predictors (*pd*) or variables (so-called $pd \gg m$ problems). Changing the loss function to include additional costs for a model with big coefficients is one way to address the stability of regression models. Penalized linear regression is a term that refers to linear regression models that apply these adjusted loss functions during training. A common penalty is to penalize a model by adding the absolute coefficient values. This is referred to as the *L1* penalty in Eqs. (3) and (4) [14].

$$\text{sum } j = 0 \text{ to } pd \text{ abs(beta } j) L1 \text{ penalty} \tag{3}$$

LASSO regularization is a penalty that can be added to the cost function for linear regression (LASSO). The weightage function (penalty) is called "lambda."

$$\text{loss} + (\text{lambda} * L1 \text{ penalty}) = \text{lasso loss} \tag{4}$$

3.4 XGB Gradient Boosting

A method that combines many base models to get optimal predictive model is called gradient boosting. In GB, weak models operate in a sequential manner. Each model attempts to correct the flaws of the previous one. This is how the boosting procedure works: Create a base model using the average of values to get one output, find the residuals by comparing the predicted and real numbers, give the erroneous forecasts greater weight, create a new model that seeks to correct the flaws in the previous one, and run the new model's predictions across the whole dataset. Finally, create a number of models, each one trying to remedy the faults caused by the preceding one. Weight the mean of all the models to get the final model [15].

4 Result Analysis and Discussion

4.1 Evaluate the Model Accuracy

See Figs. 2, 3, 4 and 5.

	precision	recall	f1-score	support
0.0	0.86	0.85	0.86	4032
1.0	0.50	0.53	0.51	1146
accuracy			0.78	5178
macro avg	0.68	0.69	0.68	5178
weighted avg	0.78	0.78	0.78	5178

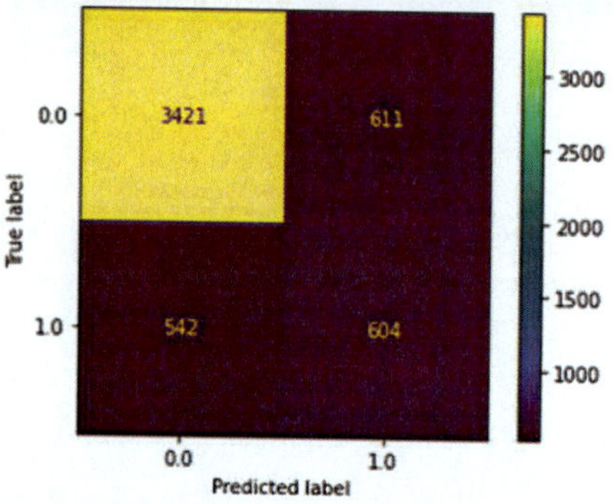

Fig. 2 Classification report and error matrix of XGBoost

	Precision	recall	f1-score	support
0.0	0.67	0 .59	0.62	4032
1.0	0.32	0.82	0.46	1146
accuracy			0.84	5178
macro avg	0.49	0.74	0.58	5178
Weighted avg	0.86	0.82	0.85	5178

Fig. 3 Classification report XGBoost with RDA and Lasso

4.2 Evaluation Metrics

To compare the two boosting models, we calculated their overall accuracy. We also generated error matrix, ROC curves, and classification report for both algorithms. We double-checked for overlap once more. At any level, there is no training and testing features did not show any overlap. ROC for each ML method employed in this study is shown and compared in Fig. 5. Error matrix (Figs. 2 and 4) is used to explain and

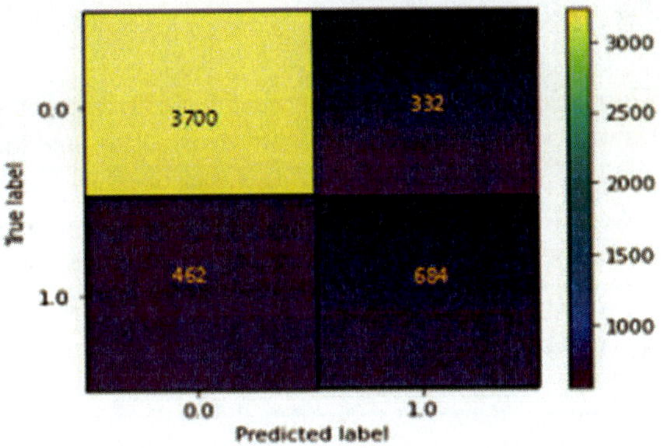

Fig. 4 Error matrix of XGBoost with RDA and Lasso

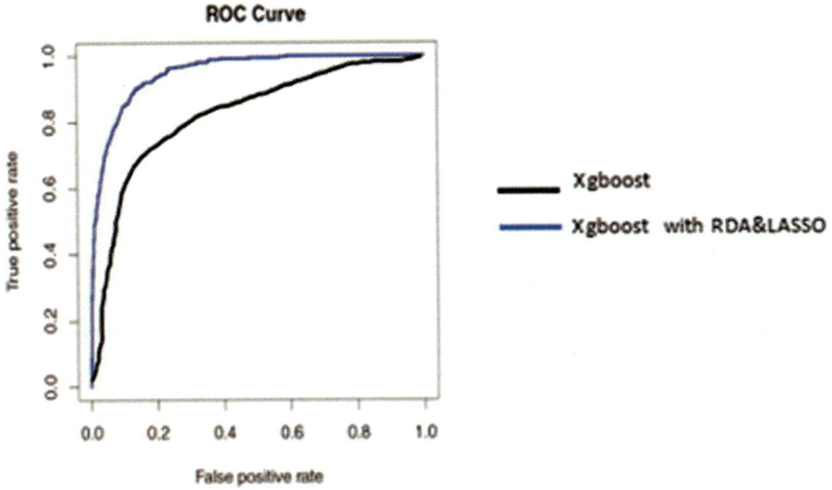

Fig. 5 ROC curve comparison

display the booster tree classifier's performance as well as to give insight into what the model misclassifies. The error matrix was used to calculate the system's TPR and TNR in Eqs. (5) and (6).

$$\mathrm{TPR} = \frac{\mathrm{TP}}{\mathrm{P}} = \frac{\mathrm{TP}}{\mathrm{TP} + \mathrm{FN}} = 1 - \mathrm{FNR} \quad (5)$$

$$\mathrm{TNR} = \frac{\mathrm{TN}}{\mathrm{N}} = \frac{\mathrm{TN}}{\mathrm{TN} + \mathrm{FP}} = 1 - \mathrm{FPR} \quad (6)$$

Condition positive (P), Condition negative (N), True positive (TP), True negative (TN), False positive (FP), False negative (FN).

4.3 Discussion

The assessment findings show that the created models are very accurate and effective. Using a boosting tree model, feature extraction, and feature selection approaches, we were able to achieve an amazing accuracy of 84.43% in predicting the risky people of COVID-19 delta variant.

5 Conclusion

This method can assist health centers and medical centers in prioritizing risky patients and reducing wait times for critical care. This research might be expanded to include additional diseases, allowing the healthcare system to respond more quickly in the event of an epidemic or pandemic.

References

1. https://www.hopkinsmedicine.org/health/conditions-and-diseases/coronavirus/a-new-strain-of-coronavirus-what-you-should-know
2. https://asm.org/Articles/2021/July/How-Dangerous-is-the-Delta-Variant-B-1-617-2
3. Casiraghi E et al (2020) Explainable machine learning for early assessment of COVID-19 risk prediction in emergency departments. IEEE Access 8:196299–196325. https://doi.org/10.1109/ACCESS.2020.3034032.141433
4. Pourhomayoun M, Shakibi M (2020) Predicting mortality risk in patients with COVID-19 using artificial intelligence to help medical decision-making. https://doi.org/10.1101/2020.03.30.20047308
5. Turabieh H, Karaa WBA (2021) Predicting the existence of COVID-19 using machine learning based on laboratory findings. In: 2021 international conference of women in data science at Taif University (WiDSTaif), pp 1–7. https://doi.org/10.1109/WiDSTaif
6. Rustam F et al (2020) COVID-19 future forecasting using supervised machine learning models. IEEE Access 8:101489–101499. https://doi.org/10.1109/ACCESS.2020.2997311
7. Alballa N, Al-Turaiki I (2021) Machine learning approaches in COVID-19 diagnosis, mortality, and severity risk prediction: a review. Inform Med Unlocked 24:100564. https://doi.org/10.1016/j.imu.2021.100564. Epub 2021 Apr 3. PMID: 33842685; PMCID: PMC8018906
8. https://www.medrxiv.org/content/https://doi.org/10.1101/2020.08.31.20185165v2
9. de Freitas Barbosa VA, Gomes JC, de Santana MA et al (2021) Heg.IA: an intelligent system to support diagnosis of Covid-19 based on blood tests [published online ahead of print, 2021 Jan 7]. Res Biomed Eng 1–18. https://doi.org/10.1007/s42600-020-
10. https://www.frontiersin.org/articles/https://doi.org/10.3389/fpubh.2020.00357/full
11. Jibril M, Algehyne EA, Sharif U, Ahmad A, Chakraborty C, Mohammed I (2021) Supervised machine learning models for prediction of COVID-19 infection using epidemiology dataset. SN Computer Science 2
12. https://towardsdatascience.com/feature-extraction-techniques-d619b56e31be
13. https://www.geeksforgeeks.org/feature-selection-techniques-in-machine-learning/
14. https://machinelearningmastery.com/elastic-net-regression-in-python/
15. https://scikit-learn.org/stable/modules/generated/sklearn.ensemble.GradientBoostingClassifier.html

Industry 4.0 Framework Using 7-layer Architecture for Smart Factory Application

L. R. Sreedhanya and Sudeep Balan

Abstract The proposed Industry 4.0 framework implementation approach provides a flexible, resilient solution for intelligent manufacturing. The 7-layer system architecture of the Industry 4.0 framework was reported based on the digitization and centralization of industrial data. An Industry IoT framework demands flexibility, scalability, interoperability, and modularity for industrial manufacturers. The system architecture features industrial real-time data capturing, cross-enterprise data sharing, standard data model transformation, management, analysis, and hybrid visualization and control. Implementation of this framework for industrial applications is directed to further research needs. This paper presents the initial motivation, current status, and future states of a sugar factory.

Keywords Industry 4.0 · IoT · SCADA · Big data · Sugar factory

1 Introduction

Industry 4.0 focuses on the digital transformation of the industrial market with smart manufacturing in discrete and process manufacturing logistics and supply chain. Industry 4.0 is a technical domain that defines the digitization of the manufacturing process. It also refers to the emergent change in the industrial value chain process in line with the Internet of Things flourishing across the corners of the world. The advanced innovative approaches and changes assist the existing industrial process in managing and organizing standard processes like prototyping, development, production, logistics, and supply. Production processes in the Cyber-Physical Production Systems (CPPS) based on intelligent products or processes are intelligent, cooperative, and self-adaptive. It enables smooth collaboration between R&D, laboratory, and production. Detailed learning and education of Industry 4.0 have triggered the use of IoT concepts, Cloud warehousing, and digitization of the plant manufacturing processes in line with the standardized methods of Industry 4.0 techniques. The use

L. R. Sreedhanya (✉) · S. Balan
Centre for Development of Advanced Computing (CDAC), Thiruvananthapuram, Kerala, India
e-mail: sreedhanya@cdac.in

M. S. Kaiser et al. (eds.), *Information and Communication Technology for Competitive Strategies (ICTCS 2021)*, Lecture Notes in Networks and Systems 401,
https://doi.org/10.1007/978-981-19-0098-3_32

of high-speed field sensors and PLCs populates a high volume of industrial data in all formats. After analysis using high-speed computational analytical tools, these data can be recycled and pushed back to the manufacturing process to decide the feedback parameters. With the implementation of Industry4.0 using the proposed framework, we aim to optimize the plant processes by achieving yield or maximizing productivity with reduced cost and less computational time.

1.1 Smart Factory

A Smart Factory, as a core idea, is a conversion from digital to physical in a reconfigurable manufacturing system. Physical systems, networking, and computations in the smart factories are cyber-physical systems (CPS). These systems do integrate with information and communication technology (ICT) components. They are autonomous systems that belong to a network or collection that makes decisions based on real-time data acquisition, historical behaviors, advanced artificial intelligence algorithms, and analytical results. Products in "Smart Factory" comprises embedded sensors and intelligent devices for data acquisition and localization of networks for managing process states, parameters, and environment conditions. They have controlling, processing, and storing capabilities. The other elements in Smart Factory include the virtual cloud data storage facility, which takes up a massive volume of data and analytical tools to compute, analyze, and measure the veracity of data and reduced latency to optimize the manufacturing processes. Data can vary from manufacturing documents, 3D models, GPS, GIS, Images, Control system, SCADA and Sensors current status data, history information, and measurement/test data. Significant features of the Smart Factory are interoperability and connectivity. Industrial IoT should establish an uninterrupted flow of real-time communication between the devices to device, Machine-To-Machine (M2M), manufacturing systems, and human–machine interface.

This paper proposes a recommendation of the components of Industry 4.0 put together in a layered structure, as shown in Fig. 1. The proposed framework follows a bottom-top level structure. All the components encompassed in Industry 3.0 was implemented and commissioned in a sugar plant industry located in India. The tools and software and hardware modules installed at the plant as per the software architecture adhered to the framework and layered architecture envisaged in Industry 3.0. This paper proposes an approach and devises a pilot installation and implementation of the framework layer mentioned below.

2 Literature Review

Numerous standardization bodies, manufacturing associations, and research groups have been energetically working toward defining the system architectures for Industry

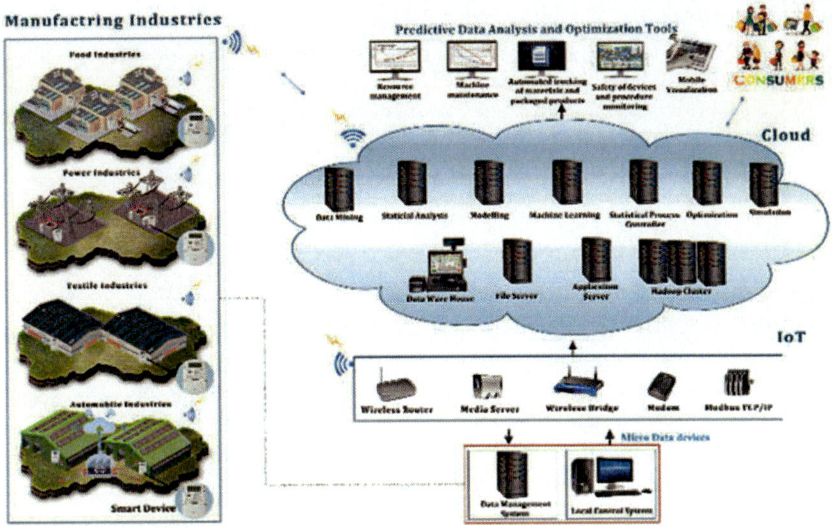

Fig. 1 Smart Factory structure

4.0 and providing possible solutions for a framework with a layered structure. One of these initiatives is the Reference Architectural Model Industrie 4.0 (RAMI 4.0) [1]. RAMI 4.0 propose a three-dimensional layer model capturing the system hierarchy, the type of information represented, and the life cycle of assets. This model shows the importance of the "Industry 4.0 component" in each layer. Another reference model is the American Industrial Internet Reference Architecture (IIRA) [2], which proposes a three-tier architecture pattern, Gateway-Mediated Edge Connectivity and Management architecture pattern and Layered Data bus pattern. Lee et al. [3] define a 5-level architecture for the understanding of Cyber-Physical Systems (CPS) with the levels indicating smart connection (I), data-to-information conversion (II), cyber (III), cognition (IV), and configuration (V). This architecture envisages close monitor and synchronization between the physical factory floor and the cyber computational space for implementations and realizations of CPS. Paulo et al. [4] propose a multi-agent-based architecture with an Enterprise Service Bus (ESB) acting as middleware between the different systems. Legacy devices are incorporated using gateways. This architecture offers a hybrid solution that combines existing solvers with multi-agent system principles. Hufnagel and Vogel-Heuser [5] capture data from various heterogeneous sources using an ESB-based architecture in the manufacturing industry. This architecture introduces a model-based idea to integrate data elements of distributed data systems and sources into one simulated database. Specific view models represent the information about data, their relations, properties, and environments. An Event-Driven Manufacturing Information System Architecture proposed by Theorin et al. [6], known as the Line Information System Architecture, LISA, enables flexible factory integration and data utilization. LISA is an event-based service-oriented architecture that offers flexibility and scalability both for lower-level and higher-level information. The Arrowhead project [7] provides a framework for interoperability between IoT devices at the service level.

This paper defines an Industry 4.0 framework using 7-layer architectures and an implementation approach in a sugar factory using 2-level digitization process.

3 Industry 4.0 Framework 7-Layer Architecture

Industry 4.0 framework describes intelligent manufacturing, with all processes connected by the Internet of Things (IoT). The framework helps create a Smart Factory using the intelligent control system, embedded software system, secure communication capability, Industrial Internet of Things (IIoT), systems of a system, enterprise systems, and visualization dashboard. This framework has 7-layer architecture, as shown in Fig. 2. They are the field sensor layer, smart electronics/intelligent control system layer, communication layer, data storage layer, analytical layer, data control/provider layer, and visualization layer.

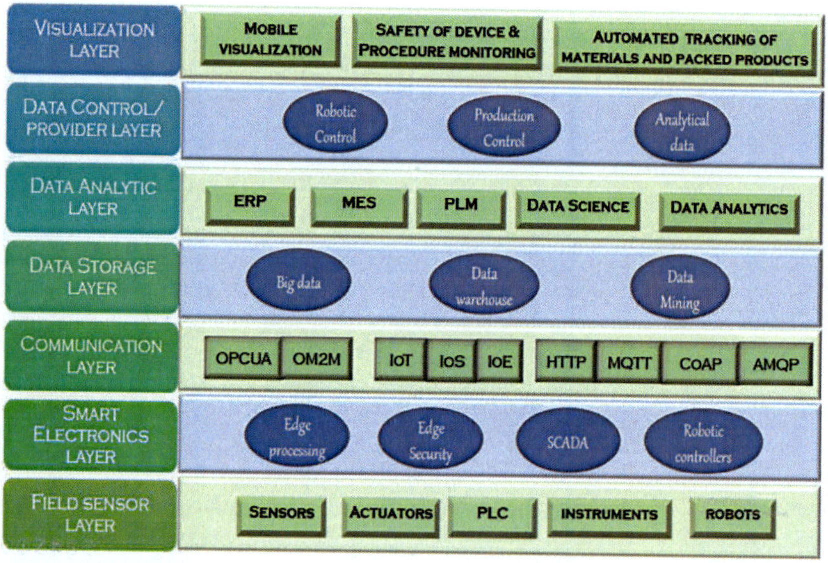

Fig. 2 Layered architecture

3.1 Field Sensor Layer

In a process control system, the real-time data sources are individual components such as a temperature sensor, level indicator, and vision sensor. Manufacturing industries use machines and robots, which have sensors and actuators that generate a massive amount of data that can be utilized for analysis and decision-making, enabling self-optimization of the process to meet custom manufacturing orders and adapt manufacturing specification in response to changing standards. The field sensor layer includes sensors, actuators, and PLC, which are the primary data capturing modules in the manufacturing industries. This layer generates measurements from the sensors, and PLC instruction for automation is heterogeneous data that is transmitted to controllers/RTUs or cloud in industrial big data analytics framework.

3.2 Smart Electronics/Control System Layer

Smart devices, Robotic controllers, Supervisory Control and Data Acquisition (SCADA) systems, and Human–Machine Interfaces (HMI) forms part of this layer. SCADA systems allow industries to control industrial processes locally or at remote locations by monitoring, gathering, and processing real-time data. SCADA Tag information is one of the data sources for IoT. Process data and real-time data are monitored through the dashboard and stored in local databases for local data management. Data collected from various smart meters and devices can better understand customer screening capabilities to use that data. Matching the smart meter/device data with additional information resources of consumers, e.g., weather conditions, would offer utilities a wide variety of data to create wise consumption-related pricing packages. IoT Gateways (or Message Oriented Middleware (MOM)) seamlessly unite edge devices into a single analytical view of the world using remote configuration, open, and secure connectivity methods such as REST, MQTT, AMQP, and OPC UA Pub-Sub.

3.3 Cloud Communication Layer

This layer must process heterogeneous data generated in the bottom layer from multiple machines or devices for inter-process communication. This layer uses a standard data model interface to process data sent and received by various subsystems in the framework. An interconnected layered data model allows for communication between each subsystem to transform data in different process conditions. This data transformation layer can communicate across CPS, IoT devices, factories and consumers (human/shop) via the Internet of Things (IoT), service-oriented computing model, Internet of People (IoP), and Internet of Everything (IoE), thereby providing the interoperability in Industry 4.0 framework. This high volume of digital format data converted to a visual format for MMI and different smart device compatible formats. A shared data model interface requires reliable, scalable, flexible, consistent, and secure data models to achieve interoperability for a real-time process. This layer should store IoT data from connected devices (manufacturers and consumers) into the extensive data system on a real-time basis, which helps in intelligent decision-making and insight for predictive maintenance.

3.4 Cloud Data Storage Layer

IoT data and applications are stored in the cloud to make it available anywhere using any web application or software. This standard data is accessible by different users, including producers, customers, operators, researchers, and programmers, through a standard interface in the framework. In the extensive data system, a tremendous amount of data is stored in big data files, characterized by four key factors known as the 4Vs of big data—Volume, Velocity, Veracity, and Variety. The volume deals with the large amounts of data procured from sensors, actuators, IoT, CPS, IoS, customers, and operations, measured in larger data units like terabytes, petabytes, and exabytes. Velocity determines data processing speed to derive insights quickly into real-time or batch or remote-control operations. Variety is about the nature of the data, whether structured, semi-structured, or unstructured, including different file types. Veracity defines the uncertainty, reliability, confidentiality, or security of data. These big data files are loaded into cluster nodes or distributed key-value and document databases depending on the size, utility of the data, computations, and significance of applications. The stored IoT data is analyzed using analytic tools like Hadoop MapReduce or Spark, Tableau, KNIME, and NodeXL.

3.5 Cloud Data Management/Analytic Layer

IoT cloud computing layers include the following: Infrastructure as a Service (IaaS) delivers computing capabilities over IoT. Platform as a Service (PaaS) models has hardware and software tools needed for IoT application development and deployment over the Internet. The Software as a Service (SaaS) model delivers IoT applications as a service. Analytics as a Service (AaaS) is the model for analyzing big data in this framework. AaaS acquires data from the cloud storage and performs data management and data control in compliance with specific manufacture requirement across the global enterprise. AaaS helps deliver accurate information to the exact place, to the right people at the correct time, like Fraud analysis, industrial fault detection, processes mining, market analysis, production analysis, and new product recommendation. MES and ERP applications are addressees planning and optimization tasks in the manufacturing processes. MES systems offer an end-to-end data flow from the ERP system up to manufacturing. This data helps the ERP systems assign activities based on the enormous real-time inflow of data rather than stored/planned information or capacities. Data analytics helps the wide-ranging assessment of data to develop models by applying numerous data analytics techniques and tools such as metric calculations, reporting, neural networks, deep learning, or machine learning. Optimization supports implementing concrete enhancement actions based on the decision to adopt the process, e.g., changing specific machine parameters.

3.6 Cloud Data Control/Provider Layer

This business layer supports business planning, supply chain management, and accounting. Analytical decisions improve efficiency, or predictive fault detection prevents downtime. The devices are interconnected at remote sites, access the enormous volume of data after analytics using high-speed analytical tools, and give an interfacing communicative layer to be accessed by the manufacturing to feedback on the data model running at the manufacturing process with the help of smart cloud robotics. Such an emergent technology aims to alleviate the complex computation and storage load and solicit services to the robots using remote and mobile devices. The communications components present in the layer allow easy access, monitor, and analysis robots and provide flexible resource management. Robotic process automation efficiently computes a vast volume of data using analytical tools and provide feedback data, which helps to improve the productivity of the process.

3.7 Data Visualization Layer

This technical priority layer addresses the need for advanced means for visualization dashboard and user interface capable of handling the endlessly increasing difficulty and scope of information to reinforce the end user in exploring and interpreting effectively. A standard and accurate demonstration are required for the machinists to operate the machine remotely. Human-To-Machine (H2M) communication required different types of visualization of data and advanced user interfaces. A hybrid visualization improves the decision-making services to recognize critical conditions that need instant attention.

4 Implementation Approach of Industry IoT Framework in a Sugar Factory Using 2-Level Digitization Process

The first-level digitalization process has been implemented in a sugar plant using our Indigenous SCADA Technology as per the layered architecture envisaged in Industry 3.0 to eliminate operator intervention. This first-level process implementation architecture consists of field sensor layer, smart electronics/control system layer, data analytical layer, and data visualization layer.

Automation area in the sugar factory includes Monitoring and Control of Auto Cane Feeding System (ACFS as shown in Fig. 3), Juice Flow Stabilization System, Pressure Reducing De-Superheating System (PRDS), 3 Element drum level Control System for 2 Boilers, Vacuum Pan Operation and Super-heated wash water for centrifugal, Monitoring the Evaporator and Juice heaters operation, Monitoring Electrical parameters of 3 MW Turbine, 500 KVA Diesel Generator, and APEB Feeders. Implementation of Industry 3.0 in the sugar factory has improved plant productivity and efficiency. It also has helped in saving electrical energy. This digitalization process in the automation system has delivered approx. 2000 data points into the framework in periodic and scheduled intervals in the range of seconds, minutes, hours, or days, which requires sophisticated data compression and analytics.

The second-level digitalization process should target the virtualization and modularization of the production process. The area of focus is the optimum control of a milling line which includes control of bagasse humidity, imbibition water, and raw juice Brix with the target of enhanced extraction of juice in the next mill and also optimization of sugar crystallization process at the vacuum pan. Using Industry 4.0 Framework, we can evaluate production decisions based upon analytics performing

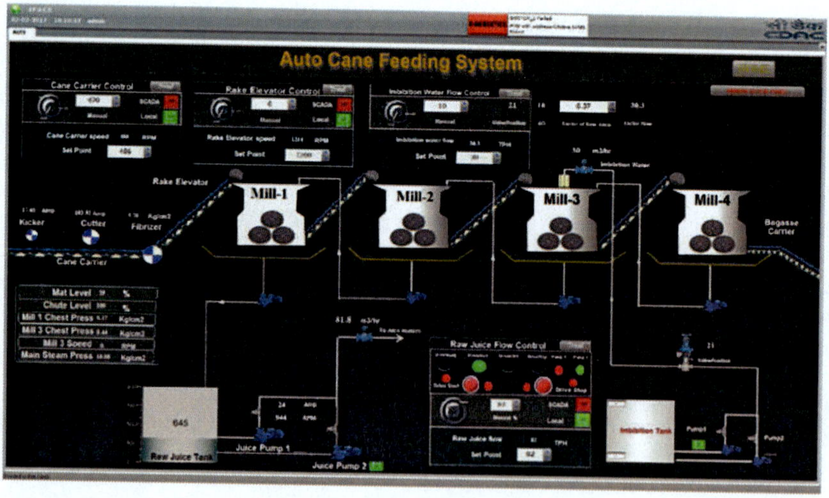

Fig. 3 HMI of ACFS

at the framework's cloud data management/analytic layer. The data visualization layer uses the visualization of products performing in their environments or used by actual people in real-time. Cloud data control/provider layer connects separate systems/processes for improved product tracking and monitoring. Smart devices and robotic controllers in the bottom layer of the framework will be integrated with the Internet of Things, cyber security, predictive data analytics, process optimization, advanced modeling, and simulation in the upper layers.

This two-step implementation approach described for the sugar factory digitalization process will help reduce the production cost, improve manufacturing productivity, promote industrial growth, change the labor force structure, and ultimately change the factory's competitiveness.

5 Challenges

The digitalization of a manufacturing industry needs essential engineering attention to make appropriate use of it. A mobile HMI or 3D digital twin is valid only if the plant is automated correctly and maintains the automated system using fully operational and calibrated instruments.

Extracting manufacturing data from different vendors is one of the significant challenges in the framework. Tag data models are introduced in the bottom layer of the framework to resolve this issue. It should define a standard data model interface to keep the data before and after data transformation in the communication layer. Tag data converted to IoT data for cloud storage and analytics, visual form for machine operators, and smart device compatible format. Modeling of industrial data is another major challenge in the framework. Integration of data from remotely located devices is essential for real-time decision-making models to appropriate actions. Integration of data is crucial and a delayed process. The decision-making models based on the Big 4 Data structure should be scalable, reliable, consistent, and secure models. The more the volume of data, more the time it takes to compute the data for analysis. Manufacturing industries must consider new IoT technologies and processes to adapt to the massive influx of data.

6 Conclusion

In this paper, we have proposed and illustrated a framework adopted for revolutionizing the process industries shortly. The proposed framework can be implemented in many process industries and technically educates the innovative digitization approaches imparted in the framework by reforming the workers as competent workers to improve the excellence and boost the internal competence of process plants.

The advantage of big data storage analytics is essential and significant in catering to the requirements of targeted markets, consumers, vendors, manufacturers, process engineers, and operators. Hence, the data is propagated from solution providers to different stakeholders like customers and all enterprise's business functions and services. This benefit of the interconnectivity of modules located at remote sites, digitization, and centralization of data play a critical role in customization. Analysis of production data and data from other sources thereby optimizes and enhances the production values of a process plant. The proposed framework layer opens up an avenue for digitization of the sugar industry where already Industry 3.0 framework was introduced on a large scale.

Smart Factory provides solutions with its automated procedures, simple setup of uncomplicated installation, and ultimately high level of scalability, enabling companies in the manufacturing industry to further enhance their efficiency and increase productivity.

References

1. DIN SPEC 91345 Reference architecture model Industrie 4.0 (RAMI4.0) (2016)
2. Industrial Internet Consortium (2017) The industrial internet of things. Volume G1: reference architecture. https://www.iiconsortium.org/IIC_PUB_G1_V1.80_2017-01-31.pdf
3. Lee J, Bagheri B, Kao HA (2015) A cyber-physical systems architecture for industry 4.0-based manufacturing systems. Manuf Lett 3:18–23
4. Leitão P, Barbosa J, Vrba P, Skobelev P, Tsarev A, Kazanskaia D (2013) Multi-agent system approach for the strategic planning in ramp-up production of small lots. In: IEEE SMC2019. IEEE, pp 4743–4748
5. Hufnagel J, Vogel-Heuser B (2015) Data integration in manufacturing industry: model-based integration of data distributed from ERP to PLC. In: IEEE 13th INDIN, pp 275–281
6. Theorin A, Bengtsson K, Provost J, Lieder M, Johnsson C, Lundholm T, Lennartson B (2016) An event-driven manufacturing information system architecture for industry 4.0. Int J Prod Res 55(5):1297–1311
7. Delsing J (2017) IoT automation: arrowhead framework. CRC Press, Taylor & Francis Group, Boca Raton

Commercial Viability of Solar Thermal Hyper Distillation of Waste Water—Sainnova, A Case Study

Pradnya Vishwas Chitrao, Pravin Kumar Bhoyar, Komal Chopra, and Rajiv Divekar

Abstract The world's fresh water resources are limited. They are constantly getting depleted by flooding, pollution, population growth, and the increasing number of uses to which water is being put. Presently available technologies for extracting pure or distilled water from brackish or saline or sewage water are costly, are energy consuming, and require continuous, expensive maintenance. As a result, promotion and fostering of technology innovation is today a major objective. This will lead to generation of distilled/clean, potable water in a cheaper, faster manner using green technology that uses less or no energy. An example of such a green technology is Sainnova. *Objective*: The researchers intend to study the commercial viability of solar thermal hyper distillation of wastewater in India that converts sewage/waste/salt water into distilled water. Very specifically, they will consider the case of Sainnova which uses solar energy technology without membranes for getting distilled water out of brackish/saline/sewage water. *Research Methodology*: The researchers will initially conduct some basic market research and will later follow it up with a detailed survey. They will also be referring to a lot of secondary sources of information. *Significance of the Study*: The research is significant as it will boost the use of green and cheap technology for extraction of distilled water and will be a sustainable and more inclusive means of acquiring pure water.

Keywords Water treatment · Self-sustainable · Renewable energy · Distilled water · Extremely economical · Patented technology

P. V. Chitrao (✉) · P. K. Bhoyar · K. Chopra
Symbiosis Institute of Management Studies (SIMS), Constituent of Symbiosis International Deemed University, Pune, India
e-mail: pradnyac@sims.edu

R. Divekar
Dr. Brig (Retd), Symbiosis Institute of Management Studies (SIMS), Constituent of Symbiosis International Deemed University, Pune, India

M. S. Kaiser et al. (eds.), *Information and Communication Technology for Competitive Strategies (ICTCS 2021)*, Lecture Notes in Networks and Systems 401,
https://doi.org/10.1007/978-981-19-0098-3_33

1 Introduction

Water is an important ingredient of life available on earth. It is essential for existence, good health, and productive livelihood. Good access to safe drinking water has multidimensional impacts on people's capacity to earn a livelihood, their physical well-being and hygiene, means of earning a living in cities as well as rural areas, the attendance of children in school, the mental health of people, and social relations. Available in large quantities, the demand for it today is becoming more than its availability [1].

India is one of the developing countries in the world that experiences a shortage of clean water as also pollution of good water resources. Once upon a time, abundant fresh water sources are rapidly depleting due to natural causes (drought, flooding, etc.) as well as human factors (like wastage, unscrupulous use for commercial purposes, negligence) leading to pollution and contamination. This has created a crisis situation that can be resolved through a technology that can make appropriate use of wastewater and will benefit society through reduction in cost of implementation.

As of now, there are available various technologies that fulfill the different requirements of different users. Easily procurable technologies in the market are technologies not controlled by the government such as Reverse Osmosis (RO) or partially permeable membrane technology for removing dirt and employing an external pressure to reverse the natural flow of pure solvent, Electro dialysis (ED) or process in which unwanted ions are extracted from solution with the help of a direct current going between two electrodes, one on each for the separation of difficult mixtures, Ion-Exchange (IE) a process for treating water in order to soften or demineralize it, Mechanical Filters (MF) or a contraption wherein a mechanical filter media within the filter unit physically catches and removes any solid waste that goes through the filter system and, after the dirt is removed, allows the cleaned water to flow back, etc.; all of them require a lot of energy as also continuous maintenance (O&M), and skilled labor for keeping the systems in good working condition. Most of these technologies usually fail in the long run due to lack of institutional support to afford their continued operation and maintenance. So in India where there is a big need for clean water and very little infrastructure and financial resources, simple technologies that require very less energy are more likely to work in the long run.

India is focusing on solar energy products through the introduction of various schemes that promote use of solar energy such as Jawaharlal Nehru Solar Mission (JLNSM2010). As per the scope of this scheme, Indian government has initiated a lot of low-cost solar projects (like solar energy parks or photovoltaic power stations, PV rooftop systems, or photovoltaic systems that have their solar panels that produce electricity fitted on top of a home or commercial structure, and solar lanterns or lighting systems consisting of an LED light bulb, solar panels, battery, charge regulator, and sometimes also an inverter and wherein the light bulb works on batteries that are charged by solar photovoltaic panel in different regions in India in order to create an eco-friendly environment and to save the reserves of fossil fuels [7].

In fact, solar energy is being used extensively to extract distilled water through use of Solar Dish Cookers (SDCs) and Solar Water Heaters (SWHs). Solar stills are built for this process of obtaining distilled water. Solar stills are basically insulated cabinets of brackish/saline water having a transparent glass cover. They heat up the water till it reaches evaporation point and then liquefy on the inside of the transparent polished surface which is positioned at a fixed slope angle [13]. It is important to find out how such commercially viable technologies are, as it will lead to sustainable and inclusive supply of clean and potable water for different purposes, and will help in healthy lives. Sainnova is a start-up that uses solar energy technology without membranes for getting distilled water out of brackish/saline/distilled water.

2 Background Review

Depletion of energy resources and waterborne illnesses brought on brackish water are two major issues of the twenty-first century [3]. Various techniques are today being used to counter the impure water issue and to make impure water drinkable. Some of them involve various processes like filtration, precipitation, sterilization, chemical treatment, etc. for removal of major impurities from impure water. Reverse osmosis is widely used for purification of water for drinking purposes; especially in households, housing societies, and offices, it involves the use of membrane. Electro dialysis, Ion-Exchange, and freezing are some of the methods that are employed to procure pure/distilled water for different household as well as business purposes.

Promotion of technology innovation is seen as a way of solving the pure water crisis in a sustainable and green manner, and which will make available pure water to all. Technology innovation involves new technologies, new management approaches, and as also methods that make existing systems more efficient [4]. One process that totally eliminates pollutants from brackish water is purification which is thermal energy-based. As per a report submitted by International Desalination Association, experts predict that purification technology will grow fastest in nations like South Africa, Jordan, Libya, Mexico, Chile, India, and China in the coming 5 years [8].

Majority of the purification techniques draw upon nonrenewable energy sources. But they put a load on the environmental resources [2]. Solar distillation is one major method of procuring potable water through the use of solar energy. A solar distillation unit having a combination of thermal collectors, photovoltaic panels, and concentrators distillates water three to six times more than conventional units.

Solar distillation is the easiest and least complicated method of procuring drinking water by using the freely available energy of the Sun. It is a better and cheaper option for all unreplenishable conditional distillation techniques. It not only removes a host of different types of pollutants in just one go, but is very easy, affordable, and eco-friendly. Solar distillation utilizes solar radiation and thus proves to be a sustainable alternative for the functioning of purification plants, especially in countries like India which come in the solar belt [5].

Solar water distillation functions on a simple and natural hydrological cycle; in this process, solar energy heats up water up to evaporation point. When the water reaches this evaporation point, the vapor spirals up and cools on the glass surface for collection. This process separates contamination like salts and heavy metals as also bacteriological life forms. Since the last hundred years, a lot of similar solar still plants and individual stills have been installed all over the world [6].

Hygienic and potable water is necessary for maintaining the health of a given community and for establishing healthy ecosystems, which are the foundation of a country's economic health, as also the health of our family members and that of the society in which we live. Water and the use of water resources, including the services employed for providing clean water, have an important role to play in all the global economies. For example, the worth of the global water market, that included controlling and cleaning up the water, was estimated at $500 billion per year in 2013 itself [9].

3 Potential

According to a World Bank report, more than 80% of the transmittable diseases in India are directly connected to water. India's population is around 1.17 billion, out of which only 15% get potable water. Most of the ground water sources have been indiscriminately utilized, leading to dramatic increase in mineral contaminants. For example, people in Rajasthan, Gujarat, and Andhra Pradesh are drinking water that has a lot of fluoride which in turn is threatening to result in mass poisoning.

Solar distillation is today frequently used to disinfect water. It can be adapted and adjusted from systems meant for one person to an entire community. The system functions well for around twenty years and does not even involve the shifting of parts. In India, the business of purifying water is transforming in a radical way in terms of technology, costing, and more enterprises offering similar services. The need for change is caused by various factors like inadequate pure potable water, poor market penetration by companies in the field of water purifiers, growing urban sprawl, and consequent increase in diseases transmitted through water. The major hurdles in this are the absence of uniform specifications and regulations for manufacturers as also low knowledge of the systems among prospective customers. The market is now also catering to a segment of customers who do not get basic amenities of the modern age like running water, electric power, and less costly products [12].

Nearly 30% of rural India does not have hygienic potable water, uninterrupted source of electric power, or also the economic capability to afford the comparatively costly versions of products that clean water. So it is necessary to provide them with solar water purifiers, which utilize solar photo voltaic systems so that they will have access to pure, safe, potable water. There is thus a lot of potential that has not been utilized in the villages.

A lot of companies are eying cleaning of water with the aid of solar energy as a prospective product for rural areas and markets that have so far not been approached.

In India, these companies have to bear in mind a few factors like the type of business model they wish to adopt as also the community's willingness to opt for paying a price for obtaining safe water for a price, in order to grow into steady, sustainable systems. The size of the profits and benefits as also the time frames within which they will be available are not fixed and many of the enterprises are still taking fledgling steps.

The Indian Government is also ready to collaborate with individual enterprises. For example, WaterHealth International (based in California) is planning to construct 225 plants for treating water in panchayat blocks in Andhra Pradesh on the basis of a build or construct and own and operate-transfer (BOOT) model as a foundation of functioning. It will obtain its profit from money that users will be charged.

The market in India has immense promise, as is evident from the reality that international companies (**Solvatten, WaterHealth, Pentair, Mörk Water Solutions**) in the water purifier segment have joined the bandwagon of renewable energy-based water purifying products and are trying to strengthen their market share [11].

The use of solar energy as an origin for commodities is expensive in comparison with other power resources, like coal or kerosene in some states. Users who are likely to use solar water purification systems face problems about paying in full the cost of the solar still as it is just not possible to pay in advance the steep capital cost [10].

4 Methodology

The researchers are faculty at Symbiosis Institute of Management Studies (SIMS, Pune) which is a constituent of Symbiosis International (Deemed University). They are collaborating with an Indian start-up company named Sainnova. The company is using green solar energy to convert waste or sewage or saline water into distilled water. The researchers are planning to conduct research at a future date on the commercial viability of the green technology of Sainnova for procuring distilled water. They will study for which segments of customers the technology will prove most beneficial.

The researchers intend to use and study the commercial viability of the patented technology that recycles ocean water/brackish/brine water into distilled water for households in rural areas or small community/hotels (Domestic and Commercial). The most important USP of the project is that the present disclosure enables recycling of wastewater by using renewable energy. Moreover, the system uses less expensive and simpler components to derive purest form of water (Distilled Water) from any type of waste. The researchers focus on field performance data of STHDW recorded during summers and winters for a period of 7–8 h which produces approximately 500–700 L of distilled water per day. The Indian Industry partners claim to reduce the cost of overall process and improve life as compared to its present competing technologies like the Electrodialysis reversal (EDR) or Reverse Osmosis (RO), Micro Filtration or Nano Filtration (NF), flocculation or Ultra Filtration (UF), Chlorine treatment, Ultraviolet Treatment, and Ozone Treatment. Furthermore, the present system can be used in space for recycling wastewater without any combustion or electricity. The processing capacity of the proposed design is multiple times faster as compared to existing competitive technology.

The yielded product (Distilled water) has many commercial uses. Besides drinking, it is utilized in automobile industries for electroplates and batteries, in pharmaceutical industries for manufacturing medicinal drugs, in soap industries, food canning, and soft drink industries, in beer brewing industries, in cosmetic industries, and many more.

5 Objectives of the Research

- Contribute to the society and help humankind get abundant drinking water wherever and whenever required in the easiest and economic way
- Study the commercial viability of the patented technology that recycles ocean water/brackish/brine water into distilled water for households in rural areas or small community/hotels (Domestic and Commercial).

6 Findings and Conclusions

Current state of the art in the field

A. **Membrane distillation (MD)**: Membrane distillation is a heat treatment or thermal method wherein absolutely pure water is obtained through evaporation. Exhaust heat or solar heat energy can be utilized to charge the plant. The structure consists of aqua phobic membrane with an aperture size of approximately 02 pm which allows minute aqua steam through while holding off liquid aqua.

B. **Reverse Osmosis (RO)**: RO requires only electrical energy, for obtaining the mechanical pressure. The need of energy grows as the saline content increases and is around 3.5 kW/m^3 for seawater. In comparison with a solar heat energy-driven MD system, the PV-RO is not so expensive. But if cost-free waste heat can be procured, the MD is an effective substitute. The RO is now already being used for various kinds of industrial waste treatment of intensive wastewater while for the MD, there is very high potential precisely in the area such as treatment of concentrated wastewater where the RO can no longer be utilized.

C. **Sainnova Technology**: As compared to the above technologies, Sainnova eliminates the membranes, filters, chemicals, and all other high-tech and expensive materials and gadgets, and its maintenance and expenses. The design mainly focuses on extraction of H_2O from any water source (even marshy water) using solar energy and delivers distilled water leaving solid waste, heavy metals, and all other impurities, which were earlier present in the water either in suspended form or dissolved form. The design uses extremely low costing but highly effective components which reduces the overall cost and efforts of manufacturing the product. Similarly, the processing capacity of the model is far higher as compared to its competing technology. The product can be used for household (domestic) purposes and can be upscaled for a larger version with more processing capacity for industrial or commercial purposes.

1. Expected deliverables

   Social:

   - Clean drinking water
   - Low-cost/cost-effective desalination method
   - Simplified and modular mobile solution for water purification.

   Industrial:

   - The system can save LDE/Fossil fuel consumption to make high heat steam and distilled water
   - Distilled water is highly in demand in Medical and pharma industry
   - Current technology of Distilled and RO water treatment process consumes lot of electricity and leave (impurity) sludge after the process adding to pollution. Sainnova has one of the most cost-effective solution for water distillation. The water will be free from BOD, COD, and heavy metals. A household system will save approximately 15 units of electricity per day (100 LPD).

2. Reasons for and benefits of cooperation

   - It will pave a new and greener way of distilled water production which industries or homes use for drinking or other utilities
   - It is most effective, simplified, and mature low-cost water distillation solution for communities
   - Cost of solar power unit is much less than industrial units
   - It will help and explore more applications related and interlinked with heat, steam and distillation
   - It will prove a breakthrough for socioeconomic scenario of industries and communities
   - It will add new sectors of industries for making our planet green
   - There have been previous research projects in India (e.g., CLIENT WaLUE) and several cooperations, e.g., IGAWE (Indio-German Alliance for Water Efficiency) and Membership in the Technical Advisory Committee for the National Urban Sanitation Plan (NUSP), India
   - Huge opportunity for implementation of cost-efficient technology for (waste)water reuse in India, Asia, and other regions through cooperation
   - Using global IEEM network and experiences from research projects in Europe, Asia, and Africa to evaluate market potential of Sainnova technology.

References

1. Boukhriss M, Zhani K, Ghribi R (2013) Study of thermophysical properties of a solar desalination system using solar energy. Desalin Water Treat 51(4–6):1290–1295
2. Gude VG, Nirmalkhandan N, Deng S (2011) Desalination using solar energy: towards sustainability. Energy 36:78–85
3. Gurung P, Sherpa TYC (2014) Freshwater scarcity and sustainable water management in the Hindu Kush Himalayan (HKH) region. Hydro Nepal 15:42–47
4. Kiparsky M, Sedlak DL, Thompson Jr BH, Truffer B (2013) The innovation deficit in urban water: the need for an integrated perspective on institutions, organizations, and technology. Environ Eng Sci 30(8)
5. Muftah AF, Alghoul MA, Fudholi A, Abdul-Majeed MM, Sopian K (2014) Factors affecting basin type solar still productivity: a detailed review. Renew Sustain Energy Rev 32:430–447
6. Sivakumar V, Sundaram EG (2013) Improvement techniques of solar still efficiency: a review. Renew Sustain Energy Rev 28:246–264
7. Solangi KH, Islam MR, Saidur R, Rahim NA, Fayaz H (2011) A review on global solar energy policy. Renew Sustain Energy Rev 15(4):2149–2163
8. Stuber MD, Sullivan C, Kirk SA, Farrand JA, Schillaci PV, Fojtasek BD, Mandell AH (2015) Pilot demonstration of concentrated solar-powered desalination of subsurface agricultural drainage water and other brackish groundwater sources. Desalination 355:186–196
9. TechKNOWLEDGEy Strategic Group 2013 (2013) Water market review: growing awareness, growing risks. 16:19. http://www.tech-strategy.com/pdf/Winter2013.pdf. Accessed 18 Sept 2013
10. Ten years of solar distillation application along the US—Mexico border. http://www.sunstar-solutions.com/10yrsStillsPaper.PDF
11. The Water Purifier Market Today. http://www.waterpurifiers.in/all-about-water-purifiers/the-water-purifier-market-today.html
12. TV Veopar Journal (2012). http://www.adi-media.com/PDF/TVJ/annual_issue/011-Water-Purifiers.pdf
13. Zarzoum K, Zhani K, Bacha HB (2013) Improving the design, modeling and simulation in dynamic mode of a solar still. Desalin Water Treat 52(34–36):6304–6314

Indian Tourist Recommendation System Using Collaborative Filtering and Deep Autoencoder

G. Ratnakanth and S. Poonkuzhali

Abstract The tourism industry's recommendation system opens the way for capitalising on visitor demand based on a prior traveller's positive or negative experience at a particular target tourist location. After weighing the advantages and disadvantages of current methods for the tourist recommendation system, this study employed collaborative filtering techniques in conjunction with deep learning approaches. In this study, a deep autoencoder is used to predict user ratings in order to give suggestions to travellers about a certain tourist attraction. The proposed recommendation system is examined and validated using two datasets comprised of user evaluations for two distinct quirky places collected from a well-known traveller's website, Tripadvisor. The first dataset contains user reviews and ratings for numerous locations in Jaipur, India, while the second dataset is based on various locations around Europe that are available in the Kaggle repository. The suggested model's performance is quantified in terms of root mean square error (RMSE).

Keywords Deep autoencoder · Recommendation · Collaborative filtering · Tripadvisor

1 Introduction

Tourism is often regarded as the most important business in a major emerging country such as India, with the potential to employ millions of people [1]. The importance of this sector is to combine several developing technologies in order to develop it in such a way that it can fulfil tourist demand at the optimal moment. The process of locating nearby required items such as hotels, tourist attractions, and destinations is a tiresome and time-consuming one. This issue can be overcome by combining

G. Ratnakanth (✉)
Anna University, Chennai, India
e-mail: rkanth.mtech@gmail.com

S. Poonkuzhali
Rajalakshmi Engineering College, Chennai, India

M. S. Kaiser et al. (eds.), *Information and Communication Technology for Competitive Strategies (ICTCS 2021)*, Lecture Notes in Networks and Systems 401,
https://doi.org/10.1007/978-981-19-0098-3_34

the Internet of Things and artificial intelligence to create dependable smart tourism requests.

Traditionally, a user would seek out his or her requirements at a remote location. This may be accomplished only through Internet browsing. As previously said, this manual method of locating more nearby or accessible items is a lengthy procedure. Thus, this may be solved by developing an automated model that places user needs in the user's hands; this is referred to as a suggestion system. The recommendation method overcomes current search obstacles in a wide variety of areas, including e-commerce [2], entertainment, and health, and is not limited to tourism requirements. A recommendation system may be designed in two ways according to popular wisdom: as a collaborative-based filtering recommendation system or as a content-based filtering recommendation system. Once again, collaborative filtering is split into memory-based and model-based recommendations. The memory-based approach is further subdivided into collaborative filtering based on user and item. The content-based filtering is determined by whether a user's previously loved product would be liked in future as well. Another widely used technique is content-based filtering. This is achieved by suggesting individuals with similar tastes through the same recommendation. The third strategy is to use a hybrid recommendation system, which combines content-based and collaborative filtering. This method capitalises on the advantages of both strategies while eliminating their disadvantages. Machine learning approaches have been used to handle prediction in recommendation systems over the last several years. Recently, machine learning was phased out in the tourist industry in favour of deep learning architectures, which have demonstrated their efficacy via experimental work using current methodologies [3]. In this sense, the recommendation system frequently incorporates deep learning techniques to enhance its effectiveness.

The following points summarises the contribution of this work:

- To scrape user evaluations for popular tourist destinations, such as things to visit in Jaipur, from traveller's websites, namely Tripadvisor.
- The data reviews were preprocessed by removing just the parameters necessary for collaborative filtering, such as user id, item id, and rating.
- The pivot table is created by organising the item ids column-by-column and user ids row-by-row. Then, each cell is summed with its associated review rating.
- Recommendation model has been built to forecast the user rating for trip suggestions by feeding the collaborative filtering input to the deep autoencoder.

The paper is organised as follows: Sect. 1 has an introduction; Sect. 2 contains a discussion of related work; Sect. 3 contains a design of the recommendation model; and Sect. 4 contains the algorithm utilised in this model. Section 4 discusses the findings received from this developed model. Finally, Sect. 5 concludes and discusses the proposed model's future development.

2 Associated Mechanisms

This section analyses the proposed deep learning-based recommendation system for tourism demand and includes the following discussion:

2.1 Recommendation System Using Deep Learning

Fessahaye et al. [4] presented a recommendation for a music playlist using both content-based and collaborative filtering techniques along with the implementation of deep neural networks. Here, the results of both filtering techniques will be fed to the deep neural network as an input in order to predict the upcoming music in a playlist as a recommendation for the user. Wei et al. [5] suggested two recommendation models for incomplete cold start items and complete start items by combining collaborative-based filtering along with deep learning models. In this work, deep neural network has been used to extract necessary content features of the item and timeSVD++ was implemented to model the temporal dynamics users, preferences and its preferred items. Wang et al. [6] presented a two-level automation model for music recommendation system using deep belief network as well as probabilistic graphical model. This model performs parallel processing of feature learning from audio and recommendation for users. Kherad et al. [7] suggested a recommendation system based on Matrix Factorisation and graph analysis methods. Apart from these techniques, deep autoencoders have been used to do initialisation of user and item latent vectors. Kiran et al. [8] proposed a hybrid recommendation system using deep learning techniques, and this model is designed in such a way to focus on to overcome the flaws arise by the traditional collaborative filtering techniques by feeding the side information into the deep learning architecture. Additionally, both the latent features and non-latent features are learnt by the matrix factorisation and embedding layer, respectively. This model is trained and evaluated using different datasets such as MovieLens 100 K, FilmTrust, Book-Crossing, and MovieLens 1 M. Lv et al. [9] proposed a sequential recommender system by combining short-term sessions and long-term sessions using recurrent neural networks along with attention mechanisms.

2.2 Tourism-Based Recommendation System

Wang et al. [10] proposed a recommendation system based on CNN along with DNN. Here, the author has preferred CNN for preprocessing of text reviews given by the user for a particular tourist site. Moreover, DNN is also implemented here to process the essential information of users as well as the items. An et al. [11] proposed a sentiment analysis-based recommendation system for tourism data. Author has utilised CNN-LSTM for applying word embedding techniques over the text reviews obtained from the user. Furthermore, by using these processed reviews, a content-based recommendation system has been designed. Abbasi et al. [12] suggested a recommendation system based on semantic clustering and sentiment classification. Author has utilised user reviews which are collected from Tripadvisor. Based on the processed reviews obtained from clustering and sentiment analysis, a context aware recommendation system has been designed. Lee et al. [13] proposed a recommendation system based on the Gated recurrent unit to process the user reviews in the form of sequential data. Apart from GRU, RNN has been used to design session-based parallel mini batch recommendation models. This model has been evaluated using the reviews given for Jeju Island.

Deep autoencoder, an efficient unsupervised algorithm, was utilised in this research to predict the review score supplied by the user or tourist. At the end of the above-mentioned relevant work, several of the major concerns associated in the collaborative filtering-based recommendation system are revealed; these are data sparsity in the employed dataset, cold start problem, and finally, difficulty in appending side characteristics as an input to the model. These concerns exist when the recommendation system is built using traditional techniques, and they are mitigated slightly by the use of deep learning techniques. The second part of this section looked at the experimental results and performance of deep learning approaches. This demonstrates how different types of neural network designs can be used to solve the aforementioned collaborative filtering challenge. However, the existing method based on neural networks almost entirely relies on supervised prediction; therefore, it fails in the majority of cases when new types of inputs are introduced into the system. To overcome this barrier, the deep autoencoder algorithm was used to create unsupervised collaborative filtering recommendation systems for tourism.

3 Framework of the Tourism Recommendation Model

Figure 1 depicts the data analytics model for the proposed framework. In this research work, a recommendation model is built for giving personalised tourist suggestions utilising an unsupervised deep auto encoder algorithm. Data collecting, data preprocessing, modelling, and recommendation system assessment are the key building components used in the architecture of this tourism recommendation system. Initially, user reviews are taken from the websites of tourists, i.e. tour advisers for well-known sites, such as those in Jaipur. Following that, only the essential parameters for collaborative filtering, such as user id, item ID, and rating, were retrieved for data review preparation. The item id is then arranged column wise, and the user id is arranged row wise to create a pivot table. The evaluation ratings for each cell are then compiled. Finally, after feeding the collaborative filtering input to the deep autoencoder, a recommendation model based on user ratings for trip suggestions was developed.

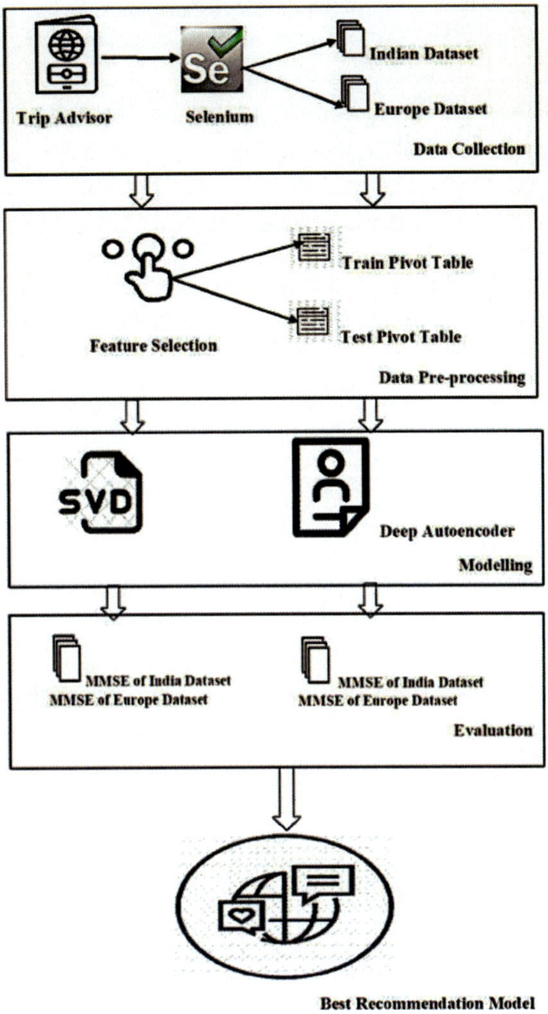

Fig. 1 Framework of the tourism recommendation model

3.1 Data Acquisition

This work is trained and tested using two databases where the first dataset is about the user reviews of the different types of places in Jaipur which is scraped in the Tripadvisor website. Another dataset is based on the user reviews of the different places in the region of Europe. This dataset also is scraped from the same Tripadvisor website. For the case of Indian dataset, the review data and its corresponding rating for performing collaborative filtering have been collected. The data has been scraped from the website using a software package called selenium which can be installed on the python environment. The column names are user id, item id and its corresponding rating. The Europe dataset was collected from Kaggle website [14].

3.2 Data Preprocessing

The pivot table was created to carry out collaborative filtering after the completion of required features from the scraped web data and the three columns were used to do this. The pivot table was created for each dataset independently, for the test and the train datasets. The pivot table was created in such a way that the column 'user id' is elucidated in a row-by-row manner, while the column 'item id' is elucidated in a column-by-column fashion. It contains the rating score that was assigned to the specific combination of user id and item id in the previous cell. The rating score will have values ranging from 1 to 5 on a scale of 1 to 5. The review score with a value of 5 indicates the highest rating, whereas the review score with a value of 0 indicates the lowest rating.

3.3 Modelling of Recommendation

The deep autoencoder model has been provided the input vectors from the collaborative filtering procedure in the form of a pivot table. In the testing dataset, this unsupervised technique is used to forecast the missed user review rating, whereas the train dataset is utilised to train the model's review score. The suggestion of a specific tourist destination can be determined based on the value by making prediction of the user review for the future data instance.

3.4 Evaluation

During the evaluation phase, the proposed recommendation model was executed by feeding the test part of the dataset as input to the neural network architecture. Typical classification metrics may be used to evaluate the performance of the deep autoencoder-based recommendation model. However, in this study, in addition to standard performance measures, the MMSE (minimum mean square error) is used to evaluate predictive models.

3.5 Algorithm

In this section, the algorithm used to design the collaborative-based deep recommendation has been elucidated and it is given below.

Deep autoencoder

The unsupervised deep autoencoder is a commonly used method for reconstructing images without degrading their original quality. Recently, deep autoencoders have garnered considerable interest for their performance on classification and prediction tasks. It also works with non-image data types such as text, alphanumeric, and numeric. The deep auto encoder's internal architecture is composed of an encoder, a decoder, and a code. Both the encoder and the decoder have the same number of hidden layers. The component 'encoder' consists of an input layer followed by 'n' hidden layers, whereas the component 'decoder' consists of an output layer followed by 'n' hidden layers. The component 'code' is used to connect the encoder and decoder components by sending the encoder's output to the decoder's input.

```
Input: User reviews dataset (Dataset_Reviews)
        Dataset_R ← Extract required features from Dataset_Reviews
    Split the Dataset_R for training and testing
        Dataset_train ← 70 percentile of Dataset_N
        Dataset_test ←30 percentile of Dataset_N
    Formation of pivot table(User_id, item_id, rating_score)
        Train_pivot ← Pivot table(Dataset_train)
        Test_pivot ← Pivot table(Dataset_test)

Initiate following parameters of deep autoencoder:
        H ← number of hidden layers
        E_f ← Encoding function
        D_f ← Decoding function
        W_i ← Weight of input parameters

        F(W) ←Train_pivot * W_i

        F(W')_i ←Train_pivot*W_i'

Opitimize the model using adam optimizer with learning rate of 0.001

End
```

Fig. 2 Pseudo for deep autoencoder

Table 1 Hyperparameters values of deep autoencoder

Hyperparameter	Value
Activation function	Selu
Dropout	0.1
Learning rate	0.001

Input: User reviews dataset ($Dataset_{pivot}$)

$Dataset_{pivot}$' ← Transpose matrix of ($Dataset_{pivot}$)

Calculate the eigen vector and eigen values of $Dataset_{pivot}$' $Dataset_{pivot}$
Using those eigen values to compute diagonal matrix ∑ of $Dataset_{pivot}$' $Dataset_{pivot}$
Compute the unitary matrix V by substituting the normalized eigenvectors of $Dataset_{pivot}$' $Dataset_{pivot}$
Compute another unitary matrix W:
W = $Dataset_{pivot}$V ∑
End

Fig. 3 Pseudo for singular value decomposition (SVD)

Figure 2 illustrates the pseudo of a deep autoencoder. Table 1 contains the values for the deep autoencoder's tuned hyperparameters.

SVD (Singular Value Decomposition)

The Singular Value Decomposition approach is a supervised technique that is well-known for its dimensionality reduction capabilities. However, in the recommendation system, the SVD is used as a collaborative approach. The resulting matrix is factorised using single value decomposition in this case. The single matrix is most likely decomposed into three matrices using the SVD algorithm. The pseudo for SVD is given in Fig. 3.

4 Result and Discussion

This recommendation model was carried out in the programming environment of Python with the use of Keras and TensorFlow as a backend. This set-up is done on a 64-bit Windows 10 machine with 4 GB RAM, 2 GHz processing rate, and a Python version 3.4.0 running on it.

Outcome of Recommendation Model

In the recommendation, classifier investigates the performance measurements based on minimum mean square error (MMSE) to evolve the best recommendation model among supervised and unsupervised algorithms. The minimum mean square error (MMSE) is a type of estimator that is used to minimise the value of the mean square error (MSE). The MMSE is calculated using the following Eq. (1)

$$\text{MMSE}(X) = E(X|Y) \tag{1}$$

where

- 'E' Minimum MSE of all predictors,
- X Estimator of the random variable.
- Y Random variable.

In Table 2, the MMSE performance metrics for SVD and Deep autoencoder methods are presented for two distinct areas of the dataset. It demonstrates that the Deep autoencoder outperforms the SVD on both datasets. For the India-based dataset, model 4 with a layer structure of (32, 30, 25, 10, 10, 25, 30, 9) outperforms the remaining three models, whereas model 2 performs well on its own for the

Table 2 Performance evaluation of SVD and deep autoencoder

Algorithm used	Different kinds of architecture	MMSE	
		Training loss	Validation loss
India-based review dataset			
Deep auto encoder	Model 1: (32, 20, 20, 9)	0.8490	0.7958
	Model 2: (32, 20, 10, 10, 20, 9)	1.1221	1.1296
	Model 3: (32, 30, 20, 10, 10, 20, 30, 9)	0.8088	0.8355
	Model 4: (32, 30, 25, 10, 10, 25, 30, 9)	**0.5101**	**0.4925**
SVD		2.0797	2.1013
Europe-based review dataset			
Deep auto encoder	Model 1: (10, 12, 12, 12)	0.0210	0.0178
	Model 2: (10, 24, 12, 12, 24, 12)	**0.0188**	**0.0135**
	Model 3: (10,30, 24, 12, 12, 24, 30, 12)	0.0397	0.0225
	Model 4: (10, 40, 24, 12, 12, 24, 40, 12)	0.0310	0.0169
SVD		0.0555	0.0656

Fig. 4 MMSE loss of deep autoencoder (India dataset)

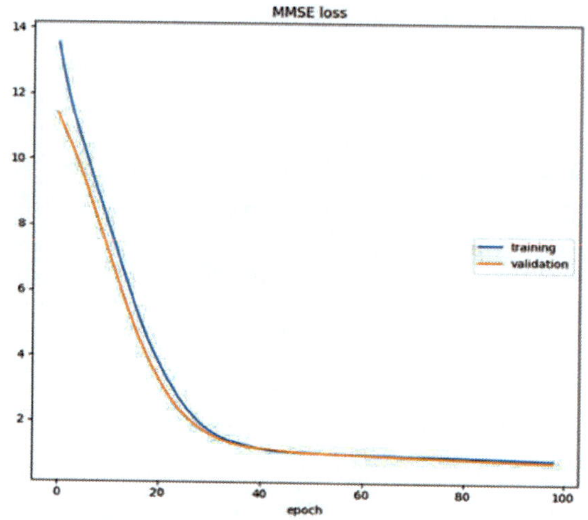

Model 1

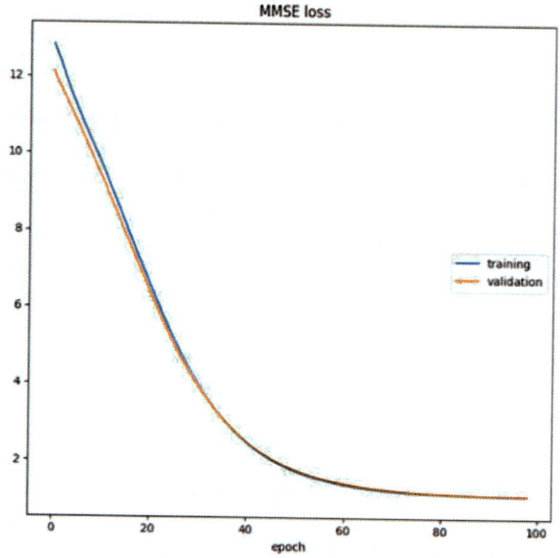

Model 2

Europe-based dataset. This demonstrates that a huge dataset just takes a few layers to get the same performance as a smaller dataset. The plots in Fig. 4 illustrate the relationship between the MMSE loss during the training and validation phases for models 1, 2, 3, and 4 with regard to the India-based dataset. Similarly, Fig. 5 depict the MMSE plot of the Europe-based dataset for deep autoencoder models 1, 2, 3,

Fig. 4 (continued)

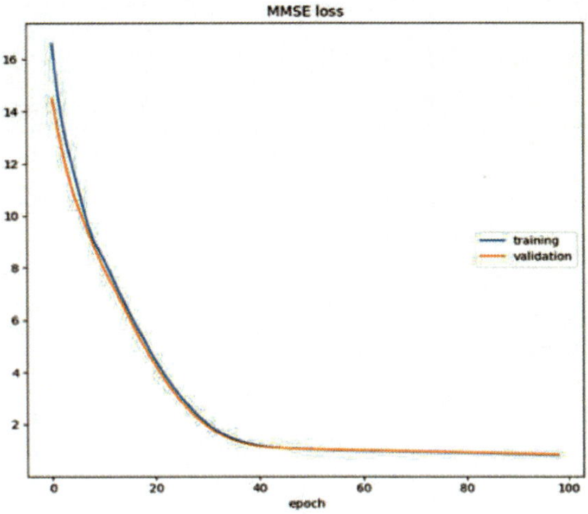

Model 3

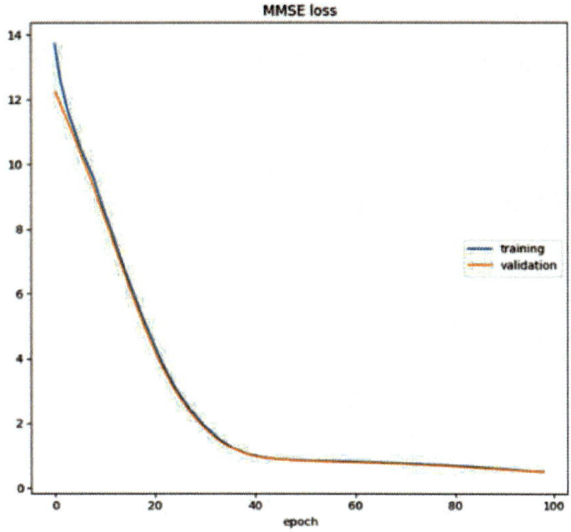

Model 4

and 4. Figure 6 illustrate the performance of the SVD method in terms of MMSE loss for India and Europe-based datasets, respectively.

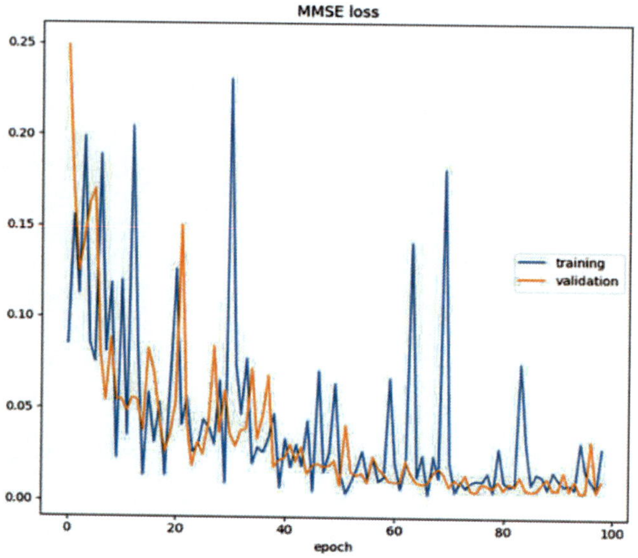

Model 1

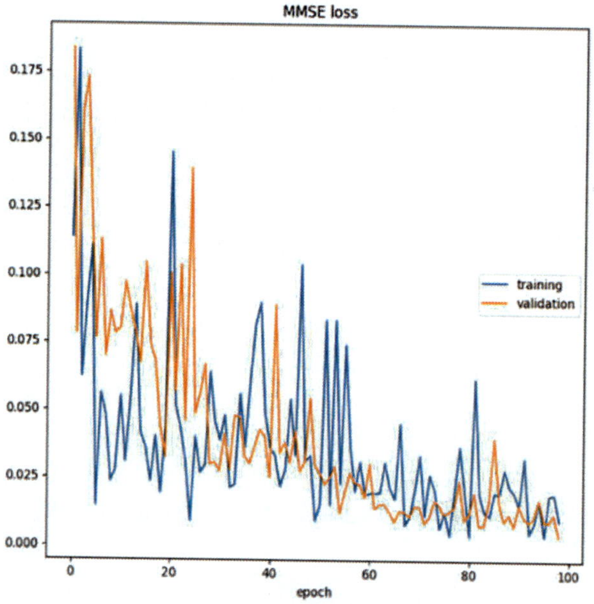

Model 2

Fig. 5 MMSE loss of deep autoencoder (Europe dataset)

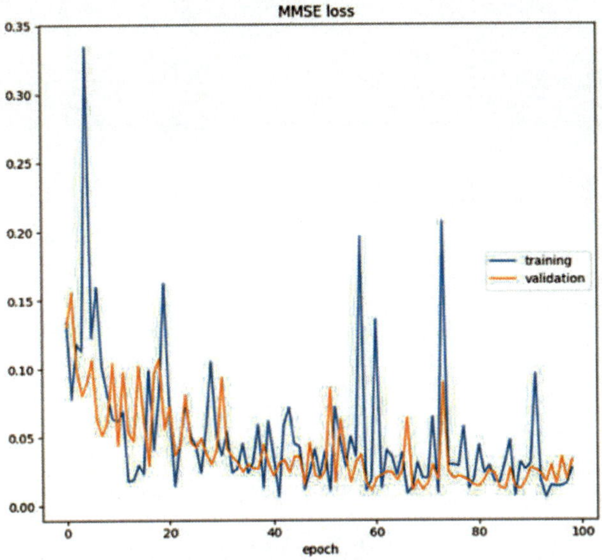

Model 3

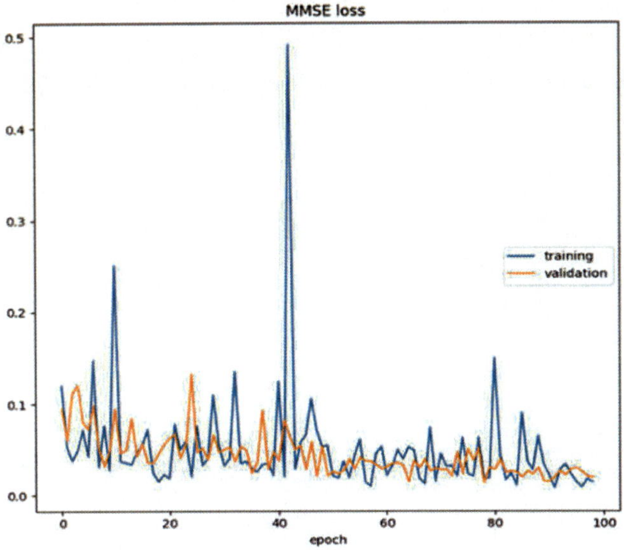

Model 4

Fig. 5 (continued)

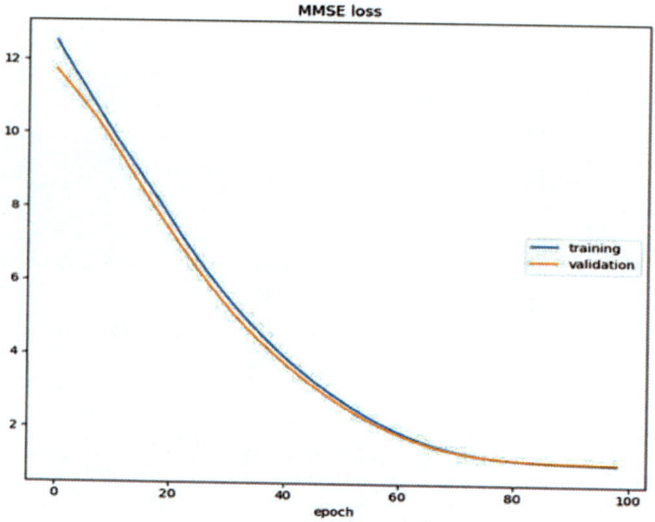

Indian Dataset

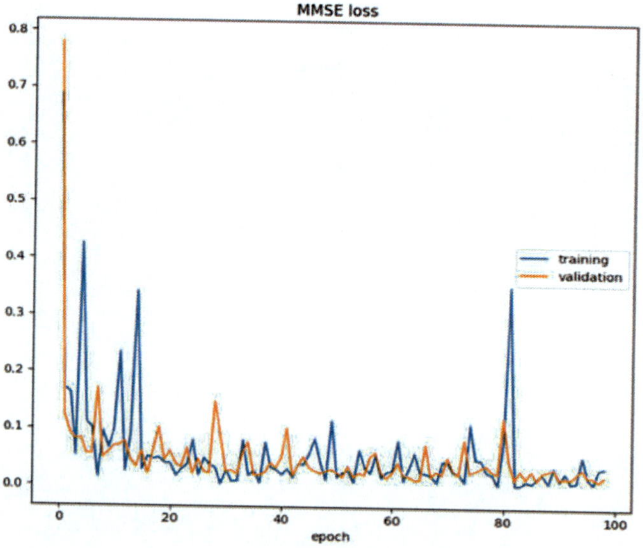

Europe Dataset

Fig. 6 MMSE loss for SVD

5 Conclusion

The recommendation system for the tourist industry, which is based on deep learning techniques, was built efficiently using an unsupervised methodology. This increases the performance of the proposed model over the standard machine learning approach, as evaluated by the MMSE. Additionally, in this study, a well-performing conventional approach called SVD (singular value decomposition) was used to conduct a comparison analysis using a deep autoencoder in order to find the best performing recommendation model. The suggested model's dependability has been assessed using both small and big datasets. As a consequence, it is discovered that deep autoencoders produce superior outcomes for both datasets. This study effort may be developed further by adding the deep autoencoder algorithm for content-based filtering and hybrid recommendation systems. The hybrid recommendation system outperforms the collaborative filtering and content-based filtering recommendation models.

References

1. Kabassi K (2010) Personalizing recommendations for tourists. Telematics Inform 27(1):51–66
2. Schafer JB, Konstan JA, Riedl J (2001) E-commerce recommendation applications. Data Min Knowl Disc 5(1):115–153
3. Ratnakanth G, Poonkuzhali S (2021) Unsupervised sentiment classification for hotel review rating using LSTM Autoencoder. In: 2nd International conference on data science and applications(ICDSA21), 10–11 April 2021, Jadavpur University, Kolkata, India
4. Fessahaye F, Perez L, Zhan T, Zhang R, Fossier C, Markarian R, ... Oh P (2019) T-recsys: a novel music recommendation system using deep learning. In: 2019 IEEE international conference on consumer electronics (ICCE). IEEE, pp 1–6
5. Wei J, He J, Chen K, Zhou Y, Tang Z (2017) Collaborative filtering and deep learning based recommendation system for cold start items. Expert Syst Appl 69:29–39. https://doi.org/10.1016/j.eswa.2016.09.040
6. Wang X, Wang Y (2014) Improving content-based and hybrid music recommendation using deep learning. In: Proceedings of the 22nd ACM international conference on multimedia, pp 627–636
7. Kherad M, Bidgoly AJ (2020) Recommendation system using a deep learning and graph analysis approach. arXiv preprint arXiv:2004.08100
8. Kiran R, Kumar P, Bhasker B (2020) DNNRec: a novel deep learning based hybrid recommender system. Expert Syst Appl 144:113054
9. Lv F, Jin T, Yu C, Sun F, Lin Q, Yang K, Ng W (2019) SDM: sequential deep matching model for online large-scale recommender system. In: Proceedings of the 28th ACM international conference on information and knowledge management, pp 2635–2643
10. Wang M (2020) Applying internet information technology combined with deep learning to tourism collaborative recommendation system. PLoS ONE 15(12):e0240656
11. An HW, Moon N (2019) Design of recommendation system for tourist spot using sentiment analysis based on CNN-LSTM. J Ambient Intell Humanized Comput 1–11
12. Abbasi-Moud Z, Vahdat-Nejad H, Sadri J (2021) Tourism recommendation system based on semantic clustering and sentiment analysis. Expert Syst Appl 167:114324
13. Lee HJ, Lee WS, Choi IH, Lee CK (2020) Sequence-based travel route recommendation systems using deep learning—a case of Jeju Island. Smart Media J 9(1):45–50

14. https://www.kaggle.com/ishbhms/travel-review-ratings

Developing Information Extraction System for Disaster Impact Factor Retrieval from Web News Data

Lakshmi S. Gopal, Rekha Prabha, and Maneesha Vinodini Ramesh

Abstract Natural and man-made disasters have had a long-term impact on our country, resulting in significant loss of life and economic benefits. India's disaster vulnerability has changed over time, emphasizing the importance of a holistic approach to disaster management. When assessing a disaster, it is important to look past the intensity scales and go deeper into the impact factors, which can help in post-disaster activities. Gathering data on hazards has become difficult due to the proliferation of knowledge available on the Internet. Using conventional natural language processing methodologies, the proposed work aims to derive data related to hazard impact factors from online news sources. Approximately, 80% of the impact factor information was retrieved effortlessly from about 200 news articles. The derived data can be used to create a severity scale that incorporates impact factors as well as intensity scales, allowing for better hazard analysis.

Keywords Information extraction · Natural language processing · Disaster management

1 Introduction

Natural and man-made disasters have been affecting our country since decades by causing massive loss in life, infrastructure, cattle, and economic gains. The holistic process of managing disasters has aided in reducing the losses of hazards and also attempts to ensure prompt decision making by the government or local authorities to provide assistance to the victims and accomplish quick recovery [1].

L. S. Gopal (✉) · R. Prabha · M. V. Ramesh
Center for Wireless Networks and Applications (WNA), Amrita Vishwa Vidyapeetham, Amritapuri, India
e-mail: lakshmisgopal@am.amrita.edu

R. Prabha
e-mail: rekhap@am.amrita.edu

M. V. Ramesh
e-mail: maneesha@amrita.edu

M. S. Kaiser et al. (eds.), *Information and Communication Technology for Competitive Strategies (ICTCS 2021)*, Lecture Notes in Networks and Systems 401,
https://doi.org/10.1007/978-981-19-0098-3_35

India is one of the most disaster-prone countries in the world due to its peculiar geo-climatic and socioeconomic conditions. Over the past decade, India has faced several disasters such as landslides, floods, droughts, and earthquakes, to mention a few. This accentuates the significance of a holistic management of disasters in our country.

Analyzing data related to existing or ongoing hazards will help in risk assessment of a locale, which is one significant phase in the disaster management cycle [2]. Generally, disaster risk is described as a combination of the severity and frequency of a hazard in a susceptible location. Other factors such as exposure and vulnerability of humans, animals, and location terrain also count. When a current hazard-related data is analyzed based on the mentioned factors, disaster risks can be examined, which aids in quick recovery actions.

To bring a comprehensive management of disasters to reality, adequate analytical data about the hazard event is required. It is also important that the data being used must be accurate and authentic. Over the years, researchers have explored web data for early disaster response, mitigation, recovery, preparedness, and other analytical experiments. The proposed work intends to analyze news data related to hazards available on the web to capture aspects that are not explicitly described in news reports in terms of severity, risk, and impact factors of a hazard event. It gives intricate details about the event such as casualties, injuries, infrastructural damages, and involvement of local or government authorities in rescue operations. Hence, analyzing news data will certainly benefit the disaster management cycle.

Analysis is rather straightforward when the data being used is in a structured form, such as a database with precise content [3]. But, news articles are unstructured, written in natural languages, and come in varying lengths. The proposed work formulates an information extraction approach to derive specific content related to disasters from news articles.

Information extraction (IE) system is a natural language processing (NLP) approach to extract 'facts' from unstructured corpus. The proposed work has developed an IE system which involves manual effort in analyzing disaster-related news articles to understand different grammar patterns in sentences which may produce hazard-relevant phrases. The phrases of interest are related to the hazard impact factors. News articles generally do not broadcast analytical details of an event, rather provide scrupulous information regarding an event. The proposed work aims in extracting information which is not explicitly described in news reports, but exhibits meticulous insights about an event.

2 Related Work

Researchers have derived terminologies and define hazard as an event that builds potential harm that threatens our social, economic, and natural capital on a community, region, or country scale. Disasters must be measured in terms of lives lost, injuries sustained, or property damaged and must be distinguished from intensity scales such as the Richter scale or an Inclinometer for an earthquake or a landslide event, respectively. The previous studies have almost exclusively focused on classification of hazards and epidemiological research such as defining a severity scale focusing on medical requirements of the injured and post-disaster trauma severity among people [4].

The impact on people, economy, and facilities has been studied by researchers to derive a scale that measures severity in terms of loss. The proposed work addresses several further questions including economic conditions of a locale into analyzing severity of a hazard as it presents the capacity of a community, region, or the nation to deal with hazard impacts.

There exist several research works that fine points the significance of web data in disaster management cycle. An important approach in mining relevant data from unstructured text includes natural language processing-based information extraction. A series of previous studies has shown various information extraction methods which incorporated natural language processing and machine learning approaches [5]. The analysis of natural language has been broken into several broad levels such as lexical, syntactical, semantic, pragmatic, and discourse analysis which extracts the meaning of a sentence grammatically which helps in defining information extraction rules.

The method of manual annotation of sentences to encode patterns that identifies the entities and relations has been widely used for domain-dependent IE systems because it generates relevant patterns only which increases the precision of extracted information. Several recent studies have experimented information extraction using the supervised machine learning approach. Although the results seem appealing, the data collection and labeling part is expensive particularly in a domain relevant scenario. Similarly, unsupervised machine learning approaches were also explored, but the performance depends upon how well the constraints are defined [6].

The proposed work attempts to build an information extraction system by using the approach of rule-based IE systems. The grammatical pattern rules are defined by manually analyzing news articles related to hazards to capture phrases or words which are relevant to hazard severity parameters.

3 Methodology

Disasters occur in all shapes and sizes, from a small-scale neighborhood fire to a large-scale flood. To comprehend the severity of a hazard, several impact factors need to be considered along with the intensity scale as it discerns the repercussions on the society. The proposed work consists of mainly two parts, viz. (i) defining a set of impact factors that can analyze severity of hazards and (ii) developing an information extraction system that focuses on capturing such impact factors by analyzing news articles published in credible web sources.

By thorough analysis of existing literature and various news articles, a set of impact factors are defined which may assist in determining hazard severity. Table 1 shows the list of impact factors to determine severity of a hazard.

Table 1 Basic impact factors to analyze severity

Impact factor	Examples
Casualty	Dead, injured, missing
Rescue operations	Initiated time, end time, rescued, authorities involved
Damages	Infrastructural, agricultural

The proposed work develops an information extraction system to capture the impact factors as mentioned in Table 1. A technique for collecting hazard-relevant news data using a web crawler was introduced in a previous study [7]. A web scraper is used to collect news articles related to hazards that occurred in India between 2018 and 2021 using this technique. With the aid of an ontology that includes keywords related to hazards, the web scraper can classify hazard-related news posts. The content of articles is analyzed in order to improve the IE framework. A news article can contain less-than-newsworthy material, such as advertising, which must be removed before further investigation. To accomplish this, a supervised machine learning model was created that predicts whether a sentence in an article is meaningful or not. Several popular ML models were experimented such as support vector machines, random forest, decision trees, and Naive Bayes, and support vector machines algorithm achieved the highest precision of around 80%. The relevant sentences are then further analyzed to extract the impact factors.

There are many methods for extracting knowledge, and the proposed work focuses on rule-based matching. A rule-based approach involves pre-defining a collection of rules that exhibit the syntactical and grammatical structure of a natural language sentence and then using them to extract knowledge from it. The overall flow of the IE system is shown in Fig. 1.

Identifying the part-of-speech of words in a sentence is the first step in developing rules. Part-of-speech is a category specified in conventional grammar for terms that fall under a particular lexicon or vocabulary based on their meaning and context in the sentence. To extract desirable information, it is also necessary to understand the syntactical structure of a sentence. The most widely used method for this analysis is to generate parse trees, where head and child entities will be generated based on the relation between the entities. Figure 2 shows a sample result from the PoS tagger and dependency parser where a landslide hazard-related sentence was given as an input. Table 2 shows samples of pattern rules of the mentioned impact factors in Table 1.

News articles contain informal phrase usages which is mapped to a similar word for easier analysis. This emphasizes the importance of using all possible synonyms of the impact factor keywords, and the proposed work does so by using the NLP tool NLTK and implementing it in Python. It is also essential to translate numbers written in natural language to numeric values in order to make pattern matching easier.

Knowledge can now be extracted using the given set of rules. Spacy's Matcher is used to do this and is a class capable of finding terms or phrases based on rules that explain the sentence structure and semantics. After the data is collected, it is saved as impact factors in the database.

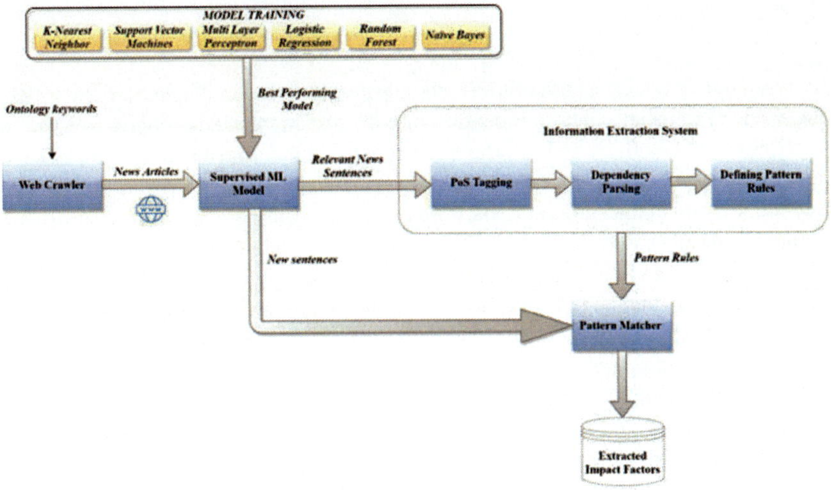

Fig. 1 Information extraction flow diagram

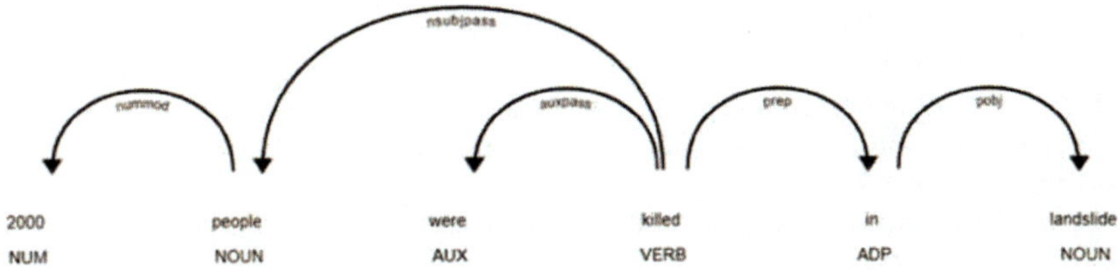

Fig. 2 PoS tagging and dependency parsing sample

Table 2 Pattern samples

Impact factor	Pattern rule	Example
No. of dead	'POS':'NUM','DEP':'nummod' 'POS':'NOUN','DEP':'nsubj' 'POS':'VERB','DEP':'ROOT'	200 people killed in a landslide
No. of injured	'POS':'NUM','DEP':'nummod' 'POS':'NOUN','DEP':'nsubj' 'POS':'VERB','DEP':'ROOT'	Five people injured after they were hit by an avalanche
Infrastructural damage	'POS':'NOUN','DEP':'pobj' 'POS':'CCONJ','DEP':'cc' 'POS':'NOUN','DEP':'conj' 'POS':'VERB','DEP':'ROOT' 'POS':'ADV','DEP':'advmod'	The states face severe infrastructural problems with bridges and roads washed away

4 Results

Extreme floods struck Kerala, India's southernmost state, on August 16, 2018. Kerala received unusually heavy rains this month, causing extreme floods in 13 of 14 districts and prompting a red alert for several days. Hundreds of people were killed or disappeared, and tens of thousands were evacuated. The rescue operations involved a number of local and federal departments, and the danger was seen as a result of global warming.

From August to September 2018, the built data scraper was used to extract news stories related to this event from various credible web sources, yielding approximately 200 news articles. The statistics of the collected data are shown in Fig. 3.

The developed system is also tested using COVID-19 news reports as a pandemic is also categorized as a disaster. The web crawler collected news reports in the duration of August 2021 specific to Kerala district as it faces the second wave with high rates of positive cases and deaths. The statistics of the data collected are shown in Fig. 3.

As can be observed from Fig. 3 that the news reports related to the flood encountered a hike toward the end of the month when the hazard was at its peak. An upsurge in disaster and preparedness-related publications was discovered through a manual examination that included rough scrutinizing. It should also be noticed that the number of articles published increased toward the end of the month, with the majority of these articles focusing on casualties and rescue missions. The COVID-19 news reports were collected and seem to be distributed uniformly.

The IE system was tested using a sample of 500 relevant news sentences out of approximately 3000 news sentences. The 500 sentences were manually assessed to achieve a gold standard comparison along with the IE system's results. To assess the system's performance, we employed precision score, which is calculated as follows:

$$\text{Precision} = \text{True Positives}/(\text{True Positives} + \text{False Positives}) \quad (1)$$

Here, true positives represent the number of correctly extracted impact factor references, and false positives represent the number of incorrectly extracted references. The IE system achieved around 74% precision which shows that false positives exist and require improvement.

Figure 4 shows the casualties of the 2018 Kerala floods based on the location. COVID-19 casualty details were also extracted district wise in Kerala in the duration of August 2021, and Fig. 5 shows the results. By analyzing the spatial distribution, we can see that Thrissur district had the maximum number of deaths during the 2018 Kerala floods. From the COVID-19 spatial distribution, it is evident that Thrissur district has maximum number of dead in the month of August. Both case studies' casualty figures were manually linked with government records [8, 9] for verification, yielding a strong correlation.

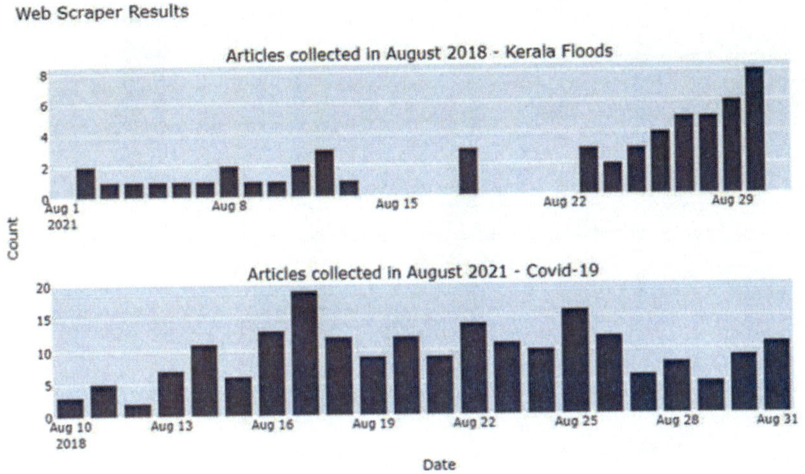

Fig. 3 Data collection statistics of 2018 Kerala floods in the month of August and 2021 COVID-19 in the month of August

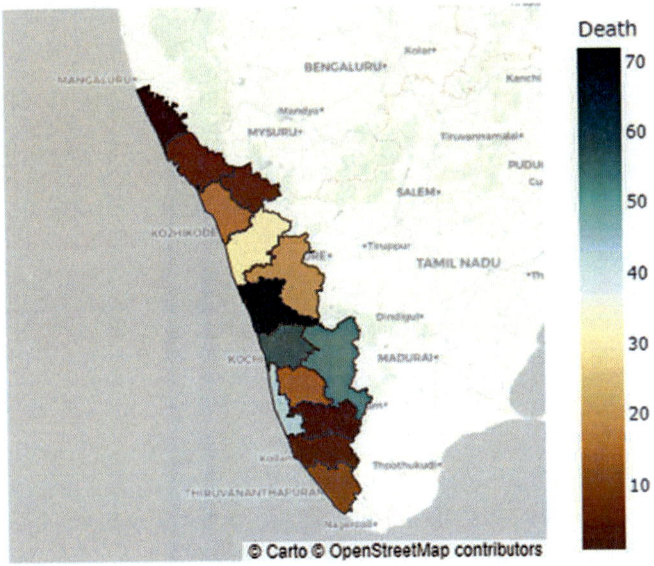

Fig. 4 Casualty statistics based on location of 2018 Kerala floods

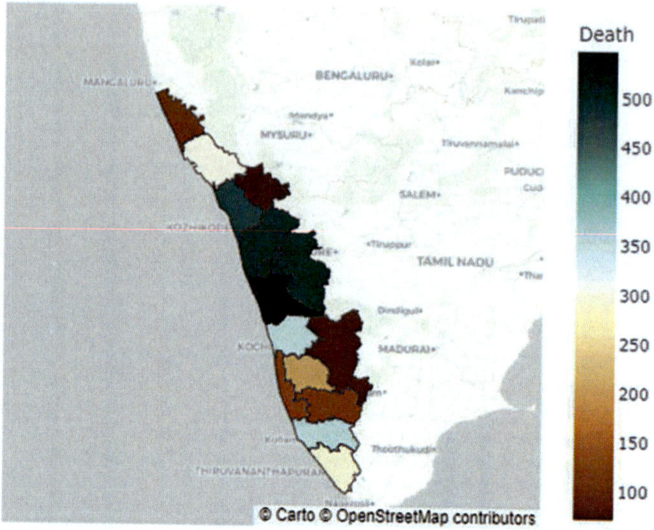

Fig. 5 Casualty statistics based on location of August 2021 COVID-19

5 Future Work

The developed knowledge extraction system complies with the study's requirements. While manually defining rules is time consuming, an automated pattern rule extraction method will be developed in the future. India is vulnerable to a variety of threats, including landslides, flooding, and earthquakes. Defining a disaster intensity scale that is applicable to all disasters would undoubtedly help with hazard analysis. In the future, more impact factors, such as a state's or country's economic conditions, may be identified in order to consider disaster loss.

6 Conclusion

The aim of the proposed research was to create an information extraction method that could extract hazard-related impact factors from an unstructured news article corpus. Patterns were established by manually analyzing news articles, and rules were specified to perform pattern matching. Except for a few exceptions, the results show that the established rules aid in the retrieval of hazard impact factors. It is also worth noting that the named object recognition model did a decent job of identifying the event locations. Although new words and article writing styles may be added in the future, a rule-based matching technique would not be optimal because it requires the work of constantly introducing new pattern rules. By implementing the future work approaches, an efficient system for extracting hazard-relevant information can be built to derive a disaster severity scale.

References

1. Vinodini RM (2014) Design, development, and deployment of a wireless sensor network for detection of landslides. Ad Hoc Netw 13:2–18
2. Nair Meera R, Ramya GR, Bagavathi Sivakumar P (2015) Usage and analysis of Twitter during. In: Chennai flood towards disaster management. Proced Comput Sci 115(2017):350–358
3. Devi GR, Anand Kumar M, Soman KP (2020) Extraction of named entities from social media text in Tamil language using n-gram embedding for disaster management. Nature-Inspired Computation in Data Mining and Machine Learning. Springer, Cham, pp 207–223
4. De Boer J (1997) Tools for evaluating disasters: preliminary results of some hundreds of disasters. Eur J Emergency Med Off J Eur Soc Emergency Med 4(2):107–110
5. Hao Z, Frank B, Riza B-N (2019) Whose story is it anyway? Automatic extraction of accounts from news articles. Inf Proc Manage 56(5):1837–1848
6. Doan A et al (2008) Introduction to the special issue on managing information extraction. ACM Sigmod Rec 37(4):5
7. Gopal LS et al (2020) Machine learning based classification of online news data for disaster management. In: 2020 IEEE Global Humanitarian Technology Conference (GHTC). IEEE
8. Dashboard G Official Kerala COVID-19 statistics. https://dashboard.kerala.gov.in/covid/deaths.php
9. Memorandum2-Floods-2018.pdf. https://sdma.kerala.gov.in/wp-content/uploads/2019/08/Memorandum2-Floods-2018.pdf

Complexity Analysis of Wireless Body Area Networks Using Soft Computing Technique

Madhu Ponakala and Rakhee

Abstract Processing the gathered information in Wireless Body Area Networks (WBANs) is crucial which consist of a few to hundreds of nodes, where coverage is an important issue in WBANs. In this article, we study the time complexity for different stages in the network, namely route discovery and route maintenance using clustering for a multi-hop WBAN nodes. We propose a soft computing technique, i.e., ANT Colony technique to improve network life, load balancing using a mechanism clustering with Breadth First Search (BFS) which avoids trapping during and ensures all nodes participate during exploration. We implemented using the network simulator OMNeT++ to execute proposed technique than conventional method.

Keywords Wireless body area networks (WBANs) · Clustering · ANT colony · Complexity · Breadth first search (BFS)

1 Introduction

Wireless Body Area Networks (WBANs) have been deployed in a variety of scenarios [1–3] and there have been few instances where they have implemented using soft computing technique for indoor hospital environment application. In this article, we extend the scope of WBANs which involves indoor patients whose vitals need to be monitored and sent to the nursing station wirelessly. Such scenario is more important in an ICU where vitals of a large number of patients need to be monitored and communicated in real time. Here, we study on different stages, namely neighbor discovery, data routing, and shortest path to provide un-interrupted monitoring of parameters with the help of sensors. The basic idea of ANT technique [6] lies on the ants communicating among the other individual nodes based on pheromone

M. Ponakala (✉)
St. Francis College for Women, Hyderabad, India
e-mail: madhus.ponakala@gmail.com

Rakhee
The University of West Indies, Mona Campus, Kingston, Jamaica
e-mail: dr.rakhee@uwimona.edu.jm

M. S. Kaiser et al. (eds.), *Information and Communication Technology for Competitive Strategies (ICTCS 2021)*, Lecture Notes in Networks and Systems 401,
https://doi.org/10.1007/978-981-19-0098-3_36

trails from one point to another point in search of food. It has well proven in many applications of Wireless Sensor Networks (WSN). The sensor nodes work ad hoc with limited resources of communication. Hence, conserving the total energy consumption of the sensor nodes along the network and finding the shortest path increases the overall efficiency of the WBAN. Efficient routing protocols [4, 5, 7] have been designed to improve these parameters and minimize the traffic load from source to destination. Further whenever a node in a network fails, various data routing algorithms have been implemented in searching for alternate paths in an adaptive fashion. Clustering using ANT technique [8] helps the network to reduce the load during data communication. Many protocols [9, 10] implemented various routing protocol using clustering technique for its scalability in WSN. It organizes its clusters with cluster head formation in order to receive and transmission of its data from all other members in its group. LEACH which has been an efficient protocol in order to address the energy consumption of nodes in the network using clustering approach [4]. The clusters formed using ANT helped to resolve this issue by continuously adapting to the network changes by using the probability of choosing the neighbor node based on pheromone value and energy in the node [3]. The route selection is based on the minimum cost function. Further, BFS technique with ANT technique is implemented to ensure the complete coverage while ants discover the route from source to destination.

2 Literature

Wireless Body Area Network (WBAN) have become a high impact on healthcare monitoring of population due to chronic diseases which requires remote monitoring continuously of the patients. Based on the IEEE 802.15.6, there are different layers of multiple tiers that participate in the communication from patient to the destination. The architecture explains the deployment of the nodes in the network which participates in sending the information from patients to the end device dedicated to receive the vital parameters without any congestion and link failure as shown in BAN architecture [2–4]. Many protocols addressed on QoS, energy efficient of the data packets in the WBAN to find the best route from source to destination [3, 4, 11]. SMART protocol provides monitoring of patients only in emergency rooms. ALARM-NET proposed a monitoring of patients residentially. Authors in proposed and energy-preserving protocol using multi-hop communication in WBAN using traditional method while choosing an intermediate node with high residual energy. “Anybody” is based on data-gathering protocol which is self-organized into clusters which reduces the number of direct transmissions to remote base station. However, displaying of real-time BAN data using traditional method has been addressed in EPR protocol but it could not address the shortest path using soft computing technique [2].

3 Motivation

Optimizing the energy consumption of the nodes using soft computing technique like ANT reduces the number of node participation when deploying and transmission of the data in WBAN. In this article, three different data processing operations, like route discovery, route maintenance, and route failure, have been addressed. For each problem, a time complexity is calculated with optimal algorithm for three operations.

4 Route Discovery

Route discovery is the process of generating shortest minimal path route from source to destination in the network. It uses forward ants and backward ants starting form tier I to tier III and same ants returning back. The forward ants are responsible for establishing the shortest path from source to destination [2], and once they find the food, they trace back to destination via same route to source.

$$\text{Modified Probability Function} = \frac{D_{ij} * \infty + P_{ij(t)\beta * \tau_i}}{\sum_{i=0}^{N} D_{ij} * \infty + P_{ij(t)} * \beta * \tau_i} \quad (1)$$

Algorithm 1: Route Discovery

```
if ( ant arrives at node i )then
  if (node i is not a cluster head )then
    if (cluster head) then
      choose cluster head neighbor using Equation 1 ;
   else
      store the ant;
      Broadcast a message regarding the cluster head to all other nodes in the clus-
ter;
  else
    if (node i==cluster head) then
     TTL--;  /* maximum number of hops of the ants*/
   if (TTL >0) then
     pick a next cluster head neighbor according to the probability function;
    choose the highest probability function and add current node ID to the routing
vector
  else
    destroy the ant;
```

4.1 Route Maintenance

Route maintenance operation is meant to maintain the path which is generated by route discovery stage. In this operation, pheromone is incremented so that shortest path is sustained and if any node does not receive any acknowledgment within a TTL, then an error message is transmitted.

4.2 Route Failure

This operation generates alternative routes in the network form BAN to NSC. Whenever it receives an error message, the pheromone value is set to zero. An alternate path exists from BAN to NSC and updates in the routing table. Once the packet reaches the BAN, it discovers new route and starts implementing ANT technique for shortest path in the network.

5 Time and Space Complexity

5.1 ANT Technique for WBAN

In this section, we analyze the time and space complexities of ANT technique applied to WBAN.

Theorem 5.1 *The time complexity of the WBAN routing using ANT technique-based algorithm is* $0(\omega * \boldsymbol{Iterations_{max}} + (d^2 * \omega))\boldsymbol{pernode}$.

Proof

$$T_1 = 0(\omega * \boldsymbol{Iterations_{max}}) = 0(20 * 500) \tag{2}$$

where ω is the number of ants, and $\boldsymbol{Iterations_{max}}$ is the maximum number of iterations for the algorithm to execute. The time complexity of computing the Euclidean distance between the nodes is as follows:

$$T_2 = (d^2 * \omega) = (d^2 * 20) \tag{3}$$

Thus, the total time complexity is

$$T_{\text{total}} = 0(\omega * \boldsymbol{Iterations_{max}} + (d^2 * \omega)) = 0(1) \tag{4}$$

Theorem 5.2 *The space complexity for WBAN routing algorithm using ANT algorithm is* $0(\omega * \boldsymbol{n})$ *per node.*

Proof Assume there are ω ants and n number of nodes distributed in the network. The space complexity will be then

$$T_{\text{space}} = 0\ (\omega * n) = 0(n) \quad (5)$$

5.2 Cluster-Based ANT Technique for WBAN

During the cluster formation in the network, the inter- and intra-cluster computation costs are minimum. The cluster rounds is equal to the size of the network. For network size N, the average number of cluster head node is 2S/N, where S is the sum of the probabilities in network size N. So the time complexity for cluster-based ANT technique is $O(d^2/N) = O(n^2)$, where d is the degree of connectivity.

5.3 Message Complexity

Theorem 5.3.1 *The complexity of the sent message count of the above algorithm is O(N)* [13].

Proof In this article, all nodes transmit one forward packet. All nodes, except the base station, send one backward to the source node. Total sent packet count is 2 N-1. Thus, the sent message complexity is O(N).

Theorem 5.3.2 *The complexity of the receive packet is* $O(\delta N)$

Proof During the broadcast communication, each ant transmits a unit of packet to its immediate neighbors to check its pheromone and energy residual. All the nodes have $\delta - 1$ backward messages. Therefore, the received message complexity is $0(\delta N)$.

5.4 Time and Space Complexities of the Cluster-Based ANT Technique Using BFS Approach

Theorem 5.4.1 *The time complexity of the cluster-based ANT technique using BFS is O(D)*

Proof Since the ants start randomly from BAN to NSC, the forward messages are transmitted until they reach the intermediate nodes in the BFS tree, where the maximum height of the tree is D; hence, the complexity is *O(D)*.

Table 1 Comparison of time complexities of proposed algorithms for WBAN

Proposed algorithms for WBAN	Time complexities
ANT technique	O(1)
Cluster using ANT technique	$0(n^2)$
Cluster using ANT with BFS algorithm	0(1)

Table 2 Simulation parameters for indoor hospital environment using ANT

Deployment	Area	21 m *21 m
	Deployment type	Variable packets
	Number of nodes	49 nodes
	Initial node energy	18,720 J
	Buffer size	32 packets
	Transmit power	−25dBm, −15dBm, −35dBm
Task	Application type	Event driven
	Max. packet size	80 K packets
	Traffic type	CBR
MAC	IEEE 802.15.4	Default values
Simulation	Time	1000 ms

Theorem 5.4.2 *The space complexity of the cluster-based ANT technique using BFS algorithm is O(D) per node*

Proof Since every node in the network must store the routing information in the form of routing table, a leaf node must hold a list with D-1 ancestors list. Thus, the space complexity of the algorithm is *O(D)* (Table 1).

6 Simulation Model

In this article, we presented the simulation of cluster-based ANT technique for data routing in WBAN for 49 nodes as shown in Table 2 [3]. Simulations were performed in OMNeT++ to show the performance of the routing protocols using ANT technique to handle the transmission of the critical information without any route failure.

References

1. Movassaghi S, Abolhasan M, Lipman J, Smith D, Jamalipour A (2014) Wireless body are networks: a survey. IEEE Commun Surv Tutorials 16(3):1658–1686

2. Ahmad A, Javaid N, Qasim U, Ishfaz M, Khan ZA, Alghamdi TA (2014) Re-attempt: a new energy-efficient routing protocol for wireless body area sensor networks. Int J Distrib Sensor Netw
3. Srinivas RMB (2016) A soft computing approach for data routing in hospital area networks (HAN). Int J Business Data Commun Netw (IJBDCN) J 12(2):16–27
4. Srinivas RMB (2016) Cluster based energy efficient routing protocol using ANT colony optimization and Breadth first search. Elsevier Proc Comput Sci 89:124–133
5. Panigrahi B, De S, Panda BS, Luk JDLS (2012) Network lifetime maximizing distributed forwarding strategies in ad hoc wireless sensor networks. IET Commun 6(14):2138–2148
6. Omolaye PO, Mom JM, Igwue GA (2017) A holistic review of soft computing techniques. J Appl Computat Mathem 6(2):93–110
7. Qu Y, Zheng G, Ma H, Wang X, Ji B, Wu H (2019) A survey of routing protocols in WBAN for healthcare applications. Sensors 19(7)
8. Moussa N, El Belrhiti El Alaou A (2021) An energy-efficient cluster based routing protocol using unequal clustering and improved ACO techniques for WSNs. Peer-to-Peer Netw Appl 14:1334–137
9. Panigrahi B, Sharma A, De S (2012) Interference aware power controlled forwarding for lifetime maximization of wireless ad hoc networks. IET Wireless Sensor Syst 2(1):22–30
10. Xiangning F, Yulin S (2007) Improvement on LEACH protocol of wireless sensor network. In: Sensor technologies and applications, IEEESensorComm 2007, pp 260–264
11. Malan D, Fulford Jones TRF, Welsh M, Moulton S (2004) CodeBlue: an ad hoc sensor network infrastructure for emergency medical care. In: Proceedings of the MobiSys 2004 workshop on applications of mobile embedded systems, pp 12–14
12. Cobo L (2015) A novel cooperative ant-based clustering approach for wireless multimedia sensor networks
13. Akram V, Dagdeviren O (2013) Breadth first search based single phase algorithms for bridge detection in wireless sensor networks. Sensors

A Novel Method of Corporation Water Distribution to Smart City with IoT Devises Based on Survey

R. Priya and S. Gavaskar

Abstract Many technologies available for water distributing system. A smart city should have facilities to meet the end user with comfort. In the existing system, there is no technical know-how to assess the usage of water to each and every house of the city. In this proposed system, the corporation has installed the know-how, viz. IoT devises, sensor devices, and Arduino for water distribution on area-wise use, and water taxes can be collected based on usage. This will ease the wastage of water based on survey.

Keywords IoT devices · Sensor devices · Arduino · Stepper motor function · Smart city water management system · Database management system

1 Introduction

At present, corporation was supplying water to all the houses to various places in a time. Nowadays, technology is increasing day by day, by which it assesses how much water is being distributed to each places. But, it is not possible to calculate how much of water they need per house. The family may contain more or less number of inmates. So, the need for water varies accordingly. In some cases, even the inmate(s) are less, they spend more water than the required minimum. Therefore, there is a need to revive the distribution of water depending on the area, the number of people reside. In a house, if supply of water is more than the required use, the balance may be kept for next day use and the supply of water to this house is to be dispensed on next day. The system was developed in such a way to avoid unnecessary use of water.

R. Priya (✉) · S. Gavaskar
Bharathiar University, Coimbatore, Tamilnadu, India
e-mail: priyaramato2016@gmail.com

M. S. Kaiser et al. (eds.), *Information and Communication Technology for Competitive Strategies (ICTCS 2021)*, Lecture Notes in Networks and Systems 401,
https://doi.org/10.1007/978-981-19-0098-3_37

2 Existing System

In existing system, water supplement is entirely depending on the time and place where the people is located. A system has to be developed which contains the information of how long water is needed for residential areas. If the area is both residential and factory, then the supply may be bifurcated accordingly. In some areas, water is now supplied on routine nature, viz. week end or week days; i.e., the drinking water is distributed once in a week and bore well water may be supplied on daily routine. Nowadays, water conservation is one of the most essential tasks. So, we should not waste water unnecessarily. Water meters are used in homes to identify the amount of water used and supplied from the corporation. In that system, (i) it is not possible to know how much water a house needs and how much water is wasted; (ii) due to lack of water, many farmers have committed suicide due to inability to farm. The water distribution is depending on the different areas, such as residential areas, agricultural lands, and industrial areas. Currently, many agricultural lands are barren due to water scarcity and other reasons. Water is distributed uniformly to agricultural and non-agricultural lands. Water supplied to non-agricultural areas is wasted unnecessarily. Sensitivity units can be attached to home appliances which measure water usage. This method is based on the WSN [1]. The water monitoring and water handling system will detect if there is a leak in the pipe, underground which helps to find and prevent leaks in the pipe and wastage of water. However, this is a very difficult and expensive method. Water scarcity can also be detected using ZigBee through the agricultural monitoring and management system. Wi-Fi is used to monitor water in the environment, and from this analysis, temperature, and pH value of the water can be determined. This method is often used to determine the water quality of water bodies, such as rivers, lakes, and ponds [2].

The new technologies of IoT are used to sensing, sharing data, and monitoring water flow management process that cannot be easily measured by an individual [3]. Water distribution system is mainly used for supply of water with sufficient quantity, and it is used to explain the usage of water and facilities [4]. Due to fast growing population, water usage is also increased. So, the water supply management system has become complex and water supply should be maintained whenever user needs. Water supply management system is defined as activity of planning, distributing and managing the water resources. Building a sufficient water supply system requires deployment of thousands of sensors and/or actuators to control water grid components such as water meter, pumps, and pressure sensors.

We have to know how much of water evaporates through heat which is called as "smart host" to monitor the water. Also, the system used for automatic water usage monitoring alerts the user when water is wasted. With this warning message, the wastage of water could be stopped. This analysis, if it can accurately determine how much water each house needs per day, then, that amount of water can be supplied to each house through the carburetor. Similarly, agricultural lands, irrespective of the industrial areas, can be prevented from wasting water completely by first analyzing how much water is required for each factory, agricultural land, and focusing on its results and distributing water. From this analysis, the proposed system is designed to avoid wasting water.

3 Proposed System

The vision of smart cities is to develop the urban people life. Based on the vision, the smart cities should be with continuous power supply, sufficient water, and best transportation, and the other important needs are designed and maintained with integrated items, sensors, and integrated networks through computerized system [5]. The proposed system has described how to control water usage. But, the corporation water supply management system (existing system) cannot find any leakage of water in pipe lines and pH level of water. The monitoring system should be developed with all facilities, and the system has to be controlled with sensors. The value of sensors can have processed with microcontroller, and the data has communicated through wired and wireless networks to computers [6].

In the proposed system, it is necessary to calculate how much water is supplied to each household on a daily basis. From the beginning of the day, a table should be generated which indicates how much water was distributed to each household. Before the next day, make a note of how much water is left over in the tank of each house. From this, corporation can calculate how much water has been used for one day. It can be done for few days. After calculating this for one month, it would be easy to know that how much water is spent daily in each house. Accordingly, water will be distributed to each household, depending on the water needs of the occupants of that home.

3.1 Methodology for the Proposed System

So far, the corporation office has been supplying water, depending on the time, as appropriate for each location. Many methods are currently being followed to avoid wasting water. However, by taking into account how much water each household needs, it is possible to reduce the need for water by 75% when supplying water to each household. In other words, the same amount of water is distributed to every house on a street. For example, let us say that approximately one house on a street is supplied with water for one hour and nearly 1000 L of water is distributed. The corporation has decided that 500 L of water per day per household is sufficient if the water supply is one day off.

- Some homes require only 350 L of water per day. The excess water is left over in the tank. It would happen daily, and if the water tank is filled, it would overflow.

- Similarly, some households require 550 L of water per day. 1100 L of water would be required for two days. However, if only 1000 L were distributed by the corporation, there would be a shortage of 100 L of water.

If it is a smart city, water should not be wasted unnecessarily. Similarly, there should be no water shortage in any house. The only way to maintain this is the corporation to know how much water each house needs daily and supply water accordingly. For that, the corporation must know the information of all the houses within the boundaries of a corporation and the extent of their water requirement. The only best way to do this is to do a survey. For a period of at least one month, all the information on how much water was distributed to each house, how much water is left, and when all the water in the tank in any house was exhausted should be mentioned in a database.

3.2 Discussion About Survey

Here is what some people said when interviewed by many to analyze this proposed system: "India has 29 states and 7 union territories. These states are divided into 718 districts. There are 38 districts in Tamil Nadu alone. Tamil Nadu is divided into 234 constituencies. Each block has several streets. There are some houses on every street. Some have noted that it is very difficult to calculate how much water each house needs."

However, if you calculate how much water each house needs every day for a period of only six months, you can make a continuous water supply to each household based on that survey. Assuming a state with a population of approximately 60 million, this would be the largest census. But, if we take each district, there will be specific assembly constituencies for them. Also, there are only about 2,12,00,000 electricity connections in Tamil Nadu. So only 2 crore 12 lakhs houses will be approximate. For example, Coimbatore has ten assembly constituencies. Coimbatore has a population of about 34.6 lakhs, according to the census. This population of 34.6 lakhs will be within these ten assembly constituencies. However, it is a very simple matter to calculate how many houses are there in each assembly constituency. Moreover, each municipal office has all the information about number of houses, streets, and houses per street and how many times a day water should be supplied. Also, there will be information about all the streets and houses in each municipality, information such as house tax and water tax for those houses, and information on which house does not have house tax and water tax. With this information, it is possible to calculate the number of houses in each municipality. Now, there is only one thing to calculate. When distributing water to every home, a sensor that detects amount of water supply in each household must be fitted to calculate how much water falls from the corporation pipe into the tank. With this, we can know the amount of water distributed. Before distributing water to each household, it is necessary to calculate how much water is left in the previously distributed water. To do this, a water level sensor must be installed in the water tank of every home.

3.3 *Survey and Analysis for the Proposed System*

Based on this methodology, the amount of water supplied from the municipality office to five houses at 6th Street in Kurinji Min Nagar under Veerakeralam Panchayat was calculated and recorded during August 2021 in a Table 1.

The information obtained from the survey taken at six houses on 6th Street, Kurinji Min Nagar, Thondamuthur Road, Vadavalli, Coimbatore, is in Table 1. Water has been distributed 11 times in the month of August 2021; i.e., water was distributed on 1.08.2021, 4.08.2021, 7.08.2021, 10.08.2021, 13.08.2021, 15.08.2021, 18.08.2021, 21.08.2021, 24.08.2021, 27.08.2021, and 30.08.2021.

Water is distributed almost every two days. Information for all days is given in Table 1. Approximately, 1100 L of water is distributed to a household each time. Accordingly, 6600 L of water is distributed to six households per day. For example, 6600 L of water has been distributed on 01.08.2021. However, 5750 L of water only spent on that day. Based on this calculation, 850 L of water has been unnecessarily over-distributed. A further 72,600 L per month (August 2021) has been distributed. But spent 63,680 L of water. So 8920 L of water has been over-distributed. Due to this excess water supply, water shortages occur in summers. The water supply should be the same every month, knowing the needs of the people. All the information in Table 1 is taken from the survey. So this information is only approximate. However, accurate measurements can be obtained through the proposed system.

However, Table 2 mentioned the water supply from corporation, water expenditure, and how much water is left, provided on 1.08.2021, 4.08.2021, and 7.08.2021, respectively; i.e., the survey for a week can be found in Table 2. Thus calculated each week, almost 800 liters of water per week remains in the water tank, without need. So the next time, the water is distributed, as always 6600 liters of water will be distributed, filling the water tank and overflowing 800 liters of water and going to waste.

Table 1 Water distribution details during August 2021

House No	House owner's Name	On 01.08.2021		On 04.08.2021		On 07.08.2021		On 10.08.2021		On 13.08.2021		On 15.08.2021		On 18.08.2021		On 21.08.2021		On 24.08.2021		On 27.08.2021		On 30.08.2021		DURING AUGUST 2021	
		Remaining	spent	Remaining	spent	Remaining	spent	Remaining	spent	Remaining	spent	Remaining	spent	Remaining	spent	Remaining	spent	Remaining	spent	Remaining	spent	Remaining	spent	Total Remaining (appro.)	Total spent (appro.)
42	R. Raja	200	900	200	900	150	950	100	1000	100	1000	100	1000	150	950	170	930	190	910	130	970	150	950	1640	10460
44	S. Palanisamy	350	750	350	750	400	700	350	750	300	800	350	750	400	700	370	730	400	700	370	730	370	730	4010	8090
45	S. Anitha	100	1000	Nil	1100	50	1050	100	1000	150	950	30	1070	Nil	1100	50	1050	30	1070	50	1050	40	1060	600	11500
47	M. Ramesh	150	950	150	950	200	900	150	950	150	950	150	950	100	1000	180	920	170	930	150	950	160	940	1710	10390
48 A	P.Karthikeyan	Nil	1100	20	1080	100	1000	Nil	1100	50	1050	100	1000	70	1030	80	1020	100	1000	50	1050	50	1050	620	11480
48	S. Murali	50	1050	50	1050	100	1000	20	1080	30	1070	50	1050	Nil	1100	Nil	1100	30	1070	10	1090	Nil	1100	340	11760
Total		850	5750	770	5830	1000	5600	720	5880	780	5820	780	5820	720	5880	850	5750	920	5680	760	5840	770	5830	8920	63680

Table 2 Water distribution on 01.08.2021, 04.08.2021, and 07.08.2021

Particulars of the house			On 1.8.2021—water distributed at 6.10 am to 7.05 am (55 minutes) with 1100 liters of water for each house		On 4.8.2021—water distributed at 6.30 am to 7.20 am (50 minutes) with 1050 liters of water for each house		On 7.8.2021—water distributed at 4.15 pm to 5.10 pm (55 minutes) with 1100 liters of water for each house	
House no.	Owner name	No. of pupils in house (Nos.)	Remaini ng water	Water spent	Remaini ng water	Water spent	Remaini ng water	Water spent
42	R. Raja	5	200	900	200	900	150	900
44	S. Palanisamy	4	350	750	350	750	400	750
45	S. Anitha	5	100	1000	0	1000	50	1000
47	M. Ramesh	3	150	950	150	950	200	950
48 A	P. Karthikeyan	4	0	1100	20	1100	100	1100
48 B	S. Murali	3	50	1050	50	1050	100	1050
			850	5750	770	5750	1000	5750

Table 3 Comparative statement for the water supply management system for one week

House no.	Average supply per week	Average remaining water per week	Average spent per week
42	1067	200	900
44	1067	350	750
45	1067	33	1000
47	1067	150	950
48 A	1067	13	1100
48 B	1067	50	1050
	6400	797	5750

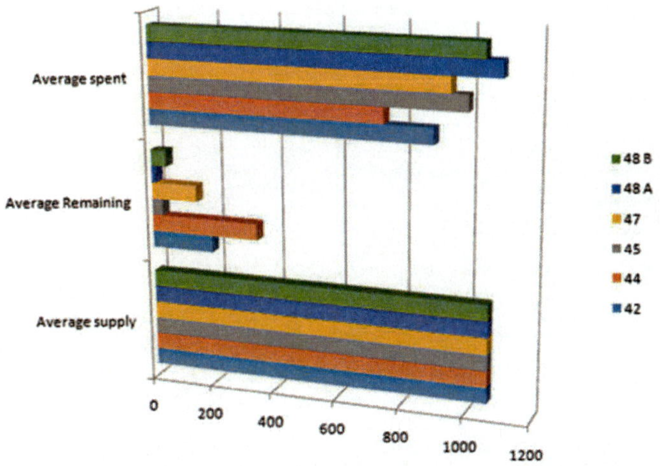

Based on Table 3, 6400 liters of water are distributed to six houses for one week only. But, 5750 liters of water are only needed for the same. 797 liters of excess water are spent unnecessarily. Based on the survey method, the corporation has supplied water of 900 liters to house no.42, 750 liters of water have supplied to house number of 44, 1000 liters of water have supplied to house number of 45, and so on. Based on this method, water only distributed based on user needs and usage. Based on this supply of water, admin can produce the water tax calculation. User can also remitted the water tax amount through online like electricity bill payment.

3.4 Algorithm of Proposed System

```
Corp(){int houseno, username, mobilenumber, watersupply,
remaining, waterspent, database.dbf, remark;
supply( );
days(){char date, watersupply, remaining, waterspent}
watertank(){int waterlevel,sensordev_level;sensor_device();}
//each 100 liter can be marked in the water tank}
define watersupply(){//water supply from corporation
if (water supply to all houses)
{if (houseno=42) then
   write: sufficient water supply to all smart cities house
read corp.supplyqty
   corp.supplyqty=900
   // do the same step to all the houses respectively
   else corp.supply(watersupply) = 1050
   end if
else    //already have some water
   read watertank.waterlevel
   if (houseno=42) then
      supply(){corp.watersupply = 900 - watertank.waterlevel}
else if(houseno=44) then
      // do the same step to all the houses respectively
else{corp.supply(watersupply) = 1050 - watertank.waterlevel}
end if}
sensor_device( ) triggered          //after reaching that level
   write: watertank.sensordev_level
   stop corp.supply();
   store data to database.dbf; end if}
database (){read houseno; corp.houseid =houseno;
sensor_device() triggered;corp.watersupply= supply(extra)
write: watertank.sensordev_level}
//after three days, there is no change in water level
{read houseno;corp.houseid=houseno;
sensor_device() triggered;
watertank.sensordev_level=watertank.nochange
return watertank.water;//return water to corporation}
verify(){if(corp.houseid=houseno)
{corp.watersupply=sum(corp.houseid(watertank.sensordev_level))
print "total water supply for month"
}else{print "check the amount of water supply"}}
//when water flow reach the value of water tank, IoT device
//sending message to user's mobile phone
int force, ledPin=7,
ATDmobileno[i] for i = 1 to no. of mobile phones;
void setup();{  Serial.begin(115200);
pinMode(ledPin, OUTPUT);}
void loop() {   if(force>700)
{ digitalWrite(ledPin, LOW);delay(1200);
    Serial.println("ATDmobileno[1];"); delay(15000);
    Serial.print("AT"); delay(3000);
    Serial.println("ATDmobileno[2];");
// dothe same statement to all the individual house
   else {digitalWrite(ledPin, HIGH);}}
// display data into the database
Corp(){
Title1 "Smart City water Management System"
Title2 "comparative statement of all house water usage"
While (table_row=fetch array(result){// Connect to database
Echo <tr>    // Retrieve all records and display them
result = SELECT * FROM server database through sensor
Echo <td> houseno<td>
Echo <td> date <td>
Echo <td> starting waterlevel <td>
Echo <td> water supplyvalue <td>
Echo <td> mobileno <td>
Echo <tr> }
days(){int houseno, mobileno, watersupply, remaining,
waterspent}

// Finding water level - algorithm
Power supply 5 V - Arudino pin for ground to GND, trigger
to pin2 and echo to pin4. Use sensor to find the water
level of tank and pass the value to Arduino to stop the
water flow supply to the house.

Waterlevel()
{   int trigpin, echopin;sensor_device( ) triggered
    write: watertank.sensordev_level; sent
    sensor_device.value to Arduino; Arduino close
    valve, stop waterflow; end if}
```

3.5 The Design of Proposed Method of Water Supply Management System

In the smart city, the corporation supply water to each house. When water flow reaches the water tank, IoT device in the water tank sends a message to user's (people residing in this house) mobile phone. IoT device also controls the water flow to the tanks of drinking water and sour water, respectively, through pH valve. The IoT device is connected with automatic charger directly, and this charger is connected with power plug of the house. But this proposed system is totally controlled by solar system. Each house tank contains valve for flow of water and sensor chip for detecting water level, when the water level touches the sensor, which is detected by sensor and will be updated to Arduino. Immediately, Arduino stops the water supply to that water tank by closing the valve. In this circuit, GSM modem is connected directly to Arduino. UNO and GSM transmitter is connected to Arduino receiver, and GSM receiver is connected to Arduino transmitter. Sensor is connected to Arduino with power supply of 5 V. Sensor ground is connected to Arduino ground. Sensor is fixed at the rim of the house tank. Transistor emitter is connected to ground, base is connected 7th pin of Arduino through 1KΩ resistor, and Transistor C is connected to solenoid valve. Solenoid valve is connected to transformer with 12v adaptor. 12 V adaptor is control the power supply based on circuit devices. Solenoid valve is connected to corporation water tank, and the other side is connected to house tank (Fig. 1).

In Fig. 2, 12 V adaptor is control the power supply based on circuit devices and converted to 5 V power supply is supplied to this circuit. Three numbers of solenoid valves are connected to corporation water tank, and the other side is connected to house water tank.

S1, S2, S3 → Water Level Detection Sensors H1, H2, H3 → House Water Tank.
C1, C2, C3 → Capacitor.
T1, T2, T3 → Transformer.
M1, M2, M3 → Solenoid Valve With Motor.
1kΩ—3 Nos. of resister

In Fig. 3, sensor is fixed in the rim of glass, and pipe is connected from solenoid vale and ended to glass. Another side of valve is connected to tank. This proposed system was tested with direct passing water to pipe and water passed through the pipe to glass. When the water touches the sensor, automatically makes a call to the user mobile phone immediately with GSM modem. SIM card is fixed under the GSM modem. So phone calls are raised from that SIM number to all. When the level of water is reaches the sensor, automatically valve is closed with the help of Arduino, and the supply of water will be stopped. All the process are controlled by Arduino.

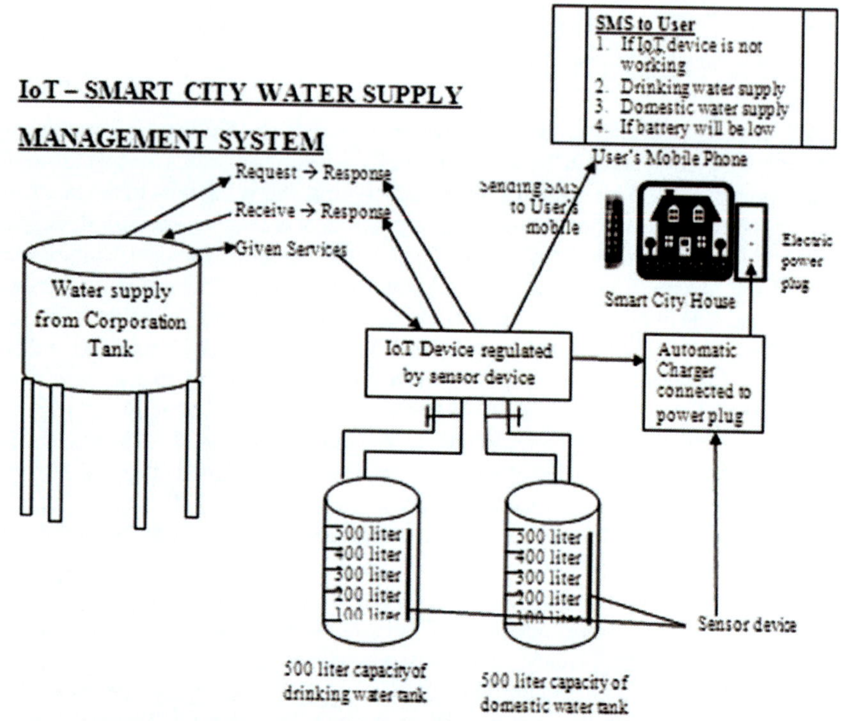

Fig. 1 Proposed design of water management system

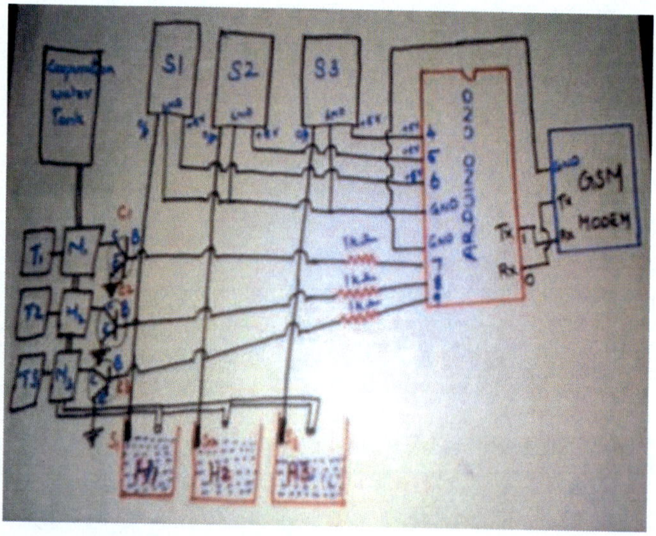

Fig. 2 Circuit diagram for proposed system

Fig. 3 Circuit connection of proposed system

4 Working Process of Proposed System

In this proposed system, sensor is fixed at the rim of glass, and pipe is connected from solenoid valve and ended to glasses (House water tank). Another side of valve is connected to corporation tank. This proposed system was tested with direct passing of water through the pipe to the glass (Figs. 4 and 5).

SIM card is fixed under the GSM modem. So phone message is raised from that SIM number to all users mobile. After that, water will be supplied to glasses. When

Fig. 4 First mobile phone with 2 Nos. of SIM card for first and second solenoid valve

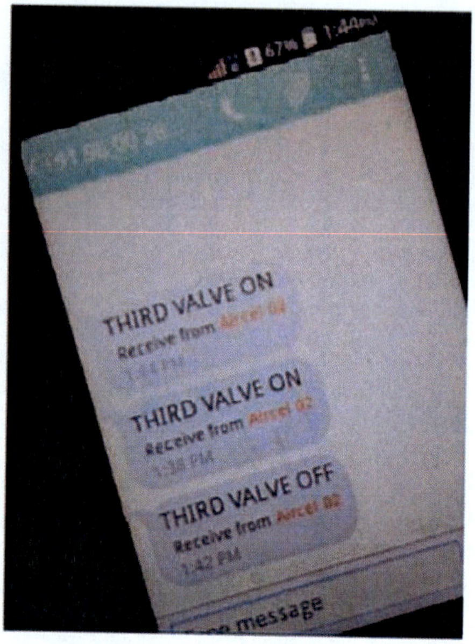

Fig. 5 Second mobile phone for third solenoid valve

the level of water reaches the sensor chip (high range), automatically valve is closed with the help of Arduino and the supply of water will be stopped. Not only this method, user can also stop the water through his mobile phone message. All the process is controlled by Arduino. Transformer is used to maintain the power supply of 5 V. For that purpose, system has been used with 12 V adapter. In this figure, the mobile one has two SIM cards. One SIM card for first house and other for second house. Another mobile for the third house. When the valve was opened, automatically messages are sent to all the SIM card (user's mobile phone). So, first mobile got two messages of "First valve ON" and "Second Valve ON" for two houses with two SIM cards. Another mobile phone got message of "Third Valve ON." After that, from the second mobile phone (for third house), a message is sent to the corporation "Third valve OFF." Immediately, the water supply will be stopped to third house. Similarly, if the user does not need water, user can also stop the water supply to his water tank through the message from his mobile phone also. In this figure, testing is made with three water tanks. When the water level reaches the sensor, automatically solenoid valve is closed and water supply is stopped by Arduino.

5 Further Enhancement

- The proposed system is tested for residential areas only. But, the system should extended to other areas such as industrial place, agricultural land, etc. In this proposed system, two sensors are needed: one is domestic water and other is drinking water. It will be extended with 2 Nos. of sensor of each house.

- The proposed system has designed to reversible process, when there is no change in house water tank level after three days. It will be also done in further work.
- Now, the system has been developed with algorithm to store the data in the server database. Based on the data in the server database, the water bill can be calculated for the individual houses.
- Water is distributed through this proposed system by calculating how much water each family needs. But in the old system, most of the water was distributed by the corporation. So, there will be more time for that. But the time will also be slightly reduced as the required amount of water will be distributed in the proposed system. However, the time for this proposed system has not yet been calculated. When developing this system, time is also calculated.

References

1. Suryadevara N.K, Mukhopadhyay S.C,"Wireless sensor network based home monitoring system for wellness determination of elderly", IEEE Sens. J. 12(6) (2012)
2. Jiang P, Xia H, He Z, Wang Z (2009) Design of a water environment monitoring system based on wireless sensor networks. Sensors 9:6411–6434
3. Balestrini M et al (2015) IoT community technologies: leaving users to their own devices or orchestration of engagement?. EAI Endorsed Transactions on IoT-10–2015
4. Sobsey MD (2002) Managing water in the home: accelerated health gains from improved water supply. World Health Organization
5. Robert E, Brookhaven H (2000) The vision of a smart city. In: 2nd international life extension technology workshop, Paris, September 2000
6. Ntambi F, Kruger CP, Silva BJ, Hancke GP (2015) Design of a water management system. INSPEC Accession Number:15615935, https://doi.org/10.1109/AFRCON.2015.7332012, September 2015
7. Arun Mozhi Devan P, Pooventhan K, Mukesh Kumar C, Midhun Kumar R (2019) IoT based water usage monitoring system using LabVIEW. Springer Nature Switzerland AG
8. Hegde N (2012) Water scarcity and security in India—in:BAIF at the Indian Science Congress (2012)
9. Hebous S, Klonner S (2014) Economic distress and farmer suicides in India: an econometric investigation" University of Heidelberg, Paper Series No. 565, May 2014
10. Whittle AJ, Allen M, Preis A, Iqbal M (2013) Sensor networks for monitoring and control of water distribution system. Hong Kong, 9–11 Dec 2013
11. Chung W-Y, Caya MVC, Chen CL (2013) Wireless sensor based monitoring and content management system for agricultural application. FedCSIS 75–78
12. Rasin Z, Abdullah MR (2009) Water quality monitoring system using Zigbee based wireless sensor network. Int J Eng Technol IJET-IJENS 09(10)
13. Ormsbee LE, Lansey KE (1992) Optimal control of water supply pumping systems. 120(2). March/April, 1994. ISSN 0733–9496/94/0002–0237,Paper No. 2646
14. NtuliAdnanAbu-Mahfouz N (2016) A simple security architecture for smart water management system. Proc Comput Sci 83:1164–1169

Neutrino Signals Throw Light on Dark Matter Searches

Sonali Bhatnagar

Abstract The question to answer is: What do we need to know about neutrinos which will lead us to dark matter searches. In this direction, the second question is why neutrinos are so important. The 2015 Nobel was given to the discoverer of neutrinos: SNO and Super-Kamiokande which was for solar neutrinos and atmospheric neutrinos; in 2016, the Breakthrough Prize went to the experiments in neutrino physics: Daya Bay, SNO, Super-Kamiokande, T2K, and KamLAND which established different aspects of neutrino oscillations. Daya Bay determined the mixing angle, SNO solved the solar neutrino problem along with KamLAND, and T2K and Super K in Japan also threw light on oscillations in neutrinos. In 2020, an article by T2K in nature on the Constraint on matter–antimatter symmetry-violating phase in neutrino oscillations has been published where in the electron–positron beam they find hints on CP violation which may help us understand matter–antimatter symmetry. Neutrinos played an important role in establishing the Standard Model, i.e., the Unified Electroweak model. Neutrinos are ubiquitous in astrophysics and cosmology, i.e., they tell about stellar evolution, supernova explosions, CMB (Cosmic Microwave Background), LSS (Large Scale Structure), BBN (Big bang Nucleosynthesis), and synthesis of heavy elements in the Universe. These studies of neutrinos give good skills to work for BSM scenarios like dark matter particles search. Neutrino masses are the first measurements of BSM physics. The future experiments as DUNE, Hyper K, JUNO, SBN at FNAL, IceCube-Gen2, nEXO, LEGEND, CMB-S4, and many more generations in this direction.

Keywords Neutrino · Solar neutrinos · Neutrino oscillations

1 Introduction

In 2021, the neutrino particle is 91 years old. It was postulated in 1930 in a letter written by Wolfgang Pauli to a conference on radioactivity held in Germany [1].

S. Bhatnagar (✉)
Department of Physics and Computer Science, Dayalbagh Educational Institute, Agra, India
e-mail: sonalibhatnagar@dei.ac.in

M. S. Kaiser et al. (eds.), *Information and Communication Technology for Competitive Strategies (ICTCS 2021)*, Lecture Notes in Networks and Systems 401,
https://doi.org/10.1007/978-981-19-0098-3_38

The concept was to explain beta decay of nuclei which were decaying into other nuclei by emitting electrons, and the energy of emitted electrons was plotted thinking a two-body decay. But to the surprise, the electrons came out in a spectrum of energies satisfying the laws of conservation of energy and momentum. This problem was thought as a fundamental property of quantum mechanics by Neil Bohr. Pauli however was more fundamental thinking the law of conservation of energy and spin-statistics should hold, postulated another particle which should be emitted in these decays calling it the neutron in analogy to proton and electron. Later when neutron was discovered two years later, this was not the particle suggested as neutron was as heavy as proton, whereas this particle had to be very light. Later in 1933, Enrico Fermi called this third particle a "neutrino" which meant "a little neutron" and developed his first theory of weak interactions and submitted it to nature. His theory required two currents interacting between proton and neutron along with the second current between electron and neutrino. Dirac equation told us the fermionic currents defining the coupling between these four fermions and write down the Lorentz invariant expressions in the weak Lagrangian of weak forces (Figs. 1 and 2).

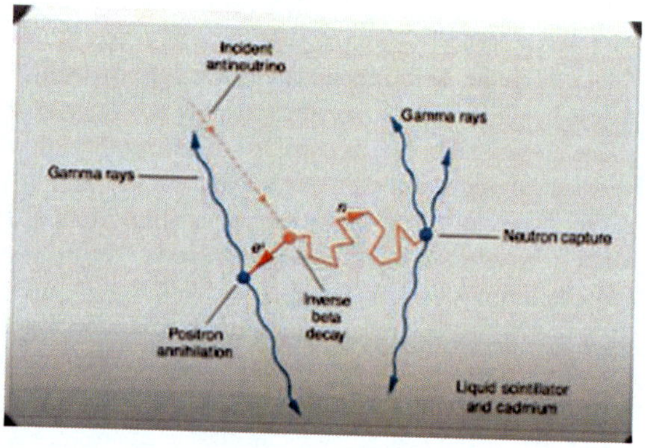

Fig. 1 Neutrino detection

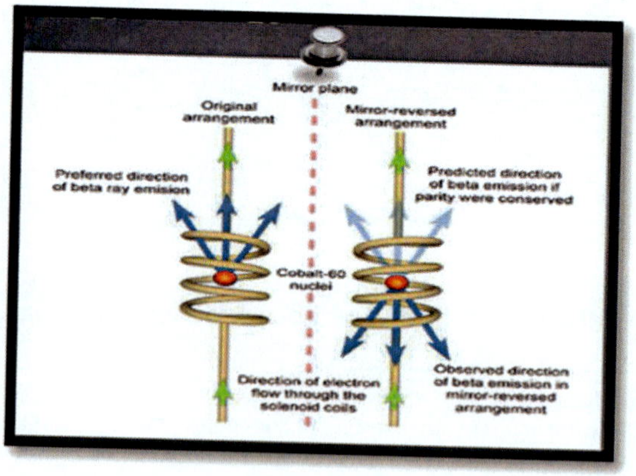

Fig. 2 Parity (Wikipedia)

The first twenty years of neutrino were lived through with the fact that it could never be detected as the coupling constant of the interaction G_F is proportional to the mass inverse square, and the cross section was proportional to Fermi constant square hence to mass inverse to power four. Putting the numbers, G_F is 10^{-5} GeV^{-2} giving the cross section of order $10^{-16} \times 0.04\ fm^2$ which is approximately $10^{-44}\ cm^2$ giving a mean free path of 10 light-years confirmed the fact that there is no practical method to detect a neutrino. The mean free path was 10^{14} cm for neutrino of energy 1 meV with a density 1 g/cm^3 of matter and for neutrinos of energy 30 meV, and nuclear density 10^{14} g/cm^3 is 0 m concluding that the neutrinos are trapped in core collapse of a supernova. The neutrinos get trapped here and slowly escape the core of SN. In 1950s, Reines and Cowan proved Hans Bethe wrong, by placing a clever detector near a nuclear reactor [2]. The experimental challenge was not only to create a big enough detector but to deal with gamma ray flashes due to reactor emissions and cosmic rays, i.e., clear the neutrino signal from other backgrounds. The governing reaction was the electron neutrino hitting a proton and creating a positron and the neutron. This positron could combine with an electron and create two gamma rays, so these gamma rays could be created by other energetic particles, cosmic rays also. This signal preference was established by filling the detector with water and cadmium chloride, water being the source of free protons and cadmium chloride has a large cross section for neutron capture creating a sequence of two gamma ray bursts within few seconds. This was the signal for the detected neutrinos.

Another experiment which throws light on our fundamental understanding of nature of weak interactions was conducted by Chein-Shiung Wu in 1956, studying parity violations in weak interactions. In this experiment, a nucleus was polarized in the presence of a magnetic field and its decay products were analyzed. The neutrinos were not observed but the electrons were observed to find that their spins were aligned preferentially with the spin of the nucleus which implied that the product of spin and momentum was dictating the distribution violated parity. At that time, Lee and Yang also published a paper for tao-theta puzzle but now we know that they are kaons. The puzzle being that one of these particles was decaying to two pions and the other one to three pions. Thus, the paradox being those two particles with same masses, decaying rates were decaying into even and odd parity final states. So, they both postulated that parity was not conserved in weak interactions giving them the Nobel Prize. Finally, in 1957 Lee and Yang proposed that parity violation was true along with CP conservation. In 1964, the $K_s{}^0$ decay gave signs of CP violation also. Next neutrinos may give us signs of this violation, but other sources are not clear. This CP violation throws light on matter–antimatter symmetry.

In 1957, another property of neutrinos was measured: left handedness. This experiment at Brookhaven national Laboratory was one of the most beautiful experiments of all times, where the neutrinos were not measured. They studied the polarizations of gamma rays to figure out how the neutrinos behaved.

The isotope of Europium ^{152}Eu was unstable with a nine-hour lifetime to electron capture a weak process and forms a new nuclei Samarium ^{152}Sm * which returns to its ground nuclei with a gamma emission. Now as visible in Fig. 3, Eu has no spin but Sm * has spin 1 and finally ^{152}Sm has no spin. The gamma rays and neutrinos account

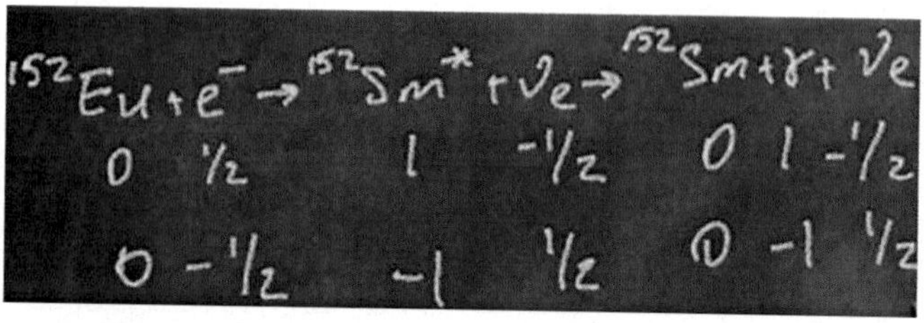

Fig. 3 The neutrino helicity

for the momentum conservation, having two options. The electron absorbed by Eu can have spin parallel or antiparallel to the direction of final neutrino, challenging the measurement of the gammas coming back-to-back with final neutrinos. To find the spin of emitted neutrino, they used resonant fluorescence to measure the gamma ray emitted by excited nuclei using gamma ray detector. The neutrinos violate parity maximally. The neutrinos were responsible for the discovery of weak neutral currents which were different from radioactivity and charged current interactions.

The sources of neutrinos are the reactors, accelerators, and sun which has a self-regulating thermonuclear reactor at its center. The sun is creating neutrinos confirmed by Homestake Solar Neutrino experiment with a groundbreaking proposal of generating the ^{37}Ar from ^{37}Cl using solar neutrinos. The theoretical prediction was of 51 atoms of argon as compared to Davis seeing only 17 in the experiment. This discrepancy took lot of time to clear out. Flavor oscillations were proposed in 1962 to understand this solar neutrino problem. The sun only produces electron neutrinos and does not have enough energy to produce muons; hence, the electron neutrinos convert or oscillate into the muon flavor on the way to earth. Oscillations implied that neutrino mass eigenstates are composed of mixture of flavor states. The neutrino mixings would resemble quark mixings if the neutrinos were massive, the amplitudes of CKM will be small along with a small third generation mixing angle. Another important phenomenon to explain the solar neutrino effect is the matter effect by finding the coherent forward scattering of neutrinos from matter. Kamiokande detector was using a better technology based on Cerenkov detectors, looking for the atmospheric neutrinos, solar neutrino problem, and proton decay. This experiment confirmed the deficit of solar neutrinos, some atmospheric neutrinos were also missing, and a month later astronomers observed the Large Magellanic Cloud supernova explosion with naked eye. Neutrinos came as the core collapse was happening, and the visible light came 2 h later. This was measured as 10^{58} neutrinos emitted in 10 s with an energy estimate of 3×10^{53} ergs which was 15% of the rest mass of the sun. This much power was possible only due to gravitational collapse of the star converting energy to rest mass of particles. Gamow in 1940 was the first to suggest that neutrinos play an important role in stellar evolution, in particular the collapse of evolved stars which was confirmed 50 years later. This was a fundamental breakthrough. In line of breakthroughs, the Super-Kamiokande conclusively showed that atmospheric neutrinos oscillate (up-down asymmetry) with maxima mixing of tau and mu neutrinos. The SNO (Sudbury Neutrino Observatory) measures independently the flux of solar neutrinos of all flavors. KamLAND observed oscillations of antineutrinos from reactors, explaining the parameters for solar data. Oscillations were measured with man-made neutrino beams like K2K, MINOS, T2K, and NOvA. Daya Bay and RENO reactor experiments measured the last mixing angle $\theta_{13.}$ Finally, neutrino oscillations had conclusive evidence: a long way from 1998 to 2000 for solar, atmospheric, and supernovae neutrinos.

2 Neutrino Oscillations

The neutrinos are chiral which implies their spin is antiparallel to their momentum whereas for antineutrinos the spin and momentum point in the same direction. This fact implies that the neutrinos behave as massless fermions and all the experiments for the creation and detection of neutrinos send the signal of neutrinos acting as massless chiral fermions. The mass signals come from oscillations which are a very delicate inference result. These particles interact only through weak interactions through the W and Z gauge bosons and not through photons or gluons which give rise to unique properties in the laboratory and astrophysics experiments. Now the fact that they oscillate gives rise to tiny masses and distinct flavor and mass eigenstates. These oscillations occur in solar neutrinos, atmospheric neutrinos, reactor neutrinos, and neutrinos created in laboratory beams in refutable ways which leads us to trying to understand them. The flavor and mass eigenstates of the neutrinos unlike the quark sector are not aligned well. In the quark sector, the up-quark emits a W boson to create a down-quark, then it turns out that it does not change to down-quark but to a superposition of d and s quark. This was first realized by Cabibbo giving us $d' = d\text{Cos}\theta_c + s\text{Sin}\theta_c$ the Cabibbo angle θ_c. Similarly, the c quark couples to $s' = -d\text{Sin}\theta_c + s\text{Cos}\theta_c$ but for the top and bottom quark, the top quark emits a W boson and converts to bottom quark. In comparison with this quark picture, the neutrinos were different as all the particles listed were eigenstates of mass whereas the neutrinos listed were the eigenstates of flavor. The Cabibbo–Kobayashi–Maskawa matrix has important contributions to CP violations, and the same was expected for neutrinos. For neutrinos, there are three mass eigenstates ν_1, ν_2, and ν_3 with measurements of the square of the mass splitting between ν_1 and ν_2, similarly between ν_2 and ν_3. The order of these states is unknown as it requires more precise information resulting in normal mass hierarchy and inverted mass hierarchy. The matrix connecting the favor eigenstates like CKM matrix is called the V_{PMNS} to the mass eigenstates. PMNS stands for Pontecorvo–Maki–Nakagawa–Sakata. This matrix is parameterized in terms of three angles with rotation between 2 and 3 states, then rotation between 1 and 3 states, and finally rotation between 1 and 2 states. To visualize it, let us go to the rest frame of this massive neutrino, which is not accurately defined as they have different masses, with three wave packets slowly shifting relative to each other. In case of two states, a simpler case, we picture it in the rest frame as two wave packets shifting relative to one another, and this particle is a superposition of two states depicted by Schrodinger equation evolution. Once derived in rest frame, this can be transformed into any laboratory frame. This gives the oscillation probability for electron neutrino to muon neutrino based on amplitude theory gives

$$P_{\nu e \to \nu\mu} = \mathrm{Sin}^2 2\theta \mathrm{Sin}^2 \Delta t/2 \text{ where } 2\Delta = \Delta m^2/2E_\nu \text{ with } \Delta m^2 = m_{\nu e}^2 - m_{\nu\mu}^2$$

This is the probability of flavor change, at time $t = 0$ start with single flavor which oscillates in appearance probability with maximum value determined by the mixing angle θ. t is length in km, mass in eV, and energy in GeV tells us that for atmospheric neutrino oscillations give the oscillation length t of order of km which is a nice scale on earth. For example, for a neutrino of energy GeV, mass difference of order EV gives the oscillation length of order 1000 km making earth a nice detector. In case of reactor with neutrinos of energy MeV, the oscillation length is of the order of 1 km showing the experiments Daya Bay where the detectors were kept a km away from the reactors to measure these neutrino oscillations. The measurements of θ_{13} from Daya Bay shown in Fig. 4 tell the histogram where the blue spectrum is un-oscillated spectrum and red is the observed spectrum which is showing the disappearance of the neutrinos. From this data on left, they re-bin it for L/E_υ in the right figure to give the expected Sin wave. As seen, this is a quantum phenomenon with a classic oscillation pattern. Since the neutrinos have masses, then they will have different arrival times depending on the different velocities. The detection and oscillation process are subject to usual quantum mechanical challenges; thus, the time of emission of neutrino has uncertainty in duration with interference effects between the wave packets corresponding to υ_1 and the wave packets corresponding to the υ_2. In case of supernova neutrinos which travel 8 kiloparsec, then by this time the wave packets corresponding to different mass eigenstates have separated in the time they reach the earth. In case of reactors, over 200 km the wave packets have

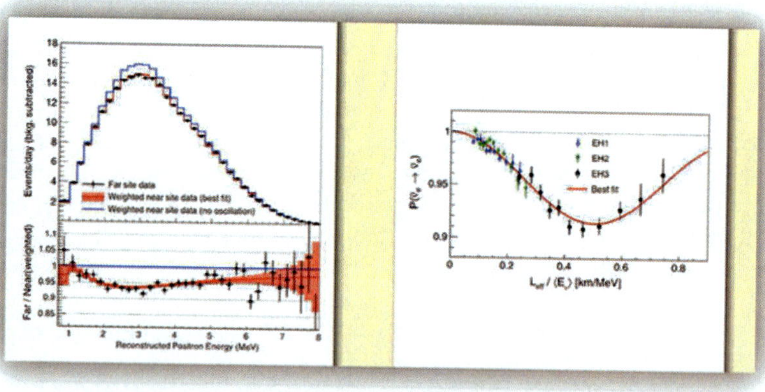

Fig. 4 Daya Bay data [3]

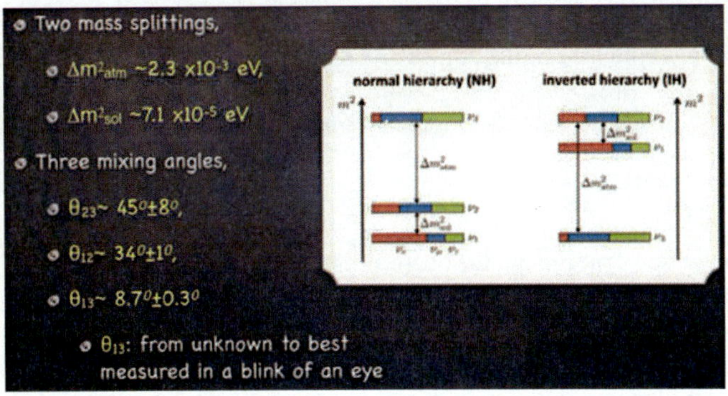

Fig. 5 Mass splitting (TASI)

not yet separated and so we see this interference phenomena. The quantum nature of oscillations can reach all kinds of paradoxes. For example, suppose neutrinos are created through pions in beam experiments (Fermi Laboratory) through a two-state process (a muon and anti-muon neutrino), in which case on knowing the energy and momentum of the pion, measured them for muon and calculate it for the muon neutrino, then from here the final neutrino state can be identified. This is a paradox.

The way out is that when the precise measurements of energy and momentum are being observed, then the position of the muon becomes more uncertain according to Uncertainty Principle, resulting in the destructive interference of phases of oscillations giving no oscillations. Summarizing, the oscillations exist due to the uncertainty principle. The oscillation phenomena only occur when there is a coherent superposition of several mass eigenstates because only then there is a relative phase buildup and can interpret that state as an oscillating composition of the electron and muon flavors. The fact that they are producing a system which is a superposition of two or more mass eigenstates rests on the uncertainty principle only. Another quantum phenomenon present here is the spin analogy. In a two-flavor system, it is a SU (2) symmetry and then any two-flavor system can be thought of as a spin. For a state, the expectation value of the polarization of spin is one way to describe the state of two-flavor system. This takes two spinors and maps them to two vector called Bloch vector, which points in some directions on the sphere of possibilities called the Bloch sphere. The time evolution of this object is given by Schrodinger equation in terms of the commutator of spin and the Hamiltonian. The Hamiltonian is the superposition of sigma Pauli matrices, and it can be solved to a graphical depiction on the Block sphere. The Hamiltonian vector on Black sphere represents oscillation. Hamiltonian graphically has two components: x-component and z component. The two basis states are νe and υμ giving the evolution as a precession. The projection on the z axis of the spin is the derived survival probabilities. This is the qubit which describes the two-state system with vacuum oscillations [4]. The present diagram of normal hierarchy and inverted hierarchy, where the color bands correspond to the flavor compositions is shown in Fig. 5. These diagrams show the atmospheric and solar splitting, with the mixing angles given with their uncertainties of measurement.

3 Solar Neutrinos

To start a particle experiment, the beam should be understood, so similarly in case of sun the beam is a thermonuclear reactor and is an average main sequence star. The Hertzsprung–Russell diagram (Fig. 6) tells us about the star's evolution with stellar luminosity on the vertical scale in log scale and temperature of the surface of the star or the color of the star on horizontal axis. On the right side of the color scale, it is red (3000 K), and on the left side, it is blue color (30,000 K), corresponding to the surface temperature of the star. On the diagram, moving along the main sequence upwards gives heavier stars in their first state of evolution while burning their hydrogen into helium. After they finish burning the hydrogen, the white stars go and occupy the region of red giant branch where the envelope expands and the core collapses until it starts burning helium. The very large stars have a different style of evolution where they go from main sequence to super giant and then perform the blue loop motion leading to the interesting astrophysics models. Our sun is also on the main sequence and is a continuous source of solar neutrinos while burning the hydrogen into helium 4. The core temperature of sun can be estimated to 1 keV based on the Schwarzschild radius. The actual temperature at the core is 15×10^6 K. In order to maintain itself, the sun needs to maintain its gas pressure and since it loses energy from the surface

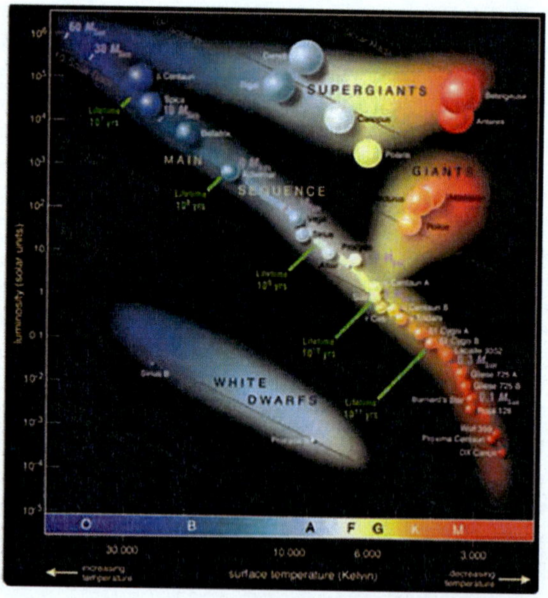

Fig. 6 H.R. Diagram [Wikipedia]

by emission of photons. The reactions are needed to replenish the heat; otherwise, the core will cool and star will compress with increasing pressure. The overall statement is that the sun shines because of quantum tunneling and shall burn for many more billion years based on the calculations before leaving the main sequence to become a red giant. The rates with which the sun burns is not related to the properties of nuclear reactions and is only related to the way energy is lost. The energy produced in the core of the sun loses out through the outer layers, mostly by photon diffusion called radiative transport. The outer 30% radiative transport becomes inefficient, leading to material boiling called convective transport. The sun as a function of the radius then about inner 70% is radiative, stable with energy flowing from hot to cold according to the temperature gradient and the outer 30% is convective due to the plasma created magnetic fields corresponding to sun spots, solar flares, and magnetic storms on earth. The information from these reactions of protons tunneling to another proton to form a deuteron is the emission of an electron neutrino, along with energy release of 0.42 meV. Thus, large number of neutrinos are produced in this reaction calling them "pp" neutrinos which could be measured by GALLEX experiments. These neutrinos were not a good source and so studying the next nuclear reaction of deuteron reacting with proton to form helium 3, photon along with energy release 5.5 MeV. Helium combining with itself forms helium 4 and 2 protons along with 12.86 meV energy. These total reactions emit quite a lot of energy 26.73 meV in the formation of helium 4 from 4 protons with only 2% of this energy went into the pp neutrinos which is the better part otherwise less energy would be left for us.

As observed, the different pp cycles of energy and no released are shown, although pp1 is the main branch. In pp2, the energy released is 1.5 meV and is occurring 17% but with no neutrinos. The main branch in which beryllium decays to ground state and captures an electron is emitting 0.862 meV. The pp neutrinos are now detectable with advanced experiments but most of the energy in pp cycle goes to non-neutrino final states. The pp3 chain gives electron neutrino with huge energy of order 15 meV, thus being responsible for the Homestake and Kamiokande neutrinos. The second main branch contributes "pep" neutrinos of energy 1.442 meV, and they have been observed. Figure 7 shows the spectrum of solar neutrinos. The black lines correspond to pp neutrinos and the ones in blue are the CNO neutrinos which are coming from the burning of hydrogen into helium using carbon as a catalyst. These neutrinos have energy above 1 meV and are still to be detected. It is observed that the conditions in the core of the sun allow the PP chain to dominate the energy production whereas the CNO reaction shown in Fig. 8 according to the Standard Model only produces 1% of the energy but going to higher mass stars the CNO reaction takes over. Finally, in the most massive stars it is the CNO cycle which is responsible for the core burning.

4 Supernova Neutrinos

At some point in the evolution of the massive star, the core reaches a critical value of the mass called the Chandrashekhar Limit where it can no longer sustain its gravity

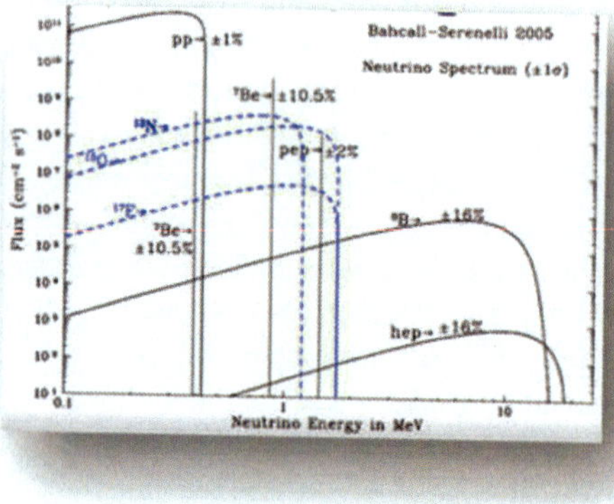

Fig. 7 Energy spectrum of solar neutrinos [5]

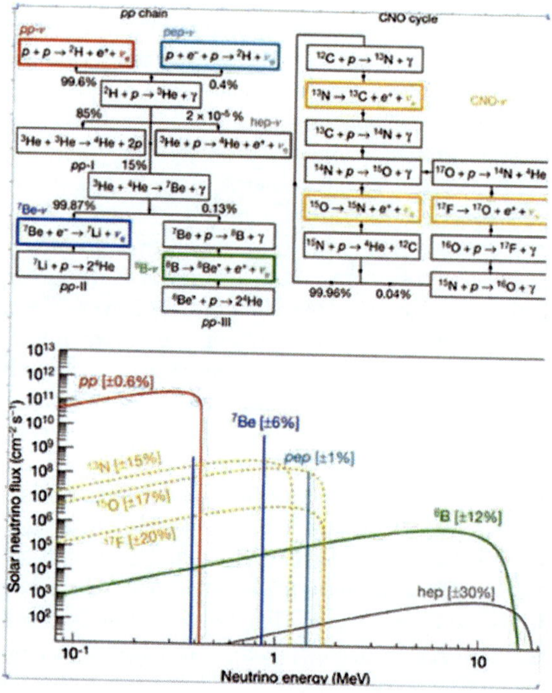

Fig. 8 Summary of solar neutrinos [5]

giving way to the core collapse. When this region collapses, the collapse is only averted when the star reaches nuclear density such that nucleons are rebound in the region, neutrinos are trapped unable to move freely out of this region. These neutrinos are providing pressure and mechanism to transport their energy through diffusion. During the collapse, a shock wave is formed inside the neutrino sphere. This is the analog of sun's photosphere where the protons are moving in stream. The shock front is formed inside the star, where it travels outward and meets the neutrinos sphere. The material falls inside at 20% the speed of light and suddenly they come across the pressure which is pulling them apart. This region where this material gets a kick, and it is this region which forms a shock wave. This shock moves outwards and breaks through neutrino sphere where it releases some of the neutrinos which are trapped resulting in a burst of the electron neutrinos trapped near the surface of the neutrino sphere. This is called neutronization burst in which the shock goes out and stalls at 200 km approximately. After the shock is revived, it explodes the outer envelope and layers come out as visible in an explosion. The visible part of the explosion gets 1% of the total energy, saying that 99% of the total energy is emitted in the neutrinos. This is important, as neutrinos determine the evolution of this object and this is because they have the smallest cross section of the known particles, in the dense stage the Fermi Sea of neutrinos and electrons is present. The electrons cannot go very far as their scattering cross section is very large, and they keep scattering and remain inside the core whereas the neutrinos scatter weakly and are most efficient in carrying the energy. The mean free path in this environment is of the order of a meter, for neutrino transport. The central object cools and turns into a neutron star. The recent supernovae explosion seen was 1987A creating a neutron star in the center, but since it had not been seen. Now the neutron star is visible 30 years later when the outer layers have cleared.

5 Sterile Neutrinos and Cosmology

In mid-1990s, the LSND experiment at Los Alamos where they created a detector close to a beam run for observing neutrinos from pion decays at rest. The baseline of this experiment was very short approximately 50 m as compared to other baselines. With mass splitting between the three active states, expect no oscillations yet the experiment claimed that electron antineutrinos were appearing from the muon neutrinos at this very short baseline. To rule this experiment out, many experiments like MiniBooNE at fermilab [6], ICECUBE experiment which had sensitiveness to atmospheric neutrinos were started. The question is still not solved. The Short-Baseline Neutrino Oscillation Program at Fermilab is going to study the appearance of different flavor of neutrinos over their long and short baselines in three detectors, one of which is still in pipeline. These sterile neutrinos if they exist have important implications for cosmology. In cosmology, the neutrinos are one of the key players just like in supernova, a lot is learnt by studying the data very carefully. There has been an evolution in this field in the last twenty years, as data in last twenty years had only one acoustic peak but now WMAP data shows three acoustic peaks which tells us about the basic composition of the Universe. With PLANCK, there is a remarkable progress in the data [7]. The oscillations convert active neutrinos to sterile neutrinos and vice versa as oscillations combine with collisions which are flavor sensitive. In collisions, the flavor leptons are subject to weak interactions, giving rise to continuous flavor transitions between active and sterile neutrinos.

References

1. www.symmetrymagazine.org/article/march-2007/neutrino-invention
2. Cowan CL, Reines F et al (1956) Detection of the free neutrino: a confirmation. Science 124:103
3. (2015) Daya bay collaboration. Phys Res Lett 115:111802
4. Richard PF et al (1957) Geometrical representation of the Schrodinger equation for solving maser problems. J Appl Phys 28(1)
5. (2018) Borexino collaboration. Nature
6. Fermilab (2018) Significant excess of electronlike events in the MiniBooNE short-baseline neutrino experiment. Phys Rev Lett 121:221801
7. Planck (2015). arXiv: 1502.01589

A Comprehensive System for Smart Homes with a Minimalist Information Security Framework

Chandan Kumar Jha, Siddhartha Sankar Biswas, and Md. Tabrez Nafis

Abstract The Internet of Things (IoT) has inexorably awakened advanced humans' existence. Aside from the advantages that this innovative technology provides to the IoT device users, there are also cyber security concerns. Traditional cyber security approaches would not work on low-power IoT devices. As a result, numerous threat detection approaches and methods that are considerate the constraints of the Internet of Things have recently been created. To detect abnormalities in network traffic time series, this research offers a threat detection strategy that combines statistical and machine learning approaches. Furthermore, this novel architecture provides a light solution for cyber security because the logic of computing is a component of the edge layer. The technique performed nicely in spans of recall, precision, accuracy, and *F*-measure, according to the findings of the test bed testing.

Keywords Machine learning · Cyber security · Statistics · Anomaly detection · Internet of Things

1 Introduction

The Internet of Things (IoT) has had a beneficial influence on many aspects of contemporary life in recent years due to its rapid expansion. However, as a result of this IoT expansion, the inescapable need for smart home security has emerged. Because IoT devices have so many unique properties, cyber security defensive approaches from other domains, such as [1] portable mobile devices and [2] business associations, are difficult to apply to IoT devices. Because (a) there is a lot of variety in IoT devices and protocols that are supported, and (b) there are limitations like low processing power and consumption, the information security community

C. K. Jha (✉) · S. S. Biswas · Md. T. Nafis
Department of Computer Science and Engineering, School of Engineering Sciences and Technology, Jamia Hamdard, New Delhi, India
e-mail: chandankmr5711@gmail.com

S. S. Biswas
e-mail: ssbiswas@jamiahamdard.ac.in

M. S. Kaiser et al. (eds.), *Information and Communication Technology for Competitive Strategies (ICTCS 2021)*, Lecture Notes in Networks and Systems 401,
https://doi.org/10.1007/978-981-19-0098-3_39

recommends an Intrusion Detection System (IDS) that a network security device initially designed to investigate security flaws in a specific programmed or machine, well suited to the nature of IoT in this approach. In the field of IoT research, the notion of edge computing has become a popular approach [3].

For the connecting and management of IoT devices-based sensors near the layer's edge, gadgets with limited computing capabilities are being used. These devices are also referred to as gateways, maybe tasked with defending against cyberattacks; however, due to hardware limitations, they can only do so to a limited level. Such functionality is essential in a number of situations, including smart homes, where personal information is exchanged and must be safeguarded from external dangers. In this paper, a low-power, low-consumption anomaly detection technique based on edge computing is suggested to improve the security of smart homes. The suggested system incorporates both machine learning and statistics techniques, such as clustering and PCA, to find out whether there are any abnormalities in an IoT network packet time series data. The work that is being proposed is based on research study conducted by Voits et al. [4], but it varies in that it includes (a) unique cluster safe zones, (b) a time series technique, and (c) various time intervals that are based on empirical variance. The goal of the research is to offer an information security solution for smart buildings with cutting edge electronics that is affordable [5, 6]. The remaining part of the paper is structured out as follows. Sections 2 and 3 address similar research in the field, whereas Sect. 4 illustrates the proposed method. In addition, Sect. 4 analyzes the outcome of the research in order to ensure that the proposed approach is valid and legitimate, and Sect. 5 highlights the findings of the research.

2 Related Work

While in the last few years, as the Internet of Things has increased in popularity, several researchers have created sophisticated techniques and network Invaders Detecting System specifications (IDS) so that the smart home security must be improved and strengthened. The four types of Invaders Detecting System identified in a recent comprehensive literature review on Invaders Detecting System in the IoT [7] are anomaly-based Invaders Detecting System, signature-based Invaders Detecting System, specification-based Invaders Detecting System, and hybrid Invaders Detecting System. Anomaly-based Invaders Detecting System tries to detect network anomalies by despairing typical network activity and then comparing it to real network behavior. Anomaly-based Invaders Detecting System [8] having the benefit of being able to identify new threats, but they have difficulties with detection accuracy Invaders Detecting System. In contrast, the Cryptographic Signature-based Invaders Detecting System is based on the assumption that hazardous network intrusion activity patterns are limited and thus they can be detected earlier. Signature-based Invaders Detecting System, in contrast, is incapable of detecting new and previously unidentified threats utilizing various assault patterns. Specification-based Invaders Detecting System are similar to Anomaly-based Invaders Detecting System in that they seek for deviations from expected behavior, but instead of utilizing machine learning techniques, they use human-developed standards [9].

Finally, Hybrid Invaders Detecting System improves performance by combining the greatest characteristics of prior Invaders Detecting Systems. In the samples below, each of the categories is represented. Cheng et al. [10] proposed a game-based zero-sum cyber intrusion defensive strategy against purposeful attacks in the case of anomaly-based IDS, and the findings indicated that this methodology might improve IoT-based network security. Furthermore, Imran et al. [11], to detect abnormalities, utilized a distributed architecture in wireless sensor networks. They discovered both global and local irregularities with Multiple hyper oblong-shaped subgroups are used, as shown by modeled and legit statistics reveal. Hamilton et al. [12] has proposed an offline Invaders Detecting System based on an Artificial Intelligence-based Neural Network approach that has been recently published, especially an Artificial Intelligence-based Neural Network with a multi-layer cellular network perceptron that has a high accuracy of over 99% in detecting Ddos flooding assaults on IoT networks. Herrmann et al. [13] have presented an energy-efficient IoT security solution based on neighbor node trust that is suited for small devices such as IoT devices. Furthermore, Li et al. [14] improved security in Software-defined IoT networks using a two-stage artificial intelligence strategy that integrated the Random Forest Classifier with Bat Algorithm. Furthermore, Chilamkurti et al. [15] built a deep learning-based distributed computing method for identifying the various types of possible cyber-attacks in the Internet of Things that outperformed the competition by achieving detection accuracy of above 99%, both standard machine learning and centralized methods were defeated. Finally, Voits et al. [4] to increase the security of smart homes using statistical and econometric ways to preventing attacks that needs the physical permission to access to the IoT devices that are being connected over a given communication network to classify IoT devices according to their network traffic. By developing the first cryptography Invaders Detecting System for IoT networks, Siddiqui et al. [16] presented a dispersed approach with low storing, connectivity, and energy usage. Similarly, Kim et al. [17] presented a lightweight surveillance system that boosts performance by replacing the usual pattern-matching methodology with methods for ancillary switching and early judgment call. Last but not least, Wang et al. [18] introduced CloudEyes, based on the bidirectional schematic structure, a cloud-based IDS. CloudEyes was shown to be both effective and practical when compared to existing Cryptography Signature-based Invaders Detecting System. The second kind is specification-based Invaders Detecting System, in which Quek et al. [19] used a Gaussian optimization algorithm in which the aggressor is seeking to mislead the defensive system by launching a variety of assaults, while the defense employs phishing mails in an attempt to capture the assailant. The testing covered both first and reoccurring cyberattacks, with the statistics showing that this technique might be beneficial for IoT information security. Umamakeswari et al. [20] proposed InDRes, a constraint-based Invaders Detecting System aimed at making current infrastructure increasingly more cost effective. Finally, Saad et al. [21] has proposed their work related to a mobile wireless sensing Invaders Detecting System for detecting node capture cyberattacks. The Sequence Probabilistic Microwave Testing has been used in the suggested method in tandem with an adjustable limit of acceptable probability of transmission errors, and simulations illustrate that it successfully detects invasions. According to improved IoT security while lowering resource use, Senouci et al. [22] combined anomaly and signature detection methods. The data revealed that resource utilization and attack detection accuracy have improved, in unlike typical anomaly detection methods. Finally, Voigt et al. [23] enhanced SVELTE by including the Count of Expected Transmissions (ETX) and regional indicators. This extension outperforms SVELTE, according to the findings. The proposed system falls within the anomaly-based category, but it differs from the rest of the relevant work in this. Because it works with attacks that require physical access, it is restricted in spans of the types of assaults it sought to establish. In addition, the technique is based on Voits et al. [4]. It is also important for normal physical attacks, but differs in that it uses the time-based discrete interval series with varying range of different discrete time intervals and analyzes the changing possible safe network zones of the IoT-based network-related node clusters rather than the continuous possible safe network zones used by Voits et al. [4].

3 Proposed Work and Research Methodology

As stated previously, there are 4 various kinds of IoT-based cyber security mechanisms [7]. The proposed system in this research paper is based on the work done by previous authors as mentioned the reference on [4], and it employs an anomalous detection mechanism in order to profit from the category's related benefit; it is also the ability to detect the possible vulnerable and unknown Internet-related and network-related cyber threats. The recommended method of the proposed system is been depicted in the given Fig. 1. It is important to note that the methodology's initial stages are concentrated on creating conventional IoT device behaviors, while the latter

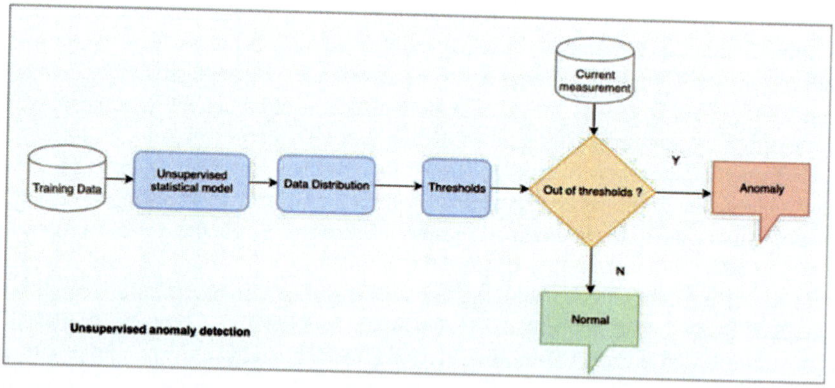

Fig. 1 To show the methodology for detecting anomalies

stages are focused on detecting anomalies. The steps of the recommended technique are discussed in detail in the following sections. The approach's initial phases focus on identifying normal IoT device activity, while the subsequent stages are focused with detecting anomalies. The steps of the proposed approach are fully discussed in the following sections. The methodology's initial input is the common Internet transportation of IoT systems and, more specifically, data about transferred datagrams. The time intervals related to immediate and long periods are then decreased from the normal operation period. This rationale is premised on the reality that only certain abnormal situations may be identified relatively instantaneously, whereas others take more time. The feature extraction procedure follows next, which effectively examines the input parameters for the multiple machine learning algorithms that follows. Table 1 displays statistics for packet headers that were sent and received by every IoT system (e.g., number of packets, packet size, event length, number of event packets, etc.). We have used features extraction approach described in to select the aforementioned criteria [4].

Table 1 Features list

Arrival time of packets for third quartile	Packets with a cross range arrival time	Maximum inter-arrival interval of data packet	
Packet count	The average amount of events data packet	Frequency of unique data packet	
Standard deviation for event packets number	Number of first quartile event packets	Number of required event packets	
Event data packet in a medium sizes	Maximum of event packet bundles	Event packets number in third quartile	
The number of intra quintile of event packets	Typical variation of event duration	Distinct event's number	
Size of average packet byte	Event's medium-length	Minimum event length	
Inter medium packets arrival time	Event length on mean	Packet size standard deviation in bytes	
Inter quartile range for each event duration		Length of the first centile incident	Bytes in a medium packet
Minimum byte packet size		Maximum event length	
Third quartile packet size bytes		Byte size of the first decile of a packet	Packet Size interquartile range in bytes
Average intra-duration of data packet		Maximum bytes each packet	Minimum packets intra-period
Cross duration of first centile package		Variance of the within period for data packet	Distinct events number

Each one of these 35 features has two or more time associated with it (one each for short distances and the other for long periods) which should be managed appropriately. As a consequence, the approach's next stage is to predict the capacity to alter' output. The rolling average method [24] is used to predict the time series output. That is just a forecasting of usual routine. The basic average methodology was selected for time series forecasting to the said possible manner, first is to respect the very low computational network capabilities of a given edge network node over a connection of associated computing devices, including a Raspberry Pi. Secondly as because of this network intrusion detection method is been used as a standardized framework for even more sophisticated network-based computational methodologies [24]. The prediction of the 70 time series is the result of the four steps mentioned previously including 35 for the long-time intervals and 35 for the short-time intervals. On the basis of these assumptions, clustered and despairing behavioral patterns will be performed accordingly [25]. Moreover, so because 35 attributes are not on the same scale, the normalizing step is necessary for the cluster to function normally in previous time spans including either long-span or short-span intervals. As a consequence, some cluster of IoT systems with reliable network behaviors may well have restricted safe areas, while others with unpredictable network activity may have larger safe areas [26].

Finally, as stated previously, the final features of the proposed Anomaly Detection method are connected to anomaly detection, including observing existing network activity of an IoT system and calculating the risk level. As a result, if an IoT device's existing Internet behavior is outside of a safe space, it is an indication of abhorrent behavior. The risk ratings are based on the deviation from the normal conduct and ranges from zero to ten, in where Zero corresponds to behavior within the safe place and ten corresponds to some of the most serious rating, according to which if the expected deviation is greater, then higher would be the score of intrusion detection. The divergence from normal behavior is calculated using the Internet-of-Things device's Euclid displacement of its clusters [27]. It is important to note that, as a consequence of PCA, IoT device behavior is characterized by major parameters, which are effectively composites of the specific characteristic specifications and network interface features, and any possible variance from the actual consistent normal representative values of a principal IoT-based networking interface component will reveal a different contrasting behavior in a given set of specifications and network interface characteristics for the Interconnected IoT devices over a given distributed communication network [28].

3.1 Measures of Validation

Veracity, fidelity, anamnesis, and F-measure are the 4 well-established performance metrics among the study on learning algorithms and were employed to put the postulated anomaly detection methodology to the trial, specifically its able to spot potential attacks [25]. The confusion matrix [Positives Correct (PC), Negatives Correct (NC),

Table 2 Experimental Internet-of-Things devices

Designator	Type	Network protocol
DS1	Weighing machine balance	Bluetooth enabled
DS2	Blood pressure gauge	Bluetooth enabled
DS3	Panic button	RF869
DS4-DS9	Movement sensor	ZigBee

Invalid Positives (IP), and Invalid Negatives (TN)] [25] is the foundation for all of these evaluation measures, and the equations for computing them are provided below:

$$\text{Veracity} = (\text{PC} + \text{NC})/(\text{PC} + \text{NC} + \text{IP} + \text{TN}) \quad (1)$$

$$\text{Fidelity} = \text{PC}/\text{PC} + \text{TN} \quad (2)$$

$$\text{Anamnesis} = \text{PC}/\text{PC} + \text{IP} \quad (3)$$

$$F\text{−Measure} = 2 * \{[\text{Fidelity} * \text{Anamnesis}]/[\text{Fidelity} + \text{Anamnesis}]\} \quad (4)$$

3.2 *Description of the Dataset*

The dataset used to test the method was taken from an actual smart home and consists of network traffic from nine IoT devices that support a variety of network protocols, including Bluetooth, RF869, and ZigBee. With Table 2, following is a list of the IoT devices that were used in the experiments:

4 Experimental Results

The simulation and experimental results are explained in this section. Before diving into the research, it is worth noting that the Televes proprietary IoT gateway was used to execute and evaluate the proposed algorithm. This network connecting portal includes an ARM Cortex-based Central Processing Unit with 4 network cores operating at the processing speed of 2.4 GHz, 2 GB of DDR4 Memory, and multiple communications port facilities. Wi-Fi, ZigBee, Bluetooth, and an internally developed connectivity in the 869 MHz frequency band are all used to attach IoT devices and appliances to the Web with providing Wired connection and UMTS wireless connectivity. The gateways, according to developers, offers memory and CPU capability similar to a Raspberry Pi 4, but with extra connectivity, authenticity offering inner memory storage space including for the windows OS, and stoutness features that for continuous operations it including a battery. Here in these gateways, both the assessment of appropriate behavior and the observation of current performance are conducted, producing the results given in Figs. 2, 3 and the respective Table 3, which denotes the 2 computing processes which are being required for the proposed algorithm's implementation on machine learning and data mining:

It should be mentioned that the proportion of the core is higher than either of the 4 cores, as well as the proportion of ram is higher than the total RAM with 2 GB size. As an outcome, the running of this embedded system demonstrates the algorithm's minimalistic design. The short time period remained 5 record minutes, while the time interval for the long duration spanned 2 h. Illustrations 4 and 5 depict IoT platform groupings with predictable behavior. Just the scores of the PCA's first two major components were utilized as in Fig. 4 to represent the clumping of IoT devices.

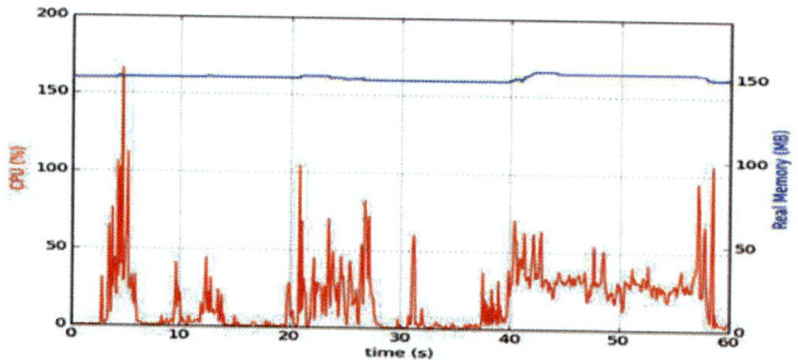

Fig. 2 Statistical processes necessitate the use of CPU and memory

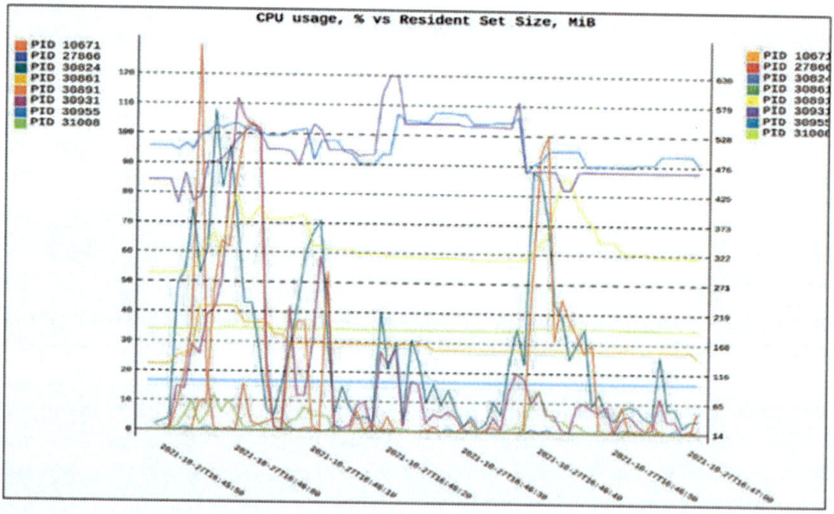

Fig. 3 Machine learning processes need a lot of CPU and memory

Table 3 Resource use by the gateway while the proposed algorithm is running

Resource	De-span nation of normal behavior		Monitoring of current behavior	
	Percentage	Value	Percentage	Value
Central processing unit utilization	12.20	0.15 GHz	3.12	0.024 GHz
Memory utilization	6.75	114 MB	6.25	118.4 MB

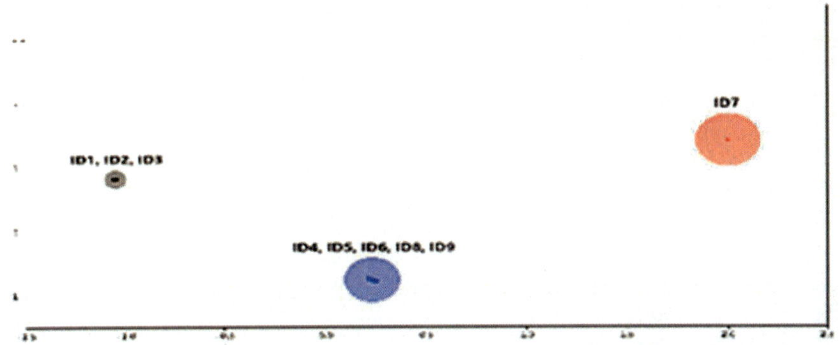

Fig. 4 IoT device clusters based on intervals of short duration

The approach used the first two primary components for the short-span evaluation and the very first 5 for the long duration evaluation. The idea that the safe zones of the clustering differ in magnitude is an important conclusion to draw through both Figures. As a response, our studies suggest that IoT systems have varying levels of network unpredictability and, as an outcome, varying levels of acceptable behavior. Furthermore, the results indicate that the long-span experiment generated more IoT device ensembles than the short-span experimental trials, and these cluster are also more compact in accordance with the safe zones with smaller widths. This is likely to have a positive long-span influence on threat intelligence.

In order to test the technique, many physical access risks linked to the Internet of Things were simulated in order to interrupt the IoT devices' normal network connections:

1. Disconnecting a device's battery.
2. Unusual traffic as a result of the emergency button.
3. Blood pressure readings obtained on a routine basis.
4. Abnormal motion that the motion detectors can capture.

When the entire threat intelligence results for both long-span and short-span analyses are reviewed, it becomes obvious that while both techniques performed admirably in spans of veracity, (For the short-span, 86.11%, and for the long-span, 94.44%.) and fidelity (Both techniques have complete fidelity, indicating that neither causes false alarms), the long-span strategy won out over the short-span strategy.

5 Conclusion

An anomaly-based technique, equipped with computational logic at the infrastructure level, is presented in this research. Deep learning, measurement, and statistical techniques also were used throughout the process. Overall, the approach achieved its objectives that included (a) identifying anomalies in IoT Internet traffic as demonstrated by several evaluation markers and (b) conforming to reduced computational restraints, as mentioned previously in the result and discussion. When compared to short-span assessment, the findings of the preceding paragraph strongly demonstrate that long-span assessment surpasses short-span analysis in identifying potential problems. These statistics show that certain cyberattacks take more time to discover and that having a long-span experiment in place improves the overall effectiveness of the approach. While not as comprehensive as a long-span strategy, short-span analysis has the advantage of identifying potential problems sooner. Finally, the presence of multiple time intervals in the proposed approach can indeed smoothly be approximated from short- and long-time intervals to more interims in spans of improving security monitoring both in spans of response time and accuracy of detecting the possible vulnerable cyber attackers onto the connected IoT devices especially when they are operating on public wireless tele communication networks or the globally available world wide web, while also keeping the contour gateway's limited calculation constrictions in thoughts.

References

1. Gelenbe E et al (2013) Nemesys: enhanced network security for seam-less service provisioning in the smart mobile ecosystem. In: Information sciences and systems. Springer, pp 369–378
2. Toupas P et al (2019) An intrusion detection system for multi-class classification based on deep neural networks. In: 18th IEEE international conference on machine learning and applications (ICMLA). IEEE, pp 1253–1258
3. Shi W et al (2016) Edge computing: vision and challenges. IEEE Internet Things J 3
4. Spanos G et al (2019) Combining statistical and machine learning techniques in IoT anomaly detection for smart homes. In: IEEE 24th international workshop on computer aided modeling and design of communication links and networks (CAMAD). IEEE, pp 1–6
5. Collen A et al (2018) Ghost—safe-guarding home IoT environments with personalized real-time risk control. In: Security in computer and information sciences. Springer International Publishing, pp 68–78
6. Augusto-Gonzalez J et al (2019) From internet of threats to internet of things: a cyber security architecture for smart homes. In: IEEE 24th international workshop on computer aided modeling and design of communication links and networks (CAMAD), pp 1–6
7. Hajiheidari S et al (2019) Intrusion detection systems in the internet of things: a comprehensive investigation. Comput Netw
8. Raza S, Wallgren L et al (2013) Svelte: real-time intrusion detection in the internet of things. Ad Hoc Networks 11(8):2661–2674
9. Sekar R et al (2002) Specification-based anomaly detection: a new approach for detecting network intrusions. In: Proceedings of the 9th ACM conference on computer and communications security

10. Chen P-Y et al (2014) Information fusion to defend intentional attack in internet of things. IEEE Internet Things J 1(4):337–348
11. Rajasegarar S, Gluhak A et al (2014) Ellipsoidal neighborhood outlier factor for distributed anomaly detection in resource constrained networks. Pattern Recogn 47(9):2867–2879
12. Hodo E, Bellekens X et al (2016) Threat analysis of IoT networks using artificial neural network intrusion detection system. In: International symposium on networks, computers and communications (ISNCC). IEEE
13. Khan ZA et al (2017) A trust based distributed intrusion detection mechanism for internet of things. In: IEEE 31st international conference on advanced information networking and applications (AINA). IEEE, pp 1169–1176
14. Li J et al (2018) Ai-based two-stage intrusion detection for software defined IoT networks. IEEE Internet Things J 6(2):2093–2102
15. Diro AA et al (2018) Distributed attack detection scheme using deep learning approach for internet of things. Future Gener Comput Syst 82:761–768
16. Amin SO et al (2009) A novel coding scheme to implement signature-based IDS in IP based sensor networks. In: International symposium on integrated network management-workshops. IEEE
17. Oh D et al (2014) A malicious pattern detection engine for embedded security systems in the internet of things. Sensors 14(12):24188–24211
18. Sun H et al (2017) Cloudeyes: cloud-based malware detection with reversible sketch for resource-constrained internet of things (IOT) devices. Softw Pract Exper 47:421–441
19. La QD et al (2016) Deceptive attack and defense game in honeypot-enabled networks for the internet of things. IEEE Internet Things J 3(6):1025–1035
20. Surendar M et al (2016) Indres: an intrusion detection and response system for internet of things with 6lowpan. In: International conference on wireless communications, signal processing and networking (WiSPNET). IEEE, pp 1903–1908
21. Gara F et al (2017) An intrusion detection system for selective forwarding attack in IPv6-based mobile WSNs. In: 13th international wireless communications and mobile computing conference (IWCMC). IEEE, pp 276–281
22. Sedjelmaci H et al (2017) An accurate security game for low-resource IoT devices. IEEE Trans Veh Technol 66(10):9381–9393
23. Shreenivas D et al (2017) Intrusion detection in the RPL-connected 6LoWPAN networks. In: Proceedings of the 3rd ACM international workshop on IoT privacy, trust, and security, pp 31–38
24. Hyndman RJ et al (2018) Forecasting: principles and practice. OTexts
25. Fawcett T (2006) An introduction to ROC analysis. Pattern Recogn Lett 27(8):861–874
26. Suthaharan S (2016) Big data analytics. In: Machine learning models and algorithms for big data classification. Springer, pp 31–75
27. Jolliffe IT (2002) Principal component analysis, 2nd edn. Springer, New York, NY
28. Kriegel H-P et al (2011) Density-based clustering. Wiley Interdisc Rev: Data Min Knowl Disc 1(3):231–240

CCC-Quality2: Cross-Cloud Service Composition Based on Quality of Clouds and Web Services for Business Applications

Kanchana Rajaram and K. Preethi Naveena Selvi

Abstract Composition of web services is a promising approach whenever it is not possible to satisfy the requirements with a single service. Combining the services deployed in a single cloud prevents to utilize the benefits offered by the other clouds. Hence, it is beneficial to compose the required services available in multiple clouds. The web services providing the similar functionality differ in their quality of service (QoS). The characteristics of the cloud platform where these services are deployed also have an impact on the quality of the services. Therefore, web services must be chosen based on their QoS as well as QoS of the deployed cloud to improve the QoS of the composite service. Towards this objective, it is proposed to compose services deployed in different clouds platforms based on QoS parameters of web services as well as cloud platforms. The proposed approach maximizes the satisfaction level offered by the composition in terms of QoS. The proposed method has been experimented with 2507 web services of QWS dataset deployed in 10 different cloud platforms. The impact on increasing the number of services deployed in each cloud as well as the number of cloud platforms available on the satisfaction level of web service selection has been studied.

Keywords Quality of service · Composition · Service level agreement · Web services · Service discovery · Utility

1 Introduction

The required functionality may be offered by a single service or a composition of multiple services when a single service cannot fulfil the requirements. The web services can be combined statically at design time or dynamically during run-time [1–4]. A dynamic web services composition [5–7] helps to automate business process functionality by developing executable workflows which can be deployed

K. Rajaram (✉) · K. P. N. Selvi
Department of Computer Science and Engineering, Sri Sivasubramaniya Nadar College of Engineering, Chennai, Tamil Nadu, India
e-mail: rkanch@ssn.edu.in

M. S. Kaiser et al. (eds.), *Information and Communication Technology for Competitive Strategies (ICTCS 2021)*, Lecture Notes in Networks and Systems 401,
https://doi.org/10.1007/978-981-19-0098-3_40

in executable environment. A template-based interleaved dynamic composition [8–11] involving web services across business organizations is based on permissible workflow templates. A web service corresponding to an activity in the workflow is selected according to the user requirements specified with respect to the results of the previous service execution. The selected service is then composted and executed.

When many identical services are available for a required functionality, it is desirable to select the service based on its quality parameters such as price, response time, availability, reliability, security and reputation. QoS-aware selection of web services improves QoS of the composite web service [12–16].

Cloud computing is a promising next-generation platform for conducting businesses [17] since it offers economic benefits to both the services providers and users. Several organizations are utilizing a cloud service provider to make use of the platform and storage in the cloud, instead of deploying them in their own local infrastructures. Web services deployed in a cloud can be selected based on their QoS in a dynamic service composition scenario [18].

With increasing business, the user requirements change frequently, and the web services in each cloud platform are usually limited to satisfy their changing requirements. Therefore, developing a quality service composition using services available in a single cloud is often challenging. In this scenario, services with missing functionalities in a cloud but are available in another cloud platform and services with better quality that are found in other cloud platforms can be used in the composition. QoS-aware composition methods supporting service invocation from different cloud environments have been investigated in [19]. The quality of web service execution largely depends on the execution environment. When web services are discovered and executed in a cloud platform that is dynamic and agile in nature, the quality of cloud platforms has an impact on the quality of web services and their composition. QoS parameters of web services such as cost, performance, availability and security depend on the corresponding QoS parameters of cloud environments. The QoS parameters related to performance of web services such as response time and throughput depend on quality parameters of the underlying cloud platform such as CPU utilization and network utilization. The existing research on quality-aware composition of web services from multiple cloud environments has not considered quality parameters of cloud platforms while selecting the services. Towards these objectives, an approach for **C**ross-**C**loud service **C**omposition based on **Quality** of cloud platforms and **Quality** of web services (**CCC-Quality2**) is proposed in this paper.

The rest of the paper is organized in the following manner: The architecture of the proposed system CCC-Quality2 and its modules are described in Sect. 2. Experimentation and results are discussed in Sect. 3. The existing literature and related works are elaborated in Sect. 4. Section 5 concludes the work.

2 Proposed Work

The proposed architecture of CCC-Quality2 is shown in Fig. 1. The architecture involves selection and composition of web services from different cloud platforms. A web service is discovered based on the requirements of the user and QoS of the web service and cloud in which it is deployed. The searching of efficient service is performed using a service registry, which contains the pool of services that are present in different cloud environments. Once the required services are found from service repository, the end point reference of those services is sent to the service composer. The service composer accesses the web services that are present in different cloud using the end point reference of the service and compose them dynamically. Finally, the results of composed service are sent back to client.

2.1 *QoS-Aware Selection of Web Services from Multiple Clouds*

The QoS-aware web service selection is based on the Utility function [20], which represents the satisfaction level with respect to the requirements of the user. The satisfaction level varies between 0 and 1, for the least and the best satisfaction, respectively. The utility function is represented by a set of QoS parameters along with weightage for each service. The utility function is defined by the following equation for web service selection (Table 1).

$$U_x = \begin{cases} \frac{Q_{\max}-Q_x}{Q_{\max}-Q_{\min}}, \text{if } y \text{ is expected to be minimum} \\ \frac{Q_x-Q_{\min}}{Q_{\max}-Q_{\min}}, \text{if } y \text{ is expected to be maximum} \end{cases}$$

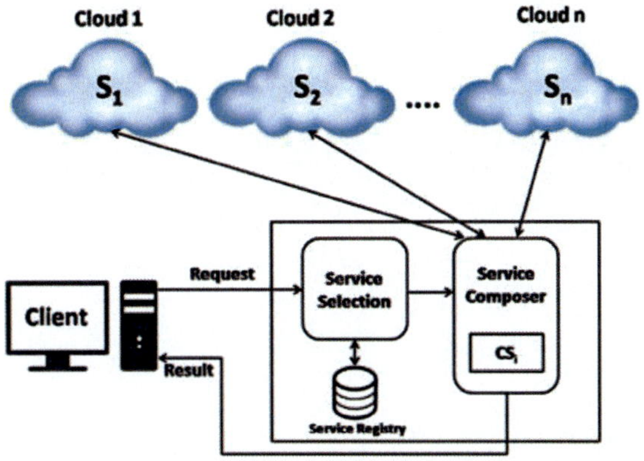

Fig. 1 Architecture of the proposed CCC-Quality2 system

Table 1 Terminologies in utility function

Notation	Explanation
Ux	Utility of QoS parameter x
Qx	Value of QoS parameter x
Wx	Weight associated with QoS parameter x
U(S)	Total utility function of a service

Table 2 QoS parameters and metrics

	QoS parameter	Metric
Web services	Security	Low, medium and high
	Cost	Rupees per hour
	Response time	Milliseconds
Cloud	CPU utilization	Percentage
	Network utilization	Data traffic in MB/ hour

$$U(S) = \sum_{x=1}^{t} (U_x * W_x) \quad (1)$$

Three QoS parameters are considered for web services, namely security, cost and response time and two QoS parameters for cloud, CPU utilization and network utilization. QoS parameters along with its metric are listed below (Table 2).

The weightage for utility function is based on user's preference. The user specifies the weightage among the web service and cloud along with the weightage of each QoS parameter for required service. For example, the weightage among web service and cloud can be 50% each, and weightage for each QoS parameters such as security, cost, response time, CPU utilization and network utilization can be 60, 20, 20, 60 and 40%, respectively. The combined utility value is calculated, and the web service with the maximum utility value is selected as the suitable service as per user's requirement.

Algorithm 1 To find the combined utility of QoS parameters of a web service and clouds.

Algorithm: CombinedUtility
Input: list of requirements for selection T,
list of QoS parameters Q,
list of weights for each QoS parameter W.
Output: Combined Utility U_s

```
Uws ← 0                  // Utility value of web service
Uc  ← 0                  // Utility value of cloud
for each web service T
  for each QoS of web service i
    if i expected to be minimum then
```

$$U_i = \frac{Q_i^{max}-Q_i}{Q_i^{max}-Q_i^{min}} * W_i$$

```
    else if i expected to be maximum then
```

$$U_i = \frac{Q_i-Q_i^{max}}{Q_i^{max}-Q_i^{min}} * W_i$$

```
        end if
    end if
    Uws=Uws + Ui
  end for
  for each QoS j of cloud where web service T is deployed
    if j expected to be minimum then
```

$$U_j = \frac{Q_j^{max}-Q_j}{Q_j^{max}-Q_j^{min}} * W_j$$

```
    else if j expected to be maximum then
```

$$U_j = \frac{Q_j-Q_j^{max}}{Q_j^{max}-Q_j^{min}} * W_j$$

```
        end if
    end if
    Uc =Uc + Uj
  end for
  UT= (Uws * Wws ) + (Uc * Wc)
end for
return UT
```

2.2 Cross-Cloud Dynamic Service Composition

Web service composer dynamically arranges multiple services into workflows supplying complex user needs. Users of the service in an application program send service request to the service composer as a query. The service composer accepts and composes the service requestor as per the request. It searches the service registry for the required web service. Service registry is the place that contains a repository of available services and allows service composer to lookup for web service. Three kinds of operations are done for consuming a web service are namely discovery of web service from service registry, binding and invocation of web service.

When a web service is invoked through web client program, it creates an object to bind with corresponding port. Then port object is created with the port binding method. Port object is used to access the methods defined in the web service. The required parameters are passed to this method as input received from the user. The output of the first web service is then passed as input to the next web service. Second web service processes the input and its output is directed to the third service and so on till the final output of composed services is obtained. Several web services can be composed to produce the output as per the requirements of the end users. The steps involved in dynamic composition of services are outlined below:

1. Submit the request for service request to the middleware where service composer is present.
2. Find a suitable service from the service registry based on the request.
3. Bind the service with the endpoint reference of the required service to establish a communication channel.
4. Establish port binding between endpoint reference and service user.
5. Send the functional parameters of the service through the port.
6. Direct the output of the web service as an input to the next requested web service.
7. The above process is continued until the user requests are satisfied.

3 Experimentation

We have considered QWS dataset that includes 2507 web services and their QoS parameter values. There are measurements for 11 quality parameters for each web service. The first nine parameters are QWS parameters measured using web service benchmark. The last two parameters represent the service name and reference to the WSDL document.

3.1 Analysis of Utility Value by Varying the Number of Services in Each Cloud

In this experiment, the size of composition is set as three services and three cloud platforms are used. Moreover, the number of services is varied from 50 to 350 which are equally distributed among three clouds. From the graph in Fig. 2, it is inferred that with an increase in the number of web services satisfying the required functionality that are deployed in each cloud, utility function value also improves. Therefore, the user satisfaction level is higher when the number of services available for selection is high.

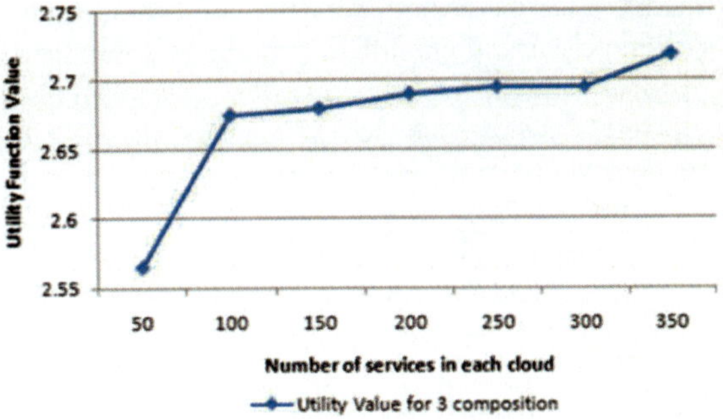

Fig. 2 Impact of number of services available on satisfaction level

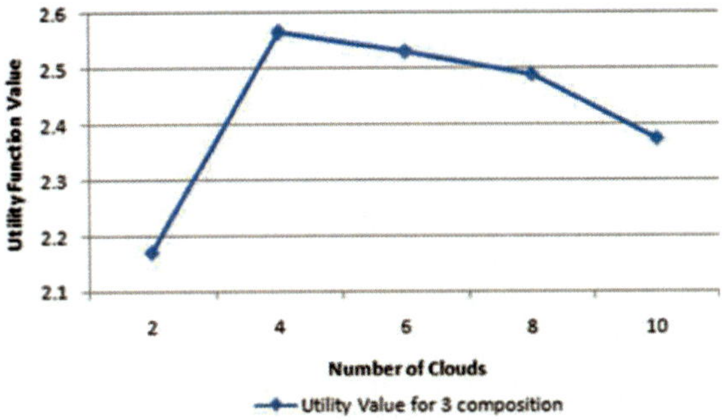

Fig. 3 Impact of the number of clouds on satisfaction level

3.2 Analysis of Utility Value for Different Number of Clouds

In this experiment, the three services are used in each compositions, and 50 web services are deployed in each cloud that are suitable for each requirement. However, the number of clouds is varied from 2 to10.

The graph in Fig. 3 shows the impact of the number of clouds where the required services are deployed, on the satisfaction level. It is observed that if the number of clouds increases beyond four, the satisfaction level starts decreasing gradually. Hence, it is necessary to find out the correct number clouds from which the services can be selected.

4 Review of Related Work

Several authors have worked with the research problem of composition of web services. Web service selection is based on the quality of services aiming to optimize the performance of composite service. Jula et al. [21] have studied and reviewed the literature on composition of web services deployed in cloud environments. They classified the methods of service composition from cloud platforms into five categories: classic and graph-based methods, combinatorial algorithms, machine-based methods, structures and frameworks and studied the different works in each of these categories. Among the QoS parameters, they found that most of the literature has considered only cost and response time of the web services.

There have been several works that select the web services deployed in a cloud based on their QoS characteristics. The problem of composing the cloud services is modelled as an influence diagram problem [22]. Influence diagrams are graphical models used to represent and solve complex decision-making problems based on uncertain information. The web services selected for composition were differentiated using non-functional properties, namely throughput, response time and cost. They refined the QoS model into continuous model involving the end users as well as cloud service providers to improve the performance. A brokerage-based cloud service architecture [23] was designed with two techniques: managing cloud service provider information using an index; selecting services by ranking the service providers and aggregating them. The cloud brokers are useful in selecting and ranking the cloud service providers based on the requirements. These works consider selecting the services from a single cloud platform.

A quality-based web service composition method was analysed for selecting web services from several cloud environments [24]. This method is exponential and cannot deliver a satisfactory result where a real-time response is required. It is a near-to-optimal solution with lesser time cost compared to traditional QoS-aware composition methods; a privacy-aware cross-cloud service composition method [25] improved the quality of a composed service by selecting the services based on the actual quality provided rather than the promised QoS values. K-means algorithm was used for filtering the suitable service, and the privacy of a cloud service is protected in a composition of services from several clouds.

Kurdi et al. [26] proposed an algorithm that reduced the number of clouds used in a composition. There is a trade-off between execution time and the number of clouds in the service composition sequence when the required services are deployed in several clouds. The method preferred to select the cloud that has the maximum number of matching services and satisfied the requirements with minimal overhead. In [27], the authors composing the tasks such as cloud service discovery, selection and composition together in an ontology and QoS-based framework to satisfy the service requests. The technique proposed in [28] involved the use of hybrid metaheuristics genetic algorithm with Tabu search that retrieved suitable web services based on their QoS parameters. In [29], web services were selected using five different multi-criteria decision-making (MCDM) methods, and they were ordered based on their QoS parameters.

The QoS parameter values of a web service that is deployed in a cloud are influenced by the characteristics of the cloud platform. None of the above works have considered the QoS characteristics of the cloud platforms while selecting the services. Moreover, most of the existing works focus on near real-time optimal solutions, while our proposed work CCC-Quality2 aims at dynamic service compositions.

5 Conclusion

Our proposed CCC-Quality2 system deals with employing quality-aware dynamic web service composition across different clouds. The web services have been selected based on the quality parameters of the web service as well as the quality parameters of the cloud platform where the services are deployed. The web service selection was made depending on the utility function that is computed based on the web service quality parameters such as response time, security and cost as well as cloud quality parameters, namely CPU utilization and network utilization. Such utility-based selection is found to deliver higher customer requirement satisfaction.

It is verified that when higher number of services are available in each cloud, the satisfaction level of the composition increases. Further, it is verified that selecting services from more than certain number of clouds will not improve the satisfaction level of the composition.

References

1. Dustdar S, Schreiner W (2005) A survey on web services composition. Int J Web Grid Services 1(1):1–30
2. Srivastava B, Koehler J (2003) Web service composition—current solutions and open problems. In: ICAPS 2003. Workshop on Planning for Web Services, 2003
3. Su X, Rao J (2004) A survey of automated web service composition methods. In: First international workshop on semantic web services and web process composition, 2004
4. Agarwal V, Dasgupta K, Karnik N, Kumar A, Kundu A, Mittal S, Srivastava B (2005) A service creation environment based on end-to-end composition of web services. In: 14th international conference on world wide web, 2005
5. Mustafa F, McCluskey TL (2009) Dynamic web service composition. In: International conference on computer engineering and technology ICCET '08, vol 2. pp 463–467
6. Tabatabaei SGH, Kadir WMNW, Ibrahim S (2008) Web service composition approaches to support dynamic e-business systems. Commun IBIMA 2
7. Raman B, Agarwal S, Chen Y, Caesar M, Cui W, Johansson P, Lai K, Lavian T, Machiraju S, Morley-Mao Z, Porter G, Roscoe T, Seshadri M, Shih JS, Sklower K, Subramanian L, Suzuki T, Zhuang S, Joseph AD, Katz RH, Stoica I (2002) The SAHARA model for service composition across multiple providers. In: Proceedings first Int'l conference pervasive computing, May 2002, pp 1–14
8. Agarwal V, Chafle G, Mittal S, Srivastava B (2008) Understanding approaches for web service composition and execution. In: Proceedings of the 1st Bangalore annual compute conference (COMPUTE '08). ACM, New York, NY, USA, Article 1, pp 8
9. Rajaram K, Babu C (2010) Template based SOA framework for dynamic and adaptive composition of web services. In: 2010 international conference on networking and information technology, Manila, 2010, pp 49–53
10. Casati F, Shan M-C (2001) Dynamic and adaptive composition of e-services. Inf Syst 26(3):143–162
11. Zeng L, Benatallah B, Ngu AHH (2001) On demand business to-business integration. In: Proceedings ninth Int'l conference cooperative information systems
12. Mecella M, Scannapieco M, Virgillito A, Baldoni R, Catarci T, Batini C (2002) Managing data quality in cooperative information systems. In: Proceedings 10th Int'l conference cooperative information systems (CoopIS)

13. Strunk A (2010) QoS-aware service composition: a survey. In: Eighth IEEE European conference on web services, Ayia Napa, December 2010, pp 67–74
14. Zeng L, Benatallah B, Dumas M, Kalagnanam J, Sheng QZ (2003) Quality driven web services composition. In: Proceedings 12th Int'l conference world wide web (WWW), May 2003
15. Liu Y, Ngu AHH, Zeng L (2004) QoS computation and policing in dynamic web service selection. In: Proceedings 13th Int'l conference world wide web (WWW), May 2004
16. Nahrstedt K, Xu D, Wichadakul D, Li B (2001) QoS-aware middleware for ubiquitous and heterogeneous environments. IEEE Comm Magazine 39(11):2–10
17. Motahari-Nezhad H, Stephenson B, Singhal S (2009) Outsourcing business to cloud computing services: opportunities and challenges. IEEE Internet Comput
18. Ylianttila M, Riekki J, Zhou J, Athukorala K, Gilman E (2012) Cloud architecture for dynamic service composition. Int J Grid High Perform Comput 4(2):17–31
19. Wu T, Dou W, Hu C, Chen J (2017) Service mining for trusted service composition in cross-cloud environment. IEEE Syst J 11(1):283–294
20. Faratin P, Sierra C, Jennings NR (2002) Using similarity criteria to make issue trade-offs in automated negotiations. Artif Intell 142(2):205–237
21. Jula A, Sundararajan E, Othman Z (2013) Cloud computing service composition: a systematic literature review. Expert Syst Appl Elsevier J 3809–3824
22. Ye Z, Bouguettaya A, Zhou X (2012) QoS-aware cloud service composition based on economic models. In: Liu C, Ludwig H, Toumani F, Yu Q (eds) Service-oriented computing. ICSOC 2012. lecture notes in computer science, vol 7636. Springer, Berlin, Heidelberg
23. Sundareswaran S, Squicciarini A, Lin D (2012) A brokerage-based approach for cloud service selection. In: IEEE fifth international conference on cloud computing, Honolulu-HI, June 2012, pp 558–565
24. Qi L, Dou W, Zhang X, Chen J (2012) A QoS aware composition method supporting cross-platform service invocation in cloud environment. J Comput Syst Sci 78(5):1316–1329
25. Dou W, Zhang X, Liu J, Chen J, Member S (2015) IEEE, HireSome-II: towards privacy-aware cross-cloud service composition for big data applications. IEEE Trans Parallel Distrib Syst 26(2):455–466
26. Kurdi H, Al-Anazi A, Campbell C, Al Faries A (2015) A combinatorial optimization algorithm for multiple cloud service composition. Comput Electri Eng Elsevier J 42:107–113
27. Modi KJ, Garg S (2019) A QoS-based approach for cloud-service matchmaking, selection and composition using the Semantic Web. J Syst Inf Technol 21(1):63–89
28. Thangaraj P, Balasubramanie P (2021) Meta heuristic QoS based service composition for service computing. J Ambient Intell Human Comput 12:5619–5625
29. Pallavi B, Aarchit J, Rahul H (2019) QoS based web service selection and multi-criteria decision making methods. Int J Interact Multimedia Artif Intell 5(4):113–121

Drone Delivery System Using GPS and Telemetry

Aman Anand, Saubhik Bandyopadhyay, Swapnil Das, Aasakti Agarwal, and Andrew Moitra

Abstract Nowadays customers and patients mostly prefer online shopping and ordering their medicines, which increases the need for the delivery person (Hayat et al. in IEEE Commun Surv Tutorials 18(4):2624–2661, 2016). The rise in online ordering has given a spike to the requirement of manpower to deliver the order to the vast majority of people located in different parts of the city and country. Drone-based delivery systems can be used to meet this requirement. The current drone delivery technology has some drawbacks like they are not economically efficient enough so that it can be implemented on a large scale, also the technology used in the current drone is not made such that it can be upgraded easily in the future. This paper is going to present a way to build a prototype of a drone delivery system using GPS and telemetry with all the necessary circuit connections. GPS is going to help drones to navigate from one point to another as well as some necessary flight modes like fail safe and return to landing position modes, and telemetry is a device which is being installed on the drone which can take various types of commands like takeoff, land, parcel drop, and data communication from the ground station. Also this paper shows the future aspects and possibility of upgrade in such a type of robotic system. Prototype shown in this paper can be upgraded to hexacopter and octacopter depending on weight lifting capacity requirement. GPS used in the prototype is precise enough for landing but if required then can be upgraded to a more advanced GPS module.

Keywords Drone · Telemetry · GPS unmanned vehicle · Parcelcopter · Quadcopter · Hexacopter · Octacopter

A. Anand (✉) · S. Bandyopadhyay · S. Das · A. Agarwal · A. Moitra
Department of Computer Science and Engineering, Amity University, Kolkata, India
e-mail: amananand972@gmail.com

M. S. Kaiser et al. (eds.), *Information and Communication Technology for Competitive Strategies (ICTCS 2021)*, Lecture Notes in Networks and Systems 401,
https://doi.org/10.1007/978-981-19-0098-3_41

1 Introduction

1.1 Background

Parcelcopter (quadcopter, hexacopter, and octacopter) is usually known as an unmanned aerial vehicle [2] which we call a delivery drone that is utilized to transport parcels, food, or some other items. This type of vehicle is one of the major parts of robotics and automation. These flying robots can be controlled remotely or can fly by itself based on the mission uploaded on the flight controller board. Drones are not only used in package transportation but also in a wide range of other tasks like firefighting, traffic monitoring, cinematography, photography, etc.

In the past few years, the concept of autonomous drone delivery systems has shown significant growth. As evidence, some companies, like Amazon, are starting a service called Amazon Prime Air. This service is designed to safely deliver the package to a given destination within 30 min using drones. The enhanced automation in such companies results in improvement in customer service by providing rapid order delivery [3]. Companies like Google, UPS, and other European companies are also experimenting with such technologies.

1.2 Problem and Scope of Work

Currently, we already have autonomous drone delivery systems that fulfill the requirement for package delivery services, and by the time they are improving as well, there are few drawbacks as well. Current drones use a GSM module for the communication of air stations and ground stations [4]. Such type of architecture is quite expensive and ultimately results in high delivery charges. This paper is an attempt to present a way to implement drone delivery systems using GPS and telemetry-based architecture. Also this paper is going to present the future scope of development in this technology, not only in delivery services but also in other fields like drone-based edge computing, 5G network feasibility, and much more. This flying robot has huge potential for solving many problems; we encounter in the development of advanced technologies which require more flexibility and speed.

2 Methodology

The entire circuitry of the drone is controlled by an Arducopter flight controller. This flight control is built on top of an Arduino Mega board. Prototype is built using a quadcopter frame, but the flight controller is capable of controlling eight motors at a time which means the same architecture can be used to build a hexacopter as well as an octocopter [5]. Apart from all these, flight controllers have one GPS port, one

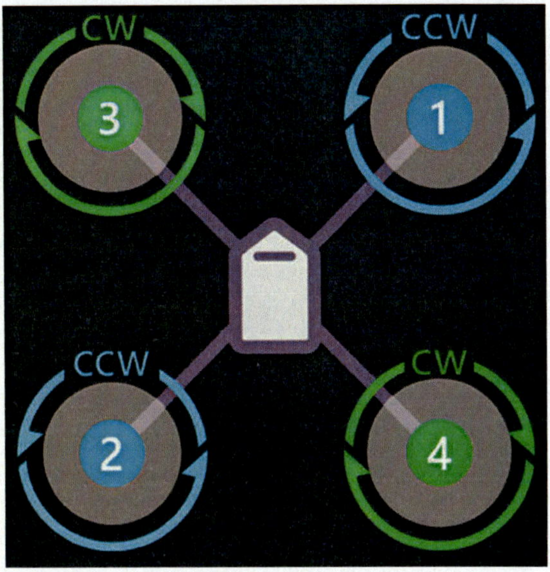

Fig. 1 Motor directions

telemetry port, and several other input output pins to control different mechanical devices and components if they are installed on drones.

Considering the quadcopter prototype, after building the frame, all four electronic speed controllers need to be soldered on the PCB (red wire to the positive terminal and black wire to the negative). Then the power module will be soldered to which will power the entire circuitry. After all the mechanical parts are installed on the frame including the flight controller board and the motors, the below steps need to be followed for the wiring part.

All the steps are written with the reference of the below figure.

Figure 1 shows the direction of all the motors and their numbers with respect to the direction of the flight controller board.

- Connect the electronic speed controller (ESC) to the output pins of the fight controller board by matching the pin number and the number of ESC given in Fig. 1.
- After connecting all the ESC to the flight controller board, we need to calibrate the accelerometer and compass.
- For the calibration process, Ardupilot mission planner software is used. Once we connected the drone via USB, first we need to upload the right firmware for the drone we built.
- When the Firmware is uploaded, then just follow the steps instructed by the software to calibrate the accelerometer and compass.

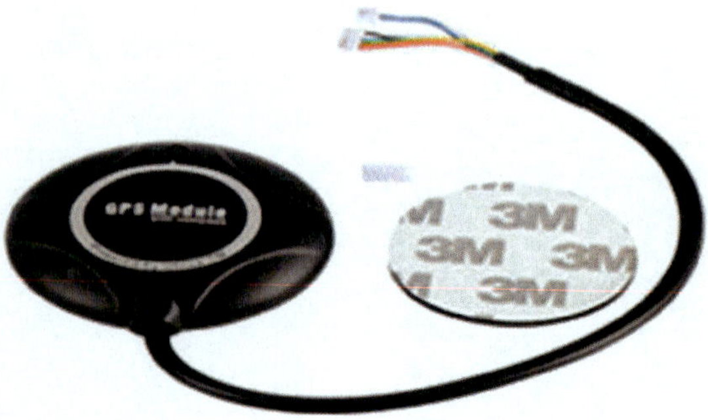

Fig. 2 GPS

- For testing purposes, it is recommended that first we should control the drone manually using remote. To control the drone with remote or transmitter, we also need the radio calibration to be done before the test flight.
- Once the testing is successful, the final step is to install the air module of telemetry on the flight controller board and ground module can be used on both smartphone, PC, laptop.

Brief about components installed on the drone:

1. ***Accelerometer***

   It is a tool used to measure the movement or acceleration of a structure with respect to its initial state. This device is responsible for the stability of the drone by controlling the rotation of propellers according to the orientation of the drone.
2. ***GPS (compass)***

   GPS stands for global pointing system. This is used to point or locate something all around the globe. It is responsible for the position tracking and the navigation of the drone in all the *x*, *y*, and *z* axis. Hence by installing the GPS or external compass on the drone, we get the access for more flight modes like altitude hold, loiter, and return to landing to position modes. Figure 2 shows the picture of the GPS module used in prototype.

3. ***Telemetry***

   Telemetry, a highly automated communications process by which measurements are made and other data collected at remote or inaccessible points and transmitted to receiving equipment for monitoring, display, and recording. Originally, the information was sent over wires, but modern telemetry more commonly uses radio transmission. Basically, the process is the same in either case. Figure 3 shows the picture of the telemetry module used in prototype.

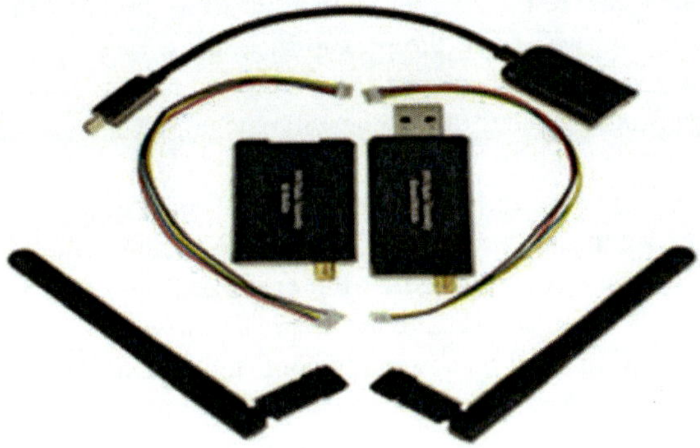

Fig. 3 Telemetry

3 Working Mechanism of Entire System

Flow Chart (Fig. 4) shown below explains in brief how the entire process of the drone delivery system works with telemetry-based architecture.

As an individual has requested some order to be delivered, the order will be received by a server available nearby the order location. Once the order is received by the server, it will do the following task:

1. First it will calculate the distance between the pickup station and delivery station and if the distance is feasible to be covered by the available drones then it will upload the location details and other package-related information on the air module telemetry of the drone [6].

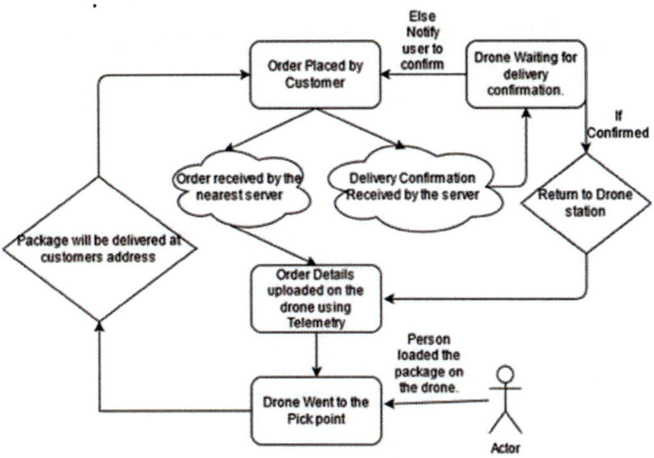

Fig. 4 Working mechanism

2. Then drone will take off straight upward at 30 m height (adjustable based on the location) and will go to the pickup location where a person will load the package on the drone.
3. Once the package is loaded, the drone will navigate itself to the delivery location [7].
4. After the arrival of the drone, the customer will pick his/her order and confirm the delivery via the app used for ordering the package.
5. When the delivery is confirmed by the customer, the drone will return to the nearest drone station. But if the customer has picked up the order and forgot to confirm then he/she will be notified to confirm the delivery.

4 Prototype Summary

Prototype is built using a quadcopter frame, 1000 kV brushless DC motor with $10 \times 4.5\,R$ propeller size capable of lifting around 400–500 gm of weight up to 10 min with 2200 mAh battery capacity. If we double the battery capacity, then we have to either compromise with weight lifting capacity or motor used in the drone. The Ardupilot APM 2.8 flight controller is used in a prototype responsible for controlling everything (motor, GPS, and telemetry) installed on the drone. This prototype is supported with four flight modes: altitude hold, stabilize, loiter, and RTL modes [8]. Once the drone arrives at the delivery or the pickup location, the drone will shift itself to loiter mode which will help in holding the position in x and y direction for smooth landing. Altitude hold mode helps the drone to stabilize or hold a certain height which is adjustable. Once the delivery is confirmed by the customer then the drone will shift to RTL mode which will navigate it to the exact location from where it was initiated [9]. Finally, the stabilize mode is just for the smooth take off by the drone at both pickup and delivery location.

Hardware used to build this entire system is flexible enough to sustain any feature upgrade in terms of hardware. Since telemetry is also involved in this entire architecture, it also leverages the power of data transfer and communication at very high speed [10]. And this technology can also be used for some other tasks like building drone edge computing architecture, traffic monitoring, and much more. Figure 5 shows the picture of prototype.

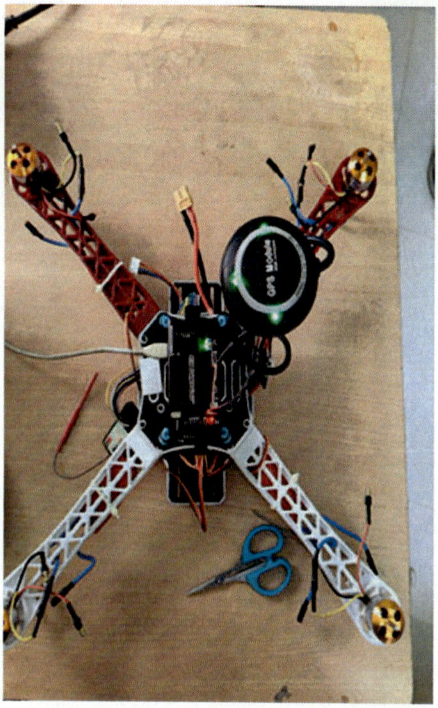

Fig. 5 Prototype

References

1. Hayat S, Yanmaz E, Muzaffar R (2016) Survey on unmanned aerial vehicle networks for civil applications: a communications viewpoint. IEEE Commun Surv Tutorials 18(4):2624–2661. https://doi.org/10.1109/COMST.2016.2560343
2. Dzul AE, Castillo P, Lozano R (2005) Modelling and control of mini-flying machines. In: AIC advances in industrial control
3. Rip MR, Hasik JM, Felt EJ (2002) The precision revolution: GPS and the future of aerial warfare. Naval Institute Press, Annapolis, p 65
4. Roberts D (2013) Construction and testing of a quadcopter
5. Kambire PS, Auti AS, Barge AA, Badadapure PR (2019) Drone delivery system
6. Fancher JC (2017) Drones for medical supplies
7. Wang E, Zhang S, Zhang Z (2012) Research on composite material UAV low-cost avionics system prototype. In: 8th International conference on wireless communications, networking and mobile computing. IEEE, pp 1–4
8. Ganesh Y, Raju R, Hegde R (2015) Surveillance drone for landmine detection. In: International conference on advanced computing and communications (ADCOM). IEEE, pp 33–38
9. Bateman J (2017) China's launching drones to fight back against earthquakes. Wired
10. Markoff J (2016) Drones marshaled to drop lifesaving supplies over Rwandan terrain. New York Times

A Smart Phone App with Geo-Spatial Information for Forecasting COVID-19 from Lung X-ray Images

K. Sujatha, N. P. G. Bhavani, K. Krishnakumar, U. Jayalatsumi, T. Kavitha, K. Senthil Kumar, V. Srividhya, and S. K. Shafiya

Abstract *Background*: Presently, the diagnosis of coronavirus-2019 (COVID-19) is a challenging task worldwide as the disease is spreading at a very faster rate when one person with the disease comes into contact with the other. Current information denotes that several people are detected with COVID-19 and the data analyst say that the rate of spread of the disease is increasing exponentially, across many countries in the world. *Novelty*: This investigation has facilitated the need for diagnosing the disease within a short duration of time by using the X-ray images of the lungs. This scheme deploys artificial intelligence like deep learning algorithms to diagnose COVID-19 among the affected people by maintaining social distancing. Real-time datasets are gathered from the government hospitals for those who are affected by COVID-19 and healthy people. Further investigation can direct the patients themselves to open the smart phone app which will record the respiratory sounds. Followed by this, the features are extracted using Discrete Wavelet Transform (DWT), where a threshold is applied to extract useful coefficients that can be used to train the deep learning neural networks using Fast Recurrent Convolutional Neural Networks (F-RCNN). The respiratory audio signals are captured to detect patients affected by coronavirus by a way of noncontact, nonintrusive approach. The results reported are

K. Sujatha (✉) · N. P. G. Bhavani · K. Krishnakumar · U. Jayalatsumi · T. Kavitha
EEE Department, Dr. MGR Educational and Research Institute, Chennai, India
e-mail: drksujatha23@gmail.com

Department of ECE, Saveetha School of Engineering, SIMATS, Chennai, Tamil Nadu, India

School of Electrical Engineering, Vels Institute of Science, Technology & Advanced Studies, Chennai, India

K. Sujatha · N. P. G. Bhavani · K. Krishnakumar · U. Jayalatsumi · T. Kavitha · K. S. Kumar · V. Srividhya · S. K. Shafiya
ECE Department, Dr. MGR Educational and Research Institute, Chennai, India

K. S. Kumar · V. Srividhya
Department of Civil Engineering, Dr. MGR Educational and Research Institute, Chennai, India

EEE Department, Meenakshi College of Engineering, Chennai, India

S. K. Shafiya
Nalla Narasimha Reddy Group of Institutions, Hyderabad, India

M. S. Kaiser et al. (eds.), *Information and Communication Technology for Competitive Strategies (ICTCS 2021)*, Lecture Notes in Networks and Systems 401,
https://doi.org/10.1007/978-981-19-0098-3_42

valued in detection of COVID-19 by using a smart phone app which is available instantly. *Objectives*: This approach seems to be an indigenous, noninvasive, and cost-effective approach that will relive the patients from trauma of undergoing the swab test and awaiting the laboratory reports, which incurs time delay. Experimental results are obtained from 20,000 samples of patients suffering from COVID-19 and also persons who are normal. This mobile phone app is effective in diagnosing the COVID-19 from the X-ray images of the lungs. Even low-income people can also use this technology. *Methods*: The effectiveness of the proposed system which uses DWT, thresholding, and deep learning algorithms resulted with a performance whose F-measure is 96–98%. The classification is carried out to classify the COVID-19-positive and COVID-19-negative cases using Fast Recurrent Convolutional Neural Networks (F-RCNN). *Expected Outcome*: A smart phone app will be developed to detect the COVID-19 by using a noninvasive and easily affordable technique. The forecasted results were in the range of 89–95% for the above said algorithms. It is significant from the above results that the severe impact of COVID-19 can be diagnosed using a noninvasive mobile phone app using X-ray images.

Keywords Artificial intelligence · Respiratory sounds · Convolutional neural network · COVID-19 · Mobile phone app · Discrete wavelet transform · Thresholding

1 Introduction

The novel coronavirus was reported in 2019 in Wuhan City, China. The WHO Director-General, Dr. Tedros Adhanom Ghebreyesus, declared that the disease was affected by the new virus named COVID-19, in February 2020. However, the first case was reported on 11th January 2020 and healthcare workers were also infected due to the rapid spread from the patients. As per report, the 11 million population of Wuhan City was restricted to travel and found to be under lockdown to avoid the drastic spread. To control the wide spreads of the COVID-19, most of the countries were under lockdown. The breaking news was that the cases continued to increase and spread to other countries such as Brazil, South Korea, India, Italy, and Iran. Thus, it leads to pandemic worldwide serious health risk, and subsequently WHO also raised the threat to the CoV epidemic to the increased rate of spread on 28th February 2020. The death rate was too high in China as compared to other countries by 13 times higher. However, the infection gets transmitted from an asymptomatic individual, which happens before the onset of symptoms. The spread crossed many countries and as of 20 June 2020, about 121,000 deaths in United States and 50,000 deaths in Brazil made a great concern by the WHO to declare the COVID-19 a pandemic. The death rate was underrated due to the confines of investigation and screening for the presence of virus. Although the lower lethal rate of COVID-19 has been reported compared to SARS and MERS epidemics, the transmissions of the SARS-COV-2 virus is widely spread than the other viruses mentioned above.

Worldwide, it has been identified that about one in five individuals are at increased risk of severe COVID-19 disease which might be due to comorbidity (Clark A et al. 2020) [1–7]. Lockdown was announced for most of the countries, and international airports were closed to avoid the spread. By mid-March, the number of cases was increased all over India and first death was reported on the same month 12th, 2020. Meantime, the disease spread throughout the national except Sikkim. In India, from 23 March 2020 lockdown was declared national level, to evade the transmission of deadly virus. During the one-month duration, there have been 135,163 deaths globally and 507 deaths in India (Available at: https://covid19.who.int/. Accessed April 17, 2020, https://www.mohfw.gov.in/ Accessed April 23, 2020).

2 Literature Survey

The current practice proposed by Department of Electronics, Goa University, Goa Taleigao, India, is the design and development of an algorithm which is used to evaluate the entire Hemoglobin content in human blood. This scheme comprises of an embedded system which uses Arduino board. Currently in hospitals, the monitoring and measurement of Hemoglobin in patients is very significant so as to trace the high and low levels of Hb during surgery in operation theater, dialysis, and donation of blood, during the testing of Stress and Intensive Care Unit (ICU). In all these places, the measurement of Hb level in the blood stream is done by an invasive approach where a sample of the blood is extracted from the patient. This system consists of a probe which is attached to the finger with one LED attached for each finger on a board with multiple chips where the wavelengths vary between 670 and 950 nm, which is built on a single silicon photo detector. Photoplethysmography (PPG) principle-based instruments with a trans-impedance amplifier are constructed to gather the signals for the finger [7–9]. This University at Goa has identified a noninvasive to monitor the level of the Hb without pain and infection less devices and instruments. This method not only facilitates a real-time monitoring but also produced outputs which are of comparable standards. This researcher has taken totally eight wavelengths which are compared with the standard values using invasive methods. The outputs are tested and then compared with the standard values obtained in the laboratory. This method is found to show increased accuracy in the measurement of hemoglobin, with minimum deviation between the results obtained from the embedded system model and laboratory results. The error difference is estimated to be 1.3778–0.3499 g/dL, and regression coefficient ($R^2 = 0.964$) is obtained during Hb analysis [10, 11].

3 Objectives of the Work

The main objectives of this work are as follows:

(i) To implement a noninvasive, remote system for detection of COVID-19 by screening the X-ray images of the lungs along with the geo-spatial information about the patients.
(ii) To relieve the patients from getting exposed to harmful radiation for longer duration during Computed Tomography (CT), X-ray imaging, and Magnetic Resonance Imaging (MRI) scanning techniques repeatedly.
(iii) To minimize the cost involved in laboratory analysis which will require the usage of chemicals and reagents.
(iv) To eliminate the biochemical hazards involved in disposal of related biomedical waste.
(v) To nullify the time delay involved in laboratory-scale analysis to identify COVID-19 and the geo-spatial information about the patients facilitates to track and monitor them consistently.
(vi) To enable a remote, online and continuous monitoring system to screen the patients affected by COVID-19 at the earlier stage with their geo-spatial information which can be prevented from spreading further by analyzing at least 10,000 samples within no matter of time.

4 Methodology

Mobile phone app for screening and detection of COVID-19 from the lung sound during respiration along with the geo-spatial information. The scheme for intelligent X-ray images of the lung diagnosis system is proposed here. The proposed scheme comprises of many stages with compatibility to share, store, and access the related database by the physicians, patients, and researchers from any part of the world unbounded by time limits along with their geo-spatial information as illustrated in Fig. 1. Such kind of flexible systems will facilitate the development of robust algorithms for diagnosis of COVID-19 without human intervention, thus preventing the spread of the disease.

Patients suffering from various types of respiratory disease develop similar symptoms as that of COVID-19 which includes cough, increase in body temperature, running nose, congestion of lungs, and difficulty in respiration. All these symptoms are related to any type of viral infection and so it becomes difficulty to differentiate coronavirus from other types of viruses responsible for other respiratory diseases. Hence, investigation has shown that extraction of wavelet coefficients from the X-ray images of the lungs will enable a noninvasive, instantaneous, early, and accurate diagnosis of the COVID-19 without encountering time delay. The need of the hour has enabled the researchers to think in a new direction to develop a smart phone app-based technology with signal and intelligent classification techniques. This state-of-the-art technology will be a freely available technology on the Google, where even a layman possessing a smart phone can download the app and use it for himself alone. As a result, each and every individual possessing a smart phone can download this app and use it for detecting the COVID-19 with less human intervention during this

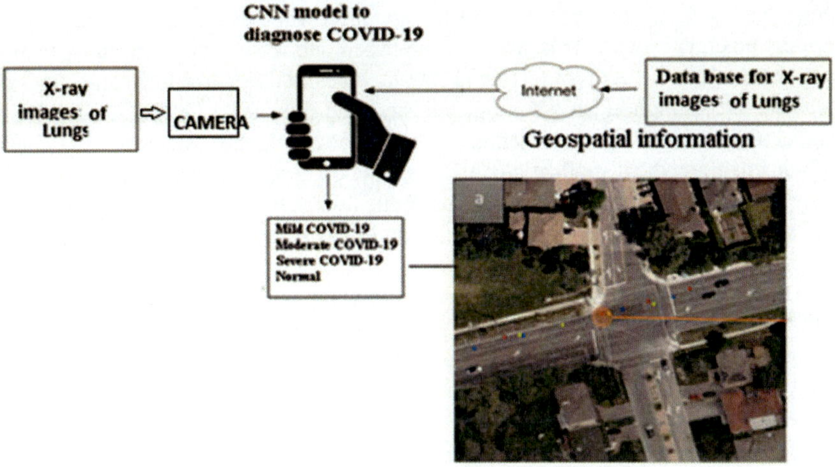

Fig. 1 Smartphone app for COVID-19 detection with geo-spatial information

pandemic. The Android API provides the best available location information based on Android 1.5 (API Level 3) location providers such as Wi-Fi and GPS (Global Positioning System) available on each smart phone. Furthermore, the API can deliver the accuracy of the reported location information determined by the available location providers. Also, this concept is easily portable and does not incur any transportation charges. The main highlight is that there is no direct contact between the patient and physician, which will control the spread of coronavirus when people come into contact. To throw more light on the technology of the proposed smart phone app is that it is absolutely free of cost and user-friendly immaterial of the knowledge possessed by the users thereby enabling early diagnosis along with the geo-spatial information about the patients, so that it becomes easy for the physicians and data analysts to track and monitor them.

5 Results and Discussion

All the organs in the human body create a unique sound during their normal functioning periods. When there arises an abnormal situation, these sounds vary which will have a unique variation in the amplitude and spectrum levels. This concept indeed finds an important application in detection of COVID-19 from the X-ray images of the lungs in Fig. 1. A correct indication of the COVID positive can be detected from the X-ray images of the lungs. This smart phone-based app can produce equivalent results as that of the swab test through laboratory analysis. This kind of smart phone app will facilitate the patients to monitor and detect the presence of coronavirus by them ensuring a remote, noncontact, noninvasive method. This smart phone app helps the patients to diagnose the COVID and infer whether the patient is in need of critical emergency care. This smart phone-based app testing can be used by health workers and can be used for testing in low-income countries like India and Nigeria, where there is high smart phone penetration and prevalence of COVID-19. This signal decomposition may not serve all applications, as in the case of certain images relating to biomedical applications. Wavelets are obtained when the signal is decomposed using the standard set of functions. Scale (or dilation) defines how "stretched" or "squished" a wavelet is which is related to frequency. The spatial and temporal coordinates of the wavelets are identified.

The noise removal is also done using the wavelet tool box. De-noising helps to improve the quality of the signal, so that appropriate and exact values of the features alone can be extracted. A high-pass filter (HPF) allows the signal portion corresponding to high frequency values alone to pass through it. The transfer function for the HPF in discrete domain is given by $1–0.99z^{-1}$. An extensive variety of features are extorted from the lung sound signal during respiration using one dimensional discrete wavelet transform. The preferred feature set serves as the input and includes the details and approximation components to train and test the F-RCNN as shown in Fig. 2.

A supervisory segmentation scheme to trace and capture the X-ray images of the lungs during respiration even in a noisy environment can be facilitated by implementing F-RCNN to detect the abnormality in the training phase by using 70% of the collected signals itself, so that an appropriate model for X-ray images of the lungs analysis can be launched. During testing of F-RCNN, only 20% of the collected sample X-ray images of the lungs are used for detection of abnormality. The remaining 10% can be used for validation after the development of the mobile phone app.

The main benefit of this method is that the proposed F-RCNN is robust enough that it is capable of restoring even the lost portion of the X-ray images of the lungs along with the geo-spatial information as in Figs. 3 and 4, respectively. The training of F-RCNN is done using Gradient Descent Rule (GDR) with optimal values of learning rate and momentum. The training includes change in weights with the filter interpretation which is derived at the first layer of the F-RCNN. The performance evaluation is done by calculating the metrics like true-positive rate (TPR), false-positive rate (FPR), false-negative rate (FNR), and accuracy. It is inferred that, by the proposed method (WT-F-RCNN) for analysis of X-ray images of lungs, the TPR value is 98.23% and 85.56% during training and validation. The overall accuracy is calculated to be nearly 96.79% from Fig. 5. This proposed technology is a wireless technology, with high computing levels so as to integrate with Internet and develop an Internet of Medical Things technology. This, MIoT technology is capable to gather, transmit, and track the medical data related to COVID-19 so as to evolve a state-of-the-art technology.

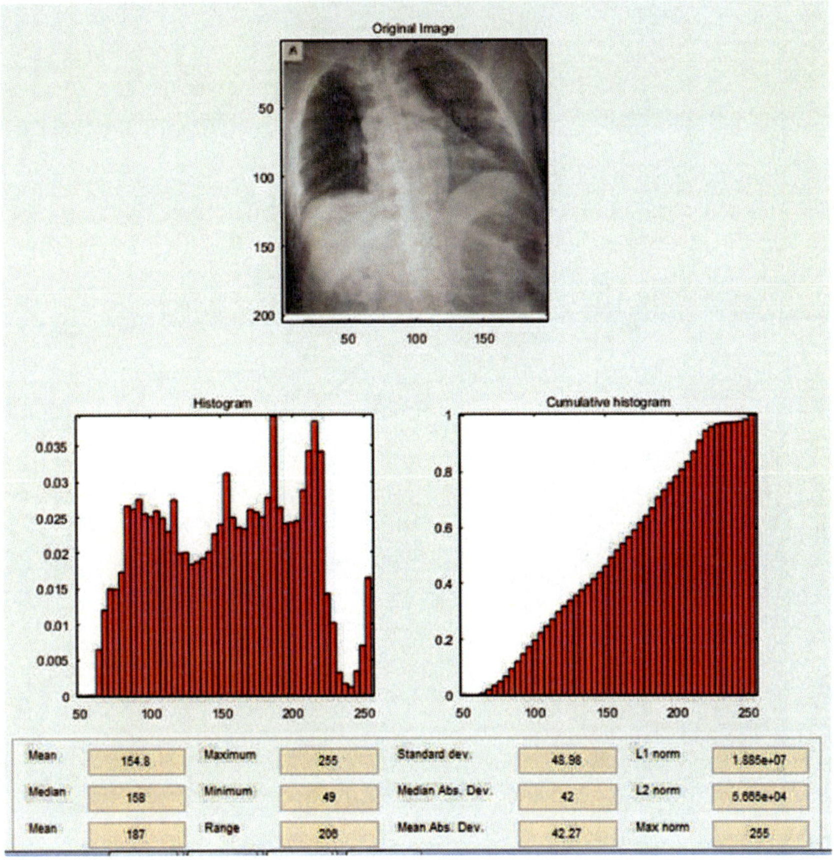

Fig. 2 Wavelet analysis for X-ray images of the lungs with COVID positive

Fig. 3 Detection of COVID-19 by F-RCNN

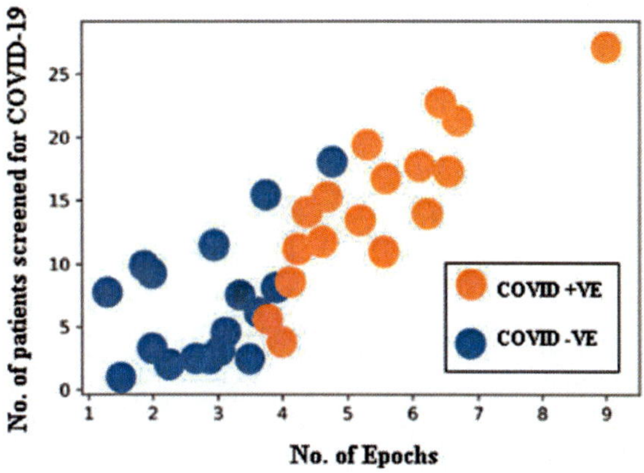

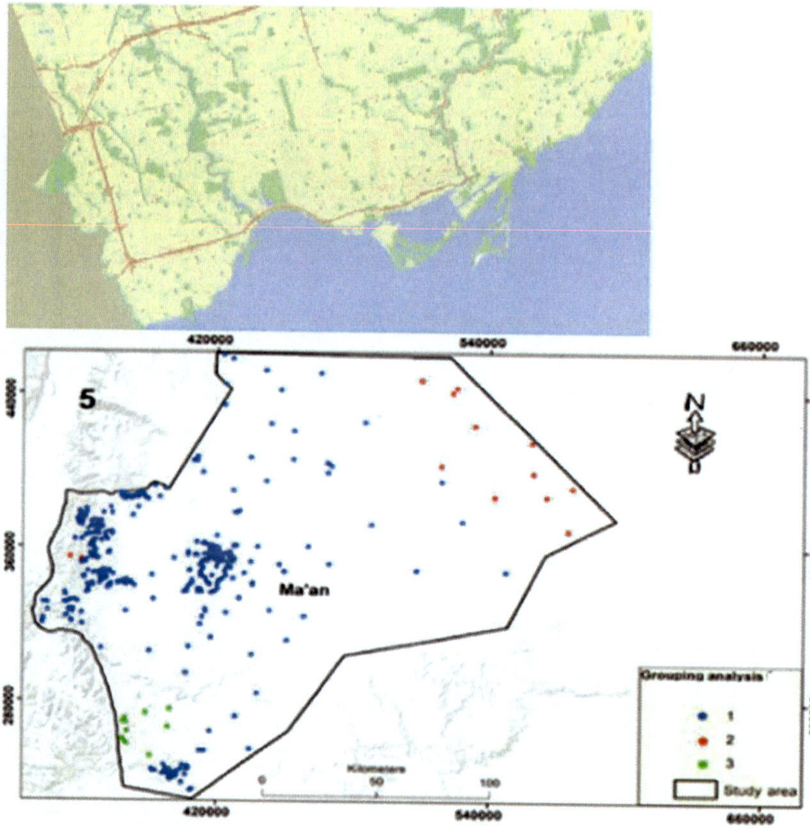

Fig. 4 Identified patient's location along with geo-spatial information on smart phone

Fig. 5 Determination of various conditions of COVID from X-ray images of the lungs using F-RCNN

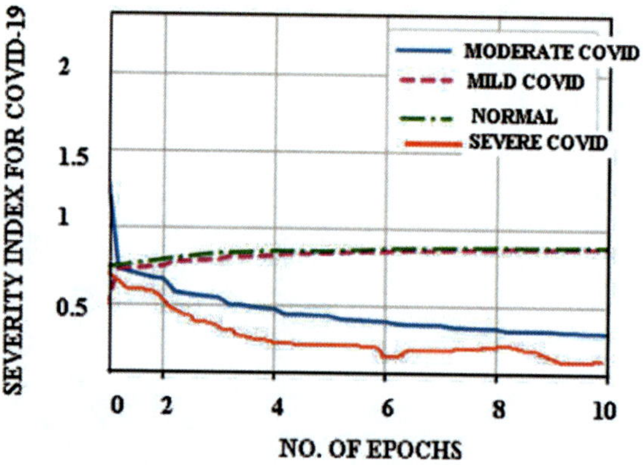

6 Conclusion

Segmentation of X-ray images of lungs along with the geo-spatial information and segmenting the homogenous regions of the images facilitates the error free identification. This noncontact type of Remote COVID Diagnostic System is used to detect the region of interest from X-ray images of lungs along with the geo-spatial information about the patients. The performance can be considerably increased even if the images are corrupted with noise. The key objective behind this work is to identify the useful component and the noisy component from the lung sounds during respiration. In future, this method of X-ray image analysis can be used for developing the database that facilitates the following applications like differentiating the diversified features present in the X-ray images of lungs and to categorize the other kinds of respiratory diseases adhering social distancing, thereby preventing the spread of coronavirus.

References

1. Di Nardo M, van Leeuwen G, Loreti A, Barbieri MA, Guner Y, Locatelli F, Ranieri VM (2020) A literature review of 2019 novel coronavirus (SARS-CoV2) infection in neonates and children. Pediatr Res
2. Wang Y, Wang Y, Chen Y, Qin Q (2020) Unique epidemiological and clinical features of the emerging 2019 novel coronavirus pneumonia (COVID-19) implicate special control measures. J Med Virol 92:568–576
3. (2019) Clinical and immunological features of severe and moderate coronavirus disease
4. Chen G, Wu D, Guo W, Cao Y, Huang D, Wang H, Wang T, Zhang X, Chen H, Yu H, Zhang X, Zhang M, Wu S, Song J, Chen T, Han M, Li S, Luo X, Zhao J, Ning Q (2020) J Clin Invest 130(5):2620–2629
5. Heimdal I, Moe N, Krokstad S, Christensen A, Skanke LH, Nordbø SA, Døllner H (2019) Human coronavirus in hospitalized children with respiratory tract infections: a 9-year population-based study from Norway. J Infect Dis 219(8):1198
6. Monto AS (1974) Medical reviews. Coronaviruses. Yale J Biol Med 47(4):234
7. Monto AS, DeJonge PM, Callear AP, Bazzi LA, Capriola SB, Malosh RE, Martin ET, Petrie JG (2020) Coronavirus occurrence and transmission over 8 years in the HIVE cohort of households in Michigan. J Infect Dis 222(1):9
8. McIntosh K, Kapikian AZ, Turner HC, Hartley JW, Parrott RH, Chanock RM (1970) Seroepidemiologic studies of coronavirus infection in adults and children. Am J Epidemiol 91(6):585
9. Nickbakhsh S, Ho A, Marques DFP, McMenamin J, Gunson RN, Murcia PR (2020) Epidemiology of seasonal coronaviruses: establishing the context for the emergence of coronavirus disease 2019. J Infect Dis 222(1):17
10. Clark A, Jit M, Warren-Gash C, Guthrie B, Wang HHX, Mercer SW, Sanderson C, McKee M, Troeger C, Ong KL, Checchi F, Perel P, Joseph S, Gibbs HP, Banerjee A, Eggo RM (2020) Centre for the mathematical modelling of infectious diseases COVID-19 working group. Global, regional, and national estimates of the population at increased risk of severe COVID-19 due to underlying health conditions in 2020: a modelling study. Lancet Glob Health 8(8):e1003–e1017
11. Yeager CL, Ashmun RA, Williams RK, Cardellichio CB, Shapiro LH, Look AT, Holmes KV (1992) Human aminopeptidase N is a receptor for human coronavirus 229E. Nature 357(6377):420

Bee Basket Pvt. Ltd—An Example of Green Entrepreneurship for a Sustainable Social System

Pradnya Chitrao, Pravin Kumar Bhoyar, and Rajiv Divekar

Abstract Entrepreneurship involves the creation of innovative organizations. Green entrepreneurs use prospects and start commercial activity for sustainable growth which is usually dramatically changing structures, devoted for a social cause, utilizing advanced technology; its product/service is usually ecofriendly and is an environmental product (Walley and Taylor in Greener Manag Int, Theme Issue Environ Entrepreneurship 38:31–43, 2002) [9]. Green entrepreneurship is thus primarily a business that uses a new innovation basically to address a societal problem and solve it completely or up to a certain extent while taking care that it is not harmful to environment. The primary aim of such a business is not just profit but alleviation of a social or community problem. In India, quite a few people are going in for green entrepreneurial ventures. Such enterprises not only try to solve social problems in an environment-friendly manner but also manage to create sustenance opportunities for many. They are also capable of creating employment opportunities and contributing to the nation's economy. The paper seeks to study the green entrepreneurial ventures in India. It seeks to find out the opportunities and challenges faced by such ventures as also how they contribute toward a sustainable social system. The research methodology employed is firstly primary research, wherein an entrepreneur venturing into such an enterprise has been interviewed. The researchers have decided to focus on one entrepreneurial venture, namely Bee Basket Pvt. Ltd as a case study. The paper also makes use of secondary sources of information. The limitation of the paper is that the researchers have restricted themselves to one Pune-based city entrepreneur though their secondary research will include enterprises all over India.

Keywords Green ventures · Technology · Social and environmental benefit · Challenges · Sustainable society

P. Chitrao (✉) · P. K. Bhoyar
Symbiosis Institute of Management Studies (SIMS), A Constituent of Symbiosis International Deemed University, Pune, India
e-mail: pradnyac@sims.edu

R. Divekar
Brig. (Retd) Symbiosis Institute of Management Studies (SIMS), A Constituent of Symbiosis International Deemed University, Pune, India

M. S. Kaiser et al. (eds.), *Information and Communication Technology for Competitive Strategies (ICTCS 2021)*, Lecture Notes in Networks and Systems 401,
https://doi.org/10.1007/978-981-19-0098-3_43

1 Introduction

Green entrepreneurship in the twenty-first century is viewed as the principle factor for putting in place an integrated and tenable economic–environment–social structure. People who combine ecological consciousness with enterprising activity are today expected to form a central factor in any step toward achieving an economy that aims at reducing environmental risks and ecological scarcities, [7]. Green entrepreneurship has the scope of being responsible in a major way for the overall transformation into a more supportable business model being as 'exemplary' results for a social metamorphosis. India also is today witnessing a growing body of entrepreneurs who wish to be environment-friendly in their business ventures.

2 Literature Review

Sustainability today is a serious concern worldwide and is a major element in determining the success and failure of enterprises. Supportable growth exercises are creating new markets, products and demand as also new business frameworks that emphasize innovation and reduce cost. Environment-friendly entrepreneurship is the shift by business persons to a supportable trade structure that is ecofriendly and which yields both monetary gains and also is concerned about protecting our planet, and making it a safe ad nice and livable place for both us and our progeny. The concept of green entrepreneurship involves both the preservation of the environment as also the promotion of entrepreneurial goals.

'Green Culture' combines the principles of ecofriendly innovative industries, environment promoting trade, environment-friendly science and environment-oriented thought processes as an integrated approach to problem-solving and growth through interdisciplinary collaboration of sections such as expansion and architecture, science and transformation, trade and pioneerdom. Green entrepreneurs try to include the environmentally safe, remunerative, and societal elements of biodiversity in a comprehensive way and to have a divergent collocating rationale as against traditional entrepreneurs [8]. While these are for-gain enterprises, they are set up with robust green (and usually community-based) standards as well [3]. Environment-friendly business is today considered essential for the creation of an integrated and tenable productive–conservational–community system. Willis et al. [10] term such environment-friendly businessmen 'disruptive innovators' who supplant and change entrenched transaction frameworks and customer expectations.

Green entrepreneurship has two perspectives, namely the return (commodity and utilities) and also the procedure (or production) of a remunerative pursuit. Businessmen can launch into a distinctly environment-friendly commerce zone like providing green and ecologically sound commodities and amenities (waste management). Or, environment-friendly businessmen can make available their commodities or amenities through an ecologically sound procedure or with the help of a set of technologies that either reduce or optimize the use of natural resources (e.g., eco-tourism). Usually, environment-friendly businessmen take into consideration both factors in their commerce plans, thereby providing for more decent employment through the use of additional impact on the environment on account of people or organizations using the end commodities and utilities (International Labor Organization). Green entrepreneurs do away with prevalent traditional techniques of production, commodities, market designs, and consumption models and in their place bring better ecofriendly commodities and amenities. They produce the forces that will impact prices and the behaviors of producers and consumers that lead to ecological progress [6]. This contrasts with a Kirznerian [4] perspective of entrepreneurship where vigilant entrepreneurs control everything from prices and inputs to end results and resource allocation, and where businessmen are driven by self-serving financial gains in a context where the environment-friendly economy is another occasion for monetary proceeds. Environment-friendly entrepreneurship has therefore been understood as bringing in new business that tries to create ecofriendly commodities or amenities or to employ sustainable resources in the interests of the community [1]. Environment-friendly businessmen are utilizing ecofriendly values as a main element of their identity and positioning them as a one-upmanship of their organization in the market place. The distinction between environment-friendly start-ups and traditional start-ups lies in the focus on ecofriendly measures of the green entrepreneurs themselves. By encouraging green entrepreneurship, it is possible to increase the weightage given to ecofriendly and supportable measures in a community's long-term plans with specific reference to sustainability and social equitability in local economies.

Many organizations in the developed nations are part of the biodiversity revolution for the last thirty years and have agreed that they can bring down deterioration of the environment and bring more profits simultaneously. The new business culture of social responsibility motivates organizations to proactively work toward supportable innovation. When organizations are environmentally responsible, it reduces society's criticism against them, and they can entice the customers by offering environment-friendly commodities and services [11]. Surveys conducted globally reveal that consumers worldwide are transforming their consumption patterns and show a preference for organizations that are socially responsible and produce ecofriendly commodities. These companies were also found to most likely report contamination prevention as a way of improving the plant. These companies also mentioned salvaging, air pollution reduction, disposal of refuse and supervision of electricity use as the different measures they were adopting in order to improve ecological balance [5].

In the present situation, ecological economic is a billion-dollar section that spans from biodegradable and environmentally safe lifestyle commodities, chemical-free farming, geothermal power, and clean technology to water and refuse administration, natural resource and land management. Today, on account of the challenge of global warming, nearly all countries are concerned about protecting the environment; consequently, a new model of business has come into existence, namely green business which is run on green philosophy. Green companies are a midway solution and compromise between two diametrically opposite prototypes of the twenty-first century, namely the commercial perspective and the environmental worldview indicating activities of corporate social responsibility [2].

3 Objectives of the Study

The researchers wished to find out how many entrepreneurs opted for green businesses in Pune. They wished to understand the challenges and opportunities available for aspiring entrepreneurs who wished to take up green business ventures. They also wanted to know the innovativeness that such ventures brought in. The study also sought to find out how such green ventures contributed toward a sustainable social system.

4 Research Methodology

The researchers have resorted to firstly primary research wherein entrepreneurs venturing into such enterprises were interviewed. They also referred to secondary sources of information.

5 Limitations of the Study

The researchers have restricted themselves to one case study of a Pune-based entrepreneur though their secondary research will include enterprises all over India.

6 Primary Research Findings

The researchers interviewed Mr Amit Godse of Pune. Famously known as the Bee Man, Mr Godse quit a highly paid job with a well-known IT company in Mumbai to start his own business of bee conservation and honey collection in mid-2014. The trigger point was the death of hundreds of honey bees in his housing society sometime in March 2013 when a beehive was destroyed by burning it. He decided to learn more about them and the art of extracting honey. He enrolled for training with the Central Bee Research and Training Institute (CBRTI) conducted by the Khadi and Village Industries Commission. He was alarmed to learn that bees were killed in large numbers in urban areas with the help of insecticides used by pest control agencies and unorganized extraction of honey that involved burning of the honeycombs in rural parts.

Mr Godse, on a trip to Malegaon in 2013 to a friend's father's organic farm, was advised to go to Mahableshwar for training in collecting honey in an organic manner. He learnt that a simple technique namely, relocation of honeycombs, could offset the problem. According to him, one can easily save bees by relocating hives from residential areas to farmlands. He has evolved a very bee-friendly technique of relocation and collection of honey. The technique involves securing a honeycomb with the help of a clip at one end after subjecting it to a few minutes of spraying of smoke and water in a controlled manner. The smoke is organic in nature made by burning coconut shell or dry leaves or cotton waste. He then cuts the honey chamber of the hive. The portion consisting of bees is placed in a net, that is, specifically designed for safe transport. He prefers to do this activity at night when all the bees are back in their hives though the courses he attended told them to remove beehives during day time. For relocation, the comb has to be attached to the pre-decided place with the clip. Godse advises not to try this technique unless you are trained.

Godse calls for separation of bees from pests and rerouting the calls for removal of hives to trained persons who use scientific techniques such as relocation. A certified trainer himself in relocation of hives and extraction of honey, he plans to set up self-help groups whom people can contact for relocation. Volunteers can be trained in relocation of beehives. Godse has already trained four people in Pune and five in Gujarat. According to environmentalists, bees improve crop quality and quantity of crops. In Godse's opinion, if all bees are wiped out from earth, then mankind will have only a few years' existence on earth. In his interview, he stated 'no bees, no pollination, no plants, no animals, no mankind.'

Amit Godse does not belong to a business family, and he did not even think of a profitable business model for his work initially. However, in 2015 he became serious about transforming his work to a sustainable, bigger business. He joined Entrepreneurship Development Program (EDP) at SIMS in Jan 2015 to learn more about building a business. Later in 2015, he was guided by deAsra foundation that helped him to prepare business plan, its proposal and apply for loans under various government schemes. After that he was accepted for the grooming program at iCreate in Ahmedabad. He learnt a lot about focusing on specific products and services, creating value and building profitable business through these programs. Initially, he wanted to do many things with the honeybees but had very little idea about actually developing a promising business around them. After these programs, he realized that he should focus on few aspects and do them well. In 2015, he also got media coverage as many local and national newspapers covered his work with honeybees. His phone number went viral after the media coverage, and he gets innumerable calls these days for relocating the beehives. Recently, he served some influential clients including divisional commissioner of Pune. He is already getting many enquiries from other cities such as Bangalore, Ahmedabad asking him if he could offer his service in those cities but he does not yet have the bandwidth to offer his service in other cities.

He started his business in 2014 bee basket. He has three partners and two employees. One of the partners, Mr Shubham Choubey is an M.B.A. From a single beehive, he harvests honey multiple times without killing them. In 2014, he managed to bootstrap. In 2015, he invested five lakhs in his business. By the year end, he got a profit of around Rs. three lakhs. In 2016, he has been getting monthly orders of around 150 kg honey. He has different kinds of floral honey, namely neem honey, karvi honey, ajwain honey, jamun honey, sunflower honey and forest honey. He trains tribals and farmers the modern method of harvesting honey. He is confident of implementing his model across all smart cities in India. He plans to start honey-based bakery products which are healthier than sugar based ones. He also plans to start e-course on bee handling and bee keeping in a green manner.

Mr Godse faces mainly two challenges: firstly, the bee population is diminishing and secondly, the mentality of farmers and tribals who only are looking for freebies. He charges around thousand to fifteen hundred rupees for handing the bee hives and relocating them on farms or farm houses and bungalow compounds. He stated that people today want bees for pollination and are they are willing to pay fees for the same. He also sells bee boxes. He has two entities: one is an NGO and the other is a private limited company, namely Bee Basket Society and Bee Basket Pvt. Ltd. The NGO trains tribals and is paid around Rs 5000/- per training. Mr Godse sells bee product through his Pvt. Ltd. Company. His business has been selected by Government for incubation center at Symbiosis Institute of Management Studies (SIMS), Pune, and he will receive a funding of Rs 6 lakhs.

Secondary Research Findings

Indians are now more and more coming up with innovative business ventures that are proving to be friendly to both clients and environment. UberHealth is one such venture which makes it possible for persons who are usually caught up in work or are staying away from their near and dear ones to look after all aspects of family health care. People can now monitor the medical records of their parents and seek a second medical opinion from doctors through UberHealth. It is the idea of Kamal Aggarwal, Ishan Jha and Ajay Pal Singh, alumni of NIT (Jalandhar). UberHealth collaborates with laboratories for taking blood samples from the patient's home and getting the doctor's recommendations on the basis of the tests' results. The company delivers both the soft copy as also the hard copy of the report as also the video of the doctor's feedback. This is a service framework of business, and so the money come from providing services to the customers.

Four Bengali youths from India worked hard to make a revolutionary product. They named it Ecotop. The commodity is a special tile that enables one to put up not only a green roof but also a tropical garden inside the bedroom if so desired. The product is a modular landscaping platform, a multilayered Lego-style tile that is for covering any plain or uneven surface in an ecofriendly manner. The two by two feet tile is portable, and leak proof, and has been designed in the USA by an Indian, tested in Switzerland by another Indian, and finally manufactured completely in India by two other Indians with the help of indigenous resources. The four friends—two from IIT Kharagpur and two others from Jadavpur University in Kolkata—collaborated on this project for one year on Cloud. Jayabrata Bhaduri, who heads Capacloud, the company that is making Ecotop, graduated from Jadavpur University in Mechanical Engineering and went on to do an MBA course from SP Jain. He left the comforts of multinational job and jumped into entrepreneurship. He now works full time for Capacloud. The other three hold full time jobs as they are yet to make profits out of their venture. The team is initially targeting the B2B segment.

Another venture Warby Parker, an online glasses retailer, is a very good instance of the interest to utilize technology for shaking up an industry that had not used innovation for many years. The founders believed strongly that their for-profit business could change the world for the better. Warby Parker's purchase-one donate-one model provides eye glasses to poor people for every pair of glasses it sells. To date, Warby Parker has given away for free more than a million pairs of glasses to poor persons. Today customers are increasingly becoming 'purpose-driven' and conscious of the effect of their buying on society and environment. So companies like Warby Parker and Toms, a manufacturer of shoes, are also making use of a similar philanthropic business model and have thus tried to acquire customers' cooperation and commitment through their social initiatives. For Warby Parker, this has supposedly resulted in a free enterprise system valuation of over $1 billion. According to Neil Blumenthal, CEO, Warby Parker, their social responsibility activities are not the main reason why customers buy spectacles from them. According to him, for customers, fashion and design are the top reasons followed by the cost of the eyeglasses and then the quality of the commodity and amenities. Social responsibility is the fourth reason. Even then, Warby Parker's social responsibility activity has made a brand personality that customers like and lean to.

Green Infra is an independent renewable power producer from India that produces electricity that is used by about one million people across four Indian states. This additional access to electricity will enable many businesses to progress and at the same time stop millions of tons of carbon dioxide emissions. The enterprise has been instrumental in creating 600 direct jobs with hundreds more supported by the supply chain. The company is actively promoting rooftop solar fittings among its portfolio companies that provide systematizations, repositories, and manufacturing complexes. Ms Ritu Kumar is Director Environmental and Social Responsibility of the venture.

Another startup called Cowism, started by Mr Chetan Raut, strives to help farmers become self-reliant and financially independent through the use of conventional, grange-based agrarian solutions. It concentrates on utilizing industrially useful strong points like native cattle for cultivation purposes in order to increase the fertility of soil, to reduce input costs, and to increase farmers' profits. Raut is a qualified engineer hailing from a farmers' family in Maharashtra's Yavatmal District. He is fully aware of the farming cataclysm that is prevalent in the region. He discovered that farmers who raised cattle incurred less input costs (chemical pesticides and fertilizers usually constitute 60% of the total input cost). This was because they used cow plop and micturate to make the land fertile. Cows, not only gave nutritive manure, but also resulted in many high-protein dairy commodities to supplement the diet of farmers' families. In addition, lacteal cattle yielded a stable source of proceeds throughout the year. Native cows' milk, in scientific terms referred to as A2 Beta Casein milk, is healthier, easier to absorb, and better for immunity building in comparison to the more easily available A1 milk.

Supported by this knowledge in the field, Raut did his Master's in Social Entrepreneurship from the Tata Institute of Social Sciences (TISS) in Mumbai in order to hone his innovative ideas. At TISS, Raut came up with the idea of Cowism for improving farming profits and increasing farmer security. Cowism began under the patronage of the Arunokalp Social Organisation for Rural Development (ASORD) in Chandrapur, Maharashtra. It enlightens local agriculturalists on the merits of cultivation along with related pursuits like raising local cows. It supports the use of natural defoliants and fertilizers like Jeevamrutha that can be made on the farm with cow dung and urine as its ingredients thereby reducing input costs. Cowism does have a potential of transforming the scenario. But it requires a big change in farming behavior because of which its effects are rather slow. So, in order to help the farmers, Raut sacrificed his big urban life and went back to his village to be with the society he wanted to empower. Cowism was born and nurtured in its initial phase in the social entrepreneurship incubation cell at TISS and was guided by DBS in preparing its business model and promotion. On account of DBS' initial investment, the team could bear the cost of obtaining and preserving assets like cows and develop the framework required for the running of its training and resource centers and research and development facilities on its farm. Cowism is essentially all about helping farmers to become self-sufficient and economically independent through the use of traditional, locally available farming solutions. It acknowledges the reliable, financial value of assets like livestock in improving soil richness, lowering input costs, raising farm earnings, and improving the health of farming households.

These are but a few of the many ventures that enterprising Indians are taking up. They are all green in nature and meant to benefit society at large.

7 Recommendations

In order to behave in a supportable or tenable manner, organizational actions cannot be restricted to just technical solutions; it must also accept modern ecofriendly responsible values, beliefs, and behaviors. Green business strategy is now not an alternative but a necessity; the future will be shaped by it.

From both the primary research and the secondary findings, it is evident that Indians have to show courage and take decisions that involve moving away from secure employment in order to concentrate all their energies and often their saved financial resources on the ventures that they set up. They need to undergo training in management aspects of a business along with gathering more information of the business and its sector that they dream of taking up. They need to enroll in Entrepreneurial Development Programs being offered by B-Schools like Symbiosis Institute of Management Studies (SIMS), Pune. They need to be supported by their family as well as government. The latter indeed is coming up with various schemes to motivate more people to go in for entrepreneurship as it has positive outcomes for the economy as also the eco-social system. Essentially, India's climate plan—termed as the **Intended Nationally Determined Contribution**—is steeped in the country's national strategy for a comprehensive and supportable approach to economic growth. Instead of being detached from it, resolving climate change is important for attaining the 17 Sustainable Development Goals which were also approved by the world's governments last year.

Green entrepreneurship is indeed a major path toward India becoming a socio-eco-environmentally friendly economy.

References

1. Allen JC, Malin SA (2006) Green entrepreneurship: a method for managing natural resources. In: International society of society and natural resources conference, Utah State University, U.S.A
2. Carlisle YM, Faulkner OD (2004) Corporate social responsibility a stages framework. Eur Bus J 1:143–152
3. Kirkwood J, Walton S (2010) How Ecopreneurs' green values affect their international engagement in supply chain management. J Int Entrep 8:200–217
4. Kirzner JM (1973) Competition and entrepreneurship. University of Chicago Press, Chicago
5. Purba R (2004) Greening production: a South-East Asian experience. Int J Oper Prod Manag 24(3):289–320
6. Schaltegger S (2002) A framework for ecopreneurship. Leading Bioneers Environ Managers Ecopreneurship Greener Managem Int 38:45–58
7. Schaper M (ed) (2010) In: Making entrepreneurs: developing sustainable entrepreneurship. 2nd ed. Gower, Farnham
8. Tilley F, Parrish B (2006) From poles to wholes: facilitating an integrated approach to sustainable entrepreneurship. World Rev Entrepreneurship Managem Sustain Developm 2(4):281–294
9. Walley EE, Taylor DW (2002) Opportunists, champions, mavericks...? A typology of green entrepreneurs. Greener Manag Int, Theme Issue Environ Entrepreneurship 38:31–43
10. Willis R, Webb M, Wilsdon J (2007) The disrupters: lessons for low-carbon innovation from the new wave of environmental pioneers. NESTA, London
11. Yoon E, Tello S (2009) Corporate social responsibility as a driver of sustainable innovation: greening initiatives of leading global brands. Competition Forum 7(2):290–294

Prediction of Employee Attrition Using Stacked Ensemble Method

Sanjay Gowdru, Suyash Kumar Dubli, Pooja Agarwal, and Bhoomika

Abstract Employee attrition is becoming a big issue in businesses. The question that many HR managers are asking is why workers quit the company. Hiring new staff, rather than maintaining existing personnel, will always take more time and money. The project's objective is to anticipate employee attrition before he or she departs the organization. The problem is predicting whether or not an employee will leave based on specific indications. Instead of focusing on a single classifier technique, the proposed study would tackle the problem utilizing a stacked classifier algorithm. The ensemble model will compute the prediction of each classification, and a new model will be constructed based on the ensemble technique used for prediction. Accurate forecasting will aid businesses in taking the required actions to reduce attrition.

Keywords Employee attrition · Machine learning · Prediction model · Stacked ensemble · Base models · Meta-model

1 Introduction

Employee attrition is described as the loss of personnel due to a natural process such as retirement, resignation, job termination, personal health, or, in simplest terms, voluntary or involuntary attrition.

Involuntary attrition is when a company initiates the exit and not the employee, or in other words, the company decides to let go of the employee. This might be due to various reasons, for example, when there is a merger and acquisitions.

There are different factors which cause employee attrition. The four factors affecting employee attrition can be divided into four categories mainly: personal motivation, professional motivation, challenges in the workplace, and poor employee-to-job fitment.

S. Gowdru (✉) · S. K. Dubli · P. Agarwal · Bhoomika
PES University EC Campus, Bangalore, India
e-mail: sanjugr7022@gmail.com

P. Agarwal
e-mail: poojaagarwal@pes.edu

M. S. Kaiser et al. (eds.), *Information and Communication Technology for Competitive Strategies (ICTCS 2021)*, Lecture Notes in Networks and Systems 401,
https://doi.org/10.1007/978-981-19-0098-3_44

Employees are viewed as the most important resource in every organization; thus, they must be treated with extreme caution. Employee turnover has recently been a major concern for all businesses, owing to its negative impact on workplace productivity and meeting organizational goals on time. Machine learning approaches are being used to help organizations anticipate attrition. Organizations may take the essential measures for staff retention or succession in a timely manner with high precision in prediction.

This paper aims to create a model which will predict employee attrition using a stacked ensemble method.

The following sections make up this document. The work/literature review in Sect. 2 is highlighted. The project's motivation and terminology are discussed in Sects. 3 and 4. The dataset description, preprocessing, feature engineering, and analysis are all presented in Sect 5. The proposed approach is shown in Sect. 6, and finally, the experimental design and results are summarized in Sect. 7.

2 Related Work

Because of the severity of its impact, voluntary employee attrition is one of the most serious problems for any firm. It may also be viewed as a defacement of the organization's intellectual capital. The literature review shows the many approaches and techniques for forecasting employee attrition suggested by various scholars.

Divyang [1] Many types of ensemble methods were explored for classification, including boosting and bagging; however, boosting algorithms performed the best, with adaptive boosting offering up to 88.8% accuracy in evaluation score. The study gives useful information on different ensemble strategies as well as the dataset.

Qutub [2] This paper discusses many machine learning algorithms that have been created to forecast employee attrition automatically and accurately. In this work, the IBM attrition dataset is used to train and evaluate machine learning models. The decision tree model, random forest model, gradient boosting model, AdaBoost, and logistic regression model are among the six machine learning models used for training and assessment.

Jain [3] In this study, we learned how the OSEMN model is used to build a typical machine learning pipeline. It employs an XGBoost-based machine learning technique to create a precision model to predict employee attrition. XGBoost is a scalable and highly reliable approach for dealing with all types of noise in large datasets and converting the data into a usable format for accurate findings.

Karande [4] In this study, we learned how to apply the random forest, SVC, and logistic regression algorithms to form an ensemble to predict employee attrition. This knowledge aided us in developing our final model.

Mani [5] They used a voting ensemble that included SVC, logistic regression, and multi-layer perceptron in their study. This research helps us understand how a simple voting ensemble is constructed and how using an ensemble technique may greatly improve a model's accuracy.

3 Motivation

Stacking takes a collection of diverse base learners, trains them in parallel, and then combines them by training a meta-model to create a prediction based on the predictions of the base models. The advantage of using stacked ensemble is that an algorithm performs better than a single model in the ensemble.

Our trained model will be able to predict employee attrition based on different attributes which are organization specific. For instance, location-specific attributes like the distance from the employee's home to workplace. Our pre-trained model will pick up these attributes from the dataset and output the result accordingly.

4 Terminologies

4.1 Gradient Boosting

Gradient boosting classifiers are a group of machine learning techniques that integrate a number of weak learning models to create a more strong prediction model. Decision trees are often used while doing gradient boosting.

4.2 AdaBoost Classifier

A meta-estimator that starts by training a classifier to the original dataset, then fits several copies of the same dataset, changing the weights of incorrectly classified cases such that future classifiers focus more on tough circumstances.

4.3 XGB Classifier

"Extreme Gradient Boosting" is what XGBoost stands for. XGBoost is used for supervised learning tasks, such as predicting a target variable yi using training data (with many features) xi.

4.4 Logistic Regression

To forecast the likelihood of a target variable, the supervised learning classification method logistic regression is utilized. There are only two classes since the aim or dependent variable is dualistic in nature. The dependent variable is a binary variable, meaning that data is recorded as 1 (representing success/yes) or 0 (representing failure/no). $P(Y = 1)$ is mathematically predicted by a logistic regression model as a function of X.

4.5 SVC

SVM is a supervised machine learning solution to address classification and regression issues. However, it is primarily used to address classification problems. Each data item is considered a system in n-dimensional space (where n is the number of features), with the feature vector being the value of the SVM algorithm over a certain position. After that, we classify the data by identifying the hyper-plane that clearly separates the two groups.

4.6 Decision Tree

A decision tree builds regression or classification models in the form of a tree structure. It divides a dataset into smaller and smaller divisions over time while also building a decision tree. The ultimate result is a tree with decision nodes and leaf nodes. Each of the two or more branches of a decision node indicates a value for the attribute under evaluation. The best predictor is represented by the root node, which is the topmost decision node in a tree.

5 Data Preprocessing and Analysis

5.1 Data Preprocessing

The dataset is acquired from IBM HR Data Analytics [7]. The dataset has 26 numerical features and 10 categorical features (including the target feature).

In order to conduct research, the data collection includes a large number of characteristics that directly or indirectly impact employee attrition. In fact, we had to first select and filter certain attributes, resulting in a dataset with a reduced number of useful attributes. We divided our data into two categories: training (80) and testing (20).

The feature engineering for this project and attribute selection is as follows:
From the dataset, we have dropped four attributes as they were constant. The attributes being employeeCount, employeeNumber, over18, and StandardHours. Because the suggested machine learning algorithm cannot handle categorical input, we must

Table 1 Attrition distribution before and after ADASYN sampling

Before sampling	
0	863
1	166
After sampling	
0	863
1	822

employ an encoding approach to convert it to numeric form. The techniques used for encoding were:

1. Ordinal encoding: Each unique category value is allocated an integer value in ordinal encoding. An ordinal encoding may be sufficient in some fields.
2. One Hot encoding: For categorical variables with no ordinal relationship, the integer encoding may be insufficient at best or deceptive at worst. Allowing the model to presume a natural ordering across categories rather than forcing an ordinal link through an ordinal encoding may result in poor performance or unexpected findings. The attributes encoded using this technique were Department, EducationField, Gender, JobRole, MaritalStaus, and Overtime.
3. Handling Imbalanced data: As shown in the figure below, the attrition distribution is highly disproportionate. If we train the model using the dataset, the recall score would be very less. So, we tried different sampling methods and found that ADASYN [8] method makes the attrition nearly 50:50 and gives better results (Table 1).

After the encoding processes, the resultant training data had 1685 rows and test data had 441 rows. The number of features was 43 (Figs. 1, 2 and 3).

5.2 *Analysis*

The research study has given the following analysis:

1. Age: Mostly, the employees had age between 30 and 40 years, 35 years being the most frequent age in dataset.
2. Distance from home: Hugely, the distance between the employees and the work place was around 3 on the scale of 1–30.
3. Monthly income: This analysis showed that in the organization, most employees are earning less money and then only few employees earn more money.
4. Mutual Info Gain: Using mutual info [9], we got the top five important features contributing to attrition as shown in fig. below. We can see that whenever the employee is worked over time, the chance of that employee quitting increases (Table 2).

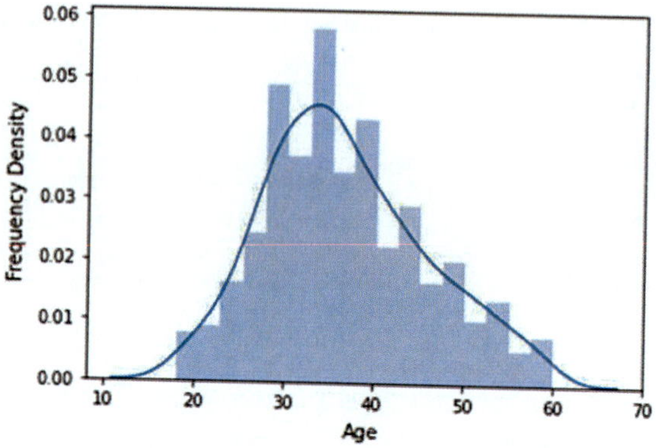

Fig. 1 Age histogram

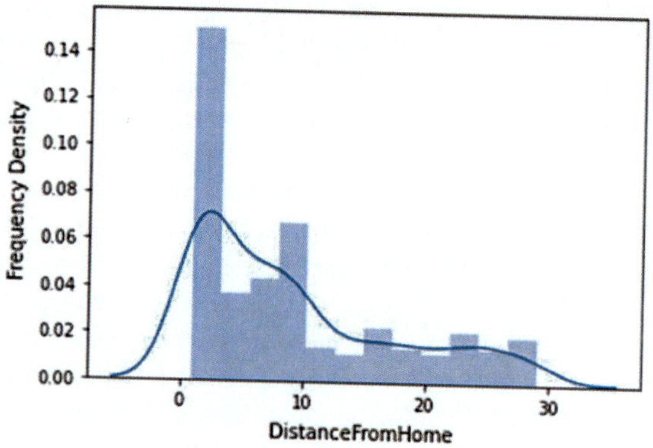

Fig. 2 Distance from home histogram

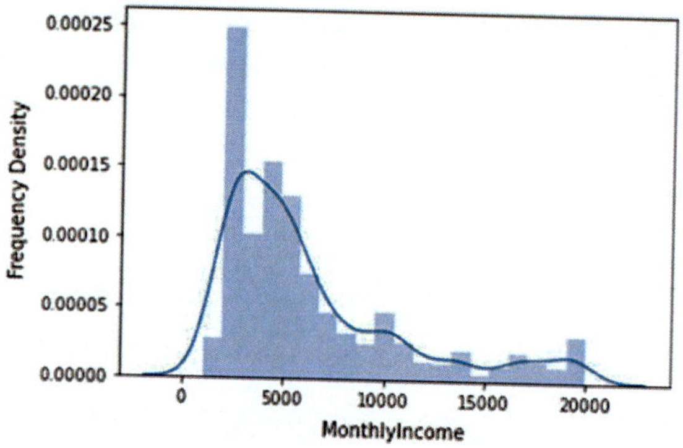

Fig. 3 Monthly income histogram

Table 2 Top five features with highest mutual information

Features	Mutual info gain
Over time	0.05
Job level	0.04
Gender	0.035
Years in current role	0.03
Daily rate	0.023

6 Methodology

Ensemble techniques combine several learning algorithms to provide greater prediction performance than any of the individual learning algorithms could provide. Bagging, boosting, and stacked ensemble methods are the three main types of ensemble techniques.

The method used in this paper is stacked ensemble. Stacking [6] uses a variety of base learners and trains them in parallel before combining them by training a meta-model to forecast based on the predictions of base models chosen. Stacking proposes a two-level approach to solve prediction problems. The "Level-0" or base learners comprise a set of heterogeneous learners and a meta-learner.

As shown in architecture diagram figure, we have used the following base learners: logistic regression, decision tree, support vector machine, Gaussian NB, gradient boosting, and its variants Adaboost and XGBoost. For meta-learner, random forest classifier is used (Fig. 4).

A detailed description how our stacked model works is explained with the help of Fig. 5. After preprocessing and splitting the data into training set and testing set, the former set is again made into K-folds. All the seven base learners used in our model are fitted on K-1 folds. Now, the trained base learners are used to make predictions on the Kth fold. Then, using the attributes given in Kth fold and also the predictions of

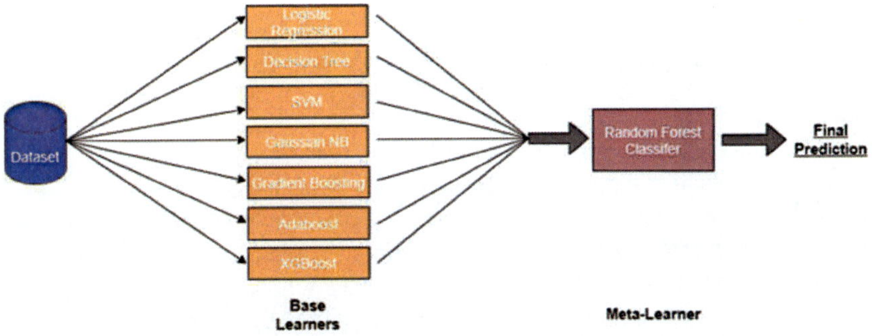

Fig. 4 Stacked ensemble model architecture

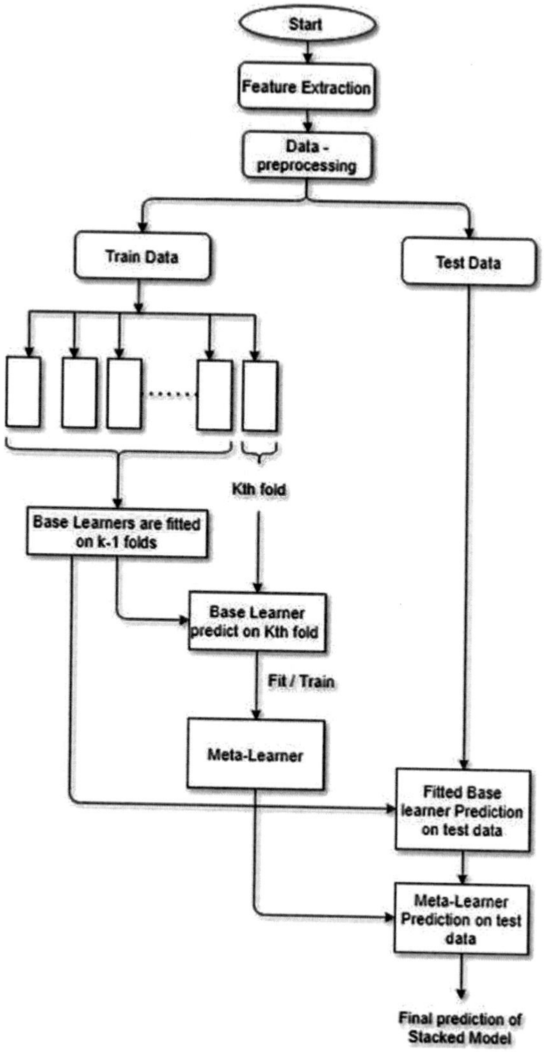

Fig. 5 Stacked ensemble in-depth diagram

base learners, the meta-model is fitted. For this, a "pass-through" hyper-parameter of stacked ensemble approach is used. Now that, both base learners and meta-learner are trained, and we will evaluate metrics of our model. First, the base learners are used to make predictions, and using these predictions, meta-learner predicts the final output. Stacking ensemble technique generalizes both voting ensemble and weighted average ensemble methods, allowing the meta-learner to learn how to optimally aggregate the predictions from contributing members. Learners are classified as experts or less experts depending on their ability to adapt to new conditions, such as out-of-sample data. This is a key contrast from basic averaging and voting, albeit it adds a layer of complexity that makes the approach difficult to execute effectively and minimize data leaking, which leads to inaccurate and optimistic results.

7 Experimental Design and Results

7.1 Experimental Design

First to handle the imbalanced dataset, we have experimented using various sampling techniques to make sure that the newly introduced synthetic data points do not introduce noise in the data and make sure to keep our model focused on both values in target feature. We have used sampling technique and trained the model on preprocessed training data using our stacking model with minimal fine tuning. As shown in Table 3, best results were given by ADASYN and SMOTETOMEK technique. Since ADASYN had better recall score, we decided to use it. The meta-learner chosen should be able to identify which base learner gets better results for certain kind of data. Since there is no procedure for selecting meta-learner, we have experimented with various popular supervised learners, as shown in Table 4, and found that random forest classifier performs the best.

Table 3 Sampling techniques and their accuracies

Sampling technique	Accuracy (%)
No sampling done	86
Borderline SMOTE	86
ADASYN	**88**
SMOTETOMEK	88
SMOTEENN	83

Table 4 Meta-model accuracies

Meta-model	Accuracy (%)
DT	86
SVC	60
LR	72
GB	83
ADA	86
DT	86
RF	88

Table 5 Hyper-parameter tuning

Models	Parameters
LR	C = 50
DT	Criterion = gini,max depth = 11 , min samples = 3, splitter = random
SVC	C = 100
GNB	Var smoothing = 0.01
GB	Learning rate = 1.2, n estimators=100
ADA	Learning rate =1 , n estimators =300
RF(meta-model)	Max depth = 3, max leaf nodes = 8, min samples split = 6

Now that, we have chosen our meta-model, and we have fine-tuned the base learners using GridSearchCV from sklearn library and the meta-model manually. Table 5 presents all the parameters used to build the final model.

7.2 *Results*

Based on the stacked ensemble model built from the Table 5, we have got the following results as shown in Tables 6, 7, and 8. We have got accuracy of 89.12 percent. From confusion matrix, Table 7, we can observe that our model accurately predicted 37 employees who were going to quit out of 71 and 356 employees who did not quit were accurately predicted. Hence, there are fewer false positives than false negatives. Four metrics are chosen chosen, other than accuracy, for evaluation: precision, recall, F-score, and support, and the respective values are shown in Table 8. Even though our F-score is 0.61, our model performed better, both in terms of accuracy and recall, than all the other models proposed by papers discussed in section related work.

Table 6 Model overall accuracy

Accuracy	89.12%

Table 7 Confusion matrix

	True No	True Yes
Predicted No	356	14
Predicted Yes	34	37

Table 8 Stacked ensemble model metrics

	Precision	Recall	F-score	Support
0	0.91	0.96	0.94	370
1	0.73	0.52	0.61	71
Accuracy			0.89	441
Macro-Avg.	0.82	0.74	0.77	441
Weighted Avg.	0.88	0.89	0.88	441

8 Conclusion

Employee attrition is a major concern for all businesses since it has a negative impact on workplace productivity and meeting organizational goals on time. Organizations must use machine learning approaches to forecast staff attrition in order to solve this challenge. Our model can help organizations to detect such employees which tend to leave. The features we considered were quite extensive and included most aspects which affected employee attrition. Better accuracy was obtained by using stacked ensemble machine learning algorithms to solve a problem than the existing model predictions.

Ensemble learning also made it possible to identify the most important factors that contribute to employee attrition. Thus, this project successfully classifies the employees who will leave the organization. Since our ensemble model uses variety of base learners, the meta-learner can easily identify the base learner which is able to learn better and give more weightage to that base model implicitly. The robustness of the model is well understood from evaluation methods.

9 Future Scope

We can develop a suggestion engine to assist companies in retaining individuals who are likely to depart the company in the near future, based on our current model. We can also build custom models depending upon the type data provided by the company.

Acknowledgements We would like to express our thanks to Dr. Pooja Agarwal, Professor, PESIT, for sharing her pearls of knowledge with us during the study process. We also owe a debt of gratitude to Dr. Arti Arya, Professor at PESIT, for her invaluable contributions.

References

1. Divyang J (2017) Evaluation of employee attrition by effective feature selection using hybrid model of ensemble methods. Diss. Dublin, National College of Ireland
2. Qutub A et al (2021) Prediction of employee attrition using machine learning and ensemble methods. Int J Mach Learn Comput 11
3. Jain R, Nayyar A (2018) Predicting employee attrition using xgboost machine learning approach. In: 2018 international conference on system modeling and advancement in research trends (SMART), IEEE
4. Karande S, Shyamala L (2019) Prediction of employee turnover using ensemble learning. In: Ambient communications and computer systems. Springer, Singapore, pp 319–327
5. Mani P. Prediction of employee retention using Cassandra and ensemble learning
6. Stacking Ensemble Machine Learning With Python. https://machinelearningmastery.com/stacking-ensemble-machine-learning-with-python/
7. IBM HR Analytics Dataset. https://www.kaggle.com/pavansubhasht/ibm-hr-analytics-attrition-dataset
8. ADASYN: Oversampling Technique. https://imbalanced-learn.org/dev/references/generated/imblearn.over_sampling.ADASYN.html
9. Mutual Info Score. https://scikit-learn.org/stable/modules/generated/sklearn.metrics.mutual_info_score.html

Transfer Learning Approach for Diagnosing Skin Cancer with Deep Convolutional Neural Network

S. Poonkuzhali, B. U. Anu Barathi, and S. Vinodhkumar

Abstract Skin cancer is the world's deadliest form of cancer. Tumours on the skin are abnormal growths of tissue that can be malignant (cancerous) or benign (harmless). The primary hurdles include categorising skin lesions such as melanoma and moles, which share a high degree of resemblance, complicating the detection and diagnosis of RGB images of skin. The automatic skin categorisation approach proposed in this study is essential for early skin cancer detection. The proposed method employs transfer learning with a pre-trained deep learning network. The Kaggle dataset is used for training and testing, and the transfer learning method RESNET50V2 is used to classify with 98.63% accuracy into benign and malignant. In terms of epochs, the suggested method RESNET50v2 outperforms existing approaches like VGG16, with a classification rate of 97.41%.

Keywords Deep learning · Transfer learning · Melanoma · Malignant melanoma · Skin cancer classification · RESNET50V2

1 Introduction

Skin cancer is one of the most lethal illnesses, with skin melanoma being the 19th most frequent kind of cancer in both men and women [1]. Globally, the prevalence of skin malignancies, including melanoma and non-melanoma, is increasing. According to data, the incidence of skin melanomas in the USA is 22.7 per 100,000 persons. According to the research, 23 new cases of melanomas of the skin were reported in

S. Poonkuzhali (✉) · S. Vinodhkumar
Department of CSE, Rajalakshmi Engineering College, Chennai, Tamil Nadu, India
e-mail: poonkuzhali.s@rajalakshmi.edu.in

S. Vinodhkumar
e-mail: vinodkumar.s@rajalakshmi.edu.in

B. U. Anu Barathi
Department of CSE, Sathyabama Institute of Science and Technology, Chennai, Tamil Nadu, India
e-mail: ambitiousanu@gmail.com

M. S. Kaiser et al. (eds.), *Information and Communication Technology for Competitive Strategies (ICTCS 2021)*, Lecture Notes in Networks and Systems 401,
https://doi.org/10.1007/978-981-19-0098-3_45

2017 for every 100,000 people, and two people died of cancer [2]. Early identification of skin cancer is critical for developing a suitable treatment method for patients.

Mishra [3] demonstrated how to automatically identify and categorise melanoma in skin lesions using machine learning techniques in order to predict the existence of melanoma. Masood et al. [4] emphasise the usefulness of computational acuity approaches for skin cancer in streamlining the data processing process and providing physicians and dermatologists with more accurate and trustworthy diagnosis.

To identify skin images using the prior approaches given by Burroniet al. and Schindewolf et al. [5, 6], the following procedures are required: processes for extensive pre-processing, segmentation and feature extraction. Noel Codella et al. [7] proposed a classification based on hybrid method for melanoma. They used classification techniques that integrated deep learning, SVM and sparse coding to obtain 93.1% classification accuracy. Ozkan et al. [8] developed a skin lesion classification system and classified the lesion as melanoma using four machine learning techniques. Additionally, they deployed artificial neural networks that were 92.5% accurate.

Bi et al. [9] proposed an approach for melanoma identification, achieving a 92% accuracy despite the fact that lesions were classed as melanoma or non-melanoma. Chakravorty et al. [10] developed a classification system for skin lesions with irregular colour and structure distributions based on the Kullback–Leibler divergence of the colour histogram and the structural similarity score. Their categorisation rate was 83%. Waheed et al. [11] classified melanoma using a machine learning model based on distinguishing aspects of skin lesions such as distinct types of colour and texture elements. Their algorithm correctly categorised data in 96%.

Premaladha et al. [12] suggested a melanoma classification system that improves image clarity by utilising a limited adaptive histogram equalisation method and a normalised method for segmenting the filtered grayscale image. The deep learning algorithm has a classification accuracy of 92.89%. Hosny et al. [13] devised a system for classifying skin lesions automatically. The process of fine-tuning is carried out with the assistance of a pre-trained deep learning network and transfer learning. A data augmentation approach is employed to augment the images, followed by AlexNet transfer learning. The technique of transfer learning resulted in a classification accuracy of 98.6%. Hosny et al. [14] suggested an autonomous skin lesion detection system based on transfer learning theory and a pre-trained deep neural network with a classification accuracy of 96.86%.

1.1 Transfer Learning Method

Weiss et al. [15] examine the application of machine learning and data mining approaches to a variety of real-world challenges. Typical machine learning techniques are based on the assumption that the training and testing data come from the same domain, have the same input feature space and have similar data propagation properties. Obtaining training records can be too expensive or impossible in some circumstances. Additionally, high-performance trainers must be developed using data from other fields that are more readily available. This is referred to as transfer learning. According to Witten Ih et al. [16], machine learning has been widely employed in a number of applications where patterns can be produced from prior knowledge (training data) in order to forecast prospective events. Shimodaira [17] defines classical machine learning as the process of training and evaluating data utilising the same input function space and data distribution. If there is a data distribution disparity between the training and evaluation sets, the predictive learner's effects will be lessened. Collecting training data that conforms to the domain of the function and the expected data propagation characteristics of the test data can be challenging and costly in some cases. Additionally, a high-performance learner for the target domain is required, which is taught from a comparable source domain. This acts as a catalyst for transfer learning.

When specific training data is scarce, the demand for transfer of learning arises. Numerous machine learning techniques have been effectively applied to transfer learning, including text sentiment classification [18] and image classification [19–21]. Deep learning has risen to prominence in a number of research fields in recent years. It is critical to understand how to transfer learning successfully utilising deep neural networks, also referred to as deep transfer learning. Deep transfer learning can be classified into four types: instance-based, mapping-based, network-based and adversarial-based deep transfer learning [22]. Navarun Das et al. [23] proposed research on oral biopsy images, including transfer learning with pre-trained deep convolutional neural networks, and pre-trained models were chosen to solve their classification problem. Arora et al. [24] showed that deep learning models on the ImageNet dataset outperformed previous techniques. MobileNetv2 surpasses with other techniques in simulations. As a consequence, the vitality of the utilised algorithm verifies the proposed study work's benefits, as it might be used as an accurate and dependable approach for assessing food quality in real-time applications. Mohammad Rahimzadeh et al. [25] proposed a neural network architecture that combines the Xception and ResNet50V2 networks. They used training techniques to develop a large number of deep convolutional networks for the classification of X-ray images.

This paper employs a transfer learning approach to categorise colour photographs of skin cancer into two types. The suggested approach offers two significant advantages over existing computer-aided categorisation systems for skin cancer. For instance, the proposed approach is applicable to any sort of image. Furthermore, the proposed method eliminates the requirement for any pre-processing. It operates directly on previously acquired skin colour photographs. The remainder of this chapter is organised into four sections. Section 2 discusses the frameworks for the proposed methods. Section 3 describes the experiment's findings, and Sect. 4 concludes.

2 Framework of the Proposed Methodology

The proposed work's design is presented in Fig. 1, which details the entire process's workflow. The skin cancer dataset was created using Kaggle's well-known datasets. The training and test datasets are both easily accessible, with the training set comprising 2637 samples and the test set containing 660 samples and can be acquired from the Kaggle website. In total, 3297 samples were used.

Following data processing, it is automatically pre-processed, and feature extraction is completed. To make the most of the few training samples, we will also augment the images with several random transformations, ensuring that our model never sees the same image twice. This avoids overfitting and encourages the model to generalise further.

The skin cancer dataset is then fed into RESNET 50V2 for analysis. RESNET50V2 is a network architecture family with a variable number of layers for image classification. The use of batch normalisation before each weight layer separates ResNetv2 from ResNetV1. ResNet50v2 with 50 layers has been introduced. We used the weights learned from training on the "Imagenet" image classification in the RESNET model. The RESNET model now has four dense layers. We used a training dataset of 2637 images, with an image size of 224 × 224 and a colour of RGB with two classes. Then, we implemented the testing dataset, which included 660 images divided into two classes: benign and malignant. Following that, we trained and tested the RESNET model with 25 epochs and obtained 99.58% accuracy.

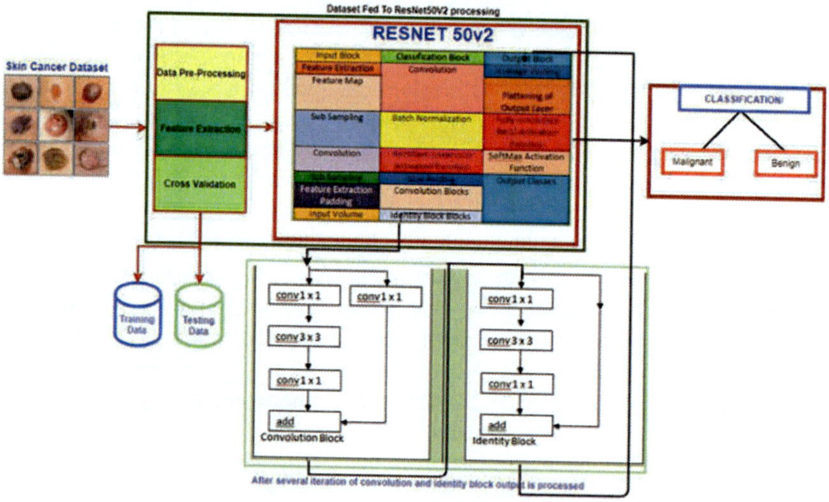

Fig. 1 Proposed system architecture

2.1 Transfer Learning—RESNET50V2

Skin cancer identification requires a significant number of images to be performed correctly and effectively. Unfortunately, no dataset with thousands of tagged photographs is readily accessible to meet this need. Transfer learning theory is critical for resolving this difficult challenge, as it enables the use of a large medical dataset such as ImageNet in the associated ResNet50 architecture.

The primary reason for adopting ResNet50 over AlexNet is because ResNet enabled the training of ultra-deep neural networks with hundreds of thousands of layers while still achieving superior performance. The ResNet architecture is the most frequently used for image recognition because it achieves the highest accuracy. It is composed of 50 convolutional layers. The ResNet50 architecture is unique in that it proposes a residual connection between prior layers and the present one. One of the most significant advantages of the ResNet50 is that it avoids bad outcomes while increasing network depth. Thus, we can raise the depth while maintaining a low training time and a high level of accuracy. The weights of the convolution layer are initiated at a modest rate to avoid a dramatic change in weight, whilst the weights of the fully connected layers are initialised randomly. The more images trained, the more accurate we get. Due to the increased number of photographs obtained during the augmentation process, the achieved accuracy is likewise increased. This transmission of knowledge enables us to determine whether the tumour is malignant or benign. The augmentation step was used to increase the number of photographs and to circumvent the labelled image constraint. This approach is quite beneficial for determining the type of skin cancer and giving more accurate results.

2.2 Augmentation Process

The primary limitation of the skin cancer dataset is the scarcity of labelled images for training and testing. By expanding the amount of photographs available for training purposes, the augmentation process contributes to resolving this issue. We employed the most frequently used augmentation methods for skin cancer datasets in this process of augmentation. Augmentation not only improves the training dataset, but also helps improve the model's accuracy, as we need to test the model on a variety of photographs that may have been cropped, rotated or blurred.

Geometric Augmentation: The input images were modified using a random combination of anticlockwise rotation, clockwise rotation, horizontal flip and vertical flip to generate new dataset samples of the ones given as input under the same label.

Colour Augmentation: This involves input images that are used to generate a series of new datasets by blurring the image and adding noise to the image. Each image can reproduce the augmented image of itself and can be directly stored in the train dataset folder.

The augmentation process consists of rotations, flips, wrap shift, adding noise and blurring the image. The number of images to be generated needs to be specified as per the need. The input images are applied until the total number of augmented images that need to be generated is met. The augmentation methods as functions are called randomly and hence the augmented images of each input dataset. The augmented images are converted into RGB and then stored along with the training dataset itself.

Initially, the training dataset contained 2637 pictures separated into two classifications, benign and malignant. Through the augmentation method, we can receive 1000 additional benign photographs and 1000 additional malignant images, which can be added to the train dataset folder to improve the model's accuracy and efficiency.

3 Results and Discussion

The performance of the suggested model RESNET50v2 is shown in Table 1. Two experiments were conducted using the same data set. The initial test was conducted using only the files from the initial dataset, without any augmentation. In the second experiment, the modified images were employed.

The proposed method's efficiency is compared to that of existing approaches [8–11, 26]. The comparative technique yields the following findings in Table 2, demonstrating that the proposed approach achieves the highest categorisation rate. The acquired categorisation rates are viewed and depicted in Fig. 2.

The study of two convolution neural networks, VGG-16 and ResNet-50v2 used to diagnose skin cancer is seen in Table 3. According to the results, the ResNet-50v2 is the highest and can achieve an accuracy of 97.41%.

Table 1 Performance measure of proposed model

Skin cancer dataset	Average accuracy (%)
Original images	98.25
Augmented images	99.63

Table 2 Comparative study of the proposed model

Classification method	Accuracy (%)
Symmetry type [10]	83
Joint reverse [9]	92
ANN [8]	92.5
SVM [11]	96
DCNN-AlexNet [26]	98.61
DCNN-proposed method RESNET50V2(Augmented Images)	**99.58**

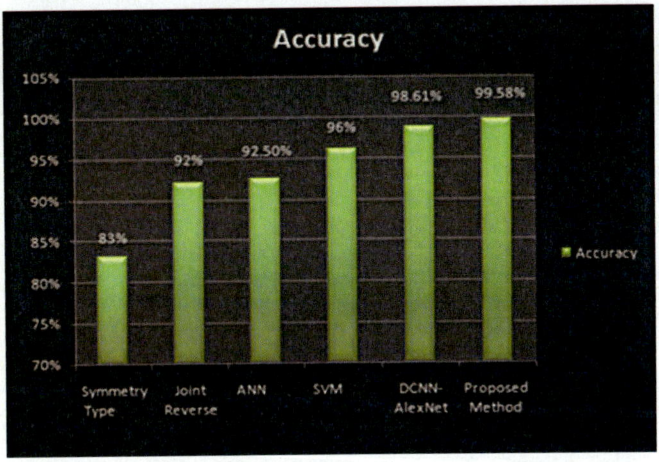

Fig. 2 Classification rates for the proposed and existing processes are depicted graphically

Table 3 Comparison with VGG16 model

Skin cancer dataset	Average accuracy (%)
RESNET50V2—8 Epochs	**97.41**
VGG16—8 Epochs	85.06

4 Conclusion

To create an effective deep neural network, a large number of labelled images are necessary. To solve this critical issue, we are using transfer learning and image augmentation to improve the performance of a pre-trained Resnet50v2 network. Skin cancer classification utilising the RESNET50v2 architecture achieved an accuracy of 98.25% when used with 2637 train dataset photographs and classified the research dataset images as benign or malignant. Following the augmentation, roughly 4637 training photographs are trained in the RESNET50 model, which classifies the testing images with an accuracy of 99.63%. Additionally, it can be demonstrated that when the model is trained using only enhanced photographs, it achieves 100% accuracy, demonstrating that it is one of the most powerful algorithms for skin cancer classification. Additionally, the RESTNET50V2 model was compared to the VGG16 model, and it was determined that the RESNET50V2 model has the maximum accuracy.

References

1. American Cancer Society: Cancer facts and figures (2018). Available: "https://www.wcrf.org/dietandcancer/cancer-trends/skin-cancer-statistics", Accessed: 12 (2020)

2. United States Cancer Statistics (2020) Data Visualiztions. Available online at "https://gis.cdc.gov/Cancer/USCS/DataViz.html", Accessed 12, Aug (2020)
3. Mishra NK, Emre Celebi M (2020) An overview of melanoma detection in dermoscopy images using image processing and machine learning, 2016. Available: https://arxiv.org/abs/1601.07843, Accessed: 12 Aug (2020)
4. Masood A, Al-Jumaily A (2013) Computer aided diagnostic support system for skin cancer: a review of techniques and algorithms. Int J Biomed Imaging 323268
5. Burroni M, Corona R, Dell'Eva G, Sera F, Bono R, Puddu P, Perotti R, Nobile F, Andreassi L, Rubegni P (2004) Melanoma computer-aided diagnosis: reliability and feasibility study. Clin Cancer Res 10:1881–1886
6. Schindewolf T, Stolz W, Albert R, Abmayr W, Harms H (1993) Classification of melanocytic lesions with color and texture analysis using digital image processing. Anal Quant Cytol Histol 15(1):1–11. PMID: 8471104
7. Codella N et al (2018) Deep learning, sparse coding, and SVM for melanoma recognition in dermoscopy images. vol 9352. LNCS, Springer, pp 118–126. Available: https://doi.org/10.1007/978-3-319-24888-2_15, Accessed: 15 Aug (2018)
8. Ozkan IA, Koklu M (2017) Skin lesion classification using machine learning algorithms. Intell Syst Appl Eng 5(4):285–289
9. Bi L, Kim J, Ahn E, Feng D, Fulham M (2016) Automatic melanoma detection via multi-scale lesion-biased representation and joint reverse classification. In: 2016 IEEE 13th international symposium on biomedical imaging (ISBI), pp 1055–1058
10. Chakravorty R, Liang S, Abedini M, Garnavi R (2016) Dermatologistlike feature extraction from skin lesion for improved asymmetry classification in PH2 database. In: 38th international conference of the IEEE engineering in medicine and biology society (EMBC), pp 3855–3858
11. Waheed Z, Waheed A, Zafar M, Riaz F (2017) An efficient machine learning approach for the detection of melanoma using dermoscopic images. In: International conference on communication, computing and digital systems (C-CODE), IEEE, pp 316–319
12. Premaladha J, Ravichandran KS (2016) Novel approaches for diagnosing melanoma skin lesions through supervised and deep learning algorithms. J Med Syst 40.4
13. Hosny KM, Kassem MA, Foaud MM (2018) Skin cancer classification using deep learning and transfer learning. In: 2018 9th cairo international biomedical engineering conference (CIBEC), Cairo, Egypt, pp 90–93. https://doi.org/10.1109/CIBEC.2018.8641762
14. Hosny KM, KassemID MA, Foaud MM (2019) Classification of skin lesions using transfer learning and augmentation with Alex-net. PLoS ONE. https://doi.org/10.1371/journal.pone.0217293
15. Weiss K, Khoshgoftaar TM, Wang D (2016) A survey of transfer learning. J Big Data 3:9. https://doi.org/10.1186/s40537-016-0043-6
16. Witten IH, Frank E (2011) Data mining, practical machine learning tools and techniques, 3rd edn. Morgan Kaufmann Publishers, San Francisco
17. Shimodaira H (2000) Improving predictive inference under covariate shift by weighting the log-likelihood function. J Stat Plan Inf 90(2):227–244
18. Wang C, Mahadevan S (2011) Heterogeneous domain adaptation using manifold alignment. In: Proceedings of the 22nd international joint conference on artificial intelligence, vol 2. pp 541–46
19. Duan L, Xu D, Tsang IW (2012) Learning with augmented features for heterogeneous domain adaptation. IEEE Trans Pattern Anal Mach Intell 36(6):1134–1148
20. Kulis B, Saenko K, Darrell T (2011) What you saw is not what you get: domain adaptation using asymmetric kernel transforms. In: IEEE 2011 conference on computer vision and pattern recognition. pp 1785–92
21. Zhu Y, Chen Y, Lu Z, Pan S, Xue G, Yu Y, Yang Q (2011) Heterogeneous transfer learning for image classification. In: Proceedings of the national conference on artificial intelligence, vol 2. pp 1304–9
22. Tan C, Sun F, Kong T, Zhang W, Yang C, Liu C (2018) A survey on deep transfer learning. In: Kůrková V, Manolopoulos Y, Hammer B, Iliadis L, Maglogiannis I (eds) Artificial neural

networks and machine learning—ICANN 2018. ICANN 2018. lecture notes in computer science, vol 11141. Springer, Cham. https://doi.org/10.1007/978-3-030-01424-7_27
23. Das N, Hussain E, Mahanta LB (2020) Automated classification of cells into multiple classes in epithelial tissue of oral squamous cell carcinoma using transfer learning and convolutional neural network. Elsevier Neural Netw 128:47–60
24. Arora M, Mangipudi P, Dutta MK (2021) Deep learning neural networks for acrylamide identification in potato chips using transfer learning approach. J Ambient Intell Human Comput. https://doi.org/10.1007/s12652-020-02867-2
25. Rahimzadeh M, Attar A (2020) A modified deep convolutional neural network for detecting COVID-19 and pneumonia from chest X-ray images based on the concatenation of Xception and ResNet50V2". Elsevier Inform Med Unlocked 19:100360
26. Hosny KM, Kassem MA, Foaud MM (2018) Skin cancer classification using deep learning and transfer learning. In: 2018 9th cairo international biomedical engineering conference (CIBEC), Cairo, Egypt, pp 90–93. https://doi.org/10.1109/CIBEC.2018.8641762

A Comprehensive Framework for IoT-Based Data Protection in Blockchain System

Dhyanendra Jain, Anupam Kumar Sharma, Prashant Singh, Amit Kumar Pandey, and Prashant Vats

Abstract Thousands of IoT devices (IoT) equipment are often linked together to share information or services in the cloud. Nonetheless, IoT security and protection are important issues because of the growing number of IoT devices from which data is obtained. Because of its distinguishing feature, which includes persistence, assessment capability, and scalability, blockchain technology notably set out to tackle security and safety concerns in IoT frameworks. This article demonstrates two unique commitments: ultralight customizable blockchain technology and scalable time-based consensus. The Ultralight extensible enables scalability by establishing an overarching networking in which greater asset devices collaborate to deal with the blockchain technology. Preparing expenses and procrastination is the time-based general agreement. The group leaders adopt a confident method to decrease the processing overhead for confirming additional squares indefinitely. Blockchain technology is increasingly been used to provide a distributed, safe, trustworthy, and autonomous framework for the energy cooperation.

Keywords IoT · Scalable lightweight blockchain · Data packets · Networks attacks · Packet sniffing

D. Jain (✉) · A. K. Sharma · P. Singh · A. K. Pandey · P. Vats
Dr. Akhilesh Das Gupta Institute of Technology and Management, Guru Gobind Singh Indraprastha University, New Delhi, India
e-mail: dhyanendra.jain@gmail.com

A. K. Sharma
e-mail: anupam.sharma@adgitmdelhi.ac.in

P. Singh
e-mail: prashant.ert@gmail.com

A. K. Pandey
e-mail: amitpandey33@gmail.com

P. Vats
e-mail: prashantvats12345@gmail.com

M. S. Kaiser et al. (eds.), *Information and Communication Technology for Competitive Strategies (ICTCS 2021)*, Lecture Notes in Networks and Systems 401,
https://doi.org/10.1007/978-981-19-0098-3_46

1 Introduction

The Blockchain technology and Internet of things: The Blockchain is a digital database containing information as a mechanism made up of informational data blocks that include the hashes of the preceding block, comparable to an ordered list that links every informational data blocks. This even assures that no changes can be made to the informational data block without disrupting all subsequent blocks. As a result of the widespread consensus, blockchains are commonly assumed to be highly secure. The IoT devices [2] is sometimes referred to as the arrangement of genuine goods termed "things" which have been inserted with coding, sensor systems, and supportive innovations to exchange and interconnect data with various frameworks and devices via the World wide web. Blockchain is a distributed ledger which has received a lot of attention from professionals and researchers across different fields due of its numerous advantages, which include decentralization, anonymity, confidentiality, and persistence [3]. As shown in Fig. 1, blockchain technology is a collection of connected squares. Every square comprises two key elements: swaps and a squared header. A swap forms the basic communication channel in the blockchain, providing multiple hubs to transmit information among each other. The theoretical foundation of transaction is depicted on the provided in Fig. 1. It must be observed that the exchange mechanism can differ slightly across blockchain deployments. The IoT technology, or Internet of things, refers toward the thousands of physical devices connected to the Internet across the world that are currently receiving and exchanging data.

1.1 The Difficulties of Using Blockchain in IoT

The blockchain-based deployments toward the IoT ecosystem are indirect and encounter the following challenges:

1. Flexibility and running costs: Every square gets transmitted to it and authenticated by every node in typical blockchain operations. The associated costs

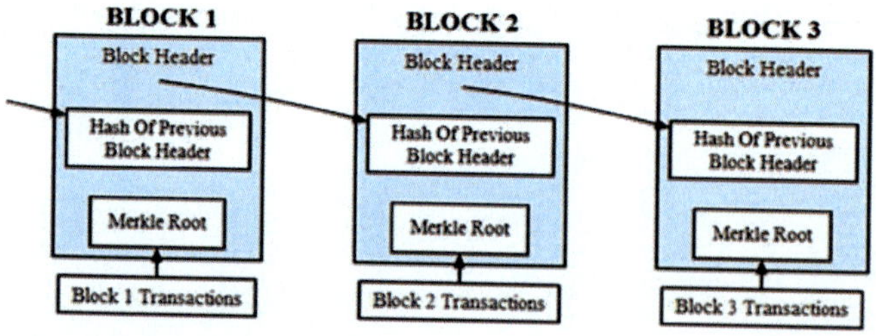

Fig. 1 The Structure of blockchain technology

are resistive because a large percentage of IoT devices has limited information transmission capabilities [4].

2. Complicated accord computations: The majority of the accord estimations used in blockchain consume large wealth of the collaborating nodes, which are far beyond the capabilities within most Remote systems.
3. Indolence: There is a significant delay associated with ensuring that a transfer is confirmed by blockchain hubs.
4. Procedures: It is determined using the rate of exchange, which is recorded every millisecond. This one has a specific value.
5. Blockchain memories imprint: A vast majority of Remote systems will undoubtedly generate a significant portion of transactions, which will overall increase the reminiscence perception at the absorbing equity gateways that maintain the blockchain.
6. Implementation specific resource requirement: IoT systems will almost certainly possess distinct capacity constraints.
7. Lack of protection: Keeping its all of customer's transactions in the blockchain might jeopardize the client's security by linking several transactions made by the same client, which eventually allows DE anonymization of user identification.
8. Compliance with said Data Protection Directive: One of the essential requirements of the Data Protection Directive [5] is the opportunity to be forgotten.

2 Proposed System

2.1 Scalable Lightweight Blockchain

Scalable Portable Blockchain expands blockchain flexibility with grouping the overarching nodes that integrate IoT systems, online cloud, plus Network Operators, and only the Member Nodes are in dealing with said blockchain. Primary administrators of blockchain comprise affirmation and competence of specific exchanges as well as square of exchanges. The exchanges being defined as that of the primary mechanism for exchanging control messages across stations. Blockchain Exchange Transferals are analyzed to generate a square that can then be attached to a blockchain to create the distributed record. Sensor nodes are maintained apart from the interchange thread. Exchanges being transmitted between the overlying stations, and information packets being routed toward its destination. This partitioning enables optimal unicast directing of information packages, resulting in reduced delay and bundling cost for transferring information. The suggested Lightweight Scalable Blockchain technology can solve the shortcomings in present blockchain configurations as suggested in blockchain release promotes dispersion by forming an overarching network wherein strong capital gadgets collaborate to engage with said blockchain. To reduce operating costs, the overlie is divided into distinct groups, and the group heads are in responsible of interacting with the blockchain network. The findings discuss a Time-based Conventional wisdom computation that reduces mining preparatory cost and delay. The group leaders use the distributed trusting technique to rationally reduce the processing expense for dynamic analysis of new squares. The Lightweight Scalable Blockchain incorporates a Consolidated Data transmission Optimization computation that ensures that such blockchain traffic does not deviate significantly from either the total interchange workload in the large corporation. Portable Scaling Blockchain technology will be investigated together in smart home environment as a delegated framework for more comprehensive IoT systems. Qualitative claims demonstrate that our technique is resistant to several potential threats. Comprehensive demonstrations indicate that package cost and latency are reduced, and blockchain scalability is increased as compared to relevant baseline methods. IoT gadgets generate, monitor, and exchange enormous amounts of protection and wellness fundamental data, as well as privacy sensitive data, and thus are intriguing foci of various digital cyberattacks [6]. Consider the following scenario: a knowledgeable interior inspector requests to talk with a Network Operator representative. The blockchain contains a record of this collaboration's exchange. In this manner, the interior regulators straightforwardly transmit temperature measurements to the Network Operator individual, who subdivides the assessments and might even issue instructions to the interior regulators to control the temperature. As shown in Fig. 2, the correspondence in Lightweight Scalable Blockchain is distinct among exchanges. Unlike communicative interactions, data packages are able to transmit data and can be guided in optimal ways across the overarching organization employing directing standards, for instance, what OSPF does [7] in the open shortest path first routing algorithm.

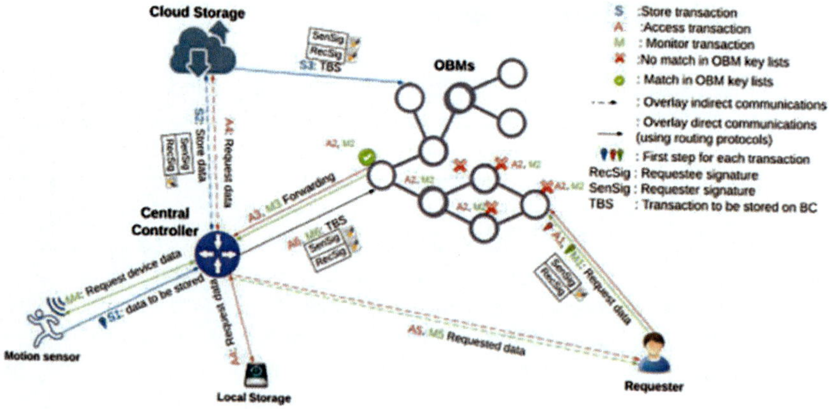

Fig. 2 To show scalable lightweight blockchain OSPF Technology

2.2 Time-Based Convergence Method

Using Lightweight Scalable Blockchain, we present temporal consensus computation over further asset-heavy solutions such as Consensus protocol. To demonstrate utter randomness among frame generating units, every overriding Block Manager must wait patiently for an erratic timeframe, called as a waiting time, before producing new block. To prevent technically called blocks administrators for repeatedly claiming to have had a limited prolonged waiting period, adjacent overlie blockchain administrators check the frequency to which an overlie frame supervisory generates new blocks at the beginning of a prolonged waiting time frame. Because the holding firm duration differs for every overlie frame administrator, an overlie frame administer may obtain additional square produced by another overlie frame supervisors that comprises some or all of the transactions that seem to be currently within the overlie frame author's group of exchange. Upon the instance, this overriding frame administrator must erase such transactions out of its lagoon because they have been stored in the blockchain by some other overriding blocks head. Allowing overlie frame controllers should wait for an erratic timeframe that lowers the amount of duplicate occludes which can be generated all the times. The most severe begin delay in the preceding organization is compensated at twice the most severe begin delayed.

3 Proposed Work

To protect the overlie vs a malicious technically called frame supervisor who may generate a huge number of spaces with fraudulent transactions, which would be known as an invaded and annexed attack, the frequency with which such technically called frame head can generate slabs is limited to quite a limited extend that only one block can really be generated over a timeframe denoted by time equally. Consolidated Data transfer monitoring alters the memorandum of understanding timeframe (Fig. 3).

Every overlie frame supervisory examines the consistency with which most other overlie frame supervisors develop frames. The checksum of such BK which will be used for the next exchanging is saved in the present exchange's yield section. This ensures anonymity since the transaction operator can alter their BK for every

Fig. 3 To show the algorithm used for overlie transaction

Algorithm 1 Transaction verification.

Input: overlie Transaction (A) Output:
True or False Requester verification:
1: if (hash (A.BK) 6= A−1.output) then
2: return False;
3: else
4: if (A.BK redeem A.signature) then5: return true;

Table 1 Depicts an illustration of a trustworthiness tables

Blockchain direct overlying evidence	Count of previously verified blocks	11	21	31	41	51
	Validation is required	82%	62%	43%	32%	23%
Blockchain indirect overlying evidence	% Of OBMs members approved block	21%	45%	62%	88%	99%
	Validation is required	83%	76%	74%	67%	42%

regulated exchange. As a result, the overlying frame heads subsequently validate this just by matching the frequency of a BK from A to output of such previous encounter in a comparable document. The sign of A, is included within the fourth column of A, is then verified using its BK within A. If there really is a multiple stage transaction, the label of those adopting an attractive node is validated using respective BKs. A is validated if the measures complete successfully. To carry out this method, the override frame administrators maintain a trust column, an example of which is shown in Table 1. Immediate evidence triumphs over indirect proof. If the overarching frame supervisors possess definite evidence regarding a square maker, a limited portion of the trades within the grid were selected to just be allowed in accordance with Table 1. Whether no instantaneous evidence is available, the technically called blocks supervisor examines to see if circular evidence is available and then picks an alternative partition of transactions dependent on the amount of many other technically called frame directors who have voiced support for the cubic generators, as shown in Table 1. Table 2 shows the several techniques that enable Scalable lightweight blockchain.

Table 2 Several techniques that enable scalable lightweight blockchain

Specifications	Employed method
Privacy	In all transactions, encryption is employed
Completeness	Each transfer employs a checksum
Availability	Make certain that only cluster members get transactions
Non-repudiation	The event provider signs the transaction
Authentication	Every node saves the original transactions

4 Research Analysis for Privacy and Security of Data Packets

4.1 Blockchain Security Reassurance

The Minimal Expandable Blockchain security reassurance, as well as the migration to quasi dissatisfaction gives the Illustration of Threat. It is common for the attacker to become the Key areas frame administrator or a hub in the Key area's hierarchy. Malicious actors can smell written communications, reject transactions, produce fraudulent transactions and squaring, disembowel various trades in an effort to make personally identifying a center line, and endorse false trades to validate conspiring nodes. The attacker can pretend to be several locations in the same way that overlie blocking administrator by signing contracts or squaring using numerous BKs to swamp the coordinate or consume up valuable capital of something like the having a priority node. Potential enemies may plan to undermine the configuration monarch's integrity by canceling the repetition with that an overlie frame controller creates new patches. Compact Expandable Blockchain technology protects against all this attack by employing appropriate defense techniques. Overlie block administrators will not transmit an exchange to their different understudies until they establish a connection with a component in their crucial regimen, (ii) Every overlie hubs does have a restriction for even the most astounding rate of exchange obtained from the environmental systems. When the threshold is reached, the essential information is updated to prevent stations from transmitting interactions to the destination hub. Furthermore, contaminating devices and key areas ports is difficult due to the usage of Discrepancies blocking administrator secret tables.

4.2 Dropping Attack

In this invasion, the Key areas blocking administrator lowers transactions from or to his batch participants in order to suppress them from the full transition. To defend against such a attack, a grouping component can switch the Key areas blocking supervisor with somebody who is associated if it notices that its transactions are not being processed.

4.3 Blockchain Tuning

The attacker publishes a fake number of squares and claims it to be the largest. As a result, all gateways accept the attacker document as the legitimate file. The suggested time-based consistency computation restricts the number of squares that a Discrepancies block controller may generate over a periodic length. It limits the number of spiteful frames which a Discrepancies block controller may append, preventing the attacker from constructing a lengthier record than that of the actual record.

4.4 Negotiating a Time Extension

In this attack, a vengeful Discrepancies block controller generates over than 1 square in each accord time. This invasion could be controlled by an aggregation of destructive nodes. Different Discrepancies block monitors will validate that because a malicious square follows the consensus criteria and may therefore be stored on the blockchain. To prevent such attack, each Discrepancies block manager saves the period from the last block produced by other Discrepancies block managers in the organization have generated. As a result, each Discrepancies block administrator has complete control over whether or not the pact criteria are followed. If the temporal distance between different squares generated by the same Discrepancies block administrator is not precisely the agreed-upon timeframe, then subsequent squares will indeed be rejected. Along with the trusted database, the timestamp signature is maintained fully updated and renewed.

4.5 Negotiating the Duration of the Hold-Up

In this attack, malicious Discrepancies block administrators conspire to generate transactions with shorter waiting up a really good chance to put forged transactions in the blockchain. Ensure that every Discrepancies block administrator waits for an indeterminate amount of time before producing subsequent squares. The adjacent Discrepancies block administrator screens the unpredictable opportunity to make sure that such an overlie frame administrator does not frequently select a shorter exposure frequency. A cluster of harmful adjacent gateways, on the other hand, may entice and continuously enable a selected node(s) to choose brief faring occurrences and thus generate blocks. The authentic Discrepancies block administrator may detect this attack throughout multiple rounds of time equally while the malicious Discrepancies block administrator mines various patches with short retention duration.

We investigate Sybil attack from the perspective of a blockchain hub and a Discrepancies block management. Figure 2 discusses the Scalable lightweight Blockchain OSPF Technology, which results in reduced bundling cost and beginning latency as compared to the conventional since, over the last, informational packages are exchanged across all organizational behavioral model blockchain (OBMBC) as opposed to the prior, wherein data packages are routed along perfect paths. As you can see, the two methods for the benchmark evolve directly as the number of OBMBC's increases. The amount of transmitting traffic generated is directly proportional to the number of OBMBCs, explaining the corresponding increase in bundling expense. Because the information packets are now transmitted, the delay in receiving information at the client application immediately increases. In contrast, with Scalable lightweight Blockchain OSPF Technology, only the package expense increases as the number of OBMBC's increases. Because the data packages are delivered directly to the client application, the setup latency is unaffected by the number of OBMBC's. These results demonstrate the sufficiency of maintaining the communicate and exchange channels independent of each other. Figure 4 depicts how the existence of a greater number of OBMBC's enhances cybersecurity substantially, albeit the effects are not as obvious. When PTV reaches 100, OBMBCs authenticate all trades in a square, resulting in a 0% attack accomplishment rate. When the PTV decreases, the attacking effectiveness rate increases. The lowest assessment of PTV that really can assure security for the companies automate used in this simulation is 60. This has been repeated for various numbers of OBMBC's to determine the smallest calculation of PTV whereby an attack can usually be recognized.

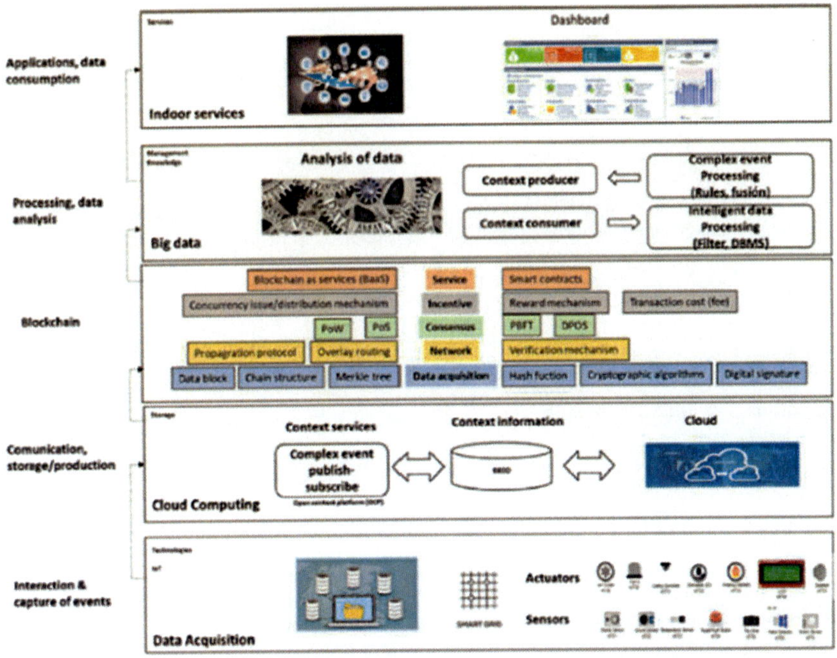

Fig. 4 To show the arrangements of integration of Blockchain technology with IoT security applications greater number of OBMBC's enhances cybersecurity substantially

5 Conclusion

The article argued that while blockchain technology is a strong invention for providing privacy and reliability in IoT, its use in the IoT paradigm offers a number of significant challenges, including: complexity, transmitting data capacity, idleness overhead cost, and scalability. We suggested a Scalable lightweight Blockchain for IoT to overcome these challenges includes an IoT amicable consensus computation that eliminates the need to settle any enigma prior to invading and occupying new sector to the database. Scalable lightweight Blockchain combines a distributed trust mechanism in which the Discrepancies block manager's data packet handling time duration and latency for passing new blocks continually decreases as they acquire confidence for one another. The distributed transfer rate the board technique modifies some conceptual model parameters to ensure that the organization's utilization is within an approved operating range. Lightweight Scalable Blockchain is exceptionally safe against a wide range of attacks, according to security testing.

References

1. "BitShares 4.0 I OSS Solutions for finance and business management based on blockchain technology," June (2020)
2. Aich S, Chakraborty S (2019) A review on benefits of IoT integrated blockchain based supply chain management implementations across different sectors with case study. In: 2019 21st international conference on advanced communication technology (ICACT), PyeongChang, Korea (South), pp 138–141
3. Abramaowicz M et al (2016) Cryptocurrency-based law. Ariz L Rev 58:359
4. Mikhaylov K et al (2016) Analysis of capacity and scalability of the lora low power wide area network technology. In: European wireless 2016, pp 1–6
5. EU-http://europa.eu/rapid/press-release-MEMO-17-1441_en.htm, 2018
6. Sicari S et al (2015) Security, privacy and trust in IoT: The road ahead. Computer_ Netw 76:146–164
7. Open Shortest Path First (R) https://www.ietf.org/rfc/rfc2328.txt. [Online; accessed 19-November-2016]
8. Celic L, Magjarevic R (2020) Seamless connectivity architecture and methods for IoT and wearable devices. Automatika 61(1):21–34
9. Xiao S et al (2017) Self-evolving trading strategy integrating internet of things and big data. IEEE Internet Things J 5(4):2518–2525
10. Amoozadeh M et al (2015) Security vulnerabilities of connected vehicle streams and their impact on cooperative driving. IEEE Commun Mag 53(6):126–132
11. Ho G, Leung D et al (2016) Smart locks: lessons for securing commodity internet of things devices. In: Proceedings of the 11th ACM on Asia conference on computer and communications security, ACM, pp 461–472
12. Kosba A et al (2016) Hawk: the blockchain model of cryptography and privacy-preserving smart contracts. In: Security and privacy, 2016 IEEE symposium on, IEEE, pp 839–858
13. Huh S et al (2017) Managing Iot devices using blockchain platform. In: Advanced communication technology, 2017 19th international conference on, IEEE, pp 464–467
14. Vivekanadam B (2020) Analysis of recent trend and applications in block chain technology. J ISMAC 2(04)

15. Aitzhan et al (2018) Security and privacy in decentralized energy trading through multi-signatures, blockchain. IEEE Trans Dependable Secure Comput 15(5):840–852
16. Sivaganesan D (2019) Block chain enabled internet of things. J Inf Technol 1(01):1–8
17. Bhalaji N (2019) QoS and defense enhancement using blockchain for fly wireless networks. J Trends Comput Sci Smart Technol (TCSST) 1(01):1–13

Nurilo is a Social Platform for the Sale and Purchase Of Framework Products Using Blockchain And Cloud Computing

Atul Kumar, Aakrut Dabhi, Mohammed Kharodawala, and Karan Khunti

Abstract A delivery chain is a gadget of companies, humans, activities, data, and assets concerned in shifting a service or product from provider to client. The management of the statistics system in most sectors drastically in healthcare, economics, meals, and education is centralized. There's small information of the item beginnings, handling or shipping travel. The challenge gets to be not as quantitative but too subjective. The biggest challenge of the supply chain remains within the traceability and information administration framework. Blockchain technology is considered to be an important development after the invention of the Internet. Blockchain technology records transactions on databases of which are encrypted and distributed out of many computer networks as digital ledger out of lines of online transactions. This technology can be used in e-commerce industries to develop platforms to buy online and sell decentralized. E-commerce industry current faces many challenges face many challenges such as security measures of e-commerce system, transparency and trust, efficient, and conventional supply chain exercises include a few middle people, believe, and execution problems. The feasibility of block chain could be utilized to disturb delivery chain characteristics for way good execution, dispersed administration, and handle computerization. Blockchain innovation can make strides in the straight forwardness of money related exchanges, upgrade the adaptability of framework operation, and realize handling robotization.

Keywords Blockchain technology · Supply chain risk management · Blockchain architecture · Supply chain · Resilience · Supply chain · Logistics · Integration applications · Literature review

A. Kumar (✉) · A. Dabhi · M. Kharodawala · K. Khunti
Parul University, Gujarat, India
e-mail: atul.kumar8844@paruluniversity.ac.in

M. S. Kaiser et al. (eds.), *Information and Communication Technology for Competitive Strategies (ICTCS 2021)*, Lecture Notes in Networks and Systems 401,
https://doi.org/10.1007/978-981-19-0098-3_47

1 Introduction

Blockchain is a creative, decentralized, and divisional "nice in elegance" innovation, which keeps up with category, honesty, also, accessibility of the relative multitude of exchanges and information [1]. That's a common, extended and dispersed record which will be useful to preserve information and exchanges sponsored by a cryptographic wallet over a distributed organization Item cancelation follows a multi-arranged cycle. Significant stages include:

(1) Recognition of the cancelation up-and-comer [1]
(2) Analysis and renewal of the recognized up-and-comers dependent on the cancelation choice trigger, like quality improvement, value decline, or client fragment shift [1]
(3) Evaluation and choice arrangement on the cancelation up-and-comer where the generally speaking authoritative goal and methodology advancement are basic rules [1].
(4) Implementation of the item cancelation choice.

In each phase of distinguishing, choosing, and carrying out item cancelation choices, dynamic data is expected to work with and support supervisors. Key cycles should be overseen in order to guarantee sound item erasure dynamic. Inability to guarantee data openness, unwavering quality, obviousness, discernibility in those cycles can undermine item cancelation hierarchical outcomes, or give problematic choices. Here innovation helps association. Blockchain innovation is one of the arising innovation's that can be troublesome to how associations work together. Blockchain can add to inventory network data the board; data that can be utilized for item erasure choices [1].

Research limitation

This effort targets joining entire composing which disseminates artefact, artefact in preview, overview documents and brief. Consider relating it with supplicates, combination and execution of distributed ledger advancement in delivery chains and coordinations. This would catch noteworthy information from bounty composing and utilizing an exact composing looking through the framework. It uncovers the most re- cent things of distributed ledger in delivery chain errands and clears the path for investigation in future. The standard for work is kept on delivery chains. The examination patterns related to this place is quicking from MMXVII and it will continue at a tremendous speed.

2 Overview of Blockchain

2.1 Features of Blockchain [2]

Cryptography—Blockchain exchanges are confirmed and dependable since of complex computations and cryptographic verification between the parties.

Immutability—The data which is stored in the blockchain cannot be modified because they are secured by hash signature and if someone tries to or modified it then that block will be dropped from the network.

Provenance—It is possible in the blockchain to find out about the from where the product came from means we are able to track the producer of that particular product.

Decentralization—The blockchain runs on the peer to peer network which is behaviorally decentralized and we are able to access the entire network at once and also due to its decentralized nature it consumes more power. Consensus algorithms like proof of work are responsible for management of these entire networks.

Anonymity—Every user on the blockchain network has a unique id which is generated by a hashing function and it provides uniqueness to each and every user of the network.

Transparency—The blockchain network is completely transparent to everyone because each block in the network is connected to the other block through the hash and no one is able to modify the information stored in the block.

3 Blockchain Technology For Supply Chain Function

Nowadays the customers are frequently questioning about the quality of product, the origin of product and also for some other details and in majority of cases even the large companies are unable to give them a satisfactory answer [3]. When this kind of case goes into judiciary then we see a conflict between governments and corporations and now the blockchain has a solution for all these problems/questions, it provides transparency into the full delivery chain and provides security of consumer's information. All the data of product are stored in block and then that block will be added into the blockchain network and broadcasted to every connected node in the network.

3.1 *Benefits of Using Blockchain in Supply Chains [3]*

Sourcing:

Acquirement is an upstream supply chain management work. Provider choice, follow-up, communication and organization advancement are a few of the most obtainment exercises. Providers can meet the materials, assets, innovation and information needs of a specific item. In the event that these items are expelled, the relationship can be debilitated. Supplier entry is reduced resulting in the removal or change of supplier. Removing a product reduces your dependence on a particular supplier. Once you have decided to reject a product, it is important to re-evaluate and re-select important suppliers. Product recall can reduce the overall consumption of resources in the supply chain. A thinner product portfolio reduces overall material supply, energy consumption and environmental impact, as discarded products are no longer produced, distributed, consumed or discarded.

3.1.1 Manufacturing [3–4]

The complexity of the product portfolio is directly related to production capacity. Discarding the product provides a new perspective for solving capacity problems. Product removal reduces portfolio complexity and frees up production capacity. Gned items free up assets and emptying materials diminishing fabricating complexity. Increasing production capacity frees up resources and improves the quality and delivery efficiency of the rest of the product. Planned waste and waste collection can reduce waste in the production process.

3.1.2 Distribution [3–4]

Dispersion and coordination exercises influence item prohibition. These exercises incorporate characterizing and overseeing conveyance channels, warehousing, bundling accomplices, affiliates, third-party coordination suppliers, and stock sorting and collection faculty. The more complex your item portfolio, the more troublesome it is to select coordinations for your organization. Eliminating products by minimizing and streamlining product portfolios results in reduced logistics partners, deliveries, warehousing and packaging requirements. Due to the compact distribution network and reduced logistics diversity, the performance of the entire supply chain in the distribution.

3.1.3 Basic Inner Workings of Blockchain Technology

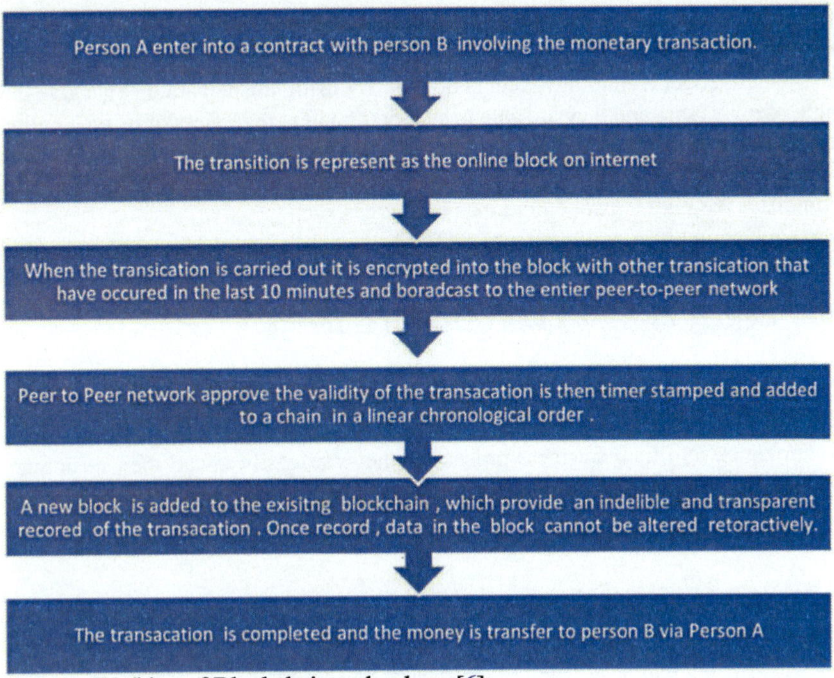

Inner Woíking of Blockchain technology [6]

4 How Blockchain Will Revolutionize E-commerce [6]

How Blockchain Will Revolutionize E-Commerce

Smart Contracts

It is a special type of programs or codes that are able to computerize some errands formulated on present rules. It can develop ventures for e-commerce by reducing the costs required to contract employees to perform some assignments that some kind of programs can computerize. This can moreover encourage stock administration. This implies that online retailers can oversee the control of stock things.

Easy to Access Receipts and Warranties

When purchasing something online the major problem people can face is losing the receipts papers. Thanks to the distributed ledger technology this problem is solved now because it can store all transaction information into one of its blocks and if someone tries to alter the information then the hash value of that particular block will change and that block will be dropped from the chain.

5 Paying Content Creators [7]

They play a major role in the e-commerce industry but nowadays one can ignore their work and most of the time they cannot get the money they deserve but by using distributed ledger technology we can do that. By using this technology we give the right of ownership and they are able to sell their content regardless of its type online and are able to earn a decent amount of money. The owner of the business will use digital wallets to pay the creators. This will reduce the time for payments and also eliminates the middleman, other kinds of charges, etc.

5.1 Loyalty Programs and Personal Offers

Customers can also benefit from personalized deals and discounts offered by retailers thanks to blockchain technology. When online merchants incorporate blockchain technologies into their business processes, they can easily award their customers 4444 redeemable reward points when they reach certain spending thresholds.

5.2 Supply Chain Monitoring [8]

With distributed ledger store operators always able to monitor which items are available and which are not and this way it offers assistance to the operators of stores. When these operators use this technology to screen the supply chain, they can anticipate merchants from providing certain items and cultivate straightforwardness all through the endeavor.

5.3 Generating Legitimate Reviews [8]

These days, we are able as it were expected that the surveys we discover online are veritable. But positive reviews might fair be produced by dealers looking to extend their turnover, while negative surveys may effortlessly be composed by competitors attempting to lean out the playing field. Luckily, blockchain innovation can offer assistance to confirm audits.

5.4 *Making Payments Without Intermediaries [8]*

The platform which uses the distributed ledger with cloud storage can take the users information and divide it into small fragments and after that it stores those fragments into blocks of distributed network and also one copy of that information stored into cloud database. They also added some kind of security to this information and also they signed these documents with digital signatures in order to provide this information uniqueness and also to provide users right of ownership. When users query this information then this platform is able to give them this information very quickly because they store two copies of this data and they directly query from storage instead of a distributed ledger network.

It does not store data, but only stores the encrypted hash value of the relevant data block, thereby providing a footprint of the data to verify its originality. We are able to store our data in blockchain but there is one issue that we cannot query our data that is stored on blockchain so if we combine cloud storage with blockchain then we will be able add querying capability to blockchain. We can store our product details or transaction details on blockchain block and after that we can encrypt that data and store it into mongo dB atlas cloud.

6 Conclusion

6.1 *An Analysis of the Corpus Case [6–7]*

Blockchain was created and released in 2008. Since then, in the research the more emphasis has been put on digital currencies such as Bitcoin, as well as, with security and protection concerns. Events of these advancements and changes in development shifted from their single currency application to several regions not long ago. During the period 2008–2015, because of the special nature of the blockchain the associated distributions were dispersed in the categories of special meetings, board reports, news- letters and were then commented again. The distributed ledger and the uses of it in different formats have pulled in consideration of the creators, analysts and specialists. Ex:- the Extraordinary Topic Meeting 2018 Global Meeting on distributed ledger " focuses on conversations about the innovative use of the blockchain, as well as skills in various fields, including identity contracts and supply chain. I guessed it.

6.2 *Analysis of the Association Schedule of Subjects*

Blockchain-related research can still be broken down into years to look at the advancement of the research topic. In this sense, an evolutionary time-line can be used to further delineate the relevance of the probe's target. It was 2016, but most

Topics	Architecture Building & System Design	Case Study	Narrative Analysis Inductive Reasoning	Algorithms & Protocols	Survey & Interview	Theory / Model Building	Literature Review	SEM	Simulation Experiment	Others
General Influence & Overview	68	47,63,73,97	1,100,119,135			117,119,131,132	19,20,98,107			26,112,132
DLT & SC	36,82,89,91,102,116	30	95	111			93			
Buisness Process Management And Quality Management	16,21,22,37,50,74		40	32		32	3			
Ownership Management	67	114		27						
Information Sharing	16,37,74,113,114,116	55		62,111					126	
Security And Privacy	91,114,116	73	90,92	62		4,117	107	78		
Buisness Model and Operation	105	80.128	1		10,18, 80,133	77,128	80,134	77,78, 79,143		
Acceptance And Adoption		115	129				20			
Supply Chain Integration		122								
Manufacturing		64,66								
Physical Distribution & Logistic	74,82,104,105,116,123	48,65,73,128,130	110		52,81	128	72,101		93	
Transaction Processing	96,123			33						
Financing , Account , Payment	34,37	130	66	33						
Governance	96		95							
Credit Evaluation	94									
Blockchain with Other Technology	68,69,70,84,85,104,127	55,97,122	60,92	71						
Traceability & Transparency	16,49,50,51,52,57,82, 84,85,87,89,94,96,103, 116,123,124,125,127	30,47,52, 58,124	110	67	18	54,132	59,72,106		57,70,126	88,132
Agriculuture and Food Application	50,51,69,84,85,86,87,94		109,110				59,106			88
Other Industry Applications	49,56,57,67,96,108	55,63,64,80	60,100,120	61,62,108					108	
Tactical Planning / Risk Analysis	76	118	75,120							

Fig. 1 Relationship [7]

blockchain conversations revolve around Bitcoin or other advanced forms of money. 'Here. As a result, the special application of SCM's blockchain is largely unexplored. The importance of the long-term points, and the fuzzy area shows new points. In the main period (2016), articles were disseminated in reduced numbers and the main topics studied were format, recognizably and simplicity of the production network. Several uses were introduced as layouts not found in local entries in the next step.

6.3 The Relationship Between of Main Research Methodologies and the Main Topics

See Fig. 1.

References

1. Kuo FF, Li CT, Shan MK, Lee SY (2012) Intelligent menu planning: recommending set of recipes by ingre-dients. In: Proceedings of the ACM multimedia 2012 workshop on multimedia for cooking and eating activities, November. pp 1–6
2. Shino N, Yamanishi R, Fukumoto J (2016) Recommen-dation system for alternative-ingredients based on co-occurrence re- lation on recipe database and the ingredient category. In: 2016 5th iiai international congress on advanced applied informatics (IIAI-AAI), July. IEEE, pp 173–178

3. Ueta T, Iwakami M, Ito T (2011) A recipe recommendation system based on automatic nutrition information extraction. In: International conference on knowledge science, engineering and management, December. Springer, Berlin, Heidelberg, pp 79–90
4. Saravanakumar K, Deepa K (2016) On privacy and security in social media–a comprehensive study. Proc Comput Sci 78:114–119
5. Lewis T (2018) Digital food: from paddock to platform. Commun Res Practice 4(3):212–228
6. Ghahtarani A, Sheikhmohammady M, Rostami M (2020) The impact of social capi- tal and social interaction on customers' purchase intention, considering knowledge sharing in social commerce context. J Innov Knowl 5(3):191–199
7. Sascha F (2012) Social Networks and the promotion of regional food culture
8. Zhou L, Zhang P, Zimmermann HD (2013) Social commerce research: an integrated view. Electron Commerce Res Appl 12(2):61–68
9. Turban E, Bolloju N, Liang TP (2010) Social commerce: an e-commerce perspective. In: Proceedings of the 12th inter national conference on electronic commerce: roadmap for the future of electronic business, August. pp 33–42
10. Chang M, Guillain LV, Jung H, Hare VM, Kim J, Agrawala M (2018) Recipescape: an interactive tool for analyzing cooking instructions at scale. In: Proceedings of the 2018 CHI conference on human factors in computing systems. pp 1–12
11. Tian F "A supply chain traceability system for food safety based on

Error Detection and Error Concealment of Medical Images Using Frequency Selective Extrapolation (FSE) Algorithm

Komal Nair, Neha Motagi, Rupali Narayankar, and P. K. Rajani

Abstract The fundamental problem of any transmission procedure is the introduction of unwanted discrepancies at the output obtained. These discrepancies cause a hindrance to the further processing of vital data present in the transmitted image. As a result, some mechanisms through which error can be identified and rectified is needed. Error is prone to occur during transmission and error detection becomes extremely important. The method for error detection through correlation function and spatial error concealment of lost image information through frequency selective extrapolation (FSE) algorithm in erroneous medical image transmission on a network is illustrated in this paper. X-ray image is used here as a medical image. This can be used for testing whether COVID, normal, or pneumonia. In this COVID scenario, the traveling is restricted and even in rural areas they cannot travel frequently. So, diagnosis can be done through the transmission of medical image taking into account the availability of a suitable transmission medium for the same.

Keywords Frequency selective extrapolation · Error concealment · COVID · Pneumonia

1 Introduction

During the transmission of medical images from a remote location, the image is bound to have errors in it due to the unreliability of the transmission medium. This leads to loss of information in the images. Such a loss of information can prove fatal for many data analytic purposes. In the field of medical research, much data analytic work is being performed on radiology samples for the extraction of vital information about many medical conditions. Thus, the need is felt for mechanisms that help counter

K. Nair (✉) · N. Motagi · R. Narayankar · P. K. Rajani
Pimpri Chinchwad College of Engineering, Pune, India
e-mail: komalnair@gmail.com

P. K. Rajani
e-mail: rajani.pk@pccoepune.org

M. S. Kaiser et al. (eds.), *Information and Communication Technology for Competitive Strategies (ICTCS 2021)*, Lecture Notes in Networks and Systems 401,
https://doi.org/10.1007/978-981-19-0098-3_48

the problem of lossy images and as a result loss of information. For the purpose of error detection, correlation function is used. The resemblance between the two photos or frames is referred to as correlation. The correlation function determines the statistical correlation between random variables separated by a temporal gap in space. The correlation values of the error image and the error-free image are different. As a result, correlation finds the incorrect image. Correlation is used to calculate the 2D correlation coefficient between A and B matrices or vectors of the same size image m X n. Error concealment (EC) is the process of concealing, congealing, or hiding faults at the output or receiver end. The digital image is first examined for any possible faults that may have happened during the X-ray image's transmission. This procedure proves extremely useful when any loss of data is intolerable or unacceptable for carrying out further processing of images to obtain valuable information from the same. In error concealment (EC) techniques like frequency selective extrapolation, the picture information of the suitably received adjacent blocks that are correctively received is gradually and collectively approximated by a weighted linear and simple set of basis functions (FSE). Extrapolation is utilized to locate the misplaced or mislaid block in FSE error concealment procedure. One of the greatest spatial error concealment techniques is the frequency selective extrapolation algorithm (FSEA). This is most beneficial for photos with a countable amount of errors or a small number of errors. Only blocks, also known as full mistakes, can be hidden with FSEA.

The error concealment algorithm is based on the spatial error concealing principle, which may be summarized in three steps.

1. Discrete linear approximation (DLA)
2. The rule of successive approximation
3. Choosing appropriate base functions

2 Error Detection

Error detection refers to processes or procedures that help us identify the unwanted discrepancies and vulnerability of any transmission medium. This step is extremely crucial when the loss of data is intolerable and the corrupted image needs immediate fixing to recover the data. In many of the latest medical researches being carried out, especially the ones using AI and machine learning, a lot of processes include processing or training the algorithm over these images. If the images are being transmitted from a remote location, it is obvious that certain noise is bound to get introduced. If the algorithm is trained over the corrupted medical images then ambiguities in the output are obtained. Thus, a mechanism for error concealment has to be followed with the first and most important step being error detection. The algorithm that is introduced in this paper uses correlation function for the detection purpose. The normalized correlation function is

$$\text{Correlation}, r = \frac{\sum_m \sum_n (A_{mn} - \underline{A})(B_{mn} - \underline{B})}{\sqrt{\sum_m \sum_n (A_{mn} - \underline{A})^2 \sum_m \sum_n (B_{mn} - \underline{B})^2}} \quad (1)$$

where

r = 2D correlation coefficient between A and matrices or vectors of the same size image or frame $m\ X\ n$.

$\underline{A}$ = is the mean of image matrix A and B is the mean of image matrix B.

The correlation with threshold is discovered via trial and error with various images and sorts of faults.

3 Error Concealment

Once the errors are detected in the medical images in the form of missing sections or blocks, an attempt is made to effectively conceal these blocks of identified errors. Frequency selective extrapolation (FSE) is an algorithm that effectively works to recover these missing patches of information. The algorithm of FSE is as follows:

(1) Discrete linear approximation: The medical image comprises of both correctly received blocks and missing blocks. The correctly received block is approximated, and this extrapolated piece is then substituted. This principle is applied because the neighboring image pieces do not differ drastically.

(2) The rule of approximating successively:

$$\begin{aligned} g^{(v)}[m,n] = \frac{1}{2MN} \sum_{(k,l)\in\mathcal{K}_v} \Big(& c^{(v)}_{(k,l)} \varphi_{(k,l)}[m,n] \\ & + c^{(v)}_{M-k,N-l} \varphi M - k, N - l[m,n] \Big) \end{aligned} \quad (2)$$

(3) Selection of apposite basis function:

$$\begin{aligned} \Delta E^{(v)}_{\mathcal{A}} = & \frac{1}{2(MN)^2} \Big(|\Delta c|^2 \sum_{m,n\in\mathcal{A}} \varphi_{u,v}[m,n] \varphi^*_{u,v}[m,n] \\ & + \mathrm{Re}\Big\{ \Delta c^2 \sum_{m,n\in\mathcal{A}} \left(\varphi_{u,v}[m,n]\right)^2 \Big\} \Big) \end{aligned} \quad (3)$$

It works by creating an approximation of the surrounding blocks of the missing patches that are correctly received and then substitutes those patches with the resulting block. In order to study about the extent to which these substitutions are happening successfully, certain parameters about the image quality needs to be studied and analyzed. Some parameters that we will be considering in our analysis mentioned in this paper are mentioned below.

1. Mean squared error (MSE):

   It measures the difference between original and corrugated pixels. MSE is given by

$$MSE(x, y) = \frac{1}{N}\sum_{i=1}^{N}(x_i - y_i)^2 \tag{4}$$

   where x_i and y_i should be of same class and N = total number of pixels.
2. Peak signal-to-noise ratio (PSNR):

   The peak signal-to-noise ratio (PSNR) between two images is evaluated in decibels. The PSNR increases as the quality of the compressed or reconstructed image improves. The two error metrics used to measure the quality of picture or video reconstruction are mean square error (MSE) and peak signal-to-noise ratio (PSNR). It is a logarithmic representation of the MSE's inverse. PSNR is measured in decibels (dB) by

$$\text{PSNR}(dB) = 10\log_{10}\left(\frac{(2^n - 1)^2}{\text{MSE}}\right) \tag{5}$$

   where *n* is number of color /gray levels in image.
3. Structural similarity index (SSIM):

As a measure of perceived visual discrepancy, SSIM employs structural distortion. However, the majority of the other methods are based on sensitivity of error. The structural similarity index (SSIM) quality assessment index is calculated using the three computation parameters of luminance, contrast, and structural. The multiplication or product of the three terms are used for overall index calculation. It is given by

$$SSIM(x, y) = [l(x, y)]^{\alpha}.[c(x, y)]^{\beta}.[s(x, y)]^{\gamma} \tag{6}$$

where

$$\left.\begin{array}{l} \text{luminance term, } l(x, y) = \frac{2\mu_x\mu_y + C_1}{\mu_x^2 + \mu_y^2 + C_1} \\ \text{contrast term, } c(x, y) = \frac{2\sigma_x\sigma_y + C_2}{\sigma_x^2 + \sigma_y^2 + C_2} \\ \text{structural term, } s(x, y) = \frac{\sigma_{xy} + C_3}{\sigma_x\sigma_y + C_3} \end{array}\right\} \tag{7}$$

where μ_x, μ_y, σ_x, σ_y, and σ_{xy} are the local means, standard deviations, and cross-covariance for images *x*, *y*. If $\alpha = \beta = \gamma = 1$ (the default for exponents) and $C_3 = \frac{C_2}{2}$ (default selection of C3).

4 Experimental Results

The error detection and concealment for normal, COVID, and pneumonia patient's X-ray are described first in the experiment results. The algorithms were implemented on MATLAB, and the results were collated.

4.1 Error Detection and Concealment

Error Detection results:

The transmitted medical images have a lot of unwanted errors in them. These errors need to be detected and fixed for the correct processing of the data. As evident from the experimental results shown below, the original image is received as shown. This image is then further converted to a gray scale image, and then the missing blocks or patches in the image is highlighted using the correlation function. This process of identifying the missing patches here, namely noised image, and then sending this image for error concealment, constitutes the entire phase of error detection (Fig. 1).

Results of error concealment phase:

Error concealment phase plays a vital role in the concealment of error blocks. There are many algorithms prevalent that accomplish the task of concealing errors. In this paper, one such algorithm called frequency selective algorithm (FSE) is used. FSE has proved extremely effective in concealing block errors by correctly approximating surrounding blocks and substituting the missing blocks with the result. This makes the corrupted image whole and complete which enables efficient processing of the recovered image.

The results of error concealment on few medical images have been shown below. A complete image quality analysis has been performed to show how the various

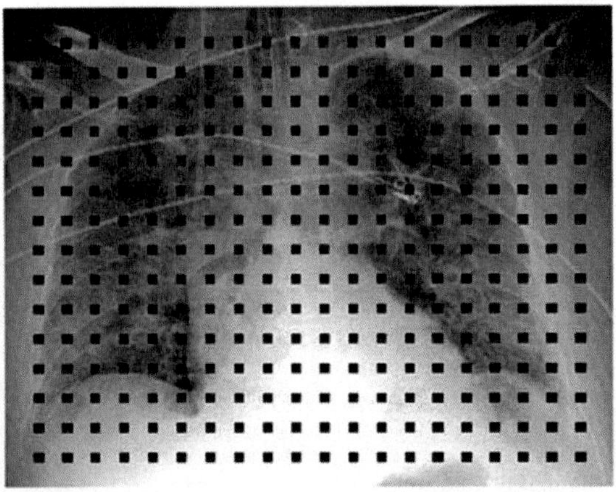

Fig. 1 Detected error blocks in COVID X-ray image

parameters change in the recovered images. A significant improvement has been observed in the image parameters after the application of FSE. During the pandemic of COVID-19 many such medical images were put to analysis for various reasons. In this paper, the medical images have been experimented with to help remove the errors in the transmitted images which will then give valuable outputs after the application of machine learning algorithms on it. Thus, this method will basically prove to be an effective aid to parties involved in data analysis activities for various research purposes.

A definite conclusion can only be drawn if the dataset is large that is to say, the sample dataset is large enough. For this reason, many radiology images have been put to test with FSE and the conclusion is that FSE definitely works to improve the quality of the erroneous image. This algorithm works patch by patch to approximate and substitute the missing patches. In images having huge missing blocks that is classified as nothing but error in the image, FSE works effectively as it is a substitution algorithm.

As will be evident in the analysis that follows, many erroneous medical images of normal, pneumonic, and COVID patients were analyzed for image quality parameters. Only a few have been desiccated and disparately arranged in a tabular fashion for quick analysis of the parameters. Individual sets have been created for these three conditions, namely normal patients, pneumonic, and COVID patients, and image quality analysis has been performed.

The conclusions show the effectiveness of FSE algorithm (Fig. 2).

Such ten radiology images have been considered for the analysis of image quality parameters. As mentioned, to come to a definite conclusion the size of the dataset has to be sufficiently large. All the images used are in.jpg format.

The conclusion can be drawn as shown below (Fig. 3):

Thus, the percentage improvement as shown is a good predictor about the quality of the recovered image. It is observed that MSE is improved by 99.49% and PSNR by 62%. The average time taken for concealing the erroneous medical images is 5.69 s (Fig. 4).

The same has been performed on the radiology samples of pneumonic patients. All the images considered for the analysis are in.jpg format (Figs. 5 and 6).

The conclusion of the analysis is

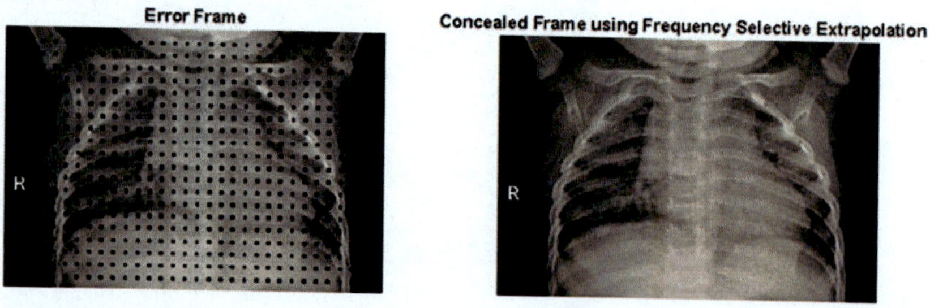

Fig. 2 Error concealment of normal patient's medical image

Normal Patient's Image Quality Analysis			
Parameter	Erroneous image parameter	Concealed Image parameter	Improvement
MSE	1752.0355	8.8797	99.49%
PSNR	10.0121dB	26.9309 dB	62%
SSIM	0.8293	0.9226	11.25%

Fig. 3 Image quality parameters

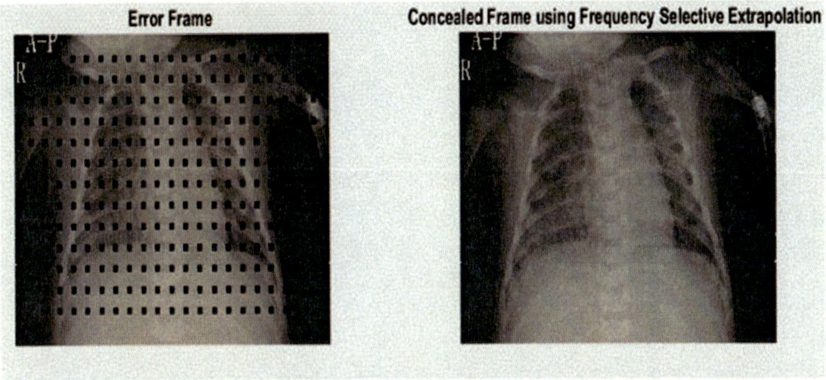

Fig. 4 Pneumonia patient's error concealment

		Erroneous Image				Error Concealed Image				
S. No.	Name of Image	MSE	Entropy	PSNR	SSIM	MSE	Entropy	PSNR	SSIM	Time taken
1	Pneumonia X-ray 1	1192.0302	0.5126	6.8952	0.8031	5.7162	0.1816	30.177	0.9164	2.519
2	Pneumonia X-ray 2	2298.0052	0.5438	4.2206	0.7882	6.231	0.2165	28.8885	0.9073	3.6643
3	Pneumonia X-ray 3	1833.4626	0.5581	5.1084	0.7986	6.6393	0.2429	29.5199	0.9179	2.595
4	Pneumonia X-ray 4	1845.5748	0.6474	6.4894	0.7322	7.2161	0.1652	39.7269	0.9271	4.388
5	Pneumonia X-ray 5	1839.3737	0.5289	5.3911	0.7362	6.2636	0.2291	30.1871	0.9354	3.4922
6	Pneumonia X-ray 6	2267.3732	0.6452	5.3782	0.8217	6.2711	0.1902	29.2719	0.8992	3.2821
7	Pneumonia X-ray 7	1842.5432	0.7822	6.3812	0.8901	6.1527	0.3812	27.281	0.9181	4.2818
8	Pneumonia X-ray 8	1829.2871	0.6382	6.2681	0.8279	8.1678	0.1828	29.7392	0.9364	3.9281
9	Pneumonia X-ray 9	1834.782	0.5884	6.2002	0.832	5.8298	0.3427	30.7491	0.9273	5.3981
10	Pneumonia X-ray 10	1878.9763	0.6387	5.2737	0.8541	6.5711	0.5797	29.6762	0.9132	3.2801
	Average Values	1866.1408	0.60835	5.76061	0.80841	6.50587	0.27119	30.52168	0.91983	3.68287

Fig. 5 Image analysis of radiology images from pneumonia patients

Pneumonic Patient's Image Quality Analysis			
Parameter	Erroneous image parameter	Concealed Image parameter	Improvement
MSE	1866.1408	6.5058	99.65%
PSNR	5.76061dB	30.5216 dB	81%
SSIM	0.8293	0.9226	12%

Fig. 6 Image quality parameters

.

It is observed that MSE improved by 99.65% and PSNR improved by 81%. It takes an average of 3.68 s to conceal the erroneous medical pictures of pneumonic samples (Figs. 7, 8 and 9).

The analysis performed on ten COVID patient's medical image is as shown (all in.jpg format)

It has been observed that MSE improved by 99.66% and PSNR by 77% followed by reduction in from 0.6944 to 0.5441 which is desired. Thus, there is a significant

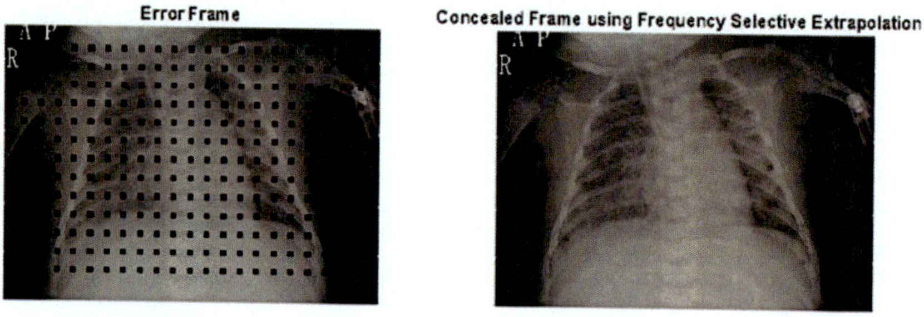

Fig. 7 COVID-19 patient's error concealment

		Erroneous Image				Error Concealed Image				
S. No.	Name of Image	MSE	Entropy	PSNR	SSIM	MSE	Entropy	PSNR	SSIM	Time taken
1	Covid X-ray 1	2519.2493	0.4553	3.9034	0.795	13.3816	0.0088	26.6511	0.9067	2.6504
2	Covid X-ray 2	1866.7867	0.9673	5.2615	0.8399	5.7908	0.9343	30.3451	0.9173	4.1955
3	Covid X-ray 3	1868.7869	0.6894	5.2075	0.8290	6.6789	0.8364	27.9675	0.9185	3.6310
4	Covid X-ray 4	1845.6238	0.6382	5.2379	0.8191	6.1827	0.5283	28.6181	0.9162	3.7193
5	Covid X-ray 5	1894.7172	0.7282	5.2387	0.8253	6.6282	0.6282	28.6722	0.9722	4.2871
6	Covid X-ray 6	2531.7191	0.8324	6.3232	0.8201	5.2973	0.5664	30.2373	0.9627	3.1797
7	Covid X-ray 7	1892.3256	0.4202	5.1879	0.8211	5.1333	0.3021	29.2718	0.8933	3.6672
8	Covid X-ray 8	2782.7989	0.6891	7.2912	0.8273	7.8791	0.5829	29.8739	0.9122	4.1801
9	Covid X-ray 9	1833.8282	0.7283	7.3872	0.8223	5.2839	0.3929	31.7391	0.1891	3.2828
10	Covid X-ray 10	1814.7478	0.7951	8.2738	0.8127	8.3582	0.6611	27.6483	0.9196	4.8179
	Average Values	2085.0584	0.6944	5.93123	0.8212	7.0614	0.5441	29.1024	0.8508	3.7611

Fig. 8 Analysis on COVID patient's medical images

Covid Patient's Image Quality Analysis			
Parameter	Erroneous image parameter	Concealed Image parameter	Improvement
MSE	2085.0584	7.0614	99.66%
PSNR	5.9312 dB	29.1024 dB	77%
SSIM	0.8201	0.8508	4%

Fig. 9 Image quality parameters

improvement in the mentioned parameters. The average time taken for concealing the erroneous medical images is 3.76 s.

The peak signal-to-noise ratio (PSNR) has been enhanced in the error-free picture, as shown in the table. The structural similarity index (SSIM) is a measure of image quality. The SSIM values vary from 0 to 1. The MSE of the error concealed image is lower, while PSNR and SSIM are higher, indicating that error concealment produces a higher-quality X-ray image. After the quality of the image is improved, it can be successfully applied for the data analytic research purposes.

5 Conclusion

This paper presents a method for error detection and concealment in this paper. Lossy images are undesirable for the analysis and extraction of useful information from these images. There is a need for effective error detection and error concealment mechanisms for the recovery of erroneous images. The proposed mechanisms can prove extremely useful as an aid in the field of medical research, where a lot of valuable information is extracted from radiology reports and any loss of information can lead to substantial gap between result datasets.

6 Future Scope

During this situation of pandemic, there has been an unprecedented increase in video consultation by doctors from remote locations for the purpose of real-time diagnosis. The proposed mechanism can be implemented to counter the problem of errors in the transmitted image to facilitate proper diagnosis and research for medical purposes. However, the above mechanism of FSE works well for block errors. The probability of eradication of random errors through FSE is quite low. Thus, with the proper use of error concealment algorithms, this problem of scattered errors can be taken care of.

References

1. Khan H, Kushwah KK, Singh S et al (2021) Smart technologies driven approaches to tackle COVID-19 pandemic: a review. 3 Biotech 11(2):50. https://doi.org/10.1007/s13205-020-02581-y
2. Kumar R, Nagpal S, Kaushik S, Mendiratta S (2020) COVID-19 diagnostic approaches: different roads to the same destination. Virus Disease
3. Chamola V, Hassija V, Gupta V, Guizani M (2020) A comprehensive review of the COVID-19 pandemic and the role of IoT, Drones, AI, blockchain, and 5G in managing its impact. IEEE Access 8:90225–90265

4. Rajaraman S, Siegelman J, Alderson PO, Folio LS, Folio LR, Antani SK (2021) Iteratively pruned deep learning ensembles for COVID-19 detection in chest X-Rays. IEEE Access 8:115041–115050
5. Rajani PK, Khaparde A (2017) Video error concealment using block matching and frequency selective extrapolation algorithms. In: 2017 the second international workshop on pattern recognition (IWPR 2017), in Nayang Technological University, Singapore during May 1–3, 2017
6. Rajani PK, Khaparde A, Ghuge A (2017) Video error concealment using moment invariance. In: 2017 international conference on computing communication control and automation (ICCUBEA), Pune, India, on 17th to 18th August 2017
7. Rajani PK, Khaparde A (2016) Comparison of frequency selective extrapolation and patch matching algorithm for error concealment in spatial domain. In: The 8th international conference on signal processing systems (ICSPS 2016), held at AUT University, Auckland, New Zealand from November 21–24, 2016

Employment of Machine Learning and Data Mining Technique in Hybrid Recommender System: A Comprehensive Survey

Indu Hariyale, M. M. Raghuwanshi, and Kavita R. Singh

Abstract In the present time of advanced world, personalization of data is generally significant for addressing user's needs more adequately, in this way expanding consumer loyalty. Recommender system has become a compelling data filtering tool that reduces data overburden for web users. Hybrid recommender systems consolidate at least two recommendation algorithms so as to increase better system performance to maintain strategic limitations from certain restrictions and issues of pure recommender systems. The article investigates the cutting edge in hybrid recommender system from the most recent decade. The paper tended to the most significant issues considered and recommendation technique incorporated to overcome. This paper examines hybridization classifications, the application areas, and the evaluation procedure and proposed future research direction. The greater part of the investigations joins collaborative filtering with another machine learning strategy. Hybrid recommender system is a hot subject in the digital world and represents newer opportunities to the researcher such as contextualizing recommendation algorithm, involving parallel hybrid algorithm, cross recommendation, and scalability.

Keywords Hybrid recommender system · Recommendation filtering · Collaborative · Context · Content · Machine learning and data mining

1 Introduction

Recommender systems are data filtering tools and method offering recommendations to items to be valuable to a user [1–3]. Recommender system is likewise depicted as a dynamic methodology for user under complex data conditions [2]. Recommender systems play a noteworthy activity in web business administrations

I. Hariyale (✉) · K. R. Singh
Computer Technology, YCCE, Nagpur, India
e-mail: induhrl@gmail.com

M. M. Raghuwanshi
Computer Engineering, GHRCEM, Pune, India

M. S. Kaiser et al. (eds.), *Information and Communication Technology for Competitive Strategies (ICTCS 2021)*, Lecture Notes in Networks and Systems 401,
https://doi.org/10.1007/978-981-19-0098-3_49

(e-commerce) [8]. Huge relationship on the world: for instance, eBay, Amazon, Facebook, LinkedIn, Netflix, etc., have amassed their own one of a kind recommender system to pick up the interests of their potential user. One most significant thought in recommender system examination is the need to consolidate at least two recommendation systems to accomplish greatest forecast precision [3]. Cutting edge recommendation methods involved characteristics and weaknesses. Examiners have chosen to consolidate strategies so as to improve precision of recommender system. Mixture recommender systems incorporate at least two recommendation systems so as to improve better execution with less of the drawbacks of any individual one [3]. Collaborative, [4] content-based, and information-based filtering algorithm are for the mostly utilized to construct recommender system. CF-based RS experiences the ill effects of issue like [3] cold start, versatility, sparsity, gray sheep, and so forth. Content-based RS endures [4] overspecialized recommendations. In hybrid recommender system, scientists have joined the collaborative filtering with some other technique which further enhance prediction accuracy. The purpose of this review paper is to interpret [5] challenges that hybrid RSs could effectively incorporate. This paper gives deep dive into machine learning [6], data mining technique included, RS approach combinations used, and hybridization categories [7]. We also saw the application area, evaluation procedures, characteristics, and metrics that were utilized. Significant contributions of the study are as follows:

- We for the most part condense recommendation filtering technique employed in RS.
- We review the challenges and problems encountered in deployment of hybrid RSs.
- We study various machine learning and data mining techniques, and their combination is used in hybrid RSs.
- We discuss hybridization categories used in hybrid RSs.
- We also throw some light on certain application areas where hybrid RSs being received.
- We study various evaluation process and metrics utilized to measure the quality of hybrid RSs.
- We focus different research gaps most promising for the future research in hybrid RS that have yet to be investigated.

2 Recommender System

Each recommender system follows a specific method so as to deliver product recommendations to user. A recommendation process consists of three stages: data collecting stage, learning stage, and prediction or recommendation stage. The available sources are the user data (demographics, name, age, etc.), the item data (keywords, genres, tag, etc.), and the user-item rating data (obtained by transaction data and explicit ratings). Data collection stage: This gathers important data of users for building a user profile model which includes user's characteristic, nature, and visited, ranked item history. Recommender system depends on various sorts of information, for example, the most advantageous explicit feedback, which incorporates, express contribution by users with respect to their choice for item [9] by surmising user inclinations through perceiving user conduct. Hybrid feedback is a blend of explicit and implicit feedback. Explicit feedback: Explicit feedback is collected in the form of user rating. Implicit feedback: Implicit feedback is gathered from user's inclinations through observing activities of users, for example, historical backdrop of buys, route visited, and web surfing time, links pursued by the user, email's content, etc. In [9], author proposed an updating technique for time-variant implicit feedback using matrix factorization and stochastic gradient descent method. Hybrid feedback: It is a combination of implicit and explicit feedback in RS so as to limit shortcomings of both. Learning stage: In this stage, learning algorithm is applied to filter and exploit the user's characteristics or features from the feedback gathered in data collection stage. Prediction stage: This stage first extracts the item feature and then prescribes it to the specific user on the basis of their test. It is straightforwardly based upon the dataset gathered (Fig. 1).

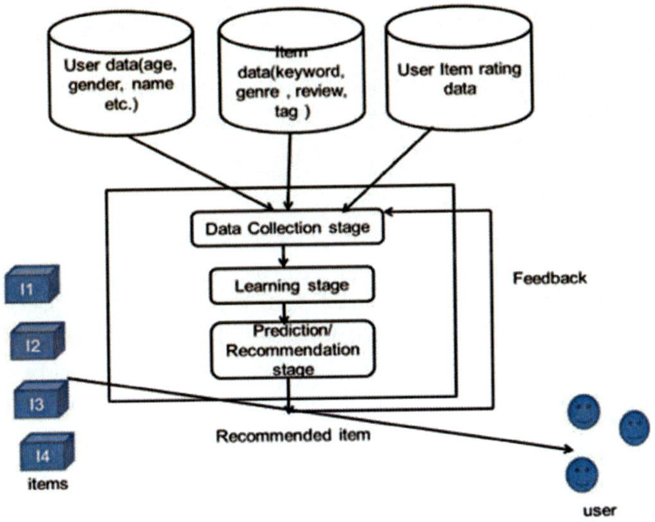

Fig. 1 Stages of recommendation model

3 Review of Literature

3.1 Recommendation Filtering Techniques

RSs are classified into two categories [10, 11]: personalized RS and non-personalized RS [11]. Non-personalized RSs do not take into account the interest of the user, and results are similar for every user. Personalized RS incorporate earlier activity of users on social network, ratings given by them, their interest, etc., to generate top N results. Six types of personalized RSs found in study are as follows: demographic, knowledge-based, utility-based, content-based filtering (CB), collaborative filtering, and hybrid RSs. Demographic filtering technique (DMF): Users' demographic features (such as, education, age, and gender) are incorporated in this technique to find out identical users [12–14]. Knowledge-based filtering technique (KBF): Features or characteristics are important aspects in knowledge-based RS. Knowledge of the items made them to be designed for special concern [15–17]. Utility-based filtering technique (UTF): Utility-based recommenders construct their recommendation based upon similarity between requirements of users and group of choices. Utility of every item is calculated to give results [18]. Content-based filtering (CB): Description of items (i.e., characteristics and attributes) is utilized in CB filtering, which further used to find equivalent user profiles [4, 6, 19, 20]. Collaborative filtering technique (CF): Earlier rating of user for specific item is utilized to predict recommendation result for the new user or item [2, 3, 8–10]. User-based CF: User-based CF search K equivalent users on the basis of same items they have bought. User-based CF [2, 10] method first compute similarity between the dynamic user and new users by looking at their ratings on a similar item and finds the neighbor of the dynamic user as stated by the similarity. It then computes item's predicted rating for the same item. Item-based CF: Item-based CF finds K similar items based on common users who have bought those items. Item-based [10, 13] CF calculates predictions by employing the similarity. It constructs an item's similarity model by recovering a dynamic user's rating for all items. On the basis of similar neighbor, top N item list is generated. It then makes forecasting of unknown rating on dynamic items as indicated by neighbor items and generates the list of recommended items. Cosine [4, 13] and [4] Euclidean Pearson, [13] and Jaccard distance are mostly employed as similarity measure.

3.2 Challenges and Issues

This subsection depicts the most widely recognized issues and challenges that experience in exploiting RSs and are viewed as significant in the RS research work.

Data Sparsity. This issue ascends from the way that user for the most part rate limited number of accessible items, particularly when the catalog is huge.

Cold Start Problem. This issue is vigorously tended in [5, 6] insufficient information present in system about new users or items.

Accuracy. Accuracy is defined as the effective anticipation of item inclinations for every user in RS.

Scalability. Scalability is very hard to accomplish characteristic which is identified with increasing count of users and items in system [7, 8].

Diversity. This is an ideal trademark that is getting consideration recently [14, 18].

Synonymy. It represents same item's inclinations which have been defined by multiple identity and section.

3.3 Machine Learning and Data Mining Technique

Machine learning utilization in recommender system brings enhancement in performance. This paper addresses dissemination of machine learning (ML) procedures and data mining (DM) used in hybrid recommender systems.

K-NN (K-Nearest Neighbors). It is a classification algorithm with a few adaptations widely used in different domain. K-NN classifier is used in collaborative RSs. It finds the K similar or neighborhood objects on the basis of their profile.

Clustering. Clustering algorithm detects categories' group on the basis of identical characteristics of data. K-means method is used for clustering which segregate whole data in the form of K cluster. [10] implements clustering by incorporating case-based reasoning and genetic algorithm. Another example is a solution to the cold start problem by blending clustering and association rules and CF technique.

Association rules. Association rule mining detects relevance in vast datasets. It is represented as rule such as X → Y; here, category of objects is represented by X and Y and is also the example of association rule mining where a hybrid model is built by combining state-of-the-art technique, association rule, clustering to handle data sparsity, and cold start problem [14].

Matrix manipulation. The strategies recognized are Principal Component Analysis (PCA), Latent Dirichlet Allocation (LDA), Singular Value Decomposition (SVD), Dimensionality Reduction, and comparable grid factorization systems [9] mostly used in RS.

Decision tree. Tree and graph design is incorporated to divine decision tree by inspecting sets of data to forecast class of datasets.

Artificial Neural Network. A composition of multiple associated neurons organized in layered and in deliberate manner is defined as Artificial Neural Network (ANN). Each associated neuron has a weight based upon the impact of neurons. Bayesian classifiers think about each quality and class name as irregular factors. For example, N is the number of feature ($X1, X2..., X$), class C is the objective of Bayesian classifier. Bayes are represented as $P(C|X1, X2,..., X)$ $P(X1, X2, ..., XN|C)$ $P(C)$. Main result is the estimated value of C which is computed using Bayes theorem [19]. Naive Bayes classifiers are powerful for separating clamor focuses including insignificant traits. Others: Link analysis Genetic Algorithms, Swarm optimization,

Convolution Neural Networks, Expert system, Deep NN, etc., need more research in hybrid RSs.

Combination of Filtering Techniques. Mostly used techniques for amalgamation have been present in the studies described below:

- Collaborative filtering along with content-based filtering (CF-CBF)
- Collaborative filtering along with other machine learning and data mining technique. Here, we have not considered CBF technique which are included in previous (CF-Others)
- Content-based filtering along with others (CBF-Others)
- Collaborative filtering with content-based filtering in addition to others (CF-CBF-Others)
- Item-based CF including user-based CF (ICF-UCF)

3.4 Categories of Hybridization

Weighted. Weighted category joins various system's consequences which further produce a recommended forecast list with blending scores of every system being used [11, 13].

Cascade. Refinement of recommendation result is done by placing the result of one strategy as input to other strategy in cascade way. Utilizing a course information-based and collective recommender is a case of course hybridization strategy. **Switching.** Exchange of one recommendation strategy's results by another one's as per the efficiency for building decent rating.

Feature augmentation. The method utilizes the ratings with data created in old R.

Mixed. Mixed hybrids consolidate recommendation aftereffects of various recommendation systems simultaneously as opposed to having only one recommendation for every item.

3.5 Evaluation Metrics

Accuracy. Portion of right recommendation results presents in complete absolute recommendation results is defined as accuracy. Root Mean Square Error (RMSE) and Mean Absolute Error (MAE) [12] along with Correlation is utilized to measure accuracy. It is computed as follows [3]

$$MAE = \frac{1}{M}\sum_{u,i}\left|X_{u,i} - R_{u,i}\right| \tag{1}$$

$X_{u,i}$: Calculated rating result of user u for item i,

$R_{u,i}$: Actual rating of user u for item *i*, *M*: Count of user.

Root Mean Square Error (RMSE). If the system has less RMSE value, it represent greater accuracy value. RMSE is calculated as:

$$RMSE = \sqrt{\frac{1}{M}\sum_{u,i}\left(X_{u,i} - R_{u,i}\right)^2} \tag{2}$$

F-measure, precision recall curve (PRC), weighted errors, precision, receiver operating characteristics (ROC), and reversal rate are other metric utilized evaluate accuracy [10]. A fragment of prescribed items which is really significant to user is defined as precision. Recall incorporated related items that are included as a fragment of prescribed items' group. They are computed as:

$$\text{Precison} = \frac{\text{TP}}{\text{TP} + \text{FP}} = \frac{\text{Correctly recommendened items}}{\text{Totally recommendened items}} \tag{3}$$

$$\text{Recall} = \frac{\text{TP}}{\text{TP} + \text{FP}} = \frac{\text{Correctly recommendened items}}{\text{Totally useful recommendened items}} \tag{4}$$

Harmonic average of precision along with recall is defined as F-Measure or F-Score,

$$\text{F—Score} = \frac{\text{2*Recall*Precision}}{\text{Recall + Precision}} \tag{5}$$

TP is true positive, FP is false positive, and FN is false negative value of items. MSE, nDCG, AUC, and execution time are other evaluation metrics. MovieLens, EachMovie, FilmTrust, motion pictures, Last.fm, clients' traits, BookCrossing, Tripadvisor, and Scrumptious are the mostly used datasets in RSs.

4 Discussion on Research Gap

Broadening or improving the proposed solutions is the most widely recognized future work the researchers mean. Widely recognized future enhancement that most researchers embraced is broadening and enhancing of recommendation solutions. Expansion of personalization and application of context aware recommender system multi criteria RS [13] need to be incorporated in hybrid system. Cross-domain RSs and Hybrid RS are a developing examination subject.

5 Conclusion

This paper skint twenty investigations given in international conference and journals deal with hybrid recommender system. Incorporation of machine learning enhances the expertise of hybrid system. Resolving cold start drawback still is a bone of contention in system. Utilizing a lot of data is going to be conceivable work course we tend to found. Parallelization of filtering technique need to be incorporated to offer higher recommendation quality. Different future work bearing can be utilizing the undiscovered deep learning, Q-learning, and blockchain in hybrid RSs to boost the computational complexness of current strategies.

References

1. Ricci F, Rokach L, Shapira B (2011) In: Introduction to recommender systems handbook. Spinger. https://doi.org/10.1007/978-0-387-85820-3_1
2. Isinkaye FO, Folajimi YO, Ojokoh BA (2015) Recommendation systems: principles, methods and evaluation. Egyptian Inform J F.O. 1–13
3. Chen R, Hua Q, Chang Y-S, Wang B, Zhang L, Kong X (2018) A survey of collaborative filtering-based recommender systems: from traditional methods to hybrid methods based on social networks. IEEE Access, pp 64301–64320
4. Lops P, de Gemmis M, Semeraro G (2011) Content-based recommender systems: state of the art and trends. In: Ricci F et al. (eds) Recommender systems handbook, Springer
5. Kumar B, Sharma N (2016) Approaches, issues and challenges in recommender systems: a systematic review. In Indian J Sci Technol 9(47). https://doi.org/10.17485/ijst/2016/v9i47/9489
6. Portugal I, Alencar P, Cowan D (2017) The use of machine learning algorithms in recommender systems. Expert Syst Appl https://doi.org/10.1016/j.eswa.2017.12.020
7. Prasad RVVSV, Valli Kumari V (2012) A categorical review of recommender systems. Int J Distrib Parallel Syst (IJDPS) 3(5): 73–84
8. Zhao X (2019) A study on e-commerce recommender system based on big data. In: 2019 IEEE 4th international conference on cloud computing and big data analytics. IEEE Press, pp 222–226
9. Lin C-Y, Wang L-C, Tsai K-H (2018) Hybrid real-time matrix factorization for implicit feedback recommendation systems. IEEE Access
10. Kumar NP, Fan Z (2015) Hybrid user-item based collaborative filtering. In: 19th international conference on knowledge based and intelligent information and engineering systems. Elsevier, pp 1453–1462
11. Khatwani S, Chandak MB (2016) Building personalized and non personalized recommendation systems. In: International conference on automatic control and dynamic optimization techniques (ICACDOT), IEEE Press, pp 623–628
12. Ghazanfar MA, Prugel-Bennett A (2010) A scalable, accurate hybrid recommender system. In: Third international conference on knowledge discovery and data mining. IEEE Press, pp 94–98
13. Kermany NR, Alizadeh SH (2017) A hybrid multi-criteria recommender system using ontology and neuro-fuzzy techniques. Elsevier, pp 50–64
14. Zare A, Motadel MR, Jalali A (2019) Presenting a hybrid model in social networks recommendation system architecture development. AI & Society, Springer, pp 1–15
15. PCRS (2017) Personalized course recommender system based on hybrid approach. In: International conference on smart computng and communication-ICSCC 2017 Kurukshetra India.Elsevier Press

16. Samın H, Azım T (2019) Knowledge based recommender system for academia using machine learning: a case study on higher education landscape of Pakistan. vol 7. IEEE Access, pp 67081–67093
17. Burke R (2013) Knowledge-based recommender systems. To Appear in the Encyclopedia of Library and Information Science. Citeceer 2013, pp 1–23
18. Guttman RH (1998) Merchant differentiation through ıntegrative negotiation in agent-mediated electronic commerce'. Master's Thesis, School of Architecture and Planning, Program in Media Arts and Sciences, Massachusetts Institute of Technology
19. Laquinta L, de Gemmis M, Lops P, Semeraro G, Filannino M, Molino P (2018) Introducing serendipity in a content-based recommender system. In: International conference on hybrid ıntelligent systems. IEEE
20. User preferences modeling using dirichlet process mixture model for a content-based recommender system (2018) Knowledge-Based Systems. https://doi.org/10.1016/j.knosys.2018.09.028

GRAB Thought Model of Consciousness for Classifying Thoughts and Identifying Mental States

R. J. Ramteke and S. S. Gharde

Abstract Consciousness is the state of ability to see, hear, and feel along with the state of realizing and noticing. Intelligence generates in the mind through consciousness or thoughts. Consciousness refers to thinking, decision-making, awareness, alert state of mind, and internal talk. Concept of arousal of thoughts and hence existence of consciousness has been experimented by developing the computational models of consciousness by research fraternity. This paper elaborates the process of consciousness generated in mind. The basis for the process is considered from the Buddhists literature Abhidhamma. Various actions are occurred in the mind while generating the consciousness. According to Buddhist psychology literature Abhidhamma, consciousness is classified into more types than available in cognitive science or psychology or neuroscience. There are 89 types of consciousness. These are broadly classified into unwholesome, wholesome, resultant, and functional. Also, there are 52 mental states which are classified into three types: ethically variable, unwholesome, and beautiful. In the proposed computational model of consciousness, actions of generating consciousness are implemented which are identified as classified thoughts and mental state or mental factor. Non-axiomatic logic is applied for generating thoughts and hence consciousness. Convolutional neural network is used for classifying generated thoughts. Experimental results are performed by collecting data from people of different ages and professions. For analyzing the results, accuracy, sensitivity, specificity, and error rate statistical evaluation parameters are used. Performance of the computational model provides 97.03% accurate results for classification of thoughts and 93.39% accuracy for identifying mental states for generated thoughts.

Keywords Consciousness model · Thought model · Abhidhamma · Cognitive science · Computational model of consciousness

R. J. Ramteke (✉) · S. S. Gharde
School of Computer Sciences, Kavayitri Bahinabai Chaudhari North Maharashtra University, Jalgaon, India
e-mail: rakeshj.ramteke@gmail.com

M. S. Kaiser et al. (eds.), *Information and Communication Technology for Competitive Strategies (ICTCS 2021)*, Lecture Notes in Networks and Systems 401,
https://doi.org/10.1007/978-981-19-0098-3_50

1 Introduction

Mere possessing physical body by human is no more than dead. Sensation and thought generation capability makes human live and conscious. This thinking ability is consciousness. Tremendous thoughts arise in mind within moment but exposed very few of them. Thought can be positive and negative. 70% generated thoughts are negative. Once you become aware of such thoughts, you can replace those negative thoughts of doubts, fear, or failure with positive one. In COVID-19 pandemic, we came across such many cases that aged people survived only because of their strong will power and positivity into their thoughts. On the contrary, we also encountered many cases that middle-aged people affected by Covid-19 got nervous, frighten, and loss hope of life. Hence, the more control over thoughts result into more joy and satisfaction as Buddha says, "We are shaped by our thoughts; we become what we think. When our mind is pure, joy follows like a shadow that never leaves."

Definition of consciousness varies through different faculty of education like science, philosophy, psychology, and neuroscience etc [1]. According to William Calvin and George Ojeman (1994), consciousness refers to focusing attention, mental rehearsal, thinking, decision-making, awareness, alert state of mind, voluntary actions and subliminal priming, concept of self and internal talk. According to Jeff Hawkins (2004), consciousness is combination of self-awareness and qualia (feelings associated with sensations but not related to sensory inputs). According to Susan Greenfield, consciousness is a dynamic process, and it changes with the development of brain [2].

Consciousness is the set of thoughts. The process of generation of these thoughts is explored by the science in the modern time, and types of consciousness are also identified by it. But, in ancient time, this human psychology of mind was analyzed and put forth in Buddhism. This is known as Buddhist psychology which was collected under Abhidhamma. It is one of the parts of Tripitaka and preaches the higher studies in Buddhism. A very striking and deeply impressive feature of the Abhidhamma is the analysis of the entire realm of consciousness. It is the first time in the history of human thought that this was undertaken so thoroughly and realistically, without admixture of any metaphysics and mythology.

This paper is organized as follows: Section 2 describes Related Work. Proposed Model is elaborated in Sect. 3. Section 4 describes Core Algorithm of GRAB Thought Model. Experimental Analysis and results is briefly described in Sect. 5. It is followed by conclusion.

2 Related Work

Concept of computational model of consciousness has been traced in 1973 in the book "Computer Models of Thought and Language" edited by Roger C. Schank and Kenneth Mark Colby [3]. The volume contains nine papers on models of natural language processing, belief systems, and memory, artificial intelligence, and modeling in psychology. They were built the model using three areas which are natural language processing, memory, and problem-solving.

In 2016, Shivani Poddar, Sindhu Kiranmai Ernala, and Navjyoti Singh from International Institute of Information Technology, Hyderabad introduced PACMAN computational model for identifying personality traits [4] of individual using mental factors described in Abhidhamma model. They applied Abhidhamma literature of Buddhism to propose a theoretical model of an individual as a stochastic finite state machine. Facebook social media data are used which as acced from myPersonality standard dataset to model a user's personality as an evolution of his/her mental states. The psycholinguistic tool, namely Linguistic Inquiry and Word Count (LIWC), is used for feature generation. They present a new dataset PACMAN and machine learning module for analysis of mental states of a user from his/her social media data. Support vector machine is used as classifier, and hamming loss is used as measure to test the results. Accuracy is found to be 90%.

In 2017, Eric Lou [5] from Stanford University, USA investigated a novel computer simulation of the mind using Buddhist theories. Proposed model is composed of the mechanics of human consciousness and cognitive processes. Consciousness, mental states, sense-based, and mind-based cognitive processes are designed and developed through this model. By utilizing timed events, individual queues for each of the five senses that are regulated by a central queue for the brain, feature and object memory, and a concurrent sense and mind architecture are implemented and demonstrated using tick-tack program. This research work is useful in modeling and mimicking human cognition.

In 2020, Paola Di Maio [6] investigated the context of Abhidhamma in neuroscience. According to Paola, it becomes easier to read investigate brain and its signals due to new technology like neuroimaging and electroencephalogram (EEG). Nobel Prize laureate in biology Francis Crick seeded that initial line of thoughts is followed by neuroscience of consciousness. It is driven by the experimentation into sensorial, perceptual organs, and by the computation of perceptual signals in the brain.

In the research performed by Banerjee and Pal [7] in 2017, arousal of thoughts in mind is demonstrated by using Z*-numbers. Z*-numbers are a logical tool for facilitating expression of self, encase subjective and objective contemplation, and self-reflection and self-conscious reactions. The purpose of their research is to identify fast and slow thinking, reactions and learning, and self-conscious decisions. Their research is performed on the observation of images and assumes sensory inputs. Emergence of thought is implemented, but mental factors in which thought arouse do not address.

In 2020, Jeste et al [8] have conceptualized the idea of artificial wisdom. They examined the constructs of human intelligence and human wisdom in terms of their basic components, neurobiology, and relationship to aging, based on published empirical literature. Also, they reviewed the development of AI with respect to the model of human intelligence. Governing principles for artificial wisdom is identified by them. This paper covers the importance of artificial intelligence with it limitations and future need of new term as artificial wisdom. While constituting artificial wisdom, they pointed out the human behavior, experiential learning, and emotions which must be part of the artificial wisdom along with intelligence. Human behavior is based on his quality of consciousness. Hence, while developing artificial wisdom, researchers should be focused on the development of consciousness as like as humans in machines.

Recently in 2021, Signaet al. [9] performed review on different models on consciousness, robot consciousness, and conscious mind. Authors also focus on Baars' global workspace theory which has recently been ranked as the most promising theory in its field but not reflected in mathematical models of cognitive science. Also, they propose simple mathematical models that are well designed for computer implementation.

3 Proposed Thought Model

This model proposed novel approach which has prime focus on these issues. Proposed model is named as Gharde–Ramteke Abhidhamma-based (GRAB) thought model. The basic framework of the GRAB thought model is prepared by considering intelligent distribution agent (IDA) cognitive life cycle. [1] Also, process of consciousness which is mentioned in Abhidhamma is crafted while devising the model of consciousness. Various terms used in the proposed GRAB thought model are borrowed from psychology. Cognitive cycle is started from incoming sensory data; cognition is established by processing on it; memory is devised to tackle the input, and to incorporate experience with the input, generation of thoughts is performed, and at the end, quality of thoughts and mental factors (mental states) is analyzed for the generated thoughts. The whole cognitive cycle is rendered as GRAB thought model. This GRAB thought computational model is divided into three stages: understanding phase, consciousness phase, and action selection phase.

Process of consciousness is mentioned in Abhidhamma. It is classified into fourteen steps for generating thoughts. These are as follows: re-birth or conception (Patisandhi), stream of being—life continuum (Bhavanga), decease (Cuti), turning toward (avajjana), seeing (Dassana), hearing (Savana), smelling (Ghayana), tasting (Sayana), touching (Phusana), receiving (sanpaticchana), investigating (santirana), determining (Votthapana), apperception (javana), registering or identifying (tadalambana).

Understanding phase includes step no. 1 to 10 till the receiving and preprocessing of inputs from the user (see Fig. 1). In consciousness Phase, step no. 11 to 14 are incorporated. In this consciousness phase, non-axiomatic logic is applied for the generation of thoughts. Steps are executed (see Fig. 2). The last phase is action selection phase where generated thoughts are classified in one of the four major types of consciousness. Working of this phase is rendered in Fig. 3. Convolutional neural network is applied for classifying thoughts and also for classifying mental factors of identified thoughts.

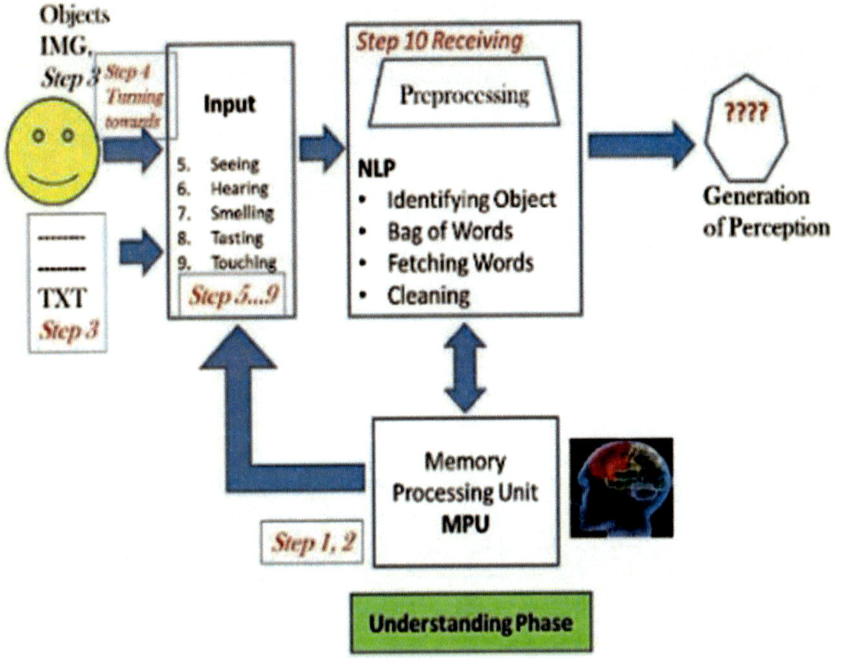

Fig. 1 Understanding phase

4 Core Algorithm of GRAB Thought Model

Core algorithm describes overall functioning of the proposed system. Samples are collected from the users by showing the images and asking three questions. These are taken as input to the algorithm. Samples are preprocessed; NAL is applied which is followed by applying CNN classifier. Training and testing are performed. Output of the proposed system is identified quality thoughts and mental factor. Algorithm for generating consciousness is given below.

Algorithm 1 Core Algorithm of GRAB Thought Model **Input:** image i and three questions, database DB.

Output: Thought T as wholesome (W), unwholesome (U), resultant (R), functional (F). Mental factor MF as ethically variable factor (F), unwholesome (U), beautiful factor (BF).

Steps:

1. Start
2. Give input (image i and questions Q)
3. Read input i text from answer of questions Q and store in database.
4. Partition of database DB in training Dtr and testing Dts data
5. Apply preprocessing

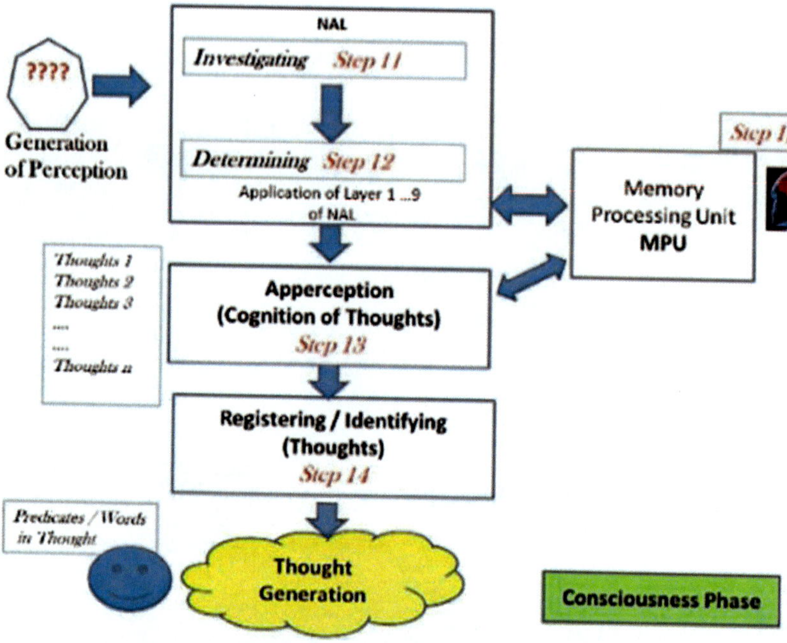

Fig. 2 Consciousness phase

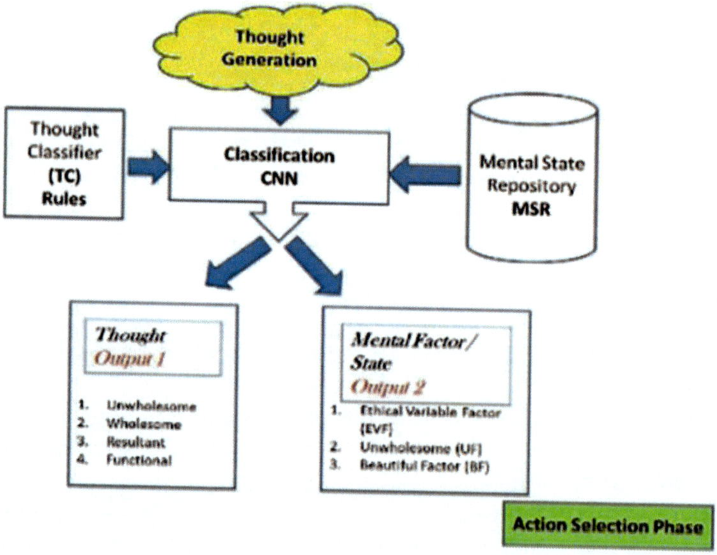

Fig. 3 Action selection phase

6. Apply non-axiomatic logic (NAL) for feature learning with length of keywords, learning duration, and number of layers
7. Apply NAL (identifying and determining thought) for generating rules for consciousness/thought
8. Apply convolutional neural network (CNN) for classification of generated thought.
9. Recognition of quality thoughts T.
10. Apply CNN for classification of mental factors.
11. Recognition of mental factor MF.
12. Testing Samples
13. GOTO 1
14. End

5 Experimental Analysis and Results

In this research, questionnaire instrument comprises multiple-choice questions and dichotomous questions. All questions are set in Google Forms for different variety of images. All images are categorized in the four broad areas such as basic shapes, heroes, places, and feelings. Samples collected through Google forms are extracted in Excel sheet and preprocessed for getting ready to use in the proposed system. Natural language processing (NLP) techniques are used for preprocessing the samples for subsequent phases of implementation of algorithm. Proposed system implemented using MATLAB 2019a. Training samples and testing samples are used randomly in the 80:20 ratios. Non-axiomatic logic is implemented which is law of thought for generating consciousness through thoughts in the proposed system. Thoughts under consciousness are classified broadly in four categories using convolutional neural network classifier.

All samples used in training are primarily distributed among four different classes based on the frequency values of words appeared in thoughts. Scores of all words under the thoughts are computed using support and confidence. It is rendered in Fig. 4. After applying non-axiomatic logic, correlation between the generated thoughts through the frequently occurred words is explored by using support and confidence digraph as shown in Fig. 5. Predicted thoughts are classified either in wholesome, unwholesome, functional, or resultant class based on the probability of frequently occurred item sets. After applying non-axiomatic logic (NAL), various values are accumulated in NAL table such as frequently occurred thoughts along with their frequency, support, and confidence.

After implementing the proposed system, results are tested by applying statistical evaluation measures where accuracy, sensitivity, specificity, and error rate parameters are used for analysis. Sensitivity measures the proportion of positives that are correctly. Specificity measures the proportion of negatives that are correctly identified. Factors like specificity and sensitivity both are equally important. Result can be very specific without being sensitive, or it can be very sensitive without being

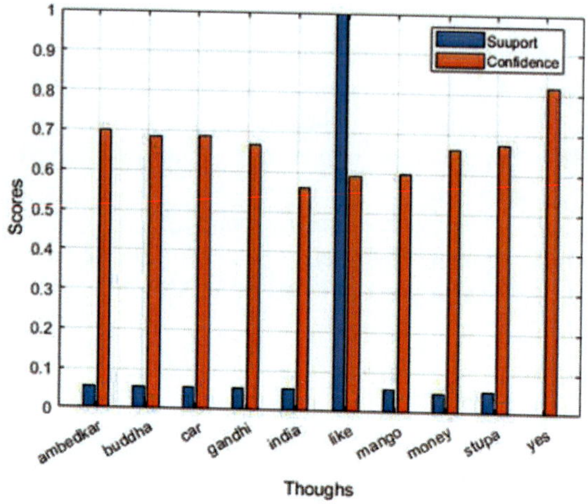

Fig. 4 Scores of thoughts

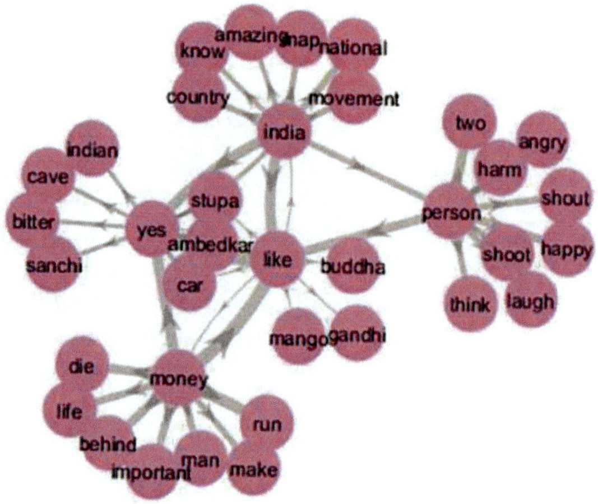

Fig. 5 Digraph of support for thoughts

specific. Hence, high sensitivity and high specificity are treated as prominent values for good results. Terms that are commonly used for computation of sensitivity, specificity, and accuracy are true positive (TP), true negative (TN), false negative (FN), and false positive (FP). Sensitivity, specificity, and accuracy are described in terms of TP, TN, FN, and FP.

Total 89 samples are randomly chosen for testing purpose. Evaluation using statistical measures is shown in Table 1 and Table 2. Among those samples, average accuracy of all classified samples is found to be 91.77%. Similarly, average, sensitivity, and specificity are computed as 90.70 and 91.63%, respectively. Average error rate (ERR) of all correctly classified samples is 0.07% which is very less. These statistical values indicate that the proposed system works efficiently.

Table 1 Statistical measures of result thought

Thought class	Accuracy	Sensitivity	Specificity	ERR
Wholesome	89.78	92.03	86.53	0.10
Unwholesome	93.30	91.47	97.14	0.07
Functional	92.24	89.34	96.76	0.08
Resultant	97.72	89.96	86.08	0.02
Average	91.77	90.70	91.63	0.07

Table 2 Statistical measures of result mental state

Mental state	Accuracy	Sensitivity	Specificity	ERR
Ethical variable factor	91.68	83.19	97.37	0.08
Unwholesome	93.40	94.97	98.72	0.07
Beautiful factor	95.09	89.88	91.99	0.05
Average	93.39	89.35	96.03	0.07

6 Conclusion

In cognitive science, research on study of mind and mental process is found to be challenging issue. Abhidhamma—the Buddhist psychology model provides suitable solution to this challenge because it is recognized by researcher's fraternity worldwide. Process of consciousness or thought generation which was discussed in Abhidhamma centuries ago is accepted by science as it was quoted by eminent researchers in the literature. Hence, proposed system of computational model of consciousness is implemented using Abhidhamma model. 89 types of consciousness are broadly categorized into four major classes', viz., wholesome, unwholesome, functional, and resultant. Non-axiomatic logic which is very much suitable algorithms for generating thoughts is applied for developing consciousness in the proposed system through the generation of thoughts. These thoughts are classified using the classifier convolutional neural network. Results are provided using statistical measures having average values 91.77% accuracy, 90.70% sensitivity, 91.63% specificity, and 0.07% error rate. Following results of statistical values prove that the proposed system of computational model of consciousness works efficiently on variety of samples.

This research paper will give new insight to the researchers of cognitive science to think about process of consciousness by considering Buddhist psychology. Mental state of the human being can also be predicted as like as generation of thoughts using Abhidhamma model. There are 52 mental states which are also known as mental factors. Implementation of computational model for classifying thoughts into one of these mental factor is the future scope of this computational model.

References

1. Zhao T, Zhu Y, Tang H, Xie R, Zhu J, Zhang JH (2019) Consciousness: new concepts and neural networks, frontier of cellular neuroscience. https://www.frontiersin.org/articles/https://doi.org/10.3389/fncel.2019.00302/full Accessed on November 20, 2020
2. Starzyk J, Prasad D (2010) A computational model of machine consciousness. Int J Mach Consciousness World Scientif Publishing Company 3(2):1–26
3. Schank R, Colby K (1973) Computer models of thought and language. Comput Sci Psychol. https://doi.org/10.2307/1421990 Corpus ID: 143599033 https://www.semanticscholar.org/paper/Computer-Models-of-Thought-and-Language-Schank-Colby/3e3725292a28029f308034c6110bd4d9e0dad4ce Author, F.: Article title. Journal 2(5), 99–110 (2016)
4. Poddar S, Kiranmai Ernala S, Singh N (2016) PACMAN: psycho and computational framework of an individual (Man), emotion and sentiment analysis. In: Fernando Sánchez-Rada J, Schuller B (eds) Proceedings, pp 10–17
5. Eric L (2017) A novel computer simulation of the mind using buddhist theories
6. Di Maio P (2020) Neuroscience meets abhidharma, buddhistdoor global, https://www.buddhistdoor.net/features/neuroscience-meets-abhidharma Article published on: 16 July 2020, Accessed on: 31 May 2021
7. Banerjee R, Pal SK (2017) A computational model for the endogenous arousal of thoughts through Z*-numbers. Elsevier J Inform Sci 405:227–258
8. Jeste D, Graham S, Nguyen TT, Depp C, Lee EE, Kim H-C (2020) Beyond artificial intelligence: exploring artificial wisdom. Int Psychogeriatrics 32:993–1001
9. Signa A, Chella A, Gentile M (2021) Cognitive robots and the conscious mind: a review of the global workspace theory. Curr Robot Rep 2:125–131. https://doi.org/10.1007/s43154-021-00044-7

An Approach to Perceive Session Hijacking in IoT Health Care

S. Sarika

Abstract Internet of Things (IoT) is a wireless network that configures itself and allows the Internet to access the real world of physical objects. IoT technologies are used in diverse platforms as they are flexible to most existing technologies in providing networked services. IoT plays a vital role in healthcare systems. It helps to improve patient care with remote monitoring and telemonitoring. IoT's heterogeneous nature has raised various privacy and security concerns. Session hijacking is one of the serious issues threatening Web application. As the number of vulnerabilities in the Web increase, session hijacking becomes a great concern to all users in the network. In session hijacking, a hacker takes control over a valid session between the user and the server and pretends to be authentic. This research paper gives an introduction to IoT and its role in healthcare systems, its security and privacy challenges, and a multi-agent paradigm to prevent session hijacking attacks.

Keywords IoT · Health care · IoT threats · Session hijacking · Multi-agent systems

1 Introduction

Internet of Things (IoT) [9] is a growing topic of conversation both in the workplace and outside of it. It is a giant network of connected things and people which collect and share data about the way they are used. The interrelated computing devices in IoT are provided with unique identifiers (UIDs). Medical IoT devices have new ways of tracking patients and facilities to optimize treatment and reduce costs. But, many of these smart devices are at security risk. The connected medical devices like Wi-Fi-enabled infusion pumps, CT scan machines, and smart MRI machines increase the attack surface of devices by sharing information.

IoT has played an important part in healthcare systems. In IoT health care, the privacy and security of the users, data, and devices are important. Three main aspects

S. Sarika (✉)
Naipunnya Institute of Management and Information Technology, Pongam, Kerala, India
e-mail: sarika@naipunnya.ac.in

M. S. Kaiser et al. (eds.), *Information and Communication Technology for Competitive Strategies (ICTCS 2021)*, Lecture Notes in Networks and Systems 401,
https://doi.org/10.1007/978-981-19-0098-3_51

of the system are confidentiality, integrity, and availability. In order to accomplish the security facets, IoT needs to achieve these requirements. In the IoT network, design elements can communicate with each other and can share data among themselves. Important information like credentials can be stolen by attackers, if there are no filtering techniques available. To meet the higher demands in terms of security, reliability, and privacy, advanced technologies and methodologies should be developed.

Session hijacking [7] is the commonly experienced Web-based threat in connected networks. In this attack, a hacker gains unauthorized access to an authorized session. Most networks and Web sites are vulnerable to this kind of attack. This paper explains the vulnerabilities in IoT healthcare systems and a novel approach using a multi-agent system to resist session hijacking.

The paper is organized as follows: Section 2 reviews the related work in this area. Section 3 explains IoT in healthcare systems and its threats. Section 4 details the proposed method using software agents to resist session hijacking. The implementation part is explained in Sects. 5 and 6 concludes the paper.

2 Related Work

Session hijacking is a type of man-in-the-middle attack which steals sensitive information of users that are transferred through the network. It is one of the attackers' favorites as it can take data from the server storage without the users' consent. In literature, different techniques have been proposed, but they still show vulnerabilities in terms of confidentiality, integrity, and authentication.

SessionShield [10] is a client-side mechanism against session hijacking where session identifiers are not used by legitimate client-side scripts. The session identifiers from the incoming HTTP traffic are detected and isolated from the browser so that they will not be available for malicious scripts. In SessionSafe [8], the authors propose a method that uses URL randomization and one-time URLs to prevent these attacks. The session ID is not used by genuine client-side scripts as the cookies are sent from another subdomain. As the malicious JavaScript code cannot access the cookie, the cookie is safe. The technique is not false proof to every type of attack, and it is difficult to implement.

In SessionLock [1], session identifiers are available only at the client side. This is possible by including fragment identifiers in the URL. The authentication of the user is performed over a secure connection. On successful authentication, the Web server sets session ID as a secure cookie and redirects the user to a URL that includes the session ID as its fragment identifier. The subsequent requests are not made over a secure connection, and thus no need to perform the cookie settings.

The concept of one-time cookies [4] prevents session hijacking by signing each user request with a session secret key securely stored in the browser. It does not require state synchronization in the Web application but uses a single, static token to authenticate each request. As each token is tied to a particular request by using a hash-based message authentication code (HMAC), the reuse of OTC tokens is not possible. In [6], D. Karis et al. have proposed a hash-based method to prevent session hijacking. In this method, when a legitimate user submits a request, the application server requests the user's login credentials for authentication. It will in turn generate a unique hash digest by concatenating the host IP address, name, platform, major and minor versions of the browser. When there is an attempt for stealing session ID, the application server calculates the hash digest of the attacker's browser details and system parameters and compares it with the hash digest generated for the legitimate user's session ID. If there is a mismatch, the attacker is denied access to the session information. Cookie-based virtual password authentication protocol [13] preserves the advantages of conventional password authentication. At the time of authentication, the cookie is stored on the client machine by the Web server. Any legitimate user can easily be authenticated to the Web server from a machine that stores the cookie. The login process for a legitimate user includes some additional computations which make it difficult for an attacker to easy login. The virtual password generated by the user will differ in different sessions of SSL. Ogundele et al. [12] proposed a method to resist session hijacking attacks in Web application management. In this paper, the user is first authenticated to the server using a user ID and password. Then, a session ID is created using a rigid algorithm and saved to cookies. If the system is inactive for a constant period of time, it is set to timeout the session. The timeout session is 15 min, which will not provide the attacker enough time to launch an attack. Table 1 shows an analysis of the existing techniques for the prevention of session hijacking.

3 IoT in Healthcare Systems and Its Threats

Health care means taking preventive or appropriate measures to enhance the well-being of an individual. This includes surgery, prescribing medication, or giving other tips/advice in the routine life of an individual. Normally, these programs are provided by a healthcare system made up of hospitals and doctors. IoT plays an important role in many fields of health care [2]. This includes:

- Tele/online consultation and remote patient monitoring system where the patients are allowed to consult the doctor through an online interface or telephone and get prescription from the doctor. This was much common during this COVID pandemic stage.
- Laboratory facilities in distant areas and the results are delivered to the hospital network at the time of consultation.
- Palliative care for bedridden patients and their remote monitoring.
- Special care clinic for senior citizens which includes nursing home and residency facility for the needed patients.
- Real-time data collection from the IP patients like blood pressure monitoring, pulse rate, and blood oxygen level.

Table 1 Comparison of existing techniques to prevent session hijacking

Method/Authors	Technology in use	Server/Client side	Advantages	Limitations
SessionShield	Intercepting Set-Cookie header	Client	Lightweight	Multiple browsers are not allowed
SessionSafe	URL randomization	Server	Malicious scripts cannot access cookie as it is sent from another domain	Difficult to implement
SessionLock	Fragment identifiers	Server	Secure connection is not required for subsequent requests	Does not resist session fixation and can be used for URLs that can be rewritten using JavaScript
Dacosta et al	One-time cookies with HMAC	Server	Reuse of session tokens are not possible	Only one session per user
D'silva et al.	Hash digest	Server	Authentication and integrity are preserved	Complexity is more
S K Sood	Virtual password authentication	Server	Difficult to execute online dictionary attack	Computational cost is more
Ogundele et al.	Session ID regeneration	Client	Secure and reliable	Re-authentication needed

IoT devices are susceptible to numerous threats. They collect measurable and analyzable data in health care. Security in IoT is paramount as it is confidential information. IoT devices are prone to many security vulnerabilities. Some of the common security challenges are data privacy threat, data manipulation, session hijacking [7], ransomware [11], and DDoS attacks [5].

3.1 Session Hijacking

Session hijacking occurs when a session is hijacked in devices connected to IoT network. It uses a valid session of a doctor, nurse, patient, or another health worker to gain unauthorized access into the system. A session refers to the communication time interval between two parties in a single system. When a user logs in to the system using a password, a session is created. The session will be valid until he logs out. Every session is associated with a session ID. This id may be stored in cookies. The anatomy of session hijacking is given below.

1. The user is logging into a system and opens a Web application in a browser.
2. A session is created by the server in the browser with a session ID.
3. The user is allowed to log in into the system once the authentication is established, and session cookies are attached to HTTP header.
4. The attacker intrudes into the system and obtains the session ID by sending a malicious script or stealing the session cookie.
5. As soon as the attacker takes control over the session ID, the original session ID is replaced by the fake ID created by the attacker.
6. Attacker forward the private details of patients to money making frauds, mafias behind organ transplantation, or any other persons who makes money through identity theft.
7. In some cases, attacker demands money from the hospital officials to give back the hospital records which is already blocked by the attacker.

4 Intelligent Agents in Resisting Session Hijacking

The use of intelligent agents in attack detection of IoT health care is a new concept. Agent technology [3] is embraced by the distributed systems due to its autonomy, collaborative behavior, scalability, and flexibility. As the number of devices connected to healthcare system using IoT increases, the security challenges of these heterogeneous devices also increase. Intelligent agents are already used in a number of e-health applications, such as diagnostics, smart emergency applications, tele monitoring, and eldercare. The proposed work explains the application of intelligent agents in session hijacking attack detection. In the proposed multi-agent system (MAS), agents in different level are deployed at the client side to resist session hijacking. The system model is depicted in Fig. 1.

The proposed method consists of three agents in two levels: monitor agent in level 1 and S1 and S2 agent in level 2. Different tasks are assigned to each agent. Monitor agent is managing the session agents in each session. When user authentication is completed, a session ID is created for the session. In order to provide message

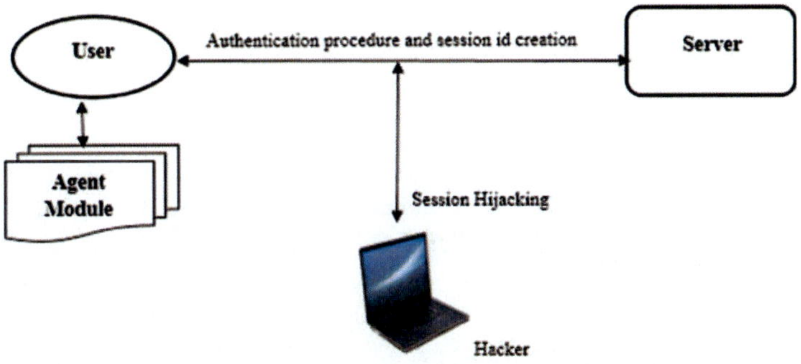

Fig. 1 System model

integrity and authentication, HMAC of session key along with username is also stored in cookie. The advantage here is, an encrypted session key. HMAC is used for security. To prevent the attacker from obtaining the session key in any case, the HMAC value is recalculated every 5 min. This is done by S1 agent. If there is a change in HMAC value after 5 min, the session is timed out. Moreover, the timestamp of the current session is also deleted. After the generation of new session ID, user has to be re-authenticated. The old session data are not deleted immediately as it may result in inconsistent state if the network is unstable. If HMAC value is same after every 5 min, the user continues with the session. The session agent S2 manages a counter and it automatically times out the page if the page does not detect any user activity for continuous 15 min. A pop-up message is also displayed to user about the page signing out activity. Operation of the system is formally explained in Algorithm 1.

Algorithm 1

```
1.Start
2.User logs into a web application in browser
3.Authentication is established with the server
4.Session id is created and it is stored with HMAC in session cookie
5.Add new TickerBehaviour to S1 agent for every 5 minutes for HMAC verification
6. If there is a mismatch in HMAC value
   Invalidate old session and create new session id
    delete_old_timestamp
7.Add a counter to S2 agent with value 900 sec
  If onmousemove not detected, then
    counter=counter-1
  else reset counter to initial value
  If counter=0 then session timeout
   Display page timeout pop up message to user
8.Repeat step 6 till session timeout
9.End
```

5 Implementation

The proposed method is implemented in Java platform. It uses JADE software framework to make use of agents. The method uses a three-tier architecture. The bottom layer represents j2se/j2ee runtime environment. The middle layer includes JADE platform, which provides multi-agent operations. The top layer is the application layer where the agent communication and task execution happen. Agents communicate and cooperate with other agents by exchanging messages and beliefs.

Figure 2 shows the session ID created after user authentication. If the user credentials are verified, the session ID and HMAC of session ID along with username are

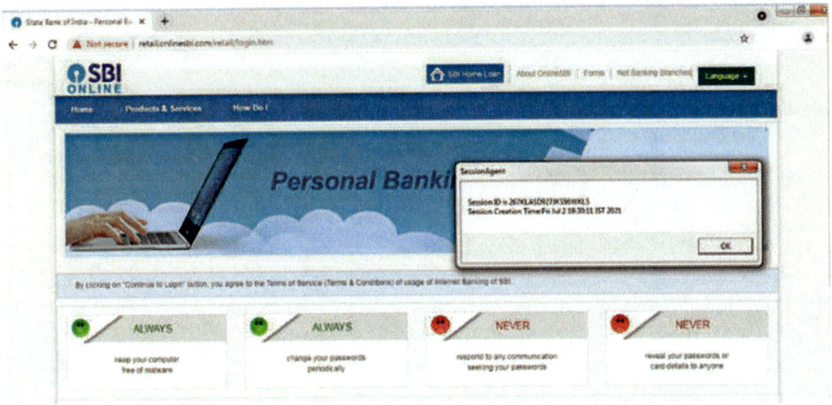

Fig. 2 Session ID created after authentication

stored in the cookie of user. Figure 3 shows the new session ID creation if a mismatch is found in HMAC after 5 min. In that case, the old session is invalidated; its timestamp is deleted, and the user is re-authenticated to the new session.

Figure 4 shows the page sign-out activity if the user has not interacted with the page for continuous 15 min. In this case, the user is given an alert message that the page is signed out due to inactivity. Here too, the user needs to be re-authenticated.

The virtue of using intelligent agents is that the attack detection can be done in parallel with different sessions. The multiple agents deployed in the application can work in concert with each other. It requires internal communication between agents. The agents in this system communicate using FIPA agent communication language. The number of internal messages exchanged between the agents is reduced to minimize complexity involved in the system.

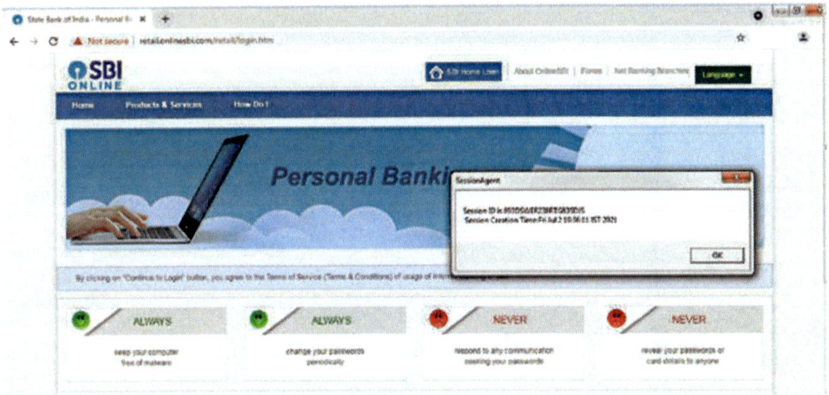

Fig. 3 Session ID re-creation after a breach is found

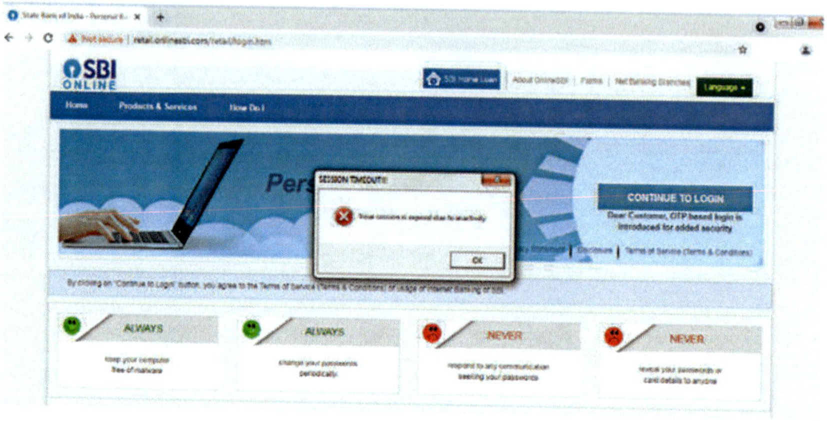

Fig. 4 Page sign-out message due to inactivity

6 Conclusion

Internet of Things is an intelligent connection of physical devices embedded with Internet connectivity, sensors, and other hardware. IoT-based healthcare infrastructures are prone to several security threats which cause malevolent actions to the devices connected. IoT threats can lead to identify theft and financial losses for the parties concerned. This paper has provided an introduction to IoT healthcare systems, the various challenges involved in IoT healthcare systems, and a novel approach using artificial intelligent agents to prevent session hijacking. Session hijacking is the most venomous form of attack in IoT systems. It causes a serious threat to the session key establishment schemes inside a healthcare networked system. The agent-based paradigm is used to monitor session duration and the hijacking of sessions by stealing session IDs. This is an ongoing research work and is expected to give better results with good accuracy.

References

1. Adida B (2008) Sessionlock: securing web sessions against eavesdropping. In: Proceedings of the 17th international conference on World Wide Web
2. Atzori L, Lera A, Morabito G (2010) In; The internet of things: a survey, computer networks. vol. 54(15), Catania, Italy pp.1–17
3. Babu BS, Srikanth K, Ramanjaneyulu T, Narayana IL (2016) IoT for healthcare. Int J Sci Res 5(2):322–326
4. Balaji PG, Srinivasan D (2010) An introduction to multi-agent systems. In: Innovations in multi-agent systems and applications-1. Springer, Berlin, Heidelberg, pp 1–27
5. Dacosta I, Chakradeo S, Ahamad M, Traynor P (2011) One-time cookies: preventing session hijacking attacks with disposable credentials. Georgia Institute of Technology

6. Donno MD et al (2017) Analysis of DDoS-capable IoT malwares. In: Federated conference on computer science and information systems (FedCSIS). IEEE
7. Jain V, Sahu DR, Tomar DS (2015) Session hijacking: threat analysis and countermeasures. In: 2015 international conference on futuristic trends in computational analysis and knowledge management
8. Karis D et al (2017) An effective method for preventing SQL injection attack and session hijacking. In: 2017 2nd IEEE international conference on recent trends in electronics, information and communication technology (RTEICT). IEEE
9. Martin J (2006) SessionSafe: implementing XSS immune session handling. In: European symposium on research in computer security. Springer, Berlin, Heidelberg
10. Nick, N et al (2011) SessionShield: lightweight protection against session hijacking. In: International symposium on engineering secure software and systems. Springer, Berlin, Heidelberg
11. O'Gorman G, McDonald G (2012) Ransomware: a growing menace. Symantec Corpora-tion
12. Ogundele, Israel O et al (2020) Detection and prevention of session hijacking in web application management. Int J Adv Res Comput Commun Eng 9(6):1–10
13. Sood SK (2011) Cookie-based virtual password authentication protocol. Inform Secur J A Global Perspect 20(2):100–111

A Graphical-Based Video Steganography for Information Security

Payal Bose, Shawni Dutta, Vishal Goyal, and Samir Bandyopadhyay

Abstract In today's digital media, data communication over the Internet increasing day by day. Therefore, the data security becomes the most important issue over the Internet. With the increase of data transmission, the number of intruders also increases. That is the reason it is needed to transmit the data over the Internet very securely. Steganography is a popular method in this field. This method hides the secret data with a cover medium in a way so that the intruders cannot predict the existence of the data. Here, a steganography method is proposed which uses a video file as a cover medium. This method has five main steps. First, convert the video file into video frames. Then, a particular frame is selected for embedded the secret data. Second, the least significant bit (LSB) coding technique is used with the double-key security technique. Third, an eight characters password verification process is applied to secure the data. Fourth, reverse the encrypted video. Fifth, the signature verification process to verify the encryption and decryption process. These five steps were followed for both the encrypting and decrypting processes.

Keywords Double-key encryption and decryption · Least significant bit (LSB) coding · Password verification · Signature verification · Video steganography

P. Bose (✉) · S. Bandyopadhyay
Lincoln University College, Petaling Jaya, Selangor D.E, Malaysia
e-mail: bosepayal91@gmail.com

S. Dutta
The Bhawanipur Education Society, Kolkata, India

V. Goyal
GLA University, Mathura, India
e-mail: vishal.goyal@gla.ac.in

M. S. Kaiser et al. (eds.), *Information and Communication Technology for Competitive Strategies (ICTCS 2021)*, Lecture Notes in Networks and Systems 401,
https://doi.org/10.1007/978-981-19-0098-3_52

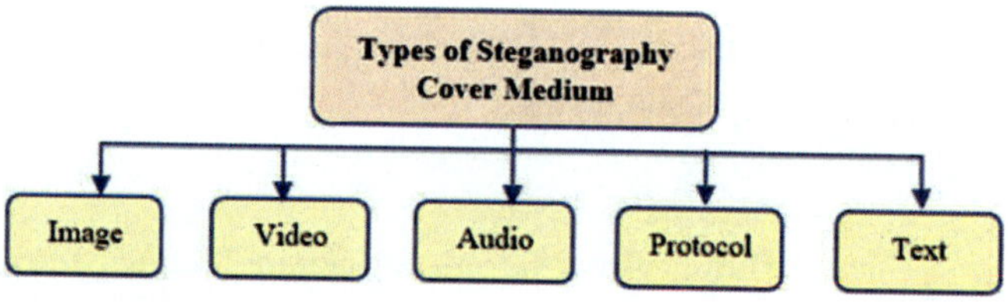

Fig. 1 Types of steganography cover medium

1 Introduction

In today's digital world, Transmission of secure data through the Internet is the biggest challenge. Therefore, for secure data transmission, two types of security techniques are available (1) Steganography and (2) Cryptography.

Steganography is the most commonly used data secure technique. It is used to conceal the secret data inside a cover medium. The cover medium embedded the secret data in a very efficient that it is too difficult to find the original data from it. In steganography, several types of cover mediums are available. Figure 1 shows the details of it.

Cryptography is a technique where a secret key is used to convert the secret data into encrypted secure data. This paper proposed a steganography technique with a video cover medium. In this technique, to secure data or information, the least signification bit algorithm (LSB) with dual-key encryption method is used to encrypt and decrypt the secret data.

2 Literature Review

Video steganography technique is the extension of the image steganography process. The most common technique for steganography is the least significant bit technique (LSB). In this technique, the least significant or the last bit of the frame holds the value of the secret message. For this reason, this technique is vulnerable and can be break very easily. Several authors proposed several steganography techniques. For hiding the secret data, the authors proposed an affine transformation technique. The data embedded into the video frame is based on the coefficient of wavelet transformation. The affine transformation is used to distribute the pixel values [1]. In today's world, for secure communication, authors describe several steganography techniques. Finally, they proposed a method that is useful and more compatible with steganography [2]. The authors survey different techniques of the steganography method and gave a detailed analysis of each process [3–5]. The authors compare the two most popular steganography techniques LSB and discrete wavelet transform (DWT) base on two factors efficiency and capacity of accepting multiple images into one cover image. They also evaluate the performance of the algorithm based on the cover image capacity, imperceptibility of data, and security [6]. To enhance security and avoid data hacking, the authors implemented an edge-based steganography technique. They used an adaptive embedded process over dual-tree complex wavelet transform (DT-CWT) method. They show that this method is significantly better than other steganography techniques [7].

3 Proposed Methodology

This experiment is the extended part of the image steganography. It has five major parts for both the encryption and decryption methods. The parts are (1) Preprocessing, (2) LSB coding, (3) Double-key protection, (4) Reverse the video with password verification, and finally (5) Signature verification. Figure 2 shows the basic structure of the process.

3.1 Preprocessing

To perform this experiment, the first step is to preprocess the input cover medium. Here, a video cover medium is used. Therefore, for preprocess, it is needed to split the input cover video into its corresponding frames. The next step is to select a particular frame among all the frames and perform the next steps [8].

3.2 Least Significant Bit (LSB) Coding

LSB is a common coding technique in steganography. It is used to embed the secret code into cover images [9, 10]. According to this technique, first, it converted all the image pixels and the secret data into their binary values. Then, it embeds each bit of the binary value of secrete data at the least significant position of pixels value. This

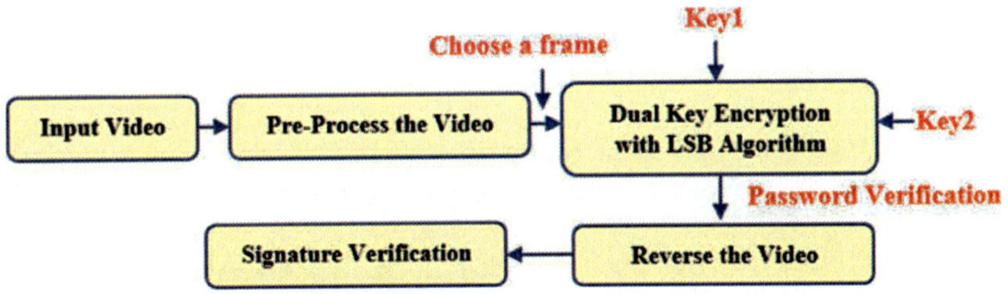

Fig. 2 Basic structure of video steganography process

is an easy method to implement and easy to decode. For an RGB frame, this process can embed a 3-bit value into each pixel.

For example, let consider the pixels of a 24-bit image—

$$\text{Pixel}_1 = (11110000 \quad 00110001 \quad 01011000)$$
$$\text{Pixel}_2 = (11100011 \quad 10110000 \quad 11011001)$$
$$\text{Pixel}_3 = (11100000 \quad 10110011 \quad 11010101)$$

Now a secret message 'A = 11,001,010' needs to embed into the above image. After applying the LSB algorithm, the message 'A' embeds into the above pixels; it is converted into unrecognizable pixels.

$$\text{Pixel}_1 = (11110001 \quad 00110001 \quad 01011000)$$
$$\text{Pixel}_2 = (11100010 \quad 10110001 \quad 11011000)$$
$$\text{Pixel}_3 = (11100001 \quad 10110010 \quad 11010101)$$

3.3 Double-Key Protection

To secure the embedding and decrypting process, in this experiment, two keys are used [10, 11]. Key1 is a 64-bit secret key, and key2 is a secret key with 128 bit or greater. These two keys help to perform the LSB algorithm to encrypt or decrypt the secret message.

3.4 Reverse the Video with Password Verification

Reverse video [12, 13] is a display mechanism. It used to invert the video sequences from end to beginning. In this experiment, this technique is used to make a secure encrypting process. When the video sequence is inverted, then the number of the encrypted frame in the video file is also changed. This process helps to hide the exact encrypted frame very securely. Equation 1 shows the calculation for finding the reverse encrypted video frame number.

$$\begin{aligned}&\text{Reverse Encrypted Video Frame No} = \\ &(\text{Total Video Frames} - \text{Actual Encrypted Video Frame No.}) + 1\end{aligned} \tag{1}$$

Finally, a password with eight characters is used for the verification process to perform this mechanism more securely. The below algorithm shows the whole verification process for encryption and decryption.

Encryption Method

- Select a particular frame from the input video file
- Start LSB coding with Dual key encryption
- Register a password (exact eight characters long)
- Reverse the encrypted video

Decryption Method

- Calculate the reverse encrypted_frame no and verify
- If correct
  - Apply password verification
  - If correct
    - Check key1 & key2
    - If ok
      - Apply LSB coding for decryption
    - else
      - Invalid!
  - else
    - Invalid!
- else
  - Invalid!

3.5 Signature Verification

In this section, a handwritten signature database is used for final verification. The database is downloaded from Kaggle. It contains more than 1000 genuine and forged handwritten signature images. Both encryption and decryption methods needed two-step signature verification [14–16] to complete the whole process. For signature verification, local binary pattern and multiclass support vector machine mechanisms are used. For signature registration, the very last two frames of reverse encrypted video are used. The below algorithm shows the verification process.

Encryption Method

- Register Sender Signature [Sender Encryption] from sender
- Register Receiver Signature [Receiver Encryption] from sender

Decryption Method

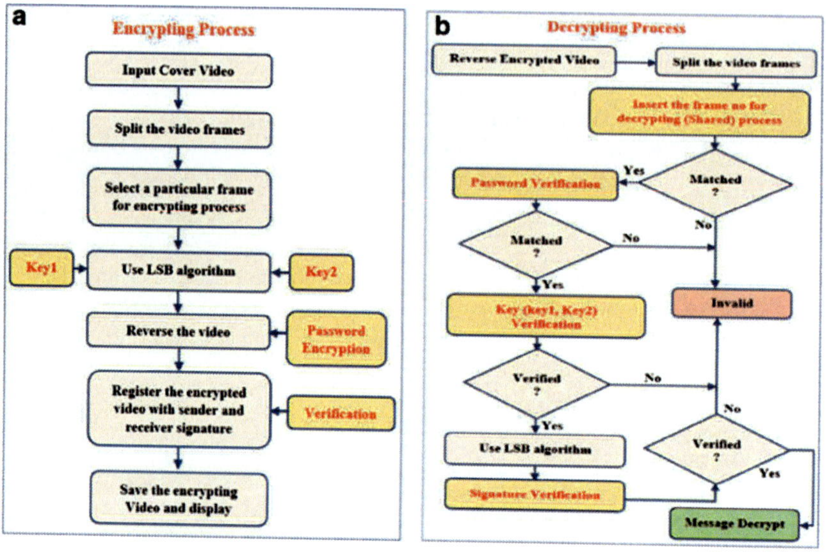

Fig. 3 **a** Encryption process, **b** Decryption process

- `Accept Receiver Signature [Receiver_Decryption] from receiver`
- `Accept Sender Signature [Sender_Decryption] from receiver`
- `If(Sender_Decryption == Sender_Encryption) and(Receiver_Decryption == Receiver_Encryption)then`
  - `Message Decrypt Successfully!`
- `Else`
  - `Invalid!`

Figure 3a–b shows the entire encrypting and decrypting process of this experiment.

4 Result and Discussions

The proposed methodology was applied on a random audio–video interleaved (.avi) video file. For implementing this process, MATLAB 2020a App designer was used. The initial interface is shown in Fig. 4a–b.

The encrypting process is shown in Fig. 5a–b. The secret key1 and 2 and secrete messages are shown in Fig. 6a–c.

The whole decrypting process is shown in Fig. 7a and b.

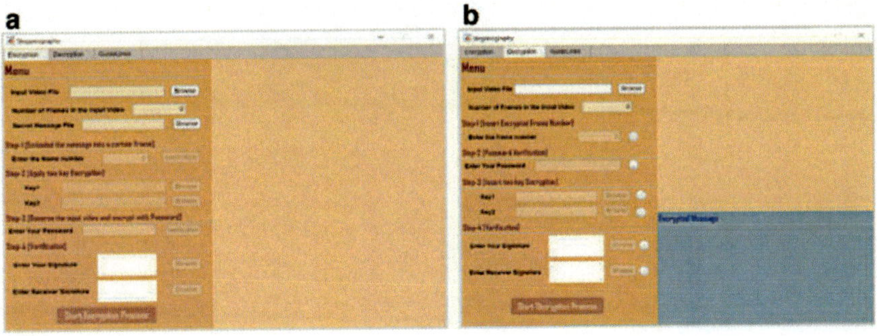

Fig. 4 Initial interface of **a** Encryption process (Left) **b** Decryption process (Right)

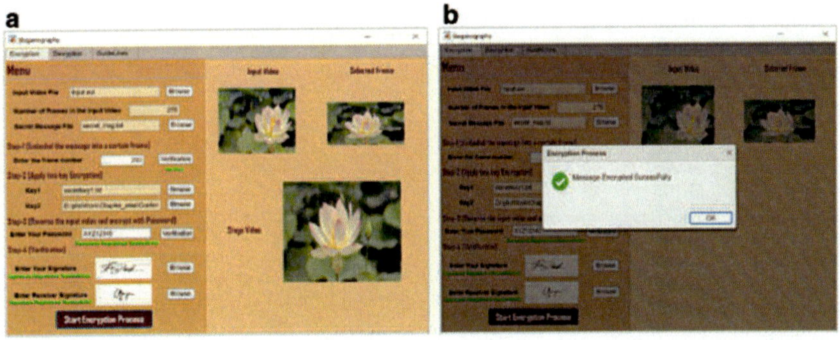

Fig. 5 **a** Different parameters setting before encryption process, **b** After encryption process

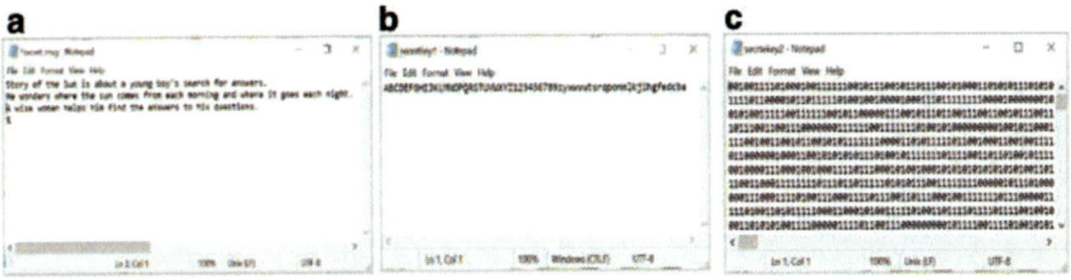

Fig. 6 **a** Secret massage, **b** Secrete Key1, **c** Secret Key2

In the decryption process, if any verification process goes wrong, the system returns the error message. Figure 8a shows the error output message of the system. The experimental result in the signature verification process is shown in Fig. 8b.

The performance analysis of this process is shown in Table 1. The accuracy comparison among the three classification models for the signature verification process is shown in Fig. 9.

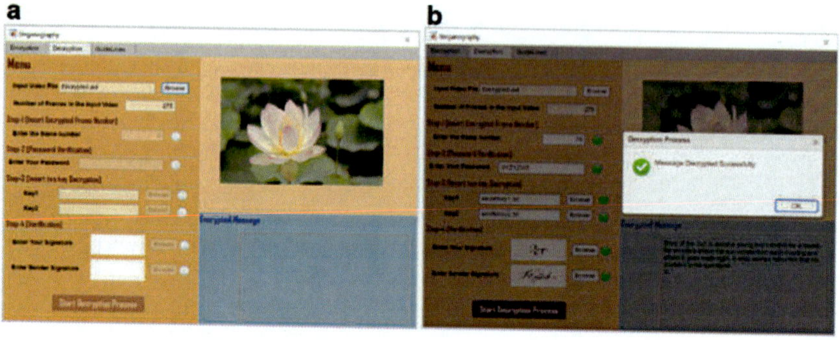

Fig. 7 **a** Parameter setting before decryption, **b** After decryption

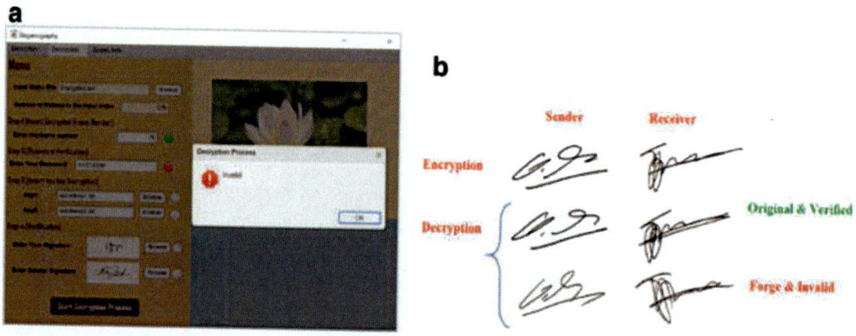

Fig. 8 **a** Decrypting process (Error Output), **b** Signature verification process

Table 1 Performance analysis of signature verification

Method		Accuracy (%)
Feature extraction	Classification	
Linear binary pattern (LBP)	Support vector machine	96.2
	Decision tree	91.2
	K-nearest neighbor	93.7

In this experiment, a special character is used in the secret message file. This special character indicates the end of the message. This experiment has some limitations; they are 1) the input and output video file must be uncompressed audio–video interleaved (.avi) video file, 2) the secret message, secrete key1 and key2 file must be in text (.txt file) format. This proposed method is suitable for confidential data sharing.

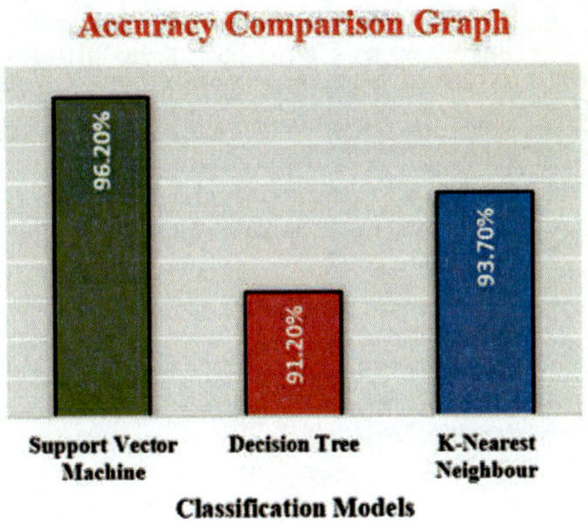

Fig. 9 Graphical representation

5 Conclusion

The aim of the proposed methodology is to create a secure steganography process with a video cover medium. The dual key, secrete key1 and key2, and the least significant bit coding are used for double protection. In this experiment, reverse video encryption method was used. For a third-party intruder, this process is one of the disadvantages to identify the correct encrypted frame number. Finally, the password verification and the signature verification process give high security throughout the whole process. The entire process shows an efficient method to encrypt and decrypt the secret data using the steganography technique.

References

1. Ramalingam M, Mat Isa NA, Puviarasi R (2020) A secured data hiding using affine transformation in video steganography. Procedia Comput Sci 171:1147–1156. https://doi.org/10.1016/j.procs.2020.04.123
2. Sindhu R, Singh P (2020) Information hiding using steganography. Int J Eng Adv Technol 9:1549–1554. https://doi.org/10.35940/ijeat.d8760.049420
3. Dhawan S, Gupta R (2021) Analysis of various data security techniques of steganography: a survey. Inf Secur J 30:63–87. https://doi.org/10.1080/19393555.2020.1801911
4. Raja Ratna S, Shajilin Loret JB, Merlin Gethsy D, Ponnu Krishnan P, Anand Prabu P (2020) A review on various approaches in video steganography. In: Lecture notes on data engineering and communications technologies. pp 626–632. https://doi.org/10.1007/978-3-030-28364-3_64
5. Al-Omari Z, Al-Taani A T (2015) A survey on digital image steganography. In: The 7th international conference on information technology. Al-Zaytoonah University of Jordan, pp 109–115. https://doi.org/10.15849/icit.2015.0016
6. Gutub A, Al-Shaarani F (2020) Efficient implementation of multi-image secret hiding based on LSB and DWT steganography comparisons. Arab J Sci Eng 45:2631–2644. https://doi.org/10.1007/s13369-020-04413-w
7. Kadhim IJ, Premaratne P, Vial PJ (2020) High capacity adaptive image steganography with cover region selection using dual-tree complex wavelet transform. Cogn Syst Res 60:20–32. https://doi.org/10.1016/j.cogsys.2019.11.002

8. Darbani A, Alyannezhadi MM, Forghani M (2019) A new steganography method for embedding message in JPEG images. In: 2019 IEEE 5th conference knowledge based engineering innovation KBEI 2019. pp 617–621. https://doi.org/10.1109/KBEI.2019.8735054
9. Saleh MA (2018) Image steganography techniques—a review paper. Ijarcce 7:52–58. https://doi.org/10.17148/ijarcce.2018.7910
10. Sumathi CP, Santanam T, Umamaheswari G (2013) A study of various steganographic techniques used for information hiding. Int J Comput Sci Eng Surv 4:9–25. https://doi.org/10.5121/ijcses.2013.4602
11. Duan X, Guo D, Liu N, Li B, Gou M, Qin C (2020) A new high capacity image steganography method combined with image elliptic curve cryptography and deep neural network. IEEEAccess 8:25777–25788. https://doi.org/10.1109/ACCESS.2020.2971528
12. Cao M, Tian L, Li C (2020) A secure video steganography based on the intra-prediction mode (Ipm) for h264. Sensors (Switzerland). 20:1–13. https://doi.org/10.3390/s20185242
13. Younus Z, Younis G (2020) Video steganography using knight tour algorithm and LSB method for encrypted data. J Intell Syst 29:1216–1225
14. Chang C-C (2021) Neural reversible steganography with long short-term memory. Secur Commun Netw 2021:1–14. https://doi.org/10.1155/2021/5580272
15. Zhou Y, Zheng J, Hu H, Wang Y (2021) Handwritten signature verification method based on improved combined features. Appl Sci 11:5867. https://doi.org/10.3390/app11135867
16. Poddar J, Parikh V, Bharti SK (2020) Offline signature recognition and forgery detection using deep learning. Procedia Comput Sci 170:610–617. https://doi.org/10.1016/j.procs.2020.03.133

Interactive Search Group Mining from Annotated Web Pages

P. Sijin and H. N. Champa

Abstract Machine learning facet is a query-dependent upper layer service. Each facet is a set of items which describe and summarizes one important aspect of a query. Faceted search allows the users to search on a facet list of a query to pick out the desired one without browsing for a long time. The proposed facet mining framework is called facet mining from Annotated Documents (FMAD) integrates keyword search by category browsing and produces an interface which has several conceptual dimensions. The FMAD is designed to achieve an interactive data summarization process by liberalizing topic identification and interpretation on user side. It is an ontology-based query-dependent facet framework which follows a series of phases from initial facet weighting to facet item ranking. The FMAD produced high-quality clusters by using a context similarity approach which provides high cluster discrimination. FMAD facets automatically extracted from document description, annotated documents, and metadata records.

Keywords Facet · Metadata · Item ranking · Datapoint

1 Introduction

Facet mining is the process of exploration and discovery within an information collection of selected choices, and it is used to summarize the knowledge and information contained in the knowledge base with high-quality structured data, large databases of text-annotated objects, semantically related feature set such as WordNet for a query [1–3]. Query reformulation and query recommendation are the methods widely used by the users to explore their intention. Direct and instant answer groups are available with faceted search, e.g., cell phone groups with display size, color, OS used. In the

P. Sijin (✉)
University Visvesvaraya College of Engineering, Bangalore University, Bengaluru, India
e-mail: psijin@gmail.com

H. N. Champa
University Visvesvaraya College of Engineering, Bengaluru, India

M. S. Kaiser et al. (eds.), *Information and Communication Technology for Competitive Strategies (ICTCS 2021)*, Lecture Notes in Networks and Systems 401,
https://doi.org/10.1007/978-981-19-0098-3_53

Table 1 Query facets for query: "Watches"

Facet name	Group items
Watch category	Men, women, kids, etc.
Watch brand	Alpina, Citizen, Favre Leuba, Movado, Victrinox, etc.
Watch color	Black, brown, pink, red, etc.
Watch type	Analog, digital, quartz, etc.

case of vague, ambiguous queries which are short in nature and noisy, facets are useful by provided with a clarified list of choice of user interest. Similar to normal and approximate query results, facets are also abundant in nature. In order to avoid the facet boom, it is possible to rank the facets and could display the top-k diversified links of choice [4–7]. Table 1 shows the facet lists to item "watch" in various categories. Facet lists to watch are crawled for four categories of watch, namely watch category, watch brand, watch color, and watch type. The user can display the query facets along with top search results, since the facets reveal the different aspects of a query, the user will get an additional set of results which may relate to the given query and could be used for approximate search process [8–11].

The quality of a facet can be measured in terms of specificity and dispersion. Specificity is the quality of belonging or relating uniquely to a particular subject. Dispersion can be defined as the action of distributing something or appearance of a facet on multiple lists. The proposed dynamic facet mining process for Table 1 data generates faceted items with high ranks on an online database by a systematic ranking process based on metadata and is given in Fig. 1.

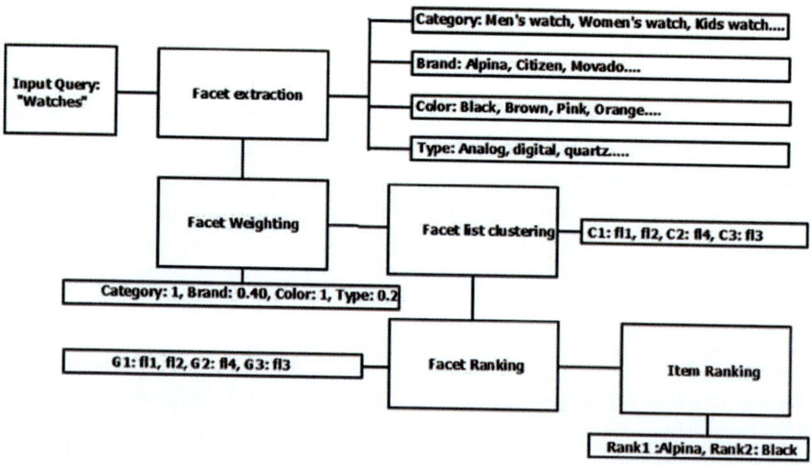

Fig. 1 Dynamic facet mining process

2 Literature Review

Takahashi et al. [12] proposed a new, user behavioral approach such as link anomaly-based detection, to detect the emergence of topics in a social network stream. The basic idea of this approach is to focus on the social aspect of the posts reflected in the mentioning behavior of users instead of the textual contents. The works by Yan and Wan [13] used a heterogeneous ranking algorithm called SRRank. A heterogeneous graph is constructed for sentences, semantic roles, and words as nodes. Using SRRank algorithm, the nodes are ranked and the nodes with the highest ranks are taken for generating summaries.

Transitional query suggestion approaches such as annotations, query logs, online summarization methods are used to design database schemas and to generate metadata attributes [8, 14–16].

Dou et al. [5] proposed a framework to list out the facets from millions of tweets called QDMiner. QDminer extracts lists from the top search results, groups them into clusters based on the features and ranks them according to how they appear in the search results.

Vandic et al. [4] proposed a framework for dynamic facet mining based on specificity and the dispersion of facet values. The proposed facet optimization algorithm ranks properties based on their importance and sorts the values within each property.

3 Methodology

The proposed facet mining process has various phases such as facet extraction, facet weighing, facet list clustering, facet ranking, item ranking. Facets are used to prepare multiple groups of semantically related queries [3].

3.1 Problem Statement

For a document Doc_i annotated with metadata M_k and a query Q_i produce facets for each Doc_i with weight $\mathrm{Doc}_{\mathrm{fi}}$, these facets are clustered and grouped to sort out top-k search items with ranks $\mathrm{WRank}_{\mathrm{ei}}$ where fi is the weight of each facet associated with that document and ei is the rank of an item e.

3.2 Facet Extraction

The FMAD performs a facet extraction of metadata of documents [6, 17, 18]. It can also able to extract the required information from web forms by applying tag

Table 2 Examples for data extraction from web page

Extraction type	Facet group	Faceted data (all data not included)
Metadata	Watch brand	Shop watches from 25+ brands like Anne Klein
Tag-based	Price range	Price = 1000–10000, 1k–10k

level extraction. The metadata of the document contains lots of information about the contents and properties related to that document. The meta properties like title, type, url, image, description, site name, sign-in client-id, sign-in cookie policy, sign-in scope are useful for achieving document categorization and data security [8], [19]. Table 2 lists out the results of metadata extraction of the query "watches" over ClueWeb dataset. After list extraction, the obtained results are listed as facets such as category, brand, color, and type.

3.3 Facet Weighing

The weight of a facet in a document is calculated based on the facet count in the document and the number of shared keywords in facet keyword lists and document. Let M_k be the metadata keyword lists for a facet. If a document accounts for the presence of more number of a particular facet in it, then that document shows the high-term frequency for that particular facet; hence, facet count is valuable for facet weight (Facet_w) calculation. In order to balance the effect of common items in a document inverted document frequency (IDFE) of that document, represented as idf_e can be considered during weight calculation and the entire equation is given in (1).

$$\text{Facet}_w = \text{Doc}_w * \text{Facet}_{\text{count}} * \frac{\text{idf}_e}{|l|} \tag{1}$$

where

$$\text{Doc}_w = \frac{N_{\text{ld}}}{|l|} \tag{2}$$

where $N_{\text{ld}} \leq M_k$ is the number of shared keywords in the facet keyword list and the document used and l is the total number of lists identified as given in (2). idf_e is calculated as in (3).

$$\text{idf}_e = \log \frac{N - N_e + 0.5}{N_e + 0.5} \tag{3}$$

where N is the total number of words in the given document and N_e is the number of terms which has a direct or semantic relation to the given item e. After facet weighing normalized value list obtained is given as (Category: 1, Brand: 0.40, Color: 1.0, Type: 0.2).

3.4 Facet List Clustering

During facet clustering list with similar intentions are grouped together. If two facet lists are related proportionally distance between them can be calculated as in (4). Quality threshold with weighted data point and web page count (WQTWC) is used to reduce the overlap of clusters.

$$d(f_{l_1}, f_{l_2}) = 1 - \frac{|f_{l_1} \cap f_{l_2}|}{\min(|f_{l_1}|, |f_{l_2}|)} \tag{4}$$

The pairwise distance obtained between the facet lists is $d(f_{l_1}, f_{l_2}) = 0.2$, $d(f_{l_1}, f_{l_3}) = 1$, $d(f_{l_1}, f_{l_4}) = 0.94$, $d(f_{l_2}, f_{l_3}) = 1$, $d(f_{l_2}, f_{l_4}) = 0.91$, $d(f_{l_3}, f_{l_4}) = 1$. WQTWC algorithm sets a cluster diameter of user choice and here it is 0.96. All the points with threshold value up to 0.96 are locked in the initial cluster as given in Algorithm 1. Algorithm 1 chooses minimum weighted list W_{min} from faceted list. Then it choses a threshold diameter $\text{Diameter}_{\text{max}}$ and randomly selected a target facet list with weight $\text{Target}_{\text{weight}}$. The pairwise distances among lists are calculated and made a core cluster for initializing the clustering. Then set $\text{Target}_{\text{weight}}$ as locus of the core cluster. After this added all faceted lists with diameter less than $\text{Diameter}_{\text{max}}$ to core cluster. The remaining faceted lists are also clustered similarly. A lookup table table is maintained for faceted list for getting website count. Based on website count performed a re-clustering operation to produce final clusters. The algorithm produced three clusters, namely C_1, C_2, C_3, are shown in Fig. 2. The faceted lists $d(f_{l_1}, f_{l_2}, f_{l_4})$ are grouped in cluster C_1. Faceted lists $d(f_{l_3})$ are grouped in cluster C_2, and faceted lists $d(f_{l_4})$ are grouped in cluster C_3. The typical WQT algorithm works on weighted list and grouped f_{l_4} in cluster C_1. The normalized website count for the given faceted lists is calculated as (Category: 1, Brand: 1, Color: 0.6, Type: 0.01). The website count for f_{l_4} is comparatively less and that increased its threshold distance above 0.96, and it is grouped in C_3.

3.5 Facet Ranking

Facet ranking is the process of grouping lists based on their hamming distance in each cluster. The weight of a group is the number of lists in that group. Here the group is defined as the lists in a cluster which have close similarity. A list grouping approach is used for this purpose. This approach set minimum weight $W_{\text{min}} = 1$ and

Algorithm 1: Quality threshold with weighted data point and web page count

Data: The facet keyword lists of various items.
Result: The pairwise distances among faceted list and facet clusters for the query.
initialization;
1. Choose a minimum weight W_{min} for facets.
2. Choose a threshold diameter $Diameter_{max}$ for facets.
3. Choose a target facet with weight $Target_{weight} \geq W_{min}$
4. Calculate the pair-wise distance among all pairs of facet lists according to (4).
5. Make a core cluster with $Target_{weight}$ as locus.
6. Add all faceted lists with weight $\leq Diameter_{max}$ to the core cluster.
7. Go to step 2 and Cluster the remaining faceted list according to their weight.
8. Calculate website count Web_{count} for each facet list.
9. Calculate the new weight by multiply (2) with Web_{count}.
10. Go to step 2 and restructure the clusters for new facets by feedbacking current cluster details.

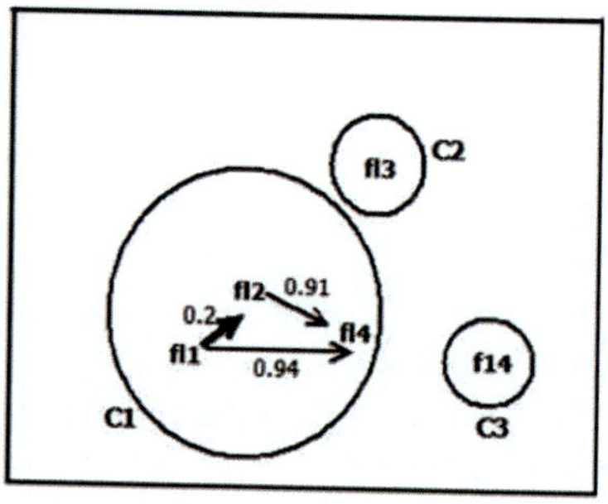

Fig. 2 Facet clustering with WQTWC algorithm

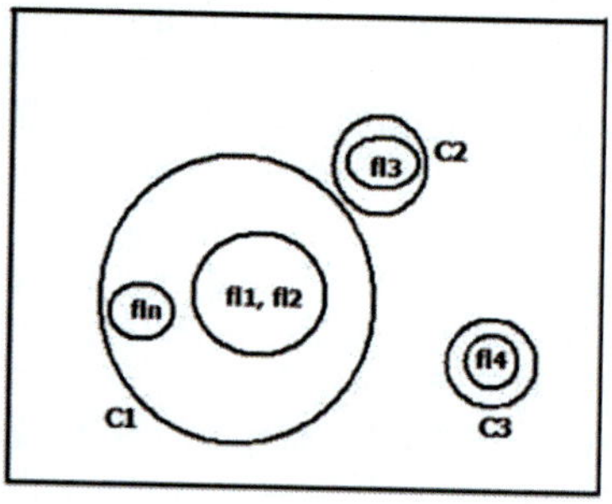

Fig. 3 During facet ranking related groups are clustered

a threshold value FLdup are identified as the similarity hamming distance, in order to form a sub-group in selected clusters. The cluster with the highest weighted lists is identified, and the highest weighted list is made as the locus of subgroups within the group. The other lists with weight less than FLdup added to the new subgroup. The same process has done to other *n* top-most clusters. The hamming distance for watch category $d(f_{l_1})$ and watch brand $d(f_{l_2})$ is calculated as in (5), and the subgroups formed in cluster C_1 are shown in Fig. 3.

$$\mathrm{HD}_{f_{l_1}, f_{l_2}} = 1 - \frac{\mathrm{hd}(d(f_{l_1}, f_{l_2}))}{\mathrm{LFP}} \quad (5)$$

where LFP is the length of fingerprint used by default it is 64, and $\mathrm{HD}_{f_{l_1}, f_{l_2}}$ is the hamming distance between two faceted lists (f_{l_1}, f_{l_2}).

3.6 *Item Ranking*

The importance of an item e in a list depends on its rank in the list and the number of lists accessing it. The rank of faceted item in a group depends on the item count in the lists of that facet. The weight of an item in a faceted list f_{l_i} is calculated as in (6).

$$\mathrm{WRank}_{\mathrm{ek}} = \sum_{f_{l_i} \in C} \frac{1}{\sqrt{\mathrm{AvgRank}}} \quad (6)$$

where average facet rank AvgRank is calculated as

$$\mathrm{AvgRank} = \frac{1}{|\mathrm{SFL}|} \sum_{l \in \mathrm{SFL}} \mathrm{Rank}_e \quad (7)$$

where SFL is the set of all list in the facet and the faceted group which contain the item *e*.

4 Performance Evaluation

4.1 *Effect of List Weighing Method*

A list which contains a term frequently in it would show a high document weight for it, but at the same time some valid facets may degrade because of less frequent appearance. If including idf_e in calculation this problem can solve. Document weight as obtained from (1) is the measure of facets which are on various lists, and it improved the facet count in lists. Figure 4 shows the effect of list weighting in facet mining.

4.2 *Effect of Search Result Quantity*

When the number of search results is increased, then the facet groups are enriched with more lists; these have more facets, and hence, the overall quality of facets is

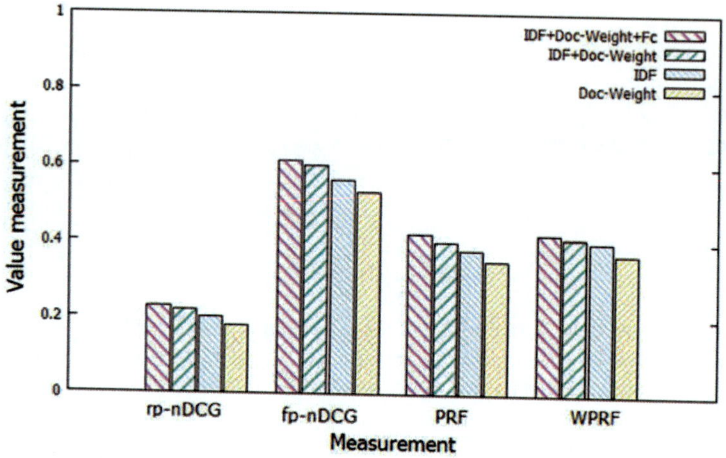

Fig. 4 Facet weight with various measurement methods

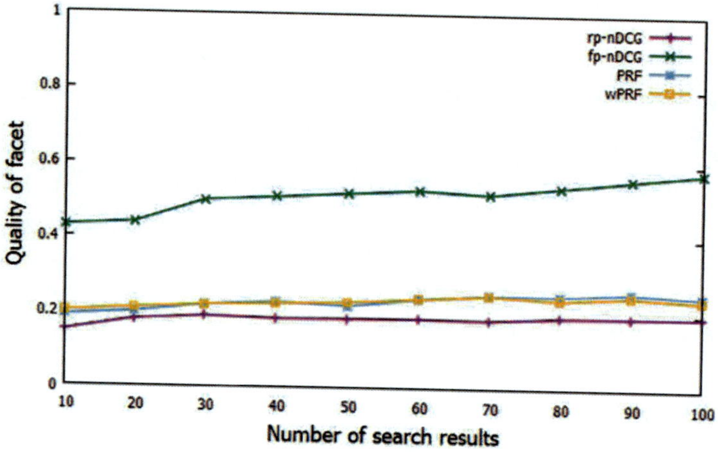

Fig. 5 Effect of search result quantity during facet mining

improved, and Fig. 5 shows this. From Sect. 3.3, it is clear that the facet count has an important role in the facet list weighing process. So if the results are more there is a possibility of their facet counts also to be more. Specific properties whose facets match many products have high impurity.

5 Conclusion

The proposed facet mining approach followed a series of phases such as facet clustering and grouping in order to organize the items, for which user is able to drill down with an intention over a navigational search platform. Because of the abundance of facet results, the FMAD followed a facet weighing and clustering scenario. Quality clusters are derived by considering disjoint facet count. Hamming distance is measured for cluster grouping by considering each facet list in a cluster. Average facet rank is calculated to list out top-k items which are specifically for query and dispersed on query dimension.

References

1. Latha KVK (2010) Facet generation framework for document retrieval. In: IEEE International conference on electro/information technology (EIT). In: 2010 International conference on advances in computer engineering, pp 602–607
2. Jiang Z, Dou Z, Wen J-R (2016) Generating query facets using knowledge bases. IEEE Trans Knowl Data Eng 29(2):315–329
3. Herdagdelen A, Ciaramita M, Alfonseca E (201) Generalized syntactic and semantic models of query reformulation. In: Proceedings of the 33rd International ACM SIGIR conference on research and development in information retrieval, pp 283–290
4. Vandic D, Aanen S, Frasincar F, Kaymak U (2017) Dynamic facet ordering for faceted product search engines. IEEE Trans Knowl Data Eng 29(5):1004–1016
5. Dou Z, Jiang Z, Hu S, Wen JR, Song R (2016) Automatically mining facets for queries from their search results. IEEE Trans Knowl Data Eng 28(2):385–397
6. Liu Q, Chen E, Xiong H, Ding CH, Chen J (2011) Enhancing collaborative filtering by user interest expansion via personalized ranking. IEEE Trans Syst Man Cybern Part B (Cybern) 42(1):218–233
7. Kim HJ, Zhu Y, Kim W, Sun T (2014) Dynamic faceted navigation in decision making using semantic web technology. Decis Support Syst 61:59–68
8. Ruiz EJ, Hristidis V, Ipeirotis PG (2014) Facilitating document annotation using content and querying value. IEEE Trans Knowl Data Eng 26(2):336–349
9. Liu J, Yan D (2016) Answering approximate queries over XML data. IEEE Trans Fuzzy Syst 24(2):288–305
10. Hahsler M, Bolaños M (2016) Clustering data streams based on shared density between micro-clusters. IEEE Trans Knowl Data Eng 28(6):1449–1461
11. Nivetha K, Ram GR, Ajitha P (2016) Opinion mining from social media using fuzzy inference system. In: International conference on communication and signal processing, pp 2171–2175
12. Takahashi T, Tomioka R, Yamanishi K (2014) Discovering emerging topics in social streams via link-anomaly detection. IEEE Trans Knowl Data Eng 26(1):120–130
13. Yan S, Wan X (2014) SRRank: leveraging semantic roles for extractive multi-document summarization. IEEE/ACM Trans Audio Speech Lang Process 22(12):2048–2058
14. Wang Z, Shou L, Chen K, Chen G, Mehrotra S (2015) On summarization and timeline generation for evolutionary Tweet streams. IEEE Trans Knowl Data Eng 27(5):1301–1315
15. Zo H, Ramamurthy K (2009) Consumer selection of e-commerce websites in a B2C environment: a discrete decision choice model. IEEE Trans Syst Man Cybernet Part A Syst Hum 39(4):819–839
16. Kules B, Capra R, Banta M, Sierra T (2009) What do exploratory searchers look at in a faceted search interface. In: Proceedings of the 9th ACM/IEEE-CS joint conference on digital libraries, pp 313–322

17. Liberman S, Lempel R (2014) Approximately optimal facet value selection. Sci Comput Prog 94:18–31
18. Vandic D, Frasincar F, Kaymak U (2013) Facet selection algorithms for web product search. In: Proceedings of the 22nd ACM international conference on information and knowledge management, pp 2327–2332
19. Kong W, Allan J (2014) Extending faceted search to the general web. In: Proceedings of the 23rd ACM international conference on information and knowledge management, pp 839–848

IPD—An Investment Portfolio Detector Using SEBI API

M. S. Karthika Devi, R. Bhuvaneshwari, K. Kavin Pragadeesh, and Baskaran Ramachandran

Abstract World at this era is growing rapidly in the financial field. Meanwhile, data storage and manipulation are also increasing. Our proposed work lies in the field of managing the investor portfolio and individual financial advisor database. The main entities are asset management companies (AMCs), individual financial advisor (IFA) database, investor portfolio, NetBanking authorities, and Security Exchange Board of India (SEBI). Investors usually invest in certain schemes offered by AMCs, and IFA helps their clients in their fruitful investment and redemption. Investors can also buy and sell individual shares in initial public offering (IPO). But, the data on the whole is not managed properly and effectively in any IPO base and with investors. So, it would be especially useful for investors and IFAs if they can view their total financial information in a single swipe. Investors should be able to view their portfolio whenever they want and their current net asset value in their device. Percentage increase in the company shares of their invested fund would yield a clear background of investor's portfolios. Financial advisors should be able to view their client's portfolio at their place anytime with the client's unique key. Same database can be extended to the mutual fund controlling authorities like SEBI and AMFI. The proposed idea can also be extended by adding NSE and BSE daily values reflected in the site. Thus, our work lies in the investor portfolio and IFA client database.

Keywords Net asset value · Initial public offering · SEBI · Asset management company

M. S. Karthika Devi · R. Bhuvaneshwari (✉) · K. Kavin Pragadeesh · B. Ramachandran
College of Engineering, Anna University, Chennai, India
e-mail: bhuvi.saakshu@gmail.com

B. Ramachandran
e-mail: baaski@annauniv.edu

M. S. Kaiser et al. (eds.), *Information and Communication Technology for Competitive Strategies (ICTCS 2021)*, Lecture Notes in Networks and Systems 401,
https://doi.org/10.1007/978-981-19-0098-3_54

1 Introduction

Financing is the process of providing funds for business activities, making purchases, or investing. The use of financing is vital in any economic system, as it allows companies to purchase products out of their immediate reach. Finance industry being a market king in the developed industry has been making its prints in almost all technology. It makes people recognize its technological advancement and helps people get used to all the financial resources. Financial institutions, such as banks, are in the business of providing capital to businesses, consumers, and investors to help them achieve their goals. One of the developing windows of finance companies is fund management and sectoring. This influences people to the core and helps them realize how it could mold or streamline life. The savings thought had been embedded in almost all people, and they were moving toward it. Many asset management companies were established and serving people with their vital strength in managing their portfolio and developing a wide and colorful future for the people. They were thinking of many ways to ease transactions and providing digital support in all the fields to people. This work also deals with a creative idea of providing an easy, simple, and digital way of viewing the portfolio and fund details of individuals all under one roof.

2 Related Works

Jaiwat [4] has proposed a simple method with fewer security and less complexity. It breaks the only higher-order accessibility of the financial resources and paves the way for the least cadre investors, financial advisors, and fund managers to view the finance market index any time. The end tabulations and results show equivalence between the tracking portfolio and the financial market index. The model predictive control (MPC) is used for finding optimal portfolios. Recalculation of the controller in every sampling time is the benefit of this model.

A new strategy for deriving the best portfolio of funds after monitoring the growth and efficiency of the market quotient was proposed by Li [1]. The author had proposed a genetic algorithm to derive at the optimized financial portfolio results in this paper. This work also exhibits the security issues in handling the funds and financial data and solved the issue using the proposed genetic algorithm. It stands strong with its globally accepted and optimized algorithm to generate a portfolio, avoid the risk, and earn the income.

Bi [2] has dealt with educational funds of Harvard University in this work and proposed mathematical methods of analyzing the funds, portfolio structure, and advises university education fund in China on their updates. This work mainly categorizes the types of funds Harvard University deals with and tabulates the profit and income with respect to each fund. A plot with year-wise funds growth has been analyzed and advises to focus on sound sectors.

The objective of enlarging or varying the field of operation of the portfolio to minimize the risk was proposed by Kedia et al. [3]. Risk factor and investment banking analysis was carried out. The idea of adoption from this work is usage of simple *k*-means cluster approach for optimal portfolio generation. The elbow method has been used for clustering, and the results were plotted. The improvisation of this work is that it could be extended for other datasets instead of considering BSE100 only.

Yu et al. [6] have studied quantitative risk management in portfolio generation since earlier times. Considering such difficulties, they have used methodologies such as neural networks (NNs) monitoring to build an economic factor-based predictive model (EFPM). Then, they combined it with the Copula-GARCH simulation model and the mean-conditional value at risk (Mean-CVaR) framework to derive an optimal portfolio. The results were tabulated and analyzed with 99% accuracy.

3 Proposed Work

The above related works have been gone through disadvantages, and further advancements have been analyzed. The clients who invest in various funds do not have a common portal to connect all their IFAs and view their fund details in what they invested. This project mainly overcomes the real-time difficulties faced by financial advisors and clients. It stands with its strong objective of providing financial portfolio details to the clients at any time with the accurate NAV of the funds in what they have invested and also providing an interface and portal for the individual financial advisors to overcome their issue of maintaining a business client base. It is a web application to view funds of clients from all the AMCs and their net asset values in a single page and allows IFAs to view their clients details in one place and add more clients and funds details. Clients would be authenticated and authorized before entering into the applications. Details about the clients and funds were to be fed to the database by the IFAs which is used for building up the portfolio pane of the clients. The NAV of each fund of the client was fetched from mutual funds API and embedded in the interface. And also, the profit and loss indications compared to the original principle are also provided. All the funds of all AMCs of the clients were given in the single page so that they need not wander between the offices for current details of the funds. This application also cares for the service of messaging between the clients and IFAs for remote communication. It is accessible from all devices. It also paves the way for connecting clients to connect from remote with IFAs. This application also promotes the security of client data as the ledger is maintained only by their trustworthy admins (IFAs) and would be shared in the outer circle. The below diagram (see Fig. 1) depicts the overall flow of the application with respect to clients and advisors.

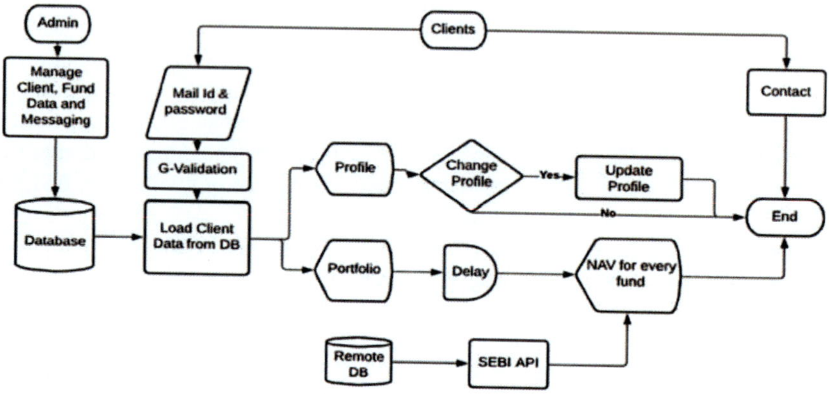

Fig. 1 Architecture diagram

3.1 Admin Portal

Admin module deals with three segments. They are clients, funds, and NAV, respectively. Each segment contains a form to enter respective information.

Id Generation. Clients and fund should have a unique identifier to maintain and fetch records from the backend database. Each client and fund will be given a unique tag to represent themselves in the common pool of admin's client list. Admin would generate unique keys for both clients and funds. This unique Id created would be given to the client which he/she uses at the time of registration. Secondary information like name, address, phone number, etc., for clients and fund name and face value for funds were collected, respectively. The management functions of the admin portal are depicted in Fig. 2.

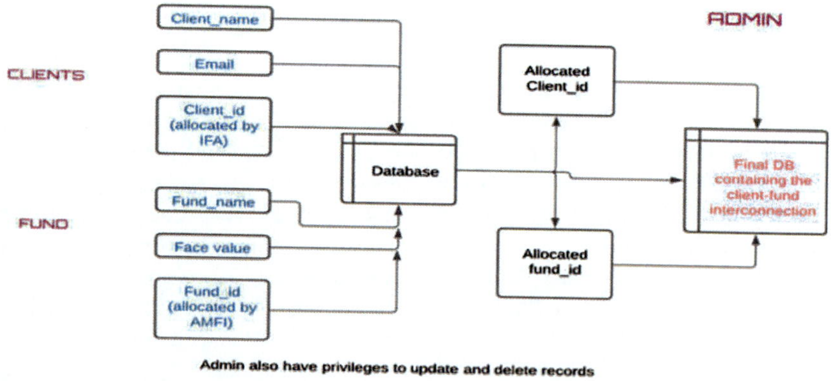

Fig. 2 System design of admin module

Client—Fund Relationship. Relationships between clients and their respective funds were established in a separate portal, and the record would be maintained separately. In case if any redemption happens, it would be easy for admins to delete the desired record alone. Figure 3a–d gives an overview of various pages in admin portal in real time.

Algorithm

1. Figure out the Entities in the application (say Clients, Contact).
2. Normalize the Database Structure.
3. Generate unique IDs for Clients and Funds (AMFI Id—for NAV fetching).

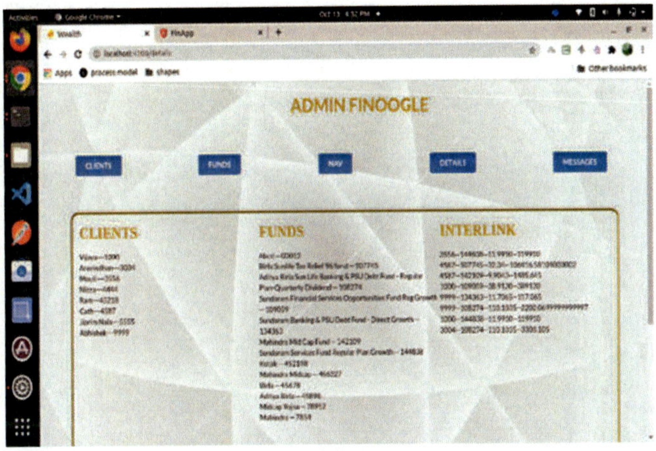

(a)

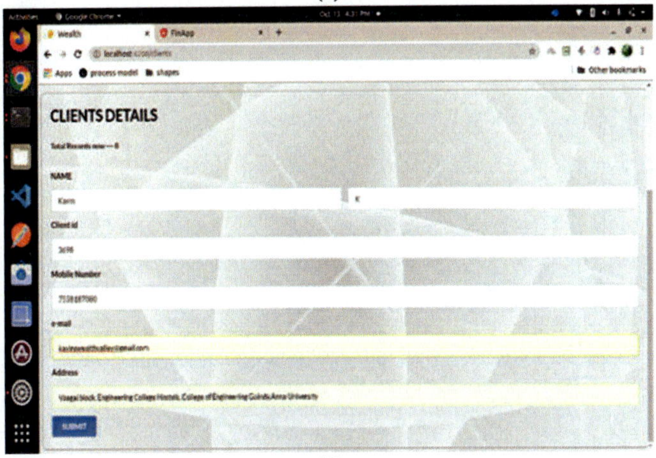

(b)

Fig. 3 **a** Admin portal—records viewing page. **b** Admin portal—client registration page. **c** Admin portal—fund registration page. **d** Admin portal—mapping clients with funds

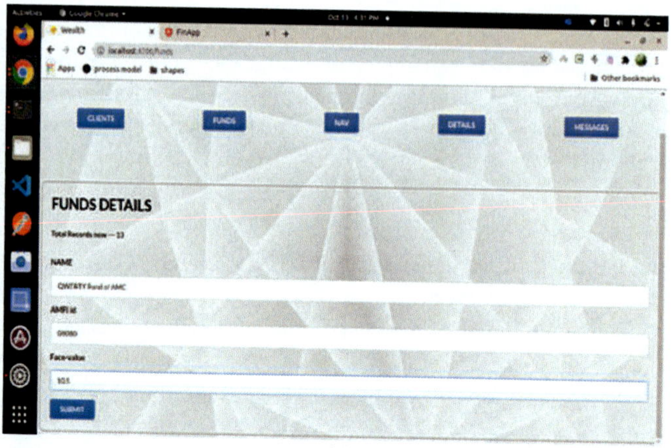

(c)

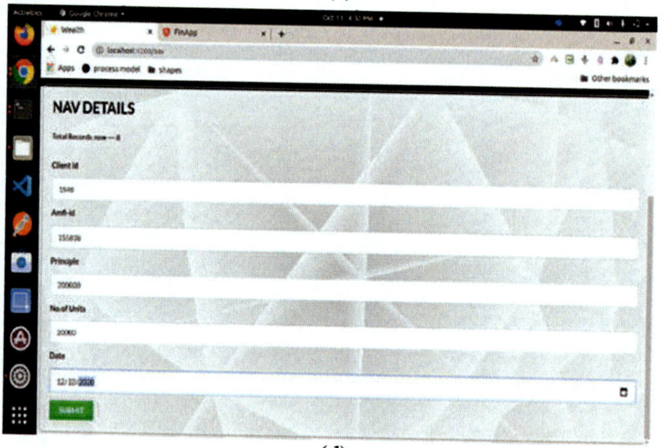

(d)

Fig. 3 (continued)

4. Build the interface for Financial Advisors for data collection of clients, funds and relationships between both.
5. Build the interface for Clients including Authentication, Profile management and Portfolio Design.
6. NAV for each fund of the client is been fetched by Mutual Funds API.
7. Fetch Fund Details like Fund name, Date of investment and Principal Amount from Database and displayed in the portfolio portal.
8. Multiply the NAV with no. of units to obtain the current amount of the fund of the client.
9. Current Capital is being compared with Principal to derive at profit or loss indication.
10. Update the profile for client information if any.

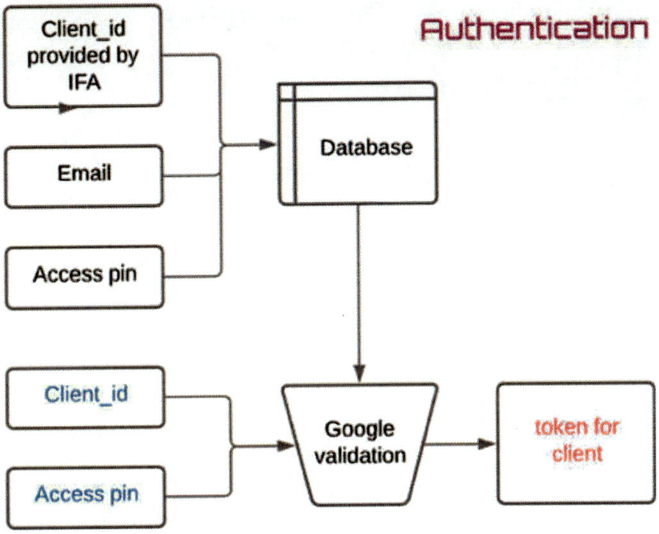

Fig. 4 System design of authentication module

3.2 *Authentication*

This module deals with the registration of clients with their Id and creating records for them in database. Clients while signing up will enter the particulars. Google Firebase authentication scheme has been used in this application and maintains the records uniquely. Clients would use e-mail and password for signing in after registration. Its process flow is explained clearly in Figs. 4 and 5 and shows the user interface.

3.3 *Portfolio Design*

Portfolio is the main and core module of the whole project. It involves fetching the correct NAV of all the funds required. It works on top of SEBI API which is hosted in RapidAPI platform, and the key is provided for the users. The manipulation of the total amount and the gain/loss percentage has been calculated in the frontend and displayed to the user. It shows all the funds of the clients when he/she logs into the application along with its day amount, date of investment, and NAV of that fund. The amount today is displayed green or red depending upon the hike or low rate at present. Figure 6 explained how clients and fund data is been processed to produce portfolio.

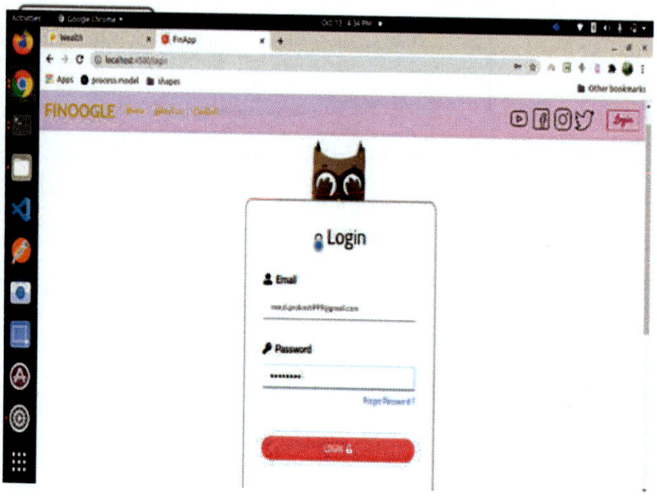

Fig. 5 Snapshots of authentication portal

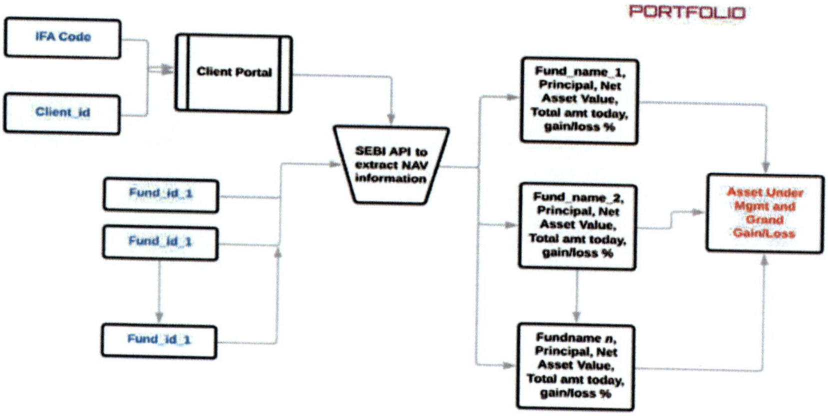

Fig. 6 System design of portfolio module

3.4 *Profile Updation*

A profile updating portal has been developed for the clients to change their profile and will be reflected in all the regions of respective client's identity. Client Id plays a vital role out here to modify details in the database.

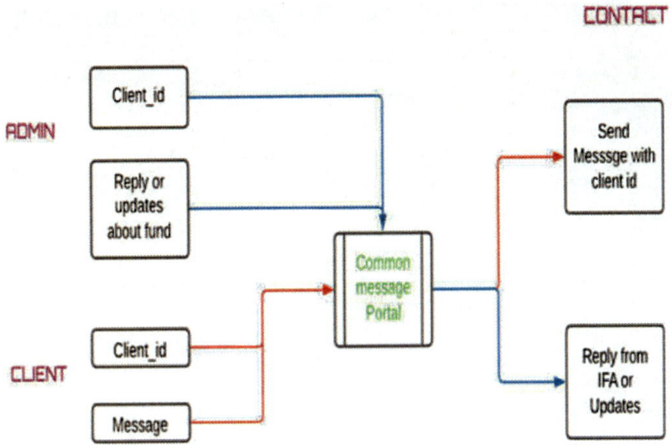

Fig. 7 System design of admin module for user message storage and retrieval

3.5 *Message Passing Interface*

This message passing portal would help clients to ping their advisors for any information. The client in remote can send any message, and it will be reflected in admin's messages pane. From there, admin can contact clients via mail or over phone. The message pane displays the date, subject, and message in both clients and message pane.

All the messages from clients would be stored in database and can be viewed anytime. Figures 7 and 8 depict the user message storage and retrieval in message passing pane along with user interface.

4 Experimental Results

In this application, the results deal with hitting third-party mutual funds API and fetching fund information along with NAV according to the requests as shown in Figs. 9 and 10. Experimental part deals only with the portfolio module.

4.1 *NAV Extraction*

NAV extraction is done with the help of mutual funds API in RapidAPI portal. Each fund would be associated with a fund Id after being registered and authorized by the Security and Exchange Board of India (SEBI) and Association of Mutual Funds in India (AMFI). This would be the Id allocated by financial advisors in admin

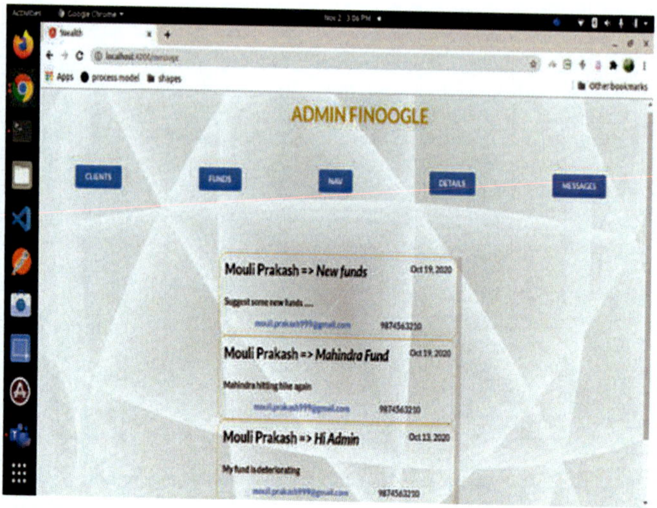

Fig. 8 Snapshot of message pane in admin portal

Fig. 9 Response object for Ex. 1 (Sundaram Mutual Fund)


```
▼[{…}]
 ▼0:
   Date: "08-Jan-2021"
   ISIN Div Payout/ISIN Growth: "INF903JA1EX7"
   ISIN Div Reinvestment: "-"
   Mutual Fund Family: "Sundaram Mutual Fund"
   Net Asset Value: "14.9520"
   Scheme Category: "Open Ended Schemes(Equity Scheme - Sectoral/ Thematic)"
   Scheme Code: "144838"
   Scheme Name: "Sundaram Services Fund Regular Plan Growth"
   Scheme Type: "Open Ended Schemes"
  ▶ __proto__: Object
  length: 1
 ▶ __proto__: Array(0)
```


Fig. 10 Response object for Ex. 2 (Mahindra Mutual Fund)


```
▼[{…}]
 ▼0:
   Date: "08-Jan-2021"
   ISIN Div Payout/ISIN Growth: "INF174V01473"
   ISIN Div Reinvestment: "-"
   Mutual Fund Family: "Mahindra Manulife Mutual Fund"
   Net Asset Value: "12.3540"
   Scheme Category: "Open Ended Schemes(Equity Scheme - Mid Cap Fund)"
   Scheme Code: "142109"
   Scheme Name: "Mahindra Manulife Mid Cap Unnati Yojana - Regular Plan - Growth"
   Scheme Type: "Open Ended Schemes"
  ▶ __proto__: Object
  length: 1
 ▶ __proto__: Array(0)
```

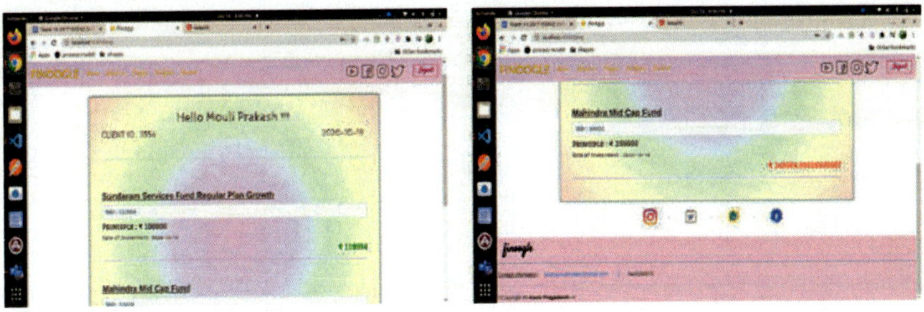

Fig. 11 Portfolio design with fund details of particular client

portal. This fund Id would be passed as the parameter get request for fetching fund information such as name, AMCs, NAV, and type.

4.2 *Interface View Generation*

Frontend for portfolio fetches fund information for clients who have been logged in other than the NAV of the fund. It has been binded to the frontend layouts.

4.3 *Results*

After NAV has been extracted from the first step, it is multiplied with the no. of units purchased to derive at the clients at present amount for a fund. Profit or loss manipulation is done by comparing with the principal amount of the clients, and it is been indicated by the color codes. Figure 11 depicts the interface output of the deduced results above. Table 1 shows the sample, then NAV detected values of some real-time funds.

5 Conclusion and Future Works

The financial services sector is a primary driver of a nation's economy. Financial services are the processes by which consumers and businesses acquire financial goods. The finance field is still growing and evergreen. Moreover, the data was getting more and more digitized and moving toward a successful marketing and finance level. So, we have also developed this app to help common people to get instant details about their portfolio. All the planned modules have been completed, and the app is successfully built for both admin and clients. The app would definitely help many finance advisors and clients to organize their data and monitor it periodically. It can

Table 1 Manipulation of current capital of funds according to NAV of funds (example values)

S. No.	Fund name	Date	Principal	No. of units	NAV	Current capital
1	Aditya Birla Sun Life Banking & PSU Debt Fund—Regular Plan-Quarterly Dividend	02/2/2020	200,000	20,000	111.087	2,221,740
2	Mahindra Manulife Mid Cap Unnati Yojana—Regular Plan—Growth	08/10/2020	120,000	12,000	12.3540	148,248
3	Mahindra Mid Cap Fund	10/18/2020	250,000	25,000	9.9428	248,570

also be hosted with the permissions and knowledge of SEBI and AMFI to make it global and accessible if certain external real-time requirements like payment gateway were met.

References

1. Li B (2019) Research on optimal portfolio of financial investment based on genetic algorithm. In: International conference on economic management and model engineering (ICEM ME). IEEE, Malacca, Malaysia, pp 497–500
2. Bi F, Wu W (2017) Research on portfolio management of university education funds. In: 4th international conference on industrial economics system and industrial security engineering (IEIS). IEEE, Kyoto, Japan
3. Kedia V, Khalid Z, Goswami S, Sharma N, Suryawanshi K (2018) Portfolio generation for Indian stock markets using unsupervised machine learning. In: Fourth international conference on computing communication control and automation (ICCUBEA). IEEE, Pune, India
4. Jaiwat P (2018) Financial market index tracking by model predictive control. In: 22nd international computer science and engineering conference (ICSEC). IEEE, Chiang Mai, Thailand
5. Li Y, Wang X, Burns CM (2017) Ecological interface design for financial trading: Trading performance and risk preference effects. In: IEEE international conference on systems, man, and cybernetics (SMC). IEEE, Banff, AB, Canada, pp 600–605
6. Yu J, Le T, Chang KC, Guharay S (2019) Fusing economic indicators for portfolio optimization—a simulation-based approach. In: 22th international conference on information fusion (FUSION). IEEE, Ottawa, ON, Canada, Canada
7. Bataev AV (2018) Analysis of the application of big data technologies in the financial sphere. In: IEEE international conference quality management, transport and information security, information technologies" (IT&M M&S). IEEE, St. Petersburg, Russia, pp 568–572
8. Kurt S, Heitz J, Bundi N, Breymann W (2018) Large-scale data-driven financial risk modeling using big data technology. In: IEEE/ACM 5th international conference on big data computing applications and technologies (BDCAT). IEEE, Zurich, Switzerland, pp 206–207

9. Zhilkina Anna N (2008) Enterprise financial management quality control. In: International conference on management science and engineering 15th annual conference proceedings. IEEE, Long Beach, CA, USA pp 1101–1104
10. Zhilkina AN (2009) Financial monitoring data in enterprise knowledge management. In: International conference on management science and engineering. IEEE, Moscow, Russia, pp 1083–1086
11. Loehr K, Khan RA (2011) Project finance—On new standards for processes in project and Program Management. In: University of applied sciences and arts and University of Kassel. IEEE, Kassel, Germany
12. Chuang L (2009) Information system design of finance sale management of SME. In: School of economics and management. IEEE, Henan Polytechnic University, pp108–111

Cloud-Based Patient Health Information Exchange System Using Blockchain Technology

Prashant Singh, Dhyanendra Jain, Anupam Kumar Sharma, Ashu Jain, and Prashant Vats

Abstract With the fast advancement of ubiquitous technologies and mobile computation, a massive quantity of individual health-related information is being created and gathered constantly at all occasions. These individual datasets include crucial data and are the property of the specific users, so they must be owned and governed by individuals. Presently, the majority of these datasets are held and managed by several wireless service suppliers, and this unified data storing poses data security vulnerabilities and impedes information sharing. Such access to patient information is significant tools for medical academic and clinical ventures. We present a theoretical architecture for revealing private perpetual clinical informational data utilizing blockchain solutions which are backed by cloud computing to exchange well-being data confidentially and responsibly in this academic work. In addition, to have control over data integrity, we develop an information quality control and assurance mechanism data mining and machine learning technologies. The proposed system's major function is to empower users to securely possess, manage, and exchange their healthcare data in a protected manner by utilizing the Data Protection Regulation compliant manner to profit from their statistics. This also allows academics and conventional data users to acquire high-quality healthcare data for scientific and business reasons efficiently and effectively.

Keywords Mobile computing · Cloud storage · Healthcare · Machine learning · Data sharing · Blockchain

P. Singh (✉) · D. Jain · A. K. Sharma · A. Jain · P. Vats
Dr. Akhilesh Das Gupta Institute of Technology and Management, New Delhi, India
e-mail: prashant.ert@gmail.com

A. K. Sharma
e-mail: anupam.sharma@adgitmdelhi.ac.in

M. S. Kaiser et al. (eds.), *Information and Communication Technology for Competitive Strategies (ICTCS 2021)*, Lecture Notes in Networks and Systems 401,
https://doi.org/10.1007/978-981-19-0098-3_55

1 Introduction

Well with increasing use of mobile data processing, smart wearable, as well as wireless perceiving, humans now use multiple kinds of personal devices, such as cellular phones, apple watches, intelligent l—band, and google glass, among others, to recognize variety of medical implementations, which including remote health monitoring [1], treatment planning [2], and senior citizens concern [3]. Such technologies generate a huge volume of healthcare data, which seem to be significant source for medical investigation and industrial businesses. Effectively communicating healthcare information can benefit all parties, especially users across the globe, patients, investigators, enterprises, and also the context of the public health service. As a personalized resource, patient information should indeed be controlled by that of the individuals directly, but in reality, it is typically held by different vendors, equipment manufacturers, or distributed across multiple medical practices [4, 5]. In generally, it creates obstacles to information exchange and jeopardizes data privacy and security since such unified data repositories and permission suppliers are appealing victims for computer hackers [6]. Because with cryptos, blockchain technology had received increasing attention in recent decades, especially in the financial sector. For instance, Cryptocurrency had first been launched in 2009 [7] and since then has allured the interest of many researchers from different fields of study [8–10] and acquired worldwide appeal owing to its advantages, like the utter lack of centralization, a presumed greater standard of pseudonyms, and dispersed convergence atop gray distributed nodes. The focus of this article is to postulate a healthcare information sharing system that relies on blockchain technology and fog storage systems that will allow people to automatically and safely share one's personal health information while also assisting investigators and infomercial system users in obtaining necessary data in an effective, straightforward, and compliant manner with information regulatory requirements such as GDPR [11].

2 Related Work

For years, researchers have been interested in leveraging private information obtained by smartphones and other mobile devices to ensure the effective and efficient management of critical and patient care services. Most individuals feel their medical information is confidential and are reluctant to disclose it owing to data protection and privacy considerations [12]. The development of blockchain technology in a health care related management proved that employing a distributed system of peer group supported by a shared database, trustworthy and fully transparent computing is feasible [2]. The research of Zyskind et al. [13] employed blockchain technology to preserve the data integrity and confidentiality by developing a system that transforms a blockchain into more of an automated authorization supervisor that would not involve reliance in a 3rd party, ensuring that individuals have and administer to their data. The study reported within [14] presented a blockchain-based application architecture called Healthcare Informational Gateway portal (HGD) that allows patients to even own, manage, and exchange their healthcare data in a precisely secured manner without invading their privacy. The research under [15] established an integrated health care delivery record keeping system termed MedRec that uses blockchain technology to manage Healthcare Information electronically. A current finding [12] offered a blueprint for one blockchain-enabled distributed healthcare information infrastructure. For many decades, researchers have been interested in evaluating large amounts of data generated by smart technology [16–20]. For example, Researchers in Zheng et al [21] employed accelerometer information from a wearable device to assess the trembling degree of individuals with Critical Motion. The researchers of Chernbumroong et al [16] detected the everyday activities of older individuals using data obtained from personal and cellular devices with modeling techniques. The investigators in Chen et al [22] used a machine learning system for Human Activity Attribution (HAR) problems and got better results. They introduced a novel healthcare information communication platform public blockchain, cloud technology, and deep learning technologies, prompted by the research stated above.

3 Proposed Methodological Framework

As shown in Mamoshina et al. [12], patient information may be separated into static and dynamic statistics. Static data relates to individual information that remains essentially unaltered throughout the patient's life, such as genotype or biometrics. Dynamic dataset provides the patient's activities throughout time, such as pulse Electro-Encephalic-Graphic (EEGs) statistics, or the condition of said individual at the moment of recording, such as blood testing dataset. The patient information may be subdivided into two categories based on the data collecting method: prolonged data and instantaneous data. The prolonged data is spread throughout timeframe and indicates the patient's condition or activities throughout that time. Prolonged data are often quantitative approach of constantly shifting data in real time. The instantaneous data, on either hand, are collected in a single metric. The instantaneous data might be either immutable section specifies or unstructured information that changes continuously. Our research is focused solely on time series, as seen in Fig. 1, because this form of information provides for the majority of information recorded by wearables and mobile phones. These are generally of high recurrence and big quantity, but that they can be saved or transmitted in the very same way that that other healthcare information could, there is very little research on the subject.

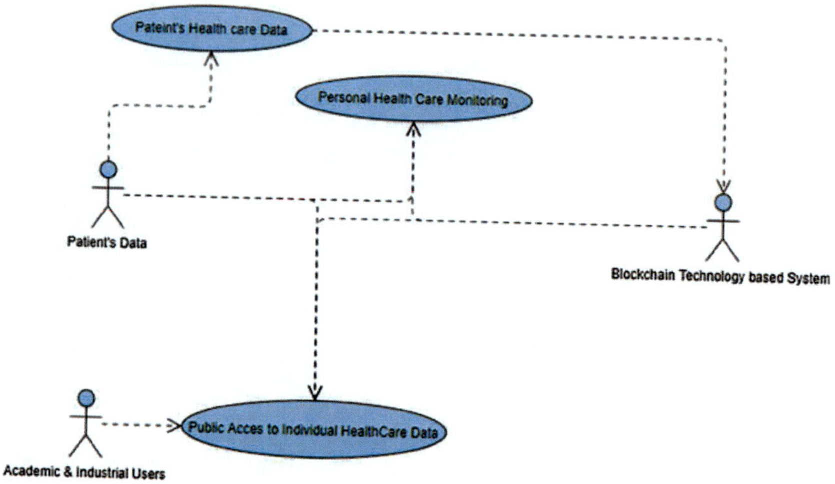

Fig. 1 Generalized use case architecture of the proposed system

4 Visual Design of the Conceptualized System

Figure 2 depicts the structure of the proposed healthcare information storage and handling system. This framework outlines three roles and responsibilities:

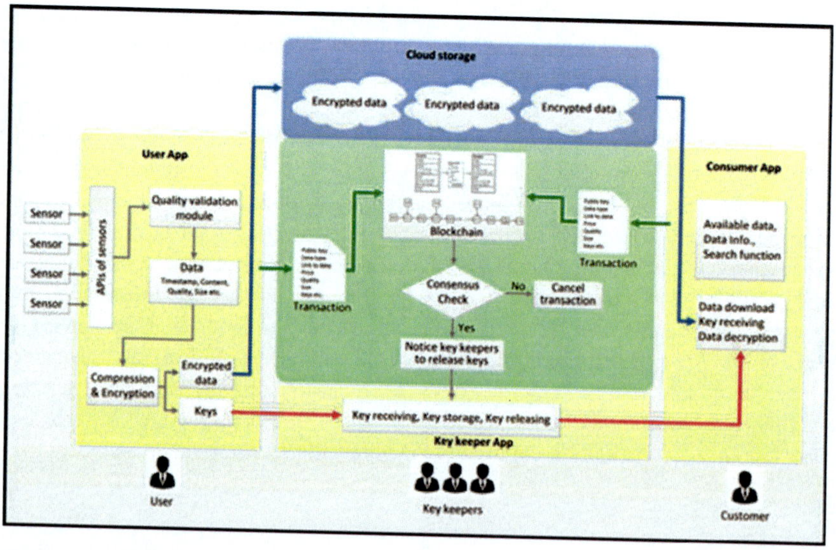

Fig. 2 The overall workflow diagram of the proposed framework

1. Subscribers: create, publish, and exchange (transfer) personalized health information in return for financial or service advantages.
2. Relevant custodians: retain the secret keys used to decrypt the encrypted once the participant uploads it and deliver the credentials to consumers when the transaction is made. Participants will be compensated financially for each transaction that is verified.
3. Visitors: to purchase user information and reward subscribers and key custodians with money or commodities.

A relevant App was created for each one of the above-mentioned functions to assist individuals in attaining their objectives.

I. Subscriber App: software that runs on a handheld computing device, such as a tablet or phone. It can communicate with various wearable technology or even other sensor systems. The Interfaces of the related sensors help with collecting information. To obtain a final grade, the obtained samples would be sent via a content verification mechanism, which would be released later. Following authentication, the verified data will now be merged with various identifying markers. After then, the aggregated information will be packed and encoded. After that the ciphertext will be stored on the Internet. The decryption key will be divided into multiple portions and handed to password custodians. After that a request will be produced and published to the blockchain network.
II. Password guardian App: operates on a physical drive connected via the Internet or even on a remote server. This is capable of receiving credential sharing from the network and securely storing information. Whenever a method for analysis transaction is confirmed, it receives a notification to transfer the matching particular establishment to the transaction's recipient.
III. Consumers Widget: runs on a community Internet-connected device or on a remote server. It can give the customers all or most of the data samples and enable them to browse for a specific study in particular. Whenever a set of data is selected, the Widget will advise the consumer to start generating an exchange for choosing to buy that data-set. After that the exchanged transactional processing data would be sent to all distributed blockchain network nodes. After that the ciphertext will be stored on the Internet. The decryption key will be divided into multiple portions and handed to password custodians. After that a request will be produced and published to the different blockchain.

A. Data integrity affirmation: As previously stated, our research solely looked at observed variables. Generally, these samples are produced by standard network node sensors. Furthermore, the collected information sequence can indeed be assessed utilizing artificial intelligence technology to check whether the data is legitimate following such confirmation structures or inspections.
B. Hardware perspective: when another gadget is associated with the client App, the equipment data of that gadget and the sensors installed in it will be gained by the client App. If a gadget or a sensor is from an approved assembling, it

is perceived as a certified equipment and the information delivered by it are dependable. Else, it will be would not associate with the App.

C. Software prospects: Through using cutting-edge AI and machine learning, the instances of time series analysis sets may be characterized with great accuracy. It is possible to view regular workouts by a service user using information collected from the wearable gadgets [21–23] of an activity tracker. Only data with return affects will be kept, and noise and insignificant data will be removed.

D. Approval for dissemination of data: the blockchain subsystem is one of the key pieces of the framework. It is used to get all the vicious circle of exchanging information. These steps allow developers to successfully develop blockchain solutions. We choose the Ethereum therefore in assessment as our application's development framework.

E. Cloud capabilities: The main reason for synchronizing dispersed storage inside the knowledge transfer architecture is that the big dataset can be answered down. In general, the unending, distinct input throughout a drawing measure is collected with a high repetition. Then, due of its replication across various hubs, blockchain is not excellent for sending out huge amount of information.

F. Data encryption: The information will be encrypted to ensure safety and reliability before the client's cloud transfer employing symmetrical key computations such as Rijn-deal ASE encrypted communication algorithm [24], in a mixture of limiting encryption layouts [12, 25–28]. Subsequently the symmetrical key for the unlocking of information is divided into multiple proposals that use the enigmatic strategy [29] of Shamir, and the key offers are then distributed across various important stakeholders at that moment in time. Based on the absolute number and the blockchain-based patient's cyber security model [12], the baseline number of key custodians is elaborated to unlock the information. For the encrypted data to be downloadable, you have to acquire both the information connectivity and authentication.

G. Crypto token: It is necessary to provide customers specific benefits when sharing their information when requesting the sharing of information. The advantage may be good governance, comparable to the control and determination of disease, although that method may not be suited to all conditions. If particular situations arise, it is more intelligent to offer a certain financial benefit.

H. General work: the communication between clientele, key patrons incorporate three pipelines as shown in Fig. 2: encoded data, blockchain exchange services and key decoding offerings. The whole procedure of work is as follows:

Step 1, User collects information, packages and scrambles using the User application. Then the information is sent to the distributed storage at this time and the key offers are applied to key staff members.

Step 2, an interchange will be made to advise several members of the transfer and sharing of information. After the client's confirmation, this transaction is entered into the blockchain through the computation of agreement and customers and validators are noticed about the exchange.

Step 3, the information authenticator will identify behavior, in the form of a communication approval, by carrying out the knowledge mining and AI calculations to ensure that information is credible in line with the regulations and necessities, and finally make the information substantially mutually compatible with the particulars.
Step 4, a customer selects data that he/she wanted to purchase, ensures that the correct validators are assured and then exchanges information to obtain them. The exchange is included into the blockchain when it has been approved by the recipient.
Step 5, the calculation of the contract will approve the exchange of information for the purchase. The trade will be supported if the customer has enough homeostasis (PHD coins) to buy the data. As the cost of data shows, the dazzling contract will be sent to particular PHD coins and the work procedure takes place at step 6. Another factor is the withdrawal of the exchange and the procedure returns to stage 3.
Step 6, essential data controllers will be notified that an exchange has been approved for the acquisition of this information. Then, important agents transmit a verified correspondence channel to your customer their key sections of the linked data collection.
Step 7, the PHD coins in the shrewd contract will be scattered across the client's and the managers' records, according to a predetermined rule.
Step 8, the customer is connected to the information and receives sufficient key offers to remove them. Then he/she genuinely wants to download and decode the information from the distributed storage. The data is ready to be used and the task is closed.

5 Information Security and User Privacy Assessment

The security of information in the proposed framework depends on the three agreements that follow.

1. The entry is restricted to the information placed on the cloud. To accept the encoded information, you must know the address and receive validation. In addition, the information is encrypted before transfer to the distributed store so as not to induce the information to leak even the disruption of capacity. The symmetric key is divided and distributed across many key managers to disassemble the information. In this way, a lone key manager's compromise would not cause the information breach.
2. The hash capability and cryptographically mark plans employed in the blockchain agreements are utilized for information dissemination measurement. Blockchain technology allows consumers to employ the public key to trade per component in blockchain and therefore, until the exchange is specifically recalled, no visible individual data is linked with this exchange.

3. A blockchain-based cryptographic framework (PKI) would be provided for the symmetric screening key to enable information transmission and competence, which allows key managers to develop valid correspondence channels for instance to encode the symmetrical data storage key using the pair of keys during the ciphering of the medical text data of a user is done in a way so that only a recipient can unbolt the desired information its emphasis is very much on the cycle of knowledge exchange.

6 Conclusion

In this article, we presented a framework for the exchange of information depending on blockchain, cloud databases and AI techniques. The program enables customers to efficiently and safely request, monitor and submit their wellness information and gains from this engagement. In this study we have initially categorized individual information on well-being into several groups, as shown by information characteristics (dynamic and static information) and information about techniques for getting information on smartwatches as well as cellular telephones. To communicate the large quantity of consistently single information using hash pointers for the capacity region, we offered several responses. In addition, by using distributed ledger technologies storage, our proposal Framework reduced the size of consistent unique wellness information. We also advised that greater size of information relating to well-being be placed in the cloud in a tumbling arrangement and the only possible way to store and distribute the information and metadata is to blockchain. Thirdly, the suggested framework to regulate the quality of information from the equipment as well as programming point of view supported by AI methods has integrated a Data Quality Certification Modules.

References

1. Son D, Lee J et al (2014) Multifunctional wearable devices for diagnosis and therapy of movement disorders. Nat Nanotechnol 9(5):397
2. Zheng X et al (2017) Continuous monitoring of essential tremor using a portable system based on smartwatch. Front Neurol 8:96
3. Gao Y et al (2015) An empirical study of wearable technology acceptance in healthcare. Ind Manag Data Syst 115(9):1704–1723
4. Varshney U (2007) Pervasive healthcare and wireless health monitoring. Mobile Netw Appl 12(2–3):113–127
5. Zhang J et al (2016) A secure system for pervasive social network-based healthcare. IEEE Access 4:9239–9250
6. Peterson K et al (2016) A blockchain- based approach to health information exchange networks. Proc NIST Workshop Blockchain Healthc 1:1–10
7. Nakamoto S (2008) Bitcoin: a peer-to-peer electronic cash system
8. Ali ST, et al (2015) Bitcoin: perils of an unregulated global p2p currency. In: Cambridge international workshop on security protocols. Springer

9. Bohme R et al (2015) Bitcoin: economics, technology, and governance. J Econ Perspect 29(2):213–238
10. Harlev MA, et al (2018) Breaking bad: de-anonymizing entity types on the bitcoin blockchain using supervised machine learning. In: Proceedings of the 51st Hawaii international conference on system sciences
11. The European Parliament (2016) Regulation (eu) 2016/679 of the European parliament and of the council of 27 April 2016 on the protection of natural persons with regard to the processing of personal data and on the free movement of such data, and repealing directive 95/46 (general data protection regulation) [GDPR]. Off J European Union 59(L119):1–88
12. Mamoshina P et al (2018) Converging blockchain and next-generation artificial intelligence technologies to decentralize and accelerate biomedical research and healthcare. Oncotarget 9(5):5665
13. Zyskind G, et al (2015) Decentralizing privacy: using blockchain to protect personal data. In: Security and privacy workshops (SPW), 2015 IEEE. IEEE, pp 180–184
14. Yue X et al (2016) Healthcare data gateways: found healthcare intelligence on blockchain with novel privacy risk control. J Med Syst 40(10):218
15. Ekblaw A, et al (2016) A case study for blockchain in healthcare: "medrec" prototype for electronic health records and medical research data. In: Proceedings of IEEE open & big data conference, vol 13, p 13
16. Chernbumroong S et al (2013) Elderly activities recognition and classification for applications in assisted living. Expert Syst Appl 40(5):1662–1674
17. Ronao CA et al (2016) Human activity recognition with smart- phone sensors using deep learning neural networks. Exp Syst Appl 59:235–244
18. Pulliam C et al (2014) Continuous in-home monitoring of essential tremor. Parkinsonism Relat Disord 20(1):37–40
19. Alsheikh MA, et al (2016) Deep activity recognition models with triaxial accelerometers. In: AAAI workshop: artificial intelligence applied to assistive technologies and smart environments
20. Ermes M et al (2008) Detection of daily activities and sports with wearable sensors in controlled and uncontrolled conditions. IEEE Trans Inf Technol Biomed 12(1):20–26
21. Zheng X, et al (2015) Detection and analysis of tremor using a system based on smart device and nosql database. In: Indus-trial engineering and systems management (IESM), 2015 international conference on. IEEE, pp 242–248
22. Chen Y. et al (2015) A deep learning approach to human activity recognition based on single accelerometer. In: Systems, man, and cybernetics (smc), 2015 IEEE international conference on. IEEE, pp 1488–1492
23. Zeng M, et al. (2014) Convolutional neural networks for human activity recognition using mobile sensors. In: Mobile computing, applications and services (MobiCASE), 2014 6th international conference on. IEEE, pp 197–205
24. Daemen J, et al (2013) The design of Rijndael: AES-the advanced encryption standard. Springer Science & Business Media
25. Shamir A (1979) How to share a secret. Commun ACM 22(11):612–613
26. Blakley GR, et al. (1979) Safeguarding cryptographic keys. In: Proceedings of the national computer conference, vol 48, pp 313–317
27. Robling Denning DE (1982) Cryptography and data security. Addison- Wesley Longman Publishing Co., Inc.
28. Desmedt Y (1992) Threshold cryptosystems. In: International workshop on the theory and application of cryptographic techniques. Springer, pp 1–14
29. Katz J, Menezes AJ, Van Oorschot PC, Vanstone SA (1996) Handbook of applied cryptography. CRC press

Detection of Faulty Steel Used in Construction Industry Using Machine Learning Algorithms

S. Poonkuzhali, P. Revathy, E. Mohana, and M. Shobana

Abstract The construction industry consumes nearly half of global steel production. Every year, nearly 3 billion tonnes of raw materials are utilized to manufacture steel products, mainly for building purposes alone. Detecting faulty steel plays a major role utilizing the quality of manufactured building products, which in turn lowers the maintenance cost. In recent days, advanced analytical tools and machine learning algorithms have gained more popularity in fault diagnosis in steel products. In this work, Naive Bayes, support vector machine, and bagging classifier are applied to detect faulty steel and perform performance evaluation in terms of precision, recall, and accuracy. Detection of fault type is used to accomplish some prevention measures with respect to occupational safety.

Keywords Construction · Faulty steel · Naive Bayes · Machine learning algorithm

1 Introduction

The construction industry is one of the oldest industries in the world, and it has recently incorporated many new technologies to groom its field. For example, the injury was predicted based on the previous recorded data. Accident prediction using machine learning based on the position of heavy loaded equipment which causes fatal accidents. Apart from accident prediction and imposition of heavy load detection, the benefits of crew productivity can also be increased using machine learning as well as deep learning techniques. This work concentrates on fault product detection, whereas steel is considered as one of the most important materials involved in the construction field. Recently, many researchers have focused their attention concentrating on the faulty steel prediction and regression to improve the product's quality. In this paper, the types of faults that occurred in steel products have been classified using machine learning techniques. This is to enhance the different kinds of treatment that can be given according to the types of fault.

S. Poonkuzhali (✉) · P. Revathy · E. Mohana · M. Shobana
Department of CSE, Rajalakshmi Engineering College, Chennai, Tamil Nadu, India
e-mail: poonkuzhali.s@rajalakshmi.edu.in

M. S. Kaiser et al. (eds.), *Information and Communication Technology for Competitive Strategies (ICTCS 2021)*, Lecture Notes in Networks and Systems 401,
https://doi.org/10.1007/978-981-19-0098-3_56

The main contribution of this work can be given as the ability to classify the type of fault exit in construction steel using Naive Bayes, SVM, and Bagging. The organization of this paper is given as Sect. 2 provides a brief explanation of the related works. Section 3 describes the working of the proposed system. Section 4 describes the result of the proposed system. Section 5 provides the conclusion and future work of the proposed idea.

2 Literature Survey

The following section summarizes the existing work in the field of study. Chatterjee et al. [1] proposed a system which showed 94.44% *F*-measure and 93.3% accuracy for the same dataset. The computational model was based on MLP. Perez et al. [2] had considered images and, using CNN, had depicted the degradation of buildings over time. The process was supported by transfer learning and optimization techniques.

Jain et al. [3] performed analysis on steel plate fault data using a decision tree model with higher accuracy. Srivastava [4] presented a comparative study of several ML models over the steel plate fault analysis issue. They had compared ensemble techniques for comparison and achieved 79.23% and 0.203 RMSE. Much new functionality has also been summarized with deep learning techniques. Zhang et al. [5] analysed the superiority and categorization of the profound learning approaches. On the classification precision, a comparison study of algorithms was conducted using the open-source Western Reserve University (CWRU) data set. Different profound learning methods have been used to diagnose the defects. For certain situations of application, detailed advice was given. Fakhr et al. [6], compared the performance of three data mining models, which were then used to classify seven faults in a steel plate.

Kholief et al. [7] also used ANN and deep auto encoder to classify the faults in the steel plates. The system was able to provide very low false rates. Hoang [8] has developed and performed a similar image-based machine learning experiment with five labels and bestowed an accuracy of 85.33%. Kazemi [9] built a model based on the CRISP-DM cycle and found the results of the C5.0 decision tree model were superior to neural models and a few other ensemble models. They achieved a training data accuracy of 95.56% and a testing data accuracy of 95.66%. Kishore Kumar et al. [10] applied various classifiers to explore the best classifier for detecting email spam using a data mining tool.

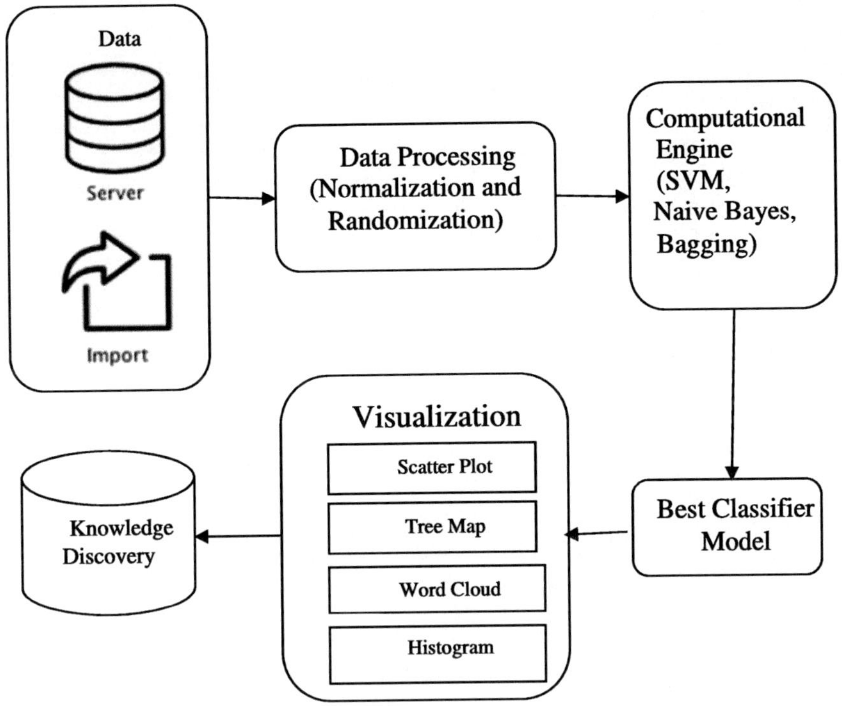

Fig. 1 Proposed system architecture

3 Architectural Design and Proposed Methodology

3.1 Proposed Methodology

The proposed framework incorporates data collection, data preprocessing, and a computational model for classification, visualization, and knowledge discovery. The proposed framework is depicted in Fig. 1.

3.2 Data Preprocessing

Data preprocessing is the process of converting the data into a suitable form for further processing. Initially, normalization is applied to the data in order to normalize the values and then it is subjected to randomization. After preprocessing, the computational model is explained in the next sub-section.

3.3 Computational Model

Classification is a supervised algorithm used to classify the data based on the class label. Various classification algorithms such as SVM, Naive Bayes, and Bagging are applied to the pre-processed dataset. Then, the best classifier for identifying the steel fault is done based on the accuracy. For a new data set, predictor accuracy is defined as how well an instance is correctly identified as a fault or fault-free steel.

Algorithm for Naïve Bayes

Algorithm 1: Naive Bayes Classification

Input: IA=(x1,x2,x3,...,xn)
Output: Class label for test set x
Data: Testing set x
/* Calculation of Maximum Posterior Probability */
while *True* **do**
  Extract the input IA with n attributes(x1, x2, x3... xn);
  Calculate the probability for separate values of TA;
  Calculate prior probability for m classes as p(Cm)
  Calculate posterior probability for m classes as
  $P(\frac{X}{C_i}) = \prod_{k=1}^{m} P(\frac{X_k}{C_i}) = P(\frac{x_1}{C_i}) * P(\frac{x_2}{C_i}) * \dots * P(\frac{x_k}{C_i})$
  Obtain the label for IA by calculating $P(\frac{C_i}{X}) = P(\frac{X}{C_i}) * P(C_i)$
  Find class with maximized $P(\frac{C_i}{X})$ maximum posterior probability and IA belongs to class $P(\frac{X}{C_i})P(C_i) > P(\frac{X}{C_j})P(C_j)$ for $i < j < m$

4 Results and Discussion

The model was designed for a Python 3.7 software environment on the Windows 8.1 Operating Systems on a Core i3 laptop with a 2.30 GHz CPU and 4 GB RAM. The classification of faults is done by using three algorithms: naive Bayes, Bagging, and support vector machine (SVM), and their performance has been compared using evaluation metrics like accuracy. Furthermore, classification metrics like precision and recall values have been discussed only for the best classifier amongst the three algorithms.

4.1 Dataset Description

Semeion, the Research Centre for Communication Sciences, supplied the dataset. The primary purpose of the data set was to properly identify the surface defect type into plates in stainless steel with six potential defect kinds (plus "other"). A total of 34 qualities are included. The input vector consisted of 20,7 parameters that explain the geometric form and outline of the defect and one output label (Type of fault). A one-hot vector containing seven defects is used to create the seven last columns. The dataset has 1941 cases labelled by different sorts of faults. The Pastry, Z Scratch, K Scratch, Stains, Dirtiness and other Faults are the seven sorts of dependent variables that signify each form of fault. In Fig. 2, the visualization of each independent variable has been displayed separately to explore the relationship that exists amongst the given variables. Table 1 illustrates the sample distribution in the given dataset.

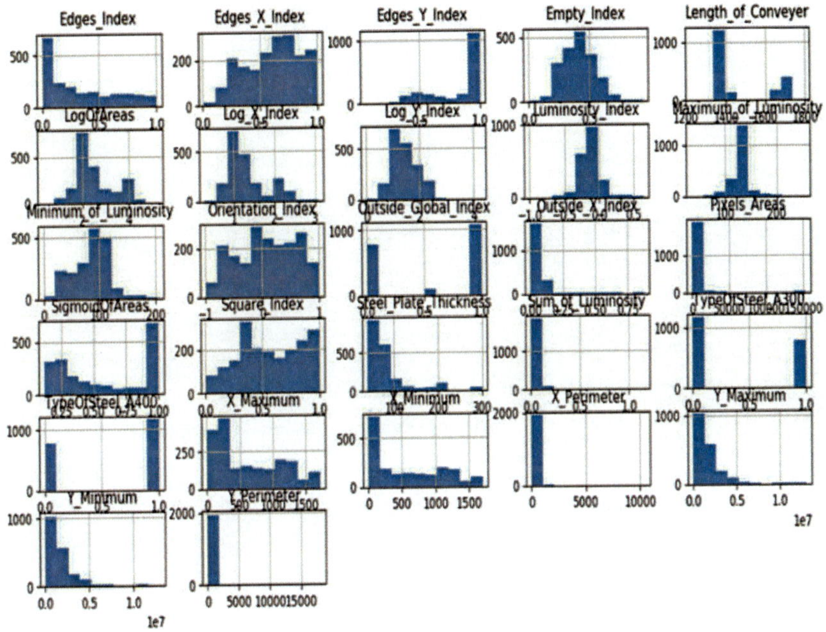

Fig. 2 Visualization of independent variables

Table 1 Sample distribution of dataset

Type of fault	Number of samples
Pastry	158
Z_Scratch	190
K_Scratch	391
Stains	72
Dirtiness	55
Bumps	402
Other faults	673

Tables 2, 3 and 4 show the confusion matrix for naive Bayes, SVM, and bagging, respectively. Table 5 illustrates the accuracy metrics for all the three classifiers. From these evaluation methods, the best classifier to classify the faults in steel can be chosen as naive Bayes. Its confusion matrix and accuracy value outperform the other two classifiers. In Table 4, more or less all the faults are classified correctly, except for pastry, z-scratch, and other faults. In order to extend the performance evaluation, the precision and recall graph for the naïve Bayes alone is given in Fig. 3. In Fig. 3, the fault type "Bumps" reaches the maximum range whereas pastry reaches the minimum range.

Table 2 Confusion matrix for Naïve Bayes

Pastry	Z_Scratch	K_Scratch	Stains	Dirtiness	Bump	Otherfaults	
150	0	1	0	0	4	3	Pastry
3	175	0	0	5	2	5	Z_Scratch
3	0	346	2	0	6	34	K_Scratch
0	0	0	69	0	3	0	Stains
2	0	0	0	49	4	0	Dirtiness
2	0	1	0	1	391	7	Bump
3	1	9	6	5	6	643	Otherfaults

Table 3 Confusion matrix for support vector machine (SVM)

Pastry	Z_Scratch	K_Scratch	Stains	Dirtiness	Bump	Otherfaults	
70	9	0	1	0	18	60	Pastry
0	163	2	0	0	9	16	Z_Scratch
1	0	361	2	0	2	25	K_Scratch
0	0	0	67	0	1	4	Stains
8	0	0	0	0	2	45	Dirtiness
4	11	2	3	0	247	135	Bump
33	25	20	6	2	130	457	Otherfaults

Table 4 Confusion matrix for bagging

Pastry	Z_Scratch	K_Scratch	Stains	Dirtiness	Bump	Otherfaults	
21	0	0	0	0	1	15	Pastry
0	47	0	0	0	2	2	Z_Scratch
0	0	71	0	0	0	2	K_Scratch
0	0	0	14	0	0	2	Stains
0	0	0	0	7	1	3	Dirtiness
1	0	0	0	0	60	18	Bump
2	1	5	1	1	20	91	Otherfaults

Table 5 Performance evaluation of classifier

Algorithm	Accuracy
Naïve Bayes	93.92
SVM	70.32
Bagging	80

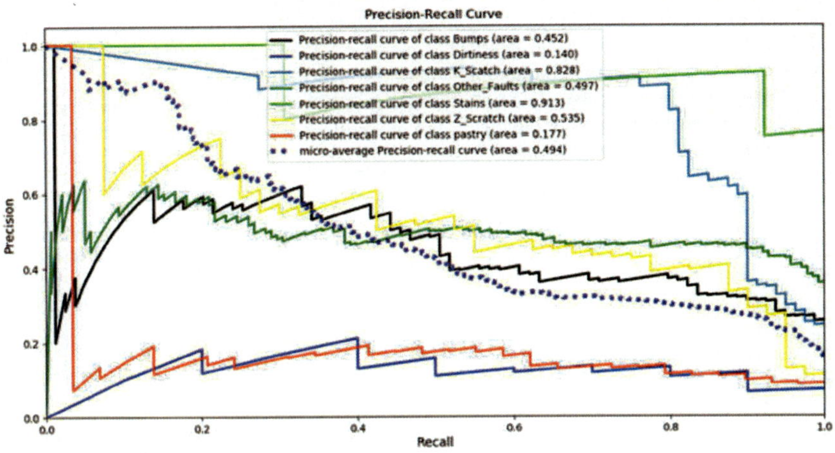

Fig. 3 Precision-recall graph for Naïve Bayes

5 Conclusion

In the proposed system, detection of faulty steel was implemented using naive Bayes, SVM, and bagging machine learning classifiers. The Naive Bayes classifier provides the highest accuracy of 93.92% when compared to SVM and bagging classifiers. In the future, deep learning algorithms can be used to enhance the accuracy further in real-time steel plants.

References

1. Chatterjee S, Sarkar S, Hore S, Dey N, Ashour AS, Shi F, Le D-N (2017) Structural failure classification for reinforced concrete buildings using trained neural network based multi-objective genetic algorithm. Struct Eng Mech 63(4)
2. Perez H, Tah JHM, Mosavi A (2019) Deep learning for detecting building defects using convolutional neural networks. Sensors 19(16):3556–3578
3. Jain S, Azad C, Jha VK (2013) Steel faults diagnosis using predictive analysis. Int J Comput Eng Appl 4(2)69–78; Author F (2010) Contribution title. In: 9th international proceedings on proceedings. Publisher, Location, pp 1–2
4. Srivastava AK (2019) Comparison analysis of machine learning algorithms for steel plate fault detection. Int Res J Eng Technol 6(5):1231–1234

5. Zhang S, Wang B (2020) Deep learning algorithms for bearing fault diagnostics—a comprehensive review. IEEE Access 8:29857–29881
6. Fakhr M, Elsayad AM (2012) Steel plates faults diagnosis with data mining models. J Comput Sci 8(4):506–514
7. Kholief EA, Darwish SH (2017) Detection of steel surface defect based on machine learning using deep auto-encoder network. In: Proceedings of the international conference on industrial engineering and operations management rabat. Morocco, pp 11–13
8. Hoang N-D (2018) Image processing-based recognition of wall defects using machine learning approaches and steerable filters. Comput Intell Neurosci 2018
9. Kazemi MAA, Hajian S, Kiani N (2018) Quality control and classification of steel plates faults using data mining. Appl Math Inf Sci Lett 6(2):59–67
10. Kishore Kumar R, Poonkuzhali G, Sudhakar P (2012) Comparative study on email spam classifier using data mining techniques. In Proceedings of the international multi-conference of engineers and computer scientists, vol 1, Hongkong

Identification of MRI-Based Adenocarcinoma Tumours with 3-D Convolutionary System

Manu Phogat, Prashant Vats, Balwant Kumar, Preeti, and Sandeep Singh

Abstract Computer-assisted adenocarcinoma diagnostics can help physicians achieve better diagnostic accuracy and efficient Ness that is not dependent on the subjective judgement and experience of the physician. Classification techniques of pancreatic tumours are subject to a partial automated test problem and disregards temporal and spatial character traits. For adenocarcinoma magnetic resonance imaging (MRT) classifications, we employed 3-dimensional variants of ResNet18, ResNet52, ResNet34, and Inception-ResNet throughout this study. We have suggested a normalised gradient descent to ameliorate the impact of imbalanced data. In order to match the impacts of the proposed gradient descent, there have been two different sets of experimental evaluation. With calibrated gradient descent, experimental findings demonstrate the effectiveness of the four methods is mainly enhanced with the exception of ResNet52's accuracy. In addition, the falsified imperfections are being diminished. ResNet18 is amongst them most efficient and attains 91% precision.

Keywords Calibrated logistic regression · Adenocarcinoma · 3-D neural convolutionary web · Magnetic resonance tomography (MRT)

M. Phogat (✉) · S. Singh
SGT University, Gurugram, Haryana, India
e-mail: manu_feat@sgtuniversity.org

S. Singh
e-mail: sandeep_feat@sgtuniversity.org

P. Vats
Fairfield Institute of Management and Technology, GGSIP University, New Delhi, India

B. Kumar
Goverment College for Girls, Datta, Hisar, Haryana, India

Preeti
Rajiv Gandhi Goverment College for Women, Bhiwani, Haryana, India

M. S. Kaiser et al. (eds.), *Information and Communication Technology for Competitive Strategies (ICTCS 2021)*, Lecture Notes in Networks and Systems 401,
https://doi.org/10.1007/978-981-19-0098-3_57

1 Introduction

A frequent cancerous pancreatic tumor is adenocarcinoma cancer. During the last several years, its incidence and fatality have substantially grown. Most standard approaches cannot function end-to-end for automatic categorization of pancreatic cancers. In the process of texture classification and afterwards explicitly modelled for classifications, Klawikowski et al. [1] employed the research instrument (IBEX). In order to make a diagnosis adenocarcinoma cancer before tumorigenesis, Ngom et al. [2] utilised a high—dimensional space quadratic device inductively coupled plasma mass. In these extracellular vesicles, Ko et al. [3] characterise the Nucleotide contents and employ a machine learning technique to produce prediction panels for specimens from diverse people with cancer. As a result of a pattern recognition, Dmitriev et al. [4] suggested to utilise combined medical information about a patient plus precise imagery knowledge from the positron emission tomography images, based on a Bayes mixture of an RF classifiers and convolution learning algorithm. The performance of the classifier of 83.6% were achieved in the suggested methodology. Above noted research, issues with the categorization of pancreatically modifiable tumours had been that these were often dependent upon the removal of handmade characteristics with the development of a predictor for differentiating malignant tumours. In this article, we utilize four distinct 3-D CNN designs, which are changes to the remnant CNN and the empirical genesis, to detect the number of structural models suited to given MRI sample. The 3-D residue CNN can deal with the above-mentioned issues. Firstly, such systems could capture morbidities and sophisticated semantic and syntactic categorization characteristics. Furthermore, remnant designs in the network ease limited training dataset's pain. We additionally provide a balanced gradient descent to prevent misleading adverse effects and alleviate imbalances in the classification. As a consequence, the 3-D MRI picture categorization uses an edge deep learning technique.

2 Related Work

3-dimensional CNN models have been implemented to identify various pancreatic disorders and their localization in earlier research. These previous systems have demonstrated the efficacy of 3-dimensional CNN model for the treatment of clinical images.

Hosseini-Asl et al. [5] have utilised the DSA-3-D-CNN (DSSA-3-D-CNN) as initialisation of the system of showcase and precise adjustment for Alzheimer's diagnostic testing in the multiple domain pictures by retraining for Convolutionary self-encoders. Kang and Liu et. al. [6] have been exploring deep CNN multi-video for identification of pulmonary nodular, incorporating 3-dimensional Beginnings and 3-dimensional Early beginnings with both Chained and Ordered Non-cyclic Graphs architectures. The amendments to the ADNI set of data again for brain MRI neural categorization were suggested by Sergey et al. [7]. Mehrtash et al. [8] introduced the SPIE Prostate-ex AAPMNCI three-dimensional neural network technique, which was designed to identify the clinically meaningful discoveries of bladder cancer in MRI scans. Oda et al. [9] presented a gross to intermediate two-stage approach, using the three-dimensional deep convolutional neural network to broadly identify a prospective region that will be utilised for definitive delineation as insight into the subsequent 3-dimensional CNN model. Xia et al. [10] proposes to solve the restricted quantity of 3-D information that is annotated whilst leveraging abundant spatial features through all three planes, a new delicate scratchy to refined volumetric pancreatic carcinoma segmentation methodology.

3 Proposed Work

3.1 Proposed Model

In the adenocarcinoma MRI-based image categorization we examine eight alternative methods using 3-dimensional CNN model. Eight modelling variants are: ResNet52, ResNet34, ResNet18, and Insomnia ImageNet with binary transversal

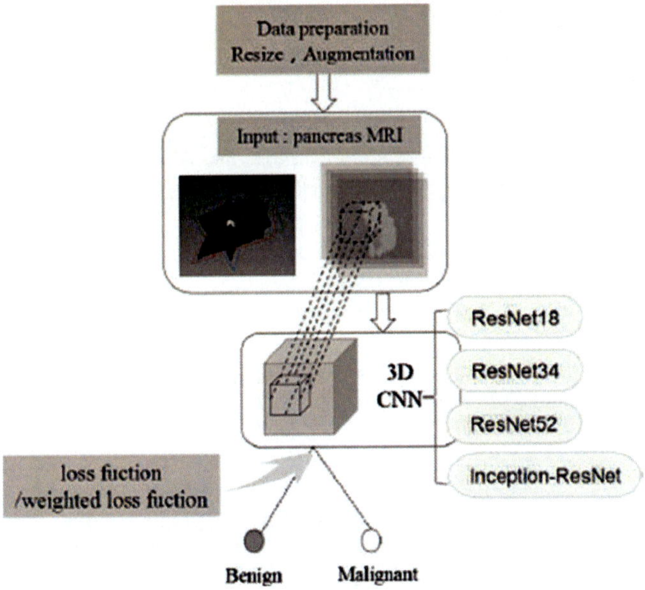

Fig. 1 To show the work model for the proposed approach

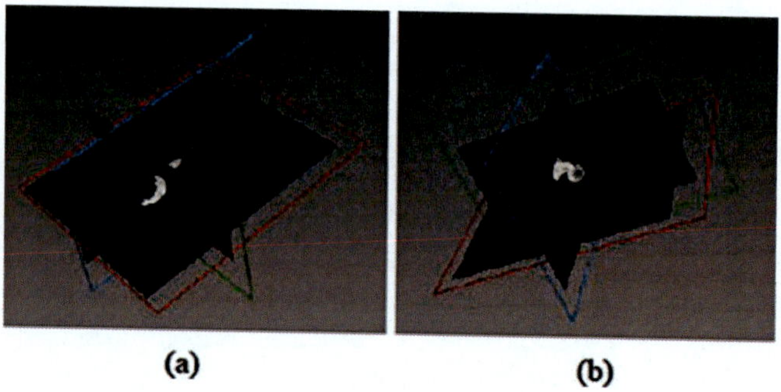

Fig. 2 3-D pancreatic. a Healthy pancreas. b Cancerous pancreatic visualisation

Table 1 MRI pictures' whilst still

322 × 261 × 63	287 × 232 × 42	510 × 510 × 112
252 × 193 × 51	253 × 190 × 46	252 × 223 × 42
510 × 510 × 70	510 × 510 × 23	510 × 510 × 82

centrepiece, accordingly the suggested weight able nonlinear function, acknowledging the impact of the width and depth of the modelling. Figure 1 shows the process of our methodology as follows:

3.2 Data Preparations

As demonstrated in Fig. 2 and their dimensions, actual Magnetic resonance photographs are recorded in the form of the newly raw rasterized data in Table 1.

Furthermore, to improve the data sample, we utilise data enrichment. Benign FMRI pictures are over double what of malignancy MRI scans in our entire dataset. In order to alter the picture sharpness in three dimensions without altering the fundamental texturing function the closest neighbour approximation technique is used. The approximated pieces are shown as Fig. 3. We apply null filling spacing to adapt the picture size 128 to 128 to 128 pixels. We may also utilise this buffering.

3.3 3-Dimensional Networks Deep Convolutional Fcn

3.3.1 Network for Three-Dimensional ResNet

In 2015, the structure of the remaining deep learning [11] topped the ImageNet competition and showed the ability of improving significantly the performance of

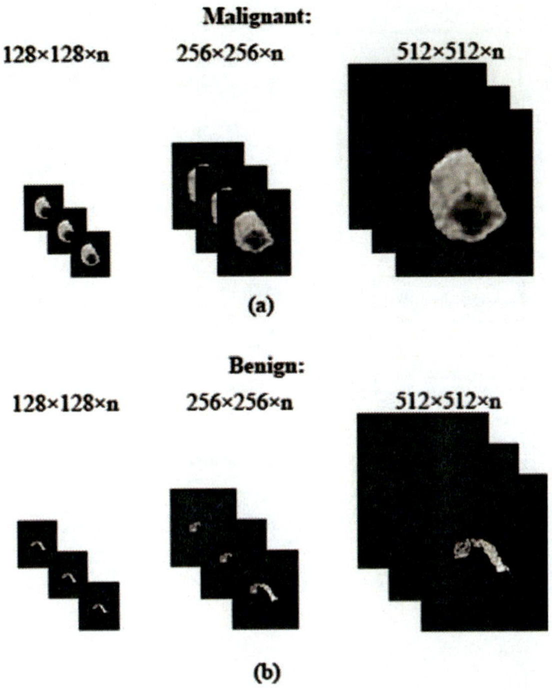

Fig. 3 Pancreas visualisation after exponential rise

the network with quick convergence. Articles have already demonstrated the findings for the categorization of the 3-dimensional brain Images picture [7]. In Fig. 4, we utilise ResNet, made up of subunits. Furthermore, illustrated in Fig. 4d, the outcome of final ResNet subsystem is forwarded to an averaging convolution layer to even further decrease this to 1 to 1*1*1*1*512 feature maps and a pooling layer with a binary output for stochastic non-linearity categorization.

3.3.2 Three-Dimensional RESNET Emergence

The output of the penultimate formation-ResNet module is transmitted, as seen in Figs. 5b, into such an intermediate pool scale to even further decrease that to 15 PART1 PART 256, accompanied by a smoothed plane with 3840 Concealed Modules and a convolution layer with digital signal for Gaussian quasi categorization. We use Adams with a computed value of 0.0001 and a lot size of 8 over 100 iterations to develop the resulting binary classifier systems.

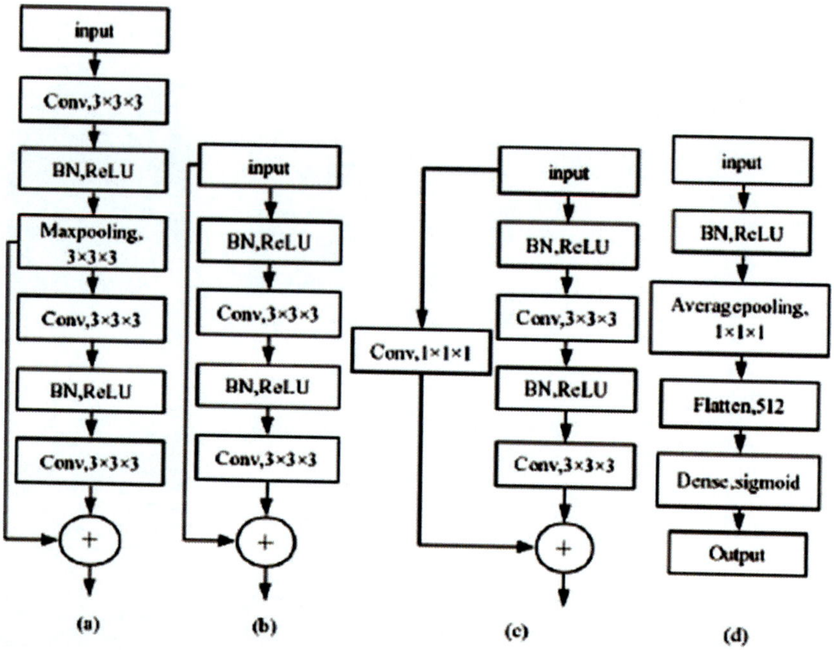

Fig. 4 Modules ResNet: **a** The initial component input. The intermediate segments are **b** and **c**. **d** The modules of outcome

3.3.3 A Weighted Indexing Error Function

In view of the class-balanced statistical model presented in the experiment. By selecting a reduced sample dimension, we provide the samples a bit more substance to diminish this problem of information asymmetry. We employ the deep learning cross entropy algorithm and adapt this to suit our model based on Holistically-Nested Edge Detection [13] which introduces a per-pixel image classification class weighting and using a three-way decision support [14] we optimise the judgement call expenses. In our approach, each time that the parameters get modified and replicated, we simply compute the number of useful classifications, then determine the weight distribution for every group. This is the formulation:

$$X1 = \mathrm{POS\_NUM}/(\mathrm{POS\_NUM} + \mathrm{NEG\_NUM}) \tag{1}$$

$$\mathrm{WGHT1} = X1/(1 - X1) \tag{2}$$

$$\mathrm{LOSSES} = -(\mathrm{W1QLOGP} + (1 - Q)\mathrm{LOG}(1 - P)) \tag{3}$$

Where

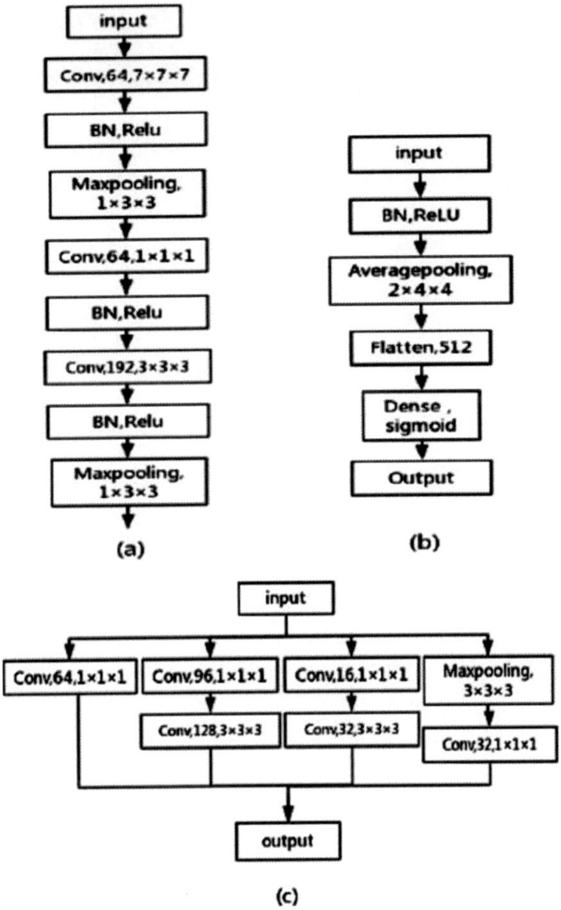

Fig. 5 **a** To show the initial component input with Inception-ResNet and is **b** the component of outcome. **c** To show the **c** the subsystem inception 3-D

POS_NUM = Positive numbered data samples for each set of Data images.
NEG_NUM = Negative numbered data samples for each set of Data images.
W1 = Weighted Coefficient for Positive Numbered Samples.
Q = Actual Value and P = Predicted Value.

4 Experimental Setup

The machine learning interface is used to execute our models through Keras. To prevent overcasting, we utilise early stops. We prepare and verify these classifiers on 173 datasets using 690 sample data. An Geforce Gtx 1080 Tis (up to 16 gb) Chip is used to speed all tests. Magnetic resonance Images from the individual patient after their consent were acquired out of this investigation. Our data set includes MRI scans of 20 regular people and 20 individuals with illness. Every individual member has several distinct MRI groups, including such improved MRI and simple MRI, examined by various means or layer thicknesses. We thus have 77 innocuous MRI scans as well as 38 cancerous MRI scans. Throughout data increase, 442 normal MRI pictures and 421 malignancy MRI scan are the statistics that we eventually used during investigation. For assessing prediction accuracy, the evaluation methods include correctness, precise, retention, area beneath receiver's functional arc (Auc Values score) and True positive rate. The true negatives (TN), true positive (TP) and false positives (FP), false negatives (FN) are computed. They were computed employing:

$$\text{Accuracy} = \left[(\text{true positive} + \text{true negatives}) / (\text{true negatives} + \text{true positive} + \text{false positives} + \text{false negatives})\right] \quad (4)$$

$$\text{Precision} = \left[\text{true positive}/(\text{true positive} + \text{false positives})\right] \quad (5)$$

$$\text{Recall} = \left[\text{true positive}/(\text{true positive} + \text{false negatives})\right] \quad (6)$$

$$F1\text{Chart} = \left[\{2 * \text{true positive}\} / \{(2 * \text{true positive}) + \text{false positives} + \text{false negatives}\}\right] \quad (7)$$

The Auc Values Rating can effectively sum up the results of the imbalanced parameters of the model and became the standard for various challenge for data analysis. In particular, this recall can inform us whether or not frequency of negative cases decreases. The follows may be learned out from equation variation (6):

$$(\text{true positive} + \text{false positives}) = \text{Number of Malignant Cells in Pancreas} \quad (8)$$

$$\text{Recall} = \{(\text{Number of Malignant Cells} - \text{false negatives})/\text{Number of Malignant Cells}\} \quad (9)$$

The FN would be lowered if the recall score grows bigger.

5 Result Analysis

As the initial specimens are imbalanced and the affirmative sample is to be given greater weight, we suggest a weighting gradient descent listed in Sect. 4 of our research paper. We perform two sets of comparative studies to demonstrate the correctness and logic of the suggested gradient descent. The influence of varied breadth and depth on outcomes is examined within every set of trials. Figure 6 and Respective Data in Table 2 demonstrate empirical studies. First, we categorise the pancreas Image data with four classifiers ResNet52, ResNet34, ResNet18 and Linear Cross entropy nonlinear function Inception-ResNet. Five criteria in Sect. 4 shall be used to analyse the outcome of classifications. Figure 6 illustrates the findings. It shows clearly that the system Initiation-ResNet delivers the greatest results, with

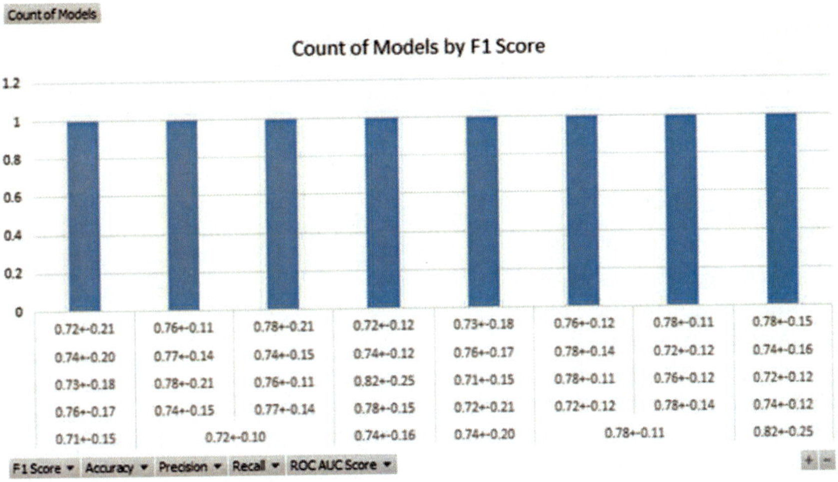

Fig. 6 Illustrates the findings examined within every set of trials

Table 2 * Indicates that this approach utilises the suggested scaled gradient descent

MODALS	ACC	PRE	RCAL	AUC -ROC score	FI score
Rees_Net18	0.78 ± 0.14	0.76 ± 0.12	0.72 ± 0.12	0.78 ± 0.11	0.78 ± 0.11
Rees_Net18*	0.77 ± 0.14	0.76 ± 0.11	0.74 ± 0.15	0.78 ± 0.21	0.72 ± 0.10
Rees_Net34	0.74 ± 0.12	0.72 ± 0.12	0.74 ± 0.16	0.78 ± 0.15	0.82 ± 0.25
Rees_Net34*	0.76 ± 0.17	0.73 ± 0.18	0.74 ± 0.20	0.72 ± 0.21	0.71 ± 0.15
Rees_Net52	0.78 ± 0.15	0.82 ± 0.25	0.74 ± 0.12	0.72 ± 0.12	0.74 ± 0.16
Rees_Net52*	0.72 ± 0.21	0.71 ± 0.15	0.76 ± 0.17	0.73 ± 0.18	0.74 ± 0.20
Inception_Rees	0.72 ± 0.12	0.78 ± 0.11	0.78 ± 0.14	0.76 ± 0.12	0.78 ± 0.11
Inception_Rees*	0.74 ± 0.15	0.78 ± 0.21	0.77 ± 0.14	0.76 ± 0.11	0.72 ± 0.10

82% efficiency and is being tracked by using ResNet18. In our observations, the categorization capability of the simulation initially drops and afterwards grows with depths of the leftover network structure which suggests the usage of the initial error function results to over compatibility although in this investigation is unworkable. We subsequently conduct out the recommended error function trials. Figure 6, which would be represented in red, displays the effects as well. Notice that its red section is an improvement in the initial outcomes. Every modelling parameter can be shown to have some degree of change, excluding the accuracy of ResNet52. The retrieval scoring around 1.0 implies that false negatives are decreased. Amongst these, ResNet18 delivers the highest results having 91% precision, trailed by ResNet34, ResNet leads in 3rd and ResNet52 has the lowest enhanced results. The findings showed a substantial increase in the suggested weighting gradient descent. The findings reveal that now the weight loss mechanism presented improves considerably with regard towards the classification performance and also demonstrate that the proper breadth and depth of the modelling for our set of data has to be investigated several times. Gastrointestinal cancer usually has different characteristics. It is hard to differentiate some features amongst malignant and benign. By using a new gradient descent, empirical outcomes are improved and also the false negatives are reduced.

6 Conclusions

In order to classify pancreas tumours, we utilised 4 networks of 3-dimensional convolutional neural network parameters ResNet34, ResNet3418, ResNet52, and original conception. Measured values and analyses indicate that ResNet18 is best positioned for classifying the innocuous from cancer in 3-dimensional MRI scans with a suggested weighting gradient descent technique, especially particular to decreasing the incidence of false—negative that really can minimise misinterpretation in real world applications. Furthermore, the breadth and depth of the connection cannot enhance categorization correctness of our tiny data sets provided we apply the suitable weighting gradient descent. Our technique is a highly autonomous way of categorising malignancy that benefits from the geospatial data in the MRI picture.

References

1. Klawikowski S et al (2016) Development of a CT radiomics based early response prediction model during delivery of chemoradiation therapy for pancreatic cancer. Med Phys 43(6):3350–3350
2. Li Y et al (2012) Diagnose the premalignant pancreatic cancer using high dimensional linear machine. In: IAPR international conference on pattern recognition in bioinformatics, pp 198–209
3. Ko J et al (2017) Combining machine learning and nanofluidic technology to diagnose pancreatic cancer using exosomes. ACS Nano 11(11):11182

4. Dmitriev K et al (2017) Classification of pancreatic cysts in computed tomography images using a random forest and convolutional neural network ensemble. In: International conference on medical image computing and computer- assisted intervention, pp 150–158
5. Hosseini-Asl E et al (2016) Alzheimer's disease diagnostics by a 3D deeply supervised adaptable convolutional network. Front Biosci 23(2016):584–596
6. Liu K et al (2017) Multiview convolutional neural networks for lung nodule classification. PLoS One 12(11):12–22
7. Korolev S et al (2017) Residual and plain convolutional neural networks for 3D brain MRI classification. In: IEEE international symposium on biomedical imaging
8. Mehrtash A et al (2017) Classification of clinical significance of MRI prostate findings using 3D convolutional neural networks. Proc Spie Int Soc Opti Eng 10134(2017):101342A
9. Roth HR et al (2018) An application of cascaded 3D fully convolutional networks for medical image segmentation. Computer Med Imag Graph Off J Computer Med Imag Soc 66(2018):90
10. Zhu Z et al (2017) A 3D Coarse-to-fine framework for automatic pancreas segmentation
11. He K et al (2015) Deep residual learning for image recognition, pp 770–778
12. Szegedy C et al (2016) Inception-v4, inception-resnet and the impact of residual connections on learning
13. Xie S et al. (2016) Holistically-nested edge detection. In: IEEE international conference on computer vision
14. Chen Y et al (2017) Three-way decision support for diagnosis on focal liver lesions. Knowledge-Based Syst 127(C):85–99

Does Stock Exchange Influence Economic Growth? An Exploratory Study from the Prospective of Gambia

Malick Jeng, Mohit Pathak, and Arti Chandani

Abstract Stock exchange helps in promoting the economic growth by increasing the opportunities for capital investments, equity financing, foreign direct investment, net export and increase employment opportunities. The Gambia being a developing country has low savings rates, unstructured companies that are family owned and ineffective financial intermediaries. Establishing a stock exchange in an under-developed economy like Gambia can help in mobilising the savings and encourage investments which may lead to economic growth. By using a sample of five African countries, whose economies were similar to Gambia before establishment of stock exchange, this study aims to investigate the effect of setting up a stock exchange on economic growth and highlight the potential benefits and challenges of setting up a stock exchange in a county like Gambia.

Keywords Capital investment · Debt financing · Economic growth · Foreign direct investment · Gambia · Net exports · Stock market

1 Introduction

In the last few couple of decades, countries all around the world witnessed a huge surge in the growth of stock markets, and emerging economies have gained heavily from this boom. As a result, various other under-developed and developing economies started focusing on establishing and developing stock market and exchanges to attract investors from all around the globe. Stock markets play an important role in the creation and growth of industry and trade by funding investments by companies that translate positively into the economy by contributing to the gross domestic product of the nations (GDP) [1]. Establishing a stock exchange (SE) and developing the stock

M. Jeng · A. Chandani (✉)
Symbiosis Institute of Management Studies, Symbiosis International (Deemed University), Pune 411020, India
e-mail: arti.chandani@sims.edu

M. Pathak
Symbiosis International (Deemed University), Pune 412115, India

M. S. Kaiser et al. (eds.), *Information and Communication Technology for Competitive Strategies (ICTCS 2021)*, Lecture Notes in Networks and Systems 401,
https://doi.org/10.1007/978-981-19-0098-3_58

market in a county is expected to simulate economy growth by boosting domestic savings and increasing the level of investment [2, 3]. According to Tessema [4] in the absence of stock markets, banks enjoy absolute monopoly to fund any short-term or long-term business needs that impose high interest rates and demand collateral. Such monopoly can be drastically reduced if stock markets are present in an economy as they can help in development of a strong and sound financial system in country. The formation of stock markets in small and under-developed economies of Africa, such as the Gambia is predicted to increase domestic savings and increase investments [5], therefore enhancing the economic growth of the country.

The Gambia is an under-developed country in Africa with low per capita GDP and its economy is characterised low savings rate, complex tax regime, unstructured companies that are individually owned, and ineffective financial intermediaries. Establishing a SE may help in boosting the economy of such countries by mobilising savings and encouraging investments which act as a source of medium and long-term funding for financing both government and private sectors. A significant surge in the mobilisation of savings could increase the savings rate. The stock market is expected to promote investment by providing individual investors in Gambia with an alternative financial instrument to meet their asset allocation and liquidity needs [6]. Gambian firms face many challenges when it comes to raising funds for their diversification and expansion, through Initial Public Offering (IPO), equity markets can be an opportunity for growing firms to raise capital at lower costs.

2 Literature Review and Hypothesis Development

Many researchers have identified the benefits of the stock market and SE and their importance to individual investors, companies and economy [7]. According to Levine and Zervos [6] countries immensely benefit from stock market exchange, as stock markets greatly influence economic prosperity. Growth of a listed company is also one of the many advantages that stock markets promote. The other merit that a stock market may provide is efficiency, healthy governance and check and balance of existing listed companies. This could be done through mounting external pressure and discipline on company's management. Stock and bond investment serves as an increment to the volume of cumulative savings because it encourages and creates incentives to invest. According to Tessema [4] stock market are more volatile investment compare to time deposits, bonds, etc. their returns are also very rewarding and can offer investors with the options to diversify across industries within the market this would enhance mitigate the bitterness to suffer sudden market crashes.

As revealed by Levine [8] countries with an edge in a better and more equipped financial and liquid market, do enjoy more rapid in the accumulation and higher gains in productivity. When firms are assured of the ability to exit hence high liquidity, there is higher tendency to invest in the financial market. One of the inclusive roles that a stock market play is that different section of the society could buy shares, become shareholder of a particular company, depending on their financial ability.

The people sharing the earnings of a business will enhance the vast gap existed within the society, and hence it bridges the large income inequalities [9]. The sense of ownership amongst investors increases when equity and distribution of enterprise earnings increase in a company [4].

Furthermore, self and external regulation within the stock market enhances an assurance that the benchmark is working resourcefully, fairly and without any fraudulent practices. The surrender of vital data on financial information of a public firms assists in preventing fraudulent and ill practices during the buying and selling of stock decisions of various investors. Those fundamentals enhance in exercising a check on a particular firm's top-level stakeholders like directors and management [9]. In securities market, Initial public offering (IPO) and further public offerings (FPO) of stocks facilitates to mobilised substantial funds for large undertakings whose outputs are in the top most demand by society. Therefore, efficient firms are compensated by access to investment funds [4]. The absence of stock market in Gambia, firms will inevitably relay mostly on their retained resources for funding their major undertakings, banks which attracts high interest rates against the companies, which ultimately affects their profits adversely. Based on the literature we present the null and alternate hypothesis of our research; H0: The stock market has no implications for the county's economic development; H1: The stock market has impacts on the county's economic development.

3 Research Methodology

3.1 Sample and Research Design

This study employs secondary data obtained from the website of World Bank [10]. The time series data cover five pre and post years of data from the year of SE establishment in the five sample countries. The sample African countries are Cape Verde, Eswatini, Rwanda, Malawi, and Mozambique. Since, these sample countries are very similar to Gambia in terms of economic growth, therefore, we will be able to estimate whether establishing a Stock Exchange will help Gambia in achieving economic prosperity or not. For data analysis we used SPSS software package and Regression analysis to estimate the association amongst the dependent and independent variables of the study.

3.2 Variables

Economic growth (EGROW) is taken as a dependent variable and is measured from one year to the next as the percentage increase in GDP. It is the amount of total value added by the country's citizens, plus any taxes and minus any subsidies. Not included in Commodity Value. FDI, net exports and Capital expenditure are taken as independent variable in our study. FDI represents the international ownership of the manufacturing facilities. It is measured as the amount of equity capital, earnings reinvestment, other long-term capital, and short-term capital as shown in the balance of payments. Net export is determined by deducting goods and services imports as a percentage of GDP from the goods and services exports as a percentage of GDP. Capital expenditure takes into consideration firms' acquisitions of new plants and equipment and is taken as a percentage of GDP for our research.

3.3 *Statistical Model*

$$\text{EGROW} = \beta 0 + \beta 1\,\text{FDI} + \beta 2\,\text{NEXP} + \beta 3\,\text{CAP} + \varepsilon \quad (1)$$

Where β_0 is intercept, ε is error term, EGROW is Economic growth, i.e. rate of change of real GDP measured in Percentage, FDI is Foreign Direct Investment as a per cent of GDP, NEXP is Net Exports of goods and services as per cent of GDP, and CAP is Capital investment as per cent of GDP.

4 Data Analysis and Results

This section presents the descriptive statistics and regression analysis to present an overview of the economic indicators of the sample countries and examine the association that exists amongst them. The aim is to identify the changes in the dependent and independent variables before and after establishment of the stock exchange. This summary of descriptive statistics presents the 10 years' data (5-year pre and 5-year post SE establishment) of the sample countries. The descriptive statistics of all the five countries before and after SE establishment is presented in Table 1.

4.1 *Regression Analysis*

For identifying the nature of association amongst the dependent and independent variables we used regression analysis on pre-SE establishment and post SE establishment data of the five sample countries separately. The regression analysis of Cape Verde, pre and post SE is presented below in Table 2. The regression analysis of pre-SE data shows GDP is positively associated with FDI, NEXP, CAP but the results are not significant. Whereas, post SE establishment, the association between GDP and FDI is positive but the association between GDP and NEXP and CAP is negative but the results are not significant.

Table 1 Summary of descriptive statistics—pre and post stock exchange establishment

Country		Pre-SE establishment				Post SE establishment			
		GDP	FDI	NEXP	CAP	GDP	FDI	NEXP	CAP
Cape Verde	Mean	7.23	4.49	−35.16	42.50	6.00	10.13	−27.78	46.34
	SD	4.92	2.47	2.05	1.98	6.36	2.73	1.85	2.24
	Min	2.23	1.60	−37.56	40.05	−1.27	6.98	−30.19	43.05
	Max	14.28	7.32	−32.54	45.09	15.17	12.67	−25.88	49.08
Eswatini	Mean	7.35	84.38	63.63	21.70	5.34	73.94	56.45	17.36
	SD	2.63	5.51	8.38	4.43	1.68	2.38	3.02	1.64
	Min	3.22	79.77	52.88	14.86	3.05	70.64	54.12	15.65
	Max	9.64	92.84	71.20	26.22	7.10	76.38	61.28	19.72
Rwanda	Mean	0.77	24.15	4.26	15.65	2.90	29.25	5.26	23.19
	SD	0.74	0.78	2.28	1.40	0.97	1.34	2.15	1.84
	Min	0.25	23.54	2.05	13.86	1.72	28.03	3.56	21.58
	Max	2.02	25.21	7.04	17.70	4.07	31.00	8.96	25.75
Malawi	Mean	0.24	42.76	21.01	20.37	1.43	37.89	24.69	13.64
	SD	1.26	13.05	13.74	5.30	1.10	3.74	4.46	1.30
	Min	−1.30	29.31	6.44	15.17	0.56	33.59	17.56	11.60
	Max	2.11	61.75	39.87	29.10	3.30	43.29	28.87	14.90
Mozambique	Mean	8.24	2.00	5.97	25.50	7.48	5.03	6.62	36.19
	SD	3.86	1.23	12.69	7.94	4.02	1.53	48.91	6.48
	Min	2.25	1.25	−13.78	18.25	1.19	2.51	−83.09	27.58
	Max	11.30	4.18	19.01	37.88	12.09	6.27	40.00	43.18

Table 2 Regression analysis—Cape Verde pre-stock exchange establishment

	Pre				Post			
	β	SE	*T* stat	*P*-value	β	SE	*T* test	*P*-value
Intercept	−7.25	46.16	−0.16	0.9	−48.34	19.74	−2.45	0.25
FDI	1.57	0.85	1.85	0.32	1.98	0.38	5.27	0.12
NEXP	1.14	0.95	1.2	0.44	−1.92	1.15	−1.67	0.34
CAP	1.12	1.03	1.08	0.47	−0.41	1.05	−0.39	0.76
Multiple *R*	0.94				0.99			
R^2	0.89				0.99			
Adjusted R^2	0.55				0.95			
SE	3.30				1.47			

The regression analysis of Eswatini, pre and post SE is presented below in Table 3. The regression analysis on data before SE shows that GDP is positively associated with FDI and NEXP, whereas GDP is negatively associated with CAP but the associations are not significant. The regression analysis on data post—SE shows that GDP is negatively associated with FDI and NEXP but is positively associated with CAP, but the results are not significant.

The Rwanda, pre and post SE regression analysis are shown in Table 4. The regression analysis shows that though the correlation is not significant, GDP is negatively associated with FDI but positively associated with NEXP for the data before SE. Whereas GDP is linked to CAP positively and significantly (at a level of 10%). Analysis of regression for post SE data reveals that GDP is positively correlated with FDI and CAP but is negatively associated with NEXP but the results are not important.

The regression analysis of Malawi, pre and post SE is presented below in Table 5. The regression analysis shows that for the data before SE, GDP is negatively associated with NEXP but positively with FDI and CAP, though the association is

Table 3 Regression analysis—Eswatini pre-stock exchange establishment

	Pre				Post			
	β	SE	*T* test	*P*-value	β	SE	*T* test	*P*-value
Intercept	−2.49	20.61	−0.12	0.92	101.13	63.71	1.59	0.36
FDI	0.01	0.23	0.07	0.96	−1.32	0.79	−1.67	0.34
NEXP	0.23	0.12	1.93	0.30	−0.45	0.39	−1.16	0.45
CAP	−0.28	0.26	−1.07	0.48	1.58	0.93	1.70	0.34
Multiple *R*	0.94				0.87			
R^2	0.88				0.76			
Adjusted R^2	0.51				0.05			
SE	1.83				1.63			

Table 4 Regression analysis—Rwanda pre-stock exchange establishment

	Pre				Post			
	β	SE	*T* test	*p*-value	β	SE	*T* test	*p*-value
Intercept	0.66	2.43	0.27	0.83	−3.02	44.47	−0.07	0.96
FDI	−0.3	0.09	−3.14	0.2	0.07	3.47	0.02	0.99
NEXP	0.07	0.03	2.02	0.29	−0.14	0.56	−0.25	0.84
CAP	0.44	0.04	10.62	0.06	0.19	2.66	0.07	0.95
Multiple *R*	1.00				0.37			
R^2	0.99				0.13			
Adjusted R^2	0.98				−2.5			
SE	0.12				1.8			

Table 5 Regression analysis—Malawi pre-stock exchange establishment

	Pre				Post			*p*-value
	β	SE	*T* stat	*p*-value	β	SE	*T* stat	
Intercept	−4.52	3.56	−1.27	0.42	−9.00	4.98	−1.81	0.32
FDI	0.1	0.07	1.46	0.38	0.24	0.2	1.18	0.45
NEXP	−0.05	0.1	−0.54	0.69	−0.18	0.16	−1.15	0.46
CAP	0.07	0.24	0.29	0.82	0.43	0.77	0.56	0.68
Multiple *R*	0.89				0.92			
R^2	0.79				0.85			
Adjusted R^2	0.16				0.39			
SE	1.16				0.86			

not significant. Regression analysis for post SE data shows that the GDP is positively associated with FDI and CAP but is negatively associated with the NEXP, but the results are not significant.

The regression analysis of Mozambique, pre and post SE is presented below in Table 6. The regression analysis shows that for the data before SE, GDP is positively associated with FDI and NEXP but negatively with CAP, though the association is not significant. Regression analysis for post SE data shows that the GDP is positively associated with FDI, NEXP and CAP, but the results are not significant.

The association between GDP and FDI in Cape Verde, Malawi and Mozambique before and after SE establishment remains positive indicating no change. Whereas in Eswatini, the relationship was positive before the SE establishment but it became negative after establishment of SE. In case of Rwanda, the relationship was negative before the SE establishment but it turns positive after establishment of SE. The association between GDP and NEXP in case of Cape Verde, Eswatini, Rwanda became negative from positive after establishment of the SE. In case of Malawi, the association remains negative for both the period, indicating no change. In case of

Table 6 Regression analysis—Mozambique Pre—Stock Exchange establishment

	Pre				Post			
	β	SE	T stat	*p*-value	β	SE	*T* stat	*p*-value
Intercept	16.99	1.02	16.69	0.04	−103.13	48.60	−2.12	0.28
FDI	1.45	0.24	6.00	0.11	8.66	3.31	2.62	0.23
NEXP	0.06	0.02	2.82	0.22	0.22	0.10	2.16	0.28
CAP	−0.47	0.04	−12.62	0.05	1.84	0.89	2.07	0.29
Multiple *R*	1				0.95			
R^2	0.99				0.91			
Adjusted R^2	0.98				0.64			
SE	0.58				2.40			

Mozambique, the association remains positive for both the period. The association between GDP and CAP in Rwanda and Malawi remains the same for both the period. In case of Eswatini and Mozambique the association becomes positive from negative after establishment of the SE, whereas in case of Cape Verde, the association turns into negative from positive after the establishment of the SE.

5 Conclusion and Recommendation

This study explored the plan of establishing a stock exchange in Gambia, for which we have examined the impact of stock exchange establishment on economic growth of sample African countries. As documented by authors that stock exchange establishment and stock market development increase the FDI, NEXP and CAP [11, 12] which increases the GDP and leads to economic growth of the country therefore, we have taken five-year data post establishment and five-year data pre-establishment of stock exchange and examined the possible effect of FDI, NEXP and CAP as proxy for SE on the GDP of Cape Verde, Eswatini, Rwanda, Malawi and Mozambique. Descriptive statistics revealed that the average GDP of countries has increased after establishment of stock exchange, except in Cape Verde and Eswatini. Whereas the regression analysis shows that the impact of FDI, NEXP and CAP on GDP are inconclusive, meaning that the establishment of stock exchange market does not influence the FDI, NEXP and CAP and therefore the GDP of the sample countries. Therefore, we accept the null hypothesis that the SE has an impact on the economic growth of the county. Our result is consistent with those of Harris [13] and Osamwonyi and Kasimu [14].

The reasons associated with this may be the lack of awareness amongst local companies and investors regarding the working of stock exchange and stock market which refrain them from listing their company and investing in the stock exchange market. Poor legal system may also be the reason, as it may not be strong enough to ensure safety of investor's funds, which may be associated with low foreign and local investments. Before establishing a stock exchange in under-developed economy, the availability of necessary regulatory agencies, financial institutions, knowledge of possible stockholders need to be addressed by the government.

6 Limitations and Scope for Future Researches

This research requires only the use of secondary data, and the sample taken to study the relationship between stock exchange and economic growth is also very small and restricted to the continent of Africa only. Future researchers can use a broader data set covering a larger time frame and including cross continent nations to study such relationship. Future researches may also collect primary data on the awareness amongst local companies regarding stock exchange and stock markets working as well.

References

1. Kapaya SM (2020) Stock market development and economic growth in Tanzania: an ARDL and bound testing approach. Rev Econ Polit Sci 5(3):187–206
2. Baier SL, Dwyer GP Jr, Tamura R (2004) Does opening a stock exchange increase economic growth? J Int Money Financ 23:311–331
3. Singh A (1997) Financial liberalisation, stock markets and economic development. Econ J 107(442):771–782
4. Tessema A (2003) Prospects and challenges for developing securities markets in Ethiopia: an analytical review. Afr Dev Rev 15(1):50–65
5. Yartey CA, Adjasi CK (2007) Stock market development in Sub-Saharan Africa: critical issues and challenges. International Monetary Fund, Washington, DC
6. Levine R, Zervos S (1998) Stock markets, banks, and economic growth. Am Econ Rev 537–558
7. Owusu EL, Odhiambo NM (2014) Stock market development and economic growth in Ghana: an ARDL-bounds testing approach. Appl Econ Lett 21(4):229–234
8. Levine R (1996) Stock markets: a spur to economic growth. Finance Dev-English Ed 33(1):7–10
9. Musonera E, Safari V (2008) Establishing a stock exchange in emerging economies: challenge and opportunities. J Int Manag Stud 3(2)
10. World Bank (2020) https://databank.worldbank.org/home.aspx. Last accessed 22 Aug 2020
11. Adam AM, Tweneboah G (2009) Foreign direct investment and stock market development: Ghana's evidence. Int Res J Financ Econ 26:178–185
12. Oseni IO, Enilolobo OS (2011) Effect of foreign direct investment and stock market development on economic growth in Nigeria (1980–2009). Europ J Bus Manag 3(12):34–42
13. Harris RD (1997) Stock markets and development: A re-assessment. Eur Econ Rev 41(1):139–146
14. Osamwonyi IO, Kasimu A (2013) Stock market and economic growth in Ghana, Kenya and Nigeria. Int J Financ Res 4(2):83

Digitization of Measuring Scales in Social Science Research Area

Mahesh Bapusaheb Toradmal, Mita Mehta, and Arti Chandani

Abstract Development in the space of computing technologies opened the doors for organizing large scale population studies on Internet where psychometric assessments are involved. This is resulting in higher speed of research, large datasets, and optimal costing to conduct research. However, mode of test conduction should not change the ethical responsibilities of researcher. Online mode of conduction should not have negative impact of it on its psychometric properties of measuring scale (MS). Over the period, Internet survey/testing became boon for people living in rural area away from physical testing facilities. During this period, we observed that Internet base survey conduction is having its own advantages and challenges. The aim of this paper is to explore and provide inputs to maintain psychometric properties of MS on Internet while taking technological advancement in this space. This paper should help while creating new research tools and/or MS like online surveys on Internet. It should help in creating complex experimental psychological models with enhanced statistical power of it. We will take help of technology advancement in the field of machine learning and artificial intelligence.

Keywords Measuring scale · Digitization · SEM · Scale selection · Factor analysis · Python programming · Machine learning

1 Introduction

Over the period it is observed that mode of test conduction did not change ethical responsibilities of the psychologist to maintain test reliability and validity. More usage of Internet to conduct psychometric assessments makes them readily available. However, survey/assessment reliability, validity, administration, item security and test taker confidentiality are more concerned while conducting online surveys. Now a day researcher may put these surveys/tests on web for people to complete it in hope

M. B. Toradmal · M. Mehta (✉) · A. Chandani
Symbiosis Institute of Management Studies, Symbiosis International (Deemed University), Pune, India
e-mail: mita.mehta@sims.edu

M. S. Kaiser et al. (eds.), *Information and Communication Technology for Competitive Strategies (ICTCS 2021)*, Lecture Notes in Networks and Systems 401,
https://doi.org/10.1007/978-981-19-0098-3_59

that they are getting results. Researcher found this as best mechanism to obtain data provided by respondent. At the same time, it is observed that available surveys/tests are designed with different goals. Few are designed for personal development and growth; few are to make important diagnostic and treatment decisions and few to benchmark cognitive ability of person. Goal of each survey should guide us on how to administer it. It is observed that some amount of motivation is required at examinee level to obtain better realistic responses from them. There are changes that individual may fake on personality inventory or cheat on intelligence test.

In spite of all these challenges, Internet testing is boon for people lives in rural area away from physical testing facilities. From ethical point of view, there are more challenges for researcher like examinee integrity, test security and challenges related to special population like disabling conditions, culturally and linguistically diversification. People with less computer familiarity may have higher cognitive load compare to those who are familiar. Test taker may not give more or required attention to the digitized instructions mentioned prior to test. Because of this, question arises on whether to use tests which are developed for paper pencil format directly as it is on digital platform. Therefore, more study should be done on how we should use online platform to conduct these tests. Equivalence and normalization should be done before using these tests.

Test construction, Environment (mode in which it will be conducted/administered), Test Analytics (The way scores are analyzed and interpreted) of test should be supported/studied before using it to assess ability of individual to perform specific activity.

In this paper we will study these three areas and try to put more light on how we could develop/use different tests on online platform to achieve desired results. This paper is outcome of systematic study of existing literature in this area. We had gone through different research work done in this area starting from 1960 to 2018. Papers published in initial years are studied to understand test construct. However, latter part of paper review is done to understand its implementation using computer.

2 Discussion

2.1 Test Construction

2.1.1 Using Existing Psychometric Scale Versus Developing New

Most of the researcher faces this dilemma of whether to use existing scale or develop new scale for their own research. Answer lays in the question itself! We should call it as "Conceptual fitment" of scale to our research objective. Any scale which is old and fulfill our research objective should be used. However, scale fitment to our samples should be verified with the help of mentioned tools in subsequent sections below. Any pilot study is good help to solve this dilemma. Another challenge researcher face is too many scales available and having question on which to use? In such a scenario, researcher should use two criteria to select most appropriate scale. First is a psychometric property of scale concerning reliability and validity and second is conceptual fitment of scale toward objective of research. In many cases, researcher might think of changing items in existing scale. In this case there are no such a black and white rule available; however, any change in the question reforming, removing of any item, adding of new item should be gone through validity and reliability check. In case if existing scale is not available then there is separate guidelines available to develop new scale. In this case, once preliminary items are derived for scale, it is guided to conduct pilot study. Goal of pilot study is to achieve an optimal balance between parsimony and comprehensive theoretical coverage of scale by removing extra preliminary items. How many numbers of items in one scale is next hard question to answer. It is said that minimum three items in scale will help better in CFA. At other side count of maximum number of items in scale is driven by complexity of the variable being measured. Larger number of items improves richness of multidimensionality of variables. These should be trade off/balanced to maximize response rate.

2.1.2 Reliability and Validity of Scale

Currently Computerized Cognitive Test (CCT) is playing important role in assessing youth potential to get admission in any educational institute. This is to find right fit candidate to undergo specific course in desired education stream. This means it is the first foundation brick for human resource supply to perform specific job identified in that field. This result is higher dependency of organizations in getting skilled motivated and passionate taskforce for their industrial growth. This should create more importance toward such CCTs. Validation of their psychometric properties is more important. As the cognitive ability of person is function of multiple traits present in examinee, we need to consider multidimensional Item Response Theory (IRT) model along with different methods to check reliability (internal and test–retest), validity (Content and construct), effect of factoring through factor validity methods. While doing this, development speed of technology must be adopted quickly by psychometric research in CCT. This should help to develop more accurate tools that really measures intended traits in examinee. IRT should be used for design, analysis and scoring of tests/questionnaires. This is to measure human behavior in online networks. This should help to classify true information vs misinformation. Psychometric scales become more trustworthy with strong reliability and validity testing on it. Following Table 1 should give us idea on different methods for reliability and validity test.

Research in psychology most of the time deal with latent variables which could not be measured directly. More precautions should be taken in this scenario while designing scales. There are multiple methods which need to be used to link latent variable to items included in the scale. Disaggregation method, partial disaggregation method, or combining all items to form one indicator called as total aggregation. Each

Table 1 Reliability and validity methods

Concept	Measure
Internal reliability	Cronbach's alpha (α) ≥ 0.75
Test–retest reliability	Pearson correlation (r) ≥ 0.80
Content validity	Content validity index (CVI) ≥ 0.80
Convergent construct validity	Pearson correlation (r) ≥ 0.30
Divergent (or discriminant) construct validity	Pearson correlation (r) < 0.20
Criterion validity	Pearson correlation (r) ≥ 0.40

of these three methods is having its own advantages and disadvantages [10]. There are multiple fit indices commonly used to support these factoring. Researcher should use these to increase creditability. Chi-Square, Relative x^2, Root Mean Square Error of Approximation (RMSEA), GFI, AGFI, RMR, SRMR, NFI, NNFI, CFI are fit indices could be used [2]. There are problems related to sample size and power of fit indices. We will try to address it in third section of discussion where we will talk about Structural Equation Model (SEM).

2.1.3 Factoring

Factor analysis allows refining and comparing solution until the most meaningful solution is arrived [1]. A strong solution with stable factors nullifies the impact of sample size; however, a large sample size is suggested to decrease sampling error [8]. Principal Component Analysis (PCA) and Exploratory Factor Analysis (EFA) are variable reduction technique. We use this when underlying latent construct cannot be directly measured and data include unreliability due to measurement error. The data Correlation matrix produced should be evaluated to measure the strength of relationships. It needs to be done before factorization of set of variables. Correlation more than 0.3 indicates enough commonality. If inter correlations are unexpectedly low, then it may result into low variance. This means, samples collected from online platform is too homogeneous and we may fail to reveal a factor or common relationship that does exist in population. Therefore, while conducting test on online platform, sample parameters need to be collected to understand sample representing which kind of population. If the determinant matrix is zero, then it is singular matrix then in this case items are not factorable. Here we need to conduct Bartlett's test of sphericity to evaluate that determinant value is significantly different than zero. This test is conducted to reject null hypothesis. "Given data/items are not factorable" is the null hypothesis for this test. Once the factors are derived and tested, factor rotation should be conducted in order to improve the meaningfulness, reliability and reproducibility of it to achieve simple structure. Oblique and orthogonal rotations of factors are two methods of rotation. Eigen value should be considered while managing factors in the model [7]. Factoring should help us to address specific, common and error variance in the model.

2.2 Environment

Beyond considering psychometric properties of scales, it is highly require understanding mode of conduction of these assessments. Arrangement of previous training task for CCT helps to set the ground before appearing for actual assessment [4]. As CCT will be conducted in absence of instructor, it is highly important to address issues arising due to this. End user issues, technical issues of CCT should be recorded. Help of IRT model is require examining the adjustment of items difficulty level to the individual participant character if require. It will also help to understand that examinee's response to one item is not influenced by the response to another item. More focus should be given in designing assessment platform to avoid measurement errors.

There are many platforms/ web applications available for researcher to conduct their survey online. Few of them are paid and few are free. Google Forms Service is one of the most famous tools of this type. This free software is designed to conduct web survey. There are many such a platforms like survey monkey, Qualtrics, Mettl, Limesurvey. However, these platforms do not have number of factions that play an important role in psychological research [6]. The psychological community do not trust the accuracy of the conclusions drown out of these tests fully. This is mainly because of variables such as technical properties of computer at examinee side and challenges with different display screens. However, quantification of this kind of technical noise is possible with the help of advancement in technology space. Modern technologies being developed made presentation of test more convenient. It helps automatic control of presentation of tests. It results in reducing errors related human factors [6]. It also helps to record performance with accuracy compared to other methods available.

While considering different technologies, it is required to define self-sufficient architecture to deliver required goal of designing psychometric testing platform. Given the fact that users (examinee and researcher both) of such platforms will increase over the period, it should have capability of horizontal scalability of application. This should result in flexibility offered. Multi-component architecture should help over monolithic software architecture. It will help in scaling and training algorithmic models involved. While having multi-component architecture, sharding and replication methods should help to make it quickly scalable. Maintenance cost of such platforms should be low compare to other available platforms. To achieve this, we should use non-blocking I/O framework like LAMP stack. At one side where we are talking about non-blocking I/O platforms to design and on other side we need to use advance technology framework using Django stack. This framework helps to represent data to researcher in required forms. Django kind of frameworks are I/O blocking. We need to find exact architecture to deliver both values to identified users to achieve goal of it. Existing tools are having limited scope for researcher to explore patterns in collected data. Analysis and reporting of findings are typically done by researchers. They utilize specialized software packages available outside of the application. Outsider tools used could help analyze only completed surveys. Additional new surveys and hypothesis are not supported on ongoing basis. At the same time these outsider tools allow researcher to manipulate data to influence results [4]. To overcome all these challenges, below is the recommended web application architecture for psychometric tests.

User Interface layer component will allow us to make this platform accessible for majority of users on different devices. Application accessibility is one of the important aspects which platform should achieve.

Business Logic Layer will manage traffic from examinee side to allow non-blocking I/O framework. This should help us to have larger scalability of platform. Loopback framework is written in Node.js (JavaScript). It is the most widely used popular language in the world according to the statistics from GitHub and StackOverflow [6].

Analytics Layer using Django framework is python based. It is freely available open-source framework. This helps rapid development of platform. Achieving interchangeable and scalable configuration items is possible. Statistical packages available in it help to do better analysis of the collected data in the application itself.

Database Layer using MySQL as relational database management system help to remove redundancy of data. Test database will collect live data which keyed by examinee. While Training data is the history data which will keep updating on regular interval from test database. This database is used to train the model to achieve higher reliability and validity for designed model. Progressive approach of retraining model should improve predictability of the model. Meta database is additional information gathered from test giver environment which will help to analyze impact of not observed variable to model (Fig. 1).

2.3 Test Analytics

While doing test data analysis, we are having many traditional ways of doing it. We could use linear regression or multiple regressions to derive model. These models are observed delivering required results where variables could be measured directly as items on measuring instrument. Studies involving psychology are more prone to have latent variables like personality traits and intelligence level and emotional intelligence to predict. This becomes more complex when one study pulls endo and exogenous variables together. If we are having reflective measurement model, then we require indicator reliability test by considering each item loading on factors. Most of traditional regression techniques assume perfect measurement; however, it is rare to achieve in psychometric assessments. SEM is best technique used to accommodate measurement errors and get unbiased estimates and to evaluate model fitment. SEM is the collection of statistical techniques that allow to set a relation between one or more independent variables, either continuous or discrete, one or more dependent, either continuous or discrete, to be examined [9].

Structured prediction methods have become central theme of many machine learning applications. Research studies where latent construct is present are more prone to use SEM. Now a days SEM is widely used in the field of psychological

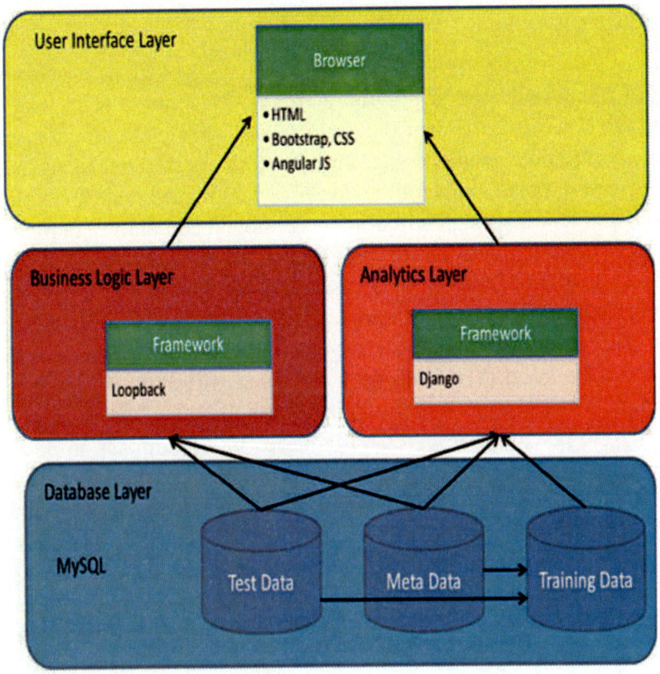

Fig. 1 Architectural design of web application

studies for developing prediction model. There are number of software available working with SEM model with limited access or need to go for paid version of it [3]. There are different implementable algorithms available for SEM. LISREL, Mplus and EQS are few of propriety and commercially distributed software. Adjustment in it from researcher perspective is not possible. OpenMx, sem, lavaan are free and open-source popular packages written in R. Python and available packages like sklearn, Semopy are open-source which are easily available with good amount of support. We found Semopy as best fit for researcher to develop integrated approach for web portal designed for psychometric tests. Semopy module named *semopy.efa* allows us to perform exploratory factor analysis by letting model know how much items we would like to combine to form factor. Two main functions available in semopy for specifying and estimating an SEM model are Model and Optimizer. There many more functions available for researcher to integrate it with existing platform to achieve desired results.

3 Conclusion

The goal of this article was to provide guideline for researcher to take advantage of technological advancement in the fiend of statistics and computation. For this we reviewed different available literature to build best settable skeletal for any psychometric portal platform. Digitization of psychometric scale is two-sided blade. Wrong usage of it will have potential to undermine the goal of undertaken research hypothesis. We recommend better comparable presentation layer of portal along with non-blocking I/O framework for business logic to make testing portal scalable. While doing this, we also recommend utilizing framework like Django using python to deliver statistical analysis edge to application. Relational database like MySQL could be good fit for this application. It will allow maintaining training data to retrain derived model and live data separately. Metadata referring user activities will help to collect unobserved variables to find its impact on derived model. We suggest using SEM for deriving model. Building matured model with higher accuracy of predictability is progressive and adaptive method. We require keeping model trained with newly collected data. This should be ongoing process. We could achieve this with few database administration activities. Python package semopy with different function available will really provide good support during building analytics part embedded in the application. This should improve hygiene of the test data by reducing human intervention in it during analysis. However, we recommend more investigation in technological space for find more better options to build SEM. Development in technology space is always going to be boon for researcher. It is required that research community catch the development speed of technology quickly in the area of psychometric research using CCT.

This study is having its limitation. We conducted this study with the available technological advancement at this time frame. One should be very careful while implementing or using these guidelines in future where technological advancement is ongoing. Researcher is having future scope of studying and integrating new test analytics methods to achieve higher predictability of model. One should keep monitoring upcoming new popular open-source packages.

References

1. Beavers AS, Lounsbury JW, Richards JK, Huck SW, Skolits GJ, Esquivel SL (2013) Practical considerations for using exploratory factor analysis in educational research. In: Practassess reseval, vol 18, no 1. University of Massachusetts Amherst, p 6. https://doi.org/10.7275/QV2Q-RK76
2. Coughlan J, Hooper D, Mullen M (2008) Structural equation modelling: guidelines for determining model fit. https://doi.org/10.21427/D7CF7R
3. Igolkina AA, Meshcheryakov G (2020) Semopy: a python package for structural equation modeling. Struct Equ Modeling 27(6):952–963

4. Jones AS, Horsburgh JS, Jackson-Smith D, Ramírez M, Flint CG, Caraballo J (2016) A web-based, interactive visualization tool for social environmental survey data. Environ Model Softw 84:412–426
5. Naglieri JA, Drasgow F, Schmit M, Handler L, Prifitera A, Margolis A, Velasquez R (2004) Psychological testing on the Internet: new problems, old issues. Am Psychol 59(3):150–162
6. Nikulchev E, Ilin D, Kolyasnikov P, Zakharov I (2018) Programming technologies for the development of web-based platform for digital psychological tools. Int J Adv Comput Sci Appl: IJACSA, 9(8). https://doi.org/10.14569/ijacsa.2018.090806
7. Raven MR (1994) The application of exploratory factor analysis in agricultural education research. J Agric Educ 35(4):9–14
8. Suhr DD (ed) (2006) Exploratory or confirmatory factor analysis? SAS institute Inc. https://support.sas.com/resources/papers/proceedings/proceedings/sugi30/203-30.pdf
9. Ullman JB (2006) Structural equation modeling: reviewing the basics and moving forward. J Pers Assess 87(1):35–50
10. Williams LJ, O'Boyle EH Jr (2008) Measurement models for linking latent variables and indicators: a review of human resource management research using parcels. Hum Resour Manag Rev 18(4):233–242

Intelligent IoT-Based Monitoring Rover for Smart Agriculture Farming in Rural Areas

Aishwarya Girish Menon and M. Prabhakar

Abstract For generations, since the first agricultural revolution in prehistoric times, human beings have depended on agricultural produces for their sustenance and survival. In the present era, man-made and natural disasters, global warming, heavy industrialization and deforestation have severely affected the quantity and quality of agricultural crop produces. As the population increases every year by millions, stable crop produce is essential for maintaining the very livelihood of these people. To improve agricultural production, a real-time monitoring system that continuously monitors crop health till the harvest season is necessary. The Arduino system proposed in this paper performs three major functions that include an area monitoring function where the system detects the presence of pests, trespassers and fire using sensors and wards off the pests or extinguishes the fire, a weather monitoring function that collects the climatic conditions during the analysis interval and updates the farmer via SMS and, a soil monitoring function that collects information regarding soil parameters and controls any soil parameter fluctuation using a microcontroller. It also provides a sprinkler system that sprays the required amount of water, insecticides, pesticides or other nutrients required to maintain crop health. The system is self-sustainable and self-maintainable as it runs on solar energy and the crop locations are pre-programmed into the system microcontroller using a method called geotagging. The system also guarantees controlled irrigation thus saving water.

Keywords Internet of Things · Agriculture · Real-time monitoring

A. G. Menon (✉) · M. Prabhakar
REVA University, Bengaluru, Karnataka, India
e-mail: aishwaryagm1999@gmail.com

M. Prabhakar
e-mail: prabhakar.m@reva.edu.in

M. S. Kaiser et al. (eds.), *Information and Communication Technology for Competitive Strategies (ICTCS 2021)*, Lecture Notes in Networks and Systems 401,
https://doi.org/10.1007/978-981-19-0098-3_60

1 Introduction

Since the Neolithic age, agriculture is one of the major factors that ensured the sustainability and survival of human beings. As time progressed humans began to shift to plant-based diets as a source of nutrition which was easier to obtain and store in large quantities rather than hunting for animals every day. In the present era, agriculture has been under extensive pressure due to an increase in man-made and natural calamities, heavy industrialization, deforestation and global warming. These issues have resulted in the extreme reduction of harvest quantity and production rate as crop health was affected. Soil erosion, acid rains, droughts and waterlogging affected soil fertility and health, making the soil unfit for growing healthy crops. The absence of a proper real-time weather forecast system in the rural areas, make it difficult for farmers to keep track of the climate continuously as they have to be physically present at the agriculture field to measure weather parameters. Sunny days with extreme sunlight have resulted in low irrigation and rainy days with heavy rains have resulted in excess irrigation due to lack of real-time weather updates in the agriculture farms. Farmers have to hire extra labor or purchase costly high-end machinery to maintain and monitor their farms which demands a large amount of the budget to go toward maintenance of the crops than toward the purchase of essential nutritional supplies for the crops.

The term Internet of Things refers to an interconnected network of sensors, physical devices and electronic devices to one another for the exchange of data, communication between each other and to gain control over connected devices without human–computer interaction. It has proven to be of great assistance to farmers as it is more cost-effective, accurate and reliable with its observations and work. The proposed system in this paper fully automates maintenance and monitoring tasks as well as the data analyses and data exchange between the proposed system and farmer's cellular device. It uses the GSM technology to exchange data between the farmer's mobile device and the system's microcontroller via SMS. The system uses GPS technology and pre-programmed location coordinates using geotagging of each crop for performing the monitoring tasks. The proposed system carries out 3 monitoring functionalities, area monitoring, soil monitoring and weather monitoring. Expensive computer systems or smartphones are not required to receive the alerts as a simple basic mobile phone would suffice to receive SMS.

2 Related Work

The agricultural field has various research works and projects based on the Internet of Things. In paper [1], the authors proposed an IoT-based system that uses a 4-layered system. System senses and collects sensor data using temperature sensor and soil moisture sensor to analyze temperature and soil moisture content. The microcontroller controls actuators and sensors. When the temperature was high, the polyhouse roof was opened and when the soil moisture content was low, the irrigation system was activated. The microcontroller transmits sensor data to cloud servers for storage using Wi-Fi. The system stores the farming data on a cloud computing platform called ThingSpeak. This data is compared with time to plot the graph. It provides graphical analysis of the sensor readings with respect to time in real-time.

In paper [2], the authors proposed a rover system with an embedded microcontroller. The system consists of a temperature sensor and a humidity sensor for measuring temperature and humidity levels, a soil moisture sensor to measure soil moisture content, a water level sensor to detect the amount of water left in the irrigation tank, a flame sensor to detect crops on fire, and a PIR sensor to detect the presence of animals on the crop field to sound the alarm. A camera module was used to capture images of crops to analyze whether the crop had any disease using image classification algorithms. If the soil moisture content is low, the soil moisture sensor signals microcontroller to activate motor pump of the irrigation unit. Temperature and humidity sensors helped to determine whether the field should be irrigated or not by analyzing the weather conditions. The results were finally sent as Tweets using the Wi-Fi network.

In paper [3], the authors proposed a system using Arduino Uno as the microcontroller. The system used the temperature sensor and the humidity sensor to detect temperature and humidity levels, a soil moisture sensor to detect the soil moisture content and an ultrasonic sensor to analyze plant growth. A pesticide and water spraying motor unit was also used for warding off pests and for irrigating the soil when moisture content was low. The collected data from the sensors were sent to an online cloud-based website called ThingSpeak accessed using the Wi-Fi network. Whenever any of the parameters fluctuated beyond the ideal range, the IoT cloud servers immediately alerted the farmer to take necessary action.

3 Proposed System

The proposed system uses ATmega2560 as its microcontroller that's embedded on Arduino Mega 2560 Rev3, to control the collection, analysis and communication of data between other devices. Two solar panels provide power using solar energy, thus making the system eco-friendly and self-sustainable. One solar panel supplies solar energy during the day while the other solar panel charges a Li-On battery pack using the charging module. The Li-On battery is used during the night when solar energy is not available. A relay module switches between the two power sources with ease.

The power and ground connections are extended using a 170-pt breadboard. The data collected and analyzed by each of the monitoring functions are stored in different CSV text files using SD card module. RTC clock module keeps track of the date and time of each interval analysis. The GSM module transmits the SMS alerts to the farmer whenever requested by the microcontroller. In case the system fails to send SMS alert at a particular interval, the microcontroller stores the data to a separate text file which is accessed later to resend the SMS as soon as connection is re-established.

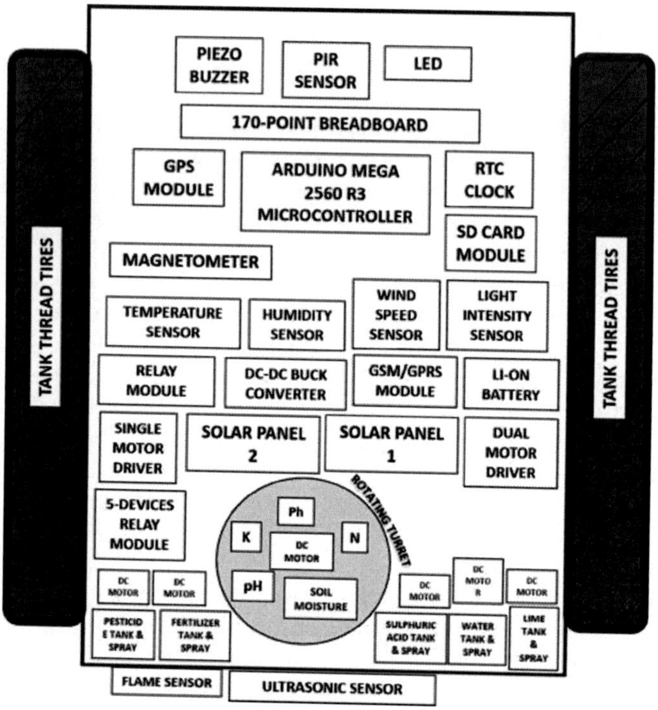

Fig. 1 Block diagram of the proposed system

The system chassis is rectangular in shape as seen in Fig. 1. The design is based on an RC car model. It uses continuous tracks such as those found on military tanks in place of tires. This makes it easier for the system to move across soft, hard, muddy, inclined or grassy terrains. The speed and motion of the tracks are controlled by two DC motors. In order to avoid trampling over the crop while approaching it, an ultrasonic sensor is placed at the rear end of the system. The continuous tracks may either be made using synthetic and rubber materials or 3-D printed chain mails. The system proposed performs 3 major monitoring tasks.

Weather monitoring is performed by humidity sensor that measures humidity level, temperature sensor that measures temperature level, light intensity sensor that measures the amount of UV radiation or sunlight available using LDR and wind speed sensor that measures the speed of wind, in the crop field. Each parameter is analyzed every alternate hour. This ensures that the farmer is aware of sudden and extreme climatic changes and thus helps the farmer take required measures to protect the crops in case of extreme climatic changes that occur rarely. The collected data from sensors are sent to the farmer via SMS after every interval analysis and stored in a CSV text file.

Soil monitoring is carried out by 5 different sensors. The pH sensor is used to measure pH levels of the soil. The soil moisture sensor measures the amount of water present in the soil using electrode probes. The concentration ratio of nitrogen, phosphorus and potassium present in the soil is analyzed using the respective sensors. The five sensors are fixed on a rotating turret that rotates with the help of a DC motor, controlled by motor driver module. The rotating turret is used to accurately place each of the sensors on the soil at a fixed spot beneath the crop. Soil parameters are analyzed every two hours. After each interval analysis the data containing the details of fluctuated parameters, rectifications performed by the system and the measures of each soil parameter along with date and time of analysis are sent to the farmer via SMS.

Area monitoring involves the detection of fire using flame sensor. In case the sensor detects fire near a crop or anywhere in the crop field, it immediately alerts the farmer through SMS and activates the water spray unit to extinguish the fire. PIR sensor detects the presence of animals or human beings in the crop field. When an animal or human trespasses into the crop field, the farmer is alerted and the microcontroller turns on flashing LEDs and a piezo buzzer that produces high-pitched sound waves which confuse animals and humans, to ward them off the field. Pests and insects emit ultrasonic waves of a particular frequency to communicate with one another. Ultrasonic sensor present on the system monitors the environment continuously to detect ultrasonic waves belonging to such frequency ranges. When it detects ultrasonic waves of the threshold frequency range, the system detects the presence of insects or pests near the crop. Once the farmer is alerted regarding this event, the system activates the piezo buzzer to produce ultrasonic waves of a particular frequency that hinders the communication between these pests and insects thereby confusing them and chasing them out of the crop field. The system navigates around the farm based on pinpoint locations on the field where the geographic coordinates of each crop are pre-programmed using geotagging, into the microcontroller. The GPS module and the magnetometer module uses these geographical coordinates to perform monitoring functions in sequential order. The magnetometer acts as a compass to identify the current position and orientation of the system.

Figure 2 represents the pre-programmed traversal path followed by the proposed system for performing monitoring tasks. The dotted lines depict the path of traversal from the crop locations $P1$–$P12$ where the system performs monitoring and sensing tasks. The arrowheads depict the direction in which the system moves. The geographical coordinates of the sample points are pre-programmed into the microcontroller. The system consists of sprinkler or spray units used to spray insecticide, fertilizer, acidifier, agricultural lime and pesticide when instructed by the microcontroller to prevent the soil parameters from fluctuating beyond permissible range. The spray units are turned on by powering a five different DC motors using a relay module to pump the liquids with pressure through five separate sprinkler nozzles.

Each liquid has its own sprinkler nozzle, DC motor and storage tank. If the soil moisture level is low, the water spray unit is activated to irrigate the soil. If it detects any pest or insect on the crop, it sprays pesticide or insecticide on the crop. If the potassium, phosphorus and nitrogen levels are low in the soil, the system sprays fertilizer on the soil to adjust the ratio level. The venturi shaped nozzle of water, acidifier, fertilizer and agricultural lime spray units are angled downwards to spray

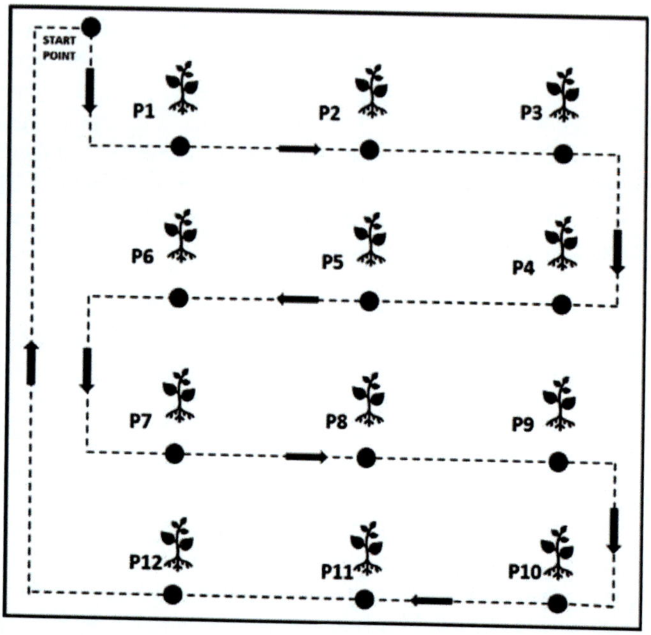

Fig. 2 Traversal path of proposed system on crop field

on the soil while the nozzles of insecticide or pesticide spray unit are angled upwards to spray the liquid directly on the crop leaves. The system generally sprays pesticide or insecticide once or twice every month on a regular basis for each crop. When the soil pH levels fluctuate to high levels, it is reduced by spraying acidifier in small quantities on the soil while low levels of pH are rectified by spraying agricultural lime mixed with water.

Figure 3 is a 3-D model that depicts how the system is modeled. The weather monitoring systems are placed at the top and at the rear end. The soil monitoring system is placed on a rotating turret at the bottom of the system. The spraying units are placed at the top with the two solar panels. The remaining electronic components are enclosed inside a protective casing. The microcontroller keeps track of the number of times each spray unit is used. If the count crosses a predefined threshold, the liquid tanks are almost empty and require refilling of the liquids. The microcontroller sends an SMS alert to the farmer stating which tank requires refilling and then to the microcontroller (Arduino Nano) fixed at the refilling station. GSM modules are used for exchanging SMS. The microcontroller at the refilling station then interprets the SMS and fills the required tank.

A separate refilling station is created that uses Arduino Nano as the microcontroller. This is placed on the roof along with a solar panel for powering it. The DC motors for filling each tank along with the nozzle units connected to supply tanks are also placed on the roof. The flooring is pressure padded. Once the system drives

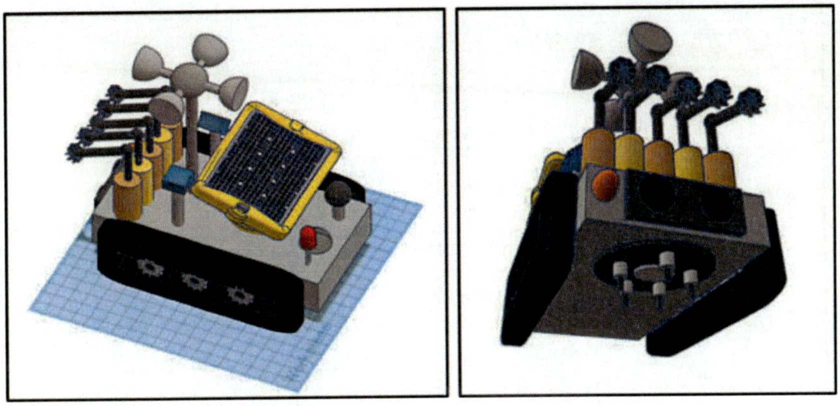

Fig. 3 3-D model of proposed system

toward the station and parks on the base, the signals from the pressure pads inform the microcontroller that the system has arrived at the station and is ready for refilling. The respective refilling units are then activated by using a 5-device relay module and DC motors to pump the liquid from supply tanks into the system's liquid tank using venturi nozzles and motors that move linearly upwards and downwards. The geographic location of the station is pre-programmed into the system. Figure 4 represents the block diagram of the refilling station. Figure 5 represents the entire process of the proposed system as a flow chart. Table 1. Provides a comparative study of proposed system.

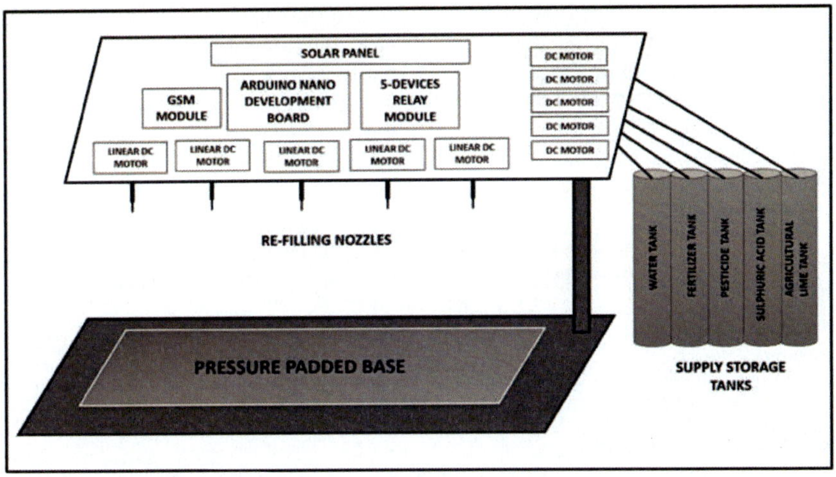

Fig. 4 Re-filling station block diagram

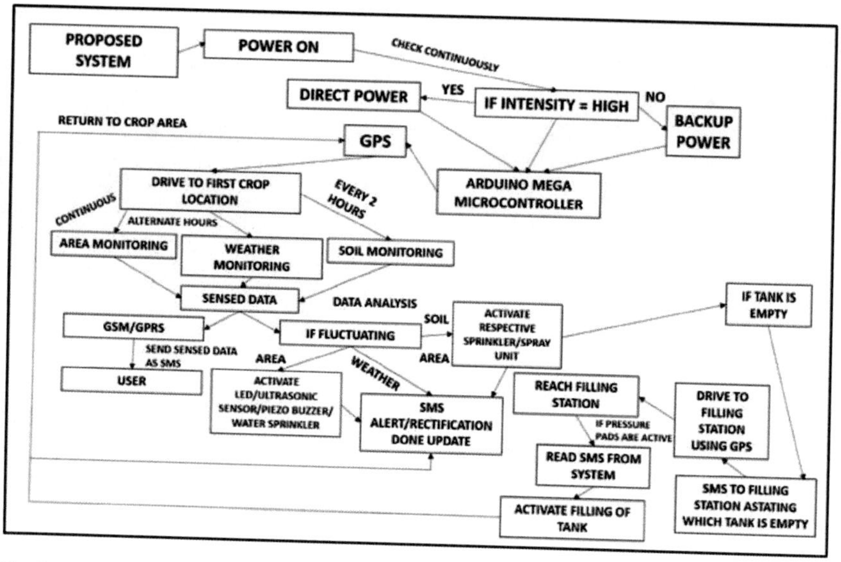

Fig. 5 Flow chart of the proposed system's function

Table 1 Comparative study between existing system and proposed system

Existing system disadvantages	Proposed system advantages
System maintenance—not automated. Power supply—from cell batteries or rechargeable batteries that must be charged manually	System maintenance—automated. Solar panels—provides uninterrupted power supply
Data transfer—using Wi-Fi network, not readily available in rural and remote areas. Unreliable and unstable	Data transfer—using GSM/GPRS mobile networking, GSM towers are available in more regions. Stable and reliable
Data is analyzed and stored in cloud servers. These servers are expensive and require stable and fast Wi-Fi connection	Data is stored in SD card and analyzed by microcontroller. There is no need for third party applications or Wi-Fi connection
Single centralized monitoring systems—Inaccurate data, distance from system varies between each crop	System performs monitoring functions separately for each crop. Data is accurate
Separate monitoring system for each crop—not affordable for farmers as it requires many hardware components	The system is a single unit that can traverse to each crop. There is no need for extra hardware
Spraying unit tanks must be manually re-filled each time	Automated filling station is available to fill the tanks automatically using Arduino and GPS

4 Results Obtained

The 3 monitoring functions has a separate set of results. To understand the expected results, a sample dataset corresponding to Tomato plant has been used in this paper.

In Fig. 6, first SMS image represents the situation when the PIR sensors detect the presence of a human or animal on field. The SMS provides the area coordinates of the region where the trespasser was detected and the date and time of the event. The second SMS image represents the situation when the ultrasonic sensors detect ultrasonic sound waves generated by insects near the crop. It provides the crop number, crop coordinates, where the insects were detected and date and time of analysis. Both the SMS alerts show rectification done by system. In Fig. 7, the first figure represents SMS sent to the farmer every alternating hour is shown which

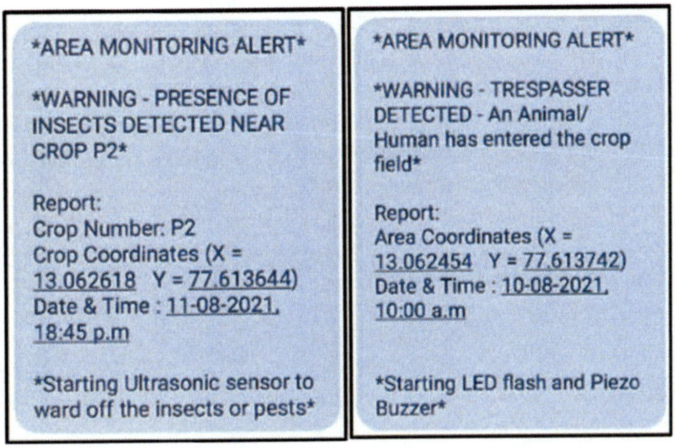

Fig. 6 SMS alerts generated for area monitoring function

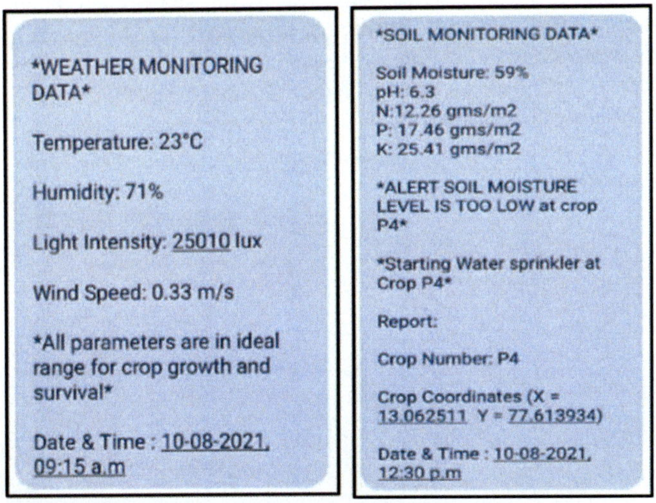

Fig. 7 SMS alert generated for weather monitoring and soil monitoring function

provides the data collected by sensors for each weather parameter, date and time of analysis and states whether each parameter is stable or has fluctuated. The second image represents SMS sent to the farmer after soil parameters are analyzed. It includes soil moisture, pH, nitrogen, phosphorus, potassium levels of the soil, date and time of analysis, and details of the parameter that fluctuated along with the crop number and coordinates. It also states the rectification performed.

Table 2 is the tabular representation containing details of the soil parameters analyzed, stored as a CSV file. Similar tables are created for weather monitoring and area monitoring as well. Figure 8 is the graphical representation of the Soil moisture parameter fluctuating for one of the interval analyses.

In the line graph, the blue line represents the values of the parameter that was recorded. The red lines represent upper and lower ideal range within which the parameter must be present for the crop to be healthy. When the blue line crosses any of the red lines acting as upper and lower bounds, it shows that the parameter fluctuated beyond the permissible range. In this case the system immediately alerts the farmer via SMS that the parameter has fluctuated.

In Fig. 8, the soil moisture level fluctuates to a lower range that is not permissible for crop number *P*4 thus activating the water spray unit to irrigate the soil beneath crop *P*4. This helps in bringing the soil moisture content back to the ideal range.

5 Conclusion

To address existing issues in the agricultural field that contributes to the sustenance of human life, the automated system proposed in this paper continuously monitors the crop health and its surroundings, collects data using sensors and analyses the data to provide automated solutions. The farmer is updated via SMS alerts using GSM technology with real-time data regarding parameter fluctuation, task performed to solve issues and general sensor data for all parameters. This method is self-sustainable and self-maintainable. It reduces the need for extra manpower and human interaction making the system cost-effective when compared to high-end machinery found in present-day markets.

Table 2 Tabular visualization of data from soil monitoring function

Crop number	Crop coordinates	Date and time of analysis	Soil parameters					Parameters fluctuated	Rectification
			Soil moisture (%)	*PH*	Nitrogen (*N*) (g/m2)	Phosphorus (*P*) (g/m2)	Potassium (*K*) (g/m2)		
*P*1	*X* = 13.062469, *Y* = 77.613737	10–08-2021, 12:15 P.M	72	6.1	11.55	17.20	25.22	Nil	Nil
*P*2	*X* = 13.062618, *Y* = 77.613644	10–08-2021, 12:20 P.M	75	6.3	11.78	17.25	25.23	Nil	Nil
*P*3	*X* = 13.062764, *Y* = 77.613579	10–08-2021, 12:25 P.M	70	5.8	12.11	17.01	25.34	Nil	Nil
*P*4	*X* = 13.062511, *Y* = 77.613934	10–08-2021, 12:30 P.M	59	6.3	12.26	17.46	25.41	Soil moisture	Water sprinkler

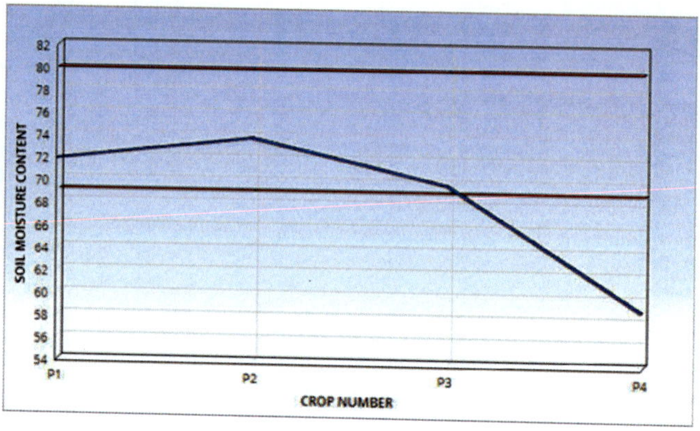

Fig. 8 Line graph representation of soil moisture parameter fluctuation

References

1. Md. Modal A, Rehana Z (2018) IoT based intelligent agriculture field monitoring system. In: 8th international conference on cloud computing, data science and engineering
2. Preksha N, Sahana A, Sushma KA, Patil SB, Nagashree (2020) Smart irrigation system using automated rover. Int J Innov Sci Res Technol 5(3)
3. Rajakumar G, Sankari MS, Shunmugapriya D, Maheswari APU (2018) Iot based smart agricultural monitoring system. Asian J Appl Sci Technol 2(2)

Wireless Sensor Network-Based Automation of Irrigation in India

Dinesh Kumar Kalal and Ankit Bhavsar

Abstract Water is a very crucial resource for any country and for developing countries like India, and it has become more important because the majority of the population relies on agricultural activity. As per studies, water demand is going to double till 2030 (Aman Kumar in Int J Res Technol 9, 9.) so the optimized use of water is need of the hour, and automation of irrigation is one of the major ways by which water resource can be used efficiently, proper water in farm field also increases the productivity and automation of irrigation reduces the human labour. This paper focusses on the automation of water used in the agricultural sector for irrigation purposes. We controlled the irrigation process using a wireless sensor network which can collect data from an agriculture farm field, and based on the soil moisture, the efficient quantity of water is supplied to the crop by this way, we can save water, soil nutrient and also reduces human labour.

Keywords WSN · Irrigation · Automation of irrigation · Sensor · Water management

1 Introduction

Agriculture has various activities like soil preparation, planting seeds in the ground, apply manure and fertilizers, provides water to crop, weeding, harvesting, and finally storage. Amongst all of these activities, irrigation which means providing water to crop for better growth and productivity is play a vital and challenging role in entire agriculture process as water is a very challenging resource.

Agriculture in a country like India has become more challenging because it holds 18% of the world population but it has access to only 4% of the total world water

---

D. K. Kalal (✉) · A. Bhavsar
GLS University, Ahmedabad, India
e-mail: dinesh.kalal@glsuniversity.ac.in

A. Bhavsar
e-mail: ankit.bhavsar@glsuniversity.ac.in

M. S. Kaiser et al. (eds.), *Information and Communication Technology for Competitive Strategies (ICTCS 2021)*, Lecture Notes in Networks and Systems 401,
https://doi.org/10.1007/978-981-19-0098-3_61

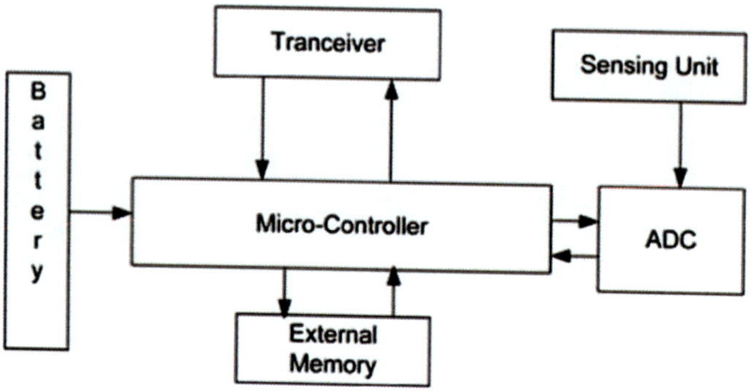

Fig. 1 Wireless sensor node

resource [1]. The agriculture sector of India is the primary sector of the Indian economy by providing 54.6% of total employment, and almost 90% of employment is related to the agricultural sector. Jain [2] but this primary sector consumes 80% of total available water resources and a very little part of it is shared by industry, drinking, energy, and other domestic use [3]. Because the maximum part of usable water is consumed by the agriculture sector only the per capita.

The intervention of technology is useful in almost every area like medical, aviation, tourism, industry, communications, education, military services, etc. The same benefits of technology can be achieved in the Indian agriculture sector also. One of the technologies focussed in this paper is using wireless sensor network (WSN) for monitoring and automation of farming activities in India [4–7]. Wireless sensor networks are a system containing various communication devices which are wirelessly connected and can be placed in a remote area as well (Fig. 1).

A wireless sensor node contains a transceiver, microcontroller, battery, and small memory. The microcontroller is used for computing tasks and assisting another component, memory is used to storing data, whilst transceivers are used for sending and receiving data [8]. Environmental data such as temperature, humidity, moisture, air quality, and pressure are received by the sensor and send this data to the server. This wireless sensor node is used to monitor and control the physical environment from a remote location.

2 Indian Irrigation and Water Availability

India is not a water-rich country as it has only 4% of the total world's fresh water, and out of this available freshwater, 80% freshwater is used for the irrigation process in agriculture. Water is a very critical resource for India as it houses 18% of the world population and about 500 million livestock. India is an agriculture-based economy where more than 55% of the population rely on agriculture for their livelihood; in the present scenario, 80% of freshwater is used in the agriculture industry only

Table 1 Demand of water in different sector [3]

Segment	Demand of water in billion cubic metres (BCM)		
	2010	2025	2050
Irrigation	550	610	815
Drinking	45	60	115
Industry	40	70	80
Energy	20	35	70
Others	55	70	115

15% by industry, and the remaining portion is used by household needs [9]. The average domestic water demand is 85 litres per capita per day(pcpd) in 2000, and it is predicted that demand will be increased to 125 (pcpd) in 2025 and 170 (pcpd) by the year 2050. On the other side, demand for water in the industrial sector also increased to 161 billion cubic metres (BCM) by 2050 [3].

Table 1 shows the assessment report of the National Commission on Integrated Water Resources Development (NCIWRD) [3].

The major source of water for irrigation, domestic, and industrial usage is groundwater as we can see in the Table 1 demand for fresh water in every sector is going to almost double by the year 2050, so proper management of water resources is need of the hour; so in this paper, we are focussing the efficient use of water resource in the agriculture sector which is in actuality a larges consumer of water resource. There are so many mechanisms for water management like automation of irrigation systems, development of decision support system for judicious utilization of rain, surface, and groundwater resources, minimizing the gap between water supply and water demand out of the above mention methods, and this paper focusses on the automation of irrigation system [10].

3 Literature Review

In this paper, we had studied the research paper [11–31] out of the listed paper, and we summarize the outcome of the following papers.

1. Muhammad et al. [11]: A natural recommendation to watering control problem applying artificial neural network controller. The suggested system is correlated with the start/ stop controller, and it is seen that start /stop controller-based system fails poorly because of its shortcomings. On the other side, the ANN-based strategy has emerged in the possible implementation of more reliable and more effective control.
2. Prisilla et al. [12]: Author of this paper mentioned water as the most critical resource available on our planet. The water resource is used in the agriculture sector for irrigation purposes. Irrigating a large area is a very complex task,

and efficient use of water is very crucial. Unfortunately in India, water management in the agriculture sector is not proper, and traditional methods like flood irrigation cause the wastage of water and degrade the soil fertility.

3. Cosmin [13]: In this paper, the author investigates the use of information technology in agriculture farms. Demands for agriculture products will be increase by 70% in the next 50 years. The most challenging part is that agricultural farms have limited resources like land, water, fertilizers, etc.
4. Jyothipriya et al. [14]: In this paper, the author gives importance to water as it is the most valuable resource to all mankind and as well as to the plants, trees. A large portion of this valuable resource is used by the agriculture sector only.
5. Swarup et al. [15]: In this paper, the author reviews the use of wireless sensors for remote monitoring of farm fields and finds the use of wireless sensor networks in the farming sector can increase the crop yield and also increases the automation of agricultural activities, which in outcomes reduces the human labour.

3.1 *Implementation of WSN-Based Irrigation*

A wireless sensor network containing a moisture sensor is deployed in the farm field, and data gathered from this sensor are passed to Arduino Uno microcontroller. Moisture value is processed, and decision for automatic on/off of water motor is taken; if moisture value is less than the prefixed value, water motor is automatically powered on, and an alert message will be sent to the farmer's mobile application from which farmer can view the status of the moisture level of farm field and the status of the water pump. The mobile application has also featured to manually on/off water pumps remotely (Figs. 2 and 3).

Fig. 2 Setup for automated irrigation

Fig. 3 Readings of moisture sensor

3.2 *Required Hardware*

3.2.1 Arduino Uno

Arduino Uno is an Atmega328 8-bit microcontroller (MCU)-based open-source, small board, it has six pins for PWM output, six analogue input pins, and 14 digital input/output pins. Arduino board can communicate with the computer system, other Arduino boards, or another microcontroller. Arduino board can be used to gather data from the sensor, process this data and also communicate this data to different devices. Arduino board can be programmed easily, and it is also user friendly and easier to handle (Fig. 4).

3.2.2 Soil Moisture Sensor FC28

Soil moisture sensors measure the volumetric water content in the soil. This makes it ideal for performing experiments in courses such as soil science, agricultural science, environmental science, horticultural botany and biology. The soil moisture sensor is one kind of sensor used to gauge the volumetric content of water within the soil. As the straight gravimetric dimension of soil moisture needs eliminating, drying, as well as sample weighting. These sensors measure the volumetric water content not directly with the help of some other rules of soil like dielectric constant, electrical resistance, otherwise interaction with neutrons, and replacement of the moisture content (Fig. 5).

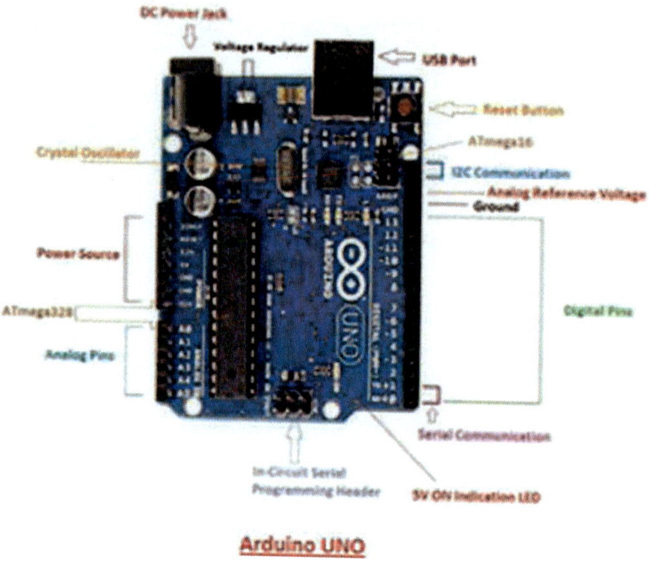

Fig. 4 Arduino Uno (microcontroller)

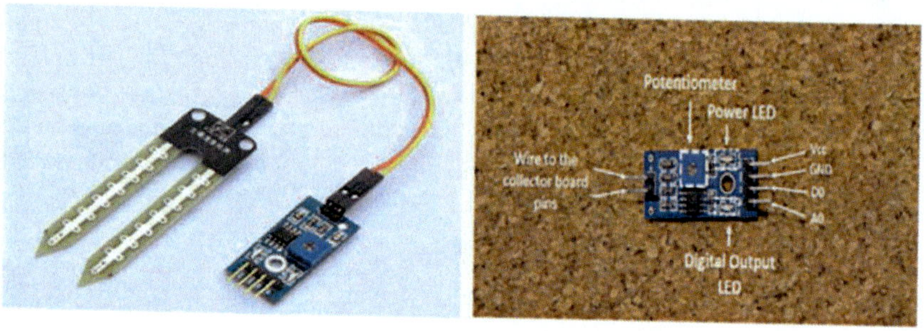

Fig. 5 Soil moisture sensor

3.2.3 Node MCU ESP 8266 Wi-Fi Module

Node MCU is an open-source firmware and development kit that helps you prototype or build IoT products. It includes firmware that runs on the ESP8266 Wi-Fi SoC from expressive systems and hardware which is based on the ESP-12 module. The firmware uses the Lua scripting language.

3.2.4 Servomotor

A servomotor is an electric device made up of linear actuator or rotary actuator which are used to rotate or push an object with a great precision. It contains a motor coupled with sensor for position feedback. Servomotor can be used as an electric water pump for automatic on/off of water supply. This motor can be attached with Arduino to operate it self using the decision based on soil moisture value.

3.2.5 12-V Asperity Type Relays

A relay is electromagnetic device which can be used as electric switch that makes circuit close and open. It can be used to on/off any electronic device, relays are generally used in control panels, manufacturing, and building automation to control power. Relays are also used for powering high-voltage devices. Panels, manufacturing, and building automation to control power. Relays are also used for powering high-voltage devices.

4 WSN-Based Irrigation Management

Wireless sensor network (WSN) technology has already proved its potential in different sectors like production, aviation, communication, health, tourism, and the aviation sector. This paper proposed WSN-based irrigation management for better productivity of farm fields and efficient use of water resources.

Wireless sensor network contains multiple sensor nodes which are interconnected and used for collecting surrounding physical and environmental data such as temperature, humidity, moisture, motion, and sound pressure. These wireless nodes are low powered and distributed in an ad hoc, decentralized fashion. The wireless sensor node can communicate with other nodes using radio signals. Wireless sensor nodes have sensing and radio transceivers, computing, and power component.

5 Conclusion

The WSN-based irrigation system has been designed using an Arduino Uno microcontroller that collects data from soil moisture sensor on getting moisture value, and it makes the decision for on/off the water motor for irrigation process to start and stop. This moisture value and status of the water motor were also passed to the farmer's mobile device over the Internet. Using the automation of the irrigation process, we can save the water resource by efficiently use of the water, and it also saves the nutrient of soil and reduces human labour.

References

1. Dhawan V (2017) Water and agriculture in India. Global Forum for Food and Agriculture (GFFA 2017)
2. Jain R (2019) J Soil Water Conser 18(4):0–0, October-December 2019 ISSN: 022–457X (Print); 2455–7145 (Online)
3. Bhat TA (2014) An analysis of demand and supply of water in India. J Environ Earth Sci 4(11). ISSN 2225-0948
4. Bhat TA (2014) An analysis of demand and supply of water in India. J Environ Earth Sci 4(11)
5. Suryadevara NK, Mukhopadhyay SC, Kelly SDT, Gill SPS (2015) WSN-based smart sensors and actuator for power management in intelligent buildings. IEEE/ASME Trans Mechatron 20(2):564–571
6. Durisic MP, Tafa Z, Dimic G, Milutinovic V (2012) A survey of military applications of wireless sensor networks. In: 2012 mediterranean conference on embedded computing (MECO). Bar, Montenegro, pp 196–199
7. Furtado H, Trobec R (2011) Applications of wireless sensors in medicine. In: 2011 proceedings of the 34th international convention MIPRO. Opatija, Croatia, pp 257–261
8. BenSaleh MS, Saida R, Kacem YH, Abid M (2020) Wireless sensor network design methodologies: a survey. J Sens 2020:9592836
9. Vellidis G et al (2008) Areal-time wireless smart sensor array for scheduling irrigation. Comput electron Agric 61:44–50
10. Upadhyaya A, Water management technologies in agriculture: challenges and opportunities. J Agri Search 2(1):7–13. ISSN: 2348-8808
11. Muhammad Umai S et al, Automation of irrigation system using ANN based controller. Int J Electric Comput Sci IJECS-IJENS 10(02)
12. Prisilla, Rooban L, Simon Vasantha Arockiam P (2012) Novel method for water irrigation system for paddy fields using ANN. Int J Comput Sci Network. ISSN: 2277–5420, Issue Date-2012–04–01
13. Cosmin (2011) Adoption of artificial intelligence in agriculture. Bull Univer Agric Sci Vet 68(1)
14. Jyothipriya AN, Saravanabava TP (2013) Design of embedded systems for drip irrigation automation. Int J Eng Sci Inven 2(4):34–37
15. Mathurkar SS, Chaudhari DS (2013) A review on smart sensors based monitoring system for agriculture. IJITEE 2(4). ISSN: 2278-3075
16. Sanjay M (2019) Portable centre pivot irrigation system with advance control through sensors. Int J Eng Res Technol (IJERT) 8(04). ISSN: 2278-0181
17. Goli KM, Madidipatla K, Sravani T (2011) Integration of wireless technologies for sustainable agriculture. IJCST
18. Priya P, Muthaiah U, Balamurugan M (2018) Predicting yield of the crop using machine learning algorithm. Int J Eng Sci Res Technol 7(1)
19. Suresh N, Nagalakshmi M, Yaswanth G, Sujatha G (2015) Real-time automization of agricultural environment for social moderization of Indian agricultural system. IJSEAT
20. Cavender-Bares K, Lofgren JB (2016) Robotic platform and method for performing multiple functions in agricultural systems. U.S. Patent US9265187B2, 23 Feb 2016
21. Bastiaanssen WGM, Molden DJ, Makin IW (2000) Remote sensing for irrigated agriculture: examples from research and possible applications. Agric Water Manag 46(2):137–155
22. Faye RM, Mora-Camino F, Sawadogo S, Niang A (1998) IEEE. An intelligent decision support system for irrigation system management
23. Rehman A-u, Abbasi A, Islam N, Shaikh Z (2014) A review of wireless sensors and networks' applications in agriculture. Comput Stand Interfaces 36:263–270. https://doi.org/10.1016/j.csi.2011.03.004
24. Gutierrez J, Villa Medina J, Nieto-Garibay A, Porta-Gándara M (2014) Automated Irrigation System Using a Wireless Sensor Network and GPRS Module. Instru Meas, IEEE Trans 63:166–176. https://doi.org/10.1109/TIM.2013.2276487

25. Rathinam DDK, Surendran D, Shilpa A, Grace AS, Sherin J (2019) Modern agriculture using wireless sensor network (WSN). In: 2019 5th international conference on advanced computing & communication systems (ICACCS), pp 515–519. https://doi.org/10.1109/ICACCS.2019.8728284
26. Khanna A, Kaur S (2020) Wireless sensor and actuator network(s) and its significant impact on agricultural domain. In: Parallel distributed and grid computing (PDGC) 2020 sixth international conference, pp 384–389
27. Sisinni E, Saifullah A, Han S, Jennehag U, Gidlund M (2018) Industrial internet of things: challenges, opportunities, and directions. IEEE Trans Ind Inf 14(11):4724–4734
28. Jayadevan J, Jasmine SM (2020) A novel architecture for internet of things in precision agriculture. Int J Appl Eng Res 15(3):204–211. ISSN 0973-4562
29. Akhil R, Gokul MS, Menon S, Nair LS (2018) Automated soil nutrient moni to ring for improved agriculture. In: ICCSP, IEEE
30. Gholap J, Ingole A, Gohil J, Gargade S (2012) Vahida attar soil data analysis using classification techniques and soil attribute prediction. Asian J Comput Sci Inf Technol
31. Aman Kumar D (2020) Application of IoT and Machine learning in agriculture. Int J Res Technol 9

Data Integrity Verification Schemes in Cloud Computing Environment: A Survey

Jyoti Rani and Rajender Nath

Abstract The popularity of cloud computing is increasing day by day. Everyone, knowingly or unknowingly, is using the services of cloud computing. Users are having very limited capability devices in terms of storage or processing, etc. To solve the storage issues, they use remote cloud storage for outsourcing their data that results in various benefits to users or reduces the burden of users. However, with these benefits, some security concerns need to be addressed that are confidentiality, integrity, and availability. Cloud data integrity means whether the user's outsourced data is intact or not. Many researchers have proposed various schemes to verify cloud data integrity. Starting with some initial work in this area, this paper reviews the most recent existing data integrity schemes based on some identified characteristics. It also provides the future research directions that can be used as a guidemap for various researchers in cloud data integrity verification area.

Keywords Data integrity verification · Cloud computing · Cloud security issues · Cloud data auditing · Outsourcing data · Cloud storage · Provable data possession · Proof of retrievability

1 Introduction

According to NIST, the definition of cloud computing is as follows: "It is a model for enabling ubiquitous, convenient, on-demand network access to a shared pool of configurable computing resources (e.g., networks, servers, storage, applications, and services) that can be rapidly provisioned and released with minimal management effort or service provider interaction. This cloud model is composed of five essential characteristics (On-demand self-service, Broad network access, Resource pooling,

J. Rani (✉) · R. Nath
Department of Computer Science and Applications, Kurukshetra University, Kurukshetra, India
e-mail: jyoti41nagwan@kuk.ac.in

R. Nath
e-mail: rnath@kuk.ac.in

M. S. Kaiser et al. (eds.), *Information and Communication Technology for Competitive Strategies (ICTCS 2021)*, Lecture Notes in Networks and Systems 401,
https://doi.org/10.1007/978-981-19-0098-3_62

Rapid elasticity, and Measured service), three service models(SaaS, PaaS, and IaaS), and four deployment models(Private cloud, Community cloud, Public cloud, and Hybrid cloud)" [21]. As the rate of generation of data is increasing day by day, devices like mobile phones, PCs, etc., are unable to handle(store/ process) data locally due to their very limited resources. Cloud storage is the best option to handle this problem, as it provides unlimited storage space with the pay-as-you-go pricing model. But, there are various security issues [23, 26, 27] related to outsourcing of data to cloud storage. Confidentiality, integrity, and availability are three core elements related to data security [5, 30, 42]. The issue of data integrity is considered in this paper. Data integrity is the protection of data from unauthorized modification or alteration, means maintaining data consistency or validity over its life cycle. In the case of cloud storage, the user stores its data into the cloud and wants the guarantee that its data will not be tempered, modified, or deleted in any case. The cloud storage server can delete rarely accessed user's data to save storage and sell the same storage to another user for more profit or there can be accidental deletion of user's data and hide this information from the user to maintain its reputation. To solve this issue, many data integrity verification schemes have been proposed by researchers with different-2 characteristics. This paper tries to survey some of them with their characteristics and cons and pros.

The basic system model which is widely used in most data integrity verification schemes is shown in Fig. 1. It consists of three entities—client, cloud storage server, and third-party auditor.

Client: Client refers to the user, data owner, or any organization which wants to outsource their data to the cloud as they are having very limited storage capacity.

Cloud storage server: Cloud storage server or simply cloud server provides a huge amount of storage space to the client for outsourcing their data.

Third-party auditor(TPA): The TPA has high capability and expertise than the client in integrity checking task. So, the client can delegate its integrity checking task to TPA to improve the efficiency and to reduce its own burden. The scheme with existence of TPA is called public scheme, otherwise called private scheme. The rest of the paper is organized as follows. Section 2 discusses some criteria to evaluate the data integrity verification schemes and also gives an overview of many existing

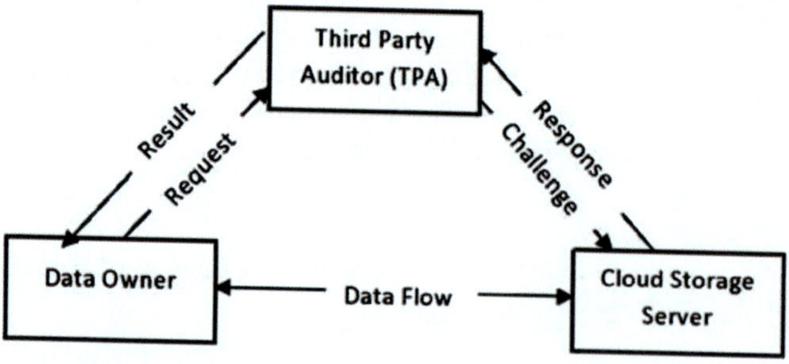

Fig. 1 Basic system model of data integrity verification scheme

schemes based on these criteria. Section 3 points out the future research directions. Finally, Sect. 4 summarizes this paper.

2 Literature Review

This section discusses some criteria to evaluate the data integrity verification schemes in cloud environment followed by an overview of the existing schemes based on these criteria.

2.1 *Evaluating Criteria for Data Integrity Verification Schemes*

After analyzing various research papers and survey papers [13, 28, 36, 41], most of the data integrity verification schemes possess all or subset of the following characteristics.

Probabilistic and deterministic schemes: The schemes which include all the file blocks for proof generation during integrity verification and give hundred percent guarantee of data integrity are deterministic schemes. On the other side, the schemes which do not include all the file blocks or a random subset of the file blocks in the proof generation(i.e., random sampling) and give a probabilistic guarantee of data integrity are probabilistic schemes. Most of the schemes are probabilistic, as deterministic schemes require more computation and communication overhead than probabilistic schemes.

PDP and POR schemes: Provable data possession(PDP) schemes can only detect the corruption, damage, or deletion to the original data by the dishonest server or any other external entity. Instead, the proof of retrievability(POR) can also recover the data loss using some correction code with detection as well.

Public and private auditing scheme: In a private auditing scheme, only the data owner can verify the data integrity, as the verification process requires some private information of the data owner. This process increases the computation and storage burden on the data owner. Consequently, to decrease the burden of the data owner, an entity named third-party auditor (TPA) is introduced, having high capability and expertise in integrity verification as compared to the data owner, which can also do the verification job in addition to the data owner. These schemes with TPA are public auditing scheme. Both types of schemes have their own merits and demerits.

Blockless verification: In earlier schemes, the user had to download all the data from the server for integrity verification. This means they did not provide blockless verification functionality. But, all the recent schemes support blockless verification which is more efficient way than earlier schemes.

Static and dynamic scheme: The schemes in which the user cannot perform insertion, deletion, or modification in data, once it is stored on the cloud, are static. These schemes are helpful in case of static data only, for example, backup or archive data, etc. On the other hand, the dynamic schemes support dynamic operations like insertion, deletion, or modification to cloud data. This means the user can verify the data integrity after dynamic operations as well. Nowadays, most of the data is dynamic in nature, which results in the demand for dynamic schemes. For providing dynamism, the data structures like Merkle hash tree, skip list, index hash table, and many more are used.
Overhead: Various types of overhead are there in the verification process which decide the efficiency of the scheme. These are as follows: computation and storage overhead on data owner (for data preprocessing, challenge generation, and verification of proof), on the cloud server(for tag or authentication information storage and proof generation), and on TPA (for challenge generation and verification of proof) in case of public auditing scheme and the communication overhead between the user, cloud server, and TPA for challenge/response transfer. The main focus of existing schemes is to minimize the burden and communication overhead on users.
Public key infrastructure (PKI) and identity-based cryptography (IDE): For privacy preservation, most of the auditing schemes have used asymmetric cryptography in which PKI and IDE are common. Earlier schemes were using PKI, but there was some drawback of PKI, i.e., complex certificate management. To overcome this problem, the auditing scheme started to use IDE, in which there is no need to manage complex certificates. Any public information of an entity can be used to create its public key.
Restriction on number of challenges: There was a restriction on data owner (in private auditing) or on TPA (in public auditing) that how many times they could challenge the cloud server for integrity verification proof. This means they could verify the integrity of remote data for a limited number of times. Later, the auditing schemes started to provide unrestricted auditing, which means data owner/TPA can challenge the cloud server any number of times for data integrity proof.
Single-copy and multi-copy auditing: Some schemes are designed to provide the auditing of a single copy of data, and others are designed to provide the auditing of multiple replicas of data in addition to the single copy. In the former, there exists only a single copy of data on cloud storage, so it needs to verify only a single copy. But in later, there exists multiple copies of data for providing high availability and reliability. Cloud server uses all the replicas of data to generate the proof and results in successful auditing if all the replicas are not modified.
Number of data owner: In traditional schemes, there could exist only one data owner of the data. But, the current remote data auditing(RDA) schemes also incorporate more than one data owner scenario. Consequently, they provide the functionality of group management like adding users in the group, revoking users from the group(also re-encrypting data so that the revoked user cannot access it), etc.
Batch auditing: It means multiple auditing requests can be handled at the same time. This attribute improves the efficiency of the scheme.
Other: Some latest schemes combine data de-duplication and blockchain technology with data integrity verification. Research is going on in these areas. Data de-

duplication means storing only one copy of data instead of multiple copies to save memory. The immutable characteristics of blockchain are utilized to record data operations.

2.2 Data Integrity Verification Schemes

This section overviews the various data integrity verification schemes classified into PDP and POR schemes. Both the schemes are further classified into static and dynamic schemes. Also some other schemes which did not fit under this classification are discussed in separate section.

2.2.1 PDP Schemes

Static PDP Schemes: The first classical PDP scheme [1] proposed by Ateniese et al. was a static scheme which was based on sampling to provide blockless integrity verification of data. It was private scheme but could be extended as public scheme with high communication cost. Two schemes S-PDP and E-PDP were discussed in that paper. Both were based on homomorphic verifiable tags for integrity verification. S-PDP was better than E-PDP in terms of security, and E-PDP was better than S-PDP in terms of efficiency. None of the dynamic operations could be performed other than append operation. This scheme provided the guidemap for other schemes.

Afterward, a private scheme [32] introduced a new entity, i.e., proxy for integrity verification of data on behalf of client when the client was unable to do the same(e.g., in prison or on the ocean going vessel, etc.) based on some conditions of warrant negotiated between them.

Morever, some special functionalities were added into the schemes like sensitive information hiding [25], privacy- preservation [3, 15, 39], data sharing with multiple users [3, 25], integrity verification of multiple copies of data [6, 31], and data deduplication [3]. Earlier schemes were not having these features. For example, they [1, 15, 25, 31, 32, 39] could support integrity verification of single copy of data only. They could not verify the integrity of multiple copies of data.

Many schemes [15, 25, 39] based on identity-based cryptography were proposed which did not need to manage complex certificates, like in public key cryptography-based schemes [32]. Most of the schemes [3, 15, 25, 31, 32, 39] were based on bilinear pairing for integrity verification of outsourced data. Other schemes [1, 6, 15] used homomorphic verifiable tags for integrity verification.

Then, a scheme was designed for handling integrity verification of some special kind of data like machine learning training's data [39]. It solved the key escrow problem and could resist against forgery and tampering attack.

Dynamic PDP schemes: The schemes which support dynamic operation on outsourced data are dynamic schemes, which means after any operation like insertion, deletion, or modification on data, the data owner is still able to verify the integrity of data. This section reviews some dynamic PDP schemes.

The scheme[2] was based on symmetric key cryptography so unsuitable for public auditing. It could verify the integrity of data for limited number of times only and also supported dynamic operations but with high overhead. To solve these problems, many dynamic PDP schemes were proposed. Different dynamic schemes [2, 8, 10, 12, 14, 17, 20, 24, 29, 37, 38, 40] used different data structures to support dynamic operations. The scheme [17] used multi-replica Merkle hash tree to support multiple-copy integrity verification. But disadvantage of this scheme was that the proof size depends on the size of data set. The scheme [24] used index matrix and data deletion matrix, [12] used tag record table, [37] used Merkle hash tree (MHT) with index logic table, [10] used auxiliary chain table, [40] used dynamic mapping hash table, and [20] used version-based MHT for data dynamics. Also, [38] was based on blockchain and [29] was based on CDH-based on ring signature as well as on vector commitment.

Most of the schemes [8, 10, 14, 17, 20, 29, 37, 40] used bilinear mapping, [12] used RSA accumulator, and [24] was based on algebraic signatures for data integrity verification. According to J. Shen et al., algebraic signatures were more efficient than bilinear pairing as well as matrix operations are more efficient than tree or table operations. Also due to RSA accumulator [12], verification could be done on low-power devices. The scheme is deterministic in nature as needed to access all blocks of a given file for proof generation. It could detect tag forgery, data deletion, replacement, and leakage attack but could not prevent them. Another scheme [29] was based on CDH-based ring signature for user authentication (addition or revocation of user to or from the group). It could resist against replace, collusion, and impersonation attack. Also, the scheme [20] enabled the data owner to share their data with multiple users. Some of the schemes [12, 20, 24, 29, 38, 40] also supported the privacy preservation during integrity verification.

Only minimum number of schemes [10, 40] supported data traceability so that client could trace their data. Instead of TPA, multiple consensus node managed the data auditing task to solve the problem of single point of failure [10]. It was specially designed for maintaining the integrity and privacy of electronic medical record (EMR) data [40].

Many schemes [8, 12] tried to reduce the computational overhead at client during initialization phase by replacing expensive operations like exponentiation with less expensive one.

2.2.2 POR Schemes

Static POR Schemes: The first POR schemes [11] which used random variable check blocks called sentinels were inserted into the original file for integrity verification and also error-correcting codes for small size error recovery. But, the problem with this scheme was that the data owner could verify the integrity of data for limited number of times only.

To address this problem, many POR schemes were proposed. The scheme [16] used the information dispersal algorithm for data recovery. Another scheme [9] was based on homomorphic authentication tags for integrity verification. A privacy-preserving scheme [4] used the blockchain technology to make the system decentralized by replacing TPA. It used the erasure code for proof of retrievability and incentive/reward mechanism to store replicas on peer node for providing proof of replication. It resists against collusion and corruption attack. The schemes [3, 4, 16] used the bilinear mapping for integrity verification of data.
Dynamic POR schemes: Both schemes [7, 33] used the modified Merkle hash tree for dynamics operations and bilinear pairing for integrity verification.

The scheme [33] also used homomorphic authenticators and BLS signatures in verification process. But, it did not include block index for authentication which was a drawback. It could be extended for multiple copies of data.

Other scheme [7] used the information dispersal algorithm for retrieval of corrupted data. This scheme was partial dynamic as it did not support insertion and deletion operation on data. It resists against replace and pollution attack.

2.2.3 Other schemes

: Some schemes which do not come fit under the above-discussed classification are discussed in this section. The scheme [22] used blockchain technology for providing a framework for integrity verification of surveillance cameras video footage. Another privacy preservation scheme [35] combined IoT and cloud computing for integrity verification of health cyber physical system using homomorphic encryption-based Merkle tree.
All of the above-discussed data integrity verification schemes are summarized in Table 1.

3 Discussion And Future Research Directions

This section discusses the open problems or research gaps in existing schemes and also provides the direction for future research. This will help the researchers to direct their research.
Multiple users with multiple data copy,ch62zhou2018data: There are various data integrity verification schemes that supported multiple copies of data and also schemes that supported multiple users. However, so far there are very few schemes proposed in the literature that supported both multiple users and multiple data copy. This area needs to be explored further.
Combining data integrity checking with data de-duplication Data de-duplication means if there are multiple copies of a given data file and then store only single copy to save storage space at cloud server. Combining data integrity checking scheme with data de-duplication can be an area for future exploration. There is also a contradiction

Table 1 Summary of various data integrity verification schemes

S. No.	Classification criteria	Schemes which satisfy the criterion
1	Probabilistic schemes	[1–4, 6–11, 14–20, 24, 25, 29, 31–34, 37–40]
	Deterministic schemes	[12]
2	PDP schemes	[1–3, 6, 8, 10, 12, 14, 15, 17, 19, 20, 24, 25, 29, 31, 32, 37–40]
	POR schemes	[4, 7, 9, 11, 16, 18, 33]
3	Static schemes	[1, 3, 6, 11, 16, 25, 31, 32] [4, 9, 15, 39]
	Dynamic schemes	[2, 7, 8, 10, 12, 14, 17, 20, 24, 29, 33, 34, 37, 38, 40]
4	Private schemes	[1, 2, 6, 11, 12, 32, 34]
	Public schemes	[7, 14, 16, 17, 24, 25, 31, 33, 37] [3, 8–10, 15, 18–20, 29, 38–40]
5	Limited challenge frequency schemes	[2, 11]
	Unlimited Challenge frequency schemes	[1, 3, 4, 6–10, 12, 14–20, 24, 25, 29, 31–34, 37–40]
6	Single-copy integrity checking scheme	[1–3, 7–12, 14–16, 18–20, 24, 25, 29, 32–34, 37–39]
	Multiple-copy integrity checking schemes	[4, 6, 17, 31, 40]
7	Batch auditing	[3, 7, 10, 14, 16, 18, 20, 24, 31, 33, 34, 38]

between POR scheme and data de-duplication that needs to be addressed, i.e., in POR schemes, multiple copies of data file or blocks are created or stored at cloud server for data recovery, conversely in data de-duplication, only single copy of data is stored to reduce the storage overhead.

Integration of data integrity and blockchain: The various properties of blockchain can be utilized in data integrity checking scheme for better or more functionalities. Even, there are few integrity checking schemes available that used blockchain technology for their operation. However, this area has not been explored fully as very few solutions have been presented so far.

Data dynamics with traceability,ch62zhou2018data: Cloud server stores the recent copy of data in case of dynamic schemes. But in some cases, we need all historic versions of data. There are various solutions available for data traceability with static schemes. However, supporting dynamic operations with traceability still needs some consideration.

Assured deletion,ch62tan2018survey: When user deletes any data at cloud, then it must be deleted permanently means there must not be any possibility for any kind of rollback to cloud server. This area could be another promising research direction in the future.

Automated remote data integrity checking [38]: Automated remote data integrity verification means data integrity checking in real time. If there is any change in

data by the cloud server or any attacker illegally, then data owner will be notified immediately. This area can be explored in future.
Dynamic POR scheme: Although many POR schemes are proposed by researchers, most of them are static or partial dynamic. There is much work to be done for fully dynamic POR schemes.

4 Conclusion

Various data integrity checking schemes are discussed with their supporting functionalities, implementation details, positive points and limitations which provides the roadmap or guidemap to various researchers who want to explore this area. So far, researchers have proposed a number of integrity verification schemes. But, there are many issues which need to be address in future. After reviewing of the existing-related literature, some open issues about the existing schemes are found, and finally, the directions for future research are discussed.

References

1. Ateniese G, Burns R, Curtmola R, Herring J, Kissner L, Peterson Z, Song D (2007) Provable data possession at untrusted stores. In: Proceedings of the 14th ACM conference on Computer and communications security, pp 598–609
2. Ateniese G, Di Pietro R, Mancini LV, Tsudik G (2008) Scalable and efficient provable data possession. In: Proceedings of the 4th international conference on security and privacy in communication networks, pp 1–10
3. Bai J, Yu J, Gao X (2020) Secure auditing and deduplication for encrypted cloud data supporting ownership modification. Soft Comput 1–18
4. Chen D, Yuan H, Hu S, Wang Q, Wang C (2020) Bossa: A decentralized system for proofs of data retrievability and replication. IEEE Trans Parallel Distrib Syst 32(4):786–798
5. Chen D, Zhao H (2012) Data security and privacy protection issues in cloud computing. In: 2012 international conference on computer science and electronics engineering, vol 1. IEEE, pp 647–651
6. Curtmola R, Khan O, Burns R, Ateniese G (2008) Mr-pdp: multiple-replica provable data possession. In: 2008 the 28th international conference on distributed computing systems. IEEE, pp 411–420
7. Fu A, Li Y, Yu S, Yu Y, Zhang G (2018) Dipor: an ida-based dynamic proof of retrievability scheme for cloud storage systems. J Netw Comput Appl 104:97–106
8. Garg N, Bawa S, Kumar N (2020) An efficient data integrity auditing protocol for cloud computing. Fut Generat Comput Syst (2020)
9. Hiremath S, Kunte RS (2020) Homomorphic authentication scheme for proof of retrievability with public verifiability. In: 2020 4th International Conference on Intelligent Computing and Control Systems (ICICCS). IEEE, pp 1017–1022
10. Huang P, Fan K, Yang H, Zhang K, Li H, Yang Y (2020) A collaborative auditing blockchain for trustworthy data integrity in cloud storage system. IEEE Access 8:94780–94794
11. Juels A, Kaliski Jr, BS (2007) Pors: proofs of retrievability for large files. In: Proceedings of the 14th ACM conference on Computer and communications security, pp 584–597

12. Khedr WI, Khater HM, Mohamed ER (2019) Cryptographic accumulator-based scheme for critical data integrity verification in cloud storage. IEEE Access 7:65635–65651
13. Li A, Chen Y, Yan Z, Zhou X, Shimizu S (2020) A survey on integrity auditing for data storage in the cloud: from single copy to multiple replicas. IEEE Trans Big Data
14. Li A, Tan S, Jia Y (2019) A method for achieving provable data integrity in cloud computing. J Supercomput 75(1):92–108
15. Li J, Yan H, Zhang Y (2020) Identity-based privacy preserving remote data integrity checking for cloud storage. IEEE Syst J (2020)
16. Li Y, Fu A, Yu Y, Zhang G (2017) Ipor: an efficient ida-based proof of retrievability scheme for cloud storage systems. In: 2017 IEEE International Conference on Communications (ICC). IEEE, pp 1–6
17. Liu C, Ranjan R, Yang C, Zhang X, Wang L, Chen J (2014) Mur-dpa: top-down levelled multi-replica merkle hash tree based secure public auditing for dynamic big data storage on cloud. IEEE Trans Comput 64(9):2609–2622
18. Liu D, Shen J, Vijayakumar P, Wang A, Zhou T (2020) Efficient data integrity auditing with corrupted data recovery for edge computing in enterprise multimedia security. Multimedia Tools Appl 79(15):10851–10870
19. Lu N, Zhang Y, Shi W, Kumari S, Choo KKR (2020) A secure and scalable data integrity auditing scheme based on hyperledger fabric. Comput Secur 92:101–741
20. Lu X, Pan Z, Xian H (2020) An integrity verification scheme of cloud storage for internet-of-things mobile terminal devices. Comput Secur 92:101–686
21. Mell P, Grance T et al (2011) The nist definition of cloud computing
22. Michelin RA, Ahmed N, Kanhere SS, Seneviratne A, Jha S (2020) Leveraging lightweight blockchain to establish data integrity for surveillance cameras. In: 2020 IEEE International Conference on Blockchain and Cryptocurrency (ICBC). IEEE, pp 1–3
23. Rao BT et al (2016) A study on data storage security issues in cloud computing. Proced Comput Sci 92:128–135
24. Shen J, Liu D, He D, Huang X, Xiang Y (2017) Algebraic signatures-based data integrity auditing for efficient data dynamics in cloud computing. IEEE Trans Sustain Comput (2017)
25. Shen W, Qin J, Yu J, Hao R, Hu J (2018) Enabling identity-based integrity auditing and data sharing with sensitive information hiding for secure cloud storage. IEEE Trans Inf Forens Secur 14(2):331–346
26. Singh A, Chatterjee K (2017) Cloud security issues and challenges: a survey. J Netw Comput Appl 79:88–115
27. Subashini S, Kavitha V (2011) A survey on security issues in service delivery models of cloud computing. J Netw Comput Appl 34(1):1–11
28. Tan CB, Hijazi MHA, Lim Y, Gani A (2018) A survey on proof of retrievability for cloud data integrity and availability: cloud storage state-of-the-art, issues, solutions and future trends. J Netw Comput Appl 110:75–86
29. Thokchom S, Saikia DK (2020) Privacy preserving integrity checking of shared dynamic cloud data with user revocation. J Inf Secur Appl 50:102–427
30. Tianfield H (2012) Security issues in cloud computing. In: 2012 IEEE International Conference on Systems, Man, and Cybernetics (SMC). IEEE, pp 1082–1089
31. Wang C, Di X (2020) Research on integrity check method of cloud storage multi-copy data based on multi-agent. IEEE Access 8:17170–17178
32. Wang H (2012) Proxy provable data possession in public clouds. IEEE Trans Serv Comput 6(4):551–559
33. Wang Q, Wang C, Ren K, Lou W, Li J (2010) Enabling public auditability and data dynamics for storage security in cloud computing. IEEE Trans Parallel Distribut Syst 22(5):847–859
34. Wei P, Wang D, Zhao Y, Tyagi SKS, Kumar N (2020) Blockchain data-based cloud data integrity protection mechanism. Fut Generat Comput Syst 102:902–911
35. Xu J, Wei L, Wu W, Wang A, Zhang Y, Zhou F (2020) Privacy-preserving data integrity verification by using lightweight streaming authenticated data structures for healthcare cyber-physical system. Fut Generat Comput Syst 08:1287–1296

36. Zafar F, Khan A, Malik SUR, Ahmed M, Anjum A, Khan MI, Javed N, Alam M, Jamil F (2017) A survey of cloud computing data integrity schemes: design challenges, taxonomy and future trends. Comput Secur 65:29–49
37. Zhao C, Xu L, Li J, Wang F, Fang H (2019) Fuzzy dentity-based dynamic auditing of big data on cloud storage. IEEE Access 7:160,459–160,471
38. Zhao Q, Chen S, Liu Z, Baker T, Zhang Y (2020) Blockchain-based privacy-preserving remote data integrity checking scheme for iot information systems. Inf Proc Manage 57(6):102–355
39. Zhao XP, Jiang R (2020) Distributed machine learning oriented data integrity verification scheme in cloud computing environment. IEEE Access 8:26372–26384
40. Zhou L, Fu A, Feng J, Zhou C (2020) An efficient and secure data integrity auditing scheme with traceability for cloud-based EMR. In: ICC 2020-2020 IEEE International Conference on Communications (ICC). IEEE, pp 1–6
41. Zhou L, Fu A, Yu S, Su M, Kuang B (2018) Data integrity verification of the outsourced big data in the cloud environment: a survey. J Netw Comput Appl 122:1–15
42. Zissis D, Lekkas D (2012) Addressing cloud computing security issues. Fut Generat Comput Syst 28(3):583–592

Automatic Brain Tumor Segmentation on Pre-operative MRI using Region Growing Algorithm

K. V. Shiny and N. Sugitha

Abstract The major issue in radiation oncology application is to detect and segment the mixed necrotic tumor tissue. Detecting and segmenting these tissues along with the neighboring vessels are a challenging issue. The normal and the abnormal tissues are used to identify the overlapping of the margins in each tissue by means of an MRI image. Various automatic seed finding methodologies have been proposed but all those suffer from various issues such as the absence of tumor or the presence of any white or gray part. However, the general purpose methods are not effective for segmenting the complex MRI image structures. Generally for the segmentation of brain tumor, region-based active contour models were employed, but the edges are not sharpen in those models, and hence, the accurate results are not obtained due to the over segmentation and under segmentation. This happens due to the initial stage of the tumors and hence is a need for a method to detect the tumor and segment it. Hence, an effective method for detecting the tumor based on MRI texture and automatic segmentation using region growing method is proposed. These approaches were employed to separate the irregular tissues from regular tissues to detect the area and to identify whether it is involved or non-involved which greatly assist the surgeon to identify the affected areas accurately. The proposed work incorporates the texture analysis and region growing algorithm which is investigated on MRI brain images to identify the boundaries of tumor in 2D MRI images.

Keywords Magnetic resonance imaging (MRI) · Segmentation · Classification · Feature extraction · Brain tumor

---

K. V. Shiny (✉)
Department of Computer Science and Engineering, Noorul Islam Centre for Higher Education, Kumaracoil, India
e-mail: kvshiny7@gmail.com

N. Sugitha
Department of Information Technology, Noorul Islam Centre for Higher Education, Kumaracoil, India

M. S. Kaiser et al. (eds.), *Information and Communication Technology for Competitive Strategies (ICTCS 2021)*, Lecture Notes in Networks and Systems 401,
https://doi.org/10.1007/978-981-19-0098-3_63

1 Introduction

In medical scenario, segmenting the tumor and brain tissues in different matters such as gray [1] and white [2] is very difficult in radiation oncology application. In specific area of radiotherapy such as spatially localized radiotherapy, contouring the brain tumor [3, 4] is a major step. These contouring are usually performed on $T1$-weighted MRI images in current scenario. By administration of the gadolinium, blood vessels, and the tumor parts, these MR images are acquired. The hyperintense areas are identified by the tumor which passes the blood brain barrier. Numerous methods for segmenting the brain tumor are identified such as single modality, multiple modalities, and priors from population atlases. Several existing methods such as perfusion imaging and diffusion imaging offer information about tumor and edema and provide low-resolution images when compared to $T1$ or $T2$ weighted image sequences. The main cause of utilizing the $T2$ MRI weighted images is to segment the edema or infiltration region which is lagging in $T1$ images.

The proposed approach integrates both region growing and edge detection techniques for segmenting magnetic resonance brain image. To handle complex brain image structures, the region growing algorithm is combined with merging methodology by initializing a basic region growing algorithm. These integrated approaches normally produce an over segmented image. Also, the boundary details are also combined to ensure the correctness of the boundaries. In this work, automatic texture-based image segmentation is used. This automatic image segmentation achieves better accuracy and efficiency in segmenting magnetic resonance brain images [5]. Also, the usage of automatic image segmentation varies from computer-aided diagnosis to studying composite structures. Segmenting complex brain image structures using common principles is difficult. Similarly, the region-based segmentation fails to achieve the expected outcome due to the lack of choosing the reasonable seed.

The region-based segmentation models are normally employed for brain tumor segmentation [30]. When compared to gradient-based techniques, the region-based models offer several advantages such as high robustness and efficiency [6, 7]. When employing 3D level-set surfaces by classical active contours, the contouring process is as good as only in initialization process since due to some factors in tumor such as they have a low-spatial proceeding with small structures. This happens because the blood cells are viewed as tumor cells which grow along with the contrast.

Various image segmentation algorithms have been proposed in recent years [5, 8]. The important among them is graph-based segmentation which supports seeding approach with graph-cuts, random walker, shorter paths, and power watersheds [3, 5, 9]. This approach overcomes the minimization factor which involves the edge weight of the graph, and it is controlled by the vertex variables or probabilities.

2 Background

2.1 Seeded Image Segmentation

In an un-directed graph with vertices v and edges e where $v \in V$ and $e \in E$, for edge e_{ij}, a weight w_{ij} is assigned between the two vertices v_i and v_j. In image segmentation process, the vertices correspond to the image pixels, and the edge weight corresponds to similarity measures. This similarity measures depend on the intensities or the features of the image. Also, each vertex is assigned an attribute x_i; which is the label probability indicator of the foreground and the background. The seeds are assigned for the foreground F and the background B by the user. Based upon the attribute value, the labels are identified as foreground and background. If the attribute value $x_i < 0.5$ is identified as foreground, the attribute value $x_i > 0.5$ is termed as foreground. Even though numerous region growing method was proposed, the seeded region growing method was the first approach proposed which accepts the image along with the seed as input. The matter which needs to be segmented is marked by the seeds, and the unallocated pixels in the border are compared, and thereby, the regions are grown iteratively. The measure of likeness is analyzed by comparing the strength value of the pixel and the mean value of the region δ. The likeness measure for all the pixels is computed, and those pixels with the minimum likeness value are assigned to the corresponding region. The process continues until all the pixels are assigned to the region [9].

2.2 Region Growing in Image Segmentation

The seeded region growing approach holds the seeds as an input. Based on the seeds, the segmentation results are analyzed. Furthermore, the noise in the image can cause the seeds to be positioned wrongly. Unlike the seeded region, the unseeded region growing approach is a customized approach which does not involve the necessity of explicit seeds. The homogeneity criteria place an important concern in region growing approach because the requirement of this process depends on the image arrangement properties which are unknown to the user. Hence, a region growing algorithm was proposed which automatically identifies the homogeneity criteria and distinctiveness from the region which is to be segmented. This approach best suits for describing the homogeneity and the outline properties of the appropriate region. By homogeneity principle, the parameters are estimated from the sample locations in the region. By means of random walk, and starting from the seed position, the regions are chosen one after another, and the homogeneity principles are updated accordingly. This method was employed in automatic and absolute segmentation technique with the pixels which has the minimum gradient length with the seed position as the region which is not segmented. This approach was experimented with CT and MR image structures and found these approach works consistent based on the model hypothesis, and the region characteristics are true. Even though the model is simple; it is strong and achieves a finite divergence degree and the expected segmentation result.

In several processes such as computer-aided analysis and treatment planning, analyzing the medical images frequently includes segmentation as the primary step for visualization or quantification [10, 11]. Several methods such as region growing, region splitting, region merging, active contours, cluster analysis methods, and watershed transformation for segmentation of CT and MR images were proposed.

Furthermore, various complex models for segmentation the images individually with Apriori information about the predictable details of the image are used. The Apriori information includes grouping the physiological or anatomical details along with the image configuration process. The segmentation process depends upon the information about the contributed model and the automation opportunity. If the contributed model is maximum, then the automation opportunity is also maximum in the segmentation technique. However, the segmentation result can be predicted easily and also the replicated information may not be accurate so that withdrawal is not feasible. The main task of the region growing segmentation approach is to minimize the replicated information and to maximize the model information. These model informations are provided in situations where information is easily provided to the client in a strong manner. Breaking the rule may cause the user to communicate and perceive the user's expectations.

The region growing segmentation process starts with the region A_1 in an iterative manner. The iterative process considers the pixel in the same way as the seeded region growing. The pixel and the seeded region growing process are analyzed such that the least amount δ is less than or greater than the threshold T. If δ is less than T, then it is added to the new region A_i; else, it is added to the new region A_{n+1}. Another alternative in this technique is pixel intensions. The test statistic is computed by using the strength of the candidate pixel and the mean and the scatter of the area. The test statistic is checked whether it is small or large. The region's mean and the scatter are computed again if the test statistic is small, and hence, the pixel is added to that region; else the pixel is discarded, and a new region is formed.

The region growing algorithm is a variant of region-based segmentation approach in medical image segmentation technique. In this process, a set n is selected with the seed points v_i; where $i = 1, 2, \ldots., n$ of voxels. The regions grow based on those seed whose neighboring voxels properties match with certain predefined criteria [12, 13]. Using this approach, a region R_i is formed. The similarity criteria to add a voxel to the new region R_i are based on the properties of the image.

In the proposed work, x is the neighbor voxel for another voxel in the region R_i. A voxel can be added to the region R_i is based upon the Euclidean distance and the threshold θ. The Euclidean distance involves the voxel x and the seed point v_i. A voxel is added to the region R_i if the Euclidean distance is less than the threshold; else it is not included in the region. Finally, the desired region is obtained by including each grown region $i = 1, 2, \ldots, n$ into region R_i.

3 Method

In this section, the segmentation framework which includes segmenting the cells and the necrotic regions in brain tumors is explained in detail. Identifying a pattern or some object in a given work space is termed as segmentation. In our work, magnetic resonance image is considered to be the workspace. The abnormalities of tissues in this image are detected. Even though the MRI image offers high-contrast image and represents every part of the image clearly, segmentation is quite difficult. Hence, the anatomical structure of these images should be observed. Also, this process tends to be time consuming and leads to incorrect results due to volume calculation incorrect segmentation process.

Moreover, the selection of the seed point by the user is eliminated, and hence, human interaction is also eradicated. We assumed that the brain tumor had grown in considerable size and shape.

The block diagram for brain tumor segmentation performed on MRI images using region growing algorithm was depicted in Fig. 1. The functions performed by each phase are explained below:

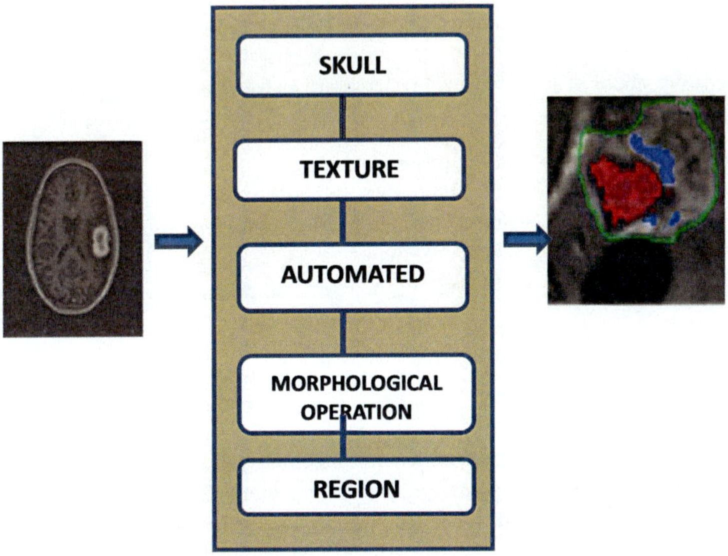

Fig. 1 Brain tumor segmentation

3.1 Input Image

The input image considered in our work is the $T2$ weighted magnetic resonance image which is used to segment the edema/infiltration region which is lacking in the $T1$ image. This work focuses on segmenting the brain tumors in $T1$ weighted magnetic resonance image with less user interaction and is able to detect fat or tumors.

3.2 Skull Part Removal

In order to achieve better segmentation results, this pre-processing step is essential. This step removes the skull, i.e., the outer part of the brain. This step mainly eliminates the non-cerebral tissues [30]. Segmenting the non-cerebral and the intracranial tissues is the main task of skull-stripping because of its homogeneity intensities. This may affect the selection of seed point product. Some assumptions to choose the gray value of skull portion range are followed.

3.2.1 Steps Involved in Skull Removal Process

The steps involved in skull removal process are as follows:

1. Identify the image size and the find the rows and columns and accumulate it in a separate variables and the process iterates.
2. Perform this iteration for all the rows and half of the columns.
3. By setting the gray value of the image to zero, the white pixels are converted into black pixels for half of the image.
4. The steps1–3 are iterated until all the rows and columns are taken.

3.3 Texture Analysis

The usual texture analysis approach cannot be employed for obtaining microscopic information in MR images visually. The texture identified in the texture analysis approach may be of different forms such as regular or irregular, soft or rough, and rude or fine. Even some complex patterns can be represented and visualized easily with respect to texture. In our work, statistical analysis of second order is carried out. This second order statistical analysis approach is employed with gray level measures such as run length measure and co-occurrence matrix. The run length measures are taken with respect to four directions such as 0, 45, 90, and 135. The image intensities range from $[\mu - 3\alpha, \mu + 3\alpha]$ during the normalization process of gray level. Through this distribution, the abnormalities in the image are identified.

3.4 Automated Seed Selection

The output of the skull removal process is considered as input in this process. We have assumed that the tumor's region has a considerable growth in its size.

3.4.1 Procedure for Finding the Seed Point

The different steps involved in finding the seed point are as follows:

Step 1: Transform the input color image into gray image.
Step 2: Next, the pixel intensities are analyzed such that,
Pixels intensities > 100
or
Pixel intensities<100 and store it in a different variables.
Step 3: The difference between the two variables is identified.
If the difference is negligible,
then goto step 4.
else
transform the image into negative and set the exterior part intensity to zero and goto step 4.
Step 4: A binary image is obtained by converting the gray image.
Step 5: Compute the maximum length and breadth of the brain and convert the pixels into a
black color in a rectangular shape by considering twenty rows and ten columns.
Step 6: An array is formed by the summation of all the rows.
Step 7: Another array is formed by the summation of all the columns.
Step 8: The array with the maximum sum is considered as the seed point.

3.5 Morphological Operation

The morphology approach is normally employed to enhance the structure and the contour of an object especially to remove the noise and to extract the object. In our work, the morphological operations are applied to enhance the object boundary and to remove the noise from the image.

Dilation and erosion are considered to be the basic morphological operations. Dilation adds pixels to the borders of objects; whereas erosion removes the borders of objects in an image. The quantity of the pixels which is added or removed is based upon the size and shape of the structuring component which is used to process the image. Based upon a rule applied to a pixel and its neighboring pixels, the pixel condition is determined in dilation and erosion operations. The dilation and the erosion operations are identified by means of the rules applied to the process the pixels.

Another important factor in processing the dilation and the erosion operations is deciding the structuring component. A matrix consisting of 0's and 1's which specify any random shape and size is termed as the structuring component. The neighborhood pixel is identified by the pixel whose value is 1.The image which is to process is always characteristically higher than two smooth or dimensional structuring elements. The pixel of interest and the pixel which is to be processed are identified by the origin. The origin is identified by thc center pixel in the structuring element. This origin identifies the pixel of interest and the pixel which is to be processed. The pixel with the value 1 identifies the neighborhood in the structuring element.

In our proposed work, we have assumed DISK shape as our structuring element. In the morphological operations of dilation and erosion, varying radii of disk structuring element is considered. Due to these varying radii of structuring element, the resultant image eliminates all the unnecessary parts in it.

The morphological operations perform various tasks in the structuring element such as positioning the origin and the center element and identifying the pixel of interest. The pixel in the border of the image is identified by the neighborhood in the structuring element which can extend the past the border of the image. From these, the values of the padding pixels can be identified, and it varies for both dilation and erosion operations.

3.6 Seeded Region Growing

The result of the morphological operations is taken as input for this stage. This approach integrates the automatic seeded region and edge detection method. This combined approach is used to identify the necrotic and enhancing tissues in the magnetic resonance images. This segmentation approach verifies whether a particular neighbor pixel should be included in a region by determining the initial seed points.

Similar to the general data clustering algorithms, this seeded region process is a repetitive process. Partitioning an image into several regions is the main objective of this approach. Based on the discontinuities between the color properties or gray levels between the regions in the border, segmentation happens in some existing segmentation techniques such as thresholding. However, region-based segmentation is employed to segment the region directly.

In our proposed approach, ten samples of brain MRI are used for experimenting with it. This approach utilizes the connected neighbor region growing method, and the results are observed.

3.6.1 Procedure for Region-Based Segmentation

The steps followed in the region-based segmentation approach are as follows:

1. Ensure that each and every pixel is inside the region, and it is complete.
   $U_{i=1}^{n} R_i = R_0$
2. Assume that there should be a clear distinct between the regions, i.e., the regions should be disjoint.
   $R_i \cap R_j = \Phi$
3. When the gray level of pixel is within the region, then the pixel may belong to a particular region.
   $P(R_i) = \text{TRUE}$ For $i=1, 2,\ldots,n$
4. By calculating the predicate, whether the two regions are identical or different is …
   $P(R_i \text{ U } R_j) = \text{FALSE}$

3.7 *Necrotic Segmentation*

Necrotic segmentation is an important step in processing the tumor regions. In this segmentation approach, quantification of necrotic regions is performed. In stereotactic radio surgery, delayed radiation necrosis occurs. This delayed radiation necrosis occurs normally for three months or more depends upon the treatment. Because of the radio surgery and the progress of the tumor, the necrosis of the tumor happens and it looks like gliomas.

Because of the different characteristics of intensities, necrotic class generally developed by means of multi-modality intensity classifiers. Figure 2 explains the segmentation process. Segmentation using single threshold and the threshold values to determine the initial seed points are depicted in Fig. 2a and b, respectively.

Our work focuses on computing the necrotic and enlarged regions in it by employing the MRI T1 weighted images. However, a hypo-intense and hyperintense

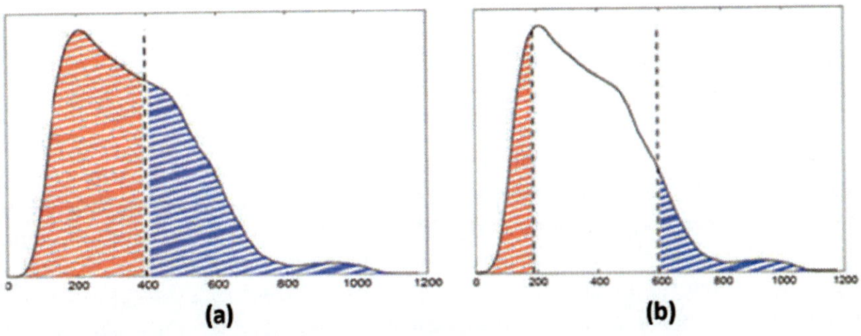

Fig. 2 **a** Single threshold-based segmentation. **b** Determining initial seeds

Table 1 Comparison of manual and automatic brain tumor segmentation

Sl No.	Analysis	Manual segmentation	Automatic segmentation
1	User interaction	More	Very less
2	Iteration	More	Less
3	Accuracy	High	Low
4	Time consuming	Yes	No
5	Shortest path problem	Exists	Reduced

regions are identified. When there is no blood flows into the seregions, it is termed as hypo-intense and its enlarged regions are termed as hyperintense. In enhancing tissues, due to the flow of blood, there will be growth in tumor tissues. The segmentation process is performed by employing the intensity threshold and by assigning certain labels based on the thresholds. If the threshold value is lower, then necrotic label is assigned to the voxels; else enhanced label is assigned.

The comparison between the manual and the automatic tumor segmentation is shown in Table 1. The analysis is performed based on several factors such as accuracy, time consumption, path, and number of iterations.

4 Results

The results can be analyzed from MRI brain images given in Fig. 3.

The Fig. 3 demonstrates the MRI images of brain with tumor cells and necrotic cells. Figure 3a–g represents the MRI images, and h–j represents the necrotic cells.

The proposed work is performed with 10 brain images, and the results are observed. It is been found that the necrotic tissues are present in the tumor, and the results are shown in Table 2.

5 Conclusion

In region-based segmentation method for identifying the necrotic and enhancing the segmentation of cell tissues in MRI brain images are performed. By means of this, abnormalities are identified. Hence, a novel approach for necrotic segmentation of MRI brain images has been proposed. The main advantage of this approach is no training sets or prior information is required. The proposed approach incorporates the features of uniform features and spatial features, and hence, the seed points are identified. The segmentation results obtained were highly accurate when compared to other segmentation methodologies. In future, it is planned to minimize the execution

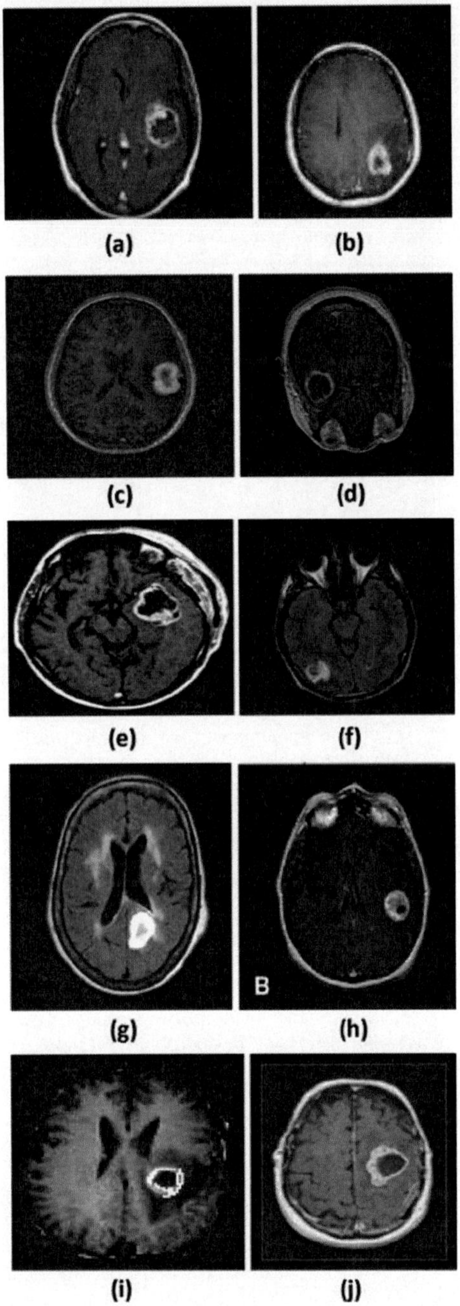

Fig. 3 **a–g** MRI of brain with tumor. **h–j** MRI of brain with necrotic inside the tumor

Table 2 Results of brain tumor segmentation

Sl No.	Image No.	Abnormality	Necrotic tissues
1	A	Present	Absent
2	B	Present	Absent
3	C	Present	Absent
4	D	Present	Absent
5	E	Present	Absent
6	F	Present	Absent
7	G	Present	Absent
8	H	Present	Present
9	I	Present	Present
10	J	Present	Present

time along with segmenting the necrotic tissues to find the accurate size of necrotic and tumor cells.

References

1. Angulakshmi M, Priya GL (2018) Brain tumour segmentation from MRI using superpixels based spectral clustering. J King Saud Univer-Comput Inf Sci
2. Rehman ZU, Naqvi SS, Khan TM, Khan MA, Bashir T (2019) Fully automated multi-parametric brain tumour segmentation using superpixel based classification. Expert Syst Appl 118:598–613
3. Soltaninejad M, Yang G, Lambrou T, Allinson N, Jones TL, Barrick TR, Howe FA, Ye X (2018) Supervised learning based multimodal MRI brain tumour segmentation using texture features from supervoxels. Comput Meth Prog Biomed 157:69–84
4. Hagargi AP, Shubhangi DC (2018) Brain tumour detection and ART classification technique in MR brain images using RPCA QT decomposition. Brain 5(4)
5. Nayak DR, Dash R, Majhi B (2016) Brain MR image classification using two-dimensional discrete wavelet transform and AdaBoost with random forests. Neurocomputing 177:188–197
6. Angelini ED, Clatz O, Mandonnet E, Konukoglu E, Capelle L, Duffau H (2007) Glioma dynamics and computational models: a reviewof segmentation, registration, and in silico growth algorithms and their clinical applications. Curr Med Imag Rev 3(4):262–276
7. Prastawa M, Bullitt E, Ho S, Gerig G (2004) A brain tumor segmentation framework based on outlier detection. Med Image Anal 8(3):275–283
8. Kaplan K, Kaya Y, Kuncan M, Ertunç HM (2020) Brain tumor classification using modified local binary patterns (LBP) feature extraction methods. Med Hypotheses 139:109696
9. Criminisi A, Sharp T, Blake A (2008) GeoS: geodesic image segmentation. Comput Vis ECC 5302:99–112
10. Biswas T et al (2009) Stereotactic radiosurgery for glioblastoma: retrospective analysis. Radiat Oncol 4(11):11
11. Gooya A, Biros G, Davatzikos C (2011) Deformable registration of glioma images using em algorithm and diffusion reaction modeling. IEEE Trans Med Imag 30(2):375–390
12. Boykov Y, Jolly M-P (2001) Interactive graph cuts for optimal boundary and region segmentation of objects in n-d images. Proc ICCV 1:105–112
13. Grady L (2006) Random walks for image segmentation. IEEE Trans Pattern Anal Mach Intell 28(11):1768–1783

14. Kalpana R, Chandrasekar P (2020) An optimized technique for brain tumor classification and detection with radiation dosage calculation in MR image. Microprocess Microsyst 72:102903
15. Hamamci A, Kucuk N, Karaman K, Engin K, Unal G (2012) Senior member, IEEE, "tumor-cut: segmentation of brain tumors on contrast enhanced MR images for radiosurgery applications. IEEE Tran Med Imag 31(3)
16. Warfield S, Zou K, Wells W (2004) Simultaneous truth and performance level estimation (STAPLE): an algorithm for the validation of image segmentation. IEEE Trans Med Imag 23(7):903–921
17. Zou KH, Warfield SK, Bharatha A, Tempany CMC, Kaus MR, Haker SJ, Wells WM, Jolesz FA, Kikinis R (2004) Statistical validation of image segmentation quality based on a spatial overlap index. Acad Radiol 11(2):178–189
18. Liu J, Udupa JK, Odhner D, Hackney D, Moonis G (2005) A system for brain tumor volume estimation via MR imaging and fuzzy connectedness. Comput Med Imag Graph 29:21–34
19. Menze B, Leemput KV, Lashkari D, Weber M-A, Ayache N, Golland P (2010) A generative model for brain tumor segmentation in multimodal images. Med Image Comput Comput Assist Intervent 13:151–159
20. Chan TF, Vese L (2001) Active contours without edges. IEEE Trans Image Process 10(2):266–277
21. Ho S, Bullitt E, Gerig G (2002) Level-set evolution with region competition: automatic 3-D segmentation of brain tumors. Proc ICPR 1:10532
22. Couprie C, Grady L, Najman L, Talbot H (2009) Power watersheds: a new image segmentation framework extending graph cuts, random walker and optimal spanning forest. ICCV 731–738
23. Sinop A, Grady L (2007) A seeded image segmentation framework unifying graph cuts and random walker which yields a new algorithm. ICCV 1–8
24. Bai X, Sapiro G (2009) Geodesic matting: a framework for fast interactive image and video segmentation and matting. Int J Comput Vis 82:113–132
25. Szeliski R, Zabih R, Scharstein D, Veksler O, Kolmogorov V, Agarwala A, Tappen M, Rother C (2008) A comparative study of energy minimization methods for Markov random fields with smoothness-based priors. IEEE Trans Pattern Anal Mach Intell 30:1068–1080
26. Vezhnevets V, Konouchine V (2005) Growcut-interactive multi-labeln-d image segmentation by cellular automata. Presented at the Graphicon, Novosibirsk Akademgorodok, Russia
27. Popovici A, Popovici D (2002) Cellular automata in image processing. In: Proceedings 15th International symposia mathematics theory networks system, pp 34–44

Cooperative Multi-Agent Nash Q-Learning (CMNQL) for Decision Building in Retail Shop

Deepak A. Vidhate and Parag Kulkarni

Abstract Based on the mentioned cooperative Nash Q-learning framework, the study presents a novel cooperative multi-agent Nash Q-learning approach for building decisions for the retail shop. The technique provides three retail stores throughout the market. Dealers can assist one another and profit from cooperative knowledge by building their own strategies that are solely focused on their goals and benefits. Assuming key hypotheses about the dealer's stock policy, reload duration, and consumer coming method, the technique is created as a Markov decision process model to build the learning algorithms. Furthermore, the research shows the results of three shop agents' cooperative reinforcement learning algorithms using Nash Q-learning over a one-year sale period. The findings of two methodologies, Nash Q-learning and joint action learning, are compared in the paper. When compared to a single-agent Q-learning approach, agents with Nash Q are to arrive at a joint optimal path. If at least one agent accepts *N*, the effect of combined agents is superior to single Q-learning.

Keywords Cooperative learning · Joint action learning · Multi-agent learning · Reinforcement learning · Nash Q-learning

1 Introduction

The retail store makes a profit by selling household items. Retailers are concerned with their sales and profits. The goal of projecting sales is to collect data from a different stores and analyze it using machine learning algorithms [1]. The relationship among customers and retailers is assessed, and changes that are expected to enhance yield are planned. Furthermore, each item's purchase history in each

D. A. Vidhate (✉)
Dr. Vithalrao Vikhe Patil College of Engineering Ahmednagar, Ahmednagar, Maharashtra, India
e-mail: dvidhate@yahoo.com

P. Kulkarni
iKnowlation Research Laboratory, Shivajinagar, Pune, Maharashtra, India

M. S. Kaiser et al. (eds.), *Information and Communication Technology for Competitive Strategies (ICTCS 2021)*, Lecture Notes in Networks and Systems 401,
https://doi.org/10.1007/978-981-19-0098-3_64

shop and department is kept track of. By evaluating these, sales may be projected, creating it easier to grasp the yield and loss that happened over the year [2]. In the Christmas season, there are greater sales in stores like clothing, footwear, and jewelry. Clothing is very popular in the summer, whereas sweaters are very popular in the winter. The season influences the goods that are bought. Sales can be anticipated for the forthcoming by looking at this historical record of purchases [2, 3]. This reveals the outcome of predicting the highest income in the retail shop. Retailers keep track on shopping habits and go to entice them with a variety of attractive schemes. Typically, the primary goal of retail shop planning is to make the most of profit by gaining considerable understanding [4, 5]. Retailers, in particular, have a difficult time predicting market conditions. Seasonal fluctuations are ignored by retailers. Furthermore, human resources are limited, and personnel are not available when they are needed. The complication of the storage management system is felt by merchants. Retailers are sometimes unconcerned about market competition or cooperation. The plans should be devised in such a way that they aid in achieving the biggest possible revenue [5, 6]. In general, the returns from the sale of a certain product are kept to forecast the greatest potential of total sales over a specified time period and in an uncertain environment. Shop and inventory forecast is important, and it is a regimented procedure at the store level [7].

2 Literature Survey

Reinforcement learning, as a model-free technique for learning in multi-agent systems, allows agents to conduct actions while they learn [8]. In most multi-agent systems, agents do not have comprehensive facts about their counterparts. The multi-agent environment is always changing as agents observe one other and adjusts their actions as a result. Techniques for Q-learning [9, 10] of reinforcement learning has been widely researched and has a solid base in the Markov model assumption. It is also rather straightforward to use, and it is appropriate for an extensive range of applications, including cellular telephone channel allocation [11], spoken dialog systems [12], and robotic control [13]. Even though the single-agent characteristics do not as it is conveyed to the multi-agent framework, its easiness of request arrives, and the approach was used in multi-agent areas like robotic soccer [14], predator–prey pursuit games [15], and Internet price bots [16]. In a multi-agent scenario, it is possible to use Q-learning in a simple method to each agent. Doing so misses two issues unique to the multi-agent approach. First, the environment contains other players who are also changing; therefore, the environment is no longer fixed. Second, other agents cause the situation to become non-stationary. Several investigations [17] have demonstrated that an agent who observes the outcome of combined acts outperforms a resulting entity who only observes its individual activities. Assume the scenario of general-sum stochastic games in [18], which extends Q-learning to multi-agent contexts. In a stochastic game, each agent's reinforcement is computed using their combined actions, and the present state and state transform are determined using the Markov process. Reinforcement of the agents in general-sum games can be randomly correlated. In a market system, the consumer and retailer share a common interest in completing a transaction, but they disagree on the price of an item. The Nash equilibrium in [19] is the essential explanation for the general-sum event. In Nash equilibrium, each agent effectively gathers accurate facts of policies of the other agents and acts logically depending on this information. For general-sum games, Nash equilibrium is the most important aspect. Most favorable Q-values are Q-values obtained in Nash equilibrium in general-sum stochastic games, and they are referred to as Nash Q-values. During recurrent episodes, the aim of learning is to calculate Nash Q-values. The agent will be able to reach Nash equilibrium and make decisions based on the Q-values it has acquired. Inferences on other agents' conduct are formed through learning about their policy needs [20]. The Nash equilibrium approach, which is based on game-theoretic origin of rationality, can be considered the most important path to form inference.

3 Nash Equilibrium

Nash equilibrium is a generally established game theory concept for two or more opponents in which each performer is supposed to be familiar with the balance policies of the other performer, and no agent has anything to achieve by merely altering its own policy. If each agent has chosen a policy and no agent may gain by altering policies, while the other actors' policies remain unaltered, the existing set of policies and payoffs is referred to as a Nash equilibrium [21, 22].

Nash Equilibrium

(G, b) be a game having n agents, where G_i is the policy group for agent i,

$G = G_1 \times G_2 \ldots\ldots\ldots\ldots G_n$ is the group of policy status.

$f(b) = (f_1(b)\ldots\ldots f_n(b))$ is its payoff value calculated at $x \in G$.

Let x_i be a policy status of agent i.

Let x_{-i} be a policy status of all agents excluding for agent i.

While every agent $i \in \{1 \cdots n\}$ selects policy x_i giving policy status $x = (x_1, x_2 \ldots x_n)$ then agent i gets value $f_i(x)$.

In Nash Equilibrium [22], the payoff based on the policy status selected, i.e., on the policy selected by an agent i as well as the policies are selected through all the other agents. A policy status $x* \in S$ is a Nash equilibrium (NE) if no individual difference in policy by any one agent is beneficial to that agent that is 1. In case of single-agent learning, forthcoming reinforcements are depended on agent's individual best policy. It is referred to Q_i^* as *Nash Q-values* for agent i [22, 23].

4 Cooperative Learning Model

The marriage ceremony is taken into account when developing the system. So many decisions must be made, starting with the venue, invitation cards, decoration, finalizing caterers, purchasing apparel, gifts, jewelry, and other stuff for the bride and groom. In addition, such periodic conditions can be implemented in the following way: When people buy clothes, they almost always go on to buy jewelry, shoes, and other accessories. Different types of shops can work together to meet customer requests. In addition, to attract clients, the retailer announces interesting ideas such as a festival offer, a discount on certain items, and "Buy one, Get one free." They can forecast and control their stock levels in those conditions. Overall number of things sold grows, resulting in higher profit margins for retailers [24]. Customers would save time and money if they could buy all of their things in one place. The mathematical notations for the above model are listed below.

1. Number of shop agents are $l = 1, 2, 3$.
2. There are $v_1 = 1/2$, $v_2 = 1/3$, $v_3 = 1/5$ indicating 50% of clients is referring shop 1, 30% of clients referring shop 2, and 20% of clients referring shop 3 [23].
3. Learning parameter = 0.2 and discount rate = 0.99.
4. Vendor has restricted store capability $I_{\max}$ and follows a stable trading strategy.
5. **States (s):** Highest inventory of each shop $= I_{\max} = i_1, i_2, i_3 = 25$. State for agent 1 become (x_1, i_1) e.g., (7, 2) that means seven client demands with two stocks in shop 1. Agent 2 state is now (x_2, i_2), and agent 3 state is (x_3, i_3). System state turns out to be **Input** as (x_i, i_i) [24, 25].
6. **Actions (a):** Action established for agent 1 is $A1 = \text{Rate } p = \{9 \text{ to } 15\} = \{9.0;$ $10.0; 10.5; 11.0; 11.5; 12.0; 12.5; 13.0; 13.5; 14.0; 15.0\}$. Action established for agent 2 is $A2 = \text{Rate } p = \{4 \text{ to } 8\} = \{4.0; 4.5; 5.0; 6.0; 7.0; 7.5; 8.0\}$. Action established for agent 3 is $A3 = \text{Rate } p = \{11 \text{ to } 14\} = \{11.0; 11.5; 12.0; 12.5;$ $13.0; 13.5; 14\}$.
7. **Output** is probable action engaged, i.e., rate in this case. State-action pair that can be simply exhibited by Q-learning, i.e., **Q(s, a)** [34]. Reward calculation is defined for output [25, 26].
8. **Reward (r):** Reward is estimated. Present state $I = (x_i, i_i)$ and future state $j =$ (x_j, i_j). Four situations are considered for reward estimation. ($v_1 = 1/2$, $v_2 =$ $1/3$, $v_3 = 1/5$)

   present state $i \to$ future state j

   Situation 1: $[x_i, i_i] \to [x_i, i_{i-1}]$ Indicate "one product is sold."

   Situation 2: $[x_i, i_i] \to [x_{i+1}, i_{i-1}]$ Indicate "one client demand fulfilled, one item sold."

   Situation 3: $[x_i, i_i] \to [x_i, i_{i-3}]$ Indicate "Three items sold—purchase one catch two."

   Situation 4: $[x_i, 0] \to [x_{i+1}, 0]$ Indicate "fresh buyer demand, zero stock are present."

With reference to above situations, reward is estimated by [27].

$$\begin{aligned}\text{Reward is } r_p(i,\ p,\ j) &= pif\ x`_1 = x_1 + 1 \cdots\cdots \text{Situation 4}\\ &= pif\ i`_1 = i_1 - 1 \cdots\cdot \text{Situation 1}\\ &= 2pif\ i`_1 = i_1 - 3 \cdots\cdot \text{Situation 2 \& 3}\end{aligned}$$

5 Results and Discussions

5.1 Results of Nash Q-Learning

Comparative study of simple Q-learning, Nash Q-learning, and joint-action learning [28]. Existing Nash Q-learning is investigated, and results for shop-agent 1, shop-agent 2, and shop-agent 3 are as follows.

Shop-Agent 1

Figure 1 depicts an annual investigation for shop-agent 1. Profit received vs months for three types of learning: simple Q-learning (without collaboration), Nash Q-learning (with collaboration), and joint-action learning are shown in the graph (with collaboration). When equated with simple Q-learning and Nash Q-learning, the joint-action leaning method yields a considerable improvement in income. The profit range for joint action learning is consistent across all months, as seen in the graph.

Shop-Agent 2

Figure 2 depicts an annual investigation for shop-agent 2. Profit received vs months for three types of learning: simple Q-learning (without collaboration), Nash Q-learning (with collaboration), and joint action learning are shown in the graph (with collaboration) [29]. The profit value derived for agent 2 using the Nash Q-learning method and joint action learning approach are nearly identical, with only a small deviation.

Fig. 1 Shop-agent 1 annual study

Fig. 2 Shop-agent 2 annual study

Fig. 3 Shop-agent 3 annual study

Shop-Agent 3

Figure 3 depicts an annual investigation for shop-agent 3. Profit received versus months for three types of learning: simple Q-learning (without collaboration), Nash Q-learning (with collaboration), and joint action learning are shown in the graph (with collaboration). For agent 3, the Joint Action leaning approach provides a consistent profit value throughout all months.

5.2 *Result Comparison for Three Agents by Nash Q-Learning*

The graph in Fig. 4 shows a comparative study within three agents for simple Q-learning and Nash Q-learning over a one-year period. It depicts the profit made by each agent over a one-year period by utilizing a simple Q-learning algorithm without

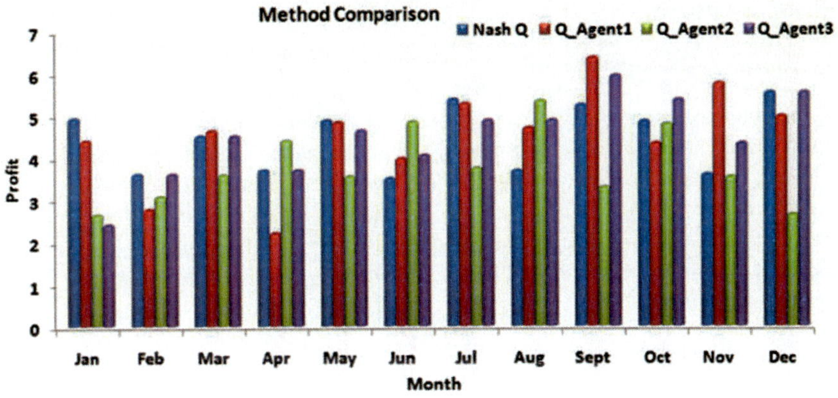

Fig. 4 Comparative study within three agents

collaboration and a Nash Q-learning algorithm with cooperation. The highest profit obtained is 6, and the lowest profit value obtained is 2. The profit obtained through basic Q-learning is greater than the profit received with Nash Q-learning over a period of a few months.

The graph in Fig. 5 shows a technique comparison between three agents for cooperative approaches, namely Nash Q-learning and joint-action learning, during one-year period. Profit obtained by employing joint-action learning by all three agents is significantly higher than the profit obtained by using Nash Q-learning. The highest profit value obtained is 7, and the lowest profit value obtained is 3. Due of cooperation, there is very little variance in each agent's month-by-month profit.

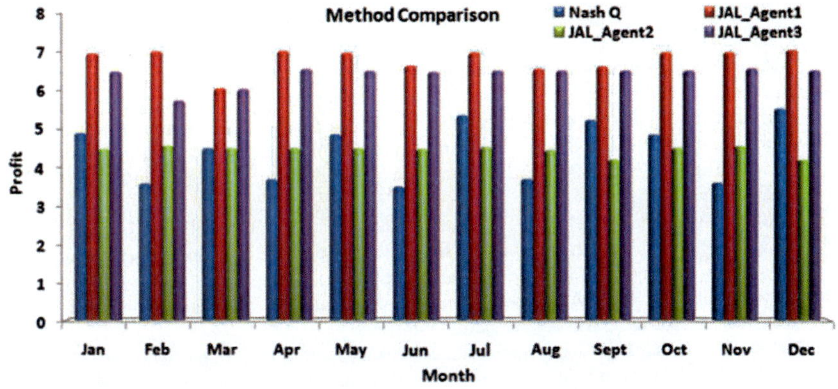

Fig. 5 Comparative study within three agents for cooperative system

6 Conclusion

For the structure in retail shop application, the paper has proposed a multi-agent Q-learning approach, i.e., Nash Q. Nash Q extends single-agent Q-learning to multi-agent situations by adjusting Q-values as agents select Nash equilibrium actions, depending on the assumption. Furthermore, the determined situation highlights that the convergence theory is described using stage games. The results of simple Q-learning, joint-action learning, and Nash Q-learning are discussed. When equated to simple Q-learning and joint-action learning, Nash Q-learning outperforms cooperative method.

References

1. Vidhate DA, Kulkarni P (2019) Performance comparison of multi agent cooperative reinforcement learning algorithms for dynamic decision making in retail shop application. Int J Comput Syst Eng, Inderscience Publishers (IEL) 5(3):169–178
2. Ten Hagen SHG (2001) Continuous state space Q-Learning for control of nonlinear systems. Ph.D. thesis, University of Amsterdam, Feb 2001
3. Vidhate DA, Kulkarni P (2019) A framework for dynamic decision making by multi-agent cooperative fault pair algorithm (MCFPA) in retail shop application. In: Information and communication technology for intelligent systems. Springer, Singapore, pp 693–703
4. Hu J, Wellman MP (2003) Nash Q-Learning for general-sum stochastic games. J Mach Learn Res 4:1039–1069
5. Vidhate DA, Kulkarni P (2017) Expertise based cooperative reinforcement learning methods (ECRLM). In: International conference on information & communication technology for intelligent system. Springer Book Series Smart Innovation, Systems and Technologies (SIST, vol 84), Cham, pp 350–360
6. Chinthalapati VLR, Yadati N, Karumanchi R (2008) Learning dynamic prices in multi-seller electronic retail markets with price sensitive customers, stochastic demands, and inventory replenishments. IEEE Trans Syst, Man, Cybern—Part C: Appl Rev 36(1)
7. Vidhate DA, Kulkarni P (2016) Innovative approach towards cooperation models for multi-agent reinforcement learning (CMMARL). In: International conference on smart trends for information technology and computer communications. Springer, Singapore, pp 468–478
8. Choi YC, Ahn HS (2010) A survey on multi-agent reinforcement learning: coordination problems. In: IEEE/ASME international conference on mechatronics and embedded systems and applications, pp 81—86
9. Vidhate DA, Kulkarni P (2017) Enhanced cooperative multi-agent learning algorithms (ECMLA) using reinforcement learning. In: International conference on computing, analytics and security trends (CAST). IEEE Xplorer, pp 556–561
10. Gosavi A (2003) Simulation-based optimization: parametric optimization techniques and reinforcement learning. Kluwer Academic Publishers
11. Vidhate DA, Kulkarni P (2016) Performance enhancement of cooperative learning algorithms by improved decision-making for context-based application. In: International conference on automatic control and dynamic optimization techniques (ICACDOT) IEEE Xplorer, pp 246–252
12. Kaelbling L, Littman ML, Moore AW (1996) Reinforcement learning: a survey. J Artif Intell Res 4:237–285
13. Walker MA (2000) An application of reinforcement learning to dialogue strategy selection in a spoken dialogue system for email. J Artif Intell Res 12:387–416

14. Vidhate DA, Pophale SS (2021) Depression scale recognition over fusion of visual and vocal expression using artificial intellectual method. Int J Comput Appl (IJCA) 183(24):16–19
15. Stone P, Sutton RS (2001) Scaling reinforcement learning toward RoboCup soccer. In: Proceeding 18th international conference on machine learning. Morgan Kaufmann, San Francisco, CA, pp 537–544
16. De Jong E (1997) Non-random exploration bonuses for online reinforcement learning. In: Collected papers from the AAAI-97 workshop on multiagent learning. AAAI Press
17. Vidhate DA, Patil A, Sarkar S (2010) Bandwidth estimation scheme for mobile adhoc network. In: International conference on business administration and information processing. Springer, Berlin, Heidelberg, pp 130–135
18. Boutilier C (1999) Sequential optimality and coordination in multi-agent systems. In: 16th International joint conference on artificial intelligence. Stockholm, pp 478–485
19. Hu J, Wellman MP (1998) Multi agent reinforcement learning: theoretical framework and an algorithm. In: Fifteenth international conference on machine learning, Madison, WI, pp 242–250
20. Vidhate DA, Kulkarni P (2016) New approach for advanced cooperative learning algorithms using RL methods (ACLA). In: VisionNet'16 proceedings of the third international symposium on computer vision and the internet, ACM DL, pp 12–20
21. Vidhate DA, Kulkarni P (2016) Enhancement in decision making with improved performance by multi-agent learning algorithms. IOSR J Comput Eng 1(18):18–25
22. Abbasi Z, Abbasi MA (2012) Reinforcement distribution in a team of cooperative Q-learning agent. In: Proceedings of the 9th ACIS international conference on artificial intelligence, distributed computing. IEEE
23. Vidhate DA, Kulkarni P (2016) Implementation of multi-agent learning algorithms for improved decision making. Int J Comput Trends Technol (IJCTT) 35(2)
24. Araabi BN, Mastoureshgh S, Ahmadabadi MN (2011) A study on expertise of agents and its effects on cooperative Q-Learning. IEEE Trans Evol Comput 14:23–57
25. Vidhate DA, Kulkarni P (2014) To improve association rule mining using new technique: multilevel relationship algorithm towards cooperative learning. In: International conference on circuits, systems, communication and information technology applications (CSCITA). IEEE pp 241—246
26. Vidhate DA, Kulkarni P (2014) Design of multi-agent system architecture based on association mining for cooperative reinforcement learning. Spvryan's Int J Eng Sci Technol (SEST) 1(1)
27. Verikas A, Lipnickas A, Malmqvist K, Bacauskiene M, Gelzinis A (1999) Soft combination of neural classifiers: a comparative study. Pattern Recogn Lett 20:429–444
28. Vidhate DA, Kulkarni P (2014) Multilevel relationship algorithm for association rule mining used for cooperative learning. Int J Comput Appl (0975–8887) 86(4):20–27
29. Vidhate DA, Kulkarni P (2014) A novel approach to association rule mining using multilevel relationship algorithm for cooperative learning. In: Proceedings of 4th international conference on advanced computing & communication technologies (ACCT-2014), pp 230–236

Real-Time Patient Monitoring System and Security Using Internet Protocols for Medical Industry

Muhammad Abul Kalam and R. Udayakumar

Abstract The Internet of Things (IoT) concepts are commonly used to connect accessible medical services and to give patients intelligent, reliable, and efficient health care. Health monitoring is one paradigm that can use the IoT benefits to enhance the lifestyle of the patient. The objective of the project was to develop a remote health monitoring system, with local sensors available to make it affordable if mass production was to be carried out. The PMS is dependent on new bus connectivity, in particular, by using the TCP/IP protocol as an interior bus rather than on a standard configuration. TCP/IP convention suite can be easily accessed at extremely low costs. Patient control and monitoring is a vital test framework for physiological signals like electrocardiographs (ECGs), breathing rates, invasive and non-invasive circulatory strain, human blood oxygen saturation (SpO2), and body temperature. PMS continuously examines important physiological signs of the patient during individual care. For healthcare personnel, observing every patient for 24 h is a bit difficult job. In hospitals, the importance of PMS is high; hence, the manufacturers of clinical equipment have central patient monitoring and tracking systems. All patient screens are connected to one server-based patient monitoring system in centralized patient monitoring systems. The utilization of networking in clinical field has become a major part for any hospital, particularly with respect to crucial issues, similar to information overflow and security issue and so on. So, we are presenting diverse set of networking topology which makes it easy for the hospital staff to manage.

Keywords Internet of Things · Patient monitoring system · Anesthetic gases · Electrocardiograph · Smart healthcare system · Sensor · IoT-based health monitoring system

---

M. A. Kalam (✉) · R. Udayakumar
Bharath Institute of Higher Education and Research, Selaiyur, Chennai, Tamil Nadu 600073, India
e-mail: makalam.cse@gmail.com

M. S. Kaiser et al. (eds.), *Information and Communication Technology for Competitive Strategies (ICTCS 2021)*, Lecture Notes in Networks and Systems 401,
https://doi.org/10.1007/978-981-19-0098-3_65

1 Introduction

Internet has become like a virtual life to every one of us in this world. People have changed the way they live, work, play, and learn. Internet provides education, money, entertainment, social networking, shopping, e-commerce, etc. The IoT is an environment where various objects can feel, communicate, and share information through a Private Internet Protocol (IP) or a public network. The connected objects regularly collect, analyze, and adopt the measures required to provide a smart analytical, planning, and decision-making network. The IOT is commonly viewed as connecting objects to and using the Internet to control or remotely monitor such objects. This term was only referred to in light of the current developments in IOT machine to market. However, as the actual IoT concept, a brilliant invisible network can be sensed, managed, and configured. The IoT-based products include embedded knowledge exchange technologies or the Internet and are estimated to be interconnected by between 8 and 50 billion devices by 2020. The entire IoT concept consists of sensors, gateways, and wireless networks, which enables users to contact and access the application. As the saying goes, "Health is wealth" it is extremely important to use innovation to improve well-being. The IoT system can ensure a safe inspection of health awareness. To outline an expert system of medical services, the sensor allows to gather customer information and to be transmitted to the cloud via Wi-Fi.

1.1 Motivation

There are few healthcare services in rural hospitals. Weak health management standard leads to healthcare problems. India's latest study data analysis states that there is a shortage of 500,000 doctors in India. The WHO determines that the doctor's ratio of patients in India will rise to 1:1000. There is a shortage of funding and management in developing countries to deal with people's problems. A common man cannot afford to check his health on a costly and regular basis. Therefore, different systems have been developed which provide a simple and reliable care unit. The system decreases the time needed for safe handling of devices.

1.2 Remote Health Monitoring

Remote health monitoring systems are an advanced technique to monitor the health of the patients from remote location through the use of IoT technology so that proper treatment at proper time be given to the patient to save the life. Along with the Internet, we also need multiple other devices like sensors which can read the health condition of the patient and transmit the patient health data to a remote computer through the IoT technology. The benefit is, it will benefit the old aged people, physically challenged people, people with disability, etc. Remote health monitoring incorporates devices like sensor, fall detection system, temperature data, and a small number of other parameters to read patients' health condition round the clock. Data were used in single parameter surveillance systems for remote detection. During design, the following characteristics of possible medical use were followed:

(a) We try to integrate the information and technology associated with the current medical practice so that we get correct and accurate information.
(b) We also consider real-time, long-term remote monitoring systems and also medical devices like wearable devices which have built in sensors.
(c) We also consider devices which are easy to use and easy to understand so that everyone can use it and follow it.

1.3 Intelligent Wireless Mobile Patient Monitoring System

Heart-related disease is on the rise and lead to sudden death. This remote health monitoring system can read the patients' health condition through a sensor and can save the patient from any type of danger. The use of wireless advancements into the current patient surveillance system is feasible. This article proposes the creation of a module that allows doctor and patient mobility by the use of an easy and familiar method that will detect anomalies in the patient's bio-signal and give a physician an SMS warning through the global mobile system (GSM) to make appropriate precautionary measures that will reduce patient critical levels. By using this device, the patient's health can be on track 24 h even without visiting a hospital. Any sudden issues can be treated at the earliest. Therefore, firsthand medical support and promoting good health are important for patient treatment and monitoring. Therefore, patients at risk must also have their heart health monitored indoors or outdoors in order to allow emergency care. Telemedicine is commonly regarded as an unavoidable part of the future of modern medical practice.

1.4 Dimensions of IoT Patient Monitoring System

The management system for patients is done through the use of IoT platform. It notifies the user if the amount of oxygen is very low by alerting the user. Figure 1 shows the principal dimensions of the IoT oxygen control system. The IoT monitoring system dimensions comprise sensors which sense the appropriate location and collect information history. The observed abnormalities do not show the fault. Changes in weather conditions may affect the behavior of the abnormality. In order to predict the danger, data from the sensor signal are analyzed and added to the algorithm. Management is based on the study; it forecasts the situation in a specific field. Management to obtain information for data processing and to equate the information obtained with past data is done appropriately and accurately. It detects frauds as well.

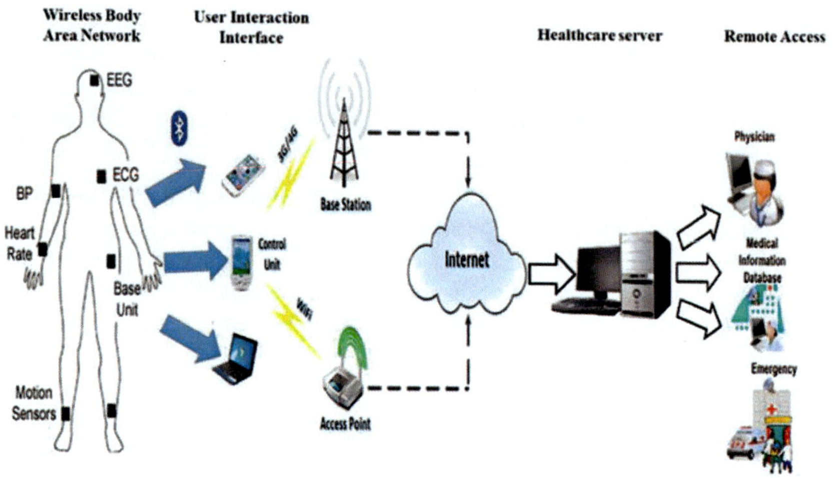

Fig. 1 Dimensions of IoT patient monitoring system

The identification is used for proper decision-making, helps to avoid and increase availability of failures. This allows data to be compared with the history of the past and alerts users.

1.5 Introduction of Patient Monitoring Systems (PMSs)

Patient monitoring is an important physiological sign monitoring device including electrocardiogram (CEG), air, invalid and non-invasive blood pressure, and human blood oxygen saturation patient monitoring systems (PMSs) (SpO2, body temperature, and others) by the use of different sensors and electrodes to receive physiological signs, including ECG electrodes, SpO2 finger monitor, blood pressure cuff, and a temperature test sample. The patient's vital physiological signs must be monitored continuously during care. Continuous technological improvements not only allow us to communicate essential physiological indications to medical staff but also simplify the measurement and thus increase the effectiveness of patients monitoring.

1.6 Classes of Patient Monitoring System in Industry

In recent years, technical advances in the field of measurement and information transmission have contributed to the patient monitoring products, comprehensive efficiency, and reliable quality. In the past, primarily, for one parameter calculation, the dominant goods produced by medical device manufacturers were used but not sufficient. Today, however, a patient monitor with many parameters is often used. Patient surveillance devices are now available in two classes in the new market.

1.6.1 Single-Parameters Monitoring Systems

Monitoring systems with single parameters can only measure a single physiological symbol. It is a pretty old technology but is still used in developing countries today, such as India, Pakistan, Bangladesh, and so on. The single-parameter monitoring systems can be manufactured and maintained at very low prices. The human body's blood pressure, the ECG (Electrocardiograph), the SpO2 sensor, etc., are all measured with a single monitoring device.

1.6.2 Multi-Parameter Patient Monitoring Systems

A patient monitoring system (PMS) is used for the transmission of vital data including electrocardiograms, respiration rates, blood pressure, and so on for various physiological essential signs of the patient. Due to the continuous advancement of PMS technology, the medical personnel are provided with the essential multiple signs of physiological measurements. The new PMS makes physiological sign measurement easier and enhances surveillance performance. PMS is very versatile now and in a single monitoring system may detect many physiological signals.

1.7 Modules Used for Data Collection

1.7.1 Electrocardiograph (ECG) Module

Electrocardiography (ECG) is a non-invasive method focused on the time-consuming interpretation of electrical heart activity. These signals or movements are captured by skin electrodes called the ECG PMS cable lead. In PMS, the ECG module contains three or five leads. The three or cinq conduits of electrodes are mounted on the skin for the sensing of physiological signals in standard configuration.

1.7.2 Non-Invasive Blood Pressure (NIBP) Module

The blood pressure monitor measures the arterial pressure caused by heart contractions which continuously shifts in the human body during the heart cycle. The following are three measured blood pressure values in PMS.

- The systematic pressures are the highest period pressure in the ventricle.

- The diastolic reading, or the bottom number, is the pressure in the arteries when the heart rests between beats. This is the time when the heart fills with blood and gets oxygen.
- The arterial pressure is the mean value of the blood pressure over the cardiac cycle.

1.7.3 Invasive Blood Pressure (IBP) Module

Arterial pressure (AP), a central venous pressure (CVP), and pulmonary arterial pressure may be controlled by blood invasion (IBP). The IBP method is developed, and the catheter is inserted in a human body part's blood vessel. The catheter port attaches to the pressure transducer directly outside the body. Inject saline into the catheter to allow the fluid to pressure into the catheter pressure of the blood within the blood vessel. The mechanical power of the blood is transformed into an electronic signal by a catheter-related pressure transducer.

1.7.4 Oxygen Saturation in Blood (SPO2) Module

The blood saturation system (SpO2) is based by means of the finger sensor and (SpO2) measuring device on blood oxygen absorption into red and infrared light. To calculate blood oxygen, the light electronic transducer is used.

1.7.5 Carbon Monoxide (CO) Measuring Module

In the patient monitoring system, the thermal dilution method is used for calculating CO. Insert the catheter through the right atrium through the pulmonary artery. Then, inject saline liquid with the catheter into the right atrium, a temperature monitor on the front of the catheter. The temperature changes, and the mixed liquid reaches into the pulmonary artery as saline fluid mixes with the blood. The sensor will then detect temperature change.

1.7.6 Gas Measuring Module

The gas (anesthetic gas) concentration is determined by its infrared absorption. An optical infrared filter and a special wavelength infrared ray are used to penetrate the gas for gas measurement. The higher levels of gas of the infrared rays shall be consumed for a given amount. This suggests that less infrared rays have entered the gas at the highest concentration of the consumed infrared gas.

1.8 *Comparison of Sensors and Technology*

(See Table 1).

2 Literature Survey

The technological infrastructure and appropriate environment required for the creation of intelligent hospitals must be created. This study shows optimizing influences, challenges, available technologies, opportunities, and the system architecture generated in the smart hospital environment using IoT technology. The necessary technical infrastructure, including intelligent computer systems and real-time big data analysis, will be splinted into five levels, and the system infrastructure, limitations, and methods needed for each layer will be specified. The study outlines the deficiencies that could lead to an intelligent design of each hospital layer and the considerations that must be taken into account to remove them. The road map should be given to managers, system engineers, and researchers who work toward optimizing the design of the intelligent hospital system [1].

Table 1 Comparison of sensors and technology

SR. No	Title of paper	Sensors used	Technology used
1	Internet Of Things (IoT)Enabled smart autonomous hospital management system—a real-world healthcare use case with the technology drivers	Various typesof sensors are used	Digital hospital
2	Remote health monitoring system fordetecting cardiac disorders	12-lead ECG probe	Mobile-based algorithm deployment, Bluetooth sensor device
3	IoT-based health monitoring via Lora WAN	*B. P*, temp., glucose	LoRa network
4	Resource-aware mobile-based health monitoring	ECG sensor	ECG processing analytics module
5	Internet Of Things (IoT)Enabled Smart autonomous hospital management system—areal-world healthcare usecasewith technology drivers	ECG, EKG	IoT-enabled data modeling

Technology plays an important part in contact, recording, and display systems and sensor devices in the healthcare field. This adapts the current trend in healthcare communications through IoT. The Internet of Things is a driver for the health sector and a popular location in a number of health applications. The microcontroller PIC18F46K22 is used as a portal for communication with the different sensors, such as sensors of temperature and pulse oximeter. The microcontroller collects sensor information and sends it via Wi-Fi to the network, thus tracking the parameters of medical treatment for doctors in real time [2].

Health monitoring is one of the most critical and technologically based programs of the last decade. People face an unexpected death problem due to multiple diseases since at the right moment patients are not being treated. The main goal was to create a safe IoT-based monitoring of patients, which enables healthcare professionals to follow up the hospital or home in order to ensure better treatment for patients through an IoT-based integrated healthcare system. A wireless healthcare monitoring system, consisting mainly of sensors, acquisitions of data, micro control units (i.e., Arduino), and programming, was developed, which provides information on a patient's physiological conditions on the Internet in real time [3].

The Internet is becoming more and more popular every day when the number of remote monitoring and pacemaker patients (especially those who need pacemakers) is increasing. In this study, a new ECG method was developed to monitor patients with pacemaker implants. Through the Wi-Fi protocol, the development system allows them to transmit data to different locations over long distances based on Internet artifacts (IoT) and transmit it to a special site display screen (www.thinger.io) in real time. The system helps ECG and can switch the temperature and magnetic field detectors through the Web site to solve the problem that your doctor can't continuously monitor patients at home and doctors in other countries/regions [4].

The government should focus more on improving the rural health care because there is still a lack of so many medical demands in the rural areas. Recently, the technology advancing with the advent of IoT was accomplished by telemedicine. But, with the latest IoT implementation, there is still space for development. No project for integrating temperature, pressure, and pulse rate sensors into a single device has been carried out in scientific research [5].

Quality of service (QoS) communications networks at local, regional, and global levels must be connected to achieve the end-to-end sharing of knowledge and to promote global ICT services. The quality of service in a network (or a country) will therefore affect service quality from start to finish. In other words, the standard should not only be viewed at national or regional level but from a global point of view. Today, individuals or corporate sectors worldwide rely on information and communication technology in their daily activities [6].

Micro-electric mechanical systems (MEMs) and microcontrollers are affordable, speed improvement, miniaturization, and power efficient in development in the integrated circuit industry. This has helped promote embedded technologies embraced by health workers. These embedded systems were also used on smartphones. Cell phones and the Internet of Things (Iot) in the most developed countries would soon embrace more internet penetration. The remote healthcare approach uses these concepts to build a social quality of life system [7].

The prospect of fast and efficient health care means the use of online telemedicine services. The systems rely on advanced wireless and wearable sensor technologies. In this paper, cost, convenience, accuracy, and data protection in a real-time heart control system will be taken into account. It is designed to provide the doctor and patients with a two-way contact interface. This study aims primarily to support the access of remote patients to state-of-the-art health facilities which would otherwise be unable because of the low physician–patient ratio. The device is subsequently checked on the android computer with wearable sensors for 40 people (ages 18–66) (i.e., smartphone under supervision of the experts) [8].

3 Design and Development

3.1 Simple Structure of the System

To assess whether a pacemaker patient's health effects are positive and normal, the sensors will be installed on the patient's body and the patient's data removed; then, the microcontroller filters and amplifies them. This information is then forwarded to the patient's personal computer or mobile phone via a Wi-Fi connection to the selected browser. This browser can be opened by the follow-up doctor together with the patient to track data at all times. The structure of the method was illustrated below. The proposed device supports patient sensors:

3.1.1 ECG sensor The ECG data are collected by the EC8232 ECG sensor from a patient's body composed of three colored electrodes. The signal is then expanded to include an operational intensifier in order to eliminate the excess frequency beyond the signal frequency range measured (0.5 and 100 Hz). The transformer is then removed from the sensor from the system machine (0–3.3 V).

3.1.2 Temperature sensor It is a small measuring unit with analog output voltage (°C) for the room temperature or body temperature. The LM35 sensor transfers ambient temperature to an analog voltage. The warm-up relationship is linearly proportional to that of the transistor LM35 voltage.

3.1.3 Heart Rate Sensor The heart level with ECG, cardiovascular measurement is performed, and the wave number (RWAVE) is calculated per minute, and the programming code for the chip is entered.

3.2 Design of the Device

With the help of the Wi-Fi protocol, the computer allows the visual display of the special site (www.thinger.io) to be moved (browser). The ECG unit with a cardiac

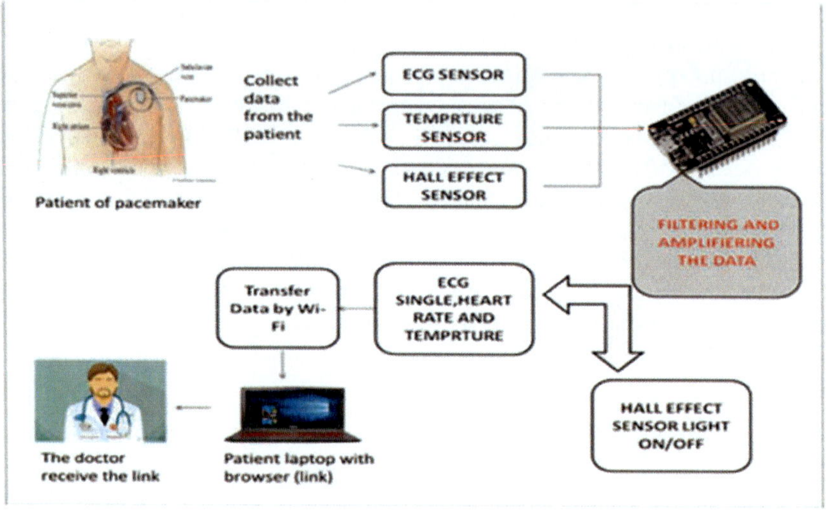

Fig. 2 Structure of the IoT-based patient monitoring system

frequency monitor, a magnet field sensor, and a temperature monitor can monitor the pacemakers in real time. A user link to the console can be created on a platform (ww.thingser.io), and later the controller can view file recovery data from the chip by showing the signal in the chart or digital counter (Microsoft Excel). The system benefits from a variety of wired options, including easy-to-work operations, reduced costs for patients and manufacturers of devices, reduced patient suffering, improved mobility, confidentiality of patient information, and other functions on the Web site (Fig. 2).

3.3 Working Flow

The machine workflow displays in the figure. The sensor results are analyzed, i.e., if suspicious behavior, an emergency plan triggered to warn the doctor about the welfare of the patient. It also eliminates critical hospital conditions.

3.4 IoT Medical Devices

Medical Web includes a network of medical and embedded devices, communication systems and complex wireless actuators with electronics, software, sensors which can be transmitted and shared via wireless. Improved IoT technology becomes an important part of the Internet connecting physical electronic systems through dynamic

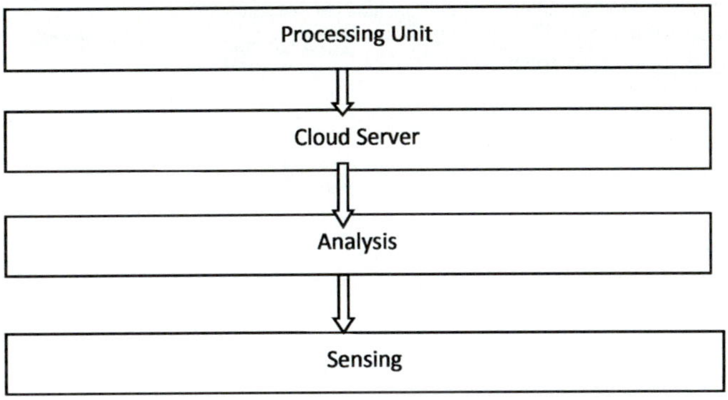

Fig. 3 Work flowchart

sensors and actuators. The IoT architecture model HCMS is shown in Fig. 3. These systems depend entirely on an integrated device to be recognized as a mini computer or microcontroller for these embedded systems. As IoT system nucleus, the microcontroller works. Depending on the device position, a number of accessories can be connected via the general input and output (GPIO). The cardiac rate, muscles, finger, image, pressure, infrared, fowl, power, humidity, the thermistor, and positioning sensors play an important role in patient life, in accordance with differing temperatures. The Internet of medical data now develop, medical devices, sanitary instruments, and their applications to improve health prevention and monitoring. The key advantages that make IoT medical devices suitable for HCMS are as follows: 1—memory, 2—power efficiency, 3—portability, 4—computational efficiency, 5—coverage, 6—durability, 7—cost, and 8—reliability (Figs. 4 and 5).

3.5 Information and Communication Technologies in HCMS

HCMS information and communication technologies play an important role in the success of IoT systems in health care. Current HCMS ICTs are categorized by heterogeneous frequencies, specifications for communications, and transmission speeds. The quality of communication for long and short distances is known as a communication standard. Licensed and non-licensed frequencies are the heterogeneous frequencies of contact. Rates are dependent on networking, sensor devices, and distribution topology for the transfer of IoT system data.

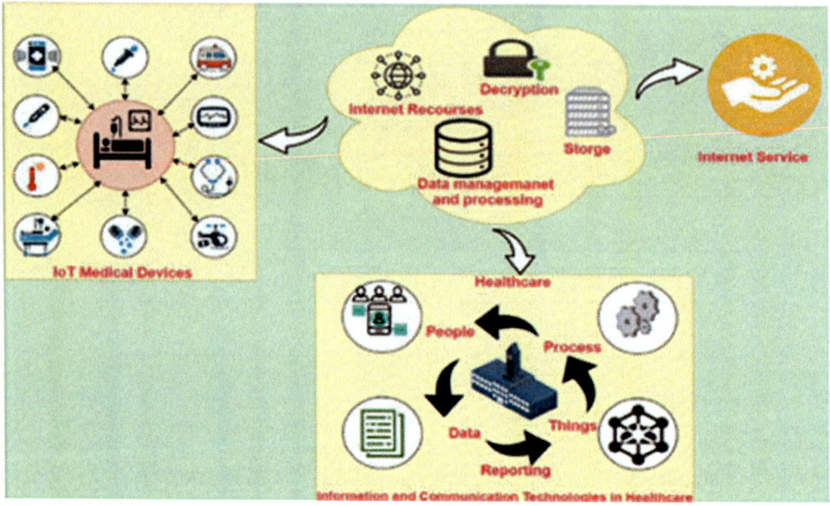

Fig. 4 Illustration of IoT and healthcare monitoring system

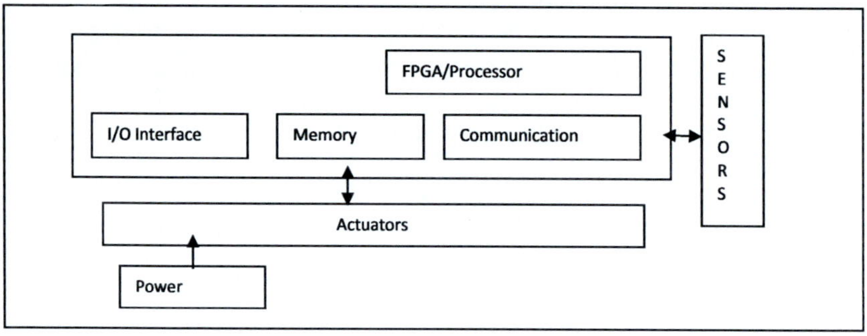

Fig. 5 IoT model for HCMS architecture

3.6 *Frequencies*

Licensed mobile network frequencies will increase traffic management efficiency, provide high levels of protection, offer a high-quality service in large areas, provide better reliability, lower infrastructure costs, and enable operators to monitor their uses. Licensed frequencies encourage long-term sustainable networking investment to ensure connectivity and productivity. The drawbacks of using licensed frequencies are subscription costs and the use of power on the IoT system. Unlicensed frequencies are less suitable for broadband IoT applications because of security problems, infrastructure costs, and interference.

3.7 *Adaptive Alarming System*

Adaptive warning device is for alerting the doctor if an emergency is involved. The alarm system is based on the algorithm threshold. The threshold values were taken from the AHA and the European Society of Hypertension study, and also after discussions with cardiologists. If average heart data deviate from the threshold value, the specified handler task generates alerts. The warning system produces a notification to inform the consumer of the abnormality. Alarm system consists primarily of data visualization, preprocessing statistics and reports. The warning system is a generalized surveillance model based on the upper and lower threshold values. The values are based on different age groups.

4 Challenges in Digital Healthcare Adoption

In automated healthcare systems using EHRs and technology, such as IoT and big data, seamless connectivity between patients and providers across a variety of health systems is expected. The Internet is also increasingly connected to various types of wearable medical technology used for real-time health surveillance. However, before creating secure, scalable, and interoperable systems through digital health care, some issues need to be addressed.

4.1 *Security and Privacy*

IoT devices can put security and privacy at the disposal of users. Unauthorized access to IoT may pose a major health and privacy risk to patients. Connected computers include the collection, merger, processing, and cloud transfers of health and mobile data. Cloning, spoofing, radio jamming, and cloud polling are vulnerable to the device layer. An attack by a man in the center leads traffic to a cloud computer directly. The use of the discovery service (i.e., universal plug-and-play) or BLE characterization requires direct communication attacks when placing and targeting IoT devices.

4.2 *Inter-Realm Authentication and Interoperability Issues*

For companies in various fields of business, developing trust in the conduct of digital medical transactions is critical. Shibboleth is a unified solution that allows entities to be authenticated both within and outside the systems. Shibboleth has been implemented with a device offering interareal authentication and was successfully validated at national level. An IDP user can authenticate and submit a service application to a service provider via a standard shibboleth framework (SP). In this sense, the IdP and SP share the user ID. Shibboleth enables a single sign-on functionality for digital health organizations, such as a federated agreement. In many areas of the digital health system, shibboleth-based systems are safe and provide good authentication. The lack of interoperability between countries on cooperation with digital health ICT infrastructure is also an important aspect of national interest as a dominant digital health problem. This is not only why ICT facilities are small or lack IT capacities, but also because they have no policy of international cooperation between countries to share sensitive health information to promote telemedicine at a distance and provide high-quality medical treatment.

5 Conclusion

The article presents a holistic study of factors dependent on IoT technology for intelligent hospital design. This technology creates a bridge through the heterogeneous network between business applications and computerized nodes. Smart healthcare systems have great opportunities for IoT technology to build and handle wearable devices, real-time exchange of information, patient tracking, and vital readings. The broad use of the Internet is aimed at the application of Internet technology to create a framework for better health through the Internet. Internet of Things will dominate the world in different areas in the future, but health care would be more beneficial. Patient health monitoring can be used extensively in emergencies because it can be tracked, registered, and maintained as a database on a regular basis. The IoT device will in future be able to be coupled with cloud storage to share the database of intensive care and treatment in all hospitals. IoT is used by active healthcare institutions in many ways such as the increase in the quality of facilities by diagnosing them and analyzing them more precisely. Locally and globally, digital healthcare adoption along with many potential solutions to enable the faster implementation of digital healthcare have been suggested. While the healthcare sector becomes increasingly involved in making more effective use of IoT and big data technology, a number of issues are needed to tackle before digital health care becomes a broad reality.

6 Future Scope

IoT will helping almost every sector, from accounting, finance, health care, manufacturing, and road safety, to home security solutions.

The Internet of Things technology-powered advanced tools will play a crucial role in various activities related to health care like:

- reducing waiting room times and tracking patients.

- ensuring the availability of the critical hardware and enhancing drug manufacturing/management.

With IoT in health care, we will do wonders, like:

- Reporting and monitoring and analysis and data assortment.
- Alerts and tracking and remote medical assistance and faster disease diagnosis.

References

1. Uslu BÇ, Okay E, Dursun E (2020) Analysis of factors affecting IoT-based smart hospital design. J Cloud Comput: Adv, Syst
2. Bhoomika BK, Muralidhara KN (2015) Secured smart healthcare monitoring system based on Iot. Int J Recent Innov Trends Comput Commun
3. Senthamilarasi C, Jansi Rani J, Vidhya B, Aretha H (2018) A smart patient health monitoring system using IoT. Int J Pure Appl Math
4. Abdul-jabbar HM, Abed JK (2020) Real time pacemaker patient monitoring system based on internet of things. In: The fourth postgraduate engineering conference
5. Saha HN, Auddy S, Pal S, Kumar S (2017) Health monitoring using internet of things (IoT). IEEE
6. Kadhim KT, Alsahlany AM, Wadi SM, Kadhum HT (2020) An overview of patient's health status monitoring system based on internet of things (IoT). Wireless Personal Commun
7. Kakria P, Tripathi NK, Kitipawang P (2015) A real-time health monitoring system for remote cardiac patients using smartphone and wearable sensors. Int J Telemed Appl
8. Ragesh GK, Baskaran K (2011) A survey on futuristic health care system: WBANs. In: International conference on communication technology and system design 2011
9. Saminathan S, Geetha K (2018) Real-time health care monitoring system using IoT. Int J Eng Technol
10. Zeadally S, Siddiqui F, Baig Z, Ibrahim A (2019) Smart healthcare Challenges and potential solutions using internet of things (IoT) and big data analytics. Potential solutions using internet of things 2019
11. Patil S, Pardeshi S (2018) Health Monitoring system using IoT. Int Res J Eng Technol
12. Naresh VS, Pericherla SS, Murty PSR, Reddi S (2020) Internet of things in healthcare: architecture, applications, challenges, and solutions. Int J Comput Syst Sci Eng
13. Vikram Gnanaraj V, Ranjana P, Thenmozhi P (2019) Patient monitoring and control system using internet of thing. Int J Innov Technol Exploring Eng (IJITEE)
14. Wang X, Le D, Cheng H, Xie C (2014) All-IP wireless sensor networks for real-time patient monitoring. J Biomed Informatics
15. Santoro G, Vrontis D, Thrassou A, Dezi L (2018) The internet of things: building a knowledge management system for open innovation and knowledge management capacity. Technol Forecast Soc Chang 136:347–354
16. Kim TY, Youm S, Jung JJ, Kim EJ (2015) Multi-hop WBAN construction for healthcare IoT. In: 2015 International conference on platform technology and service, P.P 978-1-4799-1888-1/15
17. Abdullah A, Ismael A, Rashid A, Abou-ElNour A, Tarique M (2015) Real time wireless health monitoring application using mobile devices. Int J Comput Netw Commun (IJCNC) 7(3)
18. Bhoomika, Muralitharan KN Secured smart healthcare monitoring system based on IoT. Int J Recent Innov Trends Comput Commun 3(7)

Reconnaissance for Penetration Testing Using Active Scanning of MITRE ATT&CK

S. Mayukha and R. Vadivel

Abstract In the world of penetration testing, reconnaissance is the most important and first stage of securing an enterprise. Only through offensive exploitation by the pentesters can an enterprise be secured and patched up for vulnerabilities. Vulnerabilities that would be exploited by the hackers if left unpatched or solved. The MITRE ATT&CK framework lists out all the possible techniques and tactics to exploit the vulnerabilities in a machine. This ranges from reconnaissance to impact. Any kind of hacking starts with gathering information. The more the hackers manages to get the information about the enterprise like the infrastructure of machines, open ports, active services, directory structure, etc., the more chances the hackers could succeed in exploiting the vulnerabilities and hacking the system. The pentesters can safeguard an enterprise by not hiding most of the information and not accessible to the outside world. For this, the pentesters should think like hackers with malicious intent and perform offensive attacks. Only then, the pentesters can know where the vulnerabilities are and the correct measures to patch those vulnerabilities.

Keywords Penetration testing · Hacking · MITRE ATT&CK · Reconnaissance · Active scanning · IP blocks · Vulnerability scan

1 MITRE ATT&CK

MITRE [1] ATT&CK is an acronym for MITRE Adversarial Tactics, Techniques, and Common Knowledge (ATT&CK) [2]. This framework is a collection of curated knowledge and an example of cyber adversary behaviors. This framework was created in 2013. Researchers from MITRE's Fort Meade experiment (FMX) emulated

S. Mayukha (✉) · R. Vadivel
Bharathiar University, Coimbatore 641046, India
e-mail: mayukhaselvaraj@gmail.com

R. Vadivel
e-mail: rvadivelit@buc.edu.in

M. S. Kaiser et al. (eds.), *Information and Communication Technology for Competitive Strategies (ICTCS 2021)*, Lecture Notes in Networks and Systems 401,
https://doi.org/10.1007/978-981-19-0098-3_66

both offensive and defensive behaviors of hackers and pentesters to improve post-infiltration detection of threats through behavioral analysis. By reflecting the multiple phases of a hacker's choice of platform target and attack. The listing of techniques in MITRE ATT&CK provides an overall view so that the techniques being used is understood by both defensive and offensive sides of penetration testing. This also lets the penetration testers acquire knowledge of what the attack is and how to defend it at the time of exploitation by the hackers. MITRE ATT&CK has three major verticals as well: Enterprise ATT&CK, Mobile ATT&CK, and Pre-ATT&CK™.

MITRE ATT&CK [3] is globally used by various disciplines to secure the cyber space. These disciplines include risk management, intrusion detection, security management, threat detection and intelligence, red teaming, blue teaming, and even more.

There are three main categories present in the MITRE ATT&CK framework. Tactics, techniques, and usage of attacks list.

Tactics are the short and goal-oriented attacks to achieve a specific task. It is denoted by the columns in the framework.

Techniques are the means by which that specific task is successfully completed. It is denoted by the individual cells of the framework.

Usage of attack list any real-world attack of the tactic and technique listed in the framework is documented with the information of groups that performed it and the data about how it actually took place in the real world.

The MITRE ATT&CK framework can help determine the character [4] of the hacker. This helps the pentesters to come with a fighting plan and chance to secure the enterprise in case of an attack. The framework covers a wide array of stages in penetration of a system right from reconnaissance to the impact of the attack that is created by the hackers. Figure 1 furnishes the character of the hacker to how serious the hacker is about the attack launched on the enterprise or mobile.

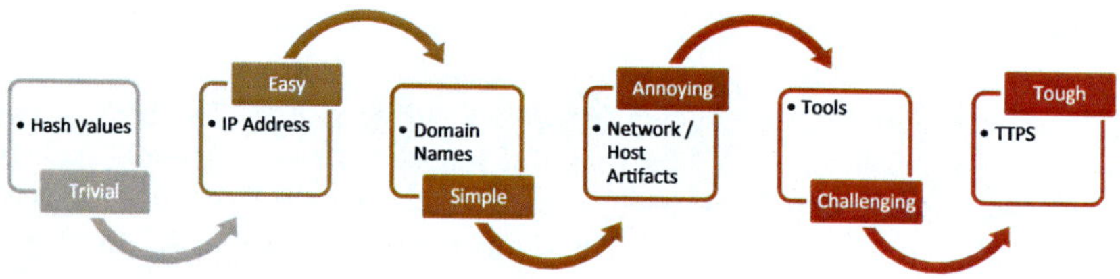

Fig. 1 Characterization of a hacker based on the level of attack launched according to MITRE ATT&CK framework

Fig. 2 Pre-ATT&CK stages

1.1 Pre-ATT&CK™

Pre-ATT&CK™ furnishes the pre-attack behavior of a hacker. This is introduced in the MITRE ATT&CK framework as the initial stages of launching an attack [5] on an enterprise. This mainly focuses on information gathering of the target. The information gathered from reconnaissance is then weaponized to use it against the enterprise. Figure 2 furnishes that the techniques with which information gathered is called as reconnaissance and using that information, weapon is created in terms of tools and payloads. Other processes such as weakness identification, target selection, priority definition, persona development, building, testing, and staging capabilities are also included in PRE-ATT&CK™.

1.2 Enterprise ATT&CK

Enterprise ATT&CK is the attack that is targeted on desktop or laptop machine environments. Machine environments such as Linux, MAC, and Windows. Such an ATT&CK is done in a controlled and sequential process. This takes a lot of patience and vigor. An attack on a machine is the foundation of MITRE ATT&CK. The framework captures real-time events to map the hacker behavior and categorize each stage of the entire attack. Some may skip few stages but not all. From the PRE-ATT&CK, where the information is gathered and weaponized is then delivered in an Enterprise ATT&CK. Then, the enterprise is exploited with payloads and tools, by which the hackers gain access to the enterprise. The next stage is where the hackers gain privilege escalation and maintain that privilege as root user to harm even more or to totally destroy the enterprise. Figure 3 furnishes the list of stages that the hacker would go with to hack the enterprise, and the number of techniques in each stage that could be used. One of the malicious hacks in entertainment history which created a huge impact for the hack of SONY in 2014.

1.3 Mobile ATT&CK

Mobile ATT&CK targets the mobile OS such as android and IOS. Almost all the stages of attack that is followed in Enterprise ATT&CK, and mobile exploiting related techniques can be followed here also as furnished in Fig. 4. When the hacker gains access to the mobile, the hacker could gain access to the microphone and camera

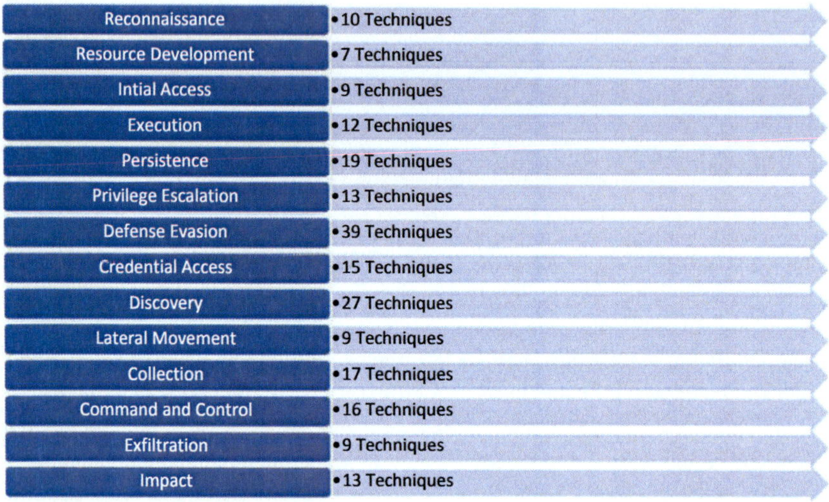

Fig. 3 MITRE-ATT&CK stages for enterprise

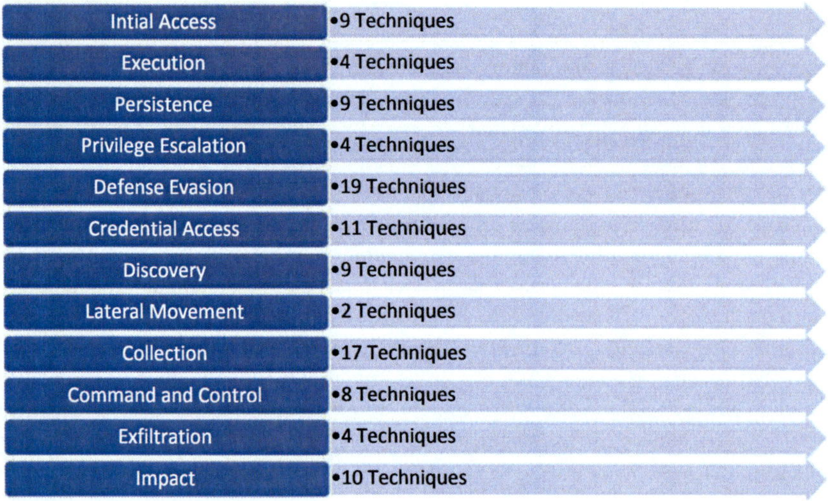

Fig. 4 MITRE-ATT&CK stages for mobile

without the knowledge of the user and start recording. One of the concepts is hot mic where the microphone of the mobile starts recording without the knowledge of the user. Hot mic is mainly used by government agencies to hack smart phones to gather intelligence of the enemy. Though there are indications now showing in the notification area if microphone or camera is in use, orange for microphone and red or green for camera depending on the operating system of the mobile. The

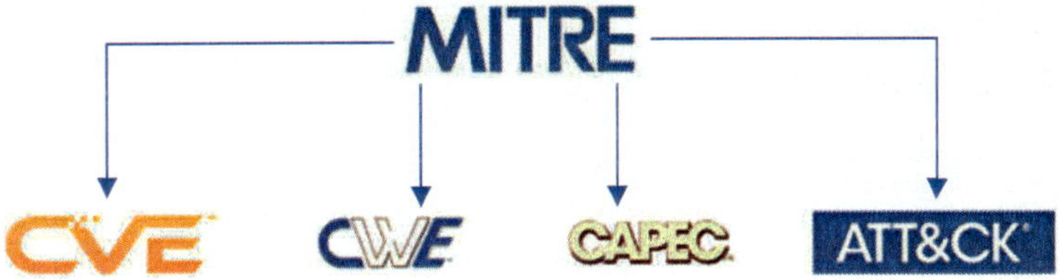

Fig. 5 MITRE supported community driven programs. Photos credit: MITRE, CVE, CWE, CAPEC, and ATT&CK

hackers are smart enough to disable the indication system and then record using microphone or camera. The pentesters can test the mobile software for vulnerabilities and exploitation and then take the necessary steps to protect the mobile device. The Mobile ATT&CK gives the pentesters a real chance to protect against the real-world attacks from various hackers with malicious intent.

In today's world of technology, any device connected to the internet is open to hack with vulnerabilities. Pentesters has to have extensive knowledge of both offense and defense mechanisms of all the attack there is. In addition to ATT&CK, to make the cyber space even safer the MITRE has three more programs as CAPEC [6] known as common attack pattern enumeration and classification, CVE [7] known as common vulnerabilities and exposures and CWE [8] known as common weakness enumeration according to Fig. 5. CAPEC is the list of common approaches or attack patterns done by hackers to exploit a cyber space. CVE is the list of vulnerabilities and exposures found in a software or an environment. CWE is the list of common weaknesses found while developing a software or a hardware. Both the lists are developed in such a way to help pentesters fix vulnerabilities and exposures and to develop weaknesses free or bug free software or hardware to avoid getting hacked or exploited by hackers with malicious intent. MITRE has a discontinued program called CPE known as common platform enumeration also.

2 Active Scanning

Active scanning is a part of reconnaissance stage in the ATT&CK framework. This is used for information gathering. Active scanning gathers information so that it can be used to target the enterprise and weaponized the information. Active scanning intrudes on the infrastructure using network traffic. This is carried out as opposed to other forms of information gathering which do not require direct intervention. Hackers may select the type of active scanning based on the type of information that is required. The active scans can be executed in multiple ways starting from a basic ICMP echo [9]. The gathered information would be the foundation for the hackers to carry out the attack by researching more on common vulnerabilities and exposures related to the specifics in the information. There are two main methods of active

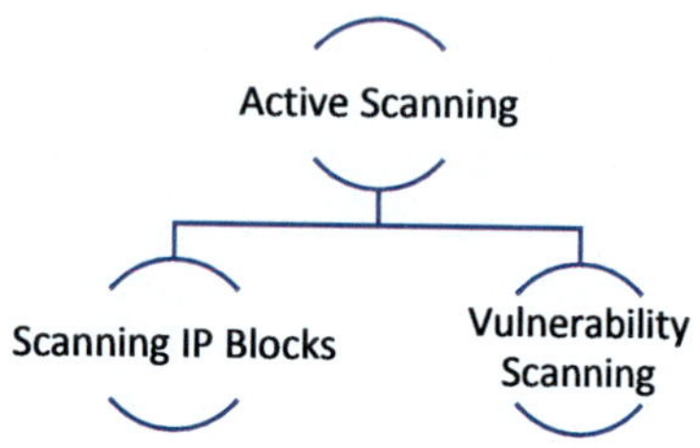

Fig. 6 Active scanning and its sub-techniques

scanning as furnished in Fig. 6. Where the active scanning is the main technique and the rest are the sub-techniques as follows.

1. Scanning IP blocks
2. Vulnerability scanning

The techniques mentioned above has various sub-techniques with which a hacker could easily attack the enterprise and exploit the vulnerability. In this paper, the tools that is used to carry out the attacks as well the prevention of how to defend the attacks is listed to support the pentesters effectively defend the enterprise. Like active scanning there are various methods using which a pentester mimicking to be hacker for offensive security could practice all the methods on the enterprise to maximize the defense system of the enterprise. This could be valuable over time and would make the enterprise a stronghold. So, offensive security is the best way to secure an enterprise in today's cyber world.

2.1 *Requirements*

In order for a pentester to carry out an active scanning on an enterprise, the pentester should have permission from the enterprise for offensive penetration testing of the infrastructure of the enterprise. Once the necessary permission is acquired by the pentester a scope of the penetration testing should be sketched out as to the limitations of intrusion done by the pentester, etc.

The techniques mentioned above have various sub-techniques with which a hacker could easily intrude the enterprise and exploit the vulnerability. In this paper, the tools that is used to carry out the attacks as well the prevention of how to defend the attacks is listed to support the pentesters effectively defend the enterprise.

2.2 *Knowledge Base*

The pentester should possess the knowledge of basic networking and penetration testing skills. The pentester should be able to understand how ping works and how a network would be configured. The network can be private or public. As long as the information gathering is within scope and with the intent of attacking the target enterprise the pentester should be able to understand the information gathered [10]. There is no use in having the information and not knowing the forward path of active scanning. Once the information is gathered the pentester should think offensively as a hacker and achieve the maximum recon possible from the infrastructure of the enterprise. The targeted enterprise can be intruded via network connections or Web site or any other means of device connected to the network. There is no limitation of what to use and what not to use. All is fair in penetration testing as long as it is not done with malicious intent.

The pentester should be able to fairly understand how the information gathered about the server or the version of the OS of the target system or the framework used in the Web site is subjected to vulnerabilities. It can easily be looked upon using CVE database. Each and every vulnerability if existed in real world is marked with an ID in the CVE database. Armed with the CVE data a pentester is able to exploit the target enterprise using tools and techniques provided by penetration testing OS [11] such as Kali [12, 13] or Parrot. It is also noted that the pentester should know how to play around the Linux distribution OS in order to execute the open source or premium tools that is available for exploitation, most of which is based on Linux OS. Pentesters should also be aware that proof of concept (POC) is required to justify their exploitation almost similar to bug bounty programs in case if there are any discrepancies from the other pentesters.

2.3 Execution Style

There are so many tools both open source and premium are available for pentesters. But just because a tool is premium it does not promise that it is the best tool available. A smart pentester should be able to understand what tools are available but at the same time which tool would perform well given the circumstances and limitations if any. For e.g., there are tools for directory finder called gobuster, dirbuster and dirb. If a pentester is to achieve greatness, the pentester should have knowledge of all three and what would be the best among three but at the same time which specific tool works better under special situations. For example, dirb is more stable than dirbuster, but at the same time, gobuster supports multithreading. So, a smart choice would be gobuster. So, the basics of active scanning to reconnaissance a network would involve few tools most probably open source such as Nmap or Zenmap, Dirb or Dirbuster, nslookup, wireshark, Aircrack-ng.

Depending on the process of the penetration of the enterprise, the pentester should choose the tools wisely. For e.g., in case of exploiting a Web site that is developed with Web site would require a WPScan rather than other tools that covers only basic operations. If the pentester has knowledge of the technology or framework used, then the way for the pentester to go is to have specific tools related to the existing technology on the enterprise which would result in even more data. This could be used to weaponized and launch the attack to exploit the enterprise. As one

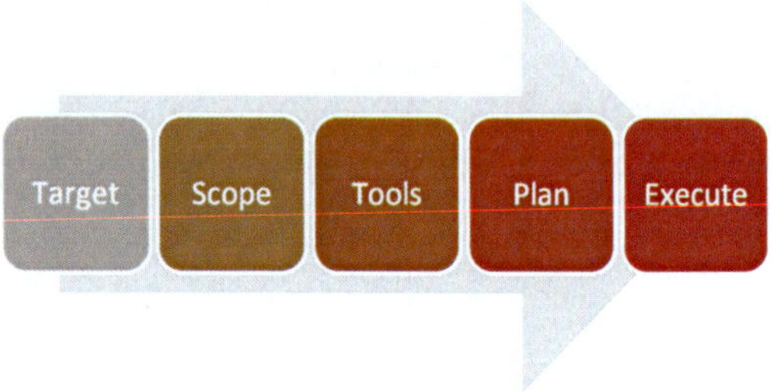

Fig. 7 Execution style of a penetration test on an enterprise

can understand, a pentester with high resilience can only master the various facets of penetration testing and can successfully establish a legit career. Figure 7 furnishes the execution style of a penetration testing. It is as important as the hack itself. This would pave the way for future security measures as well. So, it is important that there is no room for mistakes and whatever vulnerabilities are in the enterprise should be found and patched up in due course. Otherwise, there is no point in pentesting with a team of security engineers. So, the success of pentesting depends on the counter measures taken by the enterprise on advice of the pentesting team. The pentester should be strong enough to encounter vulnerabilities with all might and to adhere to the law in penetrating the enterprise which should not be harmful for the enterprise no matter what the case might be. Either the enterprise should back up all the systems before performing the pentesting or the pentesters should take note to instruct the enterprise to back up the data. In case the system goes under reset or is wiped clean due to some triggering. The backup would be the easy and safe way to get the enterprise back running with better counter measures in place to avoid such a happenstance again.

3 Scanning IP Blocks

Scanning IP blocks fall under the network mapping category and listed under reconnaissance. The main part of information gathering of a network happens here. A pentester who does not have knowledge of the infrastructure should observe and learn the infrastructure like how a hacker would do. The scanning of IP blocks can be done with a simple ping or can also use tools like Nmap and Zenmap to map the infrastructure of the target enterprise. Then, tools like dirb or dirbuster or gobuster can be used to gather information about the file structure on the network. The possibility of the gathered information leading to a successful penetration depends on the pentester's knowledge and resilience. There are so many solutions to a single problem so, the pentester should be able to think creatively to exploit the infrastructure. Adversaries may scan victim IP blocks to gather information that can be used during targeting. Public IP addresses may be allocated to organizations by block or a range of sequential addresses.

Table 1 Tools for IP blocks scanning

Application	Type	Execution
Nmap	Open Source	CLI
Zenmap	Open Source	GUI
Masscan	Open Source	CLI
Angry IP scanner	Open Source	CLI
Advanced IP scanner	Open Source	GUI

Pentesters can scan IP blocks using simple pings such as ICMP requests and responses. With that a pentester can gather an enterprise's network information like IP addresses that are actively used by the enterprise users. The pentester can gather even advanced details such as hosts, and the addresses assigned to those hosts. Another version of these scans can gather information about servers and the version of server softwares that are used to find vulnerabilities on the web. Any information about the infrastructure would help a great deal to exploit the enterprise. Table 1 describes the list of tools for IP scanning a network or to know versions of softwares that are run on the enterprise.

3.1 Procedure

First and foremost, the pentester should either open a file to dump all the outputs or have a folder specific for that pentest. The tools and procedure selected in this paper is commonly used and is not superior than the other tools. So as a pentester feel free to choose tools based on functionality. Then, input the target enterprise's IP to perform the Nmap scan. Nmap is also known as Network Mapper. With Nmap, a pentester can determine the ports associated with the IP address and the services running on each port. This also outputs the software versions running on those IP address.

3.2 Options

Nmap [14] has various options that are given for a pentester to use it for infrastructure scanning. Table 2 furnishes the flags otherwise called as options to switch what kind of scan a pentester would require. Some firewalls would block pinging itself in that case stealth scan is required to be performed. This can be achieved using the flag –sS. Optionally, Nmap scan be alternated between TCP and UDP by using the appropriate flag. Table 3 furnishes the types of options where a pentester could select the desired port to be tested on. Table4 furnishes the types of outputs that are available on the Nmap. Using these various options provided a pentester can achieve the desired results for information gathering.

Table 2 Flags for Nmap scanning

Flag	Meaning
-sS	TCP SYN Stealth scan
-sT	TCP Connect scan
-sU	UDP scan
-sF, -sN, -sX	TCP FIN, NULL, and Xmas scans
-sA	TCP ACK scan
-sW	TCP Window scan
-sM	TCP Maimon scan
-sI	TCP Idle scan
-sO	IP Protocol scan
-b	TCP FTP Bounce scan

Table 3 selecting port for Nmap scanning

Flag	Meaning
-p 22	Selecting port no. 22
-p ssh	Can be matched with multiple ports
-p 22,25,80	Connect with multiple ports
-p80-85	Hyphen to select a range of ports
-p-100,600-	Omit the first 100 ports and connect to the range from 100 to 600 ports
-p-	Omit numbers to scan the whole range
-pT:21,U:23	Separate TCP and UDP Ports can be specified
-p http*	Wildcards matching ports with same name
-p 1-102, [103-]	The range of ports is scanned only if register in nmap-services along with the higher ports

Table 4 Output flags for Nmap scanning

Flag	Meaning
-v	Verbosity level
-d	Debugging level
-packet-trace	Print every sent or received packer
-oN <filename>	Output to file
-oG <filename>	Output to grepable format
-oA <basename>	Output to all formats
-resume <filename>	Continue outputting in the file
-append-output	Output files append
-open	Hosts with open ports

4 Vulnerability Scanning

The vulnerability scanning performed by the pentester is kind of an extension from IP Block scanning. In this the pentester, with the information gathered from Nmap or the software services that are running on the target system can scan for vulnerabilities related to that software service. There are so many applications and even frameworks to scan for vulnerability and exploit the software service. One such application is the metaspoilt [15] which is actually a framework for vulnerability validation as well in addition to scanning. Burp suite is used for web applications, to find vulnerabilities and exploit them (Table 5).

4.1 Procedure

After gathering the information about softwares and services that are running on the target enterprise using Nmap, the pentester has to scan for vulnerabilities in those specific softwares. There is a database called common vulnerabilities and exposures powered by MITRE. Every vulnerability that has been proved would have a CVE Id assigned. So, a pentester has to either navigate to the CVE database or put in the keywords in the metaspoilt to get the payloads. Payloads are files that are helpful to exploit the target. Metaspoilt framework offers a way to set the target IP and the attacker IP and to select vulnerability which a pentester wants to use on the target. Metaspoilt does the rest of the work once the vulnerability is selected and the options are given. In addition to vulnerability scanning, the metaspoilt goes a step further to execute and capture initial access of the system. Vscan [16] is the exact vulnerability scanning tool where the tool scans the enterprise for vulnerabilities and report the findings in an orderly fashion.

Table 5 Tools for vulnerability scanning

Application	Type	Execution
Burp Suite	Open Source & Premium	GUI
Metaspoilt	Open Source & Premium	CLI & GUI
Vscan	Open Source	CLI
Archery	Open Source	GUI
Dagda	Open Source	CLI
OpenVAS	Open Source	GUI
Tulpar	Open Source	CLI

Table 6 Common options for vulnerability scanning

Flag	Meaning
IP_range	IP range of the vulnerability scan
Protocol	HTTP or SSH, etc.
Port	Port number

4.2 Options

The options for Vscan to scan for vulnerabilities are as furnished in Table 6. The Vscan operates with IP range or address to scan for vulnerability in a target enterprise. The protocol and the port number can be mentioned if required.

5 Conclusion

In order to penetrate an enterprise, reconnaissance is the foundation of any attack. If done right the rest of the attack would be a breeze. Active scanning plays a larger part of information gathering. Steps to prevent these kinds of scanning should be put in place in the enterprise to avoid or defend against malicious attacks. A smart pentester after understanding the working of information gathering can protect the enterprise even more better. This is why a pentester should think like a hacker and proceed with resilience to penetrate the enterprise because hackers would not quit that easily and that should be the case pentesters as well. MITRE ATT&CK is a useful guide for pentesters to observe and analyze the hacker's attacks and character and proceed accordingly. Reconnaissance should be considered the highest severity of attack and pentesters should take necessary steps in defending against it at all costs. The future scope for this study is to implement the scanning tools such as nmap, etc., and to gather the working functionality of these tools. Once the working logic is known then come up with a solution to prevent these kinds of information gathering attacks in the future would be the way to go.

References

1. MITRE Homepage. https://www.mitre.org/
2. ATT&CK Homepage. https://attack.mitre.org/
3. Xiong W, Legrand E, Åberg O, Lagerström R (2021) Cyber security threat modeling based on the MITRE Enterprise ATT&CK Matrix. Softw Syst Model
4. Munaiah N, Rahman A, Pelletier J, Williams L, Meneely A (2019) Characterizing attacker behavior in a cybersecurity penetration testing competition. In: ACM/IEEE international symposium on empirical software engineering and measurement (ESEM)
5. Mancini S, Iacono L, Hartle F, Garfinkel M, Horn D, Sullivan A (2021) Introducing the common attack process framework for incident mapping. Int J Cyber Res Edu (IJCRE)

6. CAPEC Homepage. https://capec.mitre.org/
7. CVE Homepage. https://cve.mitre.org/
8. CWE Homepage. https://cwe.mitre.org/
9. Upadhya A, Srinivas BK (2020) A survey on different port scanning methods and the tools used to perform them. academia.edu
10. McNab C (2007) Network security assessment, 2nd edn. O'Reilly, USA
11. Kali Linux Homepage. https://www.kali.org/
12. Velu VK (2017) Mastering Kali Linux for advanced penetration testing. Packt Publishing Ltd.
13. Najera-Gutierrez G, Ansari JA (2018) Web penetration testing with Kali Linux: explore the methods and tools of ethical hacking with Kali Linux, 3rd edn. Packt Publishing Ltd.
14. Nmap Homepage. https://nmap.org/
15. Metasploit Homepage. https://www.metasploit.com/
16. Vscan Github. https://github.com/xen0vas/vscan

Data Sharing Approach for Personal Health Record by Session Password and Access Key in Cloud Storage

J. Naveen John and I. Shatheesh Sam

Abstract For the safe sharing of personal health record (PHR) in the cloud, a safe sharing method is built using the proposed session password and access key. The method is secure for doctors to share the PHRs and also gain rights to the patient's medical history while maintaining their privacy. The proposed data sharing technique strengthens and secures the system, and it has been tested in a variety of circumstances. Furthermore, compared to the traditional data sharing approaches, the proposed method needs less cost, time, and memory. The proposed method utilizes less communication cost, time, and memory of 2123 bits, 21 ms, and 1234 bytes, respectively. It satisfies a variety of security properties, including tamper resistance, transparency, and decentralization.

Keywords Personal health record · Personal health information · Data sharing · Cloud server

1 Introduction

Cloud computing has established itself as a leading computing platform that enables ubiquitous and when required access to a variety of resources, like architecture, memory, and applications [1]. The cloud computing model has grown in popularity among healthcare stakeholders as a way to improve collaboration and ensure the availability of scalability and health data. As a result, cloud computing takes into account a variety of healthcare domain constraints, such as staff nurses, service providers, clinical lab employees, hospital staff, comprising of physicians, insurance

J. Naveen John (✉)
Nesamony Memorial Christian College Affiliated to Manonmaniam Sundaranar University, Abishekapatti, Tirunelveli, Tamil Nadu 627012, India
e-mail: naveenjohnmano@gmail.com

I. Shatheesh Sam
Department of PG Computer Science, Nesamony Memorial Christian College Affiliated to Manonmaniam Sundaranar University, Tamil Nadu, Abishekapatti, Tirunelveli, Tamil Nadu 627012, India

M. S. Kaiser et al. (eds.), *Information and Communication Technology for Competitive Strategies (ICTCS 2021)*, Lecture Notes in Networks and Systems 401,
https://doi.org/10.1007/978-981-19-0098-3_67

companies, pharmacy, and patients [2]. By combining these variables with the cloud, the patients can build and manage their own PHR in a shared and cost-effective health environment [3]. In particular, the PHR contains a variety of data, including private documents, previous surgeries, diagnoses, medications, test results, data from insurance companies, demographic data, patients' medical histories, and allergies discovered through patient medical issues [4].

The PHR is maintained using Web-based resources that enable patients to build and change health-related data. Health details, on the other hand, are regarded as permanent documents that can only be accessed by those with access rights. The PHR, on the other hand, allows the patient to interact more efficiently with medical providers and physicians in order to maintain updated health records, evaluate and notify about symptoms in order to make treatment and diagnosis decisions [5]. Because of the following factors, assisting the diagnosis is ineffective: (i) The patient may forgotten what happened previously, such as the medical test or medication that he or she took, as well as the accuracy of care or diagnosis that he or she received. (ii) Due to their limited experience, the patient is unable to explain the procedure or diagnosis, which impairs the current doctor's judgment. As a result, the current doctor is unable to provide correct diagnosis information. The most promising approach for resolving the above dilemma is to distribute the medical documentation with the physician so that the diagnosis can be improved. When a patient sees another doctor in the same medical facility or hospital and the patient's health records are stored in the institution, the doctors can obtain the health records straight from the local network with the patient's permission. In fact, a patient with varying symptoms can visit various doctors at separate medical institutions. In this situation, the medical center refuses the doctor's details by subscribing to the patient's personal health information (PHI) [6]. Cloud-assisted health record exchange is a vital challenge for handling healthcare information that is being built in the e-health system. These findings point to a potential solution for the e-health system's PHI exchange between hospitals. However, privacy and protection are two of the most important factors to remember in the e-health system. The cloud is regarded as a secure location for storing and handling information. If the cloud is not monitored or attacked, it can be vulnerable to failure, theft, violence, and leakage. Countermeasures, on the other hand, are built using a variety of cryptographic primitives and other techniques [7, 8]. Protection becomes a big problem in the cloud due to the centralization features of the environment. In the cloud model, various approaches are adopted to improve the safety of PHR stored on the cloud server. Integrity, transparency, audit trial, and secrecy are all ensured by the privacy protection methods. The term "audit trial" refers to the process of reviewing the use of clinical information and granting access privileges to these data.

In [4], for cloud-based PHR sharing, the secure sharing of personal health record (SeSPHR) concept is proposed. This approach maintains patient-centric oversight while maintaining PHR confidentiality. The user's access right is given after the encrypted PHR is put on an unsecured network. This system is resistant to internal attempts and implements backward–forward access prevention. But, it had an excessively high computational complexity. In [9], a privacy preserving attribute-based paradigm was created to expose the secret access-policy and allows user to exchange data with numerous parties. Every PHR communicates its aspects, by encrypting them with the group identities and access-policy. Data can be decoded by any user who belongs to the designated party.

To access the hidden pattern in cloud, an oblivious random access memory (ORAM) was developed [10]. Using shuffle correctness evidence, it ensured high effectiveness and protection of data by protecting information blocks from arbitrary change. However, it did not pay attention to data owner's engagement. A blockchain-based safe and privacy preserving sharing (BSPP) system was developed to boost the analysis in the e-health model [6]. Two forms of block chain, such as consortium and private block chain are built by devising the consent and structure of the data. In the cloud, it efficiently handles control access, reliable look for, protection, and privacy of data. However, it overlooked the keyword request and verifier election algorithms. An attribute-based encryption (ABE) system to allow two parties to securely share health records was created. By an anonymous key agreement, it gives the user the authority to produce the decryption key and mask the attributes. It successfully achieved privacy preservation but fails to preserve healthcare system's quality and security. A blockchain-based medical data exchange and security solution to improve the hospital's e-health system was developed [11]. Different factors, like transparency, forgery, and scalability, are met by this model. It developed a secure system for storing and accessing historical data from cloud. Although the method's complexity and cost of communication were less, protection was not given an importance.

The rest of the paper is organized as: The proposed method is explained in the second part. The third part describes the experimental results. Finally, the fourth part of the paper is the conclusion of the paper.

2 Proposed Work

The information security, safe information exchange, and privacy of data are the real challenges of the framework of cloud owing to the difficulty and transparency of the Internet. Thus, for protection to PHR, a strong method for sharing of data is proposed. Host registration, owner registration, user log, encryption, setup process, testing process, decryption, and download of data process are the seven stages of the proposed method. The user, owners, server of cloud, and authority are four different identities that make up the proposed method. The proposed method is depicted schematically in Fig. 1. The various stages are explained further.

A. *Host registration*

Input: Host-ID Ht_{id}, host-password Ht_{pd}

Output: Session-password Ss_{pd}

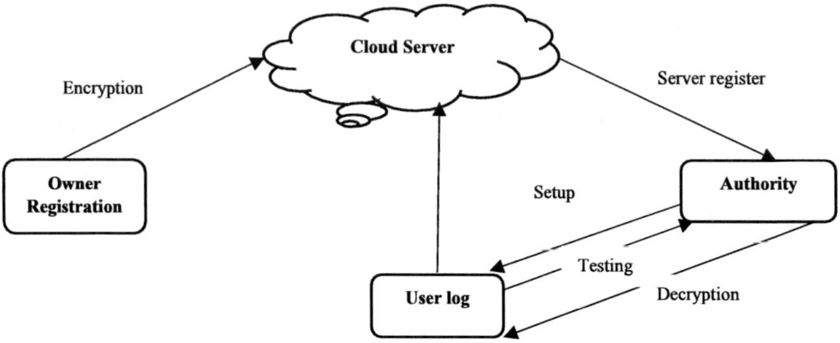

Fig. 1 Illustration for data sharing in cloud

Step 1: The cloud server is in charge of the host registration process. Primarily, server generates an ID and password Ht_{id}, Ht_{pd} and sends it to the authority.

Step 2: The authority uses the function SHA for creating session password of the server as Ss_{pd} using Ht_{id} and Ht_{pd}. The session password is created as,

$$Ss_{pd} = SHA\left(Ht_{id} \oplus Ht_{pd}\right) \tag{1}$$

Step 3: The password $Ss^{\wedge}{}_{pd}$ for the session is then saved in the cloud server. The password is then read by authority and checks with the server which contains the password of the session.

Step 4: If $Ss_{pd} = Ss^{\wedge}{}_{pd}$, the authority acknowledges the request for server's registration.

B. *Owner registration*

Input: Owner-ID Or_{id}, Owner-Password Or_{pd}.

Output: Private key Or_{prky}.

Step 1: The owner creates and sends an ID and password Or_{id}, Or_{pd} to the server to get itself registered.

Step 2: The ID and password are received by the server and stored as $Or^{\wedge}{}_{id}$ and $Or^{\wedge}{}_{id}$, respectively. The server using the ID and password saved on it generates the private key for the owner. The private key is calculated as,

$$Or_{pk} = \left(\left(ran \oplus Or^{\wedge}{}_{id}\right)|||Or^{\wedge}{}_{id}\right) \tag{2}$$

Step 3: The private key is then sent by the server to the owner, who saves it as $Or^{\wedge}{}_{prky}$. The key of the owner stored in the server is compared against key saved with the owner. If both match, the server acknowledges the request for registering the owner, and it is added to the server.

C. *User log*

Input: User-ID Ul_{id}, user-password Ul_{pd}.

Output: Private key Ul_{pk}.

Step 1: The ID and password are created as Ul_{id} and Ul_{pd} in the user log, and the user sends a request to the server. It is received by the cloud server, which records them as $Ul^{\wedge}{}_{id}$ and $Ul^{\wedge}{}_{pd}$, respectively. The server then creates a message based on the user's device number and a pseudo-number. The message is written as,

$$msn = Ul^{\wedge}{}_{de-no} \oplus \left(Ul^{\wedge}{}_{de-no}///r\right) \tag{3}$$

Step 2: The message then transmitted by the server to the user, which receives it and saves it as $msn^{\wedge}$ after which the user is verified and registered on server.

Step 3: Once the registration is successful, the user's private key is produced using security parameter, the ID $Ul^{\wedge}_{id}$, and the password $Ul^{\wedge}_{pd}$. The private key is produced by the cloud server as,

$$Ul_{pk} = Ul^{\wedge}{}_{pd} \bullet a\left(Ul^{\wedge}{}_{id} \oplus Ul^{\wedge}{}_{pd}\right) \tag{4}$$

Step 4: The user's private key is then sent to the user, who accepts and retains it in the user account. Then, owner starts the encryption step after the user log process is completed.

D. *Encryption*

Input: Input data Inp.

Output: Encrypted text Inp_{enc}.

Step 1: The owner considers an input data Inp and encrypts it during the encryption process as,

$$Inp_{enc} = Ent(a\backslash\backslash\backslash Or_{id}) \oplus \left(Inp * Or_{prky}\right) \tag{5}$$

Step 2: The owner then sends the encrypted text to the cloud server, which is retained in the server account. The user begins with the setup process once the data are encrypted.

E. *Setup*

Input: Request message Re_{ms}.

Output: Session-password Ul_{ss}.

Step 1: The server and the user are used for a new user registration method. The user composes and transmits the message request Re_{ms} to cloud server which is represented as,

$$\mathrm{Re}_{ms} = Ent\big(Ul_{id} \oplus SHA\big(Ul_{pd} \backslash\backslash Ul_{de-no}\big)\big) \tag{6}$$

Step 2: The user-generated data request message is compared to the server-stored message. If the results are same, the server approves the new user's request.

Step 3: The authority receives a message request from the server once the new user request is approved. The authority then uses the public key of the user also a random number to create the session password for the user as,

$$Ul_{ss} = SHA\big(r \backslash\backslash Ul^{\wedge}{}_{id} \oplus pky\big) \tag{7}$$

Step 4: The server receives the session password sent to it, and in the server account, it stores it as $Ul^{\wedge}{}_{ss}$.

F. *Testing phase*

Input: Authentication message Me_{ss1}, Me_{ss2}.

Output: User authentication.

Step 1: During the testing phase, the authentication messages are generated as Me_{ss1} and Me_{ss2}, which are transmitted to authority. The authenticated messages are represented as,

$$Me_{ss1} = Ent(Ul_{ss} \oplus SHA(a \backslash\backslash Ul_{id})) \tag{8}$$

$$Me_{ss2} = Ent\big(Ul_{pk} \oplus SHA(r \backslash\backslash Ul_{id-no})\big) \tag{9}$$

Step 2: The authority receives the authentication messages and is stored in the authority account.

Step 3: The user-generated authentication message is cross-checked against the authentication messages. The user is authenticated with the authority if both the messages are the same.

G. *Decryption and download phase*

Input: Data access key DA_{ky}.

Output: Original data.

Step 1: The data access key is generated with the help of ID, public key of the user along with a security parameter as by authority as,

$$DA_{ky} = r \oplus Ent\big(Ul_{id} \backslash\backslash pky\big) \tag{10}$$

Step 2: The data access key is received by user, who saves it in the user account. The data access key of user is further transmitted to server, who collects and saves in the server account.

Step 3: The data access key is checked by the cloud server, and the encrypted text is transmitted to the user if it is true. The encrypted text is then decoded using the below expression,

$$Inp = D(Inp_{enc}) \tag{11}$$

3 Examination and Results

This section describes the experimental verification and findings of the proposed process. The proposed method is compared against the existing methods like, oblivious random access memory (ORAM) [10] and attribute-based broadcast encryption (ABBE) [9], respectively. The method's performance was tested using a dataset [12].

3.1 Evaluation Metrics

The described approach's performance is assessed using measures such as cost for communication, time taken, and memory needed.

Cost for Communication (bits): The decimal numbers are used to calculate the data sent between the parties. The amount of bits available in the decimal values of transmitted data is denoted as the cost for communication.

Time Taken (ms): It is also known as runtime, and it refers to the amount of time it takes to finish a computation.

Memory (bytes): To facilitate safe data sharing, the system memory is employed to record and analyze the PHR data.

3.2 Analysis

Comparative analysis of data sharing method is elaborated in this section.

The proposed data sharing solution in cloud storage is discussed in Table 1. Figure 2 depicts a comparison of the proposed data sharing method when key length = 64, and for the file size = 100 Kb, the cost utilized for the existing methods ABBE and ORAM is 2189 bits and 2165 bits, respectively, while for the proposed method is 2123 bits. The time analyzed is 109 ms, 87 ms, and 21 ms, respectively. And, the memory used is 2012 bytes, 1687 bytes, and 1234 bytes, respectively.

Table 1 Metrics-based analysis

Metrics/Methods		Proposed Method	ORAM	ABBE
Size of File = 100 Kb	*Cost for Communication (bits)*	2123	2165	2189
	Time Taken(ms)	21	87	109
	Memory (bytes)	1234	1687	2012
Size of File = 200 Kb	*Cost for Communication (bits)*	4321	4543	4876
	Time Taken (ms)	75	132	176
	Memory (bytes)	674	1122	1876
Size of File = 300 Kb	*Cost for Communication (bits)*	6496	6876	7134
	Time Taken (ms)	100	145	198
	Memory (bytes)	1826	2345	2856

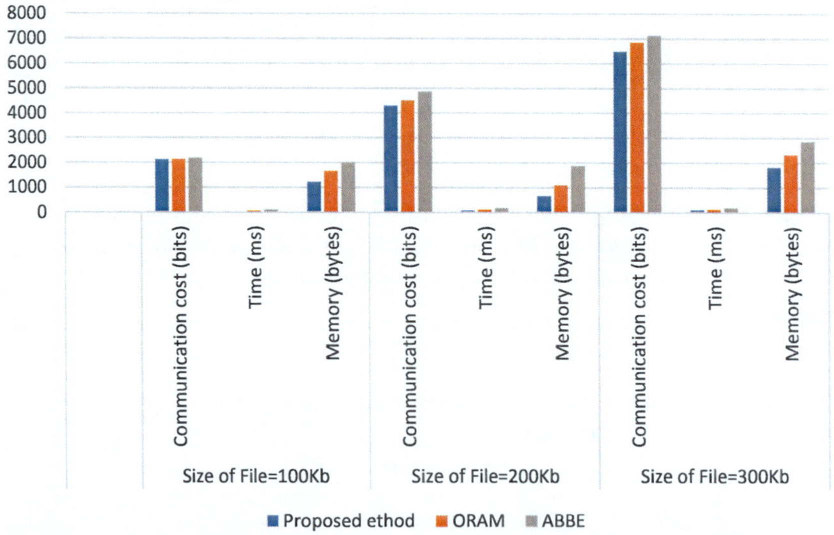

Fig. 2 Analysis for file size 100 Kb, 200 Kb, and 300 Kb, **a** communication cost, **b** time, **c** memory

When the key length = 64 and for the file size = 200 Kb, the cost utilized for the existing methods ABBE and ORAM is 4876 bits and 4543 bits, respectively, while for the proposed method is 4321 bits. The time analyzed is 176 ms, 132 ms, and 75 ms, respectively. And, the memory used is 1876 bytes, 1122 bytes, and 674 bytes, respectively.

When the key length = 64 and for the file size = 300 Kb, the cost utilized for the existing methods ABBE and ORAM is 7134 bits and 6876 bits, respectively, while for the proposed method is 6496 bits. The time analyzed is 198 ms, 145 ms, and 100 ms, respectively. And, the memory used is 2856 bytes, 2345 bytes, and 1826 bytes, respectively.

4 Conclusion

An efficient and comprehensive data sharing method is developed to provide safe sharing of PHR on cloud. Host registration, owner registration, user log, encryption, setup process, testing process, decryption, and download of data process are seven stages of the proposed method. The user, owners, server of cloud, and authority are four different identities that make up the proposed method. When the key length = 64 and for the file size = 100 Kb, the cost utilized is 2123 bits; time analyzed is 21 ms, and memory used is 1234 bytes, respectively. When the key length = 64 and file size = 300 Kb, the cost utilized is 1496 bits; time analyzed is 100 ms, and memory used is 1826 bytes, respectively. It satisfies a variety of security properties, including tamper resistance, transparency, and decentralization as compared to the existing methods. Additional entities may be used in the future to improve the performance of data sharing in cloud.

References

1. Ge C, Susilo W, Liu Z, Xia J, Szalachowski P, Liming F (2020) Secure keyword search and data sharing mechanism for cloud computing. IEEE Trans Dependable Secure Comput 1–1. https://doi.org/10.1109/TDSC.2020.2963978
2. Abbas A, Khan SU (2014) A review on the state-of-the-art privacy-preserving approaches in the e-health clouds. IEEE J Biomed Health Inform 18:1431–1441. https://doi.org/10.1109/JBHI.2014.2300846
3. Au MH, Yuen TH, Liu JK, Susilo W, Huang X, Xiang Y, Jiang ZL (2017) A general framework for secure sharing of personal health records in cloud system. J Comput Syst Sci 90:46–62. https://doi.org/10.1016/j.jcss.2017.03.002
4. Ali M, Abbas A, Khan MUS, Khan SU (2021) SeSPHR: a methodology for secure sharing of personal health records in the cloud. IEEE Trans Cloud Comput 9:347–359. https://doi.org/10.1109/TCC.2018.2854790
5. Kaelber DC, Jha AK, Johnston D, Middleton B, Bates DW (2008) A Research Agenda for personal health records (PHRs). J Am Med Inform Assoc 15:729–736. https://doi.org/10.1197/jamia.M2547

6. Zhang A, Lin X (2018) Towards secure and privacy-preserving data sharing in e-health systems via consortium blockchain. J Med Syst 42:140. https://doi.org/10.1007/s10916-018-0995-5
7. Shen Q, Liang X, Shen X, Lin X, Luo HY (2014) Exploiting geo-distributed clouds for a E-Health monitoring system with minimum service delay and privacy preservation. IEEE J Biomed Health Inform 18:430–439. https://doi.org/10.1109/JBHI.2013.2292829
8. Yang Y, Ma M (2015) Conjunctive keyword search with designated tester and timing enabled proxy re-encryption function for E-health clouds. IEEE Trans Inform Forensic Secur 1–1. https://doi.org/10.1109/TIFS.2015.2509912
9. Xiong H, Zhang H, Sun J (2019) Attribute-based privacy-preserving data sharing for dynamic groups in cloud computing. IEEE Syst J 13:2739–2750. https://doi.org/10.1109/JSYST.2018.2865221
10. Yuan D, Song X, Xu Q, Zhao M, Wei X, Wang H, Jiang H (2018) An ORAM-based privacy preserving data sharing scheme for cloud storage. J Inform Sec Appl 39:1–9. https://doi.org/10.1016/j.jisa.2018.01.002
11. Liu X, Wang Z, Jin C, Li F, Li G (2019) A blockchain-based medical data sharing and protection scheme. IEEE Access 7:118943–118953. https://doi.org/10.1109/ACCESS.2019.2937685
12. UCI Machine Learning Repository: Heart Disease Data Set. https://archive.ics.uci.edu/ml/datasets/Heart+Disease

Stock Market Prices and Returns Forecasting Using Deep Learning Based on Technical and Fundamental Analysis

Varun Vora, Manav Shah, Aryan Chouhan, and Prachi Tawde

Abstract Computer researchers and economic experts have been attracted to estimate stock values for many years. As external variables such as inflation and currency rates, socio-economic conditions, and market attitudes are regularly affected by inventory price variations, this problem is highly difficult, nonlinear, and dynamic. Neural network architectures have shown great promise toward solving this problem and have outperformed classical approaches toward stock price forecasting. In the methodology proposed, the stock price is forecasted using two unique architectures—one based on LSTM-RNNs, and the other based on transformers and time embeddings—that deal with time series sequences that are augmented by an NLP-based market sentiment-analyzing plugin that utilizes FinBERT. Stocks that are listed on the NYSE that have been considerably affected by recent developments in their respective sectors are used. Experiments show that this framework shows considerable improvement in stock price prediction compared with other standard methods in time series forecasting.

Keywords Machine learning · Deep learning · Stock price prediction · Technical analysis · Fundamental analysis · Time series forecasting · Self-attention mechanism · Long-short term memory · Recurrent neural networks · Transformers · Sentiment analysis · FinBERT

1 Introduction

The stock market is a global network that facilitates practically all significant economic transactions at a dynamic rate known as the stock value, which is based on market equilibrium. It is not surprising that the problem of stock price prediction has attracted considerable research interest over the years. There has been a significant amount of work done on ways to provide investors with a model on the basis of which an optimal trading strategy can be formed. Accurate predictions yielding superior

V. Vora (✉) · M. Shah · A. Chouhan · P. Tawde
Dwarkadas J. Sanghvi College of Engineering, Mumbai, India
e-mail: varun.vora182000@gmail.com

M. S. Kaiser et al. (eds.), *Information and Communication Technology for Competitive Strategies (ICTCS 2021)*, Lecture Notes in Networks and Systems 401,
https://doi.org/10.1007/978-981-19-0098-3_68

financial results and substantial arbitrage profit opportunities and help in hedging against market risks is a major driver of study in this area.

However, the nature of the stock market makes it incredibly challenging to predict and analyze stock market data in general. It is inherently nonlinear, dynamic, complex, nonparametric, and chaotic [1], which makes it incredibly difficult to parse useful information from and model. This nature is usually the result of macro-economic factors [2, 3] but can be influenced by micro-economic factors as well [4].

In the financial sector, the two main traditional approaches for price prediction are technical analysis and fundamental analysis. Technical analysis relies on past prices on the basis of which patterns and signals are inferred which dictate trades [5]. On the other hand, fundamental analysis focuses on the intrinsic value of companies using which reliable estimates can be made regarding share prices [6].

However, given that there is abundant data available in this field, machine learning and deep learning have been widely applied and studied toward predicting capital markets' movements [7]. The proposed method employs two unique architectures—LSTM-RNNs and transformers with time embeddings, along with a pre-trained FinBERT model for analyzing pertinent news articles and other developments that affect the stock price and style.

The key objectives and contributions of this work are:

- Develop deep learning architectures for forecasting stock prices that are better than standard econometric models.
- Use technical analysis and fundamental analysis to decipher patterns in historical data of stocks and news articles that could impact the trends of the stock prices, respectively.
- Feed material data to the optimized deep learning architectures to get effective forecasts.

2 Related Work

2.1 Technical Analysis

Econometric models like autoregressive moving average model (ARMA) and autoregressive integrated moving average (ARIMA) are some of the most common methods involved in technical analysis and their general methodology involves treating each new signal as a noisy linear combination of the previous signals as well as some independent noise factors. Pellegrini et al. [8] applies the ARIMA-GARCH model to the prediction of financial time series by introducing a generalized autoregressive conditional heteroscedastic (GARCH) model for conditional variances.

Artificial neural networks (ANN) [9, 10], fuzzy logic (FL) [11], support vector machines (SVM), [12] and particle swarm optimization (PSO) [13] are the one of the most prevalent shallow machine learning approaches utilized in today's stock forecasting model and can hence be further grouped into a class called "soft computing-based models". With the help of additional algorithms [14] showed that the machine learning method is effective and could be used to predict financial markets more accurately than economic approaches. He also demonstrated that the financial market's forecasting effect was influenced by the market's maturity and the input variable, the base forecasting time, and the forecasting interval are all factors to consider.

Researchers have also used the principle of Granger causality [15] to detect the Granger causal link in various financial time series since its inception. For example, [16] used the autoregressive integrated moving average and the Granger-causality test to analyze the connection between Hong Kong's daily stock prices, exchange rates and interest rates, and discovered sporadic unidirectional Granger causality from stock price to interest rate as well as a weak bidirectional Granger-causality between stock price and currency exchange.

Another area of research involves using tree-based ensemble approaches and deep learning algorithms to anticipate stock and stock market trends. Araújo et al. [17] employed two types of ensemble classifiers, heterogeneous and homogeneous, in light of their use of bagging and majority vote approaches. They also analyze the performance of models by looking at macro-economic variables and financial data from the Taiwan stock market. Ensemble classifiers outperformed single classifiers in terms of investment returns and prediction accuracy, according to the findings.

2.2 *Fundamental Analysis*

Due to improved data availability, deep learning has been successfully utilized for selecting and pricing securities, constructing investment portfolios, active risk management, and natural language processing (NLP) or computational linguistics has grown increasingly powerful. NLP approaches that have recently been created make it easier to capture sentiments and extract text semantics. Articles that use natural language processing techniques to forecast financial markets are forming a new field of research called natural language-based financial forecasting (NLFF).

A number of studies have been published that use text information to forecast stock movements such as [18–20]. Different types of textual elements, such as bags of words, noun phrases, named entities, and organized events, have been extracted from texts in pioneering works. Structured events from open information extraction were shown by authors [21, 22], and [23] found that capturing structured relations outperformed conventional characteristics. Structured representations for events, on the other hand, have the problem of increasing sparsity, which may limit their predictive value.

Structured representations for events, on the other hand, have the problem of increasing sparsity, which may limit their predictive value. Fader et al. [23] presented a solution to this problem by employing event dense embeddings to represent structured events. To improve event embeddings, [24] used ground truth from the knowledge graph. The effects of news attitudes on the stock market were retrieved, extracted, and studied by Shah et al. [25]. To avoid capturing redundant information in the news title and content, [26] used a two-level attention mechanism to assess the relevance of words and sentences in the news and devised a novel measurement for computing the attention weights.

To extract features, the collected articles were processed in a variety of methods. After deleting stop words, whitespace, punctuations, and numerals, [27] retrieved N-gram characteristics. They used OpenNLP to extract sentences from each document and constructed final features into a document matrix. They then utilized the SentiStrength package to detect sentiment using the Loughran and McDonald financial sentiment dictionaries.

3 Methods

3.1 Technical Analysis

The technical analysis done for time series forecasting involves hidden patterns in historical data, cyclic fluctuations, seasonality in trends, stationarity of the time-series data, and heteroscedasticity in distributions of returns from previous close prices. The common econometric models saw numerous tweaks to factor in all the mentioned characteristics, while making forecasts for future prices and returns of the stock from the last close price in the historical data for a particular forecast. Some famous examples of such transmutations in the above mentioned econometric are S-ARIMA (seasonal ARIMA), ARIMAX for including exogenous factors, and the ARIMA-GARCH model that factors conditional variances for heteroscedastic data. The ARIMA model is used as the benchmark for prediction of stock prices and returns in the datasets.

However, these models work only for time-series data with a specific set of characteristics and may not adapt to any changes in the style of the stock trend, or any idiosyncratic risk that affects the movement of the prices. This is where deep learning architectures such as RNNs come into play. Recurrent neural networks or RNNs are types of artificial neural networks in which layers of neurons are looped to form a directed graph along a time sequence. As a result, it is capable of exhibiting temporal dynamic behavior. RNNs, which are built from feed-forward neural networks, can use their internal state to process variable length sequences of inputs (memory). LSTMs are an enhancement in the original RNN architecture as they reduce susceptibility to vanishing and exploding gradient problems associated with traditional RNNs by implementing feedback connections in the looped graph of neurons. LSTMs have looked promising in predicting and forecasting entire sequences of data from historical sequences of data instead of single data points. The LSTMs have been quite popular among researchers and engineers for forecasting a stock price time series.

We have tried forecasting stock prices and returns over previous close using both vanilla and stacked LSTM with and without the inclusion of exogenous variables.

Another interesting new way of forecasting time-series data is using attention-based deep learning models, namely transformers. A transformer is a neural network that consists of an attention mechanism that coagulates the pertinent pieces of the price time series and develops expectation characteristics. The attention mechanism consists of a single-head attention and multi-head attention layer. In this case, a Time2Vector layer is used that encodes stock price data into time embeddings and feeds them to the attention mechanism layers of the transformer.

3.2 *Fundamental Analysis*

BERT is built on the notion of a transformer's attention mechanism that learns the contextual connection between words in a text. Transformers feature an encoder–decoder design in its most basic form where the input text is transformed into a tensor using an integrated algorithm and relayed ahead to the encoder. The resulting tensor is then further passed along to the decoder whose design partially resembles its counterpart. However, it contains an encoder–decoder attention layer helping to focus on the key portions of the phrase. BERT differs from a transformer in that it lacks a decoder as part of its design. In addition, roughly 15 of the words in the text are disguised at random by substituting them with tokens. BERT is taught to predict these disguised tokens as the procedure progresses.

The next obvious step was to use financial datasets to train the BERT model. Our model uses existing weights of the 'FinBERT' model, rather than random weights. FinBERT, a pre-trained NLP model, is simply a BERT model that has been trained further using huge financial corpus and then utilized for tasks involving categorization of financial sentiment. A financial sentence dataset taught the pre-trained FinBERT model utilized for further tweaking.

4 Proposed Design

Our proposed system involves providing the user with an informed decision with respect to a particular stock. It advises the user to sell, buy or hold, leveraging the best features of both worlds by concatenating the results of both technical and fundamental analysis. In cases where the movement of the stock price crosses a certain threshold value, additional information of whether the user should short or long the stock is also presented.

Our technical analysis part of the system has been implemented in two ways, the first one, using a stacked LSTM-RNN architecture and the second one, a transformer that uses time embeddings produced from a Time2Vector model. The LSTM-RNN model is a stacked neural network of four LSTM layers in a sequential model with some dropout regularization after each layer. The transformer boasts single and multi-head attention layers that receive input in the form of time embedding that are produced after passing stock data through a Time2Vector encoder model. Both are trained and tested on OHLCV data for three different stocks from the NYSE for a time period of 72 days. The end goal is to predict close prices for the next day, predict absolute returns on the next day over previous day's close and essentially, generate a strategy that can be back tested on the test set using paper trading to get a PNL out of it.

The system's basic analysis component employs a pre-trained FinBERT trained on the Thomson Reuters Text Research Collection (TRC2) dataset, which contains 1,800,370 news articles. The Financial PhraseBank dataset is used to train this model further. It is a dataset that contains polarized sentiment in financial news sentences. The dataset includes 4840 phrases from English-language financial news that were sentimentally classified as good, neutral, or negative. The dataset was cleaned initially with a custom data pre-processing module, which included converting the headlines to lowercase, removing punctuation and stop words, tokenization, stemming, and lemmatization.

The proposed LSTM-RNN design is outlined in Fig. 1 and the transformer-Time2Vec design is depicted in Fig. 2.

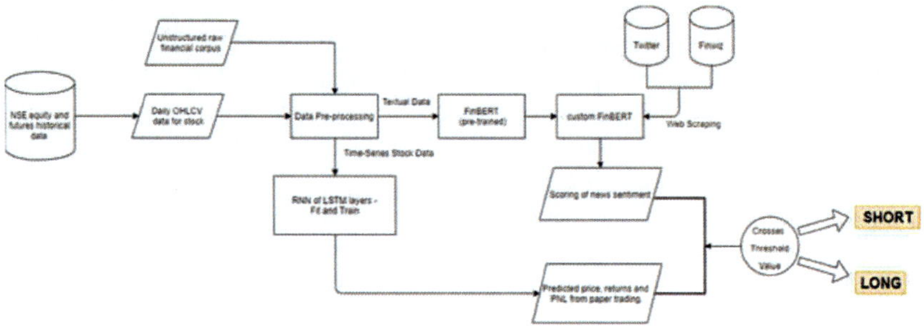

Fig. 1 Proposed design—LSTM-RNN

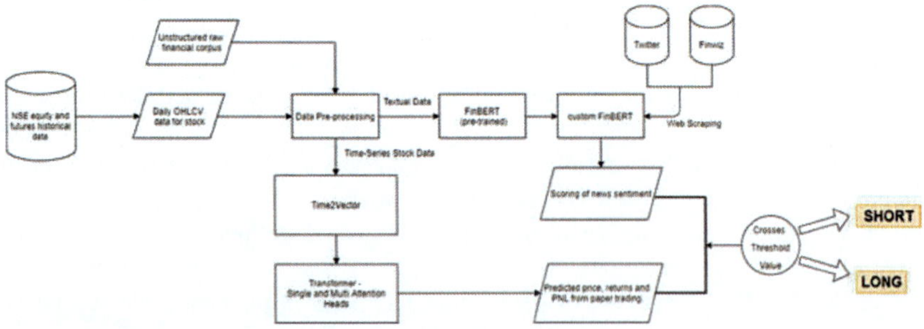

Fig. 2 Proposed design—transformers and Time2Vector

4.1 *Parameters Used in FinBERT*

- Learning Rate: $2 * 10^{-5}$
- Warm-up Proportion: 0.2 (Warm-up is a way to reduce the primacy effect of the early training examples. The learning rate is increased linearly according to the warm-up proportion.)
- Output Mode: Classification
- Discriminate: True
- Gradual Unfreeze: True

Note: The last two parameters 'Discriminate' and 'Gradual Unfreeze' determine whether to apply the corresponding technique against catastrophic forgetting.

4.2 *Parameters Used in Stacked LSTM-RNN*

- Number of Layers: 4
- Number of Units/layer: 250
- Dropout Regularization: 0.2
- Epochs: 75
- Batch Size: 18
- Sequence Size for Training Set: 15

4.3 *Parameters Used in Transformer*

- Time2Vector sequence length: 3
- Query and key dimensions: 64
- Value dimension: 64
- Number of heads: 128
- Dimensions of feed-forward layer: 64
- Batch size: 12
- Epochs: 35

5 Evaluation and Results

Analysis and evaluation of this system are carried out on three stocks of the NYSE for which data was available in the form of open-high-low-close-volume (OHLCV) time series and popular news headlines with corresponding timestamps. For all of the three stocks, the model tries to predict the closing price for the next day. Test and evaluation are carried out for a period of 72 days for which, the standard 1–1 ARIMA is used as the benchmark. Generally, machine learning and deep learning models are evaluated on the basis of certain standard metrics and performance scores which are recorded for the LSTM-RNN architecture and the transformer architecture against the scores of the 1–1 ARIMA.

But apart from that, because of the special nature of this use case, i.e., stock price forecasting, the evaluation is carried out using a popular technique used in this domain.

The obstacles while building a forecasting model for stock prices and returns are that the movement of any stock price is affected by a plethora of external factors, all of which cannot be accounted for while using deep learning architectures. The stock market prices are effectively last traded prices for any equity and are a result from a match from bid/ask quotes from two parties. So, it is not possible to be accurate every single time when it comes to predicting prices. And because of the increased volatility in most world markets, predicting exact numbers for a long time period into the future is neither effective nor pragmatic. Thus, while building a system for stock price forecasting, the aim should be to increase accuracy in terms of the number of times the direction of movement was predicted correctly, and the magnitude of change predicted. The ultimate goal of forecasting stock prices is to act on information from the model, and walk away with a nice reward. So, to evaluate how good the models are at predicting directional changes in trends or even large surges or dips for that matter, the evaluation is carried out by generating a PNL for each model by paper trading on the test set. A method for calculating PNL includes a million dollars starting capital and if the predicted return for the nest day is above or below particular thresholds the algorithm will long or short the stock. If the predicted return is not large in absolute terms, the algorithm will take no action or maybe hold its existing position. The thresholds chosen are +0.5 s and −0.5 s for going long and short, respectively where s is the standard deviation of the daily returns of the training set. This is also where FinBERT comes into play as it overrides the signal from the technical analysis algorithm when it detects a positive or negative news that can greatly affect the stock price.

The observations from the experiment are summarized in Tables 1, 2, and 3 for Dr. Reddy's Laboratories Ltd., Reliance Steel & Aluminum Co. and Wipro Ltd., respectively. Figures 3, 4, and 5 show plots of the stock price of all three, in the same order, along with their predicted prices using the three different models.

From, evaluation it is observed that LSTM is the clear winner when the metric scores are considered. But when the PNL results are considered, the transformer is clearly better at tracking directional changes or rather, the surges in returns that

Table 1 Performance metrics—RDY

Algorithm	RMSE	MAE	MAPE	PNL	Net balance
ARIMA	1.53	1	1.46	7.34	10,73,395.06
LSTM-RNN	1.02	0.78	1.15	2.68	10,26,837.13
Transformer	1.88	1.6	2.34	20.7	12,07,020.40

Table 2 Performance metrics—RS

Algorithm	RMSE	MAE	MAPE	PNL	Net balance
ARIMA	6.5	5.99	3.9	−7.47	9,25,282.04
LSTM-RNN	1.68	1.44	0.93	17.46	11,74,644.06
Transformer	4.07	3.39	2.2	21.26	12,12,597.76

Table 3 Performance metrics—WIT

Algorithm	RMSE	MAE	MAPE	PNL	Net balance
ARIMA	0.21	0.17	2.05	−8.37	9,16,257.77
LSTM-RNN	0.09	0.08	0.91	−3.74	9,62,565.38
Transformer	0.04	0.03	0.31	−3.7	9,63,046.62

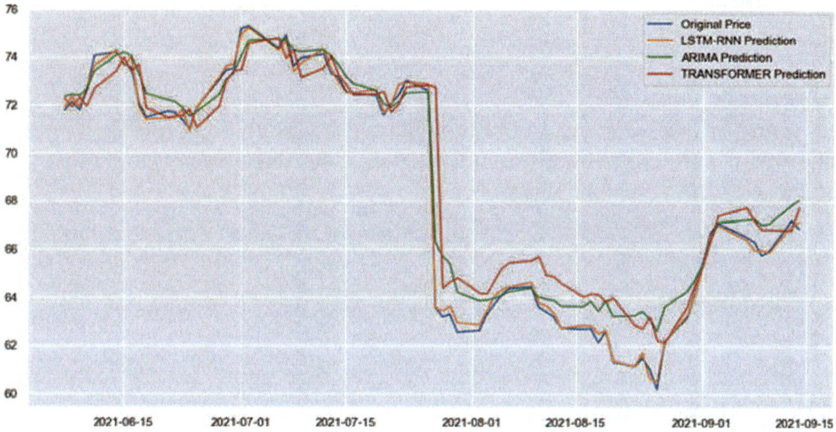

Fig. 3 Stock price prediction of RDY

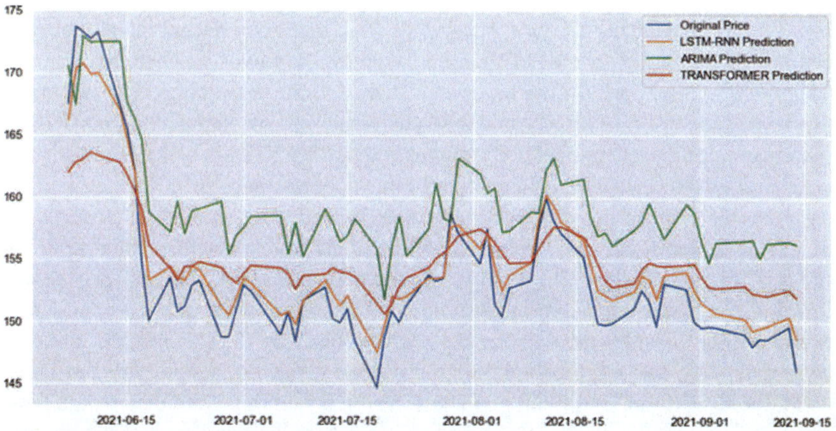

Fig. 4 Stock price prediction of RS

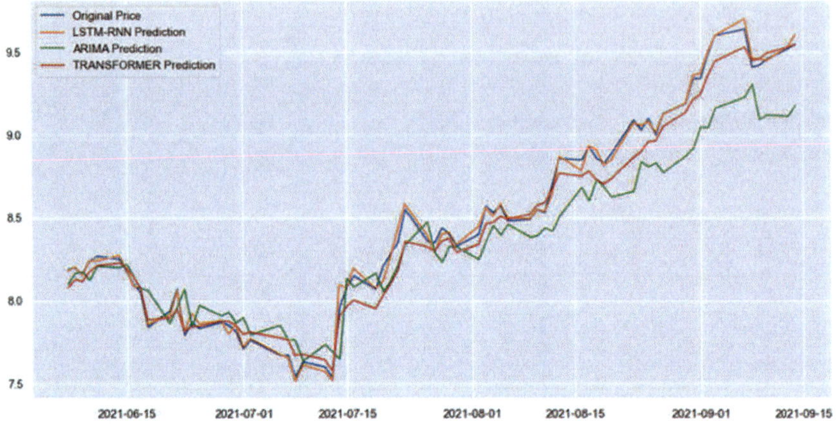

Fig. 5 Stock price prediction of Wipro

are significant and will count toward building a big profit. The FinBERT plugin augments the algorithm by giving information about directional changes that the technical signals were not indicative of.

6 Conclusion

The proposed deep learning architectures employ time series analysis and natural language processing space to forecast stock prices effectively. This work is not just about novel architectures that can revolutionize forecasting and time-series prediction, but also the way each use case and application of AI/ML should be evaluated in terms of the domain of the use case itself. Referring to this specific use case, metrics are definitely an evaluation factor, but it is the PNL that gives deeper insight into the accuracy of the model to predict changes based on the trends and patterns that are of significance and perform the best when looked at the big picture. The inference from the results of LSTM-RNN, and the transformer when compared to the ARIMA benchmark along with the FinBERT plugin is that the LSTM is better at getting the trends to almost replicate the original trend, but it does not react to directional changes as good as the transformer which does not score so well when it comes to closeness to the original stock trend. Any stock prediction model cannot provide greater accuracy without increasing the complexity of prediction operations exponentially. So, the key to optimizing a stock forecasting system is balancing accuracy and computational expense. If the residues from the model are less than half the standard deviation of the returns of the test set, the model shall provide satisfactory results. The FinBERT model scores well on giving out good news verdicts and the effect they would have on the stock price. FinBERT does this by effectively collecting the semantic characteristics of news headlines and properly classifying their sentiment. FinBERT might potentially be taught and utilized for various NLP-related tasks in the financial area, such as document clustering.

References

1. Tan TZ, Quek C, Ng GS (2005) Brain-inspired genetic complementary learning for stock market prediction. In: 2005 IEEE Congress on evolutionary computation. IEEE CEC 2005. Proceeding vol 3, pp 2653–2668. https://doi.org/10.1109/CEC.2005.1555027
2. Hondroyiannis G, Papapetrou E (2001) Macroeconomic influences on the stock market. J Econ Financ 251(25):33–49. https://doi.org/10.1007/BF02759685
3. Gopinathan R, Durai SRS (2019) Stock market and macroeconomic variables: new evidence from India. Financ Innov 51(5):1–17. https://doi.org/10.1186/S40854-019-0145-1
4. Sadeghzadeh K (2018) The effects of microeconomic factors on the stock market: a panel for the stock exchange in Istanbul ARDL analysis. Theor Appl Econ XXV:113–134
5. Menkhoff L (2010) The use of technical analysis by fund managers: international evidence. J Bank Financ 34:2573–2586. https://doi.org/10.1016/J.JBANKFIN.2010.04.014
6. Wafi AS, Hassan H, Mabrouk A (2015) Fundamental analysis models in financial markets—review study. Procedia Econ Financ 30:939–947. https://doi.org/10.1016/S2212-5671(15)01344-1
7. Shen S, Jiang H, Zhang T Stock market forecasting using machine learning algorithms
8. Pellegrini S, Ruiz E, Espasa A (2011) Prediction intervals in conditionally heteroscedastic time series with stochastic components. Int J Forecast 27:308–319
9. Wamkaya B ANN model to predict stock prices at stock exchange markets
10. Selvamuthu D, Kumar V, Mishra A (2019) Indian stock market prediction using artificial neural networks on tick data. Financ Innov 5:1–12. https://doi.org/10.1186/S40854-019-0131-7
11. Hwang H, Oh J (2010) Fuzzy models for predicting time series stock price index. Int J Control Autom Syst 8:702–706. https://doi.org/10.1007/S12555-010-0325-2
12. Madge S, Bhatt S (2015) Predicting stock price direction using support vector machines
13. El E (2016) A new particle swarm optimization based stock market prediction technique. Int J Adv Comput Sci Appl 7. https://doi.org/10.14569/IJACSA.2016.070442
14. Hsu MW, Lessmann S, Sung MC, Ma T, Johnson JEV (2016) Bridging the divide in financial market forecasting: machine learners vs. financial economists. Expert Syst Appl 61:215–234. https://doi.org/10.1016/J.ESWA.2016.05.033
15. Granger CWJ (1969) Investigating causal relations by econometric models and cross-spectral methods. Econometrica 37:424. https://doi.org/10.2307/1912791
16. Mok HM (1993) Causality of interest rate, exchange rate and stock prices at stock market open and close in Hong Kong. Asia Pacific J Manag 102(10):123–143. https://doi.org/10.1007/BF01734274
17. Araújo RDA, Ferreira TAE (2013) A morphological-rank-linear evolutionary method for stock market prediction. Inf Sci (Ny) 237:3–17. https://doi.org/10.1016/J.INS.2009.07.007
18. Schumaker RP, Chen H (2009) Textual analysis of stock market prediction using breaking financial news. ACM Trans Inf Syst 27. https://doi.org/10.1145/1462198.1462204
19. Peng Y, Jiang H (2016) Leverage financial news to predict stock price movements using word embeddings and deep neural networks. In: 2016 Conference of the North American chapter of the association for computational linguistics: human language technologies. NAACL HLT 2016—proceeding conference, pp374–379
20. Li Q, Wang T, Li P, Liu L, Gong Q, Chen Y (2014) The effect of news and public mood on stock movements. Inf Sci (Ny) 278:826–840. https://doi.org/10.1016/J.INS.2014.03.096
21. Ding X, Zhang Y, Liu T, Duan J Deep learning for event-driven stock prediction
22. Yates A, Etzioni O (2007) Unsupervised resolution of objects and relations on the web. 121–130

23. Fader A, Soderland S, Etzioni O Identifying relations for open information extraction. 1535–1545
24. Ding B, Wang Q, Wang B, Guo L Improving knowledge graph embedding using simple constraints
25. Shah D, Isah H, Zulkernine F (2019) Stock market analysis: a review and taxonomy of prediction techniques. Int J Financ Stud 7. https://doi.org/10.3390/IJFS7020026
26. Liu G, Wang X (2019) A numerical-based attention method for stock market prediction with dual information. IEEE Access. 7:7357–7367. https://doi.org/10.1109/ACCESS.2018.2886367
27. Alostad H, Davulcu H (2016) Directional prediction of stock prices using breaking news on twitter. In: Proceeding—2015 IEEE/WIC/ACM international conference on web intelligence and intelligent agent technology WI-IAT 2015, vol 1. pp 523–530. https://doi.org/10.1109/WI-IAT.2015.82

Data Encryption Basing on the Existence of Eulerian Circuits in a Group of Random Graphs

M. P. R. Murthy, G. Lalitha Devi, S. Sarva Lakshmi, and Ch. Suneetha

Abstract In cyber security, data encryption is conversion of data from readable form to coded form so that an intended and authorized person can read the original data. Information security has become an essential tool in the present digital world. Applied graph theory plays a crucial role in encryption technique as it has widespread special features equipped with easy and effective representation. The present paper explores a symmetric encryption technique using Hamiltonian circuits of weighted Eulerian graphs, circular bit shift operation, and simple logical XOR operations.

Keywords Eulerian graph · Eulerian circuit · Circular bit shift · Encryption · Decryption

1 Introduction

Secret communication of sensitive data in a disguise form with protection in a public channel is cryptography. When internet came into picture there started the era of information security. A threat of hacking and stealing the secret information in the transmission medium has become a formidable challenge. So, concealing the secret information illegible to unauthorized persons is most required aspect in communication network. Graph theory, a major part of applied mathematics has a wide role in the history of cryptography. Many complex problems in our day-to-day life can be represented and solved by graph structures as it provides simple and clear pictorial representation. The flexible properties of graph structures give an additional strength to cryptosystems. The history of cryptography shows diverse graph theory applications in the encryption process. The relationship between graph theory and cryptography is very interesting.

M. P. R. Murthy (✉) · G. Lalitha Devi · S. Sarva Lakshmi · Ch. Suneetha
Department of Mathematics, GITAM University, Visakhapatnam, India
e-mail: schivuku@gitam.edu

M. S. Kaiser et al. (eds.), *Information and Communication Technology for Competitive Strategies (ICTCS 2021)*, Lecture Notes in Networks and Systems 401,
https://doi.org/10.1007/978-981-19-0098-3_69

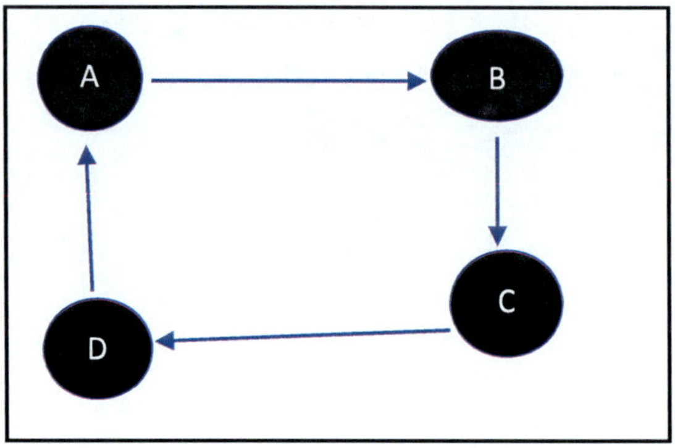

Fig. 1 Euler path 1

1.1 Graph Theory

Euler's Theorem.

A connected graph containing an Eulerian circuit is called the Euler graph [1].

1.2 Eulerian Circuit

A circuit which touches every edge of the graph exactly one and returns to the same vertex is known as Eulerian circuit. If all the vertices of the graph are of even degree, then the graph is called as Euler Graph.

1.3 Euler Path

A graph meeting every edge of the graph and coming back to the same vertex as shown in Fig. 1 is called the Euler Path.

1.4 Euler Theorem

- Euler theorem states that if there are more than two vertices, with an odd degree in a graph it does not have an Euler path.
- A connected graph having 0 or exactly two vertices with odd degree contains at least one Euler path and Euler circuit.

1.5 Obtaining Euler Path in a Graph

There are two different ways to find the existence of the Euler circuit in a graph:

1. Fleury's algorithm
2. Eulerizing a graph

1.6 Fleury's Algorithm

All the vertices of Euler's graph or circuit are of even degree. This method is erasing the edge after meeting it, from any one of the vertices the travel is started, and after finishing the travel at the edge that particular edge is erased until we reach the starting vertex [2].

1.7 Eulerizing a Graph

It is a process of changing a graph so that it contains an Euler circuit. It is in other words constructing postman or newspaper boy friendly roads in a city to save the delivery time. It is a process of identifying and pairing up the vertices with odd degree, keeping average distance as small as possible. All vertices are duplicated along with optimal path.

1.8 Euler Graph Application

Euler graphs have diverse applications in all branches of science and engineering like bioinformatics in CMOS circuit's designs to find the optimal logic gates and also to find bridge problems in the field of navigation to find out the optimal feasible solution. They are also used in Game theory, floor designs, and Chinese postman problem [3].

1.9 Euler's Theorem

A connected planar graph with P vertices and Q edges has F faces where F satisfies the solution.

$$P - Q + F = 2.$$

1.9.1 Euler's Handshaking Theorem

Sum of degrees of vertices of Euler's graph is equal to twice the number of edges of the graph [3].

$$\Sigma d\,[V(G)] = 2|e(G)|$$

1.9.2 Circular Bit Shift

A circular bit shift or bit rotation is the rearrangement of entries of a binary number to get a new binary, frequently used in cryptography to permute a bit sequence. There are two types of circular shifts: left circular shift and right circular shift. In right circular shift, most significant bit (MSB) is replaced by least significant bit (LSB), vice versa in left circular shift. When machine has rotation instructions cryptography, applications get significant speedup [13].

2 Related Work

Graph theory has many applications in cryptography. Graph theory applications are appearing in many branches of cryptography to develop encryption algorithm, key communication technique nowadays. Ustimenko [4] proposed multivariate cryptography using polynomial maps of affine space over a finite commutative ring into itself as encryption tools. Tokareva [5] established the relation between graph theory and cryptography. That paper discusses sparse graphs, Cayley graphs, and bent functions applied to mobile networks. Etaiwi [6] suggested a new complex cipher basing on cyclic graphs, complete graphs, and minimum spanning tree. Etaiwai technique of graph-based encryption has become more popular referred by many researchers. Amudha et al. [7] surveyed many graph theory applications of cryptography. That chapter proposes an encryption algorithm using Euler graphs. Yamuna et al. [8] suggested a double encryption technique using Hamilton path and complete graph. Lu et al. [9] proposed applied graph theory to image cryptography by assigning nodes and edges to images. Charles et al. [10] designed new collision resistant hash functions from expander graphs. Costache et al. [11] studied the security of a proposal for post quantum cryptography using super singular isogeny graphs. Later, Jo [12] introduced cryptographic hash functions based on Charles expander graph in post quantum cryptography. Considering the literature on graph theory applications in cryptography, the present paper explains an innovative encryption technique using weighted Euler graph.

3 Proposed Scheme

In the present scheme, a weighted Euler graph act as a secret key in communication network. Before communication, one of the users publishes a weighted Euler graph with all possible Euler circuit representations or it can be done by a trusted key management authority. An Euler circuit can be drawn in many ways using different vertices of the graph in different directions for instance consider a graph with four vertices ABCD. Euler circuit can be drawn in many ways starting from different vertices of the graph in different directions. For instance, Figs. 2 and 3 are the different Euler circuit representations of the same graph.

Similarly,

BC-CD-DA-AB, BA-AD-DC-CB, CD-DA-AB-BC.

CB-DA-AD-DC, DA-AB-BC-CD, DC-CB-BA-AD.

All the Euler circuit representations are given numbers I, II, III, IV, ...

Fig. 2 Euler path 2

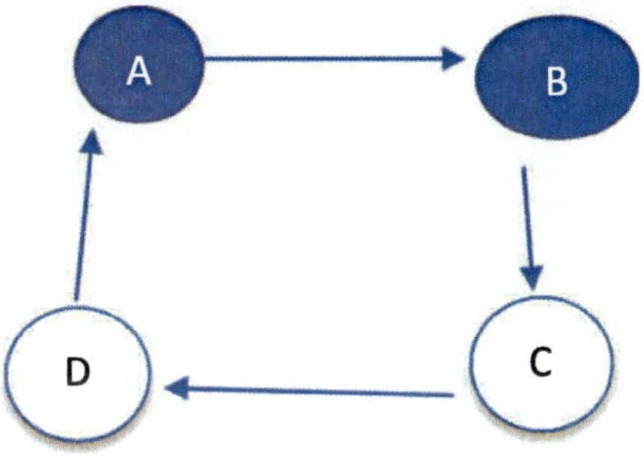

Fig. 3 Euler path 3

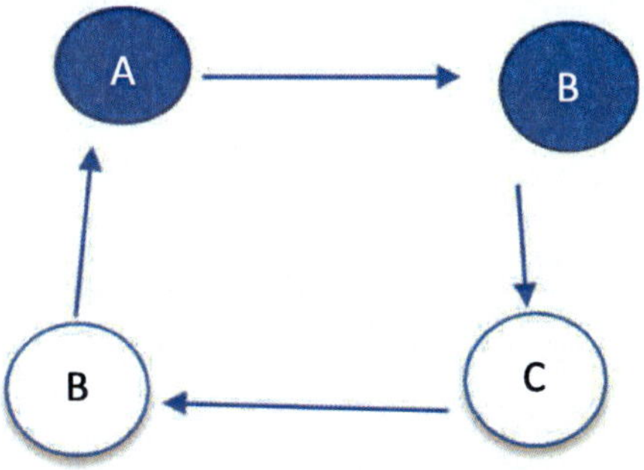

3.1 Encryption

1. The transferring message or plain text is split into plain text blocks of length equal to the number of edges of the agreed up on Euler graph. All these plain text blocks are given numbers 1, 2, 3, 4.......

   The Euler circuit representation I, II, III, IV, ... will act as encryption / decryption keys to the plain text/cipher text blocks 1, 2, 3, 4 ...... consecutively.
2. The character of all the blocks is coded ASCII binary values. The edge weights of Euler circuit representations I, II, III, IV, ... are converted to 8-bit binary numbers. If any edge weights are big numbers more than 256, then the number is reduced to mod 256.
3. The plain text characters of each block are coded to ASCLL binary equivalents

   Each binary number will undergo right circular shift operation, and the number of times equal to degree of the corresponding vertex of the key (Euler circuit representation).
4. If $M1(1)$ $M2$ (1) ....$Mn(1)$ are the characters of the first block

   $V1(I)$ $V2(I)$ ....$Vn(I)$ are degrees of vertices of first Euler circuit representation, the binary equivalents of $M1(1)$ $M2$ (2) ....$Mn(1)$ are right circular shifted p times where p is the degree of vertices $V1(I)$ $V2(I)$ ....$Vn(I)$,

   i.e., binary equivalent of $M1(1)$ is right rotated $V1(I)$ times, $M2$ (1) right rotated, $V2(I)$ times, and so on.

   For instance, if $M1(1) = 10{,}110{,}010$ and $V1(I) = 2$, it is two times right circular shifted.

1	0	1	1	0	0	1	0
0	1	0	1	1	0	0	1
1	0	1	0	1	1	0	0

5. Resulting binary numbers of the above step 3 are arranged in the form of a square matrix with order $n \times n$ where n^2 equals to the block size. Logical XOR operation is applied one each element of the matrix with its neighbouring elements in a special pattern. Each binary entry aij 0f the matrix is XORed with the neighbouring elements as.

   aij XOR ai(j + 1) XOR $a(i{-}1)\,j$ XOR ai(j−1) XOR $a(I + 1)j$ and aij is replaced by resulting new element as shown in Fig. 4.
6. Similar procedure explained in step 3–6 is applied to all the plain text blocks.

The binary numbers are converted to ASCII characters, sent as cipher text to the other party.

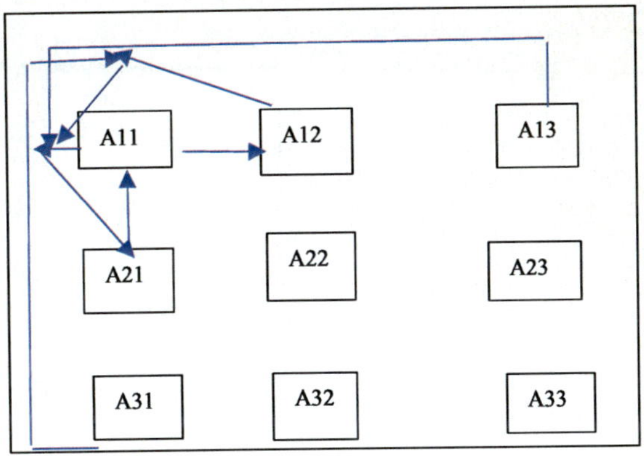

Fig. 4 Logical XOR operation

3.2 Decryption

1. The receiver after getting the cipher text first divides into cipher text blocks. As the present block cipher is symmetric decryption, starts from first block.
2. The ASCII characters are coded to equivalent binary numbers and arranged as a square matrices of order $n \times n$, where n^2 is the block size.
3. Logical XOR operation is applied on each element of the matrix starting from the last element exactly in the opposite pattern of encryption explained in the above step.
4. The output binary numbers after applying logical XOR operation are left circular shifted and the number of times equal to the degree of the vertices of the corresponding key (Euler circuit representation).

1	0	1	1	0	0	1	0
0	1	0	1	1	0	0	1
1	0	1	0	1	1	0	0

Corresponding ASCII code of all the cipher text blocks in original message.

Example If Alice and Bob want to communicate the messages, they agree upon to use an Euler graph with 12 vertices (Fig. 5) and 16 edges.

There are 32 independent Euler circuit representations in the present graph (Fig. 6) with size 16 decimal numbers each like.

AI, IB,BC,CJ,JD,DE,EK,KJ,JI,IL,LK,KF,FG,GL,LH,HA.

BI, IA,AH,HL,LG,GF,FK,KL,LI,IJ,JK,KE,ED,DJ,JC,CB, and so on.

Consider the message.

"G A Y A T R I V I D Y A P A R I S H A D D E G R E E C O L L E G"

It is divided into two blocks **G A Y A T R I V I D Y A P A R I** and **S H A D D E G R E E C O L L E G**. Degrees of vertices of first Euler circuit representation

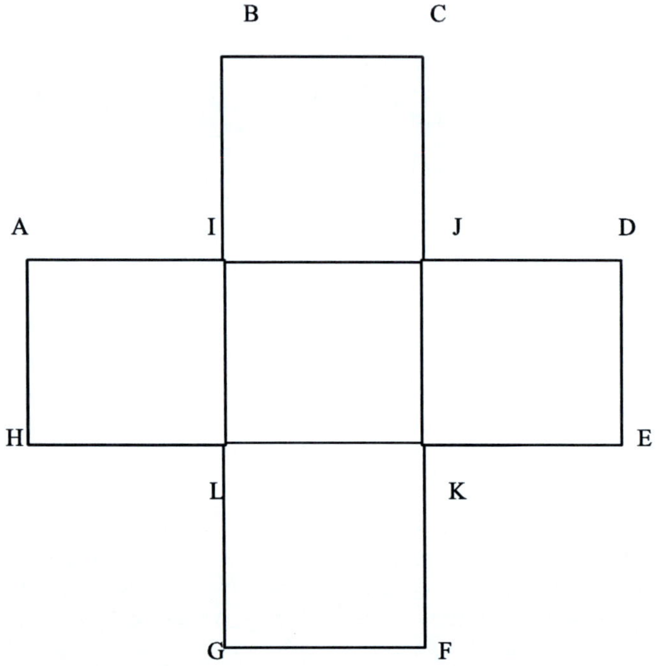

Fig. 5 Example for Euler graph

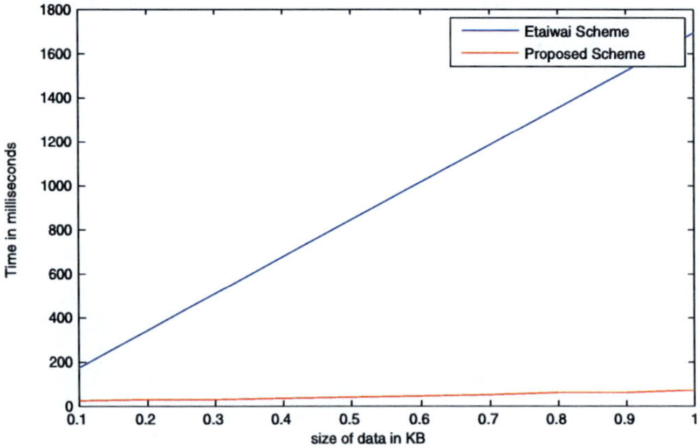

Fig. 6 Execution time graph

are [2 4 2 2 4 2 2 4 4 4 4 4 2 2 4 2]. Binary equivalents of first block characters are undergone right circular shifts equal to the degrees of vertices the outcome 8-bit binary numbers are arranged as matrix arrays. Logical XOR operation is applied on each binary as described in the encryption process to obtain the cipher, $\overline{A}$ FS Y@ M L F Y s Ị J fRS : '$y\hat{S}$. Same procedure is applied for second block also to get the cipher text.

"$\overline{A}$ FS Y@ M L F Y s Ị J fRS : '$y\hat{S}$ FF $s\dot{0}$ 1EscBZspaceBY $\overline{A}\dot{A}\dot{U}$ ENQR8".

4 Performance Analysis

Etaiwi scheme [6] has become more popular graph-based encryption technique being referred by several researchers in the history. This graph-based encryption scheme used minimal span tree and matrix multiplication algorithms. The execution time of Etawai scheme and the proposed encryption technique are implemented for different sizes of data on Intel core i5, 10th generation processor using MATLABR2011, shown in the following Table.1 and Fig. 6.

The above Table 1 and graph in Fig. 6 clearly show that the implementation time is comparatively low in our proposed algorithm. This lowers the computational overhead which is one of the required properties of a good cipher.

Besides the computational overhead, Etaiwai scheme possesses broadcast difficulty also. The cipher text size is too big. A plain text with 10 characters is mapping into 63 characters cipher text. Cipher size is calculated for different sizes of plain text, presented in the following Table 2 and graph in Fig. 7.

Table 2 and graph in Fig. 7 represent that there is abnormal nonlinearity between sizes plain text and cipher text. Even the author has concluded that there may be an

Table 1 Showing encryption bar and decryption time

Size of the data in KB	Encryption/decryption time in milliseconds	
	Etaiwi scheme	Proposed scheme
0.1	169	35
0.2	338	53
0.3	507	78
0.4	674	92
0.5	845	123
0.6	1014	141
0.7	1183	167
0.8	1352	179
0.9	1521	190
1	1691	224

Table 2 Showing plaintext bar and ciphertext characters

Plain text characters	Cipher text characters	
	Etaiwi scheme	Proposed scheme
100	625	100
200	1250	200
300	1875	300
400	2500	400
500	3125	500
600	3750	600
700	4375	700
800	5002	800
900	5625	900
1000	6250	1000

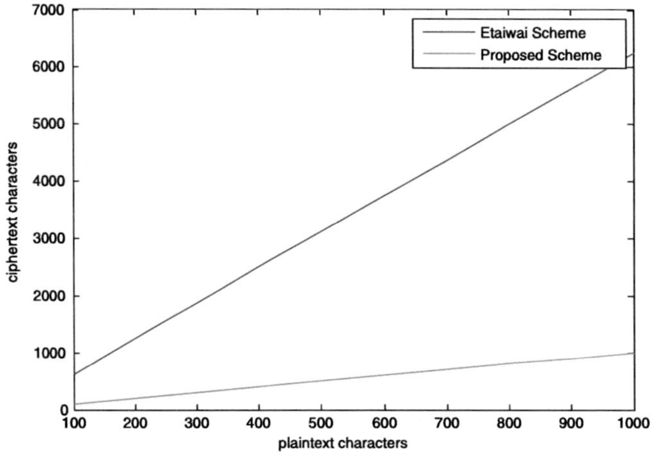

Fig. 7 Graph showing the relation between plain text and cipher text

improvement for his scheme to decrease the size of the cipher text of improving his scheme. One of his suggestions is to divide the whole message into blocks.

5 Security Analysis and Conclusions

Apart from the good performance, the block cipher proposed here achieves best security also. In the proposed scheme, a weighted Euler graph with different independent Euler circuit representations is the secret key between communicating parties.

Sharing of the secret key is either done by one of the users or by any other trusted external secret key provider agency. Cipher block chaining mode technique is applied from the second block. The scheme uses simple logical XOR operation with concatenation of preceding cipher block to the present block key (Euler circuit representation) to generate a new key for present block encryption. The robustness of the cipher depends on one of the required factors of long and strong key. As the key size of each block is the number of edge weights of Euler graph, the length of the key may be enhanced by considering the Euler graph with a greater number of vertices and edges. More the key size more the resistance of the cipher. It is hard against all types of active and passive attacks. As the repeated plain text characters are mapped to different cipher text characters, the present block cipher is free from known plain text attack linear cryptanalysis. In case, if there is same message in two different blocks, it is reflected as different cipher text characters as the encryption is done by using different keys for different blocks. This avoids the chosen cipher text attack differential cryptanalysis. All the independent Euler circuit representations will act as keys for first 32 blocks of plain text. So, a message having $32 \times 16 = 512$ characters can be communicated using one Euler graph as the secret key. If the message has more than 512 characters a new Euler graph may be considered as the secret key. So, the present cipher achieves all the required properties of a good encryption algorithm with add on good features low-computational overhead and less transport risk with same level of security as conventional graph-based cryptosystems.

References

1. Balakrishnan R, Ranganathan K A textbook of graph theory, 2nd edn. https://doi.org/10.1007/978-1-4614-4529-6
2. Deo N (2010) Graph theory with applications to engineering and computer science. Prentice Hall
3. van Steen M (2010) Graph theory and complex networks: an introduction
4. Ustimenko VA (2018) On algebraic graph theory and non-bijective multivariate maps in cryptography. In: Ganzha M, Maciaszek L, Paprzycki M (eds) Proceedings of the 2018 federated conference on computer science and information systems, ACSIS, vol 15, pp 397–405. https://doi.org/10.15439/2018F204
5. Tokareva N (2014) Connection between graph theory and cryptography. G2C2: Graphs and groups, cyclic and coverings. Sep 2014, Novosibirsk Russia
6. Etaiwi WMA (2014) Encryption algorithm using graph theory. J Sci Res Rep 3(19):2519–2527
7. Amudha P, Charles Sagayaraj AC, Shantha Sheela AC (2018) An application of graph theory in cryptography. Int J Pure Appl Math IJPAM 119(3):375–383
8. Yamuna M, Gogia M, Sikka A, Khan MJ (2012) Encryption using graph theory and linear algebra. Int J Comput Appl IJCA
9. Lu S, Manchala D, Ostrovsky R (2008) Visual cryptography on graphs. Cite Seeix COCOON, pp 225–234
10. Charles DX, Lauter KE, Goren EZ (2009) Cryptographic hash functions from expander graphs. J Cryptology 22:93–113
11. Costache A, Feigon B, Lauter K, Massierer M, Puskás A (2018) Ramanujan graphs in cryptography. arXiv:1806. 05709V2 [math.NT] 18 Dec 2018

12. Jo H, Sugiyama S, Yamasaki Y (2021) Ramanujan graphs for post-quantum cryptography. In: Takagi T, Wakayama M, Tanaka K, Kunihiro N, Kimoto K, Ikematsu Y (eds) International symposium on mathematics, quantum theory, and cryptography. Mathematics for Industry, vol 33. Springer, Singapore. https://doi.org/10.1007/978-981-15-5191-8_17
13. Pradhan D, Som S, Rana A (2008) Cryptography encryption technique using circular bit rotation in binary field. In: 2008 International conference on information processing in sensor networks (ipsn 2008). https://doi.org/10.1109/ICRITO48877.2020.9197845

Centralized Library Management System: An E-governance Approach for Improving Accessibility of Library Resources of Bangladesh

Faiyaz Ahmed, Rupok Chowdhury Protik, and Mahady Hasan

Abstract Libraries are one of the highly trusted sources of information. Bangladesh has a large number of educational institutions, and every institution has its own library. Currently, it is very challenging for people in Bangladesh to find and attain resources for their educational and research needs. Therefore, it is vital that we make this process as seamless as possible. Using a centralized library management system (CLMS) is the right step toward that simplicity. The current process of finding and attaining library resources is both time consuming and expensive. Also managing different resources is difficult by government bodies because of decentralization of the libraries. In this paper, we take a look at how other countries have implemented a solution to this problem and propose a system tailored to Bangladesh.

Keywords Centralized library management system · Library management system · E-governance · Digital library management system · Library resource · Computer science · Digital Bangladesh

1 Introduction

E-governance is increasingly getting popular around the world. With the rise of usage of Information and Communication Technology (ICT), governments are actively taking on technological projects to improve the lives of citizens. Similarly, in Bangladesh with the vision of "Digital Bangladesh" [1] many services are being digitized and are being made available to the citizens. The government is motivated to

F. Ahmed (✉) · R. C. Protik · M. Hasan
Department of Computer Science and Engineering, Independent University, Bangladesh, Dhaka, Bangladesh
e-mail: 1910740@iub.edu.bd

R. C. Protik
e-mail: 2022933@iub.edu.bd

M. Hasan
e-mail: mahady@iub.edu.bd

M. S. Kaiser et al. (eds.), *Information and Communication Technology for Competitive Strategies (ICTCS 2021)*, Lecture Notes in Networks and Systems 401,
https://doi.org/10.1007/978-981-19-0098-3_70

use ICT and E-governance to ease the problems currently faced by citizens. However, simply changing the current services and moving them online is not enough. The goal of E-governance is not just creating a digital system for the current offline solution but rather re-engineering the entire process taking the needs of the citizens into account [2].

Getting access to resources from libraries is such a problem the citizens of Bangladesh are currently facing. Libraries are very vital cornerstones of any society. Even today, libraries are one of the most reliable sources of information. Libraries are the only place where people can get access to vast amounts of resources for their education. Usually, libraries have collections of local information which is hard to find in other parts of the country. Looking up and attaining the resources is currently the biggest problem for citizens and students. The current process of attaining a resource is very time consuming and chaotic. Also due to the current pandemic, it is increasingly difficult and risky to venture out to places to attain a specific resource.

Having a centralized system that will maintain a database of every available resource in any library inside the country carries significant importance and helps everyone to search and attain specific resources that are available in any part of the country's different libraries. So, in this paper, we discuss the current problems faced by citizens in finding and attaining library resources and also take a look at other countries or localities that tried to solve this similar problem and propose a centralized E-governance approach to the library management system for tackling this issue in Bangladesh. The proposed solution will not only make it easier for the citizens to get access to the resources but also will find a way to generate revenue for further development and maintenance of the system.

2 Literature Review

Previously, different studies were done in different parts of the world including India, Australia, and Africa. Those studies discussed the implementation of an effective digital library management system. Those studies show that the importance of having a digital library management system knows no bounds. But due to different variables including the incapability of coping up with the rapid changes required, a financial plan for sustaining the managing of the library resources and expansion, and more, the projects were not successful. This paper will propose a solution that will solve the previous problems. It will also identify the way for finding out possible solutions to the problems that might arise during implementing the solution in Bangladesh.

Constructing a unified smart library and correlated issues, for instance: scrutinizing, registering, security, management, tagging, self-servicing, tracking, and identifying users is in concern for the scholars. Nisha [3] constructed an Internet of things (IoT)-based system to manage libraries based on the near-field communication (NFC) technology using the embedded NFC tags on both the books and the user cards. NFC readers can read the tags to control the operation of libraries. Users use their smartphone and handheld reader to verify the book information. On the contrary, Mrunal [4] introduced a radio-frequency identification (RFID)-based system that allows self-service actions to ameliorate the efficacy of libraries. In this case, RFID tags identify books and users to sort out the information and then send it to a PC that hosts a dedicated database. The goal of the design was to discern the benefits of RFID tags in CLMS.

Ahmad [5] analyzed the effectiveness of IoT in RFID technology at a library by engaging interviews and opinions. His research shows that the inventory control by RFID tags was achieved so easily and took less time compared to doing so by hand inventory. Coyle [6] considered RFID management. Hopkinson and Chandrakar [7] proposed RFID at Middlesex University's internal learning system.

Chang [8] delineated an initiative of IoT for Western Michigan University's internal library. He studied that the librarians were required to observe their ecosystems and to employ the IoT applications. They needed physical smart devices like sensor nodes, controllers, and cloud systems. The monitoring and management of libraries were reviewed by Srinivasan and Vanithamani [9]. They designed a book renewal system module for the library. Their system can handle returning books and issues regarding library dues—using GSM mobile technology and RFID.

Chelliaha [10] concentrated on the after-effects of the implementation of RFID for specific topics such as the processes, the technology, and the people, which are associated areas of the library. They investigated a case study of RFID application at the library of the University of Technology Sydney and discovered that the enabling of RFID ameliorated the overall process and efficiency for users. Moreover, they recognized that when RFID is applied in a creative way, it helped to build an effective process and supply chain management.

Lastly, Brian [11] recommended an IoT-based smart and secure library module which specified a process features like retrieving a book from its given location with NFC tags and Wi-Fi positioning system (WPS).

3 Problem Statement

In Bangladesh, if anyone wants to read a book or try to find resources to research on a topic, they are immediately faced with a conundrum of where to look for the resource. The first approach would be to visit the local library and query there to see if the specific resource is available or in case of researching into a topic, they are limited to the resources that are available in that local library.

In case the resources are not available there, they would need to track all libraries in proximity and either physically visit them or call them and query about the resource. There is always the possibility of either the resource not being available there or in case of querying over the phone one may not get a viable answer from the librarians. It would take a fair bit of time and physical stress to track and find the resource.

Apart from the physical and time-consuming impact, there is also a monetary impact associated with accessing resources a person would need. People usually tend to look up the resources in their locality and never tend to invest in the physical and time-consuming stress; hence, they tend to buy the resource online or through local book vendors that is money wasted if you only need the resource for a specific time period.

Students living in the rural areas where libraries are sparse and lack of transportation limit them from traveling to close-by city centers. It is very unlikely they would spend that much money just for gaining access to a resource.

Another problem for governing bodies, such as the Bangladesh National Library, is tracking local resources that are published in the country. It becomes very difficult for them to analyze and take action based on the data since the data is not readily available. For example, Chittagong public library contains handwritten parchments that the government would want to digitize for preservation.

A more centralized library management system can mostly tackle these problems we have identified. So, to summarize the problems we identified in this paper, below are the list of problems we are trying to tackle with the proposed system:

1. Physical and time-consuming difficulties faced by the people trying search for desired resources.
2. The economic impact people have to face when they instead purchase the resource due to local availability.
3. Difficulty of accessibility of resources for students living in rural areas of Bangladesh.
4. Difficulty in keeping track of resources by governing bodies to maintain locally published resources.

4 Proposed Solution

Our system brings all the education system's libraries and public libraries under one roof and provides an easy way for citizens/students to access the library resources. All they need is a national identity card and means to make payment to the system through debit cards, credit cards, or mobile wallets such as bKash [12]. The system will support integration with the courier services such as government postal service and private courier services such as Pathao [13] by integrating them through their public web APIs. In rural areas, books can be collected and submitted from the local postal office. The online platform will have a web and mobile app. The entire system is illustrated in Fig. 1.

As can be seen from Fig. 1, the resources are maintained by the libraries and universities that are under the system. The system will not incur any charge to the businesses that want to integrate themselves and their resources. Hence, they are responsible for adding and updating the resources under their label. Resource information can be uploaded by uploading MARC data by exporting them from their existing systems. All the resources provided by different suppliers will be stored in the library Management backend. Users can query by resources they need; the library management system will show them the list of available resources from different libraries and universities. The resource suppliers will have ratings on them so that it

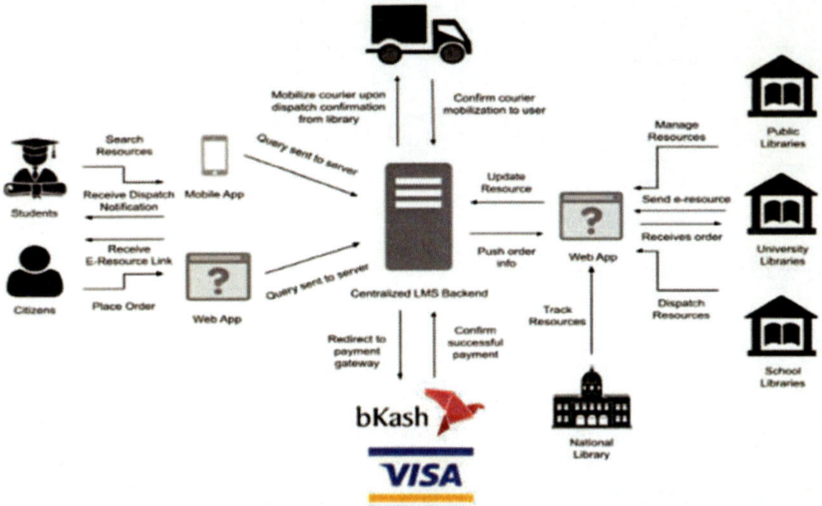

Fig. 1 Rich picture of centralized library management system (CLMS)

is convenient for the users to make a decision on where to get the available resource. Once the user selects the resource for renting and their preferred courier service, they will be redirected to a payment gateway where the user can pay by their preferred payment method. Once the payment is confirmed, an order is issued and the respective resource supplier is notified. Once the supplier confirms the order, the courier is notified to mobilize.

4.1 Planning

To implement the library management system project, we have carried out a planning exercise. The first task was to come up with a list of tasks we need to do in order to make this project successful. The detailed breakdown exercise can be summarized using a WBS diagram, as can be seen from Fig. 2.

After breaking down the work, we came up with a list of tasks and its interdependencies, and with those, we performed a planning exercise and came up with a project estimation the details of which can be found in the technical documentation [14].

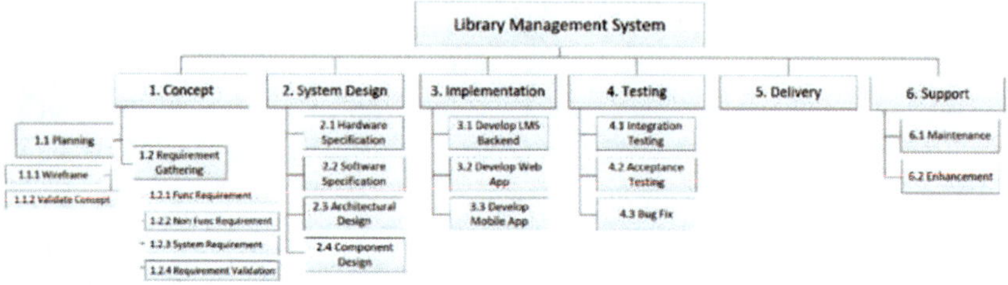

Fig. 2 CLMS work breakdown structure

4.2 Scope

To formulate an idea of how long this project will take, we mapped the above tasks to a Gantt chart as can be seen in the technical documentation [14]. By performing the project scoping, we estimate an approximate project delivery by **28 weeks**.

5 Financial Feasibility

To assess the feasibility factors which can influence the successful implementation of the CLMS project in Bangladesh, we look into the financial and the technical feasibility of the system.

Cost

Before we can determine the benefit that can be monetized from this project, we need to consider the cost of the CLMS project. We limited the scope of the project cost to the human resource required for development. There will be other costs associated with the CLMS project, and we have decided to not include those to limit the scope of our hypothesis. In order to determine the cost, we have considered the following resources, as can be seen in Table 2 in the technical document [14], would be required. We have surveyed the recent Bangladesh job sites and came up with the above human resource cost. We then take this resource and map this to our task list and come up with the total cost of developing the project as can be seen in the technical document [14]. After mapping the resource to the task, we come up with the CLMS project development cost of more than 15 lakh taka. I have also broken the cost based on top tier tasks, as can be seen from Table 1.

Benefit

For our project, our goal is to facilitate the students and citizens of Bangladesh, and hence, we are thinking of charging 50 Tk for this project service; this excludes the actual book rental fee and delivery charge. Hence with a 50 Tk service charge in mind, we perform a cost analysis, and we have considered 1000 transactions per

Table 1 Cost by top tier task

Name	Duration (in days)	Cost (in taka)
Concept	41	332,880
System design	30	111,333
Implementation	190	525,333
Testing	76	112,267
Delivery	6	35,200
Support	135	444,000
	Total Cost	*1,560,933*

Table 2 Cost–benefit analysis (in thousand BDT)

Cash flow description	Year 0	Year 1	Year 2	Year 3	Year 4	Year 5	Year 6
Development cost	1561						
Maintenance cost	0	108	108	108	108	108	108
Discount factor of 12%	1.00	0.88	0.76	0.64	0.52	0.40	0.28
Cumulative time-adjusted costs over lifetime	1561	1656	1738	1807	1863	1907	1937
Per order charge with 10% growth yearly	0	600	660	726	799	878	966
Cumulative time-adjusted benefit over lifetime	0	600	1260	1986	2785	3663	4629
Cost–benefit	−1561	−1056	−478	179	921	1757	2693

month and a 10% growth annually. Hence, the cost–benefit analysis can be seen in Table 2.

As we can see from Table 2 that the cost of the development was considered and along with it, we have considered the maintenance charge of 60,000Tk annually. Hence, we come with a yearly cumulative cost considering a discount factor 12% every year. The early cumulative cost can be calculated by multiplying the discount factor with the maintenance charge and adding to the prior year's cumulative cost.

As we can see from Table 2, the cost of the investment will be attained after approximately 2.5 years after the deployment of the project; if we plot a graph against this data, this can show the result as can be seen from Fig. 3.

6 Technical Feasibility

In technical feasibility, we will try to explore the SWOT and PESTLE analysis of the CLMS.

SWOT Analysis

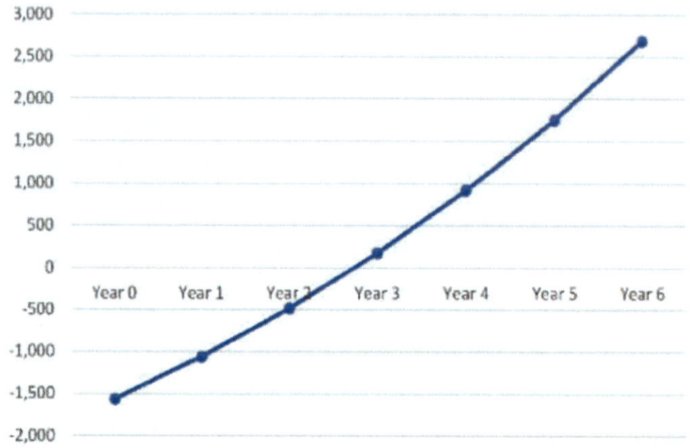

Fig. 3 Payback analysis over a 6-year period

Strength	Weakness
1.Centralized process 2. Easy governance 3. Ease of physical and time-consuming work by people trying to find resources 4. Accessibility for students in rural areas 5. In line with the Government's vision of creating a "Digital Bangladesh"	1.Possible policy changes mandating public and private libraries to integrate with this application 2. Difficult to identify all the challenges such a system would face 3. Lack of government support will weaken the success of the project 4. Reach of physical resources will be dependent on delivery service providers
Opportunities	**Threat**
1.No other system exists with similar solution 2. Good demand for a system like this 3. Bangladesh National Library is looking to implement such a system	1.Staying within the budget may be challenging 2. Maintaining data integrity 3. If not completed with proper quality, the project reputation will be hampered 4 There may be potential competitors who can outbid us in securing the government project

PESTLE Analysis

Political	Our project has a huge dependency on a policy being there for institutions and libraries to mandate the use of this software, without the mandate there might not be enough incentive for institutions to use our application to provide data
Economic	The spending capita of the general people in Bangladesh is at a rise based on the economic capability, and the price charged for acquiring the resource might be affected Bangladesh's GDP per capita is at an exponential rise; hence, it is likely there is a good possibility that this product business will be affected based on the rise and fall of the GDP

(continued)

(continued)

Political	Our project has a huge dependency on a policy being there for institutions and libraries to mandate the use of this software, without the mandate there might not be enough incentive for institutions to use our application to provide data
Social	The social aspect of our project is huge, and this will greatly impact education of the people throughout the nation and also will greatly influence the people that live in remote and rural areas of Bangladesh
Technological	One of the key features of the product will be the centralized repository for the data. Problems like big data and clustering analysis can be performed to gain insight of the topics that people are interested in This project can also give the opportunity to other libraries to analyze the error in their data, and if need be, they can use this project to correct data entry on their local system
Legal	Potential change in legislation can have a direct impact on the product. We need to be aware of potential change in policies that may positively or negatively impact our system
Environmental	Since most of our application logistics are online, the environment has little effect. However, if there are floods in rural parts of Bangladesh, then sending out resources to users will be a challenge

7 Conclusion

In this paper, we highlighted some of the current problems faced by the people in the current process of acquiring a resource. The main objective we had when we planned for the proposed solution was to make the process of finding and acquiring a library resource as seamless as possible so that citizens and students do not have to go through the hurdle they have to go through now or waste a lot of money on buying resources which could have been rented. Another objective was to allow for easy governance of locally published resources for government bodies such as the Bangladesh National Library.

Our solution tackles both problems and proposes a solution to mitigate the problem in theory. The next step of the process would be to develop a proof of concept (POC) of the proposed system and test its usage on a small scale and make adjustments to the system based on our findings. However, there are some assumptions we made in this research, and further studies are needed to check how well those assumptions hold. We strongly believe that the system we have proposed is a step in the right direction.

References

1. Bhuiyan SH (2011) Modernizing Bangladesh public administration through e-governance: benefits and challenges. Gov Inf Q 28(1):54–65
2. Bhanti P, Lehri S, Kumar N (2012) E-Governance: an approach towards the integration of higher education system in India. Int J Emerg Technol Adv Eng Website: www.ijetae.com 2(8), ISSN 2250-2459
3. Nisha P, Karande P, Desai J, Pereira S (2017) Internet of things for library management system. Int J Eng Sci Comput (IJESC) 7(4):10021–10024
4. Mrunal A, Jadhav I, Pachpande SR, Rane P (2014) A paper on RFID in library automation. Int J Adv Electron Commun Syst, Proc Int Conf Model Simul Eng Technol (ICMSET-2014)
5. Ahmad H (2016) RFID technology in libraries: a case study of Allama Iqbal library, university of Kashmir. J Indian Libr Assoc (JILA) 52(4):109–120
6. Coyle K (2005) Management of RFID in libraries. J Acad Librarianship 31(5):486–489
7. Hopkinson A, Chandrakar R (2006) Introducing RFID at Middlesex University learning resources. Program, Electron Libr Inform Syst 46(1):89–97, ISSN: 0033-0337
8. Chang A (2016) Building an internet of things environment in the library. In: The VALA2016 18th Biennial conference and exhibition, Melbourne Convention and Exhibition Centre, Melbourne, Australia from 9–11 Feb 2016
9. Srinivasan S, Vanithamani R (2013) An internet of things approach to library management and monitoring. Int J Res Eng Adv Technol
10. Chelliaha J, Soodb S, Scholfieldc S (2015) Realising the strategic value of RFID in academic libraries: a case study of the University of Technology Sydney. Aust Libr J 64(2):113–127
11. Brian AL, Arockiam L, Malarchelvi PD (2014) An IoT based secured smart library system with NFC based. Int J Emerg Technol Comput Sci Electron (IJETCSE) 11(5), ISSN: 0976-1353
12. bKash: bKash is the fastest and safest medium of financial transaction. https://www.bkash.com/
13. Patho: #1 Platform for all services. https://www.pathao.com/
14. Ahmed F, Protik RC, Ilias Md, Johra FT, Hasan M (2021) Technical paper—centralized library management system. Retrieved from https://docs.google.com/document/d/1P-ARU7K8Oo-6JGE5eS2_seuaP3iFeOpb2oYCyjSSJ9k/edit?usp=sharing

A Review on Single-Phase Integrated Battery Chargers for Electric Vehicles

K. Sreeram, Sreehari Surendran, and P. K. Preetha

Abstract To increase the adoption of electric vehicles (EVs), there is a need for extensive infrastructure. At present, the EV charging infrastructure is at its nascent stage. It is advantageous to make use of the existing 240 V AC supply for EV charging. An EV having an integrated battery charger can utilize this supply. The integrated battery chargers are classified based on the motor used, their features, and construction. This paper reviews the topologies of single-phase integrated chargers proposed for various electric vehicle motors. This paper provides a basic understanding of the need for integrated chargers, the recent developments and provides a base for additional studies.

Keywords Electric vehicle (EV) · On-board battery charger · Integrated chargers · Range anxiety · Charging infrastructure · Motors · Power factor correction

1 Introduction

Global electric vehicle (EV) penetration is increasing each year with a much faster rate of adoption, and India is the next high-potential electric vehicle market [1]. At present, faster adoption is seen in small capacity EVs because of their low cost and their utility. The sizing of the battery mostly decides the cost of an EV. To reduce the cost of an EV, we can reduce the battery's size, but the range will also get reduced. Thereby reduced battery size will cause range anxiety, which is a real problem for consumers deciding to buy an electric vehicle [2]. More availability of EV charging stations or charging points could compensate for the shortcoming of range. Unfortunately, the charging infrastructure lacks in most countries, and this lack of charging infrastructure is the main disadvantage for any emerging markets

K. Sreeram (✉) · S. Surendran · P. K. Preetha
Department of Electrical and Electronics Engineering, Amrita Vishwa Vidyapeetham, Amritapuri, India
e-mail: sreeram@am.amrita.edu

P. K. Preetha
e-mail: preethapk@am.amrita.edu

M. S. Kaiser et al. (eds.), *Information and Communication Technology for Competitive Strategies (ICTCS 2021)*, Lecture Notes in Networks and Systems 401,
https://doi.org/10.1007/978-981-19-0098-3_71

Table 1 Charging level classification

Charger level type	Chargers based on location and input	Power specification	Charging time range
Level Type–1	On-board 1-phase	1.4 kW, 12 A, 120 V AC	4–11 h
		1.9 kW, 20 A, 230 V AC	11–36 h
Level Type–2	On-board 1-phase or 3-phase	4 kW, 17 A, 240 V AC	1–4 h
		19 kW, 20 A, 400 V AC	2–3 h
Level Type–3	Off-board/external 3-phase or DC voltage	50 kW	0.2–0.5 h
		100 kW	

[3]. The utility supply at our homes and office buildings can compensate for the lack of dedicated EV charging infrastructure, but installing off-board chargers (OFBC) or external chargers at each supply socket could defeat the main purpose. This massive distribution of utility supply points could be taken advantage of if our vehicle had a battery charger installed inside the vehicle. The on-board battery charging (OBBC) technology can compensate for the lack of charging infrastructure by creating an opportunity to charge the EV from anywhere with a readily available 240 V AC socket. It has a dedicated rectifier, DC-DC converter, and power factor correction circuits, like the universal OBBC with a wide output voltage range [4]. Table 1 provides the charger level type classification.

The most readily available form of utility supply is the single-phase 240 V, which is available in homes and commercial buildings. A three-phase supply is often available only at the utility connection point of a building. Three-phase supplies in most distribution networks have phase unbalance issues [5], which can affect loads. An unbalanced supply is not desirable for EV charging. Most of the smaller houses have single-phase wiring. It is not the norm to wire three-phase sockets in homes and commercial buildings. In industrial buildings, it is also not readily available unless installed for other specific usages. This aspect makes it having a three-phase integrated charger less desirable for EV applications [6]. Therefore, this paper emphasizes the review of single-phase integrated battery chargers. These chargers can improve their utility by providing a vehicle to grid (V2G) functionality to implement smart grid operations like power quality improvement and load balancing [7]. Out of the motors used in EV. Permanent magnet synchronous motors (PMSMs) are one of the best motors for EV application followed by induction motor (IM), and switched reluctance motor (SRM) is a promising motor for future EVs with good performance and reliability [8].

The first section of this paper provides a basic overview regarding the need for EVs and its future scope. The next section reviews single-phase input integrated battery chargers based on the type of motor used in an EV followed by the conclusion.

2 Integrated Battery Charger

All single-phase charging systems are slow charging systems. Figure 1 shows the classification of various charger configurations for EV. In a simple OBBC, the existing DC-DC converter used inside the drive train of an EV is modified to work in charging mode. In the most straightforward configuration of an OBBC, a single-phase AC input is connected between the motor inverter and the modified DC-DC converter [9], thus making this topology the most cost-effective design. The modified converter even enables the regenerative capability in such systems. However, there is a disadvantage in this design because the input side filter inductor is absent. A rectifier is also needed to rectify the power supply, or else pulsating supply can cause the degradation of the battery, hence reducing its life. The converter can further be modified to provide active filter functionalities [10]. To increase the charging capacity, we need to increase the inductor size, but adding additional inductors increases the size of the system. These types of OBBC are capable of being modified for vehicle to grid (V2G) application. However, the absence of a grid side filter inductor is a demerit to its design. These filter inductors can be added but increase size and weight. Integrated charger uses electric motor parts, such as inductors for filters, to produce an on-board battery charger with minimum additional components.

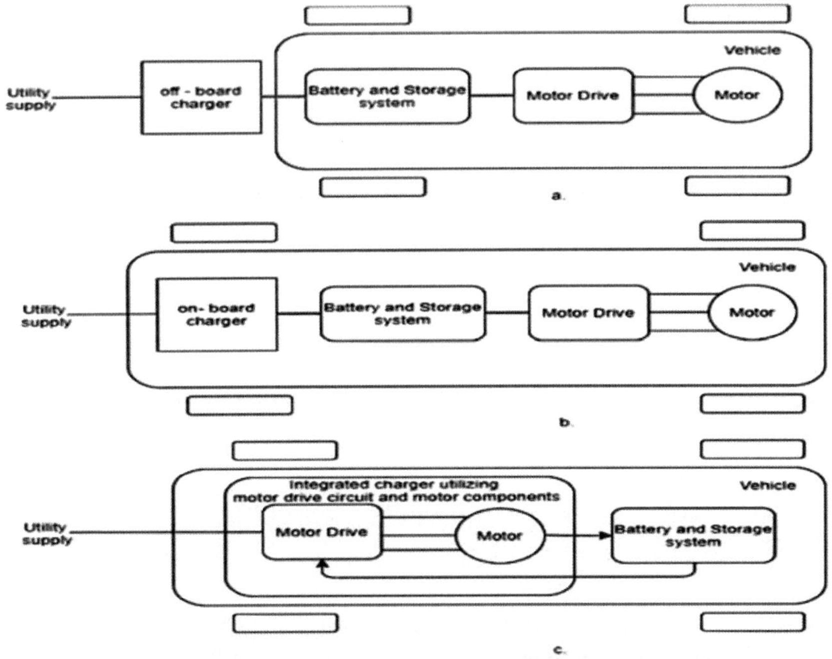

Fig. 1 Types of electric vehicle charger configuration: **a** off-board charger, **b** traditional on-board charger, and **c** integrated charger

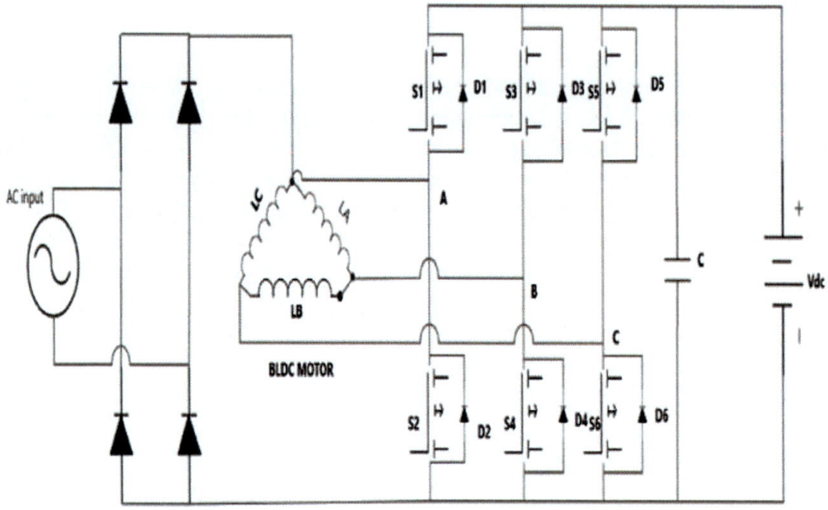

Fig. 2 Charging mode of BLDC-based integrated charger

2.1 Brushless DC Motor (BLDC)-Based Integrated Battery Charger

Brushless DC (BLDC) motor-based integrated charger is proposed in [11]. The converter acts as a three-phase voltage source inverter (VSI) during traction mode and as a front-end boost power factor correction (PFC) during charging, with minimum additional switching components. The integrated charger uses a unified control strategy where it does not need separate control for each stage. Unity power factor and maximum optimized battery charging are ensured with this control strategy. A delta connected BLDC motor-based integrated charger is developed in [12] and shown in Fig. 2. The charger topology used for charging is the two-channel interleaved boost converter topology. The motor windings act as coupled inductors during the charging mode. As the inductance changes with the rotor position, surface-mounted rotor is preferred where the winding inductance value is independent of the rotor position.

2.2 Switched Reluctance Motor (SRM)-Based Integrated Battery Charger

Switched reluctance motor (SRM)-based integrated chargers have their rotor rotate during charging. Figure 3 shows the conventional topology of a SRM drive circuit. SRM-based integrated charger having zero torque control is discussed in [13]. Due to an alternating current flowing through the motor winding, the rotor will rotate,

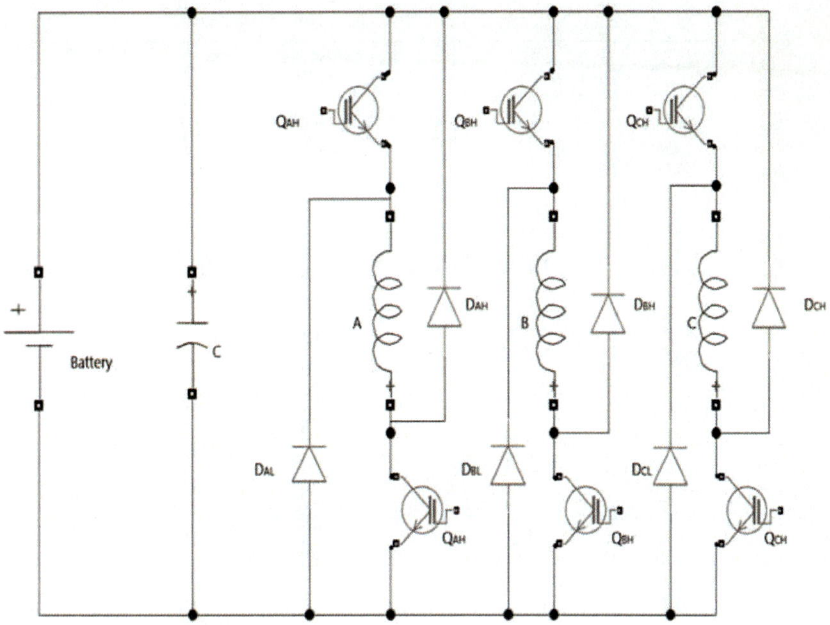

Fig. 3 Conventional SRM motor drive

causing energy loss, and mechanical wear and tear. The zero torque control prevents torque production in the rotor. The rotor will stay still while charging, thereby increasing efficiency. This charger has regenerative capabilities as well. During motoring or regenerative mode, this converter works as an asymmetrical converter. During charging, the converter works as a buck-boost converter in which the phase windings act as inductors.

A modified miller converter-based SRM drive circuit controlled by digital closed-loop control is given in [14]. This converter can operate as a buck-boost converter and control both charging and discharging of the battery. The battery also has a high-capacity capacitor connected in parallel to act as a low-pass filter for the battery charging input. In any SRM-based integrated charger, the motor winding can be used as an inductor for the power converter, reducing cost, size, and weight.

A center-tapped three-phase 6/4 pole SRM-based integrated charger is designed in [15], as shown in Fig. 4. The AC supply is given between the center-tapped points of the two phases of the motor winding with an additional relay. The topology can be configured as a full-bridge converter during charging and as a buck-boost circuit with power factor correction (PFC) capability. The PFC improves the power converter's filtering capability and efficiency, and the charging voltage is controlled using a hysteresis closed-loop control. The power converter can also operate at a wide range of battery voltage.

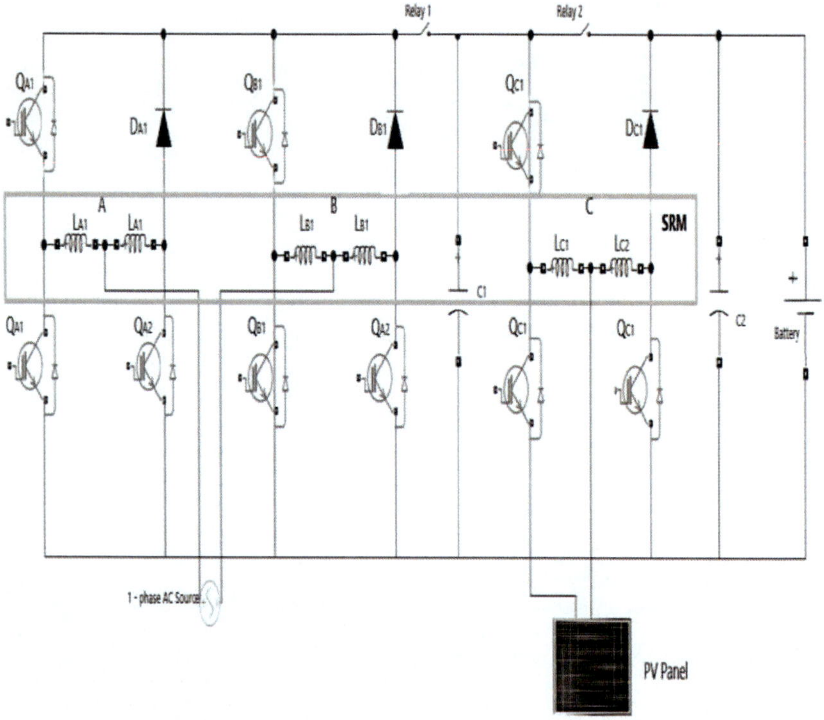

Fig. 4 SRM integrated charger topology with solar panel

Solar panel integration [16] in electric vehicles along with maximum power point tracking (MPPT) control can harness additional energy to reduce range anxiety [17] as shown in Fig. 4. The converter acts as an asymmetrical half-bridge converter during the driving mode. During charging mode, the converter will act as a bridgeless interleaved boost converter. The buck-boost converter is connected in series with the converter and has PFC functionality. The split-phase winding of the SRM motor will act as the input inductance for the bridgeless interleaved converter. During the charging mode, the parallel boost circuit will ensure that the inductor current ripples balance out each other, thus reducing total harmonic distortions and improving the power factor.

The SRM windings can be modified into a two-phase bridgeless PFC topology for charging mode, as shown in Fig. 5 [18]. The construction is much simpler and does not require extra components. A front-end buck converter is added to the asymmetrical half-bridge converter for the driving mode. The charging is done under a constant average current charging method, controlled by a double current loop control algorithm. It has regenerative charging capabilities, as well as proper input AC supply rectification. However, it is applicable only for high-voltage batteries

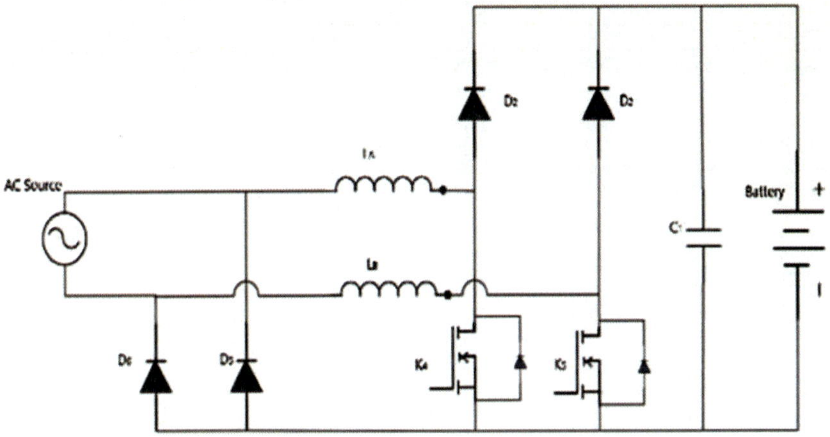

Fig. 5 SRM integrated charging mode topology as a 2-phase bridgeless PFC

because the boost converter will boost the input voltage level greater than the input AC voltage level.

2.3 Induction Motor (IM)-Based Integrated Battery Charger

The induction motor-based integrated battery charger utilizes a single-phase supply as input. A three-phase supply can also be given as input, but since a three-phase supply flows through the three-phase motor winding, it will produce a torque, causing the motor to rotate. The rotation can be prevented by using obstructing parts or mechanisms, but this will cause loss due to the torque still being produced in the motor. This problem of motor rotation is prevented in multi-phase motors, where there is zero torque production during charging. The IM-based integrated charger could be further classified based on the single or multiple motors used in the EV based on the traction drive configuration of the vehicle [19].

2.3.1 Single Motor-Based Integrated Charger

The single-phase AC supply is connected between the neutral point of the three-phase motor and the middle of the other leg that is added for the working of integrated charging [20]. Figure 6 represents the construction of that integrated battery charger based on the three-phase IM. The traction components, i.e., the motor windings and the motor inverter, will act as input filter inductors, rectifiers, and DC-DC converter for charging the battery. The motor will also not rotate during charging since the

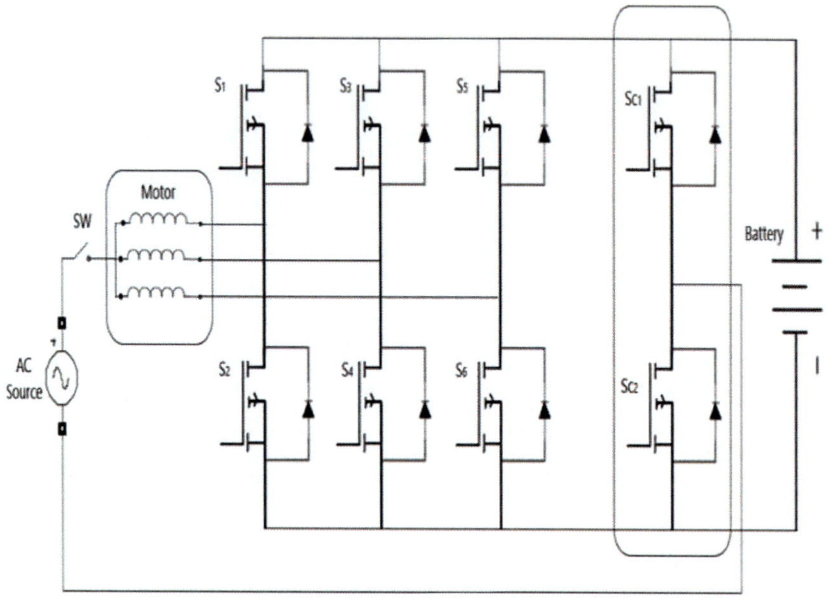

Fig. 6 Induction motor-based integrated charger

current flowing in each phase windings is equal in magnitude and of the same phase, and the converter switching will also ensure that it stays that way.

2.3.2 Dual Motor-Based Integrated Charger

In vehicles with two motors in their traction system, primarily in higher performance vehicles, the integrated charger shown in Fig. 7 is realized [21]. This two-motor design is based on an equivalent circuit of a motor having two three-phase PFC connected to the split-phase windings. The single-phase AC input is connected across the neutral points of the three-phase motor. The advantage of this setup is that it does not need additional switch legs like the single motor application, making the integrated charger's implementation more reliable, with lesser component modification and lower cost of construction. In addition, the motor does not rotate while charging since no torque is produced. This charger topology can also be applied for V2G application through unique control of the inverter switches and three-phase supply input.

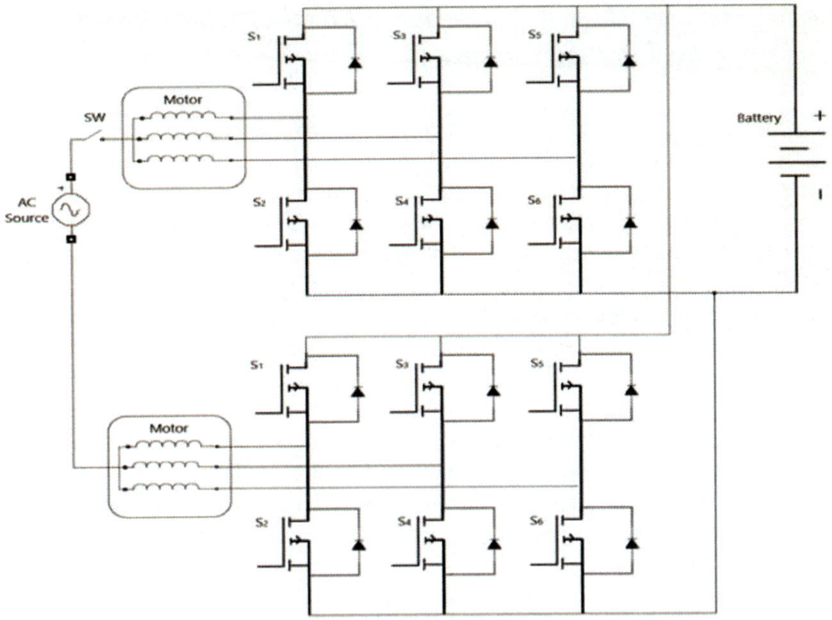

Fig. 7 Dual motor-based integrated charger

2.3.3 Four Motor/All-Wheel Drive-Based Integrated Charger

These four motors in the EV have their separate inverter driving them [22]. The AC supply is connected across the neutral points of the first two motors. The neutral point of the remaining two motors is connected to the battery. The windings of these two motors will act as the filter inductors. The inverters of the first two motors will act as the rectifier for the battery charger system, also shown in the 2-motor integrated battery charger design. The setup of the 4-motor integrated charger is shown in Fig. 8.

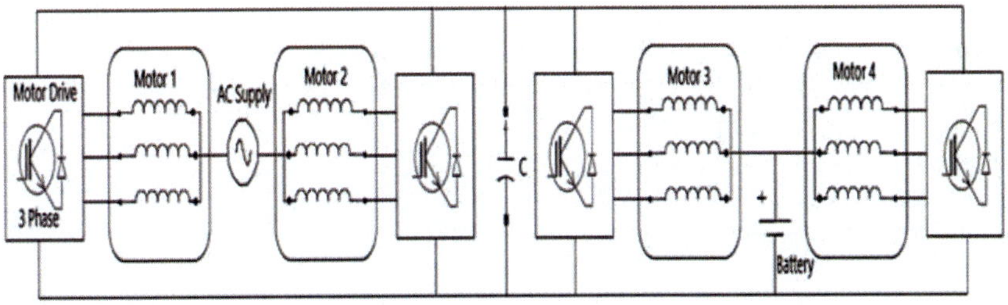

Fig. 8 Four IM integrated charger

2.4 Permanent Magnet Synchronous Motor (PMSM)-Based Integrated Battery Charger

Permanent magnet-based machines like PMSM-based integrated battery chargers are mostly multi-phase-based power drives. The three-phase supply will cause torque in the motor, which will rotate the rotor. Even though the rotation can be blocked mechanically or run at synchronous speed with a clutch-based disconnection to the drivetrain, it is still an energy loss. A two-pole three-phase PMSM with a double stator, whose windings are shifted 120 degrees apart, is used. During charging, the motor will run at synchronous speed, and the AC connected split stator windings will induce a voltage on the other winding of the inverter part. Before connecting the grid supply to the integrated charger, the battery and inverter voltages must synchronize to the grid voltage. After the successful synchronization, the grid side winding is connected to the grid. Another problem of this three-phase supply is that the battery voltage must be higher than the grid voltage because the integrated charger is in boost mode. So, vehicles having a low-voltage battery system will require a DC-DC converter to step down the voltage, or integrated charging technique cannot be applied to such vehicles [23].

A single-phase input-based integrated charger with a 5-leg inverter and 2 PMSM motors, the main motor, and an auxiliary motor is implemented in [24]. This charger has V2G functionality also. Some integrated charger topologies based on other motor topologies have also been used in PMSM-based integrated chargers. The IM-based integrated charger topology for 2-motor, as shown in Fig. 7, is applicable for 2-PMSM-based integrated charger.

Hybrid excitation flux switching (HEFS) machine is basically stator PM motors, and the drive components are divided into four parts [25]. The six-phase inverter takes care of the power delivery to the field windings and the armature windings. The phase of the AC supply is connected in parallel with the first pair of two alternative phase winding, and the neutral is connected in parallel with the other two alternative phase windings. The charging segment of the power drive is divided into two stages, as shown in Fig. 9. The first stage comprising the armature windings and the four inverter legs will act as the AC to DC converter with PFC capability, and the second stage comprising the field winding and the 6th inverter leg will act as the DC output controlling stage. Stage 1 will ensure the voltage matching with the battery. This topology does not require extra power components, has a low-torque production problem during charging, requires single-phase charging only, and can match charger output with most EV battery design voltages.

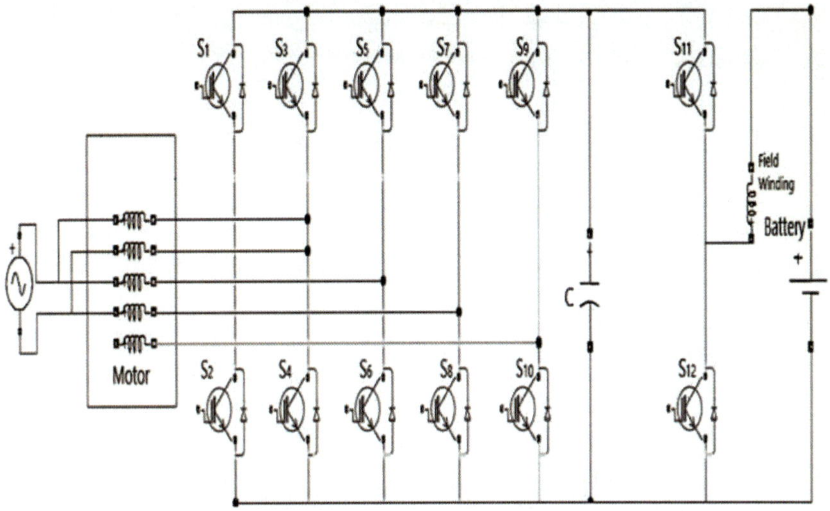

Fig. 9 Five-phase HEFS motor in charging mode

2.5 *Interior Permanent Magnet (IPM) Motor-Based Integrated Battery Charger*

Interior permanent magnet (IPM)-based traction motors are high-speed permanent magnet motors with permanent magnets in their rotor structure. The IPM has a similar electromagnetic performance to a surface-mounted permanent magnet (SPM) with a lower material cost. A typical IPM-based single-phase input integrated charger is shown in Fig. 10 with PFC capability [26]. The power factor remains at unity, and there is no need for extra filters as the application of phase interleaving will suppress the PWM ripples. The phase windings of the three-phase motor will act as a coupling inductor in the three-phase boost converter.

An isolated high-power special IPM-based bi-directional integrated charger is discussed in [27]. It uses a motor-generator concept where the motor reaches a synchronous speed before connecting the drive system to the grid to start the charging mode. A separate clutch is used to disconnect the motor from the mechanical drive train, but the rotation will still cause the motor's energy loss and mechanical degradation. The need for a different clutch system will increase the mechanical complexity.

An integrated charger based on IPM with safe charging operation and ripple suppression is given in [28]. The AC supply is directly connected in parallel to the leg and not through the 3-phase winding of the IPM. A rectifier is needed at the input because the supply is not connected to the input filter inductors. Two switches are added to change the mode from traction mode of operation to integrated battery charging operation. The rectifier used for this charger can take both single-phase

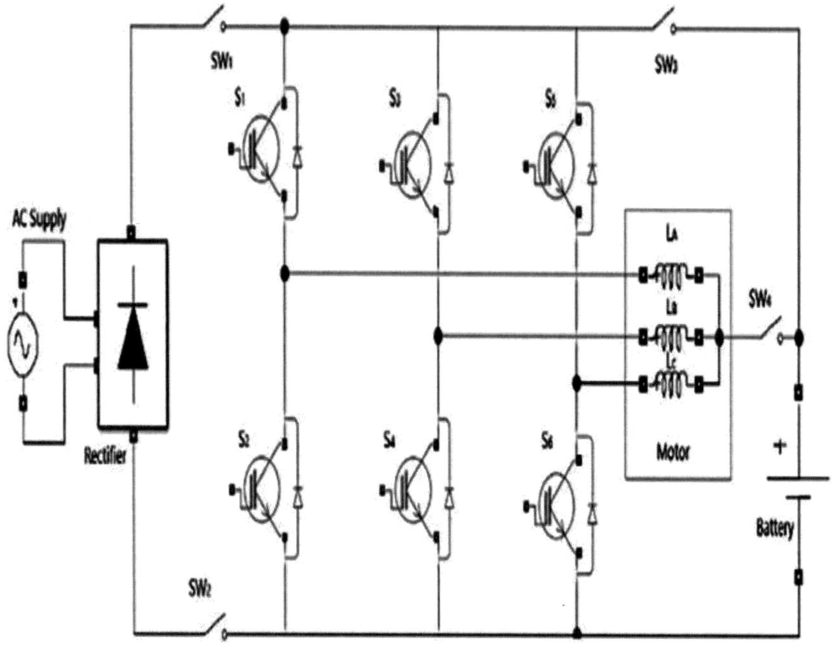

Fig. 10 IPM motor drive with integrated charging

input and three-phase input and has PFC capabilities. The charger controller ensures balanced current flow and prevents rotor operation. IPM motor-based integrated chargers can also be used with V2G functionality, with the motor stator windings acting as the filter inductors. It uses the variable frequency switching method to increase converter efficiency and reduce total harmonics distortions.

3 Conclusion

Integrated charging is important for the initial EV adoption. It will substitute for the lack of charging infrastructures. However, integrated charging is slow and cannot be a proper substitute for the performance and long-range capable passenger vehicles. Another hindrance to EV adoption is the economic problems associated with owning an EV. Larger EVs are very costly compared to conventional vehicles. Thus, for the initial period, people will substitute smaller EVs for their smaller conventional vehicles. Hence for smaller vehicles, this integrated charging technology will be the main driving force for EV adoption. Thus, smaller EVs will be more attractive because they can be charged anywhere, and for small EVs, this slow charging is as convenient as fast charging since their batteries have a much smaller capacity. Another problem that could arise is billing the socket used for charging, but using a specifically built Internet of things (IoT) smart meter inside the vehicle will solve this billing issue. In future, regular roofing sheets will be replaced by more budget-friendly PV panel embedded roofing sheets. Also, having a direct DC input feature into the existing integrated charger can enable direct input from solar panels installed on car parks to charge the battery. This paper provides a brief review of integrated chargers, with single-phase input, for EVs. With more research, integrated chargers have a vast scope as a substitute for conventional chargers, and it will attract more consumers to EV adoption.

References

1. Sreeram K, Preetha PK, Poornachandran P (2019) Electric vehicle scenario in India: roadmap, challenges and opportunities. In: 2019 IEEE international conference on electrical, computer and communication technologies (ICECCT), pp 1–7. https://doi.org/10.1109/ICECCT.2019.8869479
2. Breschi V, Tanelli M, Ravazzi C, Strada S, Dabbene F (2020) Social network analysis of electric vehicles adoption: a data-based approach. In: 2020 IEEE international conference on human-machine systems (ICHMS), pp 1–4. https://doi.org/10.1109/ICHMS49158.2020.9209373
3. Hosseinpour S, Chen H, Tang H (2015) Barriers to the wide adoption of electric vehicles: a literature review based discussion. In: 2015 Portland international conference on management of engineering and technology (PICMET), pp 2329–2336. https://doi.org/10.1109/PICMET.2015.7273259
4. Praneeth AVJS, Vincent D, Williamson SS (2019) An universal on-board battery charger with wide output voltage range for electric transportation. In: 2019 IEEE energy conversion congress and exposition (ECCE), pp 1159–1165. https://doi.org/10.1109/ECCE.2019.8912749
5. Taiwo OP, Tiako R, Davidson IE (2017) Investigation of voltage unbalance in low voltage electric power distribution network under steady state mode. In: 2017 IEEE 3rd international conference on electro-technology for national development (NIGERCON), pp 932–939. https://doi.org/10.1109/NIGERCON.2017.8281959
6. Yilmaz M, Krein PT (2013) Review of battery charger topologies, charging power levels, and infrastructure for plug-in electric and hybrid vehicles. IEEE Trans Power Electron 28(5):2151–2169. https://doi.org/10.1109/TPEL.2012.2212917
7. Turker H (2016) Review of electric motors for grid connected integrated battery chargers in electric vehicle applications. In: 2016 IEEE international conference on renewable energy research and applications (ICRERA), Birmingham, pp 1029–1033. https://doi.org/10.1109/ICRERA.2016.7884490
8. Sidharthan VP, Lingam AS, Vijith K (2015) Brushless DC motor driven plug in electric vehicle. Int J Appl Eng Res 10:3420–3424
9. Raveendran V, Krishnan NN, Nair MG (2017) Vehicle-to-grid support by electric vehicle charging stations operated at airports and metro rail stations. In: 2017 International conference on technological advancements in power and energy (TAP Energy), pp 1–6. https://doi.org/10.1109/TAPENERGY.2017.8397241
10. Solero L (2001) Nonconventional on-board charger for electric vehicle propulsion batteries. IEEE Trans Veh Technol 50(1):144–149. https://doi.org/10.1109/25.917904
11. Meher SR, Banerjee S, Vankayalapati BT, Singh RK (2020) A reconfigurable on-board power converter for electric vehicle with reduced switch count. IEEE Trans Veh Technol 69(4):3760–3772. https://doi.org/10.1109/PIICON49524.2020.9112928

12. Singh V, Tummuru NR, Rajpurohit BS (2020) Integrated battery charger with delta connected BLDC motor for electric vehicle applications. In: 2020 IEEE 9th power India international conference (PIICON), pp 1–6. https://doi.org/10.1109/PIICON49524.2020.9112928
13. Huseini SRK, Farjah E, Tashakor N, Ghanbari T (2015) Development of an integrated switched-reluctance motor drive with battery charging capability for electric vehicle propulsion system. In: 6th Power electronics, drive systems & technologies conference (PEDSTC2015), pp 579–584. https://doi.org/10.1109/PEDSTC.2015.7093339
14. Liang J, Xu G, Jian L, Lee D, Ahn J (2011) Zero-torque control of switched reluctance motor drive system with on-board charger. In: 2011 International conference on electrical machines and systems, pp 1–6. https://doi.org/10.1109/ICEMS.2011.6073540
15. Feng C, Wu J, Sun Q, Wu H, Zhang L (2019) An integrated BLIL boost converter-based switched reluctance motor drive for PEV applications with PFC charging function. In: 2019 22nd International conference on electrical machines and systems (ICEMS), pp 1–5. https://doi.org/10.1109/ICEMS.2019.8921510
16. Sreelekshmi RS, Anusree R, Raveendran V, Nair MG (2018) Solar fed hybrid energy storage system in an electric vehicle. In: 2018 9th International conference on computing, communication and networking technologies (ICCCNT), pp 1–7. https://doi.org/10.1109/ICCCNT.2018.8493846
17. Pillai PT, Sreelekshmi RS, Nair MG (2018) A novel MPP tracker control for a solar fed energy storage system in electric vehicle. In: 2018 9th International conference on computing, communication and networking technologies (ICCCNT), pp 1–6. https://doi.org/10.1109/ICCCNT.2018.8494125
18. Liang J, Xu G, Jian L, Chang M (2012) Analysis parking position of integrated switched reluctance motor drive system with on-board charger for EV. In: 2012 IEEE vehicle power and propulsion conference, pp 72–77. https://doi.org/10.1109/VPPC.2012.6422574
19. Sharma S, Aware MV, Bhowate A (2020) Integrated battery charger for EV by using three-phase induction motor stator windings as filter. IEEE Trans Transp Electrification 6(1):83–94. https://doi.org/10.1109/TTE.2020.2972765
20. Sharma S, Aware M, Tatte Y, Pandit JK, Bhowate A (2016) A split three phase induction motor for battery charging application. In: Proceedings of IEEE international conference on power electronics drives energy systems (PEDES), pp 1–6. https://doi.org/10.1109/PEDES.2016.7914515
21. Pellegrino G, Armando E, Guglielmi P (2010) An integral battery charger with power factor correction for electric scooter. IEEE Trans Power Electron 25(3):751–759. https://doi.org/10.1109/IEMDC.2009.5075276
22. Sul S-K, Lee S-J (1995) An integral battery charger for four-wheel drive electric vehicle. IEEE Trans Ind Appl 31(5):1096–1099. https://doi.org/10.1109/28.464524
23. Gao J, Jiang D, Li Q, Sun W, Qu R (2018) Current control of integrated bi-directional converter for V2G based on open winding interior PMSM. In: 2018 IEEE energy conversion congress and exposition (ECCE), pp 6703–6709. https://doi.org/10.1109/ECCE.2018.8558130
24. Haghbin S, Lundmark S, Carlson O, Alaküla M (2011) A combined motor drive battery charger based on a split-windings PMSM. In: 2011 IEEE vehicle power and propulsion conference, pp 1–6. https://doi.org/10.1109/VPPC.2010.5729071
25. Tong M, Cheng M, Hua W, Ding S (2020) A Single-phase on-board two-stage integrated battery charger for EVs based on a five-phase hybrid-excitation flux-switching machine. IEEE Trans Veh Technol 69(4):3793–3804. https://doi.org/10.1109/TVT.2020.2974019
26. Xiao Y, Liu C, Yu F (2019) An integrated on-board EV charger with safe charging operation for three-phase IPM motor. IEEE Trans Ind Electron 66(10):7551–7560. https://doi.org/10.1109/TIE.2018.2880712
27. Haghbin S et al (2010) An integrated charger for plug-in hybrid electric vehicles based on a special interior permanent magnet motor. In: 2010 IEEE vehicle power and propulsion conference, pp 1–6. https://doi.org/10.1109/VPPC.2010.5729071

28. Mukundan S, Dhulipati H, Lai C, Mukherjee K, Tjong J, Kar NC (2018) Design and optimization of traction IPMSM with asymmetrical damper bars for integrated charging capability using evolutionary algorithm. IEEE Trans Energy Convers 33(4):2060–2069. https://doi.org/10.1109/TEC.2018.2877104

Investigating Gammatone Filterbank-Based i-Vectors for Speaker Verification Task

Asmita Nirmal and Deepak Jayaswal

Abstract Most of the recent studies based on i-vector framework use conventional mel-frequency cepstral coefficient (MFCC) features. However, the main issue that is still faced with MFCC features is its sensitivity to noisy conditions. The verification performance degrades significantly in noisy conditions. In this paper, we replace the most commonly used MFCC features with new features known as gammatone cepstral coefficients (GFCCs) features. These features are built on gammatone filter bank. The speech data used in the experimentation are taken from TIMIT dataset. Our results show that combination of proposed GFCC features and conventional MFCC features in i-vector framework outperforms individual features not only for representing speaker specific information but also for making our verification system robust in noisy environment. Our experimental evaluations show that GFCC features outperforms the MFCC features with an improvement in equal error rate (EER) by 1% in matched noise conditions and 0.45% in mismatched noise conditions.

Keywords MFCC · GFCC · i-vector · Speaker verification · SNR · TIMIT

1 Introduction

Speech signal is the most natural way of communicating for the human beings. Generally, speaker's physiological and behavioral characteristics are encoded in speech signal [1]. The objective of a speaker verification (SV) system is to extract the unique voiceprint to authenticate the claimed identity of the speaker. Currently, the SV systems have been playing a crucial role in many security applications, such as: call centers, forensic verification, and banking services. The main challenge of SV systems is that it hinders precise verification in noisy conditions.

A. Nirmal (✉)
Datta Meghe College of Engineering, Navi Mumbai, India
e-mail: asmita.nirmal@dmce.ac.in

D. Jayaswal
St. Francis Institute of Technology, Mumbai, India
e-mail: djjayaswal@sfit.ac.in

M. S. Kaiser et al. (eds.), *Information and Communication Technology for Competitive Strategies (ICTCS 2021)*, Lecture Notes in Networks and Systems 401,
https://doi.org/10.1007/978-981-19-0098-3_72

Over the last few decades, many features have been developed in SV domain, of which mel-frequency cepstral coefficient (MFCC) features have been the most common one [2, 3]. MFCC features represent the speaker-specific information using the speech spectrogram [4]. They are effective for verification systems in clean speech utterances [5]. However, if these features are used for characterizing noisy speech utterances, verification performance drops substantially. Therefore, a more precise human auditory system representation built on gammatone filter bank is suggested in [6]. The new set of features extracted from this filter bank is known as gammatone frequency cepstral coefficients (GFCCs) features.

In this paper, we have implemented text-independent SV system using MFCC, GFCC features, and their combinations in the more recently used i-vector framework [7]. The i-vectors are computed using the factor analysis concept. The i-vectors use a single total variability matrix to characterize both intra-speaker variability and inter-speaker variabilities that are embedded in speech utterance. Our experiments are conducted on Texas Instruments and Massachusetts Institute of Technology (TIMIT) database [8] and noisy version of TIMIT database which is created using noise samples collected from AURORA database [9].

The remaining part of this paper is divided into following different sections. Section 2 gives the theoretical details of features that we have used in implementing our SV system. Section 3 provides experimental setup used for implementing both baseline and proposed feature extraction methods along with the database formation. Section 4 provides result analysis, and lastly, conclusions about our work are made in Sect. 5.

2 Theoretical Background

The implementation framework of baseline and proposed feature extraction is shown in Fig. 1. In this section, we give detailed background of MFCC and GFCC features followed by i-vector extraction based on MFCC and GFCC features.

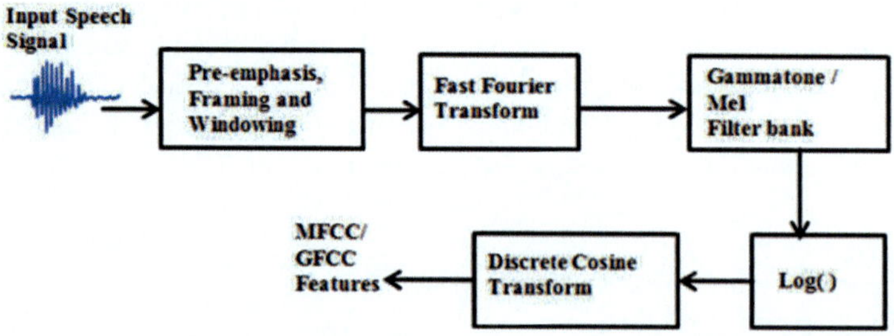

Fig. 1 Procedure to extract MFCC/GFCC features

2.1 Mel Filter Bank Cepstral Coefficient Feature Extraction

Our baseline system is implemented using conventional MFCC features. They are representative of spectral envelope of a vocal tract and primarily motivated from the source filter modeling approach of speech signal [2, 3].

The very first preprocessing step in MFCC feature extraction is pre-emphasis. It aims to emphasize the amplitude of high frequencies that are ignored during speech production. Speech signals have time varying characteristics. Therefore, after pre-emphasis, the input speech signal is segmented into short frames of length varying from 20 to 30 ms so that dynamic characteristics of speech signal are taken into consideration. After framing, hamming window is applied to minimize the spectral leakage. After that spectrum of each frame is calculated using fast Fourier transform (FFT) algorithm. Then, spectral warping is done using mel scale. As a final stage, cepstral coefficients are computed by applying discrete cosine transform (DCT) to the log magnitude of speech spectrum.

2.2 Gammatone Filter Bank Cepstral Coefficient Feature Extraction

In this paper, we replace conventional mel scale-based filter bank by gammatone filter bank [6]. From an implementation point of view, GFCC feature extraction steps are similar to that of MFCCs. To elaborate on this, GFCC simulates the human auditory system with a set of filters that are linearly set apart on the equivalent rectangular bandwidth (ERB) scale rather than mel scale. The relation between ERB scale and center frequency, f_c is given by the equation

$$\text{ERB} = 24.7(4.37\frac{f_c}{1000} + 1) \tag{1}$$

The impulse response of gammatone filter is given by the equation:

$$f(t) = At^{(N-1)}e^{-2\pi Bt}\cos(f_c t + \theta) \tag{2}$$

where θ is the phase, f_c is the central frequency of the bandpass filter, A is the constant that regulates the gain of the filter, and N the order of the filter and controls the slope of filter edges. The parameter B that controls the duration of the filter is calculated by following equation:

$$B = 1.1019\text{ERB} \tag{3}$$

Figure 1 shows the primary steps of GFCC feature extraction. It consists of pre-emphasis windowing and framing and FFT as that of MFCC. Followed by this spectral magnitude of FFT, coefficients are passed through gammatone filter bank. As a result of this, a multichannel signal is obtained. The spectral content of each of the channel is summed to calculate the energy contributed by that particular channel. After that logarithm compression of the concatenated filter bank energies is done, and DCT is applied to transform the spectral energies into the cepstral domain to find the final GFCC features.

2.3 i-Vector Feature Extraction

The identity vectors also known as i-vectors [6] are built on factor analysis concept. The i-vectors are low dimensional, fixed length representation of speech utterances regardless of their variable length. The i-vectors characterize the speech utterance as a fixed size, low-dimensional vector and that also regardless of its length. Here, the main idea is to project the intra-speaker and inter-speaker variability of speech utterance into a single subspace rather than two subspaces [10]. Both the intra-speaker and inter-speaker variability compensation is done in this low-dimensional space instead of high-dimensional super vector space. Here, every single speech utterance is represented by speaker dependent supervector, M. The factor analysis model of M is defined as, calculated by following equation:

$$M = m + Tw \tag{4}$$

where m is the speaker-independent supervector of universal background model (UBM), T is a total variability matrix that models both intra- and inter-speaker variabilities, and w is the i-vector with zero mean and unit variance Gaussian distribution.

The i-vectors are calculated from all the speech utterances of a speaker and averaged to form the enrollment i-vector. It acts as a reference probe of the speaker to be enrolled in the system. The features in the i-vector subspace include both intra- and inter-speaker variability. These variability needs to be compensated before scoring against the test trials. As shown in Fig. 2, the computational complexity at the backend is reduced by applying linear discriminant analysis (LDA) to the i-vectors [11]. Then, intra-speaker variability compensation is carried out in lower dimensional space using within class covariance normalization (WCCN) [12]. After going through preprocessing steps similar to that of training utterances, the test i-vectors are scored against enrollment i-vectors using cosine similarity measure. If the match score is greater than threshold, the decision is made to accept the claimed identity otherwise it is rejected.

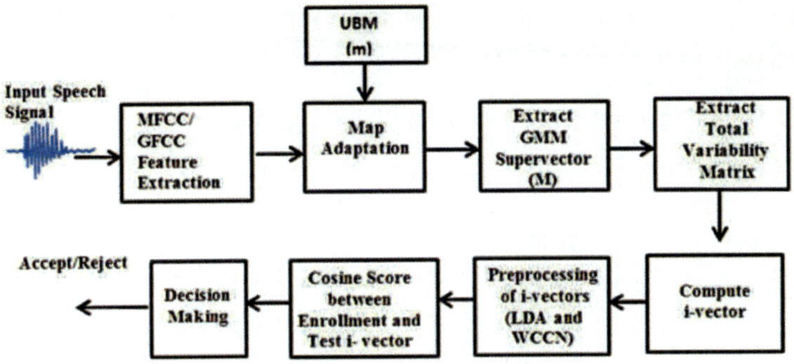

Fig. 2 Methodology to extract i-vectors based on MFCC/GFCC features

3 Experimental Setup

Our experiments are conducted on the TIMIT database. It consist of speech recordings of 630 speakers from eight dialect regions in clean environment. Each speaker has speech recordings with an average duration of around 3 s. Our experimental investigations consist of following stages: noisy dataset formation, UBM and TVM training, i-vector feature extraction, and it is preprocessing before finding match scores of target and imposter i-vectors with the enrolled i-vectors and finally computing equal error rate (EER) as a performance measure.

3.1 Noisy Dataset Formation

Original speech data of TIMIT database is enlarged with its noisy version. Noisy speech data are created by mixing noise samples obtained from the AURORA database. It contains seven different noise types. We have added clean data from TIMIT database with station, airport, and restaurant noise samples from AURORA database. Clean and noisy recordings are mixed with signal-to-noise ratio (SNR) levels varying from 0 to 15 dB. We have used data from "Train Folder of TIMIT dataset for development purpose, whereas "Test Folder" is used for speaker enrollment and evaluation purpose. The development data have a total of 36,960 obtained from 462 speakers, ten utterances per speaker, two noise types, and four SNR levels.

Speech utterances of 100 speakers randomly selected from "Test folder" are used for enrollment and testing purpose. Models are trained using 7 out of 10 utterances per speaker and tested using remaining 3 utterances. Verification trials consist of all 100 speaker model and test trial combinations, resulting into a total of 240,000 trials that includes 2400 target trials and 237,600 impostor trials. Moreover, we have evaluated our systems when GFCC and MFCC features are used individually and when fused together in the i-vector framework. Further, we have also used two different sets of noisy conditions: matched and mismatched noise conditions during training and testing phase. Performance evaluations under both of these settings show how the baseline and proposed systems perform in noisy conditions. We have elaborated the implementation procedure of SV in the following sections.

3.2 Feature Extraction

In our experimentation, we have tried to analyze the performance of GFCC and MFCC features in the i-vector framework in both matched and mismatched noisy conditions. We have used similar settings for the frame size, window type, and the hop size, number of cepstral coefficients during both the feature extraction process except the type of filter bank. At the initial stage, the pre-emphasized speech utterances are divided into short frames of 20 ms with 10 ms overlap. After that speech spectrum of hamming-windowed frame is computed using FFT. Followed by this, spectrum of the frame is multiplied with the mel filter bank and gammtone filter bank to compute corresponding MFCC and GFCC features. Total 13 coefficients obtained from each of the frame are combined with delta and delta-delta coefficients to capture the frame-to-frame dynamic variations. This combination gives a 39-dimensional feature vector for each frame. Similar processing steps are carried out for all the frames of the input speech. Thus, a feature matrix is obtained for every single speech recording of a speaker. The rows of the feature matrix indicate frames, and columns indicate coefficient.

3.3 Speaker Enrollment and Scoring Test Trials

We have used development data to train subsystems of our SV system such as: UBM, total variability matrix, and LDA and WCCN matrices. We have utilized different portions of development data to train LDA and WCCN matrices. The intention behind using different portions of data is that the aim of LDA is quite different than that of WCCN. LDA aims to model the inter-speaker variability, whereas WCCN aims to model the intra-speaker variability. So, we have used clean as well as noisy data to train WCCN so as use speech utterances with possible noise variations. For UBM and TVM training, 46,200 utterances of 462 speakers from "Train Folder" are used. While extracting the i-vectors, the number of Gaussian components of UBM and size of T matrix are varied and finally fixed to the values for which we have obtained improved performance. Experimentally, we have observed that the UBM with 512 Gaussian components and T matrix of size 400 provides better performance compared to higher values. We have used Microsoft Research (MSR) Identity Toolbox [13] for implementing front end and back of baseline and proposed feature-based systems in i-vector framework.

4 Results and Discussion

During the testing phase, the i-vector of test audio file probe is compared with all the enrollment i-vectors using the cosine similarity score. These scores are further used for system evaluation. All the scores where the speaker models are tested against the test utterance of enrolled speaker are collected into a set known as positive scores. The other scores are pooled into a set known as negative scores. A global, speaker independent, threshold are used to verify the speaker authenticity. The false acceptance and false rejection ratios are calculated using the threshold to compute the EER. We have compared system performance for all three feature extraction methods EER. Table 1 shows the EER values that are obtained for different features with SNR levels varying from 0 to 15 dB. It can also be seen in Table 1 that at low SNR levels, proposed system performs as good as baseline system. However, as the SNR levels are increased, proposed features outperform baseline features. Out of the three feature extraction methods, we have obtained the lowest EER values for MFCC and GFCC feature combination. From Table 1, it can be observed that for station noise added at the 15 dB SNR level the EERs of system corresponding to MFCC, GFCC, and their combination are 11.25%, 6.25%, and 5.31%, respectively. But, in the mismatched noise condition, EER values increase to 12.37%, 7.62%, and 7.15% for MFCC, GFCC, and combination of MFCC and GFCC, respectively. So, comparatively, there is reduction in EER of about 1% in case of GFCC and MFCC combination than that of the GFCC alone in matched conditions. Figure 3 shows comparison of all three methods in terms of average EER (%) considering all SNR levels of airport noise. The average EER is degraded from 11.75 to 13.37% corresponding to matched and mismatched conditions for GFCC and MFCC feature combination. It can be observed in the Fig. 3 that the feature extracted by gammatone filterbank can improve speaker verification performance in matched as well as mismatched noise conditions.

It can be seen in the right portion of Table 1 that there is very small improvement in the EER using GFCC features if we use the mismatched noise conditions during

Table 1 Performance evaluation with different features under matched and mismatched conditions

Type of noise	SNR levels (dB)	EER (%)					
		MFCC	GFCC	MFCC and GFCC	MFCC	GFCC	MFCC and GFCC
Airport	0	27.52	24.58	21.92	34.52	29.58	25.29
	5	22.64	12.14	9.14	24.54	13.25	10.83
	10	17.58	9.29	8.57	19.58	12.62	9.82
	15	13.21	8.28	7.38	12.21	9.12	7.23
Station	0	26.92	24.78	22.52	30.86	26.18	25.89
	5	21.14	11.12	10.63	27.12	13.35	12.55
	10	16.56	8.34	7.45	19.23	10.75	8.59
	15	11.25	6.65	5.31	12.37	7.62	7.15

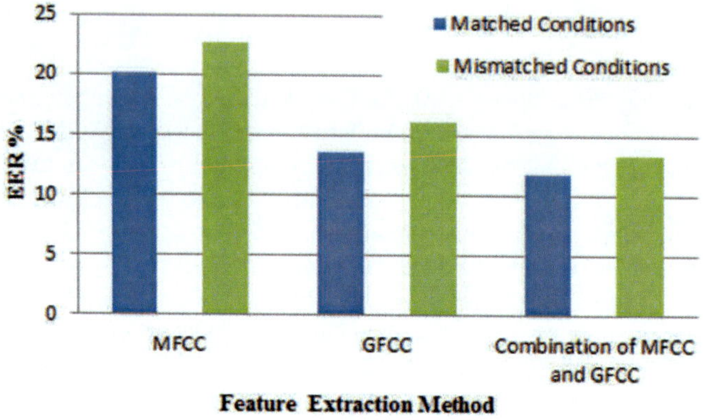

Fig. 3 Average EER graph for system trained with airport noise

testing phase. However still, the GFCC features perform better for almost all SNR levels of both matched and mismatched noise conditions compared to the MFCC features.

5 Conclusions

In this paper, we have tried to improve the EER in noisy conditions by exploring better features than the conventional MFCC feature. For doing so, we have used GFCC features that are based on human auditory system in the i-vector framework. In our experimental evaluations, we have observed that combination of GFCC and MFCC features provides best performance than that of individual MFCC and GFCC features. The improvement is significant if we use the matched noise conditions and thus shows the effectiveness of GFCC features in noisy conditions than conventional MFCC features. We have observed that speaker models trained under probable noisy conditions and right type of features can be advantageous to tackle with noisy test conditions. However, we have observed that system performance degrades under mismatched training conditions for MFCC as well as GFCC features. First way of improving could be using more noisy conditions so as to avoid training and testing mismatched conditions. Second is to use adaptive speaker modeling techniques to match with the unseen noisy conditions.

References

1. Kinnunen T, Li H (2010) An overview of text-independent speaker recognition: from features to supervectors. Speech Commun 52(1):12–40

2. Gish H, Schmidt M (1994) Text-independent speaker identification. IEEE Signal Process Mag 11(4):18–32
3. Umesh S, Cohen L, Nelson D (1999) Fitting the mel scale. In: Proceedings of ICASSP, vol 1. pp 217–220
4. Reynolds DA (1994) Experimental evaluation of features for robust speaker identification. IEEE Trans Speech Audio Process 2(4):639–643
5. Nakagawa S, Wang L, Ohtsuka S (2012) Speaker identification and verification by combining MFCC and phase information. IEEE Trans Audio Speech Lang Process 20(4):1085–1095
6. Valero X, Alias F (2012) Gammatone cepstral coefficients: biologically inspired features for non-speech audio classification. IEEE Trans Multimedia 14(6):1684–1689
7. Dehak N, Kenny P, Dumouchel P, Ouellet P (2011) Front-end factor analysis for speaker verification. IEEE Trans Audio Speech Lang Process 19(4):788–798
8. Garofolo JS, Lamel LF, Fisher WM, Fiscus JG, Pallett DS, Dahlgren NL, Zue V (1993) Timit acoustic-phonetic continuous speech corpus linguistic data consortium. Philadelphia, Technical report LDC93S1, USA
9. Hirsch HG, Pearce D (2000) The aurora experimental framework for the performance evaluation of speech recognition systems under noisy conditions. International journal of speech communication association tutorial research workshop ASR2000 on automatic speech recognition: challenges for the next millennium, Paris, France, pp 18–20
10. Dehak N, Kenny P, Dumouchel P (2007) Modeling prosodic features with joint factor analysis for speaker verification. IEEE Trans Audio Speech Lang Process 15(7):2095–2103
11. Kanagasundaram A, Vogt R, Dean D, Sridharan S, Mason M (2011) I-vector based speaker recognition on short utterances. In: Proceedings of INTERSPEECH
12. Hatch A, Kajarekar SS, Stolcke A (2006) Within-class covariance normalization for SVM-based speaker recognition. In: Proceedings of 9th international conference on spoken language processing 2006, Pittsburgh, PA
13. Sadjadi SO, Slaney M, Heck L (2013) MSR identity toolbox v1. 0: A MATLAB toolbox for speaker-recognition research. Microsoft research technical report

A Survey on In-Order 5-Stage Pipeline RISC-V Processor Implementation

Jayashree Mallidu and Saroja V. Siddamal

Abstract RISC-V is a free and open instruction set architecture (ISA) based on reduced instruction set computer (RISC) principles. RISC-V ISA enables a new phase in the field of processors through open standard association. The address of RISC-V is based on 32-bit and 64-bit variants. The essential RISC-V is a 32-bit integer instruction set defined as RV32I, which efficiently supports the operating system environments and also suits for the embedded system applications. In this paper, a survey is carried for 5-stage in-order pipeline implementation and ways to overcome pipelining hazards for structural hazards, data hazards, and control hazards on RISC-V processors. Being open-source and free, this is adopted in many commercial and academic research and projects.

Keywords RISC-V · Instruction set architecture · In-order · 5-stage pipeline · Hazards

1 Introduction

Currently, the advancement in the technology in integrated circuits has remarkably progressed to allow unique implementations of system on chip (SoC). Most of the instruction set architecture from ARM, Intel, AMD, etc., is registered and guarded by intellectual property. Very few open-source instruction set architecture like MIPS, RISC-V, and open RISC-based processors have been gaining importance in custom SoCs designs which provide a standard procedure for customization and optimization. RISC-V consists of five instruction formats. They are as follows:

- Register type (R-type)
- Immediate type (I-type)

J. Mallidu (✉) · S. V. Siddamal
KLE Technological University, Hubballi, India
e-mail: jayashree.mallidu@kletech.ac.in

S. V. Siddamal
e-mail: sarojavs@kletech.ac.in

M. S. Kaiser et al. (eds.), *Information and Communication Technology for Competitive Strategies (ICTCS 2021)*, Lecture Notes in Networks and Systems 401,
https://doi.org/10.1007/978-981-19-0098-3_73

- Store type (S-type)
- Branch type (B-type)
- Upper immediate type (U-type)
- Jump type (J-type)

Each of the above formats further has various instructions defined within. For example, R-type contains instructions like ADD, AND, and OR. These formats have their way of encoding the different information (source and destination registers, immediate values (if any), and sideband signals) (Fig. 1 and Table 1).

- opcode: This is the sideband signal which is used to decide one of the instruction format
- rd: One of the X1–X31 registers used as the destination where the result of the instruction would be stored
- rs1: One of the X1–X31 registers used as one of data sources for the instruction
- rs2: One of the X1–X31 registers used as the other data source for the instruction
- funct3/funct7: Sideband signals used for further decoding instruction within one of the instruction formats
- imm: This contains the immediate value to be processed by the instruction. Should be 0 for R-type

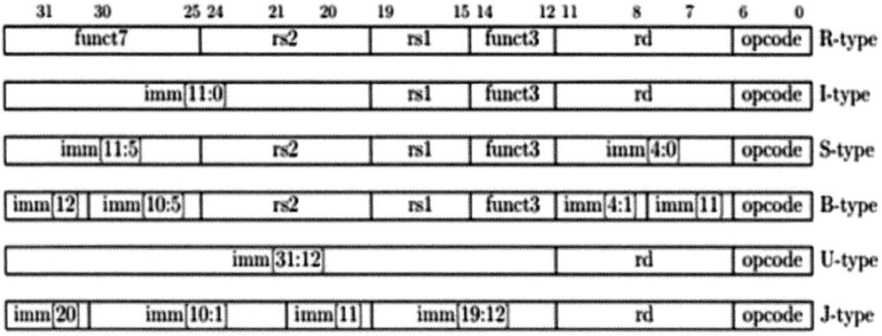

Fig. 1 RISC-V instruction formats

Table 1 RISC-V instruction opcode formats

Instruction type	Src reg 1 (RS1)	Src reg 2 (RS2)	Dest reg (RD)
R-type	Instr[19:15]	Instr[24:20]	Instr[11:7]
I-type	Instr[19:15]	–	Instr[11:7]
S-type	Instr[19:15]	Instr[24:20]	–
B-type	Instr[19:15]	Instr[24:20]	–
U-type	–	–	Instr[11:7]
J-type	–	–	Instr[11:7]

The microarchitectural design aims at having a balance between maximum clock frequency and cycle performance. In a processor, if the execution time for one instruction is one cycle, then it to have an IPC of one. Ideally, instructions per cycle is one measure that determines the processor performance. IPC can be termed as the average number of instructions executing in one clock cycle. In case of a processor that only issues one instruction per cycle comparison, this ideal IPC cannot be achieved in a pipelined processor. Certain factors introduce hazards that can force the next instruction not to execute in the following clock cycles. These hazards will result in not executing an instruction every cycle. The critical path in a processor will be enormous in any design considered, thereby decreasing the design's maximum frequency. To solve such issues and to achieve higher IPC, the concept of the pipeline is introduced. The ideal state is to have clocks per instructions (CPI) to be equal to one. But, it is quite a changeling to achieve an ideal state mainly due to memory latency, branch miss-prediction, pipeline hazards, etc.

2 Literature Survey

The authors in [1] discuss a proposed microarchitecture design supporting the base RISC-V instruction set architecture, including a pipeline stage in the design. This acts as a basis for further pipeline design space explorations taking into account of speed, area, power consumption, and implementation complexity. The work in this paper brings out the behavior of non-pipelined and pipelined implementations when the decisions for designing a specific embedded domain. The procs and corns of the basis of each aspect underlying the difference between non-pipelined and pipelined microarchitectures give proper motivation and research scope. The criterion configuration is to carry out the RV32I standard ISA in M mode. This includes designing and developing a non-pipelined, single cycle, and development of its pipelined equivalents. Such architecture leaves a lot of space for improvements. It thus can serve a criterion design for comparative analysis with more expanded solutions or maybe subject to further developments and experiments in the educational process. Based on the analysis of algorithms that are application specific and their implementation based on their penalty of clock cycles and concerning the instructions, optimal pipeline stages are defined. This paper models an efficient framework for the placement of pipeline stages.

RISC-V processor core with higher advantages of compressed instruction set extension, which is low cost, is developed in paper [2]. The authors in this paper have implemented the processor with a 6-stage in-order pipelined structure that includes the following stages: fetching of instruction (IF), decoding instruction (ID), renaming register (RR), issuing of instructions (IS), instruction execution (EX), and write backstage (WB). In-order fetching and execution of instruction according to the program order takes place. The write backstage for results here is performed in-order with implementation being out-of-order. Comparisons against existing RISC-V designs specify that the proposed architecture in this paper can approach the performance gains as much as a two-way superscalar processor at a much-reduced hardware cost. Accordingly, efficiency of the 6-stage pipelined design is much higher in terms of area and power as proposed by the authors.

In this paper [3], authors proposed an efficient five-stage pipeline that increases the system's performance. Authors proved the designing of a five-stage pipelined processor can precisely perform execution of all instructions meeting all the necessary functions, giving an enhanced performance. To conclude, the pipelined 5-stage RISC-V is implemented on FPGA. The author also proposed the method by duplicating the resources to eliminate structural hazards. Here, the memory is divided into two types: data memory and instruction memory. Loading and storing of the data take place from data memory and fetching of instruction from the instruction memory. And to solve data hazards, forwarding techniques are used in the pipeline. The architecture is designed so that data can be moved from registers in the pipeline to the next instruction where the instruction requires it. This process can minimize the usage of stalling in the pipeline.

The proposed work in [4] mainly focuses on the design and implementation of a simulator for RISC processor-based 5-stage pipeline. In this paper, different solutions for data hazards and control hazards are provided with multiple ways for fetching instructions and translations, making it distinctive. The system is divided into 5-stages pipeline depending upon the functions performed, which include instruction fetch (IM), decode with register read (REG), execute (ALU), data memory (DM), and register write (REG). The overall pipeline stage [5] can be subdivided into fetch, decode, execute, store. The authors in this paper have designed only pipeline stages without considering the hazards. The pipelining of instructions that reduce the CPI increases the speed of execution or operation and increases the overall system throughput.

In this paper [6], the authors have discussed the designing of the RISC-V instruction set and 5-stage pipeline. The analysis has been performed regarding instruction encoding, functionality, type, decoder logic complexity, pipeline and hazard detection, control flow effect, data memory access, branch instructions, execution unit hardware resources, and operating modes. The processor that is designed here includes a branch prediction unit and data forwarding unit with advanced data hazard detection. Also, ASIC implementation for the same is carried out at UMC 65 nm and 130 nm technology. The authors in [7] have designed a branch prediction technique for in-order RISC-V processors. Used for minimizing control hazards in pipelined processors. An efficient branch prediction system consists of branch target buffer (BTB) designed to achieve high performance.

3 Pipelining in RISC-V Processors

Pipelining is arranging the elements in series that the output of one element is acting as an input to another. They are, basically, used for organizing of instructions so that instructions are executed concurrently. As compared to non-pipelined architecture execution, many instructions are executed in one clock cycle. In the case of RISC-V architecture, pipelining increases the throughput of the system, and instruction per cycle also increases (IPC). In case of instruction pipelining, overlapping of fetch, decode, and execute stages of the instruction cycle takes place. Variations can be done in pipeline structure which leads to improvement in design and also increases the speed of the system (Figs. 2 and 3).

In RISC-V 5-stage in-order pipeline exists which has one step per stage that is.

1. IF: Fetching instructions from memory
2. ID: Instruction decode
3. EX: Calculate address or execution stage
4. MEM: Accessing memory operand
5. WB: Writing result back to register

Since all instructions are fixed 32 bit in RISC-V, it is easy to fetch and decode instructions in one cycle (Fig. 4).

Pipelining is the process of arranging the elements in series that the output of one element is acting as an input to other. Basically, used for organizing of instructions so that instructions are executed in concurrent fashion. As compared to non-pipelined architecture execution, many instructions are executed in one clock cycle. In case

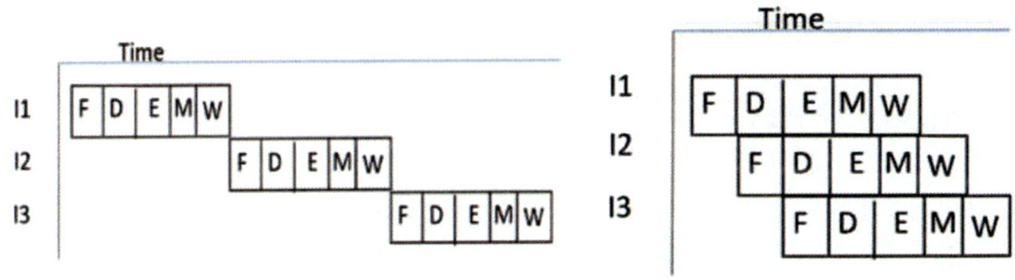

Fig. 2 Pipelined versus non-pipelined instruction execution

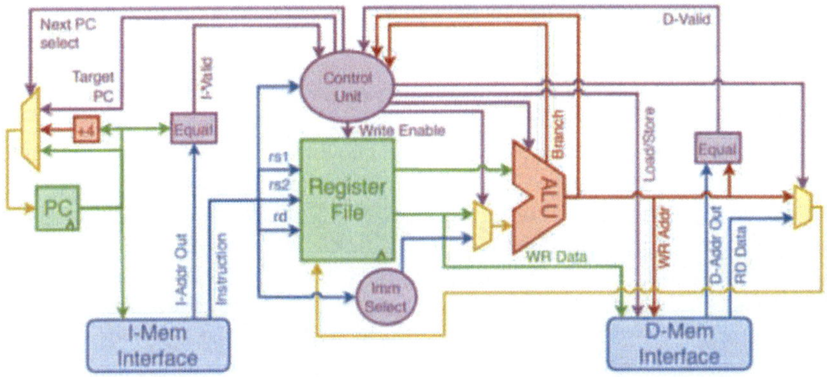

Fig. 3 Single-cycle RISC-V RV32I datapath

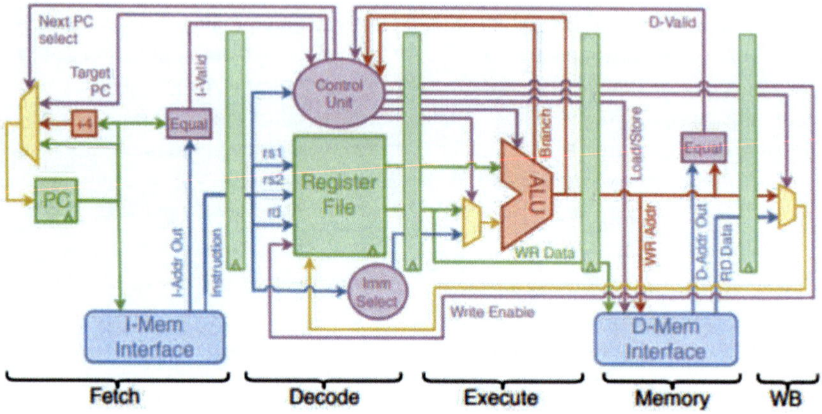

Fig. 4 Pipelined RISC-V RV32I datapath

of RISC-V architecture, pipelining is present which increases the throughput of the system and also instruction per cycle also increases (IPC). In case of instruction pipelining, overlapping of fetch, decode, and execute stages of instruction cycle takes place.

4 Pipelining Hazards

Hazards are the one the cause the starting of next instruction during next clock cycle. Different hazards are occurring in RISC-V core. They can be named as

- Structural hazards
- Data hazards
- Control hazards

4.1 Structural Hazards

Two or more instructions executing in the pipeline require access to the same resources results in structural hazards. One way to overcome structural hazards is instruction must be executed in series rather than parallel pipeline execution. But, this will not result in achieving higher IPC. So, other way to overcomes is it can be resolved by stalling new instructions till the old instruction finished their execution. Another way of overcoming structural hazards is by adding new hardware to the design.

4.2 Data Hazards

When instructions under execution show data dependency which then results in modification of data in different pipeline stages. The first method to overcome data hazards is to stall and to introduce interlocks to avoid data hazards.

4.3 Control Hazards

Control hazard arises when wrong decisions are taken in pipeline during branch predictions and therefore takes instructions into the pipeline that must consequently be rejected. Branch prediction is one way to overcome Control hazards, it is a hardware process that predicts the address of the next instruction following the branch. A combination of various quantified circuits is used to make a prediction. The branch target buffer (BTB), return address stack (RAS), and branch history table (BHT) are to be used. Such predictors are most commonly used in in-order processors. In pipelined processor, conditional branches are of a serious issue. After various cycles of instruction, the branch target address and branch direction are determined and calculated by the processor, resulting in pipeline stall. This stall leads to control hazards and thereby degrades the performance of the processor. To increase the rate of the instruction flow in case of modern processors like RISC-V, branch prediction has been implemented. Branch prediction increases the ideal rate in terms of the performance of the processor. Using this branch prediction, the processor predicts the direction, and accordingly, the instruction gets executed with the expected path.

5 Conclusion

RISC-V is the most widely used open-source ISA which is used for a wide range of applications in today's world. The efficient pipeline design and ways to over pipeline hazards which when inculcated in the design makes in more efficient. This paper provides the survey on the various pipeline architecture that is being developed along with measures to overcome pipeline hazards such as structural hazards, data hazards, and control hazards.

References

1. Khimchenko I, Christian K, Peter S (2019) RISC-V processor with configurable pipeline stage placement. In: Dortmund international research conference
2. Patsidis K et al (2018) A low-cost synthesizable RISC-V dual-issue processor core leveraging the compressed Instruction Set Extension. Microprocess Microsyst 61:1–10

3. Gao, LX, Zha HS (2013) Pipelined RISC processor design and FPGA implementation. Appl Mech Mater 336
4. Olanrewaju RF et al (2017) Design and Implementation of a five stage pipelining architecture simulator for RiSC-16 instruction set. Indian J Sci Technol 10(3):1–9
5. Mangalwedhe S, Roopa K, Kulkarni SY (2019) Low power implementation of 32-bit RISC processor with pipelining. In: Proceeding of the second international conference on Microelectronics, Computing & Communication Systems (MCCS 2017). Springer, Singapore
6. Raveendran A, Patil V, Selvakumar D (2016) A RISC-V instruction set processor-micro-architecture design and analysis. In: 2016 International conference on VLSI Systems, Architectures, Technology and Applications (VLSI-SATA)
7. Rathi CA, Rajakumar G, Kumar TA, Samuel TSA (2020) Design and development of an efficient branch predictor for an in-order RISC-V processor

Image Matching Techniques: A Review

Ramandeep Kaur and V. Devendran

Abstract The motive of this work is to distribute a review of both modern and classic area-based image matching algorithms. In the realm of machine vision, determining three-dimensional data from images is a critical issue (Joglekar and Gedam in Int J Emerg Technol Adv Eng 2(1), 2012, [1]). To match images and identification at the heart of computer, they play very significant role in our daily lives (Feature detection and extraction—MATLAB & Simulink, 2016, [2]). The motive is to behave the cameras like human eye work. In modern era, with cameras, people generate wide number of pictures, there is need to examine those pictures, which can be used for further research and to get useable results particularly for a problem given by user (Jayanthi and Indu in Int J Latest Trends Eng Technol 7:396–401, 2018, [3]). In this paper, we have described some points of which comparison for matching can be done, and along with, some different points of interest for matching are described.

Keywords Image matching · Image recognition · Feature detection

1 Introduction

Image matching methods are the procedures used to discover presence of an example inside a source picture. Matching techniques can be grouped into two classifications, for example, on the basis of features and area. In the present era, cameras are everywhere. From smart mobile phones in our hands to the cameras used in clinical science, etc. Use of computers has seen a transient ascent in the new era, with the advancement of a variety assortment of strategies to achieve some errands. These errands incorporate movement examination, scene recreation, picture rebuilding, and picture matching [4–6]. In this examination, we have zeroed in on different picture coordinating with procedures and calculations. We have looked at their exhibitions, in the end, recommending the best procedure out of the relative multitude of thought about strategies. It might happen that a portion of these calculations/strategies can work

R. Kaur · V. Devendran (✉)
Computer Science and Engineering, Lovely Professional University, Phagwara, India
e-mail: devendran.22735@lpu.co.in

M. S. Kaiser et al. (eds.), *Information and Communication Technology for Competitive Strategies (ICTCS 2021)*, Lecture Notes in Networks and Systems 401,
https://doi.org/10.1007/978-981-19-0098-3_74

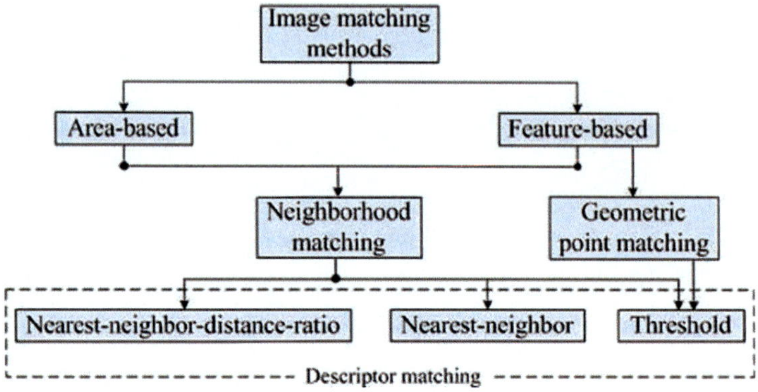

Fig. 1 Different image matching methods

well with certain informational collections, and others are not as powerful in breaking down similar informational collections. Henceforth, certain calculations end up being valuable for a particular application, while others have diverse usage. Every calculation is extraordinary from the other one, and proficiency standards contrast for each situation, despite the fact that the point of the calculation is the equivalent, for example, image matching. Image matching represents two major issues.

(a) Vague arrangements may happen, if picture coordinating is handled utilizing just nearby data.
(b) Good estimations are required.

Majority of the image matching with calculations, boundaries like interpretation, scaling, pivot, and viewpoint invariance should be regularly considered [7]. For fast and solid article acknowledgment, the sort of highlight to be utilized for coordinating and standard for best matching should be known (Fig. 1).

In Sect. 2, we have described feature detection for image matching, which covers the concept of feature-based matching in detail, using various feature extraction techniques. In Sect. 3, concept of area-based matching technique is described. In Sect. 4, a comparative study is done on the basis of different feature matching techniques. Section 5 describes the different image matching techniques like SURF, BLOB, template matching in detail. Section 6 contains the conclusion of whole study of image matching techniques.

2 Feature Recognition and Detection for Image Matching

Each image is recognized by using its kind of arrangement of features, these features are just for the one image, and this will be helpful to act as proof how images are different [7, 8]. A low-level picture preparation activity is feature ID. That is, it is frequently used as the first method on a picture, and it examines each pixel to see if there is a portion existing there [9–11]. Features for PC vision and picture preparation are figuring thoughts of picture information and making close by decisions at each picture point whether there is a picture feature of a given sort by at that point or not, and distinguishing proof refers to procedures that go for figuring thoughts of picture information and making close by decisions at each picture point whether there is a picture feature of a given sort by at that point or not.

2.1 Concept of Feature-Based Matching

Picture features are some noteworthy centers or parts of an image that can be used as a starting point for various computer calculations [12, 13]. Parts are utilized as the initial phase that determines how calculations are made; the overall calculation will always be in the same class as its component indicator. A low-level picture preparation activity is feature ID. That is, it is frequently used as the first method on a picture, and it examines each pixel to see if there is a portion existing there. If this is part of a larger computation, the estimator will evaluate the image in the district of the features on a regular basis. [2, 11, 12]. Features for PC vision and picture preparation are figuring thoughts of picture information and making close by decisions at each picture point whether there is a picture feature of a given sort by at that point or not, and distinguishing proof refers to procedures that go for figuring thoughts of picture information and making close by decisions at each picture point whether there is a picture feature of a given sort by at that point or not.

1. To choose unmistakable focuses in the pictures independently.
2. The focuses are chosen utilizing a purported interest administrator.
3. Developing a fundamental rundown of competitor sets of comparing focuses expecting a likeness measure.
4. Tracking down the last rundown of point sets predictable with an item model [14].

Choosing Distinct Points: In feature-based matching, rather than coordinating with whole pixels of a picture, just chosen focuses with specific features that are to be coordinated. The determination standard ought to satisfy the accompanying necessities: Distinctness: The focuses ought to be particular, for example, be unique in relation to adjoining focuses; for example, focuses on edges ought not be chosen if the epipolar calculation limitation is not utilized; likewise, focuses in level regions ought not be chosen.

Invariance: Level of invariance straightforwardly impacts the accuracy and the unwavering quality of the matching.

Stability: The chosen focuses ought to be required to show up in different pictures. Subsequently, the determination ought to be vigorous as for commotion. In picture grouping investigation, they chose focuses that ought to show up in long arrangements of successive casings.

Interpretability: The determination standard ought to be interpretable in some sense, for example, searching for edges, corners, or other basic, however, named highlights. This prerequisite is not fundamental from a designing perspective, yet it might be fundamental if the interest administrator is utilized for picture investigation.

Seldomness: Whereas uniqueness ensures nearby distinctness of focuses, seldomness focuses on worldwide are distinguishableness. This is fundamental in pictures with incompletely tedious examples.

2.2 *Feature Extraction*

With feature-based coordinating, features can be separated in every picture independently before coordinating with them. Nearby features are focuses, edges and lines, and locales. Bigger (worldwide) highlights are called structures. Worldwide highlights are generally made out of various neighborhood features.

Point Extraction: These features can normally remove with neighborhood administrators, regularly known as "interest" administrators. Few interest administrators recorded beneath: (1) Administrator of Moravec distinguishes focuses with high dim level fluctuations. (2) Luhmann and Ehlers utilized matrix kind element to extraction without thresholding, depending on contrast to acquire a same appropriation of the highlights. (3) Dreschler fostered an interest administrator dependent on the head and Gaussian arches along with point of central course of the force work—which compute the dim-level angles of picture f_x, f_y by utilizing edge administrator. For a 3×3 cover, f_x and f_y are registered using utilizing conditions: f_x = (wc7 + wc8 + wc9)–(wc1 + wc2 + wc3) and f_y = (wc3 + wc6 + wc9)–(wc1 + wc4 + wc7). where wci is worth of coefficients.

wc1 wc2 wc3
wc4 wc5 wc6
wc7 wc8 wc9

Compute the elements of the normal matrix

$$M = \begin{bmatrix} \sum f_x^2 & f_x f_y \\ \sum f_x f_y & \sum f_y^2 \end{bmatrix}$$

(a) Weight, wg, is related to size of the error ellipse, with proportional to contrast
det M = determinant of M;
tr M = trace of M.

(b) Roundness r, it is belonged to shape of error ellipse, which means the length of semi-minor and semi-major ellipse axes; $0 < r < 1$

Threshold criteria.

- wg(x, y) = wg(x, y) if $f(x, y)$ >= r min, wg(x, y) >= wgmin, else wg (x, y) = 0

- Non-maxima suppression by using wg(x, y) within window size
- Prediction of accurate point coordinates

5. Hannah uses an operator which selects points with steep autocorrelation function of the gray levels in all directions

Edge Extraction: Edges can be parted from administrators that recognize edges. Among all, which two commonly known are,

- Gradient administrators, i.e., depends on computation of neighborhood force inclination with utilization of cover as a rule with odd measurements.
- The zero intersections of Laplacian, those are expression points of the power work $f(x, y)$

$$\nabla^2 f = \frac{\partial^2 f}{\partial x^2} + \frac{\partial^2 f}{\partial y^2} = 0$$

The Laplacian depends on second-request subsidiaries, consequently is exceptionally touchy to clamor, and subsequently is joined with a smoothing activity, for example, convolution with a Gaussian channel, and it is classified "Laplacian of a Gaussian".

District Extraction: Its extraction of locales begins with the location of few extremely "same areas" of picture, for example, areas with power varieties under a specific limit, then maximally becomes yet for degree that they remain adequately homogeneous. Multi-fix measure dependent on general calculation fills in as follows.

1. Three patches are picked in the main picture. An inquiry is led in the up-and-comer rundown of every of three patches for a mix of competitors which struct a triangle of same shape, size, direction to the patches within main picture.
2. On the off chance that, no appropriate arrangement of up-and-comers can be discovered. Among all of patches, one patch in sync 1 is supplanted, which tends to Stage 2 rehash.
3. Stages 2 and 3 will rehash till reasonable triangles will framed.
4. By using initial two effectively coordinated with patches, different patches are handled thus from the principal picture, with the up-and-comer list being contrasted as in sync 2 by track down the one with the right math.
5. Then, the triangles are thought about along the direction, and distance of their edges is viewed as comparative, when edge direction is around 50.

SIFT [9], A Transform Scale Invariant Features Lowe released an ad libbed type of work done by him under the distribution "Scale Invariant Feature Transform computation" shortly after introducing a calculation on include recognition from finished photographs [11]. This describes a method for identifying remarkable invariant highlights that can be used to compare different locations, directions, viewpoints of an object. He portioned his method into four parts:

(1) Scale space extrema identification: A distinction of-Gaussian capacity is used to detect likely revenue foci that are invariant to direction and scale.

(2) A careful model is used to characterize area and scale at each competitor region and then heaps of their steadiness and central issues are chosen.
(3) Orientation task: Each central issue region is assigned at least one direction based on nearby picture inclination bearings. As all future computations will be compared to the allotted ones, this ensures that changes in direction, size, and area remain invariant.
(4) Key point descriptor: The nearby picture slopes are estimated and transferred into a outlook which withstands substantial degrees of light and neighborhood form twisting in the space around each central issue at the assigned scale. These steps are carried out in a slipping order, with only the most important concerns advancing to the next level. In any event, this is a costly cycle; hence, a revamped version of SIFT's descriptor, "Guideline Component Analysis SIFT (PCASIFT)", was introduced.

Regardless, it ends up being less specific than SIFT. Lindeberg introduces the "Scale Invariant Feature Transform," which is an important methodology (Lindeberg, 2012). This highlight locating technique in the SIFT administrator can be thought of as a scale-versatile mass recognition strategy in which the detected masses with relevant scale levels are found from scale space extrema of the scale-standardized Laplacian.

B. Robust Features (SURF) picture coordinating with calculation was developed with the goal of ensuring speed in the following areas: discovery, representation, and coordination. Speeded-Up Robust Features (SURF) indicator was given to this calculation. SURF, in contrast to other popular methods, uses a hessian network to greatly accelerate. Its descriptor characterizes a spread of Haar wavelet reactions around the interest point, which depends on fundamental images to save calculation time. SURF is certain to execute faster thanks to the low-dimensionality descriptor (64 measures) and the "Fast-Hessian" indication. C. Highlights from the Accelerated Segment Test (A.S.T.) (FAST) FAST is a lone element-based algorithm developed by Trajkovic and Hedley (1998) and used for this correlation.

Edward, in any case, spread the execution used for the correlation (Rosten and Drummond, 2006). Corner, which is identified by a circle covered an application pixel in the FAST identifier, which is a wedge-type locator. It works by taking a 16-pixel circle, and if n bordering pixels are above or below a limit value, t, the corner is chosen as competitor pixel. In any case, Rosten and Drummond used an AI-based indicator to extend this calculation. Where any residual identifiers use a computation to distinguish corners, this technique creates a classifier on the model and then applies it to a picture. On how a corner should behave, the classifier can be prepared. As a result, the indication is significantly faster than alternative component locators.

D. Other tests of a comparable nature various people have carried out investigations to assess various images in conjunction with calculations. They were convincing in relating the majority of the calculations. However, their work does not provide us with the photographs they used, the framework specifics, or even merely final pictures that were used for the analysis. The research aims to augment existing assessments by allowing for the quick identification of the cycle that will perform well under specified conditions. The report also looks at the accuracy of the calculations when it comes to identifying single item in a scene.

3 Concept of Area-Based Matching

Area [15]-based matching here and there called connection like techniques or layout coordinating, combine the component identification venture with the coordinating with part. These strategies manage the pictures without endeavoring to recognize notable articles. Windows of predefined size is utilized for the assessment of correspondence. Territory-based calculations can be additionally partitioned as following classes.

- Cross-relationship based
- Fourier based
- Mutual data based
- Optimization techniques (reenacted annealing)

3.1 The Principles of Area-Based Algorithms

At some picture scale, there is a level of proximity between two perspectives on a comparable scene. The perspectives become more comparable as the scale becomes coarser. If a view is divided into smaller subdistricts, the number of highlights for a given subarea decreases and the subarea looks more like its corresponding subdistrict in the opposite view. Comparability assessments based on territory can be applied to subdistricts or squares to identify the most likely correspondence between similar subregions from two different views.

On the off chance, that comparability is characterized by spot item score of two vectors $A = (x_1, y_1)$ and $B = (x_2, y_2)$ where all canisters are non-negative, then their speck item score will consistently increment if the vectors are coarsened by incorporating container substance [12]. The spot item is given by $AB = x_1x_2 + y_1y_2$: (1) Coarsening A and B by container coordination currently gives two additional terms

$$C = x_1 + y_1$$

$$D = x_2 + y_2$$

$$C.D = x_1y_1 + y_1y_2 + x_1y_2 + y_1x_2$$

$$C.D = AB + x_1 y_2 + y_1 x_2$$

(2) Because the coarser speck item has more words, it will always be larger than the first. For maintaining consistency restrictions and needs such as surface perfection, a proper similarity measure should be used for region-based sound system techniques. With the use of territory-based methodologies, the concept of cross-connection work and a pursuit space is also given. Closeness measurements got from likelihood thickness that works hypothetically offer best arrangement albeit practically speaking approximations; for example, Euclidean distance and spot item measurements have been used [1].

4 Comparative Study

See (Table 1).

Table 1 According to the findings of the comparison investigation, the feature-based correspondence technique is the best

Parameters	Area-based matching	Feature-based matching
Matching strategy	Every point which has to be matched in ABM, it is center of short window which consists of pixels in a reference image, which is statistically compared to other images which consist of similar-sized pixel window	Instead of matching all pixels in a picture, FBM matches just selected locations with specific attributes
Convergence, speed, and versatility	When compared to FBM, ABM exhibits less convergence, speed, and adaptability	FBM outperforms image correlation in terms of speed and versatility and outperforms LSM in terms of convergence range, speed, and adaptability
Initial estimates	For the unknown parameters in ABM, very good beginning values are usually required	FBM does not necessitate extremely accurate first estimates
Sensitivity	The convergence radius of LSM is extremely tiny. Even if cross-correlation is more powerful, the similarity criterion must be met	When it comes to the quality of approximate values, feature-based matching algorithms are less sensitive
Occlusions	When it comes to occlusions, ABM is very sensitive	Because the mathematical model does not have to be exactly fulfilled, FBM approaches are also less sensitive to occlusions

5 Image Matching Techniques

In this paper [3], we have considered three picture coordinating strategies for execution examination.

(A) Blob identification strategy
(B) Template coordinating with
(C) SURF include extraction

(A) Blob Detection Technique: Blob area procedures focus on finding districts that contrast in various properties, for instance, splendor or concealing, appeared differently in relation to incorporating regions, in an advanced picture. Mass is a district of a picture where a couple of properties are consistent or around steady and all of the focuses in a mass can be considered in some sense to resemble each other [12, 13]. The following is a calculation for the mass recognition process that we used: This calculation is performed by capturing a picture from a webcam or other camera in one of the specified organizations, such as MJPG 1280 × 720, MJPG 160 × 120, etc. (Fig. 2).

(B) Template Matching [3]: It is a significant level of machine-type technique which recognizes segments on a picture that match a predefined design. Calculation is: (i) A picture having writings (may be in debased structure) or articles is taken as info and changed over into dark-scale picture. (ii) It is gone through Gaussian channel to smoothen the messed up edges and commotion. (iii) It is gone through other pre-handling channels, e.g., widening, clamor pixel expulsion step, thresholding, and so on (writings or items in white). (iv) Every one of the different white areas is set apart as various items and tallied, edited to its base size. Jumping box is made around each item. (v) From that point onward, the item district is resized to the size of formats,

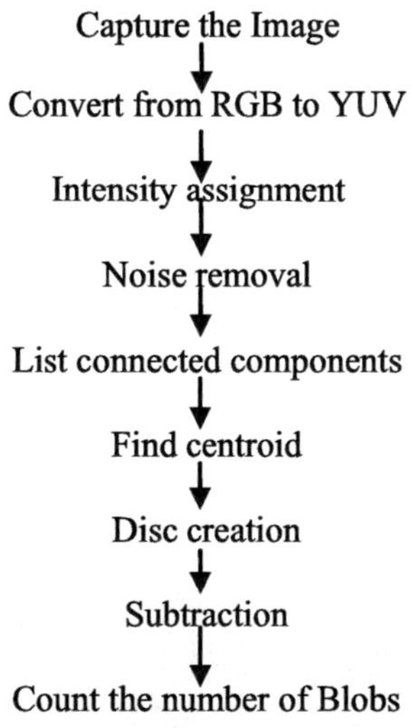

Fig. 2 Blob detection algorithm flow diagram

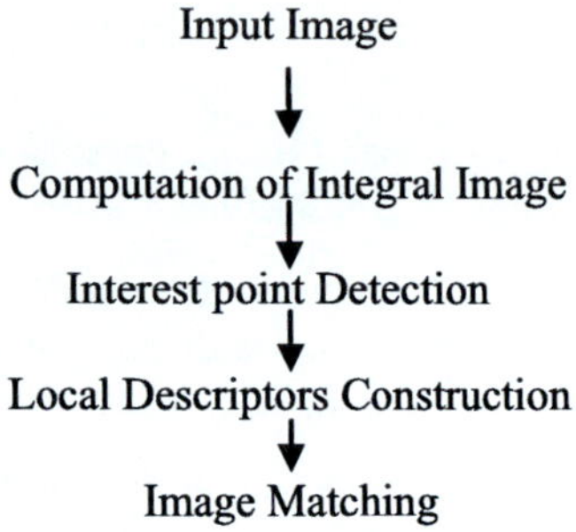

Fig. 3 SURF algorithm flowchart

and afterward, each article is contrasted with every one of the layouts pre-saved in a framework. (vi) corr2 (layout {1, n}, burn) is a capacity that figures the relationship in the format picture and the test object picture. The format picture which has the most noteworthy relationship coefficient is set apart as distinguished item or text. (vii) Also, every one of the writings or articles is analyzed, and the outcomes are put away in a content document which is shown toward the finish of program. The importance of detecting all information picture areas where the layout picture item is available is addressed by format matching algorithms. The client may (or may not) want to perceive the turned or scaled occurrences, depending on the unique situation nearby [8, 15, 16].

(C) Extraction of SURF: SURF is powerful adjacent element indication, first introduced by Herbert Bay in 2006, which can be employed in PC vision tasks such as item recognition. The SIFT description pushes it quite hard. SURF's regular version is a few times faster than SIFT, and it is more powerful against different picture alterations. SURF is based on a large number of 2D Haar wavelet reactions and nicely employs fundamental images. It makes an estimate at the determinant of the Hessian mass locator, which may be calculated quickly using a simple graphic (three number tasks). It utilizes the amount of Haar wavelet reaction near the point of interest for highlighting [6, 12, 13] (Fig. 3).

6 Conclusion

The feature-based coordinating strategy is the ideal technique for image matching with issue, and it upholds the least squares coordinating with calculation by giving dependable and precise estimated values if essential. The accuracy of this FBM methodology has been demonstrated to be appr. 1/3 of a pixel in the event that the focal point of the chosen windows is utilized as highlight focuses and the pictures show sufficient surface. Precision continuously tasks are requested by an ever-increasing number of utilizations. Both territory-based and include-based calculations stroll toward this target. However, when thick difference guide and straightforward calculations for less computational intricacy are required, territory-based calculations beat include-based calculations. Area-based strategies might be relatively more precise, on the grounds that they consider an entire neighborhood around the focuses being dissected to build up correspondences. The review presumes that every one of the various methodologies had its overall merits and burdens, reliant upon nature of correspondence issue.

References

1. Joglekar J, Gedam SS (2012) Area based image matching methods—a survey. Int J Emerg Technol Adv Eng 2(1)
2. Feature detection and extraction—MATLAB & simulink. In.mathworks.com, 2016
3. Jayanthi N, Indu S (2018) Comparison of image matching techniques. Int J Latest Trends Eng Technol 7(3):396–401
4. Sankowski, D, Nowakowski J (eds) (2014) Computer vision in robotics and industrial applications. World Sci
5. Kagami S (2010) High-speed vision systems and projectors for real-time perception of the world. In: 2010 IEEE computer society conference on computer vision and pattern recognition-workshops
6. Grimson WEL, Mundy JL (1994) Computer vision applications. Commun ACM 37(3):45–51
7. Ghafoor A, Iqbal RN, Khan S (2003) Robust image matching algorithm. In: EC-VIP-MC 2003, 4th EURASIP conference focused on video/image processing and multimedia communications, Zagreb, Croatia, 2–5 July 2003
8. Briechle K, Hanebeck UD (2001) Template matching using fast normalized cross correlation. In: Aerospace/defense sensing, simulation, and controls. International Society for Optics and Photonics
9. Babri UM, Tanvir M, Khurshid K (2016) Feature based correspondence: a comparative study on image matching algorithms. (IJACSA) Int J Adv Comput Sci Appl 7(3)
10. Zhou Y, Jiang G, Lin Y (2016) A novel finger and hand pose estimation technique for real-time hand gesture recognition. Pattern Recogn 49:102–114
11. Lowe DG (2004) Distinctive image features from scale-invariant keypoints. Int J Comput Vision 60(2):91–110
12. Verma NK et al (2016) Object matching using speeded up robust features. In: Intelligent and evolutionary systems. Springer International Publishing, pp 415–427
13. Andersson O, Marquez SR (2016) A comparison of object detection algorithms using unmanipulated testing images: comparing SIFT, KAZE, AKAZE and ORB
14. Babbar G, Bajaj P, Chawla A, Gogna M (2010) Comparative study of image matching algorithms. Int J Inf Technol Knowl Manage 2(2):337–339
15. Jia D et al (2016) Colour FAST (CFAST) match: fast affine template matching for colour images. Electron Lett
16. Local feature detection and extraction—MATLAB & Simulink. In.mathworks.com 2016
17. Bay H, Tuytelaars T, Gool LV (2006) Surf: speeded up robust features. In: European conference on computer vision, Springer, Berlin

Model-Based Design Technique to Monitor Vehicle Exhaust Emissions

Venkatesh Mane, P. Vishal, P. Nikita, Nalini C. Iyer, and K. Ashwini

Abstract The era of electronics began very early in automobiles but was rejected or discontinued mainly because of the size and cost of electronics by the consumers and original equipment manufacturers as vehicle population day by day started ramping up. This became the biggest hurdle for developing countries as most of the countries in the world have to import the required fuel, and most importantly, these vehicles on the road started polluting the environment. The air pollution caused by vehicles has been a huge impact on the health of every creature on this planet. They forced the government entities to put forth stringent emission norms for the original equipment manufacturers and suppliers. This was the main reason for electronics to enter into automobiles because the conventional (mechanical) method to control the fuel economy and exhaust emission was inadequate/inefficient. Another biggest reason for the sudden rush of electronics into automotive was the emerging semiconductor era. As a very large-scale integration of the transistors into a single chip began. That is, the system on chip development drastically reduced the size and cost of the electronics being used. The electronics in an automobile started controlling the mechanical components or even replaced many conventional mechanical components. The era of electronics began with controlling fuel economy and emission, but now modern electronics are not just limited to these objectives. Now, modern CARs use up to 100 electronic control unit (ECU). ECUs are used for every look and corner of the vehicle such as controlling the engine, braking, steering, suspension,

V. Mane (✉) · P. Vishal · P. Nikita · N. C. Iyer · K. Ashwini
KLE Technological University, Hubballi, India
e-mail: mane@kletech.ac.in

P. Vishal
e-mail: vishalbps@kletech.ac.in

P. Nikita
e-mail: nikhita_patil@kletech.ac.in

N. C. Iyer
e-mail: nalini.c@kletech.ac.in

K. Ashwini
e-mail: ashwinikamate@kletech.ac.in

M. S. Kaiser et al. (eds.), *Information and Communication Technology for Competitive Strategies (ICTCS 2021)*, Lecture Notes in Networks and Systems 401,
https://doi.org/10.1007/978-981-19-0098-3_75

infotainment, telematics, safety, and many more functions, which goes up to 2500 functions with 10 million lines of code. The paper majorly focuses on vehicle emission monitoring and updating the status of each of the vehicles on the cloud server using an IOT application. The paper's second objective is to focus on the model-based design using Simulink. The conventional method has been code-based design, but the majority of the automotive industries are focusing on model-based design because of its versatile advantages. If the vehicle under monitoring is crossing the specified limit of the emission standards, then the controller monitors the condition and updates the emission status of the vehicle on the cloud. It gives the broader picture of the corrupt vehicles to the government entities, and necessary actions can be taken.

Keywords Simulink · Model-based design · ECUs · Emission · IOT · Arduino · Raspberry-pi

1 Introduction

The vehicle emission is monitored and controlled by engine management ECU. Controlling the injection and ignition system, the engine management ECU controls the emission. According to the stoichiometry ratio, to burn one kg of fuel, there must be 14.7 kg of air required for optimum combustion. This is the ideal condition, but this ratio is never met because of varying load conditions [1, 2]. With the help of various sensors in place, engine management ECU senses the condition of the engine, driver demand, and environmental parameters; the engine management ECU always tries to control the air–fuel ratio. The ideal air–fuel ratio is 14.7:1. If the fuel quantity is more than the specified figure, then the air–fuel mixture is called a rich mixture; otherwise, it is termed a lean mixture, where the air quantity is more than the specified.

Whatever the vehicle may be, the first question that crosses our mind is that, what is the vehicle's mileage? To achieve good fuel economy, engine management ECU tries to keep the air–fuel ratio as lean as possible, but in certain conditions such as cold start or hard acceleration, engine management ECU cannot operate the engine in the lean mixture because during the cold start, the engine cinder will be very cold, and to start the vehicle, rich mixture is must [3, 4].

Rich mixture during these conditions is bound to produce exhaust particles which can harm the environment and pollute the quality of the air. To counter this, there are various systems available in places such as exhaust gas recirculation (EGR), catalytic converter, secondary air injection system, and electronic fuel evaporation control system. Basically, the engine management system controls the emission by sensing the various engine parameters.

Most important is the oxygen sensor, which informs the air–fuel ratio. If enough oxygen is present after combustion, then the oxygen sensor signals the ECU that

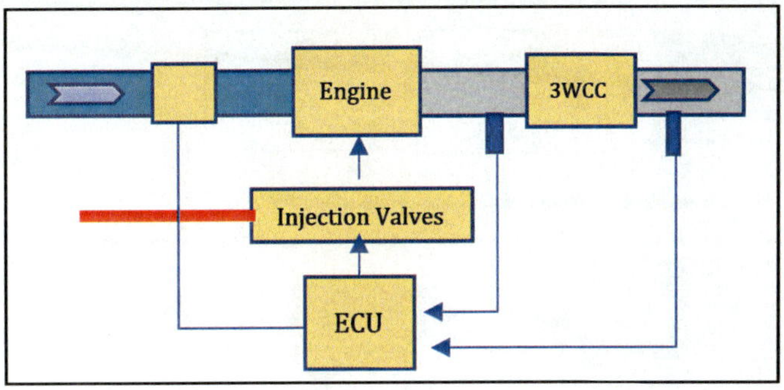

Fig. 1 Closed-loop control system

the mixture is burnt completely, and it was a lean mixture. Otherwise, it was a rich mixture [5, 6].

Typically, there are oxygen two sensors in the system, as shown in Fig. 1.

The first oxygen sensor tells the about the mixture of air–fuel, lean or rich. The second sensor tells about the performance of the catalytic converter. For example, if the mixture formed was rich due to cold start conditions, then the exhaust particles will be generated. This is sensed by the first sensor, but since it was intended to have rich mixture, engine management ECU cannot do anything that is the reason of using the catalytic converter which converts these harmful exhaust particles such as CO, HC, and NOx to N_2, CO_2, and H_2O which are not so harmful. So, the second oxygen sensor informs ECU about the performance of the catalytic converter whether it was able to convert the harmful particle to less harmful gases.

If the second oxygen sensor fails or the catalytic converter fails to oxidize and reduce the exhaust particle, then it would be violating the emission norms.

The paper presented here discusses informing the user about the failure of the system and emission limit crossing of the vehicle. If the user does not take the necessary action within a stipulated period of time, then with the help of any IOT application, the information is sent to the webserver monitored by the government entities.

In the next section, methodology adopted has been discussed.

2 Methodology and Result

2.1 System Architecture

The paper presented here basically focuses on the carbon monoxide (CO) and hydrocarbon (HC) levels of the exhaust gas. If the CO level exceeds the limit, the application monitors the same and updates this information to the webserver.

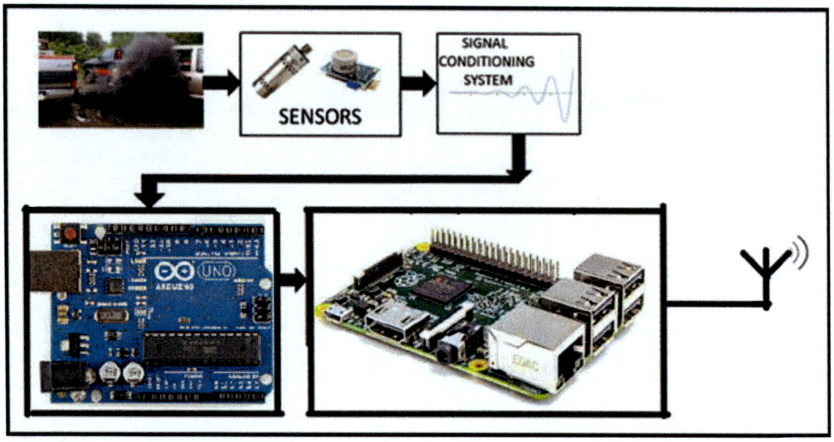

Fig. 2 System architecture

As shown in Fig. 2, the CO and HC levels are sensed through the sensors MQ-7 and IR600, respectively.

MQ-7 sensor is used to measure the carbon monoxide content, and IR6000 sensor is used to measure the hydrocarbon content of the vehicle. These sensors are fixed at the end of the exhaust pipe of the vehicle. The sensor senses the incoming emission level and this transfers the data into the Arduino board, which contains the ATP328 microcontroller. If it crosses the threshold level, the controller sends the information about this to the "Regional Transport office" (RTO) through a Wi-Fi module available with Raspberry-pi board, and the RTO office keeps track of the information using database.

2.2 *Modeling the Algorithm to Acquire the Sensor Data Using Simulink*

As shown in Fig. 3, the sensor data is acquired by the Arduino-UNO device. The model is designed with the help of Simulink using MBD techniques. The MATLAB/Simulink provides this facility to model the algorithm required to acquire the sensor data from Arduino.

Simulink support package for Arduino hardware helps to acquire the sensor data directly without writing any C code. This data can be tested and verified very quickly using Simulink toolsets. The acquired sensor data is then sent to the Raspberry-pi hardware, which has a built-in Wi-Fi module.

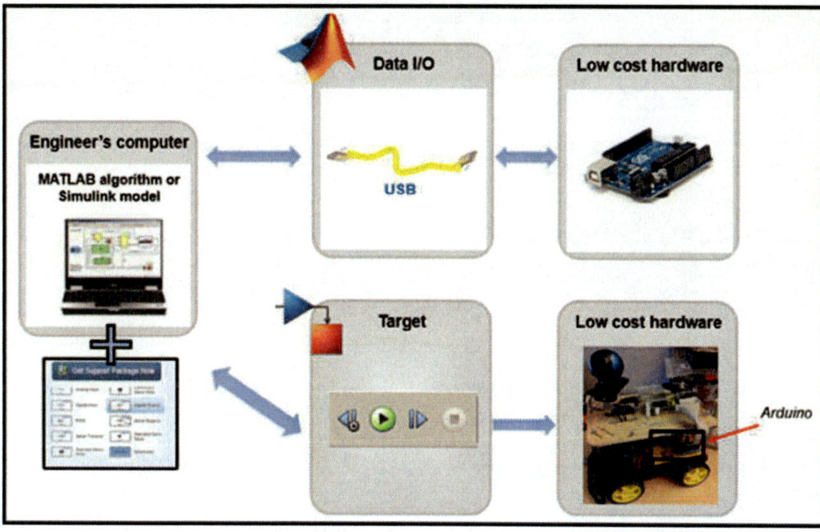

Fig. 3 Simulink and Arduino interface

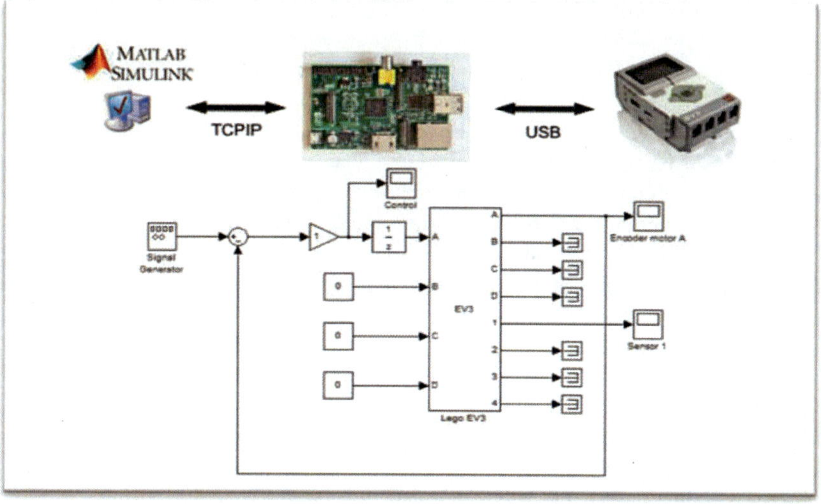

Fig. 4 Simulink and Raspberry-pi interface

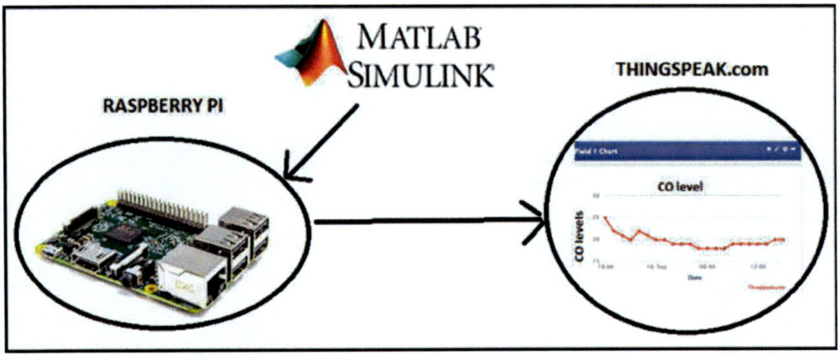

Fig. 5 Simulink, Raspberry-pi, and Thingsspeak server

2.3 Modeling the Algorithm to Acquire the Sensor Data Using Simulink

As shown in Fig. 4, the Raspberry-pi is programmed to receive the sensor data from Arduino is modeled in Simulink.

The Simulink support package for Raspberry-pi provides various toolsets to interact with Raspberry-pi and program it without writing any C code. Using model-based design technology, Raspberry-pi uploads the sensor data to the webserver [7, 8].

There is a plugin available in the Raspberry-pi support package of the Simulink, which can directly upload the sensor data received from Arduino to the web and can be easily monitored by government entities such as RTO [9, 10].

D. Simulink communication with Thingsspeak.com.

As shown in Fig. 5, the Simulink model has been developed Raspberry-pi hardware to receive the input sensor data from the Arduino, and the data is uploaded to the webserver THINGSPEAK.com.

References

1. ITU report on internet of things executive summary: www.itu.intlinternetofthings/
2. Jianli P, Paul S, Jain R (2011) A survey of the research on future internet architectures. IEEE Commun Mag 49(7):26–36
3. Communication Surveys and Tutorials 8(4):48–63, December 2006
4. Albert H, Kravets R, Gupta I (2007) Building trees based on aggregation efficiency in sensor networks. Ad Hoc Netw 5(8):1317–1328
5. Pattanashetty VB, Iyer NC, Dinkar A, Gudi S (2017) Self-coordinating bus route system to avoid bus bunching, In: ICDECT 2016, vol 2. pp 523–532
6. Pattanashetty VB, Iyer N (2015) Smart driving assistance using zigbee. No. 2015-28-0105, SAE technical paper

7. Shet R, Iyer NC, Jeppu Y (2020) Fault tolerant control system for autonomous vehicle: a survey. J Adv Res Dyn Control Syst 12(8):813–830
8. Pillai PS, Shet R, Iyer NC, Punagin S (2022) Modeling and simulation of an automotive RADAR. In: Joshi A, Mahmud M, Ragel RG, Thakur NV (eds) Information and Communication Technology for Competitive Strategies (ICTCS 2020). Lecture notes in networks and systems, vol 191. Springer, Singapore
9. Iyer NC et al (2021) Virtual simulation and testing platform for self-driving cars. In: Fong S, Dey N, Joshi A (eds) ICT analysis and applications. Lecture notes in networks and systems, vol 154. Springer, Singapore
10. Shet RM, Iyer NC, Nissimagoudar PC, Kulkarni A, Abhiram J, Amarnath SK (2021) Motion control and sensor fault diagnostic systems for autonomous electric vehicle. In: Fong S, Dey N, Joshi A (eds) ICT analysis and applications. Lecture notes in networks and systems, vol 154. Springer, Singapore

Author Index

M. S. Kaiser et al. (eds.), *Information and Communication Technology for Competitive Strategies (ICTCS 2021)*, Lecture Notes in Networks and Systems 401,
https://doi.org/10.1007/978-981-19-0098-3